Bioinformatics for Plant Research and Crop Breeding

Bioinformatics for Plant Research and Crop Breeding

Edited by
Jen-Tsung Chen
National University of Kaohsiung
Kaohsiung, Taiwan

WILEY

This edition first published 2024

Registered Offices
John Wiley & Sons, Inc., 111 River Street, Hoboken, NJ 07030, USA
John Wiley & Sons Ltd, The Atrium, Southern Gate, Chichester, West Sussex, PO19 8SQ, UK

For details of our global editorial offices, customer services, and more information about Wiley products visit us at www.wiley.com.

Wiley also publishes its books in a variety of electronic formats and by print-on-demand. Some content that appears in standard print versions of this book may not be available in other formats.

Library of Congress Cataloging-in-Publication Data
Names: Chen, Jen-Tsung, editor.
Title: Bioinformatics for plant research and crop breeding / edited by Jen-Tsung Chen, National University of Kaohsiung, Nanzih District, Kaohsiung, TW.
Description: Hoboken, NJ : Wiley, 2024. | Includes index.
Identifiers: LCCN 2024010213 (print) | LCCN 2024010214 (ebook) | ISBN 9781394209934 (hardback) | ISBN 9781394209941 (adobe pdf) | ISBN 9781394209958 (epub)
Subjects: LCSH: Plant genetic engineering–Data processing. | Agricultural informatics. | Plant breeding–Data processing. | Crop improvement–Data processing.
Classification: LCC SB123.57 B565 2025 (print) | LCC SB123.57 (ebook) | DDC 631.5/233–dc23/eng/20240401
LC record available at https://lccn.loc.gov/2024010213
LC ebook record available at https://lccn.loc.gov/2024010214

Cover Design: Wiley
Cover Image: © T Tstudio/Shutterstock

Set in 9.5/12.5pt STIXTwoText by Straive, Pondicherry, India

M WEP297176 220724

Contents

List of Contributors

Mohamed Addi
Laboratoire d'Amélioration des Productions Agricoles, Biotechnologie et Environnement (LAPABE), Faculté des Sciences, Université Mohammed Premier Oujda, Morocco

Harindra Singh Balyan
Department of Genetics and Plant Breeding, CCS University, Meerut Uttar Pradesh, India

Partha Pratim Behera
Department of Plant Breeding and Genetics, Assam Agricultural University Jorhat, Assam, India

Prateek Ranjan Behera
Department of Plant Pathology, School of Agriculture, Gandhi Institute of Engineering and Technology University Rayagada, Odisha, India

Tanmaya Kumar Bhoi
Forest Protection Division, ICFRE-Arid Forest Research Institute, Jodhpur Rajasthan, India

Ritik Digamber Bisane
Department of Genetics and Plant Breeding, Institute of Agricultural Sciences, Banaras Hindu University (BHU), Varanasi, Uttar Pradesh, India

Kolluru Viswanatha Chaitanya
Microbiology and Food Science Technology Department, GITAM School of Science GITAM University, Visakhapatnam Andhra Pradesh, India

Sona Charles
Bioinformatics and Integrative Genomics Facility, Division of Crop Improvement and Biotechnology, ICAR-Indian Institute of Spices Research, Kozhikode Kerala, India

Jyoti Chaudhary
Department of Genetics and Plant Breeding, CCS University, Meerut Uttar Pradesh, India

Ritu Chauhan
Amity Institute of Biotechnology Amity University Uttar Pradesh, Noida Uttar Pradesh, India

Himanshushekhar Chaurasia
Mechanical Processing Division, ICAR-Central Institute for Research on Cotton Technology, Mumbai, Maharashtra, India

Rahul Lahu Chavhan
Vilasrao Deshmukh College of Agricultural Biotechnology, Vasantrao Naik Marathwada Krishi Vidyapeeth Latur, Maharashtra, India

Jen-Tsung Chen
Department of Life Sciences
National University of Kaohsiung
Kaohsiung, Taiwan

Manish D
Department of Biotechnology, Dayanand
Sagar College of Engineering, Bangalore
Karnataka, India

Siarhei A. Dabravolski
Department of Biotechnology Engineering
Braude Academic College of Engineering
Karmiel, Israel

Ritwika Das
Division of Agricultural Bioinformatics
ICAR-Indian Agricultural Statistics
Research Institute, New Delhi, India

Sourav Datta
Plant Cell and Developmental
Biology Lab, Department of Biological
Sciences, IISER Bhopal, Bhauri
Madhya Pradesh, India

Abhijeet Subhash Deshmukh
Vilasrao Deshmukh College of
Agricultural Biotechnology, Vasantrao
Naik Marathwada Krishi Vidyapeeth
Latur, Maharashtra, India

Vinay Kumar Dhiman
Department of Basic Sciences
Dr. Yashwant Singh Parmar University
of Horticulture and Forestry, Solan
Himachal Pradesh, India

Vivek Kumar Dhiman
Department of Biotechnology, Dr. Yashwant
Singh Parmar University of Horticulture
and Forestry, Solan, Himachal Pradesh
India

Amine Elbouzidi
Laboratoire d'Amélioration des
Productions Agricoles, Biotechnologie
et Environnement (LAPABE), Faculté des
Sciences, Université Mohammed Premier
Oujda, Morocco

Jemaa Essemine
National Key Laboratory of Plant
Molecular Genetics, CAS Center
for Excellence in Molecular Plant
Sciences, Chinese Academy of Sciences
Shanghai, China

Mohammad Samir Farooqi
Division of Agricultural Bioinformatics
ICAR-Indian Agricultural Statistics
Research Institute, New Delhi, India

Tinku Gautam
Department of Genetics and Plant
Breeding, CCS University, Meerut
Uttar Pradesh, India; Agriculture and
Agri-Food Canada, Swift Current
Research and Development Centre
Swift Current, Canada

Pushpendra Kumar Gupta
Department of Genetics and Plant
Breeding, CCS University, Meerut
Uttar Pradesh, India; Murdoch's
Centre for Crop and Food Innovation
Murdoch University, Perth, Australia;
Borlaug Institute for South Asia (BISA)
Ludhiana, Punjab, India; Chaudhary
Charan Singh University, Meerut
Uttar Pradesh, India

Akansha Guru
Department of Genetics and Plant
Breeding, Janta Vedic College
Chaudhary Charan Singh University
Baraut, Uttar Pradesh, India

Vidya Ramesh Hinge
Vilasrao Deshmukh College of Agricultural Biotechnology, Vasantrao Naik Marathwada Krishi Vidyapeeth Latur, Maharashtra, India

Stanislav V. Isayenkov
Institute of Agricultural and Nutritional Sciences, Martin Luther University Halle-Wittenberg, Halle (Saale) Germany; Department of Plant Food Products and Biofortification, Institute of Food Biotechnology and Genomics NAS of Ukraine, Kyiv, Ukraine

Ruchi Jakhmola-Mani
Proteomics and Translational Research Lab, Centre for Medical Biotechnology Amity Institute of Biotechnology Amity University Uttar Pradesh, Noida Uttar Pradesh, India

Ulhas Sopanrao Kadam
Division of Life Science and Division of Applied Life Science (BK21 Four), Plant Molecular Biology and Biotechnology Research Center, Gyeongsang National University, Jinju, Gyeongnam-do Republic of Korea

Debojyoti Kar
Plant Cell and Developmental Biology Lab, Department of Biological Sciences, IISER Bhopal, Bhauri Madhya Pradesh, India

Deepshikha Pande Katare
Proteomics and Translational Research Lab, Centre for Medical Biotechnology Amity Institute of Biotechnology Amity University Uttar Pradesh, Noida Uttar Pradesh, India

Merlin Lopus
Environmental Sciences and Sustainable Engineering Centre, Indian Institute of Technology, Palakkad, Kerala, India

Nagrani Mahajan
Vilasrao Deshmukh College of Agricultural Biotechnology, Vasantrao Naik Marathwada Krishi Vidyapeeth Latur, Maharashtra, India

Prasanta Kumar Majhi
Department of Genetics and Plant Breeding, Odisha University of Agriculture and Technology (OUAT), Bhubaneswar Odisha, India

Poulami Majumder
Department of Biotechnology, Maulana Abul Kalam Azad University of Technology, Kolkata, West Bengal, India

Prabina Kumar Meher
ICAR-Indian Agricultural Statistics Research Institute, New Delhi, India

Manoj Kumar Mishra
Department of Biotechnology, Faculty of Engineering, Rama University, Kanpur Uttar Pradesh, India

Suma C. Mogali
Department of Genetics and Plant Breeding, University of Agricultural Sciences, Dharwad, Karnataka, India

Ashutosh Mukherjee
Department of Botany, Vivekananda College, Kolkata, West Bengal, India

Dola Mukherjee
Amity Institute of Biotechnology, Amity University, Kolkata, West Bengal, India

Sneha Murmu
Division of Agricultural Bioinformatics
ICAR-Indian Agricultural Statistics
Research Institute, New Delhi, India

Nageswara Rao Reddy Neelapu
Biochemistry and Bioinformatics
Department, GITAM School of Science
GITAM University, Visakhapatnam
Andhra Pradesh, India

Potshangbam Nongdam
Department of Biotechnology, Manipur
University, Imphal, Manipur, India

Aniket Pandey
Department of Computer Science, Nguyen
Engineering Building, Fairfax, VA, USA

Bharati Pandey
Animal Biotechnology Division, ICAR-
National Dairy Research Institute
Karnal, Haryana, India

Himanshu Pandey
Department of Agriculture, Khalsa
College, Amritsar, Punjab, India

Shree Prakash Pandey
College of Pastoral Agriculture Science
and Technology, Lanzhou University
Lanzhou, China

Vibha Pandey
Molecular Biology and Biotechnology
Division, CSIR-National Botanical
Research Institute, Lucknow, Uttar
Pradesh, India

Ajinkya Bharatraj Patil
Computational Evolutionary Genomics
Lab, Department of Biological Sciences
IISER Bhopal, Bhauri, Madhya
Pradesh, India

Prachi Pattnaik
Department of Horticulture, Institute
of Agricultural Sciences, Banaras Hindu
University (BHU), Varanasi
Uttar Pradesh, India

Angamba Meetei Potshangbam
Department of Biotechnology, Manipur
University, Imphal, Manipur, India

Asit Kumar Pradhan
Social Science Division, ICAR-National
Rice Research Institute, Cuttack
Odisha, India

Upendra Kumar Pradhan
ICAR-Indian Agricultural Statistics
Research Institute, New Delhi, India

Mingnan Qu
Jiangsu Key Laboratory of Crop Genetics
and Physiology, Agricultural College of
Yangzhou University, Yangzhou, China

Dhananjay Raturi
Proteomics and Translational Research
Lab, Centre for Medical Biotechnology
Amity Institute of Biotechnology,
Amity University Uttar Pradesh, Noida
Uttar Pradesh, India

Aryadeep Roychoudhury
Discipline of Life Sciences, School of
Sciences, Indira Gandhi National Open
University, New Delhi, India

Pradeep Ruperao
Center of Excellence in Genomics and
Systems Biology, International Crops
Research Institute for the Semi-Arid
Tropics, Patancheru, Telangana, India

Ipsita Samal
Department of Entomology, Sri Sri
University, Cuttack, Odisha, India

Santanu Samanta
Post Graduate Department of Biotechnology, St. Xavier's College (Autonomous), Kolkata
West Bengal, India

Soumya Sharma
Division of Agricultural Bioinformatics
ICAR-Indian Agricultural Statistics Research Institute, New Delhi, India

Devendra Singh
Faculty of Biotechnology, Institute of Biosciences and Technology, Shri Ramswaroop Memorial University
Barabanki, Uttar Pradesh, India

Kalpana Singh
Department of Genetics and Plant Breeding, CCS University, Meerut
Uttar Pradesh, India; Department of Bioinformatics, College of Animal Biotechnology, Guru Angad Dev Veterinary and Animal Sciences University, Ludhiana, Punjab, India

Rishita Singh
Proteomics and Translational Research Lab, Centre for Medical Biotechnology
Amity Institute of Biotechnology
Amity University Uttar Pradesh, Noida
Uttar Pradesh, India

Sadhana Singh
Department of Biochemistry, School of Sciences, U.P. Rajarshi Tandon Open University, Prayagraj
Uttar Pradesh, India

Lopamudra Singha
Department of Genetics and Plant Breeding, Institute of Agricultural Sciences, Banaras Hindu University (BHU), Varanasi, Uttar Pradesh, India

Sonali
Proteomics and Translational Research Lab, Centre for Medical Biotechnology
Amity Institute of Biotechnology
Amity University Uttar Pradesh, Noida
Uttar Pradesh, India

Mohamed Taibi
Laboratoire d'Amélioration des Productions Agricoles, Biotechnologie et Environnement (LAPABE), Faculté des Sciences, Université Mohammed Premier
Oujda, Morocco; Centre de l'Oriental des Sciences et Technologies de l'Eau et de l'Environnement (COSTEE), Université Mohammed Premier, Oujda, Morocco

Kusala Vanam
Department of Pharmaceutics, CMR College of Pharmacy, JNTU, Hyderabad
Telangana, India

Nagarjun Vijay
Computational Evolutionary Genomics Lab
Department of Biological Sciences, IISER
Bhopal, Bhauri, Madhya Pradesh, India

Dipti Jayvantrao Wankhade
Vilasrao Deshmukh College of Agricultural Biotechnology, Vasantrao Naik Marathwada Krishi Vidyapeeth
Latur, Maharashtra, India

Zhan Xu
State Key Laboratory for Conservation and Utilization of Subtropical Agro-Bioresources, South China Agricultural University, Guangzhou, China

Preface

Over the decades, in the fields of plant biology and crop breeding, a huge amount of sequencing data and omics-related information involving genomics, transcriptomics, proteomics, interactomics, and emerging integrative multiple omics, have been produced and inevitably continue to increase dramatically. In the meantime, bioinformatics tools have to be organized and upgraded fast to be capable of analyzing the resulting big raw data, which might reveal new aspects of biological mechanisms and networks as well as acquire the valuable genetic information of agronomic traits to aid in breeding programs of crops. Bioinformatics tools are powerful in solving biological problems by predicting novel functions of genes, protein-to-protein interactions, and biological networks. Nowadays, these tools are indispensable, and the knowledge of bioinformatics/computational biology/systems biology is increasingly prosperous; therefore, a timely summary of the emerging literature is necessary, which can provide a practical handbook for researchers and can also advance the future exploration of these tools and related research topics.

This book summarizes the current technical achievements in bioinformatics and presents their emerging applications in the research fields of plant science and agribiotechnology for crop improvement. Also, it presents multiple omics-assisted molecular breeding which aim to the development of high-yield, high-quality, and stress-tolerant crops. In the topic of genetic mapping for interest genes, bioinformatics tools have been upgraded in a way to facilitate the identification of quantitative trait loci (QTLs), which might advance the efficiency of gene tagging, marker-assisted selection, and the understanding of genetics of complex traits and the identification of genetically related DNA markers. In the post-genomics and currently multiple omics era, a vast amount of genomic resources can be available in public databases. Meanwhile, machine learning algorithms came into being powerful enough to deal with a multilevel gene annotation process for predicting gene functions to fulfill crop improvement.

In modern times, the world has to face multiple crises that arise from changing climate and increasing human population and, consequently, serious environmental pollutants and food shortages. Therefore, agricultural biotechnology and molecular breeding are expected to solve the problems by mitigating the stress responses of crops and thereby supporting the global zero hunger of Sustainable Development Goals (SDGs) by the United Nations. For decades, plant scientists have started to develop climate-resilient agriculture by obtaining stress-tolerant crops using strategies to coordinate modern molecular tools; undoubtedly, bioinformatics has to be an essential component due to its multilevel contributions to functional genomics-based crop improvement, particularly for designing future crops against changing climate. Each year, the global agricultural production suffers significant losses due to diseases caused by pests and pathogens, which leads a decrease in crop yield and quality. Unfortunately, under changing climate conditions, these damages may become more significant, meaning that a greater loss of food is foreseeable. Using bioinformatics tools together with multiple omics approaches, plant scientists are capable of deeply dissecting the interaction between plants and pests or pathogens, and the resulting knowledge might greatly benefit the development of crop protection.

In plant research, it is getting more and more critical to explore the knowledge of noncoding RNAs and one of the most powerful tools, bioinformatics, not only for genome-wide identification of interest RNAs and targets but also to be used for predicting their functions and interactions with players of cellular processes as well as the development of epigenomics. This book provides methods and applications of a critical type of emerging noncoding RNAs, namely microRNAs, with an emphasis on the molecular machinery for regulating plant growth and stress responses that can be unrevealing using bioinformatics tools.

Secondary metabolism research is thought to be critical for understanding plant stress responses and bioactivities of compounds derived from plants. The use of bioinformatics tools, such as molecular docking and computational simulation, can advance the prediction of molecular-level interactions concerning the system of ethnopharmacology and drug discovery. Also, this book presents how bioinformatics tools including machine learning and artificial intelligence can be applied to better understand the impact of secondary metabolites in plant defense systems, which can make considerable contributions to high-throughput metabolomics-related plant stress physiology and the development of stress-resilient agriculture.

This book presents recent advances and applications of bioinformatics/systems biology/multi-omics in plant biology and crop breeding based on an in silico platform acquired multi-omics big data, involving functional analysis, interaction network, regulatory mechanisms, and comparative analysis. The book would be an ideal reference for students, teachers, researchers, and experts in the field of plant science and crop breeding.

The editor is grateful for the instructions and help from the publisher. The editor is also extremely grateful for insightful and valuable chapters contributed by scientists and experts in the field.

Jen-Tsung Chen
National University of Kaohsiung, Kaohsiung, Taiwan

1

Bioinformatics as a Powerful Tool to Foster Plant Science Research and Crop Breeding Through Its Involvement in a Multidisciplinary Research Activity

Jemaa Essemine[1,#,], Zhan Xu[2,#], Jen-Tsung Chen[3,*], and Mingnan Qu[4,*]*

[1] *National Key Laboratory of Plant Molecular Genetics, CAS Center for Excellence in Molecular Plant Sciences, Chinese Academy of Sciences, Shanghai, China*
[2] *State Key Laboratory for Conservation and Utilization of Subtropical Agro-Bioresources, South China Agricultural University, Guangzhou, China*
[3] *Department of Life Sciences, National University of Kaohsiung, Nanzih District, Kaohsiung, Taiwan*
[4] *Jiangsu Key Laboratory of Crop Genetics and Physiology, Agricultural College of Yangzhou University, Yangzhou, China*

1.1 Introduction

Globally, the elimination of hunger and ensuring food safety worldwide constitutes a key preoccupation of all farmers around the world. Ensuring food security for the global population and eliminating malnutrition from society has never been so difficult, particularly in the context of decreasing cultivated land and water reservoirs (or points) scarcity, an ever-increasing global population, and global warming due to climate change ascribable to an enhanced pollution rate and the huge CO_2 gas release owing to human activities and industries.

Notably, crops provide various usable products including food, feed, fuel, and several important consumable elements for human life, such as drugs (in the case of tobacco), participating thereby enormously to society's needs. In a recent census and prediction, the worldwide population is expected to increase and reach 9.6 billion by 2050, and the global demand for crops will increase by 100–110% compared with 2005 (Tilman et al. 2011; Chen et al. 2019). To feed a rapidly ever-increasing population in the face of global warming due to climate change, decline of fertile land, and scarcity of usable water for agriculture, agricultural stakeholders and scientists intensify research to unveil new methods and approaches for innovations in crop breeding technology to promote (or foster) agriculture production and maintain sustainable agronomy for a promising productivity in quality and quantity.

The quickly evolving technologies and equipment may play an eminent role in large (or grand) cultures and function as a crucial growth engine in the plant science field. However, the various breeding methods including, cross-, mutation-, and transgenic breedings constitute nowadays the key strategies for crop amelioration and promotion in modern

[#] These authors contributed equally to this work
* Corresponding authors

Bioinformatics for Plant Research and Crop Breeding, First Edition. Edited by Jen-Tsung Chen.

agriculture. Thus, to fulfill that a huge work needs to be done, in which bioinformatics plays a paramount importance. For instance, bioinformatics renders handling a phenotyping big data, some sequences templates design and primers design easy and rapid tasks. Furthermore, bioinformatics constitutes, actually, the carrying vehicle in different plant science domains, especially, molecular and genetic ones. Nowadays, bioinformatics can be useful in several plant research fields such as molecular biology, epigenetics, synthetic biology (synbio), and computational biology.

Bioinformatics can be defined as a computer-assisted science aiming to deal (or handle) with a huge mass of genomic data. In the last decade, bioinformatics has been defined as a new research arena (area or discipline) encompassing advanced algorithmic methods and computational tools to compute, simulate, or sweep high-dimensional (big data) biological data (Oliver et al. 2015a,b). Bioinformatics is a multidisciplinary science that adopts state-of-the-art computational approaches in multifaceted research science to interpret and disentangle any ambiguity or invisible issue at the biological big data scale using information technology (IT) and computer science (Alemu 2015). Furthermore, bioinformatics merges the intelligence of computers, sophisticated programs and software, and statistical methods with the retrieved (or gained) bio-insights from the life sciences field to unravel (uncover) and decipher any raised biological problem. It is a recently emerged area of applied science and IT, rich with valuable insights and statistical approaches. This interdisciplinary science field (bioinformatics) relies on the application of systematic developmental features and the use of IT strategies and smart solutions to handle biological information by adopting various scenarios and ways to solve any life science-related problem, such as biological data fetch, collection and storage, mining, database sweeping, analyses and interpretation, modeling and product design (Degrave et al. 2002; Xiong 2009; Jayaram and Dhingra 2010).

Bioinformatics has numerous applications in different fields. Thus, it offers the scientific community various tools and approaches to foster big data mining and analysis. As mentioned above, bioinformatics has become pivotal in various research topics, including molecular biology, genetics, genome-wide association studies (GWAS), plant breeding, and crop improvement. Bioinformatics represents a key new theme of biological sciences that allows scientists to comprehend and manage available big data obtained following sequencing results (Ojo and Maxwell 2010). Bioinformatics splits into two sub-divisions: computational techniques and databases. The application of these techniques (or tools) and databases in creating biological valuable insights helps better understand the structure and function of living organisms (Xiong 2009). The bioinformatics tools improvement (renovation or promotion) includes developing suitable software for sequence, structural, and functional analyses, as well as building, organizing, and storing biological databases and worthy insights. These tools can be used in several areas of plant genomic and molecular biology, including molecular sequence analysis, molecular structural analysis, and molecular functional analysis (Tramontano 2009; Xiong 2009).

This chapter aims to document the necessity and bright future of bioinformatics in the different plant science domains (or sectors), especially in biology, and the impacts of that on plant breeding and crop improvement in yield and quality. This piece of work shed light on the huge (or substantial) progress observed lately in various plant biotechnology fields such as genetics, genomics, trait and gene discovery, transcriptomics, metabolomics, epigenomics, and analytical tools. This report constitutes an update for the scientific community, readers, and other stakeholders in global agriculture on the valuable news and progress in the different agricultural biotechnology areas.

1.2 Bioinformatics as a Powerful Tool for Big Data Analysis in Plant Science

In the plant science research field, to deal with big data analysis, we can perform multistep integrative bioinformatics analyses, including function prediction of single nucleotide polymorphisms (SNPs) in case of GWAS, gene ontology (GO), Kyoto encyclopedia of genes and genomes (KEGG) enrichment analysis, and protein–protein interaction (PPI). Thus, bioinformatics constitutes the backbone of molecular biology- and plant genetic-related research field. For deep investigating gene function, GO and pathway enrichment analyses can be performed using appropriate software including R package-clusterProfiler (Yu et al. 2012). Thus, based on the biological ontologies including GO and KEGG annotations, the enrichment analysis of the gene set can be fulfilled with statistical methods and visualized with the clusterProfiler package (Yu et al. 2012).

1.3 Role of Bioinformatics in Trait Mapping

Mainly, there exist three major approaches to trait mapping, namely simply linkage, linkage disequilibrium (LD) of GWAS, and joint linkage association mapping (JLAM, Varshney et al. 2018). The analyses for these three mapping categories are bioinformatic-based techniques and/or tools. Thus, bioinformatics combines biology, mathematics, statistics, and computer science to solve a puzzling biological enigma and/or a delicate living problem. Bioinformatics means the use of IT and other computational features to treat biological data that helps in deciphering (or decoding) the plant genomes. The employment of different bioinformatics tools in biological research allows informaticians and scientists to store, retrieve, analyze, annotate, and visualize the results and permit well comprehension of the biological living system and, eventually life science, obviously and fully. This will promote animal and plant health care against contagious disease and pathogen invasion and attack, respectively. Therefore, combining bioinformatics approaches and genome sequencing results permits the assembly and ligation of short- and long-reads, then compared to the reference gene, which may help scientists in disentangling the molecular mechanisms of such a disease and/or plant resistance to pathogen-attack or other kinds of abiotic stress.

1.4 Bioinformatics in Molecular Biology

Bioinformatics plays a major role in the molecular plant research field. In this regard, it has been reported that to treat (or handle) an escalating quantity of genomic insights, bioinformatics tools are needed to mine, retrieve, and analyze the DNA sequences from various organisms (Yang et al. 2008). In molecular biology, a huge set of databases can also be mined and treated with very specific, sophisticated, and well-developed software and/or algorithmic approaches. For instance, characterization of DNA sequence identity (or homology), gene seeking (or search) and retrieval, coding sequence region identification, structural and functional analyses of genomic sequences, etc., can be fulfilled (or accessible) by using various bioinformatics features and/or methods and software packages (Sensen 2005). Besides, in comparative genomics, bioinformatics could play a crucial and pivotal role in defining (or assessing) the genomic structural and functional relationship between the different biological species (Rubin et al. 2000). Bioinformatics constitutes a successful and promising new discipline in the molecular biology field for investigating epigenomics data. Hence, to generate these epigenomic data which can be managed, mined, and stored, using various bioinformatics tools (algorithmic approaches) and software, variegated, advanced, and sophisticated technologies and fostered (or boosted) analytical approaches seem to be essential (or required) for the researchers and scientific community.

1.5 Role of Bioinformatics in Genetic Variation

Bioinformatics is a very important and essential tool in genetic dissection and biostatistics-genetic related analyses such as quantitative traits-loci (QTL), LD, and *P*-value such as in GWAS (Agrama et al. 2010). Bioinformatics is a pivotal element in genetic statistical analysis to determine the degree of relatedness between parental lines, the accuracy and reliability of the obtained results as genetic sampling should be carried out, either in human or plant phenotyping screening, on a large population size endowed with a substantial natural genetic variation (Qu et al. 2020a,b). The genetic variation relies mainly on thc population size and genetic heritability factor between the members (humans) or lines (crops) of the used population for such genetic investigation of a desired trait. For instance, in GWAS authors used a mixed linear model (MLM) method which takes into consideration the population structure and, therefore, minimizes false positive rates (FPR) compared to a general linear model, GLM (Zhou and Stephens 2012). Thus, the GEMMA software considers both population structure (Q) and genetic kinship (K) (Turner 2014). In this regard, a well genetically structured population ensures reliable and trustworthy genetic variation investigations of desired, agronomic or agricultural, traits with promising and fruitful outcomes that can be important for plant breeding programs and eventually in crop improvement strategies.

1.6 Bioinformatics in Genome-wide Association Studies (GWAS)

Both bioinformatics and computational biology are known to be involved in GWAS analysis. Briefly, GWAS analysis is widely used to evaluate (estimate or assess) the statistically significant genomic variants related to a particular physiological trait or phenotype (Tam et al. 2019). In this context, bioinformatics setting parameters help in designing primers, predicting genes, and constructing sequences. In GWAS, bioinformatics can be utilized to design the protospacer-adjacent motifs (PAMs), single guide RNA (sgRNA), trans-activating crRNA (or TracrRNA), which is a different RNA type that interacts with the CRISPR (cr) RNA to form the dual guide (g) RNA in type II and subtype V-B CRISPR-Cas systems (Faure et al. 2019; Chen et al. 2019). The sgRNA can be designed using bioinformatics and computational biology through the ensuing professional and specific websites (http://chopchop.cbu.uib.no/; https://design.synthego.com/#/). Usually in GWAS, the DNA repair damage in the case of the HDR (Homology-directed repair) relies mainly on a synthetic exogenous DNA template sequence or with the help of the sister chromatid information. The repair template can be a sister chromatid or an exogenous one (template), such as an exogenous DNA or a ssDNA (single-strand DNA) containing the aimed sequence modification to be inserted in the break site (Chen et al. 2019). The GWAS analysis is important in disentangling (or deciphering) the orchestrated genetic and molecular mechanisms monitoring certain traits. It remains so much difficult and tedious to carry out GWAS analysis for scientists with low or very limited bioinformatics knowledge and skills. These limitations (or challenges) are owing (or ascribable) to the complexity of installing some pGWAS professional key files and packages, intricacies (or complexities) of commands and codes, or the quantity of genomic data required for triggering pGWAS analysis (Falola et al. 2023). The limitations may also due to the availability of computer numbers (no granted funds to purchase computer machines) to run bioinformatics and biostatistics analyses; especially for the researchers of countries in development.

1.7 Implication of Bioinformatics in "Omics"

Bioinformatics represents an eminent and complex component of omics, offering the required tools to manage and deal with huge biological data sets and interpret them into their accurate biological function in service of "omics," which can help in understanding certain biological phenomena. Omics encompasses proteomics, genomics, transcriptomics, metabolomics, and epigenomics. Thus, bioinformatics plays a prominent role in these research fields. It represents a suitable and paramount technique to interpret "omics" data sets and to gain (or collect) valuable and important biological knowledge (Mochida and Shinozaki 2010). From the sequence analysis to the gene's characterization, clustering of concerned sequences and investigating the evolutionary relationships using phylogenetics imputation and permutation operations are major tasks of bioinformatics. It also encompasses the characterization or determination of the cellular and functional annotations of genes, proteins, and active sites of the spatial (3D) protein structure in the cell (Rao et al. 2008). For the omics data sweeping and management, the first step is the determination and retrieval of sequences from different database repositories (NCBI, TAIR, Rap-db, Phytozome, PlantGDB, SwissProt) developed for bioinformatics-updated applications to deal with protein structure and interactions (PPI), DNA, RNA, and expressed sequence tag(EST) sequences (Rahman et al. 2016).

1.8 Bioinformatics in Computational Biology and Evolutionary Studies

The evolutionary relationship among plant species and/or groups of organisms is known as phylogenetics. Thus, botanists and/or taxonomists fetch the evolutionary relationship between species and groups through different anatomical methods that take tremendous time and effort to collect the required information to construct phylogenetic trees. However, using bioinformatics, phylogenetic trees can be built relying on the sequence alignment using various powerful approaches and techniques. Notably, different algorithmic methods and the newly created and advanced software can be used to construct the phylogenetic tree which can be varied according to the various evolutionary lineages. Bioinformatics sweeps, computational biological analysis, and retrieval of available data information of the concerned species from the repositories databases websites, such as the plant comparative genomics portal website (https://phytozome-next.jgi.doe.gov/), are key and collective tasks and events to construct a phylogenetic tree.

1.9 Role of Bioinformatics in Transcriptomics

First of all, transcriptomics is the investigation of all sets of RNA messenger molecules in the cell (Kumar and Chordia 2017). Transcriptomics is known (or recognized) also as the transcripts expression profiling where DNA microarray can be utilized to assess the expression level of mRNA in a well-defined (or -targeted) cell or organ. The microarray smart tool provides a huge quantity of data (big data); however, a simple run produces thousands of data points (or values) and each experiment needs several (hundreds if not more) sweepings (imputations or runs). Analyzing big data can be achieved (or fulfilled) with the aid (or assistance) of several specific software packages. Thus, bioinformatics plays a key role in transcriptome analysis where the mRNA transcript levels might be assessed (Aplin and Singh 2008). To our knowledge, RNA sequencing (RNA-seq) is also subscribed to belong to the transcriptomics field. It is evaluated using the recent emerging technology, next-generation sequencing (NGS), to probe (or estimate) the abundance and amount of RNA in a given tissue or cell at a well-determined instant (time) and conditions (normal or stressful). This method is employed to analyze the transcriptome reshuffling in the cell transcript levels under challenging conditions to determine the degree of resistance and resilience of certain plant species under climate changes.

1.10 Implication of Bioinformatics in Next-generation Sequencing (NGS) Analysis

Bioinformatics makes easier, faster, and more trustworthy the data combination and incorporation, interpretation, management, and model prediction, as well as dealing with huge amounts of data (big data or massive data) generated by high-throughput methods and technologies (Ko et al. 2018). Thus, Park et al. (2017) revealed that with the employment of bioinformatics approaches, scientists can determine the accurate insertion positions and copy numbers of tDNA, genomic reorganizations, and the missed backbone fragments or sequences, which were confirmed by the southern-blot findings and analyses (Park et al. 2017). Vlk and Řepková investigated the paramount role of bioinformatics in NGS analysis and also, emphasized the application of NGS in plant breeding (Vlk and Řepková 2017). In this context, NGS improved the sequencing coverage rate and paved the way for new tools and techniques to genotype simultaneously and substantially high SNP numbers. Thus, NGS recently discovered sequencing tools and RNA-seq are powerful and valuable methods to assess the gene expression level and/or the transcripts abundance in the plant cell. These approaches (NGS and RNA-seq) played an important and paramount role in crop improvement, plant breeding, and characterization of biomarkers (enzymes, TFs, and genes) of interest. These biomarkers might be involved in conferring plants the defense mechanisms against biotic and abiotic stresses. In accordance, Kamei and co-workers (2016) created and thereafter improved a method that permits molecular plant breeders, even with low or with no deep (or extended) bioinformatics skills, to competently (performantly or professionally) analyze and interpret *de novo* transcriptome data from any crop (Orphan Crop Browser; http://www.bioinformatics.nl).

The NGS has become famous, nowadays, as a very effective and promising tool for identifying crucial genes in large cultures, including crop plants, such as genes responsible for coping with abiotic stress. Several molecular techniques and advanced technologies such as genome analyzer, the applied biosystems SOLiD system, Roche/454 FLX, and the Illumina/Solexa have developed and become commercially available lately for the NGS analysis (Agrawal et al. 2015). The NGS might be employed to carry out whole-genome sequencing (WGS), identify, sort, and filter then isolate the transcription factors (TFs) binding sites, and expression of the noncoding RNA and a targeted re-sequencing (Tiwary 2015). Investigating the complete methodologies and equipment involved in NGS is out scope of this book chapter; however, to gain more insights about the NGS methods and techniques of sequencing, readers and researchers are addressed (or recommended) to consult the study performed by Egan et al. (2012) as well as the report of Ray and Satya (2014).

1.11 Implication of Bioinformatics in Metabolomics

Metabolite profiling is a fast-expanding methodology that is utilized for phenotyping and disease analysis in plants. It has also quickly emerged as an essential approach in functional gene annotation and a valuable tool for better understanding the cellular response to biological conditions (Schauer and Fernie 2006). However, the metabolomic analysis relies on various web-based platforms such as XCMS (https://xcmsonline.scripps.edu) and METLIN (http://metlin.scripps.edu) (Smith et al. 2005; Tautenhahn et al. 2012).

The former (XCMS) is an online open access (freely accessible) bioinformatics repository gate that makes the quick and safe storing of raw big data and then guides and helps users in data visualizing, managing, and retrieval (Tautenhahn et al. 2012). The second (METLIN) constitutes another online database repository (or gate) that can be utilized in different investigations (or reports) associated with plant metabolic dissection (or screening) in response to various environmental stresses (biotic and abiotic). This web-based platform helps investigate metabolites in plants under various conditions. Retrieving data and filtering from this metabolomic storing gate or platform is not tedious or does not need advanced bioinformatics skills to treat, manage, organize, annotate then package the raw data to be well-stored, clear, and easily accessible for a next calling and/or retrieval (Smith et al. 2005). Thus, several computational web-based platforms, statistical software, and online bioinformatics methods, encompassing sweeping, imputation, permutation, recall, filter, and retrieve, constitute the best heavy weapons for a skilled bioinformatician to design then establish or build the metabolic profiling of such a plant species under such an environmental condition (control and stressful conditions). Therefore, the incorporation (or combination) of powerful and easy-handling (or executing) computational and bioinformatics techniques with "omics" approaches ideally elaborate new metabolic profiling and pathways networks for the benefits of plant breeding programs and thereby crop improvement in yield and quality.

1.12 Bioinformatics and Epigenetics

Given the cell's tiny structure and the numerous and significant complexity of factors and/or pathways involved in the epigenetic regulation mechanisms, the computational and bioinformatics tools are regarded as key and pivotal elements in deciphering (disentangling) these mechanisms at the genome level and, subsequently, sort (or find) out the way (or fashion) by which the epigenetic level may sense the environmental clues, including biotic and abiotic stresses, and thus intervene in modulating the plant responses to the external stimuli (Angarica and del Sol 2017). Notably, computational, permutational, and even imputational analyses in bioinformatics have boosted largely the analytical methods and become powerful and valuable approaches to consolidate our investigations and enrich our skills and knowledge weapon to tackle a well-defined and complicated issue, which is in this case understanding the role of bioinformatics in the epigenetic regulatory mechanisms and analyze the epigenomic data (Yang and Lee 2004).

Thus, bioinformatics tools are very important for the histone modifications at the epigenetic level. For instance, bioinformatics tools and algorithmic approaches such as ACME, HGMM, ChIPOTle, and HMMTiling are key methods to deal with ChIP-on-chip microarray data and peak callers. However, the most bioinformatics challenge for the ChIP-seq data is a rapid and precise mapping of a huge amount (thousands to millions) of short reads, sequenced and identified to be appropriate for the regions bearing a specific histone modification, to the reference genome sequence. In this regard, several aligners were developed to solve the issues of mapping short reads, including Bowtie (Langmead et al. 2009) and BLAT (Kent 2002), among numerous others. Furthermore, for ChIP-seq data, the alignment of short reads to the reference sequence, and quality control (QC) of sequencing data, remains a bioinformatics defiance and preoccupation. Besides, the bioinformatics analysis of the histone posttranslational modification is strictly specific to the used mass spectrometry approach (Hirst and Marra 2010; Olsen and Mann 2013; Moradian et al. 2014).

In this regard, it has been reported as well that bioinformatics plays a crucial role in analyzing the genome-wide DNA methylation assays. The DNA methylation process can be determined experimentally through bisulfite sequencing (Deng et al. 2009; Booth et al. 2012), bisulfite microarrays (Bibikova et al. 2011; Schumacher et al. 2006), and enrichment methods, including MeDIP-seq and MethylCap-seq (Weber et al. 2005; Down et al. 2008; Serre et al. 2010).

In brief, bioinformatics helps in different manners in the field of epigenetics. So, it can aid in diagnosing and/or deciphering the biological abnormalities responsible for the diseases caused by the epigenome and, accordingly, the therapies can be evaluated, then the required tools for the epigenetic data analysis and development of methods for analyzing and understanding the large epigenome data sets, which may help in disentangling the epigenetic regulatory mechanisms underlying such a biological process. Eventually, with the help of bioinformatics tools, the diseases can be diagnosed, and then therapies might be well-assessed, thereby the appropriate drugs (chemical compounds) can be properly designed, synthesized, registered, and ultimately assigned to the patient.

1.13 Involvement of Bioinformatics in Synthetic Biology

In this section, we may define synthetic biology (Synbio) as ensuing; Synbio aims to imagine, predict, and design biologically based parts (enzymes and regulatory processes), novel pathways (devices), and systems (chassis) as well as reengineering existing, natural biological phenomena for the discovery of new drugs from plants such tobacco and the benefits of humankind. Thus, the Synbio approach can be divided into three categories (or groups); including design, build, and test the synthesized element such as drug or other entities (chemical compounds). Hence bioinformatics plays a major role in the different Synbio steps; especially, in the design one (step). The design encompasses collecting detailed information (gaining insights) about enzymes, pathways, or regulatory elements; however, the building step contains DNA optimization and assembly, and then eventually the test that may rely on checking the produced entities (natural products or chemical compounds) such as drugs (Carbonell et al. 2016) for disease killer such as cancer. However, the integration of these three steps (design, build, and test) remains challenging actually, because the hard connection and lack of automated tasks (or events) between the bioinformatics tools renders their (tools) adoption by the scientific community out of reach, if not even impossible. With the advent and expansion use of bioinformatics tools, Synbio has become an accessible topic for scientists and algorithms developers. Thus, the rapid development of Synbio relies strongly on the use and combination of databases, bioinformatics, and computational biology (Cai et al. 2023).

1.14 How Can Bioinformatics Promote Plant Biotechnology?

Bioinformatics helps scientists decipher certain biotechnology issues and can foster basic research. Notably, defining the whole plant's genome sequence may pave the way for deeper molecular and biotechnological accurate studies related to this organism (plant). For instance, worldwide scientists and scientific communities analyzed and debated (or discussed) gene expression levels and transcript abundance in crops including rice and wheat, for the sake of identifying crucial genes or biomarkers responsible for the tolerance of crops to severe environmental conditions (heat, drought, etc.). Therefore, combined bioinformatics and biotechnological techniques may lead to the creation of new varieties through genome editing approaches or improve the yield quality and quantity of the already existing varieties by designing, for example the plant architecture to change the leaf angle to the best position for a higher sunlight collection, which may eventually lead to an improvement in the plant organic matter and thereby the plant productivity. The interaction phenotype–genotype is the key process in unveiling any regulatory mechanism in the plant. For instance, scientists can identify the molecular basis for disease resistance and stress tolerance in the plant, and thus design the appropriate methods that can boost (or foster) the plant's resilience and adaptation to certain biotic and/or abiotic constraints. Researchers also utilize bioinformatics and biotechnology to generate new plants characterized by an elevated fruit quality, or with the ability to survive under extreme environmental conditions. Besides, bioinformatics can be used to harness genetic diversity. Thus, through phylogenetic tree analysis and by defining which plants are evolutionarily closely related based on the clades and nodes relationship of the constructed phylogenetic tree, scientists can determine the sexually compatible plant species with preferred features (such as longer stalks for rice, larger grains for barley, corn, and wheat). Further, bioinformatics can be utilized, as well, to engineer or uncover new techniques and methods to assess gene function and structure. Hence, the scientific community unveiled microRNAs (miRNAs), a member of genes in small sequence fragments, in plants. These small RNA biomarkers regulate different molecular aspects along the plant lifecycle and mainly exert tight control on its (plant) growth and development processes. These regulatory biomarkers (miRNAs) target certain DNA sequences and somehow control certain genes from being active or silent. For instance, mutations in miRNAs can be responsible for dramatic damage such as abnormal floral development, or even plant death.

1.15 Bioinformatics Use in Biotic and Abiotic Stress Management

Understanding how plants, especially crops, react to environmental stress is a pivotal task to recognize and determine the adaptation strategies adopted by the plant to cope with such stress kind and thus ensure safe growth and development by escaping (or avoiding) the deleterious effects of severe environmental constraints, either biotic or abiotic (Cho et al. 2019; Costa and Farrant 2019). Thus, bioinformatics is very promising and accurate for investigating and dissecting the plant transcriptome wealth in response to biotic and abiotic stresses. Besides, the use of bioinformatics methods and/or approaches in deeply investigating crop genomes may promote the agricultural research area and boost the researchers' efforts in dealing with any challenge at the dry level (computational side) before passing or executing the wet part (bunch work) of the experiment. The abovementioned can also simplify (make easy) the scientific community tasks in defining the promising biomarkers (genes or TFs) by checking the whole genome from different species and elucidating their function on the crops (Cho et al. 2019). This transgenic experiment was performed by Perveen et al. (2020), where authors observed that the overexpression of maize TF mEmBP-1 enhances photosynthesis, biomass, and yield in rice (Perveen et al. 2020) in their manuscript authors used a TF from C_4 to enhance the performance, photosynthesis, yield in a C_3 plant, which is rice (crop plant). Furthermore, bioinformatics serves as a powerful and successful tool to deal with and understand the various scenarios of plant tolerance and/or avoidance of different environmental scourges (İncili et al. 2022). Hence, the mechanisms of plant tolerance to the surrounding stressful conditions including both types of stress (abiotic and biotic) have been extensively investigated, but the results might be stronger and very promising if authors can raise the challenge, redirect their thinking based on renovation to pave new ways in their agricultural research activities by integrating the predictive models, discover analytical strategies, bootstrap methods of computational analysis to benefit from the artificial intelligence (AI), which relies hugely on the power of bioinformatics (Aditya-Shastry and Sanjay 2020). Thus, bioinformatics can be involved in all types of biotic and abiotic stresses to manifest in plant protection and avoid detrimental effects of severe environmental conditions.

1.16 Bioinformatics for the Investigation of Plant Resistance to Pathogens

Nowadays, among the main difficulties (or constraints) in modernized agriculture, to ensure an ever-increasing worldwide population food safety and healthy nutrition and avoid malnutrition, are the hypersensitivity of large cultures including crops to the pathogen attack and/or invasion which may cause a huge loss in the culture ascribable to fast diseases widespread that may eventually entrain a disaster owing to a partial or total culture field, resulting thereby in population malnutrition and even starvation. The investigation of plant–pathogen interaction impacts on the crops plays a prominent role in better understanding the origin of plant diseases, including pathogen identification, and disease resistance, and trying to disentangle the molecular mechanisms and find suitable treatments (Nishad et al. 2020). Hence, plants protect themselves through an orchestrated, in a concerted manner, complex defense system (physiology, biochemistry, and molecular) against a variety of pathogens, such as insects, bacteria, fungi, and viruses. Plant–pathogen interaction is a complex and tough phenomenon manifested by the appearance of new chemical compounds or biomolecules secreted by the pathogen organism or by the plant tissues as the first layer of defense, where these molecules can take various formats such as enzymes, proteins, and sugar polysaccharides. However, their way (or manner) of action, such as signal detection and transduction, may occur through pattern recognition receptors (PRRs) located mainly within the plant cells and tissues (Gupta et al. 2015; Abdulkhair and Alghuthaymi 2016; Andersen et al. 2018; Dong and Ronald 2019). In addition, bioinformatics is recognized to be widely used in biology known by its multifaceted research fields including plant pathology. In this context, bioinformatics can be applied to characterize and expect (or imagine) the "effector" proteins that can be generated by the pathogens to be able to invade the host plant. To discover new approaches to monitor the disease caused by the pathogens, researchers should understand the genetic causes and progressively the complex molecular mechanisms standing behind the pathogen infection, including characterizing the type of pathogen, its way (or fashion) of action on the plant cells and the procedure of signals transfer between cells. Thus, WGS methods using NGS permitted the sequencing and characterization of a large number of pathogens and their respective effectors which allowed scientists and researchers to collect big genetic data that may constitute the cornerstone or the strong pillar to diagnosis (or unravel) the pathogen-related disease and accordingly find the suitable pesticide to cancel its effects on the host plant. Therefore, bioinformatics methods and approaches have been invented to well dissect both the host plant genome and those of pathogen, and effectors and deeply understand the mechanistic inter-specific interactions to disentangling the disease infection mechanisms, pathogenic targets, and all expected factors to be potentially involved in the plant pathology (Dong et al. 2021).

1.17 Bioinformatics in Crop Breeding and Improvement

Bioinformatics tasks include (or encompass) the handling and treatment of sequence information. The latter implies the ensuing strategies; characterization of genes in DNA sequences from different cells, characterizing families of linked sequences and the prediction of an appropriate model, matching and/or aligning similar sequences using BLAST (Basic Local Alignment Sequence Tool) of the National Center for Biotechnology Information (NCBI; https://www.ncbi.nlm.nih.gov/), then produce phylogenetic trees to investigate the evolutionary relationships. Besides, obtaining the whole genome's genes and proteins based on a given sequence of amino acids and predicting the active and vacant sites in the protein structures available for the drug molecules binding or docking (Rao et al. 2008). Furthermore, the analysis of phenotype–genotype interactions based on a particular agronomic or physiological trait(s) is a key feature in crop breeding programs (Hu et al. 2022). Notably, bioinformatics represents a key tool of the molecular and genomic field to pave the way toward precious and promising results and to foster the analysis through applying beforehand a computational analysis and prediction of essential biomarkers such as genes and transcriptions factors (TFs) that might potentially be involved in the target project before passing ahead to the detailed sequencing, functional and structural analyses to save time and experimental costs.

An ever-green landscape for a healthy population and an intensive and sustainable agricultural production constitute urgent challenges in response to global warming due to climate change, pollution owing to a massive CO_2 release, and the ever-increasing worldwide population. Hence, the innovations and discoveries in the omics-based research domain promote the multifaceted and in a state of the art plant research activities. In this concern, the combined and/or incorporated "omics" approaches help clarify the molecular system of a plant species under such conditions that can be utilized to improve plant productivity and/or yield quality. However, the genomics strategy, particularly comparative genomics, paves the way to understanding the gene functions and their biological properties for such plant species. Therefore, bioinformatics can alleviate the wet-lab costs and economize the time of investigation by predicting some genes, and pathways that can potentially play eminent roles in plant breeding programs. Thus, bioinformatics tools are mostly and extensively used to engineer and/or predict new techniques and approaches for an enhanced or improved plant yield quantity and quality (Mochida and Shinozaki 2010).

Consequently, bioinformatics is mainly employed for crop breeding and improvement to promote (or enhance) yield and quality (Gaurav et al. 2021; Mu et al. 2022). Via an accurate and deep bioinformatics analysis of the genes, using specialized algorithmic methods and software, associated with seed germination, seedling size (length, width, and weight), reproductive yield, and artificial interference with respective genes, large culture, including crops (rice, wheat, barley, maize, soybean, etc.), can be further ameliorated (Mascher et al. 2019). Among the striking bioinformatics uses in plant breeding programs, it can help in determining the ideal leaf angle for the maximum photosynthetic rate, to generate new species with a preferred leaf angle to collect more sunlight which might be converted to chemical energy useful for electron transfer and synthesis of dry matter. This might eventually help in better accumulating organic matter in the leaf and better sap transportation through the leaf and the whole plant, and ultimately higher plant yield and productivity.

1.18 Bioinformatics Impacts on Plant Science

Mainly bioinformatics tools help manage massive data resulting from genomic sequencing experiments (Kushwaha et al. 2017) using different sequencing platforms such as PacBio (Pacific biosciences), known for its highly accurate long-read sequencing, using next NGS technologies. The trustworthy and valuable bioinformatics techniques are required to analyze, manage, retrieve, and mine data and thus strengthen our ability and boost our skills to handle, analyze, and interpret the concerned manner complex biological systems. For instance, to study the genome structure and regulation, bioinformatics spans various sectors, including databases on regulatory sequences such as enhancers, repressors, transposons, and promoters, the regulation of gene expression; analysis and recognition of genomic sequences; gene structure prediction; a massive sequencing data management and. The incorporation (integration) of bioinformatics would certainly promote plant science research activities and thereby may pave the way for crop improvements in different features including the determination of crucial biomarkers (genes or TFs) via genetic screening, amelioration of analysis methods of gene expression and/or transcript abundance levels, promotion of analysis and interpretation approaches and functional genomics assessment tools, engineer certain agrochemicals relying on the analysis of the signal perception and transduction flux to detect and even identify targets, and use of plant genetic wealth to maintain the genetic diversity for the benefits of agricultural species and promote the so-called sustainable agronomy and developed agricultural strategies.

The plant growth rate might be assessed based on the available collected physiological and phenotypic data. This enables generating a virtual plant model that can preciously expect (and even recognize) the plant growth patterns and the results of interactions with diseases or pests (Kim et al. 2022).

1.19 Application of Bioinformatics in Plant Breeding Programs

In addition to generating mutations such as indel (insertion–deletion) or base editing in the functional sequences (coding or gene body), assessing (or probing) the gene expression or the transcript abundance level constitutes a suitable, if not the appropriate, experimental technique for determining gene function and its abundance level and may hugely facilitate the plant breeding strategies and/or implementations. Gene expression might be influenced at different processing levels, including the transcription, the mRNA amplification and/or polymerization, or its (mRNA) translation. Besides, the gene regulatory process may also happen at the translational stage. Thus, the ORFs (open reading frames) are well recognized as *cis*-regulatory factors endowed with negative influences on translation and nonsense-mediated mRNA decay (Von Arnim et al. 2014). Bioinformatics sweeping and imputation revealed that uORFs are largely abundant among the plant mRNAs. For instance, more than 35% of total Arabidopsis mRNAs harbor at least one putative uORF, and targeting uORFs with antisense oligonucleotides enhances (or improves) the translation effectiveness (yield or efficacity) of the primary ORFs (Liang et al. 2016).

So far, we can say that all bioinformatics-based analysis plant science research fields, such as metabolomics, genetics, epigenetics, and Synbio may contribute to various extends to agriculture science, such as crop breeding and genome editing for crop improvement, enhancing grain yield. This may occur by unveiling new genes or promoting certain agronomic and/or physiological traits, elucidating their unknown and novel metabolic pathways. We can deduce that bioinformatics plays a master role in plant science and constitutes the main pillar of all successes and fruitful results in this field.

1.20 Conclusion

As mentioned above and well displayed in the summarizing graph of Figure 1.1, bioinformatics plays crucial roles in various plant science fields. Thus, the contribution of bioinformatics in the development of plant functional genomics has become very obvious. However, with the advent and expansion of big data, bioinformatics faced enormous challenges and limitations during its application in plant science and crop improvement. To meet the challenges, Bioinformaticians need tedious work and huge perseverance. Accordingly, unveiling

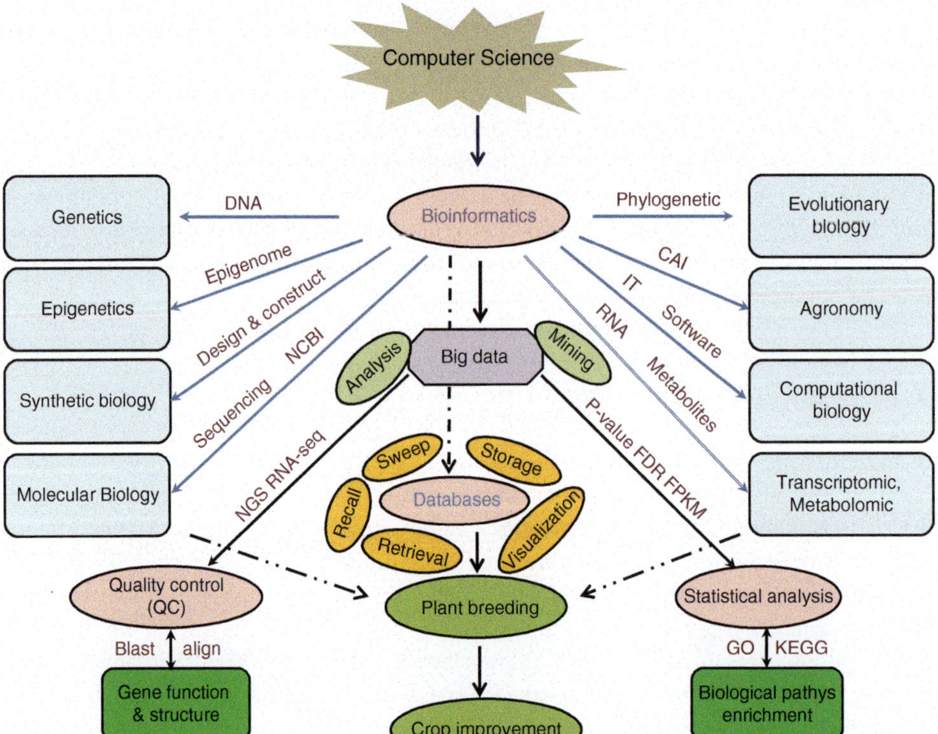

Figure 1.1 Schematic diagram summarizing the possible applications of bioinformatics in plant science and its impacts on plant breeding programs and the fruitful repercussions on crop improvement (yield and quality). Abbreviations: IT, information technology; NGS, next-generation sequencing; QC, quality control; GO, gene ontology; KEGG, Kyoto Encyclopedia of genes and genomes; FDR, false discovery rate. FPKM, fragments per kilobase of transcript per million mapped reads; CAI, computer-assisted irrigation; NCBI, National Center for Biotechnology Information; Biological pathys, biological pathways.

and developing more sophisticated and powerful bioinformatics tools and approaches, based on IT discipline, would help manage and integrate, in a concerted manner, all the available bio-insights resources and provide accurate support and robust pillars for effective breeding strategies and plant resistance analysis. Hence, well comprehend the whole genome, gene function, and structure, the discovery of genes, construct design and insertion in the genome, deployment of diagnostic elements, such as in CRISPR-Cas9 (gene insertion or deletion, indel), epigenetics regulatory events including, methylation, and acetylation) and Synbio (design biomarker and its incorporation in the gene), is fundamental and prerequisite to mediate breeding program for the sake of improving crop yield and quality to ensure food security for an ever-growing worldwide population. All these data sets require robust and effective database management systems and accurate bioinformatics tools for visualizing, retrieving, storing, and even mining the information that may eventually influence and/or guide the genetic crop improvement strategies.

References

Abdulkhair, W.M. and Alghuthaymi, M.A. (2016). Plant pathogens. In: *Plant Growth*, 1e (ed. E.C. Rigobelo). InTechOpen https://doi.org/10.5772/65325 https://www.intechopen.com/chapters/52387.

Aditya-Shastry, K. and Sanjay, H.A. (2020). Hybrid prediction strategy to predict agricultural information. *Applied Soft Computing* 98: 106811.

Agrama, H.A., Yan, W., Jia, M. et al. (2010). Genetic structure associated with diversity and geographic distribution in the USDA rice world collection. *Natural Science* 02: 247–291.

Agrawal, P.K., Babu, B.K., and Saini, N. (2015). Omics of model plants. In: *Hand Book of Plant Omics: The Omics of Plant Science* (ed. D. Barh, M.S. Khan, and E. Davies), 1–32. New Delhi: Springer https://doi.org/10.1007/978-81-322-2172-2_1.

Alemu, K. (2015). The role and application of bioinformatics in plant disease management. *Advances in Life Science and Technology* 28: 28–33.

Andersen, E.J., Ali, S., Byamukama, E. et al. (2018). Disease resistance mechanisms in plants. *Genes* 9 (7): 339. https://doi.org/10.3390/genes9070339.

Angarica, V.E. and del Sol, A. (2017). Bioinformatics tools for genome-wide epigenetic research. In: *Neuroepigenomics in aging and disease, advances in experimental medicine and biology* (ed. R. Delgado-Morales), 4–512. Springer International Publishing AG https://doi.org/10.1007/978-3-319-53889-1_25.

Aplin, J.D. and Singh, H. (2008). Bioinformatics and transcriptomics studies of early implantation. *Annals of the New York Academy of Sciences* 1127: 116–120.

Bibikova, M., Barnes, B., Tsan, C. et al. (2011). High density DNA methylation array with single CpG site resolution. *Genomics* 98 (4): 288–295.

Booth, M.J., Branco, M.R., Ficz, G. et al. (2012). Quantitative sequencing of 5-methylcytosine and 5-hydroxymethylcytosine at single-base resolution. *Science* 336 (6083): 934–937.

Cai, P., Liu, S., Zhang, D. et al. (2023). SynBioTools: a one-stop facility for searching and selecting synthetic biology tools. *BMC Bioinformatics* 24: 152. https://doi.org/10.1186/s12859-023-05281-5.

Carbonell, P., Currin, A., Jervis, A.J. et al. (2016). Bioinformatics for the synthetic biology of natural products: integrating across the Design–Build–Test cycle. *Natural Product Reports Royal Society of Chemistry* 33: 925–932. https://doi.org/10.1039/c6np00018e.

Chen, K., Wang, Y., Zhang, R. et al. (2019). CRISPR/Cas Genome Editing and Precision Plant Breeding in Agriculture. *Annual Review of Plant Biology* 70 (28): 1–31. https://doi.org/10.1146/annurev-arplant-050718-100049.

Cho, K.T., Portwood, J.L., Gardiner, J.M. et al. (2019). MaizeDIG: maize database of images and genomes. *Frontiers in Plant Science* 10: 1050. https://doi.org/10.3389/fpls.2019.01050.

Costa, M.C.D. and Farrant, J.M. (2019). Plant resistance to abiotic stresses. *Plants (Basel)* 8 (12): 553. https://doi.org/10.3390/plants8120553.

Degrave, W., Huyn, C., Roos, D. et al. (2002). *Bioinformatics for Disease Endemic Countries: Opportunities and Challenges in Science and Technology Development for Health*. IUPA.

Deng, J., Shoemaker, R., Xie, B. et al. (2009). Targeted bisulfite sequencing reveals changes in DNA methylation associated with nuclear reprogramming. *Nature Biotechnology* 27 (4): 353–360.

Dong, O.X. and Ronald, P.C. (2019). Genetic engineering for disease resistance in plants: recent progress and future perspectives. *Plant Physiology* 180 (1): 26–38. https://doi.org/10.1104/pp.18.01224.

Dong, A.Y., Wang, Z., Huang, J.J. et al. (2021). Bioinformatic tools support decision-making in plant disease management. *Trends in Plant Science* 26: 953–967.

Down, T.A., Rakyan, V.K., Turner, D.J. et al. (2008). A Bayesian deconvolution strategy for immunoprecipitation-based DNA methylome analysis. *Nature Biotechnology* 26 (7): 779–785.

Egan, A.N., Schlueter, J., and Spooner, D.M. (2012). Applications of next-generation sequencing in plant biology. *American Journal of Botany* 99: 175–185.

Falola, O., Adam, Y., Ajayi, O. et al. (2023, 2023). SysBiolPGWAS: simplifying post-GWAS analysis through the use of computational technologies and integration of diverse omics datasets. *Bioinformatics* 39 (1): btac791. https://doi.org/10.1093/bioinformatics/btac791.

Faure, G., Shmakov, S.A., Makarova, K.S. et al. (2019). Comparative genomics and evolution of trans-activating RNAs in Class 2 CRISPR-Cas systems. *RNA Biology* 16 (4): 435–448. https://doi.org/10.1080/15476286.2018.1493331.

Gaurav, K., Arora, S., Silva, P. et al. (2021). Population genomic analysis of *Aegilops tauschii* identifies targets for bread wheat improvement. *Nature Biotechnology* 40: 422–431.

Gupta, R., Lee, S.E., Agrawal, G.K. et al. (2015). Understanding the plant-pathogen interactions in the context of proteomics-generated apoplastic proteins inventory. *Frontiers in Plant Science* 6: 352. https://doi.org/10.3389/fpls.2015.00352.

Hirst, M. and Marra, M.A. (2010). Next generation sequencing based approaches to epigenomics. *Briefings in Functional Genomics* 9: 455–465.

Hu, J., Chen, B., Zhao, J. et al. (2022). Genomic selection and genetic architecture of agronomic traits during modern rapeseed breeding. *Nature Genetics* 54: 694–704.

Incili, Ç.Y., Arslan, B., Çelik, E.N.Y. et al. (2022). Comparative bioinformatics analysis and abiotic stress responses of expansin proteins in Cucurbitaceae members: Watermelon and melon. *Protoplasma* 260 (2): 509–527.

Jayaram, B. and Dhingra, P. (2010). *Bioinformatics for a Better Tomorrow*. Hauz Khas, New Delhi: Indian Institute of Technology.

Kent, W.J. (2002). BLAT-The BLAST-like alignment tool. *Genome Research* 12: 656–664.

Kim, T., Lee, S.H., and Kim, J.O. (2022). A novel shape-based plant growth prediction algorithm using deep learning and spatial transformation. *IEEE Access* 10: 37731–37742.

Ko, G., Kim, P.G., Yoon, J. et al. (2018). Bioinformatics workflow system for the analysis of massive sequencing data. *BMC Bioinformatics* 19: 43.

Kumar, A. and Chordia, N. (2017). Role of bioinformatics in biotechnology. *Research & Reviews in BioSciences* 12 (1): 116.

Kushwaha, U.K.S., Deo, I., Jaiswal, J.P., and Prasad, B. (2017). Role of bioinformatics in crop improvement. *Global Journal of Science Frontier Research: D Agriculture and Veterinary* 17 (1). Double Blind Peer Reviewed International Research Journal Publisher: Global Journals Inc. (USA). Online ISSN: 2249-4626.

Langmead, B., Trapnell, C., Pop, M., and Salzberg, S.L. (2009). Ultrafast and memory-efficient alignment of short DNA sequences to the human genome. *Genome Biology* 10: R25.

Liang, X.H., Shen, W., Sun, H. et al. (2016). Translation efficiency of mRNAs is increased by antisense oligonucleotides targeting upstream open reading frames. *Nature Biotechnology* 34: 875–880.

Mascher, M., Schreiber, M., Scholz, U. et al. (2019). Genebank genomics bridges the gap between the conservation of crop diversity and plant breeding. *Nature Genetics* 51: 1076–1081.

Mochida, K. and Shinozaki, K. (2010). Genomics and bioinformatics resources for crop improvement. *Plant & Cell Physiology* 51: 497–523. https://doi.org/10.1093/pcp/pcq027.

Moradian, A., Kalli, A., Sweredoski, M.J., and Hess, S. (2014). The top-down, middle-down, and bottomup mass spectrometry approaches for characterization of histone variants and their posttranslational modifications. *Proteomics* 14 (4–5): 489–497.

Mu, H., Wang, B., and Yuan, F. (2022). Bioinformatics in plant breeding and research on disease resistance. *Plants* 11: 3118. https://doi.org/10.3390/plants11223118.

Nishad, R., Ahmed, T., Rahman, V.J., and Kareem, A. (2020). Modulation of plant defense system in response to microbial interactions. *Frontiers in Microbiology* 11: 1298. https://doi.org/10.3389/fmicb.2020.01298.

Ojo, O.O. and Maxwell, O. (2010). Incorporating bioinformatics into biological science education in Nigeria: prospects and challenges. *Infection, Genetics and Evolution* https://doi.org/10.1016/j.meegid,11.015.

Oliver, G.R., Hart, S.N., and Klee, E.W. (2015a). Bioinformatics for clinical next generation sequencing. *Clinical Chemistry* 61 (1): 124–135. https://doi.org/10.1373/clinchem.2014.224360.

Oliver, G.R., Hart, S.N., and Klee, E.W. (2015b). Bioinformatics for clinical next generation sequencing. *Clinical Chemistry* 61 (1): 124–135. https://doi.org/10.1373/clinchem.2014.224360.

Olsen, J.V. and Mann, M. (2013). Status of large-scale analysis of post-translational modifications by mass spectrometry. *Molecular & Cellular Proteomics* 12 (12): 3444–3452.

Park, D., Park, S.H., Ban, Y.W. et al. (2017). A bioinformatics approach for identifying transgene insertion sites using whole genome sequencing data. *BMC Bioinformatics* 17: 67. https://doi.org/10.1186/s12896-017-0386-x.

Perveen, S., Qu, M., Chen, F. et al. (2020). Overexpression of maize transcription factor mEmBP-1 increases photosynthesis, biomass, and yield in rice. *Journal of Experimental Botany* 71 (16): 4944–4957. https://doi.org/10.1093/jxb/eraa248.

Qu, M., Essemine, J., Li, M. et al. (2020a). Genome-wide association study unravels LRK1 as a dark respiration regulator in rice (*Oryza sativa* L.). *International Journal of Molecular Sciences* 21: 4930. https://doi.org/10.3390/ijms21144930.

Qu, M., Essemine, J., Xu, J. et al. (2020b). Alterations in stomatal response to fluctuating light increase biomass and yield of rice under drought conditions. *The Plant Journal* 104: 1334–1347. https://doi.org/10.1111/tpj.15004.

Rahman, M., Shaheen, T., Rahman, M. et al. (2016). Bioinformatics: a way forward to explore “plant omics”. In: *Bioinformatics-updated features and applications* (ed. I.Y. Abdurakhmonov), 203–230. InTechOpen https://doi.org/10.5772/64043.

Rao, V.S., Das, S.K., Rao, V.J., and Srinubabu, G. (2008). Recent developments in life sciences research: role of Bioinformatics. *African Journal of Biotechnology* 7: 495–503.

Ray, S. and Satya, P.N. (2014). Next generation sequencing technologies for next generation plant breeding. *Frontiers in Plant Science* 5: 367. https://doi.org/10.3389/fpls.2014.00367.

Rubin, G.M., Yandell, M.D., Wortman, J.R. et al. (2000). Comparative genomics of the eukaryotes. *Science* 287 (5461): 2204–2215.

Schauer, N. and Fernie, A.R. (2006). Plant metabolomics: towards biological function and mechanism. *Trends in Plant Science* 11 (10): 508–516. https://doi.org/10.1016/j.tplants.2006.08.007.

Schumacher, A., Kapranov, P., Kaminsky, Z. et al. (2006). Microarraybased DNA methylation profiling: technology and applications. *Nucleic Acids Research* 34 (2): 528–542.

Sensen, C. (ed.) (2005). *Handbook of Genome Research: Genomics, Proteomics, Metabolomics, Bioinformatics, Ethical and Legal Issues*. Wiley.

Serre, D., Lee, B.H., and Ting, A.H. (2010). MBD-isolated genome sequencing provides a high-throughput and comprehensive survey of DNA methylation in the human genome. *Nucleic Acids Research* 38 (2): 391–399.

Smith, C.A., O’Maille, G., Want, E.J. et al. (2005). METLIN: A metabolite mass spectral database. *Therapeutic Drug Monitoring* 27: 747–751.

Tam, V., Patel, N., Turcotte, M. et al. (2019). Benefits and limitations of genome-wide association studies. *Nature Reviews. Genetics* 20: 467–484. https://doi.org/10.1038/s41576-019-0127-1.

Tautenhahn, R., Patti, G.J., Rinehart, D., and Siuzdak, G. (2012). XCMS Online: a web-based platform to process untargeted metabolomic data. *Analytical Chemistry* 84: 5035–5039.

Tilman, D., Balzer, C., Hill, J., and Befort, B.L. (2011). Global food demand and the sustainable intensification of agriculture. *Proceedings of the National Academy of Sciences of the United States of America* 108: 20260–20264.

Tiwary, B.K. (2015). Next-generation sequencing and assembly of plant genomes. In: *Hand Book of Plant Omics: The Omics of Plant Science* (ed. D. Barh, M.S. Khan, and E. Davies), 53–64. New Delhi: Springer https://doi.org/10.1007/978-81-322-2172-2_1.

Tramontano, A.L. (2009). Bioinformatics. In: *Encyclopedia of Life Sciences (ELS)*. Chichester: John Wiley & Sons, Ltd https://doi.org/10.1002/97804700 15902.a0001900.pub2.

Turner, S.D. (2014). qqman: An R package for visualizing GWAS results using Q-Q and Manhattan plots. *BioRxiv*.

Varshney, R.K., Thudi, M., Pandey, M.K. et al. (2018). Accelerating genetic gains in legumes for the development of prosperous smallholder agriculture: integrating genomics, phenotyping, systems modelling and agronomy. *Journal of Experimental Botany* 69: 3293–3312. https://doi.org/10.1093/jxb/ery088.

Vlk, D. and Řepková, J. (2017). Application of next-generation sequencing in plant breeding. *Czech Journal of Genetics and Plant Breeding* 53: 89–96. https://doi.org/10.17221/192/2016-CJGPB.

Von Arnim, A.G., Jia, Q., and Vaughn, J.N. (2014). Regulation of plant translation by upstream open reading frames. *Plant Science* 214: 1–12.

Weber, M., Davies, J.J., Wittig, D. et al. (2005). Chromosome-wide and promoter-specific analyses identify sites of differential DNA methylation in normal and transformed human cells. *Nature Genetics* 37 (8): 853–862.

Xiong, J. (2009). *Essential Bioinformatics: Introduction and Biological Databases*. USA: Cambridge University Press http://www.cambidge.org/catalogue.

Yang, H.H. and Lee, M.P. (2004). Application of bioinformatics in cancer epigenetics. *Annals of the New York Academy of Sciences* 1020: 67–76.

Yang, J.Y., Yang, M.Q., Zhu, M.M. et al. (2008). Promoting synergistic research and education in genomics and bioinformatics. *BMC Genomics* 9 (1): I1.

Yu, G., Wang, L.G., Han, Y., and He, Q.Y. (2012). ClusterProfiler: an R package for comparing biological themes among gene clusters. *OMICS* 16: 284–287. https://doi.org/10.1089/omi.2011.0118.

Zhou, X. and Stephens, M. (2012). Genome-wide efficient mixed-model analysis for association studies. *Nature Genetics* 44: 821–824.

2

Bioinformatics for Molecular Breeding and Enhanced Crop Performance: Applications and Perspectives

Rahul Lahu Chavhan[1], *Vidya Ramesh Hinge*[1], *Dipti Jayvantrao Wankhade*[1], *Abhijeet Subhash Deshmukh*[1], *Nagrani Mahajan*[1], *and Ulhas Sopanrao Kadam*[2]

[1] *Vilasrao Deshmukh College of Agricultural Biotechnology, Vasantrao Naik Marathwada Krishi Vidyapeeth, Latur, Maharashtra, India*
[2] *Division of Life Science and Division of Applied Life Science (BK21 Four), Plant Molecular Biology and Biotechnology Research Center, Gyeongsang National University, Jinju, Gyeongnam-do, Republic of Korea*

2.1 Introduction

2.1.1 Bioinformatics and Its Application in Plant Breeding

In recent decades, the term 'bioinformatics' has gained significant recognition within various realms of biological science research. As advancements in biology continue to unfold, the explosive growth in biological data has necessitated a more systematic, computerized approach for collecting, storing, organizing, managing, and analyzing this vast amount of information generated in experiments across diverse fields (Tan et al. 2022). Bioinformatics, emerging as an interdisciplinary field, provides essential tools and techniques for efficiently organizing and categorizing biological data into databases (Lesk 2019).

According to its definition, bioinformatics is an interdisciplinary field that combines computer science, mathematics, statistics, and biology to analyze and interpret biological data, especially large-scale data produced by numerous biological studies and experiments. The main goal of this approach is to derive significant patterns, insights, and information from biological data, which includes gene expression profiles, protein structures, DNA sequences, and more. Bioinformatics comprises several components, including – first, the collection, organization, and storage of large data in databases; next, the development of software tools and algorithms for interpreting biological data; and finally, the utilization of biological data and software to interpret the results (Baxevanis 2004). Bioinformatics has played a vital role in various biological domains, offering a wide range of data types, including nucleotide and amino acid sequences, protein domains and structures, and expression patterns from many organisms. Bioinformatics plays a pivotal role in various fields of biology, such as genetics, genomics, proteomics, drug discovery, evolutionary biology, and personalized medicine. It empowers researchers to make sense of the vast amounts of biological

Bioinformatics for Plant Research and Crop Breeding, First Edition. Edited by Jen-Tsung Chen.

data generated in modern biology and contributes to advancements in our understanding of life sciences (Nikalje et al. 2023, Netla et al. 2023; and Shinde et al. 2023a,b).

2.1.2 Applications of Bioinformatics in Plant Breeding

Crop improvement through various plant breeding strategies is based on repeated rounds of phenotypic selection (PS) and crossbreeding, leading to enhanced genotypes via genetic recombination (Osei et al. 2014). Currently, a vast number of crop genome sequences are available and accessible, enabling every gene and genetic variation to be identified, contributing to agronomic traits. Genome sequencing of crop populations provides an in-depth understanding of agronomic variation at the gene level. Genomics-based breeding advancements have allowed the identification of genetic variations in crop species, making genomics a prominent tool in crop breeding, including mapping quantitative trait loci (QTL) and conducting genome-wide association studies (GWAS) (Mohammadi et al. 2020). Genomic selection (GS) is an alternate strategy that uses genetic polymorphisms across the genome to eliminate the requirement for repetitive phenotyping in breeding cycles (Rahim et al. 2020).

Processing and analyzing massive genomic datasets with the help of bioinformatics is essential for gaining a functional understanding of plant genomes. A cross-disciplinary approach is required for 21st-century plant breeding to address breeding challenges and improve crop productivity effectively. Genomic and bioinformatics are critical in improving crop cultivar production when combined with innovative greenhouse technologies that accelerate plant development (Mu et al. 2022). However, integrating various data outputs for breeding is significantly hampered by the enormous amounts of genotypic and phenotypic data. To overcome this challenge and meet breeding targets, it is essential to integrate phenotypes, genomics, and bioinformatics tools and resources into breeding pipelines (Hu et al. 2018). This chapter will delve into the extensive applications of bioinformatics in plant breeding in more detail.

2.2 Data Management and Integration

The expanding realm of bioinformatics hinges entirely on effectively managing the vast amount of data generated through NGS platforms and other high-end experiments like microarray, transcriptome profiling, and mass spectrometry (MS). This substantial data is systematically stored in various databases and harnessed by researchers.

2.2.1 Data Storage and Retrieval

A database is a computerized, well-structured collection of interconnected data, offering a consistent method for searching, adding, and updating information. These databases house organized, enduring data. A database management system (DBMS) facilitates the systematic analysis and retrieval of data by interacting with users, other programs, and the database itself. Databases dedicated to biological information are commonly known as bioinformatics databases. They encompass data gathered from published literature, computational analysis, and scientific experimentation. Biological databases can be categorized based on several criteria, including the type of data they store (raw data: primary and processed data: secondary; sequence, structure, nucleotide, protein, pathway, metabolite, taxonomic, image, restriction enzyme, literature, etc.), the purpose they serve, and the organism(s) they focus on (Mahmud et al. 2021).

Databases can be classified into four categories based on their structure: text file, flat file, object-oriented, and relational databases. They can also be grouped into general and specialized databases based on the type of information they contain and typically categorized according to the kind of data they house in primary, secondary, and composite databases; see Table 2.1 for a detailed list (Kumar and Dudley 2007).

Table 2.1 Genomic selection methods.

Sr. No.	Genomic Selection Method and Models	Features	References
1.	Genomic best linear unbiased prediction (GBLUP)	It estimates the genetic value of an individual as a linear combination of marker effects, assuming a multivariate normal distribution. It is computationally efficient and easy to implement	Zhang et al. (2015)
2.	Bayesian techniques with various basis functions, such as the Bayesian Lasso, Bayesian Ridge Regression, Bayesian Sparse Linear Mixed Model, and Markov Chain Monte Carlo techniques	These methods utilize Bayesian statistics to estimate marker effects and predict genetic values. These methods incorporate prior knowledge and can handle complex genetic architectures	Tibshirani (1996), Friedman et al. (2010)
3.	Deep learning (DL) methods/machine learning (ML) approach	These techniques include neural networks, which can be modified for genomic selection, convolutional neural networks (CNNs), and recurrent neural networks (RNNs). These models enhance prediction accuracy by capturing complex patterns in genomic data	Montesinos-López et al. (2021)
4.	RR-BLUP: Ridge regression–best linear unbiased prediction	stabilizes the estimation of marker effects by adding a penalty to the squared size of the coefficients, mainly when multicollinearity is present. It helps raise the accuracy of predictions	Wang et al. (2018)
5.	Genomic-Enabled Prediction Using Kernel Methods e.g. Reproducing kernel Hilbert space (RKHS)	Kernel methods use nonlinear transformations of the marker data to capture complex relationships between markers and phenotypes	Wang et al. (2018), Neves et al. (2012)
6.	Support vector machine	SVM is a supervised learning technique that is nonparametric and utilized in regression analysis and classification. SVM considers the fitting and complexity of training samples while adhering to the structural risk minimization principle	Wang et al. (2018), Maenhout et al. (2007)

2.2.2 Data Integration and Analysis

Data integration and analysis in bioinformatics involve the amalgamation and processing of diverse biological data to glean meaningful insights and make informed decisions (Morbia et al. 2023). Data integration revolves around the fusion of various types or sources of biological data to construct a unified and comprehensive perspective. This integration proves invaluable for studying biological systems holistically and uncovering hidden information that might remain concealed when examining individual datasets. Key elements of data integration encompass:

1) **Data cleaning and preprocessing:** Eliminating errors, inconsistencies, and noise from raw data, standardizing formats, and ensuring data quality.
2) **Data fusion:** Combining information from various sources, such as clinical data, proteomics, metabolomics, and genomics, to provide a thorough understanding of biological systems.
3) **Normalization:** Adjusting data to account for measurement variations and biases, ensuring accurate comparisons across different datasets.
4) **Ontology and metadata integration:** Applying controlled vocabularies and metadata to standardize data representation and enhance compatibility and integration across diverse datasets.
5) **Semantic integration:** Incorporating semantic relationships and mappings between different data types to enable more meaningful analysis.

Bioinformatics data analysis revolves around extracting meaningful patterns, rclationships, and insights from integrated biological data. It employs an array of statistical, computational, and machine-learning techniques to decipher complex biological information (Wang et al. 2019b). The steps in data analysis encompass:

1) **Exploratory data analysis (EDA):** Summarizing, visualizing, and comprehending the general characteristics and trends within the data to guide further analysis.
2) **Statistical analysis:** Applying statistical methods to identify significant differences, correlations, and patterns within the data.
3) **Machine learning and predictive modeling:** Training machine learning (ML) models to predict outcomes, classify biological samples, or identify patterns and relationships within the data.
4) **Clustering and pattern recognition:** Grouping similar data points to identify clusters or patterns within the data, unveiling biological subtypes or relationships.
5) **Pathway and network analysis:** Analyzing biological pathways and networks to understand the interconnections between genes, proteins, and metabolites.
6) **Functional enrichment analysis:** Determining the biological functions and processes significantly enriched within a set of genes or proteins.
7) **Comparative genomics:** Comparing genomes across species to comprehend evolutionary relationships and pinpoint conserved elements or variations.
8) **Time-series analysis:** Analyzing data collected over time to explore dynamic processes, such as gene expression changes.
9) **Association and correlation analysis:** Investigating relationships between different biological variables to identify potential associations or dependencies.

Data integration and analysis in bioinformatics are iterative processes often requiring interdisciplinary collaboration among biologists, statisticians, computer scientists, and domain experts. These processes are paramount for extracting meaningful insights from the available biological data and advancing our comprehension of intricate biological systems.

2.2.3 Data Visualization and Interpretation

Data visualization and interpretation in bioinformatics involve transforming complex biological data into visual representations that are easily understandable and meaningful. Effective visualization facilitates the identification of patterns, trends, and relationships within the data, aiding in informed decision-making and hypothesis generation. The data visualization and interpretation techniques in bioinformatics are involved.

1) **Graphs and charts:** Bar charts, line plots, scatter plots, histograms, and pie charts to represent numerical and categorical data.
2) **Heatmaps:** Color-coded matrices to display relationships and patterns, commonly used for gene expression analysis or pathway visualization.
3) **Networks and graphs:** Representing biological interactions, pathways, and relationships using nodes and edges to visualize complex systems.
4) **Circos plots:** Circular visualization to display relationships between genomic features such as genes, mutations, and chromosomal positions.
5) **Geographical maps:** Visualization of spatial data, helpful in analyzing disease distribution or population genetics.
6) **3D Visualization:** Representing molecular structures or complex biological landscapes in three dimensions.
7) **Parallel coordinates:** Displaying multidimensional data, enabling the visualization of relationships between multiple variables.
8) **Venn diagrams:** Overlapping circles to illustrate the intersections and differences between data sets.

The commonly used Visualization Tools and Software includes

1) **Tableau:** A versatile tool for creating interactive and dynamic visualizations from various data sources.
2) **R (ggplot2, Shiny):** A powerful statistical computing language with packages like ggplot2 for creating customizable plots and Shiny for building interactive web applications.
3) **Python (Matplotlib, Seaborn, Plotly):** Widely used for data visualization, offering various libraries for generating different types of plots.
4) **D3.js:** A JavaScript library for creating interactive and dynamic visualizations for the web.
5) **Cytoscape:** A tool for visualizing and analyzing biological networks.
6) **Jupyter notebooks:** Allowing integration of code, visualizations, and narrative text to create shareable documents.

The data Guiding Principles for Effective Data Visualization are

1) **Clarity and simplicity:** Keep visualizations clear, simple, and easy to interpret by avoiding unnecessary complexity.

2) **Use of color:** Utilize color strategically to convey information, highlight trends, and differentiate data categories.
3) **Interactivity:** Incorporate interactive elements to allow users to explore the data and gain deeper insights.
4) **Annotations and labels:** Provide appropriate labels, titles, and annotations to convey the context and meaning of the visualizations.
5) **Contextualization:** Present visualizations about the biological question or hypothesis being investigated.

Steps Involved in Data Interpretation

1) **Pattern recognition:** Identify recurring patterns and trends in the data that could provide biological insights.
2) **Correlation and association analysis:** Analyze relationships between variables to understand potential connections or dependencies.
3) **Comparative analysis:** Compare datasets or biological entities to discover differences, similarities, or evolutionary relationships.
4) **Biological context integration:** Interpret the visualized data in the context of known biological knowledge and hypotheses.
5) **Hypothesis generation:** Formulate new hypotheses based on the observed patterns, which can guide further experiments and investigations.

Data visualization and interpretation are crucial steps in the bioinformatics workflow, enabling researchers to effectively explore and communicate complex biological data, ultimately contributing to a deeper understanding of biological systems and phenomena.

2.3 Genomic Resources for Plant Breeding

Genomic resources are a diverse array of tools, databases, technologies, and information crucial for enhancing the efficiency and precision of plant breeding endeavors. These invaluable resources empower breeders to gain deeper insights into the genetic foundations of traits, expedite breeding programs, and cultivate superior crop varieties with desired attributes. The advancement of breeding programs for sustainable farming is largely dependent on genomic resources, which make it possible to develop high-nutrition crop varieties, tolerant of biotic stress, and resistant to abiotic stress (Chavhan et al. 2017, 2018, 2023; Mahalle, et al. 2023; Hinge et al. 2021).

2.3.1 Plant Genome Sequencing and Annotation

Plant genome sequencing and annotation represent pivotal processes in modern biology, involving the decryption of the entire genetic information (genome) of a plant species and the subsequent annotation of genomic elements to comprehend their functions. Currently, around 600 complete plant genome sequences are available in public repositories (Kersey 2019). Recent advancements in sequencing technologies are consistently improving the quality of genomic assemblies. The largest plant genome sequence available to date is a gigantic 28 gigabytes in size (Danilevicz et al. 2020). This burgeoning wealth of genome sequencing data is significantly impacting breeding, conservation, and identification efforts. Notable plant genome databases include PlantsDB and Phytozome, and a comprehensive list of whole plant genome sequences can be accessed from the genome databases (Table 2.2). These resources also encompass databases for repetitive elements within plant genomes.

The guidelines for plant genome sequencing and annotation are rigorous and encompass a multifaceted process that relies on various tools and databases (Jung et al. 2020). The procedure for genome sequencing and annotation (Figure 2.1). Due to highly repetitive sequences, high heterozygosity, and polyploid genomes, the process of plant genome sequencing and annotation is inherently complex (Kong et al. 2023). However, an array of bioinformatics tools is constantly evolving and being updated to assemble raw data and generate long reads during plant genome sequencing. Notable tools in this realm include Falcon, miniasm, Flye, Hinge, CANU, wtdbg, Shasta, Wengan, HiFi reads, HiCanu, hifiasm, Hi-C reads, LACHESIS, 3D-DNA, and ALLHiC, among others (Zhou et al. 2022).

Plant breeders gain access to fully assembled genome sequences only following their structural and functional annotation. Structural annotation is crucial in unraveling gene function and its evolution across various taxa, relying on precise gene models derived from computational programs and experimental evidence. Functional annotation, however, depends on the sequence's similarity to known genes or proteins, enhancing our understanding of gene function on a genomic scale. The synergy of these two annotation types results in accurate annotation and a deeper comprehension of gene and genome evolution across diverse plant species. As the availability of plant genome data continues to rise, the significance of comparative annotation is amplified, and ongoing advancements in methodologies will further refine genome annotation in the future (Bolger et al. 2018). The comprehensive list of bioinformatics tools and software used for functional genome annotation is given in Table 2.2.

2.3.2 Transcriptome Analysis and Gene Expression Profiling

Transcriptome analysis and gene expression profiling in plant genomes are indispensable tools for gaining a molecular-level understanding of a plant's functional elements and processes. These techniques offer critical insights into genes' regulation, expression, and interactions in response to various conditions and stimuli. They involve characterizing transcriptional activity, encompassing both coding and noncoding components, and simultaneously profiling thousands of genes and transcripts. This approach deepens our comprehension of cellular function and can uncover broad, coordinated trends in gene expression that may remain hidden in more targeted assays (Lowe et al. 2017).

A snapshot of the genes and transcripts that are actively expressed is provided by gene expression profiling, which measures the amounts of gene expression under various circumstances. It offers a means to study gene regulation, gene function, and the impact of various stimuli on gene expression. Various methods are available for gene expression profiling, including gene expression microarrays, quantitative real-time polymerase chain reaction (qRT-PCR), and next-generation sequencing (NGS)-based RNA sequencing (RNA-Seq) (Lowe et al. 2017).

Table 2.2 List bioinformatics Databases and Tools used in Genomics, proteomics, and transcriptomics.

Database/Software Tool	Details	Web Link
Genome Databases		
1. GenBank	The National Center for Biotechnology Information (NCBI) manages a comprehensive database of genetic sequences, including DNA, RNA, and protein sequences	http://www.ncbi.nlm.nih.gov/genbank/
2. Ensembl Plants	Genomic database of 133 plant species (genome sequence, gene models, functional annotation)	https://plants.ensembl.org/index.html
3. NCBI Genome Database	A collection of genomic sequences, maps, domains, and functional annotations maintained by NCBI	https://www.ncbi.nlm.nih.gov/genome/
4. DNA Data Bank of Japan (DDBJ)	The Bioinformation and DDBJ Center facilitates data exchange and analysis for life science researchers and advances science	https://www.ddbj.nig.ac.jp/index-e.html
5. European Molecular Biology Laboratory (EMBL) nucleotide sequence database	a comprehensive database of nucleotide sequencing information from around the world, including raw sequencing data, sequence assembly information, and functional annotation	https://www.ebi.ac.uk/ena/browser/home
6. Genome Size in Asteraceae Database (GSAD)	GSAD is a comprehensive catalog of genome size data for the Asteraceae family. The genome sizes of 1,555 species are now available	https://www.asteraceaegenomesize.com/
7. ArrayExpress	Functional genomics data from microarray and sequencing platforms are archived here	www.ebi.ac.uk/arrayexpress
8. International Nucleotide Sequence Database Collaboration (INSDC)	The INSDC is a long-running foundational initiative run by DDBJ, EMBL-EBI, and NCBI INSDC includes the entire data pipeline, from raw reads to alignments and assemblies to functional annotation, and is enhanced with contextual information about samples and experimental configurations	https://www.insdc.org/
9. UniProt	A comprehensive resource for protein sequence and functional information, including protein sequence data, protein families, domains, and more	https://www.uniprot.org/
10. Expression Atlas EMBL-EBI	Open resource platform for gene and protein expression information. Search across 66 species, 4424 studies, 158,401 assays	www.ebi.ac.uk/gxa/home
11. Gene Expression Omnibus	Functional genomics data repository	www.ncbi.nlm.nih.gov/projects/geo

12. BarleyGenes	Gene and RNA-seq database for Barley	https://ics.hutton.ac.uk/barleyGenes/
13. Chrysanthemum Transcriptome Database	Database of chrysanthemum's gene and transcripts	http://www.icugi.org/chrysanthemum
14. Cottongen	Genomics, genetics, and breeding database for cotton	https://www.cottongen.org/
15. AutoSNPdb	An annotated single nucleotide polymorphism database for crop plants	http://autosnpdb.appliedbioinformatics.com.au/
16. *Brassica* genome gateway	This site hosts Brassica genome databases.	http://www.brassicagenome.net
17. *Brassica rapa* genome database Brassica Database (BRAD)	Newly released genome sequences of Brassiceae species, genomic data of most other Brassicaceae	http://brassicadb.cn/
18. Graingenes	GrainGenes is an online platform that provides a curated and peer-reviewed central repository for data relevant to small grains research communities	http://wheat.pw.usda.gov/
19. Gramene	Comparative plant genomics for crops and model organisms	http://www.gramene.org/
20. International Crop Information System (ICIS)	Applications for storing and analyzing phenotypic and pedigree data are included in the International Crop Information System, an open-source project being developed by a global community of crop researchers	http://www.icis.cgiar.org
21. Legume Information System (LIS)	LIS combines research techniques with genetic and genomic data from numerous legume species	https://www.legumeinfo.org/
22. MaizeDIG	Genotypic-phenotypic database for maize	https://maizedig.maizegdb.org/
23. MaizeGDB	Genetic and genomics database for maize	https://www.maizegdb.org/
24. MaizeMine	Archive to access literature, genomic, interaction, and proteomic data for maize	https://maizemine.rnet.missouri.edu/maizemine/begin.domine/begin.do
25. Maize sequence database	Genomic sequence information of maize	http://www.maizesequence.org/
26. Oryzabase	The goal of the comprehensive rice science database was to compile as much information as possible on everything from contemporary genomics to classical rice genetics, as well as on basic research and trending themes. The goal of the comprehensive rice science database was to compile as much information as possible on everything from contemporary genomics to classical rice genetics, as well as on basic research and trending themes	http://www.shigen.nig.ac.jp/rice/oryzabase/

(Continued)

Table 2.2 (Continued)

Database/Software Tool	Details	Web Link
27. Panzea	"Biology of Rare Alleles in Maize and its Wild Relatives" was the name of the NSF-funded project Panzea	http://www.panzea.org/
28. Phytozome	Phytozome enables precise and informative comparative genomics research by integrating a sizable number of plant genomes into a single resource and carrying out thorough and consistent annotation and analysis	http://www.phytozome.net/
29. PlantsDB	a database framework for comparative plant genome research	http://mips.helmholtz-muenchen.de/plant/genomes.jsp
30. PlantGDB	A resource for plant comparative genomics	http://www.plantgdb.org/
31. *PTGBase*	An integrated database for the study of plant tandem repeats. 39 plant species are currently represented in 54,130 tandem repeat gene clusters with 129,652 genes in PTGBase	https://ngdc.cncb.ac.cn/databasecommons/database/id/912
32. The Plant rDNA Database	An online database called the Plant rDNA database contains details on the numbers, locations, and architectures of ribosomal DNA signals for 2770 different plant species (4948 entries)	https://www.plantrdnadatabase.com/
33. The Plant Ontology	promote consistent usage of plant-specific terminology in annotation and enable cross-database querying	http://www.plantontology.org/
34. Plaza	an access point for plant comparative genomics, centralizing genomic data produced by different genome sequencing initiatives	http://bioinformatics.psb.ugent.be/plaza/
35. Rice Genome Annotation Project	data on rice genome sequencing and annotations is provided by a National Science Foundation initiative	http://rice.plantbiology.msu.edu/
36. The applied bioinformatics group	The examination of genetic and genomic diversity, the development of molecular genetic markers, and second-generation sequencing and assembly	http://appliedbioinformatics.com.au/projects/ssrtaxonomy/php/
37. SOL Genomics Network (SGN)	biological information about Solanaceae species and their close relatives is available in the Clade-oriented database (COD), which includes information on chromosomes, genes, phenotypes, and accessions	http://solgenomics.net/

38. SoyBase	Soybean Genetics and Genomics Database, USDA-ARS. For soybeans, Glycine max, and allied species, genetic and genomic data are available in this database	http://soybase.org/
39. TAGdb	This allows researchers to find matched read sequences that share identity with a query sequence that has been submitted These tags can then be utilized to create oligonucleotide primers for PCR amplification of the target genomic region	http://flora.acpfg.com.au/tagdb/
40. CR-EST The Crop Expressed Sequence Tag database, CR-EST	CR-EST is a publicly accessible online resource that provides access to crop EST project sequencing, classification, clustering, and annotation data at the IPK	http://pgrc.ipk-gatersleben.de/cr-est/
41. The Triticeae Repeat Sequence Database (TREP)	Database of Triticeae (wheat, barley, and close relatives) genomic repeat sequences	http://wheat.pw.usda.gov/ITMI/Repeats/
42. VvGDB	Database for grape genome	www.plantgdb.org/VvGDB/
43. Plant Promoter Database	databases of plant promoter sequences and transcription start sites of different plant species identified by experimentation	https://ppdb.agr.gifu-u.ac.jp/ppdb/cgi-bin/index.cgi
44. PLEXdb	Database for plant expression that includes phenotypic, structural genomics, and gene expression profile data sets	http://www.plantgdb.org/prj/PLEXdb/
45. Pomamo	Archive of potato sequences, literature, maps, and tools	https://www.gabipd.org/projects/Pomamo/
46. PRGdb	Archive of pathogen receptor genes of various plant species	http://prgdb.org/prgdb/
47. Rice Expression Database	Repository of gene expression profiles of rice from RNA-seq data on tissues spanning an entire range of rice growth	http://expression.ic4r.org/
48. Rice expression profile Database	Repository of gene expression profiles of rice from microarray analysis	http://ricexpro.dna.affrc.go.jp/
49. SolGenomics Network	Database and tool of genomics and genetics approach for tomato and some other plant species such as eggplant	https://solgenomics.net/gem/experimental_design.pl?id=2
50. TAIR	Database of genetic and molecular biology data for *Arabidopsis thaliana*	https://www.arabidopsis.org/
51. The Rice Annotation Project Database	Archive of rice genome sequence, structure, and function	https://rapdb.dna.affrc.go.jp/
52. Tomato Functional Genomics Database	Large-scale tomato functional genomics data sets can be stored, queried, mined, analyzed, visualized, and integrated using this comprehensive resource	http://ted.bti.cornell.edu

(Continued)

Table 2.2 (Continued)

Database/Software Tool	Details	Web Link
Metabolomics Database		
1. Metabolights	www.ebi.ac.uk/metabolights/	Metabolomics database (cross-species, metabolite structures, biological role)
Pathway Database		
1. Reactome	Pathway interaction repository	www.reactome.org
2. RiceCyc	Database of known and predicted biochemical pathways from rice	http://pathway.gramene.org/gramene/ricecyc.shtml
3. PlantCyc	Metabolic pathway reference database from over 350 plant species	http://plantcyc.org
4. WikiPathways	WikiPathways is an open scientific platform for biological pathways that are submitted, maintained, and used by the research community	https://www.wikipathways.org/
5. MaizeCyc	Containing enzyme catalysts, proteins, and other metabolites	https://maizecyc.maizegdb.org/
6. MapMan	Large datasets (such as gene expression data from Arabidopsis Affymetrix arrays) can be displayed on metabolic pathways or other process diagrams using the user-driven MapMan application	https://mapman.gabipd.org/mapman
7. BIGG	BiGG Models is a knowledgebase that reconstructs genome-scale metabolic networks. BiGG Models combines over 70 previously published genome-scale metabolic networks into a single database using a set of standardized identifiers known as BiGG IDs	http://bigg.ucsd.edu/
8. ModelSEED	ModelSEED is a resource for the reconstruction, exploration, comparison, and analysis of metabolic models	https://modelseed.org/
9. MetRxn	MetRxn is a knowledgebase that integrates information from BRENDA, KEGG, MetaCyc, Reactome.org, and 44 metabolic models into a single unified data source	http://metrxn.che.psu.edu.

10. UM-BBD	The goal of UM-BBD is to give information about microbial enzyme-catalyzed biotechnology reactions. The reactions examined are for basic nature knowledge, biocatalysis leading to specialty chemical manufacturing, and biodegradation of environmental contaminants. Individual processes and metabolic pathways are described, together with information on the beginning and intermediate chemical components, organisms that transform the molecules, enzymes, and genes	http://www.pdg.cnb.uam.es/cursos/Leon2002/pages/software/DatabasesListNAR2002/summary/117.html
11. Rhea	Enzyme and transporter annotation in UniProtKB is done according to the standards set by Rhea, an expert-curated knowledgebase of chemical and transport reactions of biological significance. Reaction participants are described by Rhea using the chemical dictionary ChEBI (Chemical Entities of Biological Interest)	https://www.rhea-db.org/
12. BRENDA	Comprehensive enzymatic and metabolic data that has been retrieved, updated, and assessed from primary literature is available in the BRENDA protein function database	https://www.brenda-enzymes.org/
13. GeneCards: The Human Gene Database	All annotated and predicted human genes are covered in detail and are easily navigable with GeneCards, an integrative database. Gene-centric data, comprising genomic, transcriptomic, proteomic, genetic, clinical, and functional information, is automatically integrated by the knowledgebase from around 150 web sources	https://www.genecards.org/cgi-bin/carddisp.pl?gene=MOS
14. Beilstein/crossfire	Information about compounds, reactions, structures, and properties can be found in the database	https://www-jmg.ch.cam.ac.uk/cil/beilstein.html
15. ATLAS of Biochemistry	A comprehensive database of all potential biochemical processes for research purposes in metabolic engineering and synthetic biology	https://lcsb-databases.epfl.ch/pathways/atlas/
16. MINE	Compared to generic chemical databases, which yield synthetic substances that are not relevant, MINEs enhance metabolomics peak identification	https://minedatabase.mcs.anl.gov/#/home
17. Chemical Translation Service	A web-based application to help standardize metabolomic reports is the Chemical Translation Service. In brief, databases and publications related to metabolism have distinct database identifiers or even simple names, which prevent searches from operating on the same database or study simultaneously	http://cts.fiehnlab.ucdavis.edu

(Continued)

Table 2.2 (Continued)

Database/Software Tool	Details	Web Link
18. EcoCyc	EcoCyc is a research database that contains information about the bacterium Escherichia coli K-12 MG1655. The EcoCyc project curates its genome, as well as transcriptional regulation, transporters, and metabolic pathways, using literature.	https://ecocyc.org/
19. AraCyc	AraCyc is a tool for visualizing Arabidopsis thaliana biochemical pathways	https://www.arabidopsis.org/biocyc/
20. PlantPathMarks database (PPMdb)	PPMdb, or the PlantPathMarks database, is an extensive, web-based, interactive, and user-friendly center for pathway-based markers found in plant genomes	http://ppmdb.easyomics.org/
21. Plant MetGenMAP	A visualization and analysis software called Plant MetGenMAP makes it possible to see the profile data in the context of biochemical pathways and to find altered pathways and enriched GO keywords from gene expression and/or metabolite profile data	http://bioinfo.bti.cornell.edu/cgi-bin/MetGenMAP/home.cgi
RNA Analysis Tool		
1. qTeller	Comparative RNA-seq expression tool	https://qteller.maizegdb.org/
Chemical Compound Database		
1. hEBI	Archive of molecular entities that focuses on small chemical compounds	www.ebi.ac.uk/chebi/
2. PubChem	Archive of chemical information of various chemical compounds	http://pubchem.ncbi.nlm.nih.gov/
Mass Spectrum Database		
3. ReSpect for Phytochemicals	Database for phytochemicals MS spectra data and literature	http://spectra.psc.riken.jp/menta.cgi/index
4. Metlin	Comprehensive MS/MS database	http://metlin.scripps.edu/

Resistance Analysis		
1. Disease Resistance Analysis and Gene Orthology (DRAGO2)	Annotate resistance genes	http://prgdb.org/prgdb/drago2
Networking and Interaction Analysis		
1. PathoPlant	Database of plant-pathogen interactions and pathogenicity-related elements	www.pathoplant.de/
2. AraNet	Tool for estimating probabilistic functional gene networks in Arabidopsis thaliana	www.inetbio.org/aranet/
3. PLANET	An analytical and visual platform for the co-function networks of photosynthetic organisms	http://aranet.mpimp-golm.mpg.de/
Posttranslational Modification Prediction Tools		
1. PHOSFER	PHOSFER predicts the phosphorylation sites in soybean proteins by employing a revolutionary machine-learning technique	https://bio.tools/phosfer
2. PhosPhAt	An Arabidopsis thaliana phosphorylation site database and a phosphorylation site predictor exclusive to that plant	https://phosphat.uni-hohenheim.de/
3.PlantPhos	Finding substrate-specific phosphorylation sites in plants by maximal dependence decomposition	https://bio.tools/plantphos
4. Musite	Musite is a stand-alone Java application that predicts the phosphorylation sites of proteins both generally and kinase-specifically	https://www.musite.net/
Software and Tools in Plant Bioinformatics		
1. The Bio-Analytic Resource	Plant bioinformatics tool repository (gene expression, mapping, molecular markers, and genomics)	http://bar.utoronto.ca/
2. SSR Primer	An online tool that makes it possible to find SSRs in submitted DNA sequences in real time	http://flora.acpfg.com.au/ssrprimer2/
3. BLAST (Basic Local Alignment Search Tool)	A crucial tool for finding homologous regions and comparing sequences when searching for sequence similarity	https://blast.ncbi.nlm.nih.gov/

(*Continued*)

Table 2.2 (Continued)

Database/Software Tool	Details	Web Link
4. Bioconductor	An open-source software that focuses on the R programming language for the analysis and understanding of genomic data	https://www.bioconductor.org/
5. IGV (Integrative Genomics Viewer)	A high-performance visualization tool for interactive exploration of large, integrated genomic datasets	https://igv.org/
6. Cufflinks and Cuffdiff	Tools for assembling transcriptomes, measuring gene expression, and analyzing differential expression with RNA-seq data	https://pubmed.ncbi.nlm.nih.gov/26519415/
7. Variant Effect Predictor (VEP):	Annotates genetic variants and predicts their effects on genes, proteins, and regulatory regions	https://www.ensembl.org/vep#:~:text=VEP%20determines%20the%20effect%20of,Transcripts%20affected%20by%20the%20variants
8. GATK (Genome Analysis Toolkit)	Software designed to analyze high-throughput sequencing data, especially for genotyping and variant discovery	https://gatk.broadinstitute.org/hc/en-us
9. Bowtie and Bowtie2	Tools for quickly and efficiently aligning sequencing reads to a reference genome	https://bowtie-bio.sourceforge.net/bowtie2/manual.shtml#:~:text=The%20chief%20differences%20between%20Bowtie,faster%20and%2For%20more%20sensitive.
10. STRING (Search Tool for the Retrieval of InteractingGenes/Proteins)	protein–protein interaction prediction database and tool that helps in elucidating the functional connections between genes and proteins	https://www.denbi.de/services/276-string-search-tool-for-the-retrieval-of-interacting-genes-proteins#:~:text=STRING%20(Search%20Tool%20for%20the,9%2C643%2C763%20proteins%20from%202%2C031%20organisms.
11. SLAM	Finding, aligning, and annotating genes (human-mouse homology identification)	http://bio.math.berkeley.edu/slam
12. Mauve	Multiple alignments of rearranged genomes	https://darlinglab.org/mauve/mauve.html

13. MGA	Multiple Genome Aligner	https://www.ensembl.org/info/genome/compara/multiple_genome_alignments.html#:~:text=Multiple%20alignments%20are%20calculated%20between,conservation%20scores%20and%20constrained%20elements.
14. Mulan	Local multiple alignments of genome-length sequences	https://bio.tools/mulan
15. Sequerome	Profiling sequence alignment data with major servers/services	https://www.bioinformatics.org/sequerome/wiki/Main/HomePage
16. AVID	Pairwise global alignment with whole genomes	https://ouci.dntb.gov.ua/en/works/9GE853J4/
17. SIBsim4 / Sim4	Software made to match a genomic sequence to an expressed DNA sequence while allowing for introns	https://www.expasy.org/resources/sim
18. Shuffle-LAGAN	Pairwise global alignment of completed genome regions	http://lagan.stanford.edu/
19. ACT (Artemis Comparison Tool)	Synteny and comparative genomics	http://www.icgeb.res.in/whotdr/reading/ACT.pdf
Protein Analysis Tools		
1. Expasy	Developed by SIB Groups, it is an extensible and integrative portal that offers access to more than 160 databases and software tools supporting a variety of life science and clinical research domains, including systems biology, medical chemistry, evolution and phylogeny, genomics, proteomics, and structural biology	https://www.expasy.org/
2. Clustal w	An extensively used method for aligning any number of homologous protein or nucleotide sequences is called Clustal W	https://www.genome.jp/tools-bin/clustalw
3. InterProScan	InterProScan is the software package that scans sequences against InterPro's member database signatures	http://www.ebi.ac.uk/interpro/search/sequence/
4. PfamScan	Used to search a FASTA sequence against a library of Pfam HMM	https://www.ebi.ac.uk/Tools/pfa/pfamscan/
5. Pratt	PRATT is a tool to discover patterns conserved in protein sequences	https://www.ebi.ac.uk/Tools/pfa/pratt/

(Continued)

Table 2.2 (Continued)

Database/Software Tool	Details	Web Link
6. HMMER3 hmmscan	HMMER is used to create sequence alignments and search sequence databases for homologs. It employs techniques with profile hidden Markov models, which are probabilistic models (profile HMMs)	https://www.ebi.ac.uk/Tools/pfa/hmmer3_hmmscan/
7. Phobius	A combined transmembrane topology and signal peptide predictor	https://www.ebi.ac.uk/Tools/pfa/phobius/
8. RADAR	RADAR (Rapid Automatic Detection and Alignment of Repeats) is a tool to identify complicated repeat structures	https://www.ebi.ac.uk/Tools/pfa/radar/
9. HMMER3 phmmer	Hmmer3 phmmer searches one or more query sequences against a sequence database	https://www.ebi.ac.uk/Tools/pfa/hmmer3_phmmer/
10. RCSB PDB	The worldwide Protein Data Bank (PDB) archive of 3D structure data for large biological molecules (proteins, DNA, and RNA) is housed at RCSB PDB (RCSB.org), the US data center for this archive. This data is crucial for basic biology research as well as health, energy, and biotechnology education	https://www.rcsb.org/
11. MMDB	Also known as the Entrez Structure database, the Molecular Modeling DataBase (MMDB) is a database of three-dimensional biomolecular structures determined through experimentation	https://www.ncbi.nlm.nih.gov/Structure/MMDB/mmdb.shtml
12. SWISS-MODEL	This fully automated protein structure homology modeling server can be accessed through the Expasy web server or the DeepView application (Swiss PDB-Viewer)	https://swissmodel.expasy.org/
13. I-TASSER	A hierarchical method called I-TASSER (Iterative Threading ASSEmbly Refinement) is used for structure-based function annotation and protein structure prediction	https://zhanggroup.org/I-TASSER/
14. BLAST	An essential tool for sequence similarity searching, it is used to compare sequences and identify homologous regions	https://blast.ncbi.nlm.nih.gov/Blast.cgi
15. CDART	Based on domain architecture – the sequential arrangement of conserved domains in proteins – the Conserved Domain Architecture Retrieval Tool (CDART) searches the NCEI Entrez Protein Database for similarities	https://www.ncbi.nlm.nih.gov/Structure/lexington/lexington.cgi

16. PFAM	The Pfam database contains evolutionary units that have been conserved. It can be used to investigate a protein's domain(s) and other functional areas, providing information about the protein's role	http://pfam.xfam.org/
17. TIGRFAM	TIGRFAMs is a carefully curated library of protein families with a primary emphasis on sequences from bacteria. It includes gene symbols, protein family names, Enzyme Commission (EC) numbers, hidden Markov models (HMMs), multiple sequence alignments, Gene Ontology (GO) terminology, descriptive language, cross-references to related models in TIGRFAMs and other databases, and references to relevant literature	http://www.jcvi.org/cgi-bin/tigrfams/index.cgi
18. PANTHER	Protein analysis through evolutionary relationships	http://pantherdb.org/
19. SMART	A biological database called Simple Modular Architecture Research Tool (SMART) finds and examines protein domains inside protein sequences	http://smart.embl-heidelberg.de/
20. EggNOG	The eggNOG database is a database of biological information hosted by the EMBL	http://eggnogdb.embl.de/#/app/home
21. INTERPROSCAN	A resource for sequence analysis (nucleotide and protein sequences), InterProScan integrates several protein signature recognition techniques into one program	https://www.ebi.ac.uk/interpro/search/sequence-search
22. CDD	The CDD protein annotation library comprises a set of thoroughly annotated multiple sequence alignment models for both full-length and ancient domains	http://www.ncbi.nlm.nih.gov/cdd/
Functional Annotation Tools		
1. TMHMM	Based on the hidden Markov model (HMM), TMHMM is a popular bioinformatics tool for predicting the transmembrane helices of integral membrane proteins	http://www.cbs.dtu.dk/services/TMHMM/
2. TOPCONS	A web server called TOPCONS estimates membrane protein topology by consensus	http://topcons.net/
3. TargetP	TargetP predicts the subcellular location of eukaryotic proteins	http://www.cbs.dtu.dk/services/TargetP/

(Continued)

Table 2.2 (Continued)

Database/Software Tool	Details	Web Link
4. Plant-mPLoc	A Top-Down Strategy to Augment the Power for Predicting Plant Protein Subcellular Localization	http://www.csbio.sjtu.edu.cn/bioinf/plant-multi/
5. AtSubP	A highly accurate subcellular localization prediction tool for annotating the Arabidopsis thaliana proteome	http://bioinfo3.noble.org/AtSubP/
6. Predotar	Predotar for prediction of organelle targeting sequences	https://urgi.versailles.inra.fr/predotar/predotar.html
7. PHOSFER	PHOSFER predicts the phosphorylation sites in soybean proteins by employing a revolutionary machine-learning technique	http://saphire.usask.ca/saphire/phosfer/index.html
8. PhosPhAt	A database of phosphorylation sites in Arabidopsis thaliana and a plant-specific phosphorylation site predictor	http://phosphat.uni-hohenheim.de/phosphat.html
9. PlantPhos	Using maximal dependence decomposition to identify plant phosphorylation sites with substrate site specificity	http://csb.cse.yzu.edu.tw/PlantPhos/Predict.html
10. Musite	Musite is a stand-alone Java application that may be used to predict the phosphorylation sites of proteins both generally and specifically at kinases	http://musite.net/
11. TAIR/Protein Interaction Data	The Arabidopsis Information Resource (TAIR) is a library of genetic and molecular biology data for Arabidopsis thaliana, a model higher plant	https://www.arabidopsis.org/download/index-auto.jsp?dir=%2Fdownload_files%2FProteins%2FProtein_interaction_data
12. Arabidopsis Predicted Interactome and Arabidopsis Interactions Viewer	Protein–protein interactions in Arabidopsis are predicted using a proteome-wide structure	ftp://ftp.arabidopsis.org/home/tair/Proteins/Protein_interaction_data/Interactome2.0/ or http://bar.utoronto.ca/interactions/ cgi-bin/arabidopsis_interactions_ viewer.cgi
13. IntAct	IntAct offers a free, open-source database system and molecular interaction data analysis tools	http://www.ebi.ac.uk/intact/
14. AtPIN	Arabidopsis thaliana protein interaction network	http://atpin.bioinfoguy.net/cgi-bin/atpin.pl

15. ANAP	ANAP was created for Arabidopsis protein interaction integration and network-based studies to facilitate functional protein network analysis	http://gmdd.shgmo.org/Computational-Biology/ANAP
16. M.I.N.D	Membrane-based Interactome Network Database	https://associomics.dpb.carnegiescience.edu/Associomics/Home.html
17. PPIM	A Protein-Protein Interaction Database for Maize	http://comp-sysbio.org/ppim/
18. PRIN	A predicted rice interactome network	http://bis.zju.edu.cn/prin/
Tools for Functional Analysis of Plant Genomes		
1. Blast2GO	Blast2GO is a bioinformatics platform for high-quality functional annotation and analysis of genomic datasets	https://www.blast2go.com/
2. KAAS	KAAS (KEGG Automatic Annotation Server) uses BLAST or GHOST comparisons against the manually curated KEGG GENES database to give functional annotation of genes. The result includes KO (KEGG Orthology) designations and KEGG pathways constructed automatically	https://www.genome.jp/kegg/kaas/
3. GhostKOALA	The automatic annotation systems BlastKOALA and GhostKOALA work with genome and metagenome sequences, assigning KO (KEGG Orthology) to each gene to describe its function and reconstructing KEGG pathways, BRITE hierarchies, and KEGG modules to infer the organisms or ecosystem's high-level functions	https://www.kegg.jp/ghostkoala/
4. Mercator	Mercator pipeline for automated sequence annotation	https://mapman.gabipd.org/app/mercator
5. TRAPID	An online program called TRAPID makes de novo transcriptome analysis quick, dependable, and easy to use	https://bio.tools/trapid

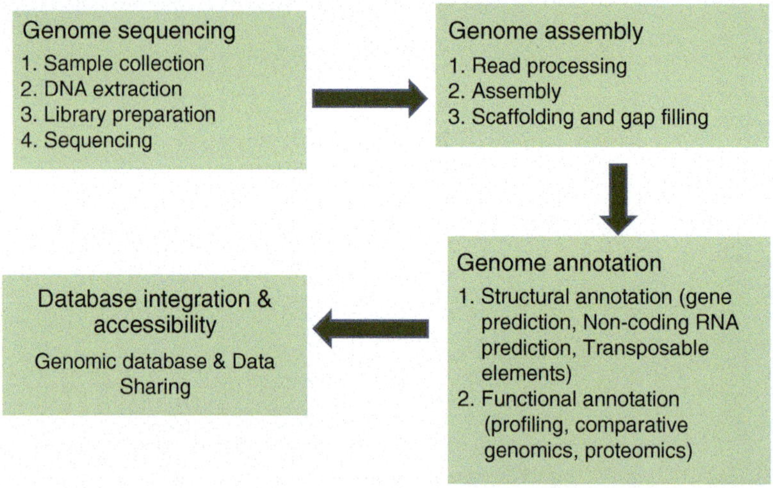

Figure 2.1 Flowchart of plant genome sequencing and annotation.

NGS-based RNA-Seq has transformed transcriptome analysis by allowing genome-wide gene expression levels to be characterized in a single experiment. This approach offers a broad dynamic range, can detect known and novel features, and can be applied to virtually any species. RNA-Seq provides a comprehensive view of the coding transcriptome, facilitating the discovery of alternative transcripts, gene fusions, allele-specific expression patterns, and more (Thompson et al. 2019).

Researchers typically begin by isolating RNA from their target cells or tissues to conduct transcriptome analysis and gene expression profiling. Subsequently, this RNA is converted into complementary DNA (cDNA) and then sequenced using NGS platforms. The resulting sequence data can be analyzed to identify differentially expressed genes, alternative splicing events, and other transcriptional features of interest. Transcriptome analysis yields direct, real-time insights into gene expression profiles and is widely used for identifying functional genes within biosynthetic pathways.

2.3.3 Proteome Analysis and Protein–Protein Interactions

"Proteome analysis" refers to thoroughly investigating the proteins expressed in a tissue, cell, or organism. Its primary goal is identifying and quantifying the entire set of proteins in a biological sample (Koh et al. 2012). Proteome analysis offers insight into protein expression patterns, posttranslational modifications, protein–protein interactions (PPIs), and protein functions (Yakubu et al. 2018). The advent of high-throughput proteomics is made possible by the remarkable success of genomics and bioinformatics, which aid in identifying the protein-coding regions of potential genes (Menschaert and Fenyö 2017).

Comprehensive proteomic studies have been undertaken for several crops, examining proteomes from various tissues at various developmental stages. These studies have reported varying numbers of proteins in different crops, such as 18,210 to 13,029 proteins in 30 Arabidopsis tissues and 17,862 proteins in 23 maize tissues (Bassal et al. 2020). Similarly, large proteome reference maps are available for bread wheat (15,779 proteins from 24 tissues), tomato fruit (7738 proteins), and the woody perennial model plant poplar (11,689 proteins from four tissues), encompassing 25% of the expected poplar protein space (Mergner and Kuster 2022). PPIs represent the physical connections between two or more proteins within a cell. These interactions are critical in various biological activities, such as signal transduction, gene control, and metabolic pathways. Analyzing PPIs aids in comprehending the organization and dynamics of protein networks, identifying novel protein functions, and unraveling disease mechanisms (Hu et al. 2021). Various methods are available for studying PPIs, including yeast two-hybrid assays, co-immunoprecipitation, affinity purification coupled with MS, and protein microarrays (Wang et al. 2022). These techniques enable researchers to identify and characterize protein complexes, map interaction networks, and explore the functional relationships between proteins. Researchers often construct "protein interaction networks" that depict the connections between proteins in analyzing PPIs. These networks can be visualized and analyzed using bioinformatics tools and software platforms like Cytoscape (De Las Rivas and Fontanillo 2010).

The "transcriptome," of all the "omics" technologies, gives the most detailed insights for real-time gene expression profiles and is the most often used procedure for detecting gene function in biosynthetic pathways. Based on a reference genomic sequence or de novo, transcriptomes can be constructed without any additional data. The genome sequence of the same organism, a closely related species, or closely related transcriptome sequences may make up the reference sequence. Through hybrid-seq transcriptome profiling analysis, unigenes – valid and diverse candidates for the study of chemical skeleton modification and secondary metabolism evolution – can be obtained (Wang and Huo 2022). The annotation of unigenes in the Kyoto Encyclopedia of Genes and Genomes (KEGG) pathway database, containing information about enzyme nomenclature, catalytic activity, and other functional parameters, allows for the grouping of unigenes into different categories, such as metabolic processes, cellular processes, catalytic activities, and more. This grouping aids in systematically evaluating the genes' biological functions in various pathways (Wang and Huo 2022).

2.3.4 Metabolome Analysis and Metabolic Pathway Reconstruction

Metabolome analysis and metabolic pathway reconstruction are foundational techniques within metabolomics. Metabolomics entails a comprehensive exploration of small molecules, also known as metabolites, present in biological systems. These metabolites encompass small molecules like amino acids, sugars, lipids, and organic acids, serving as substrates, intermediates, and products in metabolic reactions. Metabolome analysis strives to identify and quantify these metabolites, offering insights into the metabolic state of a biological system. This analytical approach aids in understanding the biochemical changes associated with diverse physiological and pathological conditions.

Metabolome analysis entails identifying, quantifying, and characterizing the entire spectrum of metabolites in a biological sample. Its overarching goal is to capture the dynamic responses of metabolites to various stimuli or conditions (Chen et al. 2022). The process of metabolite analysis involves several key steps:

Sample preparation: This step involves collecting and preparing biological samples, such as leaves, flowers, or tissues, at specific developmental stages. The aim is to extract metabolites while preserving their chemical integrity.

Metabolite extraction: Various techniques, including organic solvent extraction and solid-phase extraction, are employed to isolate a wide range of metabolites from the biological samples.

1) **Analytical techniques:** Analytical techniques such as MS, nuclear magnetic resonance (NMR) spectroscopy, and chromatography (e.g. liquid chromatography and gas chromatography) are used to analyze the extracted metabolites.
2) **Data processing and analysis:** The data obtained from the analytical techniques are processed and analyzed to identify and quantify metabolites. This often involves the use of specialized software and databases.
3) **Metabolite annotation:** The detected metabolites are annotated by comparing experimental data with reference databases. This process helps determine the identified metabolites' chemical identity and biological roles.

2.3.4.1 Metabolic Pathway Reconstruction

Metabolic pathways are a series of enzymatic events within a cell and serve as substrates for subsequent reactions. The reconstruction of metabolic pathways involves integrating data from various sources, including metabolomics data, genomic data, biochemical knowledge, and bioinformatics tools. The ultimate objective is to craft a comprehensive and precise depiction of the metabolic network, encompassing the reactions, enzymes, substrates, products, and regulatory elements involved (Yamanishi et al. 2015). This endeavor aids in grasping the functional relationships between metabolites and their roles in cellular processes. The process of analyzing metabolic pathways involves three essential steps: prediction, design or reconstruction, and optimization (Shah et al. 2021).

Knowledge-driven objectives (KDO) and data-driven objectives (DDO) are the two primary methods for pathway reconstruction (Viswanathan et al. 2008). Knowledge-driven reconstruction relies on extensive domain knowledge, be it detailed pathway information, cell type specifics, disease contexts, or system insights. Karp and collaborators have developed the pathologic pathway software, which reconstructs metabolic pathways by mapping functional annotations on the reactions and pathways in the MetaCyc collection (Karp et al. 1999; Paley and Karp 2002).

In addition to KDO and DDO methods, there are additional techniques that are not based on reference pathways but use ordinary differential equations (ODEs) or probabilistic inference techniques like graphical models and Bayesian networks (Jansen et al. 2003; Friedman 2004; Werhli et al. 2006; Zhao et al. 2012). Moreover, a combination of existing pathway knowledge and Bayesian probabilistic graphical models can be utilized (Qi et al. 2014). ML has recently been applied to optimize metabolic pathways owing to its excellent modeling capabilities (Cheng et al. 2023). Tools like DeepEC have been developed to predict enzyme commission (EC) numbers based on protein sequences (Zhou et al. 2018). Furthermore, ML can be integrated into Design–Build–Test–Learn (DBTL) cycles to enhance the exploration of design space (Cheng et al. 2023).

Reconstructed metabolic pathways yield valuable insights into an organism's metabolism, offering potential drug targets, elucidating disease mechanisms, optimizing metabolic engineering strategies, and enhancing biotechnological processes.

2.3.4.2 Metabolic Pathway Databases

Various public repositories have been used to identify, record, and characterize a large number of metabolic pathways according to their functions. Notable repositories include KEGG (Ogata et al. 1998; Kanehisa et al. 2019), MetaCyc (Caspi et al. 2018), and BioCyc (Karp et al. 2019). In addition, several other metabolite databases encompass a wealth of information about biochemical reactions and pathways. These include BIGG (King et al. 2016), BRENDA (Scheer et al. 2010), ModelSEED (Henry et al. 2010), MetRxn, Rhea (Alcántara et al. 2012), UM-BBD (Gao et al. 2010), MOS (Sutter et al. 2011), Beilste Crossfile, in addition to databases containing hypothetical metabolites and reactions, like MINE, ATLAS of Biochemistry, Chemical Translation Service, and UniChem, provides an easy-to-use web application for converting metabolite IDs between various databases, organism-specific GSMs, or knowledge bases (as in EcoCyc, AraCyc, and HumanCyc) (Wang et al. 2017b) (Table 2.1).

Both metabolome analysis and metabolic pathway reconstruction are pivotal in unraveling the mechanisms underlying biological systems, encompassing cellular metabolism, disease progression, and drug responses. These approaches yield invaluable insights for biomarker discovery, drug development, personalized medicine, and a comprehensive knowledge of biological processes at the systemic level (Chen et al. 2022).

2.4 Application of Bioinformatics, Genomics, and Proteomics in Crop Improvement and Breeding

2.4.1 Bioinformatics Databases and Tools in Genomic Research

Bioinformatics databases and tools are crucial in genomic research, allowing for the efficient analysis, storage, and interpretation of massive amounts of genomic data (For the list of databases and tools, see Table 2.2). The availability of plant genome data in public databases has been greatly increased with the introduction of NGS and other advanced sequencing methods. These technologies are invaluable for identifying genomic variations and predicting protein structure and function (Kyriakidou et al. 2018; Mathur 2018). Moreover, GWAS are instrumental in pinpointing loci and allelic variations linked to valuable traits, streamlining crop modification and enhancement (Claros et al. 2012). In essence, the evolving applications of bioinformatics in plant biotechnology empower researchers to gain a comprehensive and structured understanding of economically significant plants.

2.4.2 Genetic Diversity Study and Estimation of Phylogenetic Relationships

Bioinformatics is pivotal in examining genetic diversity and estimating phylogenetic relationships across various organisms. Genetic diversity pertains to the variation in the genetic composition of individuals within a population or species. Grasping this diversity holds immense importance in multiple biological applications, including biology, conservation, medicine, and more.

The total genetic variation present in various genotypes, strains, clones, or populations within a species is referred to as genetic diversity. Genetic diversity can be assessed using various data sources, including pedigree, qualitative and quantitative traits, isozymes, and DNA markers. The primary method for estimating genetic diversity is genetic distance (GD), which measures the genetic dissimilarity of two organisms concerning variations in their DNA sequences and gene frequencies. Genetic similarity between two entities is equal to 1 minus the GD. GD can be estimated using diverse data types such as morphological (qualitative and quantitative data) and molecular markers (e.g., RAPD, ISSR, SSR, RFLP, AFLP, Isozymes, etc.; Upadhyay et al. 2010a,b). GD is also evaluated among different populations.

Subsequently, GD-based clustering and phylogenetic analysis are carried out through distance-based methods (e.g., UPGMA and clustering) and model-based approaches (e.g. Principal Component Analysis, Principal Coordinate Analysis, Multidimensional Scaling [MDS], etc.). A list of software tools employed for phylogenetic analysis is given in Table 2.2. Phylogenetic relationships entail understanding the evolutionary history and relatedness among various organisms or genetic sequences. These investigations play a pivotal role in deciphering the evolutionary connections within a network of molecules or organisms. They assist in predicting undiscovered functionalities, tracing gene movement, and establishing genetic affinities. These valuable insights form the basis for constructing a phylogenetic tree, a tool that organizes living entities according to their degree of similarity, with more remarkable similarity leading to closer proximity to the tree.

Multiple methods, including distance-based approaches, parsimony techniques, and likelihood methods, can be harnessed to create a phylogenetic tree (Morbia et al. 2023). Genetic diversity and phylogenetic analysis serve several essential purposes, such as germplasm conservation and identifying diverse lines suitable for hybridization. They show the extent of genetic variation among crop varieties, which is important for avoiding closely related germplasm lines in hybridization programs. Utilizing closely related lines could restrict the genetic diversity of the resulting varieties.

Furthermore, these analyses make it possible to choose core collections suited explicitly for particular breeding applications and systematically classify germplasm accessions. Genetic diversity estimation and phylogenetic analysis also prove indispensable in introducing desirable genes and alleles from diverse germplasm into elite crop germplasm (Dar et al. 2019; Loera-Sánchez et al. 2019).

2.4.3 Genomic Selections and their Application in the Breeding Program

Crop improvement includes the creation of new crop varieties that possess desired characteristics such as high yield, high quality, resistance to disease, and tolerance to stress (Kim et al. 2022). These improvements can be achieved through conventional breeding methods such as selection, hybridization, and mutation breeding or through biotechnological approaches like genetic engineering and genome editing.

Genomics plays a pivotal role in understanding genetic variation, leading to enhanced crop breeding efficiency and genetic improvements in crop species (Bevan et al. 2017). Genomics involves the study of genomes' structure, function, and evolution – the complete genetic information set within an organism. GS a promising approach introduced by Bernardo (1994) and Meuwissen et al. (2001), utilizes molecular genetic markers to devise novel breeding programs and construct marker-based genetic evaluation models (Bhat et al. 2016). This approach enhances the genetic gain of complex traits in terms of time and cost (Hu et al. 2018). In order to overcome the drawbacks of marker-assisted selection (MAS) in breeding programs, GS predicts the breeding values of offspring by linking their traits with high-density genetic marker scores (Atanda et al. 2021). In GS, breeders employ a plant's genetic makeup and phenotypic data (measurable traits) to "train" a model and predict the outcome of a cross, termed "genomic estimated breeding values (GEBV)" (Liabeuf et al. 2018). This approach offers the ability to forecast cross outcomes without planting and physically measuring traits. GS represents an advancement over MAS, aiming to employ genome-wide markers to estimate the effects of all loci and compute a GEBV for comprehensive and reliable selection. The GS process comprises three main stages: model training, GEBV estimation, and selection of promising individuals based on predicted GEBV. Evaluation of prediction accuracy typically involves correlating predicted GEBV with empirically observed values or utilizing cross-validation techniques. Figure 2.2 depicts the fundamental process of GS. While GS reduces time and costs compared to traditional breeding and increases genetic gain, its accuracy hinges on several factors, including training and breeding population size and diversity, trait heritability, marker density and linkage, and more. GS also presents challenges, such as gene frequency changes and epistatic interactions, impacting GEBV estimates. Additionally, GS may entail higher implementation costs compared to classical breeding methods. Selecting appropriate models for GS is a pivotal step in the process of choosing crops based on their GEBV utilizing genetic markers (Meuwissen et al. 2001). As indicated in Table 2.3, a

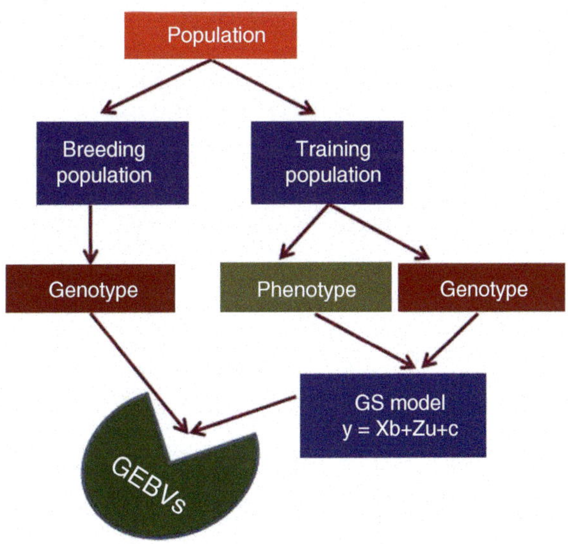

Figure 2.2 A representative scheme of the genomic selection process.

Table 2.3 Statistical tools for implementing genomic selection.

Sr. No.	Tool	Description	Year	Link
1.	Sparse Kernel Methods (SKM) R library	A tool for fitting sparse kernel models for genomic prediction and genome-wide association studies	2023	GitHub-brandon-mosqueda/SKM
2.	GenomeStudio	A software for analyzing data from Illumina genotyping arrays	2008	GenomeStudio Software Downloads (illumina.com)
3.	AxiomAnalysisSuite	A software for analyzing data from Affymetrix genotyping arrays	2012	Axiom Analysis Suite Software 4.0 | Thermo Fisher Scientific - IN
4.	Genomic Selection Toolkit (GST) R package	A tool for genomic selection using different prediction methods and cross-validation schemes	2019	CRAN - Package GSelection (r-project.org)
5.	Genomic Best Linear Unbiased Prediction (GBLUP) method	A genomic prediction method that uses a linear mixed model and a genomic relationship matrix.	2008	–
6.	rrBLUP R package	A mixed models-based tool for genomic prediction and variance component estimation.	2011	https://cran.r-project.org/package=rrBLUP
7.	BGLR R package	A tool for Bayesian generalized linear regression and genomic prediction	2014	https://cran.r-project.org/package=BGLR
	iPat web tool	A tool for imputing and predicting agronomic traits using genomic data	2018	
8.	MTGS R package	A tool for multi-trait genomic selection using a Bayesian approach with different prior distributions and covariance structures	2019	https://cran.r-project.org/package=MTGS
9.	STGS R package	A tool for single-trait genomic selection using a Bayesian approach with different prior distributions and covariance structures	2019	https://cran.r-project.org/package=STGS
10.	GMStool	A tool for determining the best marker set for predicting a continuous phenotype based on a genome-wide association study	2020	https://github.com/JaeYoonKim72/GMStool
11.	solGS	An online tool for estimating an individual's genomic estimated breeding value using a Ridge-Regression Best Linear Unbiased Predictor model	2014	http://nextgencassava.org/ http://cassavabase.org/solgs
12.	BWGS	A genomic selection web tool and its implementation in a wheat breeding program	2020	https://cran.r-project.org/package=BWGS

Table 2.3 (Continued)

Sr. No.	Tool	Description	Year	Link
13.	GenSel	A software for genomic selection and GWAS analysis for linear and categorical traits	2009	https://github.com/austin-putz/GenSel
14.	GSelection	An R package for genomic selection using the integrated model framework	2019	https://cran.r-project.org/package=GSelection
15.	lme4GS	A R package for selecting bandwidth, fitting linear mixed models with user-defined covariance structures, and genomic prediction	2011	https://github.com/perpdgo/lme4GS

number of instruments and software programs have been created to evaluate genomic prediction and apply GS. One widely used GS model is best linear unbiased prediction (BLUP), which assesses marker effects and predicts complex traits (Habier et al. 2013). However, BLUP is best suited for traits with additive genetics and may not be suitable for those with epistatic or nonadditive genetics, contributing to genetic variation in quantitative traits Cooper et al. 2014). Therefore, it is recommended not to employ simple linear or parametric models for such traits. In contrast, nonparametric and semiparametric techniques can capture both additive and epistatic effects in GS (Sao et al. 2020). The authors referred to studies by Budhlakoti et al. (2020c) for detailed comparisons of various methods in different scenarios. Lately, semiparametric and sophisticated strategies for GS have been applied (Tanaka 2018; Budhlakoti et al. 2020a, b; Majumdar et al. 2020; Tanaka 2020; and Mishra et al. 2021). Factors such as marker diversity, density, training population size and composition, number of QTL, and trait heritability affect GS accuracy. GS has much potential for several uses, including yield-focused breeding programs, germplasm enhancement, and hybrid breeding (Gidamo 2022). In Table 2.4, you can find a list of general GS methods used in crop plants, and Table 2.1 provides a list of statistical tools employed in GS.

For a detailed understanding of the methodology behind GS methods, readers can refer to sources such as Wang et al. (2018), Crossa et al. (2017), and Montesinos López et al. (2022). GS has been shown to significantly increase the selection gain per year per cycle in breeding programs compared to PS in various crops, including maize (Vivek et al. 2017; Môro et al. 2019), soybean (Smallwood et al. 2019), and several others (Robertsen et al. 2019). Nonetheless, genomics applications in crop improvement present several challenges and limitations, including *High cost and complexity* – Genomic technologies demand substantial investments in infrastructure, equipment, personnel, and data analysis. They also involve intricate experimental designs, protocols, and bioinformatics pipelines that must be tailored for each crop species and trait (Meena et al. 2020). *Ethical and social issues* – the use of genomics in crop improvement can raise ethical and social concerns among consumers, farmers, regulators, and other stakeholders. This may encompass objections to genetically modified or edited crops, concerns about intellectual property rights, and adherence to biosafety regulations (Meena et al. 2020). *Phenotypic validation* – genomics applications in

Table 2.4 List of software for phylogenetic analysis.

Sr. No.	Name of Database/ Tools	Link/URL	Use/Role
1.	PhyML, AxML	https://molbiol-tools.ca/Phylogeny.htm	Tools for constructing phylogenetic trees based on sequence data, aiding in evolutionary studies
2.	PHYLIP	https://evolution.genetics.washington.edu/phylip.html	A free suite of tools called PHYLIP is used to infer phylogenies
3.	MEGA	https://www.megasoftware.net	The desktop program known as Molecular Evolutionary Genetics Analysis (MEGA) is meant for the comparative study of homologous gene sequences from various species or multigene families, with a focus on deducing patterns of DNA and protein evolution as well as evolutionary relationships
4.	UniFrac	http://bmf2.colorado.edu/unifrac/index.psp	Comparing microbial communities with the use of phylogenetic data
5.	XplorSeq	http://vent.colorado.edu/phloware	DNA sequence compilation, management, and phylogenetic analysis
6.	RAST	http://metagenomics.nmpdr.org	Automatic phylogenetic and functional analysis of genome
7.	AMPHORA	http://bobcat.genomecenter.ucdavis.edu/AMPHORA	Phylogenetic analysis of single gene or whole genomes
8.	STRUCTURE	https://web.stanford.edu/group/pritchardlab/structure.html	The software package is used for model-based grouping of entities
9.	DARwin	https://darwin.cirad.fr/	A software program called DARwin was created for phylogenetic and diversity analysis using evolutionary dissimilarities
10.	GENE-E	https://software.broadinstitute.org/GENE-E/	GENE-E is a matrix analysis and visualization platform that facilitates data exploration visually. Among its many other tools are heat maps, clustering, filtering, charting, and marker selection
11.	SPSS	https://www.ibm.com/products/spss-statistics	SPSS Statistics is software for advanced statistical analysis
12.	MOLPHY Molecular Phylogenetics	https://sbgrid.org/software/titles/molphy	The tool is based on the maximum likelihood method for phylogenetic analyses
13.	Jalview	http://www.jalview.org/development/Version-Archive	It helps in the refinement of multiple performed alignments

crop improvement require validation through phenotypic evaluations conducted under various environmental conditions and agronomic practices. Standardized procedures and protocols are still necessary for prototyping because it is still a labor-intensive and time-consuming practice (Meena et al. 2020). To address challenges, it is crucial to integrate genomics applications in crop improvement with conventional breeding methods and other biotechnological approaches to achieve optimal results (Meena et al. 2020). Additionally, genomics should be complemented by other omics disciplines, such as transcriptomics, proteomics, metabolomics, epigenomics, and microbiomics, which can offer a deep insight into the molecular processes underlying complex features (Bevan et al. 2017).

2.4.3.1 Role of Genomic Selections (GS) in Identifying Genes and Proteins Associated with Desirable Traits

Creating climate-resilient crop varieties using conventional methods is labor-intensive and resource-demanding, with limited genetic improvement (Lush 1943). Improving phenotyping and broadening the genetic base are essential to address this challenge.

In crop breeding initiatives, GS has emerged as a feasible approach to enhance genetic gain, given the current state of extreme weather events (Yuan et al. 2019). However, Genotype Environment Interactions (GEIs) present a considerable obstacle to the successful application of GS in crop plants. GEI leads to variations in the ranking of genotypes for complex traits across different environments, resulting in reduced heritability and predictive accuracy (Boer et al. 2007; Cooper 1999).

To mitigate the impact of GEI, breeders group diverse environments into similar clusters, reducing environmental errors and enabling a more accurate determination of the true genetic potential. This process utilizes data from multi-environment trials (METs) and controlled environments (van Eeuwijk et al. 2018; Xu 2016). It is worth noting that lines from different years exhibit lower predictive power than those from the same year (Belamkar et al. 2018).

Nonadditive genetic effects like dominance and epistasis may impact the use of GS in crop breeding. While incorporating nonadditive variables into GS models can improve genetic response and prediction accuracy, it may also reduce additive variance and narrow-sense performance. By considering both additive and dominant effects in GS models, the accuracy of assessing genetic variance components for complex traits in hybrid populations can be enhanced (dos Santos et al. 2016; Cantelmo et al. 2016).

Epistasis, particularly in traits like rice flowering time (Ahsan et al. 2019) and corn oil, protein, and starch contents (Wang et al. 2015), plays a crucial role. For example, a GS analysis with 280 wheat accessions incorporated two-locus epistatic effects to improve prediction accuracy for certain traits (Wang et al. 2012). As the number of loci increases, epistatic variance becomes more pronounced, while dominance variance remains unaffected (Mäki-Tanila and Hill, 2014). Thus, it is essential to account for dominance variance in models, especially for clonally propagated crops and hybrid populations (Denis and Bouvet 2013; Hill and Mäki-Tanila 2015).

2.4.4 SNP Discovery and SNP Genotyping

SNPs are the most common type of variation in DNA sequence between alleles (Morgil et al. 2020). In plant genetics and breeding, they have played a pivotal role in advancing our comprehension of genetic diversity and crop enhancement (Morgil et al. 2020). The utilization of advanced biotechnological approaches, such as NGS, has led to the discovery and application of SNPs in numerous plant species. Here are some key insights into SNP discovery, genotyping, and their applications in plant breeding:

1) **SNP discovery:** SNP discovery involves identifying single nucleotide variations within the DNA sequence of a genome (Morgil et al. 2020). This exploration is made feasible through techniques like NGS, which offers a rapid and cost-effective means of pinpointing SNPs (Morgil et al. 2020).
2) **SNP genotyping:** SNP genotyping refers to the procedure for ascertaining the specific SNP alleles present in an individual or population (Morgil et al. 2020). Various genotyping techniques, including microarrays and high-throughput sequencing, have been developed to deliver precise and efficient determination of SNP genotypes (Mammadov et al. 2012).
3) **Applications of SNPs:** SNPs find extensive utility in various realms, including genetic diversity analysis, MAS, QTL mapping, association studies, and GS (Mammadov et al. 2012). These applications facilitate the enhancement of plant breeding and genetic research endeavors.

2.4.5 Breeding with Marker Assistance to Improve Crops

Marker-assisted breeding (MAB) is a valuable technique in crop improvement, harnessing the power of molecular markers to streamline the selection of favorable plant traits. This approach empowers breeders to identify and cultivate plants with desired characteristics more efficiently, thereby expediting the breeding process (Hasan et al. 2021). The foundational steps integral to MAS are detailed as follows:

Selection of parents with desired morphological traits: The journey commences with the judicious selection of parental plant stock boasting the sought-after morphological attributes. This entails the identification of individuals bearing the desired traits, underpinned by genetic diversity. The DNA sequences specifically linked to target traits (e.g., disease resistance, yield, quality) are identified and correlated with molecular markers, often SNPs.

Development of breeding populations: Once the parental candidates are identified, the breeding populations, typically encompassing F1 and F2 generations, are meticulously crafted through the crossbreeding of these chosen progenitors. This crosses the boundary to engender genetic diversity within the population (Jeon, et al. 2023).

Plant selection with the help of markers for desired features: The key part is figuring out how the identified molecular markers relate to the desired characteristics. This correlation is meticulously scrutinized across a population of plants. Markers demonstrating a robust association with the coveted traits are cherry-picked and seamlessly integrated into the breeding program. Breeders subsequently leverage these markers to enhance the selection of plants harboring the desired attributes throughout the breeding process.

Marker validation: The resultant plant populations are nurtured and scrutinized to assess the manifestation of the traits of interest in real-world scenarios, whether in the field or in controlled conditions. The collection of phenotypic data serves to corroborate the presence of the desired traits at a molecular level.

Correlation with morphological traits: Grounded in the insights gained from phenotypic evaluation, the breeding process is further honed by fine-tuning marker selection and breeding strategies. This iterative cycle perpetuates, driving ongoing enhancement of the traits of interest.

The spectrum of MAB encompasses a rich tapestry of applications. These include evolving phylogeny, harnessing molecular markers for assessing breeding materials, verifying cultivar purity, estimating genetic diversity, and aiding in parental selection. Moreover, it extends to the study of heterosis, identification of genomic regions under selection, support for introgression and backcrossing, enabling marker-assisted pyramiding, facilitating early generation selection, and incorporating combined MAS. Although overarching similarities exist across these domains, molecular markers have seamlessly integrated into the traditional framework of plant breeding, frequently either supplanting or complementing conventional PS (Hasan et al. 2021).

MAS is particularly invaluable for fortifying disease resistance, a characteristic that is often tested by new pathogen strains. Molecular markers assist in selecting intricate traits to measure, subject to specific environmental conditions, recessive, or underpinned by multiple genes or QTL (Xu and Crouch 2008). Multiple studies have chronicled the efficacy of MAS in tandem with PS in elevating disease resistance across a gamut of crops, such as beans (Miklas et al. 2006), rice (Bustamam et al. 2002), wheat (Wilde et al. 2008), and maize (Zhao et al. 2012). MAS plays a pivotal role in identifying and transferring genes associated with stress tolerance and response, such as an stress-responsive NAC [No Apical Meristem (NAM) (SNACs) Arabidopsis ATAF1/2 and Cup-shaped Cotyledon2 (CUC2)] and dehydration-responsive element binding (DREBs), which govern the expression of stress-responsive genes in rice (Todaka et al. 2012). By heightening root characteristics and bolstering water consumption efficiency, MAS has been instrumental in rendering rice more resilient to drought (Steele et al. 2006).

However, the adoption of MAS is encumbered by an array of challenges, including costs, accuracy, and complexity, which are currently being navigated (Dita et al. 2006; LeDeaux et al. 2006; Henkrar and Udupa 2020). Pursuing enhanced crop breeding remains a dynamic journey, with MAB marking a pivotal waypoint in this evolutionary path.

2.4.6 Association Mapping Studies in Plants

In plant genetics and genomics, association mapping (AM), often referred to as linkage disequilibrium (LD) mapping, emerges as a powerful technique for pinpointing genetic variations linked to specific traits or phenotypes within a diverse population of plants. This approach empowers researchers to unveil the genetic foundations of plant phenotypes, leading to more informed and effective breeding strategies. AM equips plant breeders with the insights needed to make judicious decisions in trait enhancement, selection of superior genotypes, and the development of cultivars boasting desired attributes for agricultural and other applications (Ibrahim et al. 2020).

Association mapping, abbreviated as AM, is a method encompassing the evaluation of QTL by considering a multitude of polymorphisms. It plays a pivotal role in identifying alleles and novel genes, shedding light on the enigmatic world of complex traits. Its remarkable high-resolution capabilities set AM apart from linkage analysis, capitalizing on the structure of LD. Central elements of successful AM encompass marker density, population characteristics, sample sizes, and the rigorous consideration of population structure. The choice of germplasm, the quality of genotypic and phenotypic data, and the application of sound statistical analysis are all pivotal components of association studies.

AM stands as a potent force for enhancing the genetic profile of crops. Nevertheless, it is imperative that ongoing research endeavors refine computational and statistical methods and seamlessly integrate AM with gene annotation and functional analysis. The development of user-friendly statistical tools and the enrichment of genetic resources emerge as compelling imperatives. Furthermore, exploring rare alleles and variants plays a critical role in advancing AM studies.

A proposed approach, known as joint linkage AM, seeks to bolster QTL mapping and address the limitations inherent in AM. Future strides in this domain hinge on the amalgamation of GWAS with functional genomics. This synergy promises to expedite the discovery of new potential genes and QTL easily (Gupta et al. 2019). In summary, AM stands as a dynamic frontier in plant genetics and genomics, a frontier with the potential to usher at the dawn of a precision age and innovation in crop breeding and improvement.

2.4.7 Mapping of QTL

QTL mapping is a valuable genetic technique employed to pinpoint specific genomic regions associated with traits or phenotypes influenced by the action of multiple genes. This method revolves around identifying genetic markers tightly linked to the trait of interest. To embark on the QTL mapping journey, several prerequisites must be met:

1) **Development of an appropriate mapping population:** A critical initial step involves the creation of a well-suited mapping population to serve as the foundation for QTL analysis.
2) **Utilization of a high-density marker linkage map:** A species-specific linkage map containing densely distributed markers is imperative. This map is the navigational tool that guides the journey through the genome.
3) **Accurate phenotypic evaluation:** A reliable and meticulous phenotypic assessment of the target trait is paramount. These assessments provide the foundation for identifying the genomic regions associated with the trait.
4) **Selection of adequate software packages:** The arsenal of QTL mapping relies on suitable software packages equipped for detecting and precisely mapping QTLs. These software packages are essential to extract valuable insights from genetic data.

Linkage mapping is the approach that detects and meticulously maps each of the QTLs governing the target trait. This mapping is achieved within relatively short confidence intervals, enabling a fine-tuned understanding of the genomic regions influencing the trait. Subsequently, the markers flanking these QTL regions can be harnessed for MAS, including recombinant selection for the specific QTL of interest. For even greater precision, higher-resolution QTL mapping is employed, enabling the localization of a QTL within a minimal confidence interval, often less than 1 centiMorgan (cM). This significantly facilitates the cloning of genes located within these QTL regions.

Bioinformatics offers a multitude of software tools tailored for QTL mapping. Some notable ones include MapMaker/QTL (Lincoln et al. 1993), PLABQTL, QTL Cartographer (Wang et al. 2015), MapManagerQT/QTX, R/QTL (Broman et al. 2019), R/QTLBIM (Yandell et al. 2007), QTL Express (Seaton et al. 2002), FlexQTL (Bink et al. 2008), MCQTL, and QGene, among others. These software solutions play a pivotal role in enabling researchers to embark on the exciting journey of QTL mapping, unlocking the genetic secrets hidden within the genome.

2.4.8 Linkage Mapping of Molecular Markers and Oligogenic Traits

Linkage mapping of molecular markers in the context of oligogenic traits is a crucial approach within the field of genetics. It serves the purpose of unraveling the inheritance patterns and pinpointing the genomic locations of genes that shape oligogenic or complex traits. As the term suggests, oligogenic traits are influenced by a relatively small number of genes, often interacting in ways that don't adhere to the standard additive model.

"SNPs" being a prime example, molecular markers offer a practical means to label and track specific positions in the genome. The process of linkage mapping entails the analysis of a genetically diverse population. Researchers examine the trait of interest and the genotypes at various molecular marker locations. This comprehensive approach enables the study of the inheritance patterns and the association of these markers with the trait within different families or populations.

By observing how markers and traits co-segregate within these populations, researchers can make inferences about the areas of the genome attached to the expression of the oligogenic trait. This process enables us to create a genetic map of the trait's genomic influences. By identifying and understanding these regions through linkage mapping, we can further our comprehension of the genetic architecture behind these complex traits. Additionally, this information opens the door to identifying potential candidate genes that may play a role in the expression of the trait. Ultimately, this approach advances our understanding of the genetic factors underpinning oligogenic traits, offering valuable insights into their inheritance and genomic locations (Delvadiya et al. 2022).

2.4.9 Role of Genomics Tools such as MAS, GWAS, and Genomic Selections in Identifying Genes and Proteins Associated with Desirable Traits

By leveraging DNA-based markers, finding genes and proteins linked to desired traits through MAS is an effective method employed in plant genetics to choose plants exhibiting coveted traits, such as high yield, resistance to diseases, drought tolerance, or superior quality. An extensive exploration of MAS in crops was conducted by Francia and colleagues back in 2005. These DNA-based markers are essentially specific DNA segments that are linked to particular genes or regions of the genome that govern a specific trait. What sets them apart from phenotypic traits, which can be influenced by environmental factors and are only observable at certain stages of plant growth, is that DNA-based markers can be identified at any point and remain unaltered by environmental variables. Consequently, using DNA-based markers can significantly enhance the efficiency and accuracy of plant breeding.

For plant breeders to effectively employ MAS, they require two fundamental components: a genetic map and a marker–trait association (MTA). A genetic map is a visual representation illustrating the relative positions and distances between DNA-based markers along the chromosomes of a given plant species. This map is invaluable in helping plant breeders pinpoint the genes or QTLs that influence a specific trait. QTLs denote those regions of the genome encompassing one or more genes governing quantitative traits, like yield or plant height.

The effectiveness of MAS primarily hinges on several key factors, including the availability of a genetic map featuring genetic markers for the trait(s); a robust correlation between the trait and the marker(s), the necessity for ample recombination between the markers and the rest of the genome; the capability to swiftly and cost-effectively analyze numerous plants. Furthermore, MAS is influenced by the proximity of the marker to the target gene. Three scenarios can be identified: The marker is situated directly within the gene. In this best-case scenario, the gene must be cloned. The genes and markers are closely positioned. This is a favorable situation, although to prevent recombination, the marker, and gene must be in close proximity. The trait(s) involves QTLs. QTLs cover significant genomic areas, making this a more intricate case. At least two markers should be in close proximity to or within the QTL region.

A MTA pertains to a statistical connection between a DNA-based marker and a trait of interest. This association implies that the marker is either positioned within or in close proximity to the gene or QTL, influencing the trait. A MTA also aids plant breeders in estimating the size and direction of the gene's or QTL's impact on the trait.

Simple sequence repeats (SSRs), amplified fragment length polymorphisms (AFLPs), random amplified polymorphic DNAs (RAPDs), restriction fragment length polymorphisms (RFLPs), and SNPs are some examples of DNA-based markers. Each marker type boasts advantages and disadvantages concerning cost, availability, reliability, polymorphism, and informativeness.

Numerous methods are available for establishing MTAs, such as linkage and AM. Linkage mapping serves as the foundation, relying on the analysis of marker and trait co-segregation in a population generated from a cross between two parents displaying variations in the trait of interest. While linkage mapping can pinpoint extensive genomic regions containing the gene or QTL of interest, achieving high resolution demands a substantial number of individuals and markers. On the other hand, AM is built on the analysis of marker–trait relationships within a population of unrelated individuals exhibiting natural variations in the trait of interest. AM can identify smaller genomic regions containing the gene or QTL of interest, but it necessitates a large number of markers and a genetically diverse and structured population.

Once the marker–trait relationship is established, plant breeders can harness MAS to select plants carrying beneficial alleles for the trait of interest. This involves employing molecular techniques like polymerase chain reaction (PCR) or hybridization. The benefits reaped by plant breeders through MAS include: the ability to trace advantageous alleles across generations and incorporate them into elite cultivars, the capability to identify superior individuals from among segregating progenies based on their genetic makeup, and the potential to sever the link between favorable alleles and undesired loci that could negatively affect other attributes.

When one gene or a significant QTL determines a trait, a single genomic segment transferred from a donor parent to a recipient parent can significantly improve the trait. For backcross programs, MAS expedites the recovery of the recurrent parent by reducing the number of backcross generations required to restore the recurrent parent's genome and selecting individuals with a more robust recovery of the recurrent parent's genome and less donor parent genome beyond the target region. Eliminating linkage drags refers to the undesired transfer of loci linked to the target region.

In scenarios where a trait is governed by numerous QTLs, transferring different genomic segments from various donor parents to a recipient parent can result in cumulative enhancements to the trait. For traits with low to moderate heritability, MAS can enhance the likelihood of selecting superior genotypes by: improving the accuracy and precision of selection for characteristics that are hard to assess or affected by external factors. Reducing the resources required to achieve selection goals for traits with low to moderate heritability, especially when both the selection aim and intensity are high. Combining information from multiple traits and environments to make selections based on overall performance.

MAS has succeeded considerably in improving simple traits regulated by a few genes, such as disease resistance and tomato quality. Although there are some instances of its success in complex traits governed by multiple QTLs, there are challenges and opportunities for improving its efficiency. Particularly for traits with low to moderate heritability, MAS is more effective than PS alone as it enhances selection gains by using markers. It can also reduce resource requirements when there is significant selection pressure to achieve a goal for a trait with low to moderate heritability.

2.4.9.1 Genome-wide Association Studies in Identifying Proteins and Genes Linked to Desired Traits

GWAS identify relationships between allele or genotype frequencies and trait status. This comprehensive process entails various steps for effective execution. Initially, the traits to be assessed are carefully selected, and an appropriate study population is chosen. The population's size, as well as its genetic and trait variance, is taken into consideration. To facilitate this, SNP arrays, imputation, and whole-genome sequencing are deployed. Subsequently, association tests are applied to zero in on the genomic regions exhibiting significant genome-wide associations with the variance of the target phenotype. In many cases, meta-analysis is employed to enhance the statistical power in the quest for associations.

The pioneering GWAS conducted by Klein and colleagues in 2005 unearthed a variant of the Complement Factor H gene, firmly linked to age-related macular degeneration. Over the past 15 years, GWAS has proven to be a potent tool for unraveling the genetic underpinnings of variation in a wide array of complex phenotypes, encompassing human and animal diseases and physiological and agronomic traits in plants using many software statistical packages as listed in Table 2.5. Substantial efforts have been made to address the challenge of population structure, minimizing the risk of spurious associations and erroneous discoveries. Techniques like mixed linear models (MLM) have been deployed to rectify the inflation of minor genetic effects and mitigate bias stemming from population structure. Innovative methods, including haplotype- and gene-based GWAS, have been developed to alleviate these issues, assuming a normal distribution for the phenotype and marker effects.

In plant genetics, GWAS serves as a potent tool for dissecting complex traits influenced by a myriad of genes and environmental factors. This approach has been widely employed

Table 2.5 The software packages for performing GWAS.

Software	Description	Features	Link
TASSEL	Software for linkage, evolution, and association analyses of traits	• Performs GWAS using GLM and MLM • Analyzes population structure and LD • Supports GBS data and visualization tools	[TASSEL] http://www.maizegenetics.net/tassel
GenStat	A statistical software for analyzing data from genotyping arrays	• Performs GWAS using GLM or MLM with population structure correction • Defines significance threshold and LD decay- Calculates SNP effects and G × E interaction	[GenStat] https://www.vsni.co.uk/software/genstat/
PLINK	A software for analyzing large-scale data sets of phenotypes and genotypes	• Performs analyses for population stratification, basic association tests, meta-analyses, gene-based tests, and epistasis screening • Provides graphical images and tables for GWAS and LD results	[PLINK] http://zzz.bwh.harvard.edu/plink/
GAPIT	An R package for genomic selection and prediction integrated tool	• Handles large data sets and reduces computational time • Includes MLM, P3D, and EMMA methods • Illustrates GWAS results by plots and tables with various parameters	[GAPIT] https://www.r-project.org/

across various crop species and attributes, encompassing domestication, stress tolerance, yield, and flowering time. For instance, studies by Zhu et al. (2008), Ersoz et al. (2007), Liu and Yan (2019), Sukumaran and Yu (2014), and Varshney et al. (2012) have harnessed GWAS to explore and understand various plant traits. A recent example by Wang et al. (2022) delves into the intricate interactions between genes and microbes that influence foxtail millet traits. By analyzing a staggering 827 cultivars and 257 microbes associated with six traits, their study unveiled that plant genes such as FLS2 and bHLH35 interact with these microbes. This suggests that modifying microbes could be a promising avenue for enhancing foxtail millet yield. Furthermore, researchers have conducted GWAS for various plant traits, including metabolites and panicle architecture, utilizing high-throughput phenotyping techniques. These efforts, as exemplified by studies from Chen et al. (2019), Zhou et al. (2019), and Spindel et al. (2018), showcase the versatility of GWAS in advancing our understanding of plant genetics.

GWAS can validate existing loci or discover novel ones associated with valuable traits. These insights can be harnessed for applications in crop breeding, genetic engineering, and genome editing, as demonstrated in studies by Gao (2021), Wang et al. (2016), and

Zhang et al. (2018). GWAS encounters several challenges, including population structure issues and rare alleles that can obscure the identification of genuine loci (Korte and Farlow 2013). A practical illustration of this challenge can be found in a study conducted on maize, where only a single gene (ZmCCT) associated with flowering time was discovered within a diverse panel of subjects (Yang et al. 2013). The influence of population structure on traits like flowering time can result in the omission of true loci when applying corrections during GWAS (Liu and Yan 2019; Huang and Han 2014). To tackle these challenges, two key approaches have been proposed: employing novel methods for detecting rare alleles (Zhu et al. 2011; Kaakinen et al. 2017) and utilizing intentionally designed populations to balance alleles and manage structural variations (Buckler et al. 2009; Dell'Acqua et al. 2015; Wen et al. 2019).

Another GWAS limitation pertains to the phenomenon where incorrect loci exhibit stronger associations than accurate ones, implying that authentic genes may be located quite a distance from GWAS peaks. Arabidopsis has observed this issue (Atwell et al. 2010; Kerdaffrec et al. 2016). Such spurious associations are termed synthetic associations and arise due to the linkage between common markers and rare variants (Liu and Yan 2019; Dickson et al. 2010). Notably, some factors involve multiple alleles within a single gene (Yano et al. 2016). As a potential solution, methods grounded in haplotypes or genes can be employed to identify such scenarios. However, it is crucial to acknowledge that current haplotype-based mapping has its limitations (Hayes 2013) and presents challenges when applied in plant genetics (Liu and Yan 2019). Improving the efficacy of haplotype analyses holds promise for advancing our comprehension of the genetics and functionality of complex traits.

2.4.10 Study of Entire Genomes and Identification of Desirable Genetic Traits

Recent advancements in computational and sequencing tools have empowered plant breeders to conduct in-depth explorations of plant genomes, comparing them across various individuals, populations, and species. These breakthroughs now offer the potential to unveil the genetic diversity, evolutionary patterns, and adaptive strategies of plants while also facilitating the discovery of novel genes and alleles associated with advantageous traits. Various techniques can be harnessed to undertake comprehensive studies of plant genomes, including *Whole-Genome Sequencing (WGS):* WGS allows for the complete decoding of an organism's genome. Beyond the functional annotation and gene expression analysis, WGS provides a holistic perspective on a plant's genome's genetic variation and structural composition. Additionally, it aids in detecting structural variations, such as insertions, deletions, inversions, or translocations, which may significantly impact a plant's phenotype (Daniell et al. 2016). *Pan-Genome Analysis:* This technique serves to compare the genomes of diverse individuals or populations within a species or closely related groups. A pan-genome comprises a core genome, consisting of genes shared by all entities, and a dispensable genome, housing genes present in some entities but absent in others. Apart from pinpointing genes unique to particular populations or ecological niches, pan-genome analysis provides insights into estimating the genetic diversity and divergence within a species or group of species (Ding et al. 2018). *Genotyping by Sequencing (GBS):* GBS entails sequencing reduced-representation libraries (RRLs) from multiple individuals to generate a profusion of molecular markers. RRLs are crafted by fragmenting genomic DNA using restriction enzymes and then selecting a subset of fragments for sequencing. GBS, characterized by its cost-efficiency and high throughput, delivers extensive marker coverage across the entire genome. It finds applications in genetic mapping, QTL analysis, GWAS, GS, population genetics, and phylogenetics, among others (Mas-Gómez et al. 2022).

2.4.11 Genomics, Transcriptomics, Proteomics, and Metabolomics Studies in Marker and Trait Identification

Genomics is the science of exploring an organism's entire genome, focusing on discovering and analyzing all its genetic sequences (Pocket 2006). This field encompasses deciphering genomic sequences, investigating gene functions, comparing genes across diverse organisms, and even generating three-dimensional structures of proteins. In the realm of crop agriculture, genomics serves as a valuable tool for deepening our comprehension of plant genomes, pinpointing genes of agronomic significance, and fostering the development of more nutritious and environmentally sustainable food sources (Pocket 2006). 'Transcriptomics' delves into the intricate patterns of gene expression, involving the analysis of an organism's complete set of RNA molecules, collectively known as the transcriptome. It provides insights into the mechanisms underlying how genes are transcribed into RNA and how RNA is edited into messenger RNA (mRNA), which conveys vital information to the ribosome for the synthesis of proteins (Pocket 2006). "Proteomics" explores the dynamic products of proteins and their interactions within a cell. It complements genomics and transcriptomics by shedding light on the functions and interplay of proteins encoded by genes. "Metabolomics" centers on the study of small molecules known as metabolites, which represent the end products of various cellular processes. It furnishes comprehensive information about an organism's metabolism and aids in unraveling the metabolic pathways underpinning various traits (Zenda et al. 2021). These distinct "omics" disciplines play pivotal roles in the identification of markers and traits. Transcriptomics examines gene expression patterns, whereas genomics provides a panoramic view of the entire set of genetic instructions. Proteomics investigates the dynamic protein products and their interactions, and metabolomics unveils the intricacies of an organism's metabolism (Pocket 2006). By harmonizing data from these diverse "omics" sciences, researchers gain a more comprehensive understanding of diseases, phenotypes, and traits of interest (Rattray et al. 2018).

2.4.12 Genome Editing Application in Crop Improvement

Conventional breeding has historically relied on natural or induced mutations, which occur infrequently and randomly in plant populations, requiring labor-intensive efforts to incorporate desired traits into elite crops. However, genome editing, an advanced molecular biology technique, offers a precise and targeted alternative to these random mutagenesis approaches (Gao 2015). Utilizing sequence-specific engineered endonucleases (SSNs), genome editing techniques have effectively altered plant genomes (Gaj et al. 2013). These technologies encompass various methods, including meganucleases or homing endonucleases, zinc finger nucleases (ZFNs), transcription activator-like effector nucleases (TALENs), and type II clustered regularly interspaced short palindromic repeat (CRISPR)/CRISPR-associated protein (Cas) systems (Sufyan et al. 2023). These genome-editing tools produce precise double-strand breaks in the DNA that are mostly repaired by the homology-directed repair (HDR) or nonhomologous end-joining (NHEJ) pathways. NHEJ is a common technique for inducing frameshift mutations at particular genomic loci, however, it is error-prone and frequently results in minor insertions or deletions at the desired locus. Conversely, the HDR pathway, a template-directed repair process, allows the insertion of custom sequences into the genome or the replacement of existing genomic sequences.

Genome editing technologies can alter the epigenome and regulate gene expression levels in addition to genetic mutations and gene substitution (Chen et al. 2019). Integrated with conventional breeding, this accelerates the incorporation of desired traits and reduces costs. Plants with genome editing that lack exogenous DNA can be created by genetically separating or crossing genes to eliminate genetic components that are necessary for genome editing from the genome (Mao et al. 2019). Of the genome-editing methods, CRISPR systems hold a unique advantage. Their potential for multiplexing enables the simultaneous editing of multiple target sites. By harnessing this capability, researchers have assembled multiple single-guide RNAs (sgRNAs) into a single Cas9/sgRNA expression vector, allowing distinct promoters to drive each sgRNA. Targeting a single gene with multiple sgRNAs can enhance editing rates, especially in crops with low transformation or editing efficiencies (Zhang et al. 2016, Li et al. 2018). Various tools have been developed for CRISPR-guide RNA design and off-target prediction, such as CRISPOR, CRISPick, Cas-Designer, CHOPCHOP, and many others. Several of these tools use ML models, like CRISPRscan, Azimuth, DeepSpCas9, DeepHF, AttnToCrispr_CNN, CRISPR-ONT, DeepSpCas9variants, DeepCpf1, DeepBaseEditor, and DeepPE. In fact, over 200 bioinformatics tools are available for designing CRISPR guide RNAs, in silico off-target detection, and genome editing in plants (Nakamae and Bono 2022; Naeem & Alkhnbashi 2023).

Genome editing can create new crop varieties from wild plants by modifying a few genes in them (Zhang 2019). Editing diploids is easier than editing polyploids, as the latter require editing more alleles at once for gene knockout. Polyploids possess multiple genomes that adapt and evolve more rapidly, which can also hide harmful mutations and enable new gene functions (Sattler et al. 2016).

2.5 Challenges and Future Directions

2.5.1 Interpretation and Management of Large Genetic Data as a Challenge

The field of plant genome research has produced a vast amount of data with the introduction of NGS technology. Furthermore, every month, millions of gigabytes of plant sequences are added to different public databases (Hale et al. 2021). Consequently, the need for continuously and rapidly upgrading these databases has become pressing. The process of updating must occur concurrently across all databases. The data generated must be employed to update secondary pathway databases. The validation of the actual pathway enzyme among potentially numerous homologous enzymes presents a significant challenge. Furthermore, problems with glycosylation, posttranslational modifications, and protein production limit the in vivo confirmation of gene activity in plant species. Only 20% of the genome and transcriptome data for over 600 plant species have been combined and thoroughly annotated (Kersey 2019). To address this, researchers have turned to ML techniques, providing a systematic approach to integrate and analyze multi-omics data and metabolic networks (Moore et al. 2019). However, challenges persist, necessitating the development of further statistical methodologies for analyzing vast datasets. Experimentation and sample preparation should be thoughtfully designed, taking into consideration the specific characteristics of plant and microbial chemical production.

2.5.2 Need for Advanced Methods and Novel Bioinformatics Tools

One of the biggest challenges in crop database building is the integration of annotated genome sequences, gene functions, interaction networks, and trait phenotypes. Relevant data are scattered throughout multiple databases with varying coverage, quality, and data formats. To tackle this, intelligent data mining tools such as KnetMiner (Knowledge Network Miner) must be developed, which searches for links and concepts across biological information networks, expediting the discovery of connections between traits and genes (Hu et al. 2018).

The ongoing development of integrative crop databases, coupled with advancements in data mining techniques, empowers breeders to gain deeper insights into complex traits and identify candidate genes associated with these traits. Effective management and the proper utilization of extensive research datasets require the implementation of Data Management Analysis and Decision Support Tools (DMASTs). Maintaining pedigrees, managing phenotypic data, conducting reliable breeding experiments, storing and retrieving high-throughput genotypic data, and carrying out various analytics are just a few of the many tasks for which these tools are invaluable.

These analytical techniques include trial analysis, geographical modifications, GEBV prediction, MTA identification, and selection index calculations. Additionally, by combining information from multiple research streams, DMASTs are essential for comprehending trait dynamics, gene activities, interactions, biology, genotype-by-environment interaction (GxE), and other aspects contributing to crop improvement programs (Rathore et al. 2018). There is a pressing need to develop such tools to effectively utilize the vast data available in plant genomics and transcriptomics, with implementation in plant breeding.

2.5.3 Potential for Bioinformatics to Contribute to Sustainable Agricultural Development

Plant genomics, transcriptomics, proteomics databases, and bioinformatic tools are pivotal in contributing to sustainable agricultural development. Bioinformatics plays a critical role in crop improvement by analyzing genetic data to identify beneficial traits and their associated genetic markers. This information aids breeders in developing crops with improved yield, disease resistance, drought tolerance, enhanced nutritional value, and improved soil nutrient cycling. Furthermore, bioinformatics plays a significant role in assessing the impact of climate change on crops and developing adaptation strategies. It involves modeling the effects of temperature, precipitation, CO_2 levels, and other factors on plant growth and development. For instance, bioinformatics can predict the phenology and yield of crops under different climate scenarios. Bioinformatics also contributes to the study of biodiversity by analyzing genetic and genomic data from various plant and animal species. Understanding biodiversity is crucial for conservation efforts and for identifying valuable genetic resources for crop improvement. New algorithms, tools, and platforms for handling large and complex omics datasets are needed to enhance the efficiency and accuracy of data analysis and integration. For instance, bioinformatics can help integrate genome, transcriptome, proteome, and metabolome data to uncover the molecular mechanisms underlying plant physiology and metabolism.

2.6 Conclusions

In the present era of Artificial Intelligence (AI) and ML technologies, bioinformatics is confronted with both remarkable opportunities and formidable challenges in its applications within the realm of plant breeding and crop improvement. The emergence of new bioinformatics tools contributes to the integration of existing bioinformatics resources, significantly expediting progress in improving crops and plant breeding. Food production systems are under tremendous strain in the current landscape, marked by a burgeoning global population, abrupt climate changes, and persistent reduction of arable land. There exists an urgent necessity to augment food production from the available agricultural land by cultivating high-yielding genotypes that can withstand both abiotic and biotic stressors (Vassilev et al. 2005). GS, MAS, GWAS, and phenomics have provided invaluable technical support to breeders, enabling them to develop novel, high-yielding varieties that exhibit greater adaptability to various environmental conditions. This, in turn, has led to a continuous enhancement of seed replacement rates.

Bioinformatics is a potent conduit through which data from diverse research domains converge to bolster the concurrent development and enrichment of molecular insights into crop plants. It encompasses data collection, organization, integration, storage, and sharing, thus facilitating multidisciplinary collaborations, expanding resources, and disseminating standardized methods. All these efforts are directed toward augmenting crop yield (Cao et al. 2020). Bioinformatics spearheads a revolution in plant breeding and crop improvement, providing critical information and tools necessary for augmenting yield and developing genotypes that can withstand both abiotic and biotic stress. The bioinformatics domain is constantly evolving, and we hold a profound optimism regarding its continued integration into plant breeding and crop enhancement endeavors. The future of crop improvement stands to benefit significantly from this dynamic field.

References

Ahsan, A., Monir, M., Meng, X. et al. (2019). Identification of epistasis loci underlying rice flowering time by controlling population stratification and polygenic effect. *DNA Research* 26 (2): 119–130.

Alcántara, R., Axelsen, K.B., Morgat, A. et al. (2012). Rhea—a manually curated resource of biochemical reactions. *Nucleic Acids Research* 40 (D1): D754–D760.

Atanda, S.A., Olsen, M., Burgueño, J. et al. (2021). Maximizing efficiency of genomic selection in CIMMYT's tropical maize breeding program. *Theoretical and Applied Genetics* 134: 279–294.

Atwell, S., Huang, Y.S., Vilhjálmsson, B.J. et al. (2010). Genome-wide association study of 107 phenotypes in *Arabidopsis thaliana* inbred lines. *Nature* 465 (7298): 627–631. https://doi.org/10.1038/nature08800.

Bassal, M., Abukhalaf, M., Majovsky, P. et al. (2020). Reshaping of the *Arabidopsis thaliana* proteome landscape and co-regulation of proteins in development and immunity. *Molecular Plant* 13: 1709–1732.

Baxevanis, A.D. (2004). Information retrieval from biological databases. In: *Bioinformatics: A Practical Guide to the Analysis of Genes & Proteins*, 3e (ed. A.D. Baxevanis and B.F.F. Ouellette), 123–124. New York: John Wiley & Sons; ISBN: 0471 478784; 540pp.; 2004; $79.95.

Belamkar, V., Guttieri, M.J., Hussain, H. et al. (2018). Genomic selection in preliminary yield trials in a winter wheat breeding program. *G3: Genes, Genomes, Genetics* 8: 2735–2747. https://doi.org/10.1534/g3.118.200415.

Bernardo, R. (1994). Prediction of maize single-cross performance using RFLPs and information from related hybrids. *Crop Science* 34: 20–25. https://doi.org/10.2135/cropsci1994.0011183X003400010003x.

Bevan, M.W., Uauy, C., Wulff, B.B. et al. (2017). Genomic innovation for crop improvement. *Nature* 543 (7645): 346–354.

Bhat, J.A., Ali, S., Salgotra, R.K. et al. (2016). Genomic selection in the era of next generation sequencing for complex traits in plant breeding. *Frontiers in Genetics* 7: 221.

Bink, M.C.A.M., Boer, M.P., Ter Braak, C.J.F. et al. (2008). Bayesian analysis of complex traits in pedigreed plant populations. *Euphytica* 161: 85–96.

Boer, M.P., Wright, D., Feng, L. et al. (2007). A mixed-model quantitative trait loci (QTL) analysis for multiple-environment trial data using environmental covariables for QTL-by-environment interactions, with an example in maize. *Genetics* 177 (3): 1801–1813. https://doi.org/10.1534/genetics.107.071068.

Bolger, M.E., Arsova, B., and Usadel, B. (2018). Plant genome and transcriptome annotations: from misconceptions to simple solutions. *Briefings in Bioinformatics* 19 (3): 437–449.

Broman, K.W., Gatti, D.M., Simecek, P. et al. (2019). R/qtl2: software for mapping quantitative trait loci with high-dimensional data and multiparent populations. *Genetics* 211 (2): 495–502.

Buckler, E.S., Holland, J.B., Bradbury, P.J. et al. (2009). The genetic architecture of maize flowering time. *Science* 325 (5941): 714–718. https://doi.org/10.1126/science.1174276.

Budhlakoti, N., Rai, A., and Mishra, D.C. (2020a). Effect of influential observation in genomic prediction using LASSO diagnostic. *Indian Journal of Agricultural Sciences* 90: 1155–1159.

Budhlakoti, N., Rai, A., and Mishra, D.C. (2020b). Statistical approach for improving genomic prediction accuracy through efficient diagnostic measure of influential observation. *Scientific Reports* 10: 1–11. https://doi.org/10.1038/s41598-020-65323-3.

Budhlakoti, N., Rai, A., Mishra, D.C. et al. (2020c). Comparative study of different non-parametric genomic selection methods under diverse genetic architecture. *Indian Journal of Genetics and Plant Breeding* 80: 395–401. https://doi.org/10.31742/IJGPB.80.4.4.

Bustamam, M., Tabien, R.E., Suwarno, A. et al. (2002). Asian rice biotechnology network: improving popular cultivars through marker-assisted backcrossing by the NARES. Poster presented at the International Rice Congress. September 16–20, Beijing, China.

Cantelmo, N.F., Von Pinho, R.G., and Balestre, M. (2016). Genomic breeding value prediction for simple maize hybrid yield using total effects of associated markers, under different imbalance levels and environments. *Genetics and Molecular Research* 15: gmr.15017232.

Cao, Y., Geddes, T.A., Yang, J.Y.H., and Yang, P. (2020). Ensemble deep learning in bioinformatics. *Nature Machine Intelligence* 2: 500–508.

Caspi, R., Billington, R., Fulcher, C.A. et al. (2018). The MetaCyc database of metabolic pathways and enzymes. *Nucleic Acids Research* 46 (D1): D633–D639. https://doi.org/10.1093/nar/gkx935.

Chavhan, R.L., Hinge, V.R., Kadam, U.S. et al. (2017). Sequence Analysis of coat protein and molecular profiling of sunflower necrosis virus (SNV) strains from Indian subcontinent. *Journal of Plant Biochemistry and Biotechnology* 27: 28–35. https://doi.org/10.1007/s13562-017-0412-z.

Chavhan, R.L., Hinge, V.R., Kadam, U.S. et al. (2018). Real-time PCR assay for rapid, efficient and accurate detection of *Paramyrothecium roridum* a leaf spot pathogen of *Gossypium* species. *Journal of Plant Biochemistry and Biotechnology* 27: 199–207. https://doi.org/10.1007/s13562-017-0431-9.

Chavhan, R.L., Sable, S., Narwade, A.V. et al. (2023). Multiplex molecular marker-assisted analysis of significant pathogens of cotton (*Gossypium* sp.). *Biocatalysis and Agriculture Biotechnology* 47: 102557.

Chen, F., Song, Y., Li, X. et al. (2019). Genome sequences of horticultural plants: past, present, and future. *Horticulture Research* 6.

Chen, Y., Li, E.M., and Xu, L.Y. (2022). Guide to metabolomics analysis: a bioinformatics workflow. *Metabolites* 12 (4): 357.

Cheng, Y., Bi, X., Xu, Y. et al. (2023). Machine learning for metabolic pathway optimization: a review. *Computational and Structural Biotechnology Journal* 21: 2381–2393. https://doi.org/10.1016/j.csbj.2023.03.045.

Claros, M.G., Bautista, R., Guerrero-Fernández, D. et al. (2012). Why assembling plant genome sequences is so challenging. *Biology (Basel)* 1 (2): 439–459. https://doi.org/10.3390/biology1020439.

Cooper, M. (1999). Concepts and strategies for plant adaptation research in rainfed lowland rice. *Field Crops Research* 64 (1–2): 13–34. https://doi.org/10.1016/ S0378-4290(99)00048-9.

Cooper, M., Messina, C.D., Podlich, D. et al. (2014). Predicting the future of plant breeding: complementing empirical evaluation with genetic prediction. *Crop and Pasture Science* 65 (4): 311–336.

Crossa, J., Pérez-Rodríguez, P., Cuevas, J. et al. (2017). Genomic selection in plant breeding: methods, models, and perspectives. *Trends in Plant Science* 22 (11): 961–975.

Daniell, H., Lin, C.S., Yu, M., and Chang, W.J. (2016). Chloroplast genomes: diversity, evolution, and applications in genetic engineering. *Genome Biology* 17: 1–29.

Danilevicz, M.F., Fernandez, C.G.T., Marsh, J.I. et al. (2020). Plant pangenomics: approaches, applications and advancements. *Current Opinion in Plant Biology* 54: 18–25.

Dar, A.A., Mahajan, R., and Sharma, S. (2019). Molecular markers for characterization and conservation of plant genetic resources. *Indian Journal of Agricultural Sciences* 89 (11): 1755–1763.

De Las Rivas, J. and Fontanillo, C. (2010). Protein–protein interactions essentials: key concepts to building and analyzing interactome networks. *PLoS Computational Biology* 6 (6): e1000807.

Dell'Acqua, M., Gatti, D.M., Pea, G. et al. (2015). Genetic properties of the MAGIC maize population: a new platform for high definition QTL mapping in *Zea mays*. *Genome Biology* 16 (1): 167. https://doi.org/10.1186/s13059-015-0716-z.

Delvadiya, N.A., Mehta, D.R., and Nandha, A.K. (2022). Construction of linkage mapping and identification of quantitative trait loci (QTL) for grain size and related traits in bread wheat (*Triticum aestivum* L.). *The Pharma Innovation* 11 (5): 1399–1405.

Denis, M. and Bouvet, J.M. (2013). Efficiency of genomic selection with models including dominance effect in the context of *Eucalyptus* breeding. *Tree Genetics & Genomes* 9: 37–51. https://doi.org/10.1007/s11295-012-0528-1.

Dickson, S.P., Wang, K., Krantz, I. et al. (2010). Rare variants create synthetic genome-wide associations. *PLoS Biology* 8 (1): e1000294. https://doi.org/10.1371/journal.pbio.1000294.

Ding, W., Baumdicker, F., and Neher, R.A. (2018). panX: pan-genome analysis and exploration. *Nucleic Acids Research* 46 (1): e5–e5.

Dita, M.A., Rispail, N., Prats, E. et al. (2006). Biotechnology approaches to overcome biotic and abiotic stress constraints in legumes. *Euphytica* 147: 1–24.

van Eeuwijk, F.A., Bustos-Korts, D., Millet, E.J. et al. (2018). Modelling strategies for assessing and increasing the effectiveness of new phenotyping techniques in plant breeding. *Plant Science* 282: 23–39. https://doi.org/10.1016/j.plantsci.2018.06.018.

Ersoz, E.S., Yu, J., and Buckler, E.S. (2007). Applications of linkage disequilibrium and association mapping in crop plants. In: *Genomics-Assisted Crop Improvement: Vol. 1: Genomics Approaches and Platforms*, 97–119. Dordrecht: Springer Netherlands.

Friedman, N. (2004). Inferring cellular networks using probabilistic graphical models. *Science* 303: 799–805. https://doi.org/10.1126/science.1094068.

Friedman, J., Hastie, T., and Tibshirani, R. (2010). Regularization paths for generalized linear models via coordinate descent. *Journal of Statistical Software* 33 (1): 1.

Gaj, T., Mercer, A.C., Sirk, S.J. et al. (2013). A comprehensive approach to zinc-finger recombinase customization enables genomic targeting in human cells. *Nucleic Acids Research* 41: 3937–3946.

Gao, C. (2015). Genome editing in crops: from bench to field. *National Science Review* 2: 13–15.

Gao, C. (2021). Genome engineering for crop improvement and future agriculture. *Cell* 184 (6): 1621–1635.

Gao, J., Ellis, L.B., and Wackett, L.P. (2010). The University of Minnesota biocatalysis/biodegradation database: improving public access. *Nucleic Acids Research* 38 (suppl_1): D488–D491.

Gidamo, G.H. (2022). Genomic selection: a faster strategy for plant breeding. In: *Case Studies of Breeding Strategies in Major Plant Species* (ed. H. Wang). IntechOpen.

Gupta, P.K., Kulwal, P.L., and Jaiswal, V. (2019). Association mapping in plants in the post-GWAS genomics era. *Advances in Genetics* 104: 75–154.

Habier, D., Fernando, R.L., and Garrick, D.J. (2013). Genomic BLUP decoded: a look into the black box of genomic prediction. *Genetics* 194 (3): 597–607.

Hale, I., Ma, X., Melo, A.T. et al. (2021). Genomic resources to guide improvement of the Shea tree. *Frontiers in Plant Science* 12: 720670.

Hasan, N., Choudhary, S., Naaz, N. et al. (2021). Recent advancements in molecular marker-assisted selection and applications in plant breeding programmes. *Journal of Genetic Engineering and Biotechnology* 19 (1): 1–26.

Hayes, B. (2013). Overview of statistical methods for genome-wide association studies (GWAS). *Genome-wide Association Studies and Genomic Prediction* 149–169.

Henkrar, F. and Udupa, S. (2020). Marker assisted selection in plant breeding. Moroccan Journal of. *Agricultural Sciences* 1 (5).

Henry, C.S., DeJongh, M., Best, A.A. et al. (2010). High-throughput generation, optimization and analysis of genome-scale metabolic models. *Nature Biotechnology* 28 (9): 977–982.

Hill, W.G. and Mäki-Tanila, A. (2015). Expected influence of linkage disequilibrium on genetic variance caused by dominance and epistasis on quantitative traits. *Journal of Animal Breeding and Genetics* 132 (2): 176–186. https://doi.org/10.1111/jbg.12140. PMID: 25823842.

Hinge, V.R., Chavhan, R.L., Kale, S.P. et al. (2021). Engineering resistance against viruses in field crops using CRISPR-Cas9. *Current Genomics* 22 (3): 214–231. https://doi.org/10.2174/1389202922666210412102214.

Hu, H., Scheben, A., and Edwards, D. (2018). Advances in integrating genomics and bioinformatics in the plant breeding pipeline. *Agriculture* 8 (6): 75.

Hu, L., Wang, X., Huang, Y.A. et al. (2021). A survey on computational models for predicting protein–protein interactions. *Briefings in Bioinformatics* 22 (5): bbab036.

Huang, X. and Han, B. (2014). Natural variations and genome-wide association studies in crop plants. *Annual Review of Plant Biology* 65: 531–551. https://doi.org/10.1146/annurev-arplant-050213-035715. Epub 2013 Nov 20. PMID: 24274033.

Ibrahim, A.K., Zhang, L., Niyitanga, S. et al. (2020). Principles and approaches of association mapping in plant breeding. *Tropical Plant Biology* 13: 212–224. https://doi.org/10.1007/s12042-020-09261-4.

Jansen, R., Yu, H., Emili, A. et al. (2003). A Bayesian networks approach for predicting protein–protein interactions from genomic data. *Science* 302: 449–453. https://doi.org/10.1126/science.1087361.

Jeon, D., Kang, Y., Lee, S. et al. (2023). Digitalizing breeding in plants: a new trend of next-generation breeding based on genomic prediction. *Frontiers in Plant Science* 14: 1092584.

Jung, H., Ventura, T., Chung, J.S. et al. (2020). Twelve quick steps for genome assembly and annotation in the classroom. *PLoS Computational Biology* 16 (11): e1008325.

Kaakinen, M., Mägi, R., Fischer, K. et al. (2017). MARV: a tool for genome-wide multi-phenotype analysis of rare variants. *BMC Bioinformatics* 18 (1): 110. https://doi.org/10.1186/s12859-017-1530-2.

Kanehisa, M., Sato, Y., Furumichi, M. et al. (2019). New approach for understanding genome variations in KEGG. *Nucleic Acids Research* 47 (D1): D590–D595. https://doi.org/10.1093/nar/gky962.

Karp, P.D., Krummenacker, M., Paley, S., and Wagg, J. (1999). Integrated pathway-genome databases and their role in drug discovery. *Trends in Biotechnology* 17 (7): 275–281. https://doi.org/10.1016/s0167-7799(99)01316-5.

Karp, P.D., Billington, R., Caspi, R. et al. (2019). The BioCyc collection of microbial genomes and metabolic pathways. *Briefings in Bioinformatics* 20 (4): 1085–1093. https://doi.org/10.1093/bib/bbx085.

Kerdaffrec, E., Filiault, D.L., Korte, A. et al. (2016). Multiple alleles at a single locus control seed dormancy in Swedish *Arabidopsis*. *eLife* 5: e22502. https://doi.org/10.7554/eLife.22502.

Kersey, P.J. (2019). Plant genome sequences: past, present, future. *Current Opinion in Plant Biology* 48: 1–8.

Kim, S.H., Bahk, S., Nguyen, N.T. et al. (2022). Phosphorylation of the auxin signaling transcriptional repressor IAA15 by MPKs is required for suppression of root development under drought stress in Arabidopsis. *Nucleic Acids Research* 50–18: 10544–10561.

King, Z.A., Lu, J., Dräger, A. et al. (2016). BiGG Models: a platform for integrating, standardizing and sharing genome-scale models. *Nucleic Acids Research* 44 (D1): D515–D522.

Koh, G.C., Porras, P., Aranda, B. et al. (2012). Analyzing protein–protein interaction networks. *Journal of Proteome Research* 11 (4): 2014–2031.

Kong, W., Wang, Y., Zhang, S. et al. (2023). Recent advances in assembly of plant complex genomes. *Genomics, Proteomics & Bioinformatics*.

Korte, A. and Farlow, A. (2013). The advantages and limitations of trait analysis with GWAS: a review. *Plant Methods* 9 (1): 1–9.

Kumar, S. and Dudley, J. (2007). Bioinformatics software for biologists in the genomics era. *Bioinformatics* 23 (14): 1713–1717.

Kyriakidou, M., Tai, H.H., Anglin, N.L. et al. (2018). Current strategies of polyploid plant genome sequence assembly. *Frontiers in Plant Science* 9: 1660. https://doi.org/10.3389/fpls.2018.01660.

LeDeaux, J.R., Graham, G.I., and Stuber, C.W. (2006). Stability of QTLs involved in heterosis in maize when mapped under several stress conditions. *Maydica* 51: 151–167.

Lesk, A. (2019). *Introduction to Bioinformatics*. Oxford University Press.

Li, R., Li, R., Li, X. et al. (2018). Multiplexed CRISPR/Cas9-mediated metabolic engineering of γ-aminobutyric acid levels in *Solanum lycopersicum*. *Plant Biotechnology Journal* 16: 415–427.

Liabeuf, D., Sim, S.C., and Francis, D.M. (2018). Comparison of marker-based genomic estimated breeding values and phenotypic evaluation for selection of bacterial spot resistance in tomato. *Phytopathology* 108 (3): 392–401.

Lincoln, S. E., Daly, M. J., & Lander, E. S. (1993). Mapping genes controlling quantitative traits using MAPMAKER/QTL version 1.1: a tutorial and reference manual. Whitehead Institute for Biomedical Research Technical Report.

Liu, H.J. and Yan, J. (2019). Crop genome-wide association study: a harvest of biological relevance. *The Plant Journal* 97 (1): 8–18. https://doi.org/10.1111/tpj.14139. Epub 2018 Dec 17. PMID: 30368955.

Loera-Sánchez, M., Studer, B., and Kölliker, R. (2019). DNA-based assessment of genetic diversity in grassland plant species: challenges, approaches, and applications. *Agronomy* 9 (12): 881.

Lowe, R., Shirley, N., Bleackley, M. et al. (2017). Transcriptomics technologies. *PLoS Computational Biology* 13 (5): e1005457. https://doi.org/10.1371/journal.pcbi.1005457.

Lush, J.L. (1943). *Animal Breeding Plans*, 2e. Charleston, SC: Bibliolife DBA of Bibilio Bazaar.

Maenhout, S., De Baets, B., Haesaert, G., and Van Bockstaele, E. (2007). Support vector machine regression for the prediction of maize hybrid performance. *Theoretical and Applied Genetics* 115: 1003–1013.

Mahalle, R.M., Bosamia, T.C., Chakravarty, S. et al. (2023). De novo mining and validating novel microsatellite markers to assess genetic diversity in *Maruca vitrata* (F.), a Legume Pod Borer. *Genes* 14 (7): 1433.

Mahmud, M., Kaiser, M.S., McGinnity, T.M., and Hussain, A. (2021). Deep learning in mining biological data. *Cognitive Computation* 13: 1–33.

Majumdar, S.G., Rai, A., and Mishra, D.C. (2020). Integrated framework for selection of additive and nonadditive genetic markers for genomic selection. *Journal of Computational Biology* 27: 845–855. https://doi.org/10.1089/CMB.2019.0223.

Mäki-Tanila, A. and Hill, W.G. (2014). Influence of gene interaction on complex trait variation with multilocus models. *Genetics* 198 (1): 355–367. https://doi.org/10.1534/genetics.114.165282.

Mammadov, J., Aggarwal, R., Buyyarapu, R., and Kumpatla, S. (2012). SNP markers and their impact on plant breeding. *International Journal of Plant Genomics* 2012.

Mao, Y., Botella, J.R., Liu, Y., and Zhu, J.K. (2019). Gene editing in plants: progress and challenges. *National Science Review* 6 (3): 421–437. https://doi.org/10.1093/nsr/nwz005.

Mas-Gómez, J., Cantín, C.M., Moreno, M.Á., and Martínez-García, P.J. (2022). Genetic diversity and genome-wide association study of morphological and quality traits in peach using two Spanish Peach germplasm collections. *Frontiers in Plant Science* 13: 854770.

Mathur, M. (2018). Bioinformatics challenges: a review. *International Journal of Advanced Science and Research* 3 (6): 29–33.

Meena, M.R., Kumar, R., Chinnaswamy, A. et al. (2020). Current breeding and genomic approaches to enhance the cane and sugar productivity under abiotic stress conditions. *3 Biotech* 10: 1–18.

Menschaert, G. and Fenyö, D. (2017). Proteogenomics from a bioinformatics angle: a growing field. *Mass Spectrometry Reviews* 36 (5): 584–599.

Mergner, J. and Kuster, B. (2022). Plant proteome dynamics. *Annual Review of Plant Biology* 73: 67–92.

Meuwissen, T.H.E., Hayes, B.J., and Goddard, M.E. (2001). Prediction of total genetic value using genome-wide dense marker maps. *Genetics* 157: 1819–1829.

Miklas, P.N., Kelly, J.D., Beebe, S.E., and Blair, M.W. (2006). Common bean breeding for resistance against biotic and abiotic stresses: from classical to MAS breeding. *Euphytica* 147: 105–131.

Mishra, D.C., Budhlakoti, N., Majumdar, S.G., and Rai, A. (2021). Innovations in genomic selection: statistical perspective. 101–111. https://ssca.org.in/media/9_Spl_Proceedings_2021_006072021_Dwijesh_Mishra_Final.pdf.

Mohammadi, M., Xavier, A., Beckett, T. et al. (2020). Identification, deployment, and transferability of quantitative trait loci from genome-wide association studies in plants. *Current Plant Biology* 24: 100145.

Montesinos López, O.A., Montesinos López, A., and Crossa, J. (2022). *Multivariate Statistical Machine Learning Methods for Genomic Prediction* (p. 691). Springer Nature.

Montesinos-López, O.A., Montesinos-López, A., Pérez-Rodríguez, P. et al. (2021). A review of deep learning applications for genomic selection. *BMC Genomics* 22: 1–23.

Moore, B.M., Wang, P., Fan, P. et al. (2019). Robust predictions of specialized metabolism genes through machine learning. *Proceedings of the National Academy of Sciences of the United States of America* 116 (6): 2344–2353.

Morbia, I., Dubey, R., and Mathur, S. (2023). Review on applicability of bioinformatics in current research and database management. *SSR Institute of International Journal of Life Sciences* 9 (2): 3195–3205.

Morgil, H., Gercek, Y.C., and Tulum, I. (2020). Single nucleotide polymorphisms (SNPs) in plant genetics and breeding. In: *The Recent Topics in Genetic Polymorphisms* (ed. M. Çalışkan, O. Erol, and G.C. Öz), 825–400. IntechOpen.

Môro, G.V., Santos, M.F., and de Souza Júnior, C.L. (2019). Comparison of genome-wide and phenotypic selection indices in maize. *Euphytica* 215: 76. https://doi.org/10.1007/s10681-019-2401-x.12.

Mu, H., Wang, B., and Yuan, F. (2022). Bioinformatics in plant breeding and research on disease resistance. *Plants* 11 (22): 3118.

Naeem, M. and Alkhnbashi, O.S. (2023). Current bioinformatics tools to optimize CRISPR/Cas9 experiments to reduce off-target effects. *International Journal of Molecular Sciences* 24 (7): 6261.

Nakamae, K. and Bono, H. (2022). Genome editing and bioinformatics. *Gene and Genome Editing* 100018.

Netla, R.V., Shinde, H., Kumar, G. et al. (2023). A comparative analysis of single-cell transcriptomic technologies in plants and animals. *Current Plant Biology* 35-36: 100289.

Neves, H.H., Carvalheiro, R., and Queiroz, S.A. (2012). A comparison of statistical methods for genomic selection in a mice population. *BMC Genetics* 13 (1): 1–17.

Nikalje, G.C., Srivastava, A.K., Shelake, R.M. et al. (2023). Profiling of conserved orthologs and miRNAs for understanding their role in salt tolerance mechanism of *Sesuvium portulacastrum* L. *Molecular Biology Reports* 50: 9731–9738.

Ogata, H., Goto, S., Fujibuchi, W., and Kanehisa, M. (1998). Computation with the KEGG Pathway Database. *Biosystems* 47 (1): 119–128. https://doi.org/10.1016/S0303-2647(98)00017-3.

Osei, M.K., Asante, M.D., Agyeman, A. et al. (2014). Plant breeding: a tool for achieving food sufficiency. In: *Sustainable Horticultural Systems: Issues, Technology and Innovation* (ed. D. Nandwani), 253–274. Springer.

Paley, S.M. and Karp, P.D. (2002). Predictions for *Helicobacter pylori*. *Bioinformatics* 18 (5): 715–724. https://doi.org/10.1093/bioinformatics/18.5.715.

Pocket, K. (2006). *No. 15.'Omics' Sciences: Genomics, Proteomics, and Metabolomics*. ISAAA.

Qi, Q., Li, J., and Cheng, J. (2014). Reconstruction of metabolic pathways by combining probabilistic graphical model-based and knowledge-based methods. *BMC Proceedings* 8 (6): 1–10. https://doi.org/10.1186/1753-6561-8-S6-S5.

Rahim, M.S., Bhandawat, A., Rana, N. et al. (2020). Genomic selection in cereal crops: methods and applications. In: *Accelerated Plant Breeding, Volume 1: Cereal Crops* (ed. S.S. Gosal and S.H. Wani), 51–88. Cham: Springer.

Rathore, A., Singh, V.K., Pandey, S.K. et al. (2018). Current status and future prospects of next-generation data management and analytical decision support tools for enhancing genetic gains in crops. In: *Plant Genetics and Molecular Biology* (ed. R.K. Varshney, M.K. Pandey, and A. Chitikineni), 277–292. Cham: Springer.

Rattray, N.J., Deziel, N.C., Wallach, J.D. et al. (2018). Beyond genomics: understanding exposotypes through metabolomics. *Human Genomics* 12 (1): 1–14.

Robertsen, C.D., Hjortshøj, R.L., and Janss, L.L. (2019). Genomic selection in cereal breeding. *Agronomy* 9 (2): 95.

dos Santos, J.P.R., Pires, L.P.M., de Castro Vasconcellos, R.C. et al. (2016). Genomic selection to resistance to *Stenocarpella maydis* in maize lines using DArTseq markers. *BMC Genetics* 17: 86. https://doi.org/10.1186/s12863-016-0392-3.

Sao, R., Sahu, P.K., Sharma, D. et al. (2020). Comparative study of radio-sensitivity and relative biological effectiveness of gamma rays, X-rays, electron beam and proton beam in short grain aromatic rice. *Indian Journal Of Genetics And Plant Breeding* 80 (04): 384–394.

Sattler, M.C., Carvalho, C.R., and Clarindo, W.R. (2016). The polyploidy and its key role in plant breeding. *Planta* 243: 281–296.

Scheer, M., Grote, A., Chang, A. et al. (2010). BRENDA, the enzyme information system in 2011. *Nucleic Acids Research* 39 (suppl_1): D670–D676.

Seaton, G., Haley, C.S., Knott, S.A. et al. (2002). QTL Express: mapping quantitative trait loci in simple and complex pedigrees. *Bioinformatics* 18 (2): 339–340.

Shah, H.A., Liu, J., Yang, Z., and Feng, J. (2021). Review of machine learning methods for the prediction and reconstruction of metabolic pathways. *Frontiers in Molecular Biosciences* 8: 634141.

Shinde, H., Dudhate, A., Sathe, A. et al. (2023a). Gene coexpression analysis identifies genes associated with chlorophyll content and relative water content in pearl millet. *Plants* 12 (6): 1412.

Shinde, H., Dudhate, A., Kadam, U.S., and Hong, J.C. (2023b). RNA methylation in plants: an overview. Frontiers. *Plant Science* 14: 1132959.

Smallwood, C.J., Saxton, A.M., Gillman, J.D. et al. (2019). Context-specific genomic selection strategies outperform phenotypic selection for soybean quantitative traits in the progeny row stage. *Crop Science* 59 (1): 54–67.

Spindel, J.E., Dahlberg, J., Colgan, M. et al. (2018). Association mapping by aerial drone reveals 213 genetic associations for *Sorghum bicolor* biomass traits under drought. *BMC Genomics* 19 (1): 1–18.

Steele, K.A., Price, A.H., Shashidhar, H.E., and Witcombe, J.R. (2006). Marker-assisted selection to introgress rice QTLs controlling root traits into an Indian upland rice variety. *Theoretical and Applied Genetics* 112: 208–222.

Sufyan, M., Daraz, U., Hyder, S. et al. (2023). An overview of genome engineering in plants, including its scope, technologies, progress and grand challenges. *Functional & Integrative Genomics* 23 (2): 119.

Sukumaran, S. and Yu, J. (2014). Association mapping of genetic resources: achievements and future perspectives. In: *Genomics of Plant Genetic Resources: Volume 1. Managing, sequencing and mining genetic resources*, 207–235. Dordrecht: Springer https://doi.org/10.1007/978-94-007-7572-5_9.

Sutter, J., Li, J., Maynard, A.J. et al. (2011). New features that improve the pharmacophore tools from Accelrys. *Current Computer-Aided Drug Design* 7 (3): 173–180.

Tan, Y.C., Kumar, A.U., Wong, Y.P., and Ling, A.P.K. (2022). Bioinformatics approaches and applications in plant biotechnology. *Journal of Genetic Engineering and Biotechnology* 20 (1): 1–13.

Tanaka, E. (2018). Simple robust genomic prediction and outlier detection for a multi-environmental field trial. arXiv preprint arXiv:1807.07268, 1–25.

Tanaka, E. (2020). Simple outlier detection for a multi-environmental field trial. *Biometrics* 76: 1374–1382. https://doi.org/10.1111/BIOM.13216.

Thompson, M.G., Valencia, L.E., Blake-Hedges, J.M. et al. (2019). Omics-driven identification and elimination of valerolactam catabolism in *Pseudomonas putida* KT2440 for increased product titer. *Metabolic Engineering Communications* 9: e00098.

Tibshirani, R. (1996). Regression shrinkage and selection via the lasso. *Journal of the Royal Statistical Society, Series B: Statistical Methodology* 58 (1): 267–288.

Todaka, D., Nakashima, K., Shinozaki, K., and Yamaguchi-Shinozaki, K. (2012). Toward understanding transcriptional regulatory networks in abiotic stress responses and tolerance in rice. *Rice* 5 (1): 1–9.

Upadhyay, A., Kadam, U.S., Chacko, P.M. et al. (2010a). Microsatellite analysis to differentiate clones of Thompson Seedless grapevine. *Indian Journal of Horticulture* 67 (2): 260–263.

Upadhyay, A., Kadam, U.S., Chacko, P.M. et al. (2010b). Microsatellite and RAPD analysis of grape (*Vitis* spp.) accessions and identification of duplicates/misnomers in germplasm collection. *Indian Journal of Horticulture* 67 (1): 8–15.

Varshney, R.K., Paulo, M.J., Grando, S. et al. (2012). Genome wide association analyses for drought tolerance related traits in barley (*Hordeum vulgare* L.). *Field Crops Research* 126: 171–180.

Vassilev, D., Leunissen, J., Atanassov, A. et al. (2005). Application of bioinformatics in plant breeding. *Biotechnology and Biotechnological Equipment* 19: 139–152.

Viswanathan, G.A., Seto, J., Patil, S. et al. (2008). Getting started in biological pathway construction and analysis. *PLoS Computational Biology* 4 (2): 16. https://doi.org/10.1371/journal.pcbi.0040016.

Vivek, B.S. et al. (2017). Use of genomic estimated breeding values results in rapid genetic gains for drought tolerance in maize. *Plant Genome* 10 (1–8): 11.

Wang, N. and Huo, Y.X. (2022). Using genome and transcriptome analysis to elucidate biosynthetic pathways. *Current Opinion in Biotechnology* 75: 102708.

Wang, D., Salah El-Basyoni, I., Stephen Baenziger, P. et al. (2012). Prediction of genetic values of quantitative traits with epistatic effects in plant breeding populations. *Heredity (Edinb).* 109 (5): 313–319. https://doi.org/10.1038/hdy.2012.44.

Wang, X., Yang, Z., and Xu, C. (2015). A comparison of genomic selection methods for breeding value prediction. *Science Bulletin* 60 (10): 925–935.

Wang, H., Xu, X., Vieira, F.G. et al. (2016). The power of inbreeding: NGS-based GWAS of rice reveals convergent evolution during rice domestication. *Molecular Plant* 9 (7): 975–985.

Wang, M., Mao, Y., Lu, Y. et al. (2017a). Multiplex gene editing in rice using the CRISPR-Cpf1 system. *Molecular Plant* 10: 1011–1013.

Wang, L., Dash, S., Ng, C.Y., and Maranas, C.D. (2017b). A review of computational tools for design and reconstruction of metabolic pathways. *Synthetic and Systems Biotechnology* 2 (4): 243–252.

Wang, X., Xu, Y., Hu, Z., and Xu, C. (2018). Genomic selection methods for crop improvement: current status and prospects. *The Crop Journal* 6 (4): 330–340.

Wang, D., Zhang, C., Wang, B. et al. (2019a). Optimized CRISPR guide RNA design for two high-fidelity Cas9 variants by deep learning. *Nature Communications* 10 (1): 4284.

Wang, X., Williams, C., Liu, Z.H., and Croghan, J. (2019b). Big data management challenges in health research—a literature review. *Briefings in Bioinformatics* 20 (1): 156–167.

Wang, S., Wu, R., Lu, J. et al. (2022). Protein-protein interaction networks as miners of biological discovery. *Proteomics* 22 (15-16): 2100190.

Wen, Y.J., Zhang, Y.W., Zhang, J. et al. (2019). An efficient multi-locus mixed model framework for the detection of small and linked QTLs in F2. *Briefings in Bioinformatics* 20 (5): 1913–1924. https://doi.org/10.1093/bib/bby058.

Werhli, A.V., Grzegorczyk, M., and Husmeier, D. (2006). Comparative evaluation of reverse engineering gene regulatory networks with relevance networks, graphical gaussian models and bayesian networks. *Bioinformatics* 22: 2523–2531. https://doi.org/10.1093/bioinformatics/btl391.

Wilde, F., Schön, C.C., Korzun, V. et al. (2008). *Theoretical and Applied Genetics* 117: 29.

Xu, Y. (2016). Envirotyping for deciphering environmental impacts on crop plants. *Theoretical and Applied Genetics* 129: 653–673. https://doi.org/10.1007/s00122-016-2691-5.

Xu, Y. and Crouch, J.H. (2008). Marker-assisted selection in plant breeding: from publications to practice. *Crop Science* 48 (2): 391–407.

Yakubu, R.R., Weiss, L.M., and Silmon de Monerri, N.C. (2018). Post-translational modifications as key regulators of apicomplexan biology: insights from proteome-wide studies. *Molecular Microbiology* 107 (1): 1–23.

Yamanishi, Y., Tabei, Y., and Kotera, M. (2015). Metabolome-scale de novo pathway reconstruction using regioisomer-sensitive graph alignments. *Bioinformatics* 31 (12): i161–i170.

Yandell, B.S., Mehta, T., Banerjee, S. et al. (2007). R/qtlbim: QTL with Bayesian interval mapping in experimental crosses. *Bioinformatics* 23 (5): 641–643.

Yang, Q., Li, Z., Li, W. et al. (2013). CACTA-like transposable element in ZmCCT attenuated photoperiod sensitivity and accelerated the postdomestication spread of maize. *Proceedings of the National Academy of Sciences* 110 (42): 16969–16974.

Yano, K., Yamamoto, E., Aya, K. et al. (2016). Genome-wide association study using whole-genome sequencing rapidly identifies new genes influencing agronomic traits in rice. *Nature Genetics* 48 (8): 927–934.

Yuan, Y., Cairns, J.E., Babu, R. et al. (2019). Genome-wide association mapping and genomic prediction analyses reveal the genetic architecture of grain yield and flowering time under drought and heat stress conditions in maize. *Frontiers in Plant Science* 9: 1919. https://doi.org/10.3389/FPLS.2018.01919/FULL.

Zenda, T., Liu, S., Dong, A. et al. (2021). Omics-facilitated crop improvement for climate resilience and superior nutritive value. *Frontiers in Plant Science* 12: 774994.

Zhang, F. (2019). Development of CRISPR-Cas systems for genome editing and beyond. *Quarterly Reviews of Biophysics* 52: e6.

Zhang, Z., Erbe, M., He, J. et al. (2015). Accuracy of whole-genome prediction using a genetic architecture-enhanced variance-covariance matrix. *G3: Genes, Genomes, Genetics* 5 (4): 615–627.

Zhang, Z., Mao, Y., Ha, S. et al. (2016). A multiplex CRISPR/Cas9 platform for fast and efficient editing of multiple genes in Arabidopsis. *Plant Cell Reports* 35: 1519–1533.

Zhang, S., Chen, X., Lu, C. et al. (2018). Genome-wide association studies of 11 agronomic traits in cassava (*Manihot esculenta* Crantz). *Frontiers in Plant Science* 9: 503.

Zhao, Y., Chen, M.-H., Pei, B. et al. (2012). A Bayesian approach to pathway analysis by integrating gene-gene functional directions and microarray data. *Statistics in Biosciences* 4: 105–131. https://doi.org/10.1007/s12561-011-9046-1.

Zhou, Y., Li, G., Dong, J. et al. (2018). MiYA, an efficient machine-learning workflow in conjunction with the YeastFab assembly strategy for combinatorial optimization of heterologous metabolic pathways in Saccharomyces cerevisiae. *Metabolic Engineering* 47: 294–302.

Zhou, Y., Srinivasan, S., Mirnezami, S.V. et al. (2019). Semiautomated feature extraction from RGB images for sorghum panicle architecture GWAS. *Plant Physiology* 179: 24–37. https://doi.org/10.1104/pp.18.00974.

Zhou, Y., Zhang, J., Xiong, X. et al. (2022). De novo assembly of plant complete genomes. *Tropical Plants* 1 (1): 1–8.

Zhu, C., Gore, M., Buckler, E.S., and Yu, J. (2008). Status and prospects of association mapping in plants. *The Plant Genome* 1 (1).

Zhu, C.S., Li, X.R., and Yu, J. (2011). Integrating rare-variant testing, function prediction, and gene network in composite resequencing-based genome-wide association studies (CR-GWAS). *G3* 1 (3): 233–243.

3

Multi-omics: An Advanced Bioinformatics Approach for Crop Improvement in Agriculture

Vinay Kumar Dhiman[1], *Devendra Singh*[2], *Vivek Kumar Dhiman*[3], *and Himanshu Pandey*[4]

[1] *Department of Basic Sciences, Dr. Yashwant Singh Parmar University of Horticulture and Forestry, Solan, Himachal Pradesh, India*
[2] *Faculty of Biotechnology, Institute of Biosciences and Technology, Shri Ramswaroop Memorial University, Lucknow-Deva Road, Barabanki, Uttar Pradesh, India*
[3] *Department of Biotechnology, Dr. Yashwant Singh Parmar University of Horticulture and Forestry, Solan, Himachal Pradesh, India*
[4] *Department of Agriculture, Khalsa College, Amritsar, Punjab, India*

3.1 Multi-omics: A Boon to Crop Improvement

The analysis and probing of entire large-scale data of scientific fields for measuring biomolecules in a living entity such as cells or tissues include "omics." Omics techniques have an advantage over traditional techniques because they allow a current analysis of several genes, proteins, and metabolites and require standardization of protocols, independent result validation for comparison, and deducing the changes in biological systems (Yang et al. 2021). The integration of such an approach in crop science to improve crop genetics in different crop varieties for favorable yield, quality, and stress tolerance is essential. Agricultural practices for commercial crops are increasing day by day, and the food security of staple food such as grain crops is needed along with the development of other important crops (Pandey et al. 2021).

Over the past few decades, sequencing technologies have undergone significant advancement and reform. Improved crop genome sequences as a result of technological reforms, the assortment, and the setup of databases for assisted identification of core functional genes collectively can help improve the expression of beneficial genetic traits, its recessive ones. The integration of information attained from experimental observations of biological processes with automated models of computation for interpretation of biological data, as in bioinformatics, can help in prediction or even the development of new ways to approach crop genetics. The study of genomes is the primary emphasis of genomics, while the investigation of the structure and function of the transcriptome, which is comprised of both coding and noncoding RNA molecules, is the primary focus of transcriptomics

Bioinformatics for Plant Research and Crop Breeding, First Edition. Edited by Jen-Tsung Chen.

(Chauhan et al. 2023a,b). On the other side, proteomics is concerned with the study of proteins, including posttranslational changes, as well as the regulatory mechanisms that are connected with those proteins. When performed in a unified fashion, metabolomics, which is a great instrument, allows the study of a wide variety of metabolites, which helps in the discovery of complex networks that are involved in the process of stress tolerance (Dhiman et al. 2022a,b; Singh et al. 2022a; Das et al. 2022; Kashyap et al. 2023). The integration of all omics approaches could serve as a foundation to improve crop genetics and breeding for greater yield and resistance to stresses. Crop improvement has made extensive use of next-generation sequencing (NGS) technologies, which are extensively high-throughput techniques capable of rapidly testing large arrays of biological samples through parallel sequencings, such as DNA nucleotides, without ring cloning methods (Scossa et al. 2021; Dai and Shen 2022; Chauhan et al., 2023a,b)

Different omics technologies involved in crop improvement are named based on their interface with hereditary biomolecules such as DNA, RNA, proteins, metabolites and pathways, disease-related gene interactions, etc. (Table 3.1).

The use of single-omics technology is often restricted to specific areas of the research, and the approach for result interpretation often gets limited in the study field. Single-omics are, therefore, more precise for a specific research study and analysis. Whereas multi-omics techniques integration combines different omic technologies to give biological context to genetic variations, novel pathways, and biomarkers. A more holistic approach to the analysis of biological data interpretations can be viewed through multi-omics, for example, e in the detection of specific genetic diseases or mutations. To improve and preserve our crops, integration of such technologies for a better understanding of the genetic and phenotypic characteristics of plants as well as the physiological effect of climate on growth, development, and metabolism (Muthamilarasan et al. 2019). Identification of quantitative trait locus (QTL) or gene of interest with traits and their exploitation for crop improvement is the goal of omics technology. The varieties with favorable traits are selected for high yield, good quality, resistance against stresses, and other agronomic traits for many important crops such as rice, wheat, maize, and other valuable crops, including vegetables, fruits with the use of omics and breeding strategies (Figure 3.1; Table 3.2).

Table 3.1 Important classes of Omics technologies.

Study Level	Omics Technologies
DNA (Gene)	Genomics
RNA (Transcript)	Transcriptomics
Protein (Proteome)	Proteomics
Metabolite (Metabolome)	Metabolomics
Ions (Ionome)	Ionomics
Phenotype (Phenome)	Phenomics
Collective/integrated multiple omics	Multi-omics/panomics

Figure 3.1 Omics in crop improvement.

3.2 Genomics: Unlocking the Crop Genome

The field of genomics is often referred to be a ground-breaking science that has fundamentally changed the way scientists conduct their research. Concurrently, the postgenomic era has arrived, which makes use of the vast amounts of genomic data that are already accessible. Because of recent breakthroughs, it is now technically simpler to modify and analyze biological systems. This has given researchers the ability to answer concerns using methods that are both practical and objective. The emergence of innovative technologies and the rapid development of new genomic techniques have paved the way for the conduct of a single experiment that can analyze an individual's whole genome. Nucleotide sequence repositories are now seeing an increase in value as a result of the transformation of a large quantity of

Table 3.2 Application in Omics technologies.

Stage	Process	Omics Technology Used	Purpose
Genotyping	Genotyping individuals or populations	Genotyping-by-sequencing (GBS)	Obtain genetic variation data
GWAS	Genome-wide association studies	SNP arrays, Next-generation sequencing (NGS)	Identify genetic markers associated with traits
Marker-assisted selection (MAS)	Marker identification and selection	SNP markers	Select individuals with desired traits
Genomic selection (GS)	Prediction of breeding values	Genome-wide marker data, phenotypic data	Estimate genetic potential for multiple traits
Transcriptomics	Gene expression analysis	RNA sequencing (RNA-seq)	Identify genes associated with traits
Proteomics	Protein analysis	Mass spectrometry	Identify proteins related to desirable traits
Metabolomics	Metabolite profiling	Liquid chromatography–mass spectrometry (LC-MS)	Identify metabolic pathways associated with traits

genomic data into data pertaining to functional genes. The identification of the exact sequence and position of all of an organism's genes is the first step toward gaining an understanding of the activities of a biological system, both on its own and as a cohesive whole. This step is necessary in order to make progress toward this goal. Enhancing crop productivity through genetic advancements remains essential, yet current progress needs to meet the requirements for sustainability and food security. Plant genomics plays a vital role in enhancing crops by identifying genetic variations that enhance performance and improve the efficiency of breeding practices. Both approaches are crucial due to the lengthy process of developing new varieties and the necessity of identifying fresh genetic sources (Gill et al. 2014).

Genomics deals with the study of the genetic makeup or genotype of an organism. The focus of genome study is to deal with the structural and functional organization, their evolution, gene mapping, gene editing, and epigenomics. Therefore, genomics is crucial for studying genetic variations; in crop breeding, genomics can play a vital role in improving crop genetics. Genetic mapping of genes through physical maps providing physical distances for nucleotide basis in the chromosome and locating the genes with qualitative traits in plant biology is provided by structural genomics. Structural genomics deals with genome sequences, identification of DNA sequences, assembly, organization, and management of sequences (Thottathil et al. 2016). It is the basis for the genomic analysis of any organism for the development of its sequence maps. Moreover, the technique can determine the protein structures encoded by genes in the 3-dimension form. In functional genomics, gene function is determined based on structural genomics by reconstruction of sequences to find out the function of genes. It provides information on the function and expression of functional genes. In epigenomics, the effect of environment on hereditary changes such as histone deacetylation and DNA or RNA methylations can be studied at the genomic level providing functional aspects of genes under stress. RNA interference (RNAi), mutagenesis, mass spectrometry, genome annotation, and other methods are frequently used to understand the function of genes and proteins. RNAi inhibits the gene expression transcriptionally and posttranscriptionally using proteins to resist pathogens. RNAi uses dsRNA precursor to suppress specific sequences, thereby silencing the desired gene; in crops, it can be used to improve disease resistance and crop development (Kamthan et al. 2015). Metagenomics is involved in using mutation for the modification of traits. In pangenomics, the genome part, which is common or variable in crops, is exploited (Golicz et al. 2016). Comparative genomics is involved in studying the functioning and structural aspect of genetic information for similar and dissimilar traits within genes using molecular tools. Such comparative analysis studies can provide direct evidence for variation in plants due to chromosomal changes across a narrow and wider range of taxa (Talukdar and Sinjushin 2015).

3.2.1 Plant Genome Databases

Genome databases provide access to genomic sequences, gene annotations, functional information, comparative genomics, and other related resources for researchers studying various aspects of plant genetics and genomics. Plant genome databases serve as repositories for storing and organizing genomic data, including DNA sequences, gene annotations, and other relevant information. These databases ensure that the data is readily accessible to researchers worldwide and contain reference genomes, which serve as standard templates for comparative genomics and functional analysis. These references enable researchers to compare and analyze the genomes of different plant species, identify conserved genes or regions, and gain insights into plant evolution and diversification (Table 3.3).

Plant genome databases play a crucial role in crop improvement efforts by providing valuable resources and information that aid in the development of improved crop varieties. Important features of plant genome database in crop improvement are:

a) They provide genetic information, which is valuable for crop breeders and geneticists in identifying candidate genes and genetic variations that contribute to desirable traits such as yield, quality, stress tolerance, and disease resistance.
b) It enables researchers to compare the genomes of different plant species, including related crops and model plants. Comparative genomics helps identify conserved genes, regulatory elements, and evolutionary relationships. By studying these similarities and differences, breeders can transfer knowledge and genetic resources from well-studied crops or models to target crops, accelerating crop improvement efforts.
c) By utilizing genetic markers from plant genome databases, breeders can efficiently select plants with improved traits and speed up the breeding process. These markers allow breeders to use marker-assisted selection (MAS), a technique that enables the selection of plants with desired traits at an early stage based on their genetic profiles.
d) The pool of database facilitates gene discovery by identifying genes associated with specific traits of interest and studying their functions and regulatory mechanisms. This knowledge is crucial for understanding the underlying molecular mechanisms of traits and designing targeted approaches for crop improvement.

Table 3.3 List of Plant genome databases.

Plant Genome Database	Focus	Website	References
The Arabidopsis Information Resource (TAIR)	*Arabidopsis thaliana*	www.arabidopsis.org	Berardini et al. (2015)
Phytozome	Comparative genomics	phytozome.jgi.doe.gov	Goodstein et al. (2012)
Gramene	Grasses	www.gramene.org	Ware et al. (2002)
The Plant Genome Database Japan (PGDBj)	Various plant species	https://biosciencedbc.jp/en/database/pgdbj/	Nakaya et al. (2017)
Sol Genomics Network (SGN)	Solanaceae family plants	solgenomics.net	Fernandez-Pozo et al. (2015)
Legume Information System (LIS)	Legume plants	legumeinfo.org	Dash et al. (2016)
Oryza Genome Annotation Project (RAP-DB)	Rice	rapdb.dna.affrc.go.jp	Sakai et al. (2013)
The Plant Genomics Resource (PGR)	Various crop plants	www.plantgdb.org	Duvick et al. (2007)
PopGenIE	*Populus trichocarpa* and related *Populus* species	popgenie.org	Sjödin et al. (2009)
Medicinal Plant Genomics Resource (MPGR)	Medicinal plants	medicinalplantgenomics.msu.edu	Lange (2016)
PhytoMine	Various plant species	phytozome.jgi.doe.gov/phytomine	Choi (2019)
MaizeGDB	Maize	www.maizegdb.org	Lawrence et al. (2008)
SoyBase	Soybean	soybase.org	
Wheat Genome Database (IWGSC)	Wheat	www.wheatgenome.org	Appels et al. (2010)
CottonGen	Cotton	www.cottongen.org	Yu et al. (2021)
Tomato Genomic Resources (TGR)	Tomato	solgenomics.net/tomato	Emmanuel and Levy (2002)

Potato Genome Sequencing Consortium (PGSC)	Potato	solanaceae.plantbiology.msu.edu/pgsc_download.shtml	Visser et al. (2009)
Banana Genome Hub	Banana	banana-genome-hub.southgreen.fr	Droc et al. (2022)
Coffee Genome Hub	Coffee	coffee-genome-hub.southgreen.fr	
The JGI Plant Genomics Resource (PhytoPortal)	Various plant species	phytoportal.jgi.doe.gov	Markowitz et al. (2007)
PlantGDB	Various plant species	www.plantgdb.org	Dong et al. (2004)
Brassica Database (BRAD)	Brassica crops	brassicadb.org	Cheng et al. (2011)
Grape Genome Database (VitisGen)	Grapevine	www.vitisgen2.org	
Cacao Genome Database (CocoaGenDB)	Cacao	www.cacaogenomedb.org	
Medicinal and Aromatic Plants Database (MAPD)	Medicinal and aromatic plants	mapd.kazusa.or.jp	Shirasawa et al. (2014)
Jatropha Genome Database (JatrophaGDB)	Jatropha	Jatropha genome. Sin. pl	Singh et al. 2016a,b
Eucalyptus Genome Integrative Explorer (EucGenIE)	Eucalyptus	eucgenie.org	Nascimento et al. (2017)
Carrot Genome Database (CarrotDB)	Carrot	apiomics.hzau.edu.cn/carrotdb	Xu et al. (2014)

Source: Adapted from Kersey (2019).

e) Plant genome databases provide critical information for genetic engineering and genome editing approaches in crop improvement. By leveraging the knowledge of gene sequences and functions, researchers can use techniques like CRISPR-Cas9 to precisely modify or edit specific genes, leading to the development of crops with enhanced traits such as increased yield, improved nutritional content, or enhanced resistance to pests and diseases (Prasad et al. 2023).
f) Role in multi-omics: Plant genome databases integrate diverse genomic and functional data, including transcriptomics, proteomics, and metabolomics. This integration allows researchers to perform systems biology analyses, uncovering complex regulatory networks and metabolic pathways in crops. Such knowledge aids in understanding the interactions between genes and traits, providing insights for crop improvement strategies.

3.2.2 Functional Genomics

Functional genomics is a wide approach that aims to understand the relationship between genotype and phenotype. It enables to identify coordinated functioning of particular genes involved in a process. An unidentified gene's function may be deduced by analyzing its sequence, structure, or location in the genome (on a chromosome) with the help of databases that store information on genes found in several organisms such as plants, animals, etc. However, in order to correctly identify the function of genes, it is very necessary to have an understanding of the part that each gene plays in the complex coordination of all activities that take place inside a cell. When investigating a cell, a number of different features may be taken into consideration. These aspects can range from the activities of DNA and RNA to the functions of proteins, as well as higher-level components such as the interactions between proteins, biochemical pathways, cell metabolism, and the overall phenotype of the cell and the organism (Bunnik and Le Roch 2013). Natural variation and functional disruptions are the two primary categories that may be used to classify different approaches to functional genomics. Comparing changes in DNA, RNA, proteins, or other cellular components across time, under varying physiological situations, or across varied cell types and body areas in multicellular animals is an example of natural variation (Werner 2010; Brunetti et al. 2013).

3.2.2.1 Functional Genomics Using Forward and Reverse Genetics

Forward genetics is a classical genetics approach to the identification and characterization of the genes associated with a particular phenotype. This approach involves identifying a relevant phenotype from a population of mutants, followed by genetic mapping and cloning of the genes responsible for the phenotype. Thus, it is a phenotype-to-gene that relies on the induction of mutations through mutagenesis, commonly referred to as phenotype-driven mutagenesis. On the other hand, reverse genetics approaches indeed rely on sequence information obtained from genome and transcript profiling projects. These approaches aim to gain insights into the function of genes by selecting specific mutations and studying the resulting phenotypic effects. Thus, it is a genotype to phenotype that relies on the induction of mutations through mutagenesis. These approaches allow researchers to selectively mutate specific genes and study the resulting phenotypic changes, providing a means to uncover gene function and its contribution to biological processes (Figure 3.2) (Abbai et al. 2017; Edwards and Batley 2010).

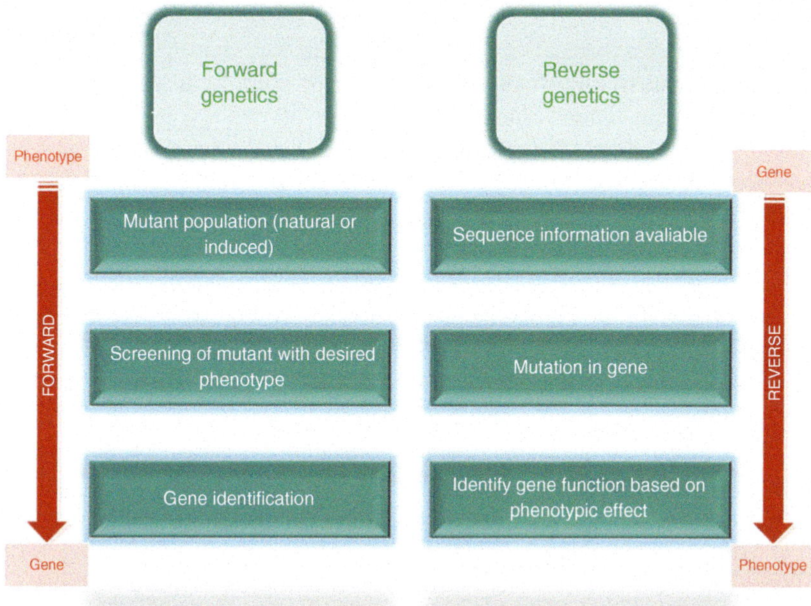

Figure 3.2 Role of Functional genomics in forward and reverse genetics.

Functional genomics tools in forward and reverse genetics approaches:

a) **Positional cloning (map-based cloning):** This approach involves genetic mapping to progressively narrow down the genomic region containing the mutant locus. It relies on linkage disequilibrium and requires a large number of polymorphic markers. Advances in sequencing projects have made positional cloning more reliable and efficient.
b) **Candidate gene approaches:** This approach involves selecting potential candidate genes based on understanding the phenotypic disorder. This can be achieved by studying similar mutants in related plants or by comparative sequence analysis of multiple mutant alleles. RNA-seq, a high-throughput sequencing technique, can also be used to identify candidate genes.
c) **Expression analysis with microarray:** Microarray technology allows for high-throughput gene expression analysis. It can help identify genes that are differentially expressed in mutants compared to wild-type plants. This approach reduces the number of candidate genes for further investigation.
d) **Exome sequencing:** Exome sequencing selectively targets and sequences the coding regions of the genome. It is a cost-effective alternative to whole-genome sequencing and can help identify genetic variations in coding regions that may be responsible for the mutant phenotype.
e) **Insertional mutagenesis:** Insertional mutagenesis involves the insertion of transposon elements or T-DNA with selectable markers into the genome. It allows for the rapid identification of mutant genes by identifying the insertion sites. However, this approach is limited to model organisms that are amenable to transformation.

f) RNA-mediated interference (RNAi) is a powerful tool used in reverse genetics to study gene function. It is a process in which small RNA molecules, known as small interfering RNA (siRNA) or microRNA (miRNA), inhibit gene expression by targeting specific mRNA molecules for degradation or blocking their translation into proteins.
g) Chemical mutagenesis and TILLING (Targeting Induced Local Lesions IN Genomes) are two commonly used approaches in forward genetics for generating and identifying mutations in genes of interest. However, it is also used in reverse genetic approaches to induce mutation.
h) Virus-induced gene silencing (VIGS) is a reverse genetics approach used to study gene function by suppressing the expression of target genes using viral vectors. It is particularly effective in plants and some invertebrate systems.
i) Fast-neutron mutagenesis is a powerful method used in forward genetics to induce random mutations in the genome of an organism using high-energy neutrons. It is particularly useful for creating large-scale mutations, including deletions, insertions, and rearrangements, which can lead to a wide range of phenotypic variations.

3.2.2.2 Functional Genomics in Relation to DNA

The understanding of functional genomics has contributed to our understanding of plant biology, genetic diversity, and crop improvement. The depth of plant functional genomics projects specifically related to plants has enlightened our understanding of providing valuable resources and knowledge for addressing challenges in agriculture, sustainability, and food security. Various consortium projects focused on plant functional genomics are as follows:

The 1001 Genomes Project: The 1001 Genomes Project focuses on studying the genomic variation in the model plant species *Arabidopsis thaliana*. The project aims to sequence and analyze the genomes of a large number of Arabidopsis accessions from diverse geographic locations. By characterizing genetic variation and its functional implications, the project contributes to our understanding of plant adaptation, development, and responses to environmental challenges.

Plant Pan-Genomes Project: The Plant Pan-Genomes Project focuses on investigating the pan-genomes of various crop species, including rice, maize, and wheat. A pan-genome refers to the complete set of genes and genetic variations within a species. By analyzing multiple accessions or varieties of a crop species, the project aims to identify the core genome (genes shared among all individuals) and the dispensable genome (genes that vary among individuals). This research helps to understand the genetic diversity and potential traits for crop improvement.

International Wheat Genome Sequencing Consortium (IWGSC): The IWGSC is a collaborative effort to sequence and analyze the wheat genome. Wheat is a complex and economically important crop, and the project aims to generate a high-quality reference genome for different wheat varieties. By understanding the structure and function of the wheat genome, the IWGSC project facilitates breeding efforts, crop improvement, and understanding of wheat biology.

Plant Reactome Project: The Plant Reactome Project is focused on developing a curated knowledge base of plant metabolic pathways and signaling networks. The project integrates genomic, transcriptomic, and metabolomic data to create comprehensive and interconnected pathway networks for various plant species. This resource aids in understanding plant metabolism, growth, development, and responses to environmental stimuli.

Phenomics Infrastructure for Excellence in Plant Science (PhenomeENAL): PhenomeENAL is an infrastructure project that focuses on plant phenotyping. It aims to develop standardized and high-throughput methods for measuring and analyzing plant traits, such as growth, development, and responses to biotic and abiotic stresses. By linking phenotypic data with genotypic information, the project helps in understanding the genetic basis of complex traits and supports crop breeding and improvement efforts.

3.2.2.3 Functional Genomics in Relation to RNA and Protein

The ultimate goal of genomics is to understand the function of a gene, the expression of RNA transcript level (study of transcriptomics), and resulting protein interactions (proteomics) decide the fate of an organism, including its growth and development. The functions of expressed RNA and proteins are studied using the following techniques and processes:

a) **Transcriptome Analysis:** RNA provides valuable insights into gene expression patterns and regulation. Transcriptome analysis, such as RNA sequencing (RNA-seq), allows researchers to profile the complete set of RNA molecules (transcripts) in a cell or tissue. It helps identify differentially expressed genes, alternative splicing events, and noncoding RNA species.
   - **RNA Sequencing (RNA-Seq):** RNA-Seq is a widely used technique for transcriptome analysis. It involves sequencing the RNA molecules in a sample and generating a comprehensive profile of gene expression. RNA-Seq allows the identification of expressed genes, alternative splicing events, and noncoding RNA molecules. It provides quantitative information on gene expression levels and can be used to compare gene expression across different samples or conditions.
   - **Differential Gene Expression Analysis:** Transcriptome analysis often includes comparing gene expression levels between different samples or conditions. Statistical methods are applied to identify differentially expressed genes, which are genes that show significant changes in expression between the compared groups. This analysis helps identify genes involved in specific biological processes, developmental stages, or responses to environmental stimuli.
   - **Gene Co-expression Network Analysis:** Gene co-expression network analysis aims to identify groups of genes that show similar expression patterns across samples. By examining the correlations in gene expression, co-expression network analysis helps in identifying genes that may function together in specific biological pathways or processes. This approach provides insights into gene regulatory networks and functional modules in plants.

b) **RNA Interference (RNAi):** RNAi techniques, such as siRNA or short hairpin RNA (shRNA), are employed to selectively silence gene expression. By suppressing the translation or stability of target mRNA molecules, RNAi allows researchers to investigate the functional consequences of gene knockdown or knockout.

c) **RNA Localization and Dynamics:** Studying the localization and dynamics of RNA molecules provides insights into spatial and temporal regulation of gene expression. Techniques like fluorescence in situ hybridization (FISH) and single-cell RNA sequencing (scRNA-seq) enable the visualization and characterization of RNA molecules within cells and tissues.

d) **Proteome Analysis:** Proteomics techniques aim to comprehensively study the complete set of proteins (proteome) within a cell, tissue, or organism. Mass spectrometry-based methods, such as shotgun proteomics or targeted proteomics, enable the identification, quantification, and characterization of proteins (Dhiman et al. 2021). This information is valuable for understanding protein expression levels, posttranslational modifications, protein–protein interactions, and functional protein networks.

   - **Protein Extraction:** Proteome analysis starts with extracting proteins from plant tissues. Various extraction methods are employed to isolate proteins while preserving their native structure and maintaining their abundance and functionality.
   - **Gel Electrophoresis:** Gel electrophoresis techniques, such as SDS-PAGE (sodium dodecyl-sulfate polyacrylamide gel electrophoresis) or 2D-PAGE (two-dimensional polyacrylamide gel electrophoresis), are commonly used to separate proteins based on their size, charge, or both. This separation allows the visualization and quantification of individual protein bands or spots.
   - **Mass Spectrometry (MS):** Mass spectrometry is a key technique used in proteome analysis. It involves ionizing proteins and measuring their mass-to-charge ratios. Different types of mass spectrometry, such as MALDI-TOF (matrix-assisted laser desorption/ionization time-of-flight) or LC-MS/MS (liquid chromatography–mass spectrometry/mass spectrometry), are employed for protein identification, quantification, and characterization.
   - **Protein Identification and Quantification:** Protein identification is typically performed by matching the mass spectrometry data against protein databases using search algorithms. Quantitative proteomics methods, such as label-based or label-free approaches, are used to determine relative protein abundances between different samples or conditions.

e) **Functional Annotation:** Functional annotation of proteins involves assigning putative functions to identified proteins. This is done by comparing the identified proteins against existing protein databases, gene ontology (GO) annotations, and pathway databases. Functional annotation helps in understanding the biological processes, metabolic pathways, and protein–protein interactions in plants.

f) **Protein–Protein Interactions:** Investigating protein–protein interactions provides insights into protein function and complex formation. Techniques like yeast two-hybrid (Y2H) assays, co-immunoprecipitation (Co-IP), and affinity purification coupled with mass spectrometry (AP-MS) help identify and characterize protein interaction networks.

g) **Protein Localization and Dynamics:** Techniques such as immunofluorescence microscopy, live-cell imaging, and super-resolution microscopy allow researchers to visualize and track the localization, trafficking, and dynamics of proteins within cells and tissues. This information contributes to understanding protein function in various cellular processes.

h) **Systems Biology:** Integration of transcriptomic and proteomic data enables a more comprehensive understanding of gene function and regulatory networks. Systems biology approaches integrate multi-omics data, including RNA and protein measurements, to model and predict cellular behavior and response to stimuli.
i) **Functional Validation:** Functional genomics studies often involve validating findings at both the RNA and protein levels. Confirmation of gene expression changes or protein abundance by techniques such as qPCR, Western blotting, or immunohistochemistry strengthens the evidence and provides a more complete understanding of gene function.

3.2.3 Structural Genomics

The term "structural genomics" focuses on elucidating the three-dimensional structures of crucial biological macromolecules, with a primary emphasis on proteins (Chen and Yiu 2014). By following these steps, structural genomics can contribute to crop improvement by providing insights into the three-dimensional structures and functions of proteins. This information helps in understanding protein interactions, identifying potential targets for crop improvement, rational design of molecular markers, and engineering proteins for desired traits. These are the steps that are used in structural genomics:

a) **Genomic DNA Sequence:** This step involves obtaining the complete DNA sequence of the crop's genome, typically using advanced sequencing technologies. The genomic sequence provides information about the entire set of genes present in the crop.
b) **Identification of Protein-coding Gene:** Using bioinformatics tools and computational analysis, the genomic DNA sequence is analyzed to identify regions that code for proteins. This step involves predicting and annotating protein-coding genes within the crop's genome.
c) **Protein Expression in Bacteria:** Once the protein-coding genes are identified, the corresponding genes are cloned into bacterial expression vectors. These vectors are then introduced into host bacteria, such as *Escherichia coli* (*E. coli*), which are commonly used for large-scale protein production. The bacteria are cultured, allowing for the expression and accumulation of the target proteins within the bacterial cells.
d) **Experimental Determination of Protein Structure and Function:** The expressed proteins are purified from the bacterial cells using techniques like chromatography. Experimental methods such as X-ray crystallography, nuclear magnetic resonance (NMR) spectroscopy, or cryo-electron microscopy (cryo-EM) are employed to determine the three-dimensional structure of the purified protein. Additionally, functional assays are conducted to investigate the biochemical or biological activities of the protein.
e) **In Silico Prediction:** In parallel with experimental methods, computational approaches are employed to predict protein structures and functions based on the genomic and protein sequence information. These in silico predictions use methods like comparative modeling, threading, and ab initio modeling. In silico predictions are particularly useful when experimental determination of protein structure and function is challenging or time-consuming.

3.2.4 Comparative Genomics

The discipline of comparative genomics examines the similarities and differences in the sequences of genomes found in different species and, in some cases, across individuals of the same species. This field highlights how genome comparisons have greatly contributed to genome annotation. This is especially true in terms of gene prediction and the discovery of conserved regulatory elements. In addition, by using comparative genomics, it is feasible to recognize minute differences between various genomes. This opens the door for the investigation of areas of the genome that code for proteins and the effect that these regions have on the structure and function of proteins. These kinds of comparison investigations make it possible to locate and describe important regulatory components that are included inside DNA (Talukdar and Sinjushin 2015).

Since the evolution of genome sequencing technologies, many major crops and model plants have been sequenced. That sequence provides a comprehensive view of the crop's genetic information, allowing researchers to identify genes and regulatory elements associated with specific traits. Identification of genetic markers, such as single nucleotide polymorphisms (SNPs) and simple sequence repeats (SSRs), allows the selection of specific traits in breeding programs with greater accuracy and efficiency (Varshney et al. 2009). The studies on evolutionary history and domestication using phylogenomic analyses help to identify certain genes responsible for enhancing desirable traits in crops, such as larger fruits, improved yield disease resistance, drought tolerance, pest resistance, and so on. This information is invaluable for understanding the genetic basis of agronomic traits and identifying candidate genes for crop improvement.

3.2.5 Epigenomics

Epigenomes are chemical modifications that are reversible and present in the DNA or the histones that bind the DNA. Changes in gene expression may occur as a result of these adjustments even while the underlying genetic sequence remains unchanged. Epigenetic alterations might be tissue-specific, induced by environmental factors, connected with the development of illnesses, or inherited from one generation to the next. Epigenetic changes can also be associated with the transmission of diseases. It is possible for the epigenomes of several cell types within an organism to have large degrees of diversity. High-throughput assessments of epigenetic alterations largely fall into two categories from a biochemical point of view. These categories include DNA cytosine methylation (at CpG sites) and different modifications of particular histone proteins (acetylation) in chromosomes (known as histone marks). For example, in a study of high-yield rice cultivars, researchers found hypermethylation of genes related to grain filling, resulting in increased grain size and yield potential (Shin et al. 2020). Researchers observed increased histone acetylation at stress-responsive genes in maize, leading to enhanced expression and improved drought tolerance (Yu et al. 2018). In addition, RNA editing may be used as an additional mechanism for epigenetic modifications to gene expression. These modifications can largely be identified via the use of transcriptomic approaches. A study showed that m1G modifications on stress-responsive mRNAs influenced their stability and translation, resulting in improved stress tolerance in both rice and Arabidopsis plants (Wang et al. 2017).

3.3 Metabolomics: Profiling the Crop's Metabolic Processes

The collection of all of the metabolites that are found in a biological sample is referred to as the metabolome. This comprises signaling molecules like hormones and exogenous compounds like medications and their by-products, as well as metabolic intermediates involved in numerous biochemical processes like carbohydrates, lipids, amino acids, and nucleic acids. The metabolome is dynamic and may fluctuate within an organism or among individuals of the same species as a result of variables such as changes in nutrition, stress, physical activity, pharmaceutical effects, and illness (Singh and Agarwal 2021, 2022; Singh et al. 2021, 2022b, c, d). These causes can all cause changes in the metabolome. Methods such as mass spectrometry and NMR spectroscopy are used often for the purpose of measuring the constituents of the metabolome, which includes the lipidome. An understanding of the link between crop genomes and phenotypes under certain physiological and environmental settings may be gained by integrating functional genomics with other omics methods.

Plant metabolomics is a rapidly advancing field that has significantly benefited from large-scale studies of metabolites. The availability of comprehensive databases of global metabolites and their correlations with metabolic pathways has opened up new possibilities for understanding plant biochemistry and genetic regulation. Integrating metabolomics data with other omics approaches, such as genomics and proteomics, has proven to be invaluable in improving genome annotations and identifying novel metabolic pathways. One of the prominent applications of metabolomics in plant research is the use of metabolic-based quantitative trait loci (mbQTL) mapping. This method allows researchers to identify specific genetic variants that influence metabolic traits in plants. By utilizing mbQTLs, researchers can gain insights into the genetic basis of stress-responsive traits and uncover regulatory circuits governing metabolic processes. The approach is analogous to traditional genetic QTLs, proteomic QTLs, and epigenetic QTLs, which all focus on identifying genetic loci associated with specific traits.

In the context of plant breeding, metabolomics-based genome-wide association studies have become essential tools for detecting genetic variations linked to metabolic traits (Sharma et al. 2023). By analyzing the genetic diversity present in populations of crops, researchers can identify regions of the genome that are associated with desirable metabolic traits (Sunya et al. 2022). This information can then be used to facilitate metabolomics-assisted crop breeding, where breeders aim to develop new plant varieties with improved metabolite profiles and enhanced agricultural traits. In summary, the integration of metabolomics with other omics approaches and the use of metabolomics-based QTL mapping and metabolomics-based GWAS have significantly expanded our understanding of plant metabolism and genetics. These powerful techniques hold great promise for discovering novel metabolic pathways, improving genome annotations, and advancing crop breeding efforts to enhance agricultural productivity.

Improving the quality of crops through the metabolic engineering of phytohormones involves enhancing specific traits that contribute to better nutritional value, flavor, appearance, and overall consumer appeal (Negi et al. 2021). Different phytohormones play distinct roles in determining various aspects of crop quality, and targeted manipulation of these hormones can lead to desirable improvements. Metabolic engineering can be used to enhance crop quality:

- **Improving Nutritional Content:** Phytohormones can influence the synthesis and accumulation of various nutrients in plants. For example, cytokinins have been associated with promoting nutrient uptake and assimilation. By enhancing the expression of genes involved in cytokinin biosynthesis or signaling, researchers can potentially increase the levels of essential nutrients, such as vitamins, minerals, and antioxidants, in crops (Sharma et al. 2022).
- **Enhancing Flavor and Aroma:** Phytohormones like jasmonic acid and ethylene play crucial roles in the production of volatile compounds responsible for the characteristic flavors and aromas of fruits and vegetables. Metabolic engineering can be used to modulate the expression of genes related to these hormones to enhance the production of desirable flavor and aroma compounds, thereby improving the taste and sensory experience of the crop.
- **Extending Shelf Life:** Ethylene is a hormone known to promote fruit ripening and senescence. By engineering ethylene biosynthesis or signaling pathways, researchers can regulate the timing of fruit ripening and extend the shelf life of harvested produce, reducing postharvest losses and improving the marketability of the crop.
- **Optimizing Texture and Appearance:** Phytohormones like gibberellins are involved in cell elongation and growth. By manipulating gibberellin levels, researchers can influence the texture and size of fruits and vegetables, leading to improvements in appearance and consumer acceptance.
- **Reducing Anti-nutritional Factors:** Some phytohormones can influence the synthesis of compounds that have anti-nutritional effects, such as certain phytates or oxalates. By engineering these hormone pathways, it may be possible to reduce the levels of these compounds, thereby increasing the bioavailability of nutrients and improving the nutritional value of the crop.

It is essential to note that metabolic engineering for quality improvement should be carefully designed and evaluated to avoid unintended negative effects on other aspects of plant biology or nutritional composition. Furthermore, extensive testing and regulatory oversight are necessary to ensure the safety of genetically engineered crops intended for human consumption. The potential benefits of crop quality improvement through metabolic engineering are promising, but it is crucial to strike a balance between innovation and safety in agricultural biotechnology.

3.4 Phenomics

Phenomics deals with organizing phenotypic data from appropriate genomes that result from environmental interactions of organisms' genotypic characteristics. Phenomics is the study of the large-scale collection and analysis of phenotypic data, whereas the phenome is a catalog providing the measurements. It studies how the genetic instructions from a single gene or the genome translate to phenotypic traits of an organism on a large scale. The method employs sampling of large-scale data for genetic variation in the population; these genotypes are assayed for different traits using high throughput procedures standardized for accuracy. As the conditions are monitored properly, the phenotypic data is captured along with other related data defining the conditions. Hence, these analyses should reveal genotype–phenotype relationships and correlations between unrelated phenotypes (Pasala and Pandey 2020).

Phenomics application can be applied under controlled conditions in laboratories for a set of populations to study dynamic environmental changes, as in agriculture with monoculture practices. Its role in applied biology is crucial as its relevance in crop development, higher yield, complex environmental interaction including stress tolerance, hormone signaling and metabolism, and interaction with other organisms. Setting up experimental designs specific to the expression of phenotypes can provide standardized models for analysis following agricultural practices (Zhao et al. 2019). Linking crop phenotyping with genomics studies for crop improvement using multi-omics data can be studied in a series of steps (Yang et al. 2020):

i) **Crop Population Studies Using Comprehensive and High-capacity Phenotyping Platform:** This platform allows for the efficient and accurate measurement of various phenotypic traits in large crop genetic populations. It likely uses advanced technologies and automation to handle a high volume of samples.
ii) **Identification of Elite Phenotypic Traits:** The platform can capture the dynamic changes in phenotypic traits, which are characteristics or observable features of the crops. This data is valuable for understanding how the crops respond to different conditions and treatments over time.
iii) **Sorting of Phenotypic Data:** This refers to the collected information on the observable traits of the crops, including growth patterns, yield, disease resistance, and other agronomically relevant characteristics.
iv) **Multi-omics Data:** This includes various types of omics data that can be linked to the phenotypic trait of the crops. By combining phenotypic data and omics-based strategies, elite traits in crops can be easy to detect at genomic, transcriptomic, and proteomic levels.
v) **Mapping:** QTL mapping and GWASs (genome-wide association studies), in combination with statistical techniques, s are used to identify the genetic regions associated with specific phenotypic traits. By correlating phenotypic data with genomic information, researchers can pinpoint genes that contribute to desirable traits.
vi) **Mining QTL/Genes:** This process involves identifying and extracting the relevant QTL or genes responsible for specific phenotypic traits of interest. This information is crucial for understanding the genetic basis of crop traits and breeding better varieties.
vii) **Genetic Transformation Techniques:** Once promising QTL/genes are identified, genetic manipulation techniques such as CRISPR/Cas can be employed to introduce these beneficial genes into other crops or varieties. This process aims to enhance desirable traits and improve overall crop genetic potential.

3.5 Ionomics

Ionomics is the study of inorganic components within an organism, i.e., mineral nutrients and other trace elements collectively called ionome. Ionomics provides valuable insights into the uptake, transport, and utilization of essential mineral nutrients by plants (Huang and Salt 2016). By studying the elemental composition of different plant tissues, researchers can identify variations in nutrient levels among different genotypes or species. This information helps in the development of nutrient-efficient crop varieties and the optimization of fertilization practices to enhance nutrient uptake and utilization. It plays a crucial role in biofortification studies, which aims to enhance the nutritional quality of crops by increasing the concentration of essential minerals and trace elements (Carvalho and Vasconcelos 2013; Negi et al. 2021). Through ionomics analyses, researchers can identify genotypes or accessions with naturally high levels of specific minerals. These genotypes can then be used in breeding programs to develop crop varieties with enhanced micronutrient content, addressing micronutrient deficiencies in human diets. Ionomics for plant stress tolerance and resilience can be studied by observation of elemental responses of plants to various stress conditions, such as drought, salinity, heavy metal toxicity, and nutrient deficiencies (Singh et al. 2013, 2022e, f; Yadav et al. 2023). By understanding how plants accumulate, redistribute, and detoxify elements in response to stress, researchers can identify genes and pathways involved in stress tolerance. This knowledge aids in breeding crops with improved stress resilience and adaptation to challenging environments (Pita-Barbosa et al. 2019). Ionomics data, including elemental composition profiles, can be integrated with genetic mapping techniques to identify QTL associated with elemental accumulation and distribution. These QTLs can be used to develop molecular markers for MAS, enabling breeders to select plants with desired elemental traits more efficiently. Ionomics studies also contribute to environmental monitoring by assessing elemental contamination in soil and plants. By analyzing the elemental composition of plants growing in polluted environments, researchers can identify plants with natural tolerance or hyperaccumulation capabilities for specific elements. Such plants can be used in phytoremediation strategies to remove or mitigate environmental contaminants (Singh et al. 2016a,b).

3.6 Omics-Assisted Breeding: Accelerating Crop Improvement

Pan-genome studies aim to explore the genetic diversity and genomic variations within a species or across related species (Table 3.4). These studies provide insights into the core genome (genes shared by all individuals), dispensable genome (genes present in some individuals), and accessory genome (genes unique to specific individuals).

3.7 Conclusion and Future Perspectives

Multi-omics has become a potent and promising platform for crop breeding and agronomic practices. By combining genomes, transcriptomics, proteomics, metabolomics, and high-throughput phenotyping data, scientists have better-understood crop attributes' intricate

Table 3.4 Pan-genome studies conducted in some major crops.

Crop	Species	Number of Accessions/ Genotypes Studied	Key Findings	References
Rice	*Oryza sativa*	3010	Identified a core set of genes shared among all accessions. Revealed 29 million SNPs, 2.4 million indels, and 90,000 structural variations. Discovered 10,000 novel genes	Wang et al. (2018)
Maize	*Zea mays*	26	Uncovered 103,000 pan genes found in large proportion in all accessions. Highlighted extensive genetic diversity within the species. Identified genes associated with various features pan-gene complement, homoeologs retention post-polyploidy, abundance of transposable elements, NLR disease resistance gene copy number, and methylation profiles	Hufford et al. (2021)
Soybean	*Glycine max*	26	Identified over 14.6 million SNPs and 12.7 million small insertions and deletions. Genetic variations were detected in agronomic trait-associated genes	Liu et al. (2020)
Wheat	*Triticum aestivum*	10	Revealed extensive genetic diversity in the pan-genome. Identified 140K genes with an average of 128K genes present across in each cultivar. 36 million SNP were identified	Montenegro et al. (2017)
Brassica crops	*Brassica rapa, Brassica oleracea*	199 (*B. rapa*); 44 (*B. oleracea*)	Explored the genomic diversity within different *Brassica* species. 2.2 and 3.8 million high-quality SNPs; 303 and 417K indels were identified in *B. rapa* and *B. oleracea* Identified genes associated with traits such as oil content, flowering time, and disease resistance. Investigated the evolution and domestication of Brassica crops	Cheng et al. (2016)

molecular relationships. This comprehensive approach has identified genes, regulatory networks, and biological processes that affect yield, stress tolerance, and nutritional quality. Omics-based selection and identification of candidate genes for particular phenotypes have helped genetic manipulation and MAS of enhanced crop types. It has expedited breeding by lowering the time needed to choose suitable candidates for field testing. Thus, farmers may profit from crop varieties with higher yields, tolerance to biotic and abiotic challenges, and nutritional content. The multi-omics strategy for crop enhancement has a bright future as technology advances. High-throughput phenotyping and genomics technologies will allow the gathering of large volumes of data at lower prices and with better accuracy. This will improve agricultural traits and genetic knowledge. Integration of AI and machine learning techniques will help multi-omics datasets analyze and recognize patterns. These methods may detect complicated gene connections and predict crop performance in varied environments. Functional genomics and gene editing (CRISPR-Cas9) technologies may augment the multi-omics approach by accurately modifying target genes to give desired features. Gene editing allows remarkable crop improvement speed and precision. Production of biofortified and climate-resilient crops can be achieved in the future by improving stress tolerance genetics and crop metabolic pathways. This will address global food security and nutritional issues. Open data sharing and worldwide cooperation will boost multi-omics-based crop improvement and cross-disciplinary research.

References

Abbai, R., Subramaniyam, S., Mathiyalagan, R., and Yang, D.C. (2017). Functional genomic approaches in plant research. *Plant Bioinformatics* 215–239.

Appels, R., Akhunov, E., Alaux, M. et al. IWGSC: PhySICal MaPPInG Standard ProtoColS WorkShoP Plant and animal Genome Meeting, San Diego, CA, USA Tuesday, 12 January 2010. Edited by WJ Raupp, Department of Plant Pathology, Kansas State University, Manhattan, KS 66506-5502 USA. Facilities during manuscript editing were provided by the Plant Pathology Department and the Wheat Genetic and Genomic Resources Center, Kansas State University. (2010): 5.

Berardini, T.Z., Reiser, L., Li, D. et al. (2015). The Arabidopsis information resource: making and mining the "gold standard" annotated reference plant genome. *Genesis* 53 (8): 474–485.

Brunetti, C., George, R.M., Tattini, M. et al. (2013). Metabolomics in plant environmental physiology. *Journal of Experimental Botany* 64 (13): 4011–4020.

Bunnik, E.M. and Le Roch, K.G. (2013). An introduction to functional genomics and systems biology. *Advances in Wound Care* 2 (9): 490–498.

Carvalho, S.M. and Vasconcelos, M.W. (2013). Producing more with less: strategies and novel technologies for plant-based food biofortification. *Food Research International* 54 (1): 961–971.

Chauhan, D., Singh, D., Pandey, H. et al. (2023a). Impact of transcription factors in plant abiotic stress: a recent advancement for crop improvement. In: *Plant Transcription Factors*, 271–286. Academic Press.

Chauhan, D., Pandey, H., Prasad, G., and Singh, D. (2023b). Role of next-generation sequencing technology in crop improvement. In: *Smart Breeding: Molecular Interventions and Advancements for Crop Improvement Publisher*. Francis and Taylor.

Chen, Y.W. and Yiu, C.P. (ed.) (2014). *Structural Genomics: General Applications*. Humana Press.

Cheng, F., Liu, S., Wu, J. et al. (2011). BRAD, the genetics and genomics database for Brassica plants. *BMC Plant Biology* 11: 1–6.

Cheng, F., Wu, J., Cai, C. et al. (2016). Genome resequencing and comparative variome analysis in a *Brassica rapa* and *Brassica oleracea* collection. *Scientific Data* 3 (1): 1–9.

Choi, H.K. (2019). Translational genomics and multi-omics integrated approaches as a useful strategy for crop breeding. *Genes & Genomics* 41: 133–146.

Dai, X. and Shen, L. (2022). Advances and trends in omics technology development. *Frontiers in Medicine* 9: 911861.

Das, R., Deb, P., Pandey, H. et al. (2022). Botanical synthesis of silver nanoparticles (Agnps) and its antifungal effect against *Alternaria porri* causing purple blotch of onion: an in vitro and natural epiphytic study. *Journal of Agriculture and Food Research* 10: 100390. https://doi.org/10.1016/j.jafr.2022.100390.

Dash, S., Campbell, J.D., Cannon, E.K. et al. (2016). Legume information system (LegumeInfo.org): a key component of a set of federated data resources for the legume family. *Nucleic Acids Research* 44 (D1): D1181–D1188.

Dhiman, V.K., Chauhan, V., Kanwar, S.S. et al. (2021). Purification and characterization of actinidin from *Actinidia deliciosa* and its utilization in inactivation of α-amylase. *Bulletin of the National Research Centre* 45: 1–9.

Dhiman, V.K., Rana, N., Dhiman, V.K. et al. (2022a). Effect of rhizobial isolates and nitrogen fertilizers on nursery performance, nodulation behavior and nitrogenase activity of *Dalbergia sissoo* Roxb. seedlings. *Plant Stress* 4: 100080.

Dhiman, V.K., Singh, D., Pandey, H. et al. (2022b). Mycoremediation of agricultural waste for the cultivation of edible mushroom. In: *Sustainable Management of Environmental Contaminants: Eco-friendly Remediation Approaches*, 471–482. Cham: Springer International Publishing https://doi.org/10.1007/978-3-031-08446-1_18.

Dong, Q., Schlueter, S.D., and Brendel, V. (2004). PlantGDB, plant genome database and analysis tools. *Nucleic Acids Research* 32 (suppl_1): D354–D359.

Droc, G., Martin, G., Guignon, V. et al. (2022). The banana genome hub: a community database for genomics in the Musaceae. *Horticulture Research* 9: uhac221.

Duvick, J., Fu, A., Muppirala, U. et al. (2007). PlantGDB: a resource for comparative plant genomics. *Nucleic Acids Research* 36 (suppl_1): D959–D965.

Edwards, D. and Batley, J. (2010 Jan). Plant genome sequencing: applications for crop improvement. *Plant Biotechnology Journal* 8 (1): 2–9.

Emmanuel, E. and Levy, A.A. (2002). Tomato mutants as tools for functional genomics. *Current Opinion in Plant Biology* 5 (2): 112–117.

Fernandez-Pozo, N., Menda, N., Edwards, J.D. et al. (2015). The Sol Genomics Network (SGN)—from genotype to phenotype to breeding. *Nucleic Acids Research* 43 (D1): D1036–D1041.

Gill, S.S., Gill, R., Tuteja, R., and Tuteja, N. (2014). Genetic engineering of crops: a ray of hope for enhanced food security. *Plant Signaling & Behavior* 9 (3): e28545.

Golicz, A.A., Batley, J., and Edwards, D. (2016 Apr). Towards plant pangenomics. *Plant Biotechnology Journal* 14 (4): 1099–1105.

Goodstein, D.M., Shu, S., Howson, R. et al. (2012). Phytozome: a comparative platform for green plant genomics. *Nucleic Acids Research* 40 (D1): D1178–D1186.

Huang, X.Y. and Salt, D.E. (2016). Plant ionomics: from elemental profiling to environmental adaptation. *Molecular Plant* 9 (6): 787–797.

Hufford, M.B., Seetharam, A.S., Woodhouse, M.R. et al. (2021). De novo assembly, annotation, and comparative analysis of 26 diverse maize genomes. *Science* 373 (6555): 655–662.

Kamthan, A., Chaudhuri, A., Kamthan, M., and Datta, A. (2015). Small RNAs in plants: recent development and application for crop improvement. *Frontiers in Plant Science* (6): 208.

Kashyap, P., Shirkot, P., Das, R. et al. (2023). Biosynthesis and characterization of copper nanoparticles from *Stenotrophomonas maltophilia* and its effect on plant pathogens and pesticide degradation. *Journal of Agriculture and Food Research* 100654. https://doi.org/10.1016/j.jafr.2023.100654.

Kersey, P.J. (2019). Plant genome sequences: past, present, future. *Current Opinion in Plant Biology* 48: 1–8.

Lange, B.M. (2016). Online resources for gene discovery and biochemical research with aromatic and medicinal plants. *Phytochemistry Reviews* 15: 489–510.

Lawrence, C.J., Harper, L.C., Schaeffer, M.L. et al. (2008). MaizeGDB: the maize model organism database for basic, translational, and applied research. *International Journal of Plant Genomics* 2008.

Liu, Y., Du, H., Li, P. et al. (2020). Pan-genome of wild and cultivated soybeans. *Cell* 182 (1): 162–176.

Markowitz, V.M., Ivanova, N.N., Szeto, E. et al. (2007). IMG/M: a data management and analysis system for metagenomes. *Nucleic acids research.* 36 (suppl_1): D534–D538.

Montenegro, J.D., Golicz, A.A., Bayer, P.E. et al. (2017). The pangenome of hexaploid bread wheat. *The Plant Journal* 90 (5): 1007–1013.

Muthamilarasan, M., Singh, N.K., and Prasad, M. (2019). Multi-omics approaches for strategic improvement of stress tolerance in underutilized crop species: a climate change perspective. *Advances in Genetics* 103: 1–38.

Nakaya, A., Ichihara, H., Asamizu, E. et al. (2017). Plant genome database Japan (PGDBj). *Plant Genomics Databases: Methods and Protocols* 45–77.

Nascimento, L.C., Salazar, M.M., Lepikson-Neto, J. et al. (2017). EUCANEXT: an integrated database for the exploration of genomic and transcriptomic data from *Eucalyptus* species. *Database* 2017: bax079.

Negi, S., Singh, D., Pandey, H., and Singhal, R.K. (2021). Biofortification for high Fe and Zn in various Poaceae crops by using different molecular breeding and biotechnological approaches. *Plant Physiology Reports* 1–11.

Pandey, H., Singh, D., Das, R., & Pandey, D. (2021). Precision farming and its application. *Smart Agriculture Automation Using Advanced Technologies: Data Analytics and Machine Learning, Cloud Architecture, Automation and IoT*, 17-33.

Pasala, R. and Pandey, B.B. (2020). Plant phenomics: High-throughput technology for accelerating genomics. *Journal of Biosciences* 45 (1): 111.

Pita-Barbosa, A., Ricachenevsky, F.K., and Flis, P.M. (2019). One “OMICS” to integrate them all: ionomics as a result of plant genetics, physiology and evolution. *Theoretical and Experimental Plant Physiology* 31 (1): 71–89.

Prasad, G., Chauhan, D., Pandey, H. et al. (2023). Recent advances in CRISPR-Cas for climate-resilient horticulture in fruits crops. In: *Climate-Resilient Agriculture: A Molecular Perspective* (ed. A. Singh and S. Pandey). Apple Academic Press.

Sakai, H., Lee, S.S., Tanaka, T. et al. (2013). Rice Annotation Project Database (RAP-DB): an integrative and interactive database for rice genomics. *Plant and Cell Physiology* 54 (2): e6.

Scossa, F., Alseekh, S., and Fernie, A.R. (2021). Integrating multi-omics data for crop improvement. *Journal of Plant Physiology* 257: 153352.

Sharma, A., Pandey, H., Manpoong, C. et al. (2023). Biometric analysis, biosynthetic pathway and multipurpose uses of buckwheat local varieties in Namsai district of eastern Himalayas in India. *Trends in Food Science & Technology* .

Sharma, S.K., Singh, D., Pandey, H. et al. (2022). An overview of roles of enzymatic and non-enzymatic antioxidants in plant. *Antioxidant Defense in Plants: Molecular Basis of Regulation* 1–13. https://doi.org/10.1007/978-981-16-7981-0_1.

Shin, D., Lee, S., Kim, T.H. et al. (2020). Natural variations at the Stay-Green gene promoter control lifespan and yield in rice cultivars. *Nature Communications* 11 (1): 2819.

Shirasawa, K., Isobe, S., Tabata, S., and Hirakawa, H. (2014). Kazusa Marker DataBase: a database for genomics, genetics, and molecular breeding in plants. *Breeding Science* 64 (3): 264–271.

Singh, D., Chetia, H., Kabiraj, D. et al. (2016a). A comprehensive view of the web-resources related to sericulture. *Database* 2016: baw086.

Singh, D. and Agarwal, V. (2021). Screening of antimicrobial, anti-quorum sensing activity and cytotoxicity of origanum oil against Gram-positive and Gram-negative bacteria. *Biomedicine* 41 (3): 599–603.

Singh, D. and Agarwal, V. (2022). Herbal antibacterial remedy against upper respiratory infection-causing bacteria and in vivo safety analysis. *Vegetos* 35 (1): 264–268. https://doi.org/10.1007/s42535-021-00281-3.

Singh, D., Dhiman, V.K., Pandey, H. et al. (2022a). Crosstalk between salicylic acid and auxins, cytokinins and gibberellins under biotic stress. In: *Auxins, Cytokinins and Gibberellins Signaling in Plants* (ed. T. Aftab), 249–262. Cham: Springer International Publishing.

Singh, D., Pandey, H., and Singh, V. (2022c). Natural products that target cancer stem cells. In: *Handbook of Research on Natural Products and Their Bioactive Compounds as Cancer Therapeutics*, 169–186. IGI Global https://doi.org/10.4018/978-1-7998-9258-8.ch008.

Singh, D., Pandey, H., Dhiman, V.K. et al. (2022b). Secondary metabolite engineering for plant immunity against various pathogens. In: *Metabolic Engineering in Plants*, 123–143. Springer Nature Singapore: Singapore.

Singh, D., Pandey, H., Dhiman, V.K. et al. (2022e). Assessment of environmental impacts of metal/metalloid pollution on plants. In: *Metals Metalloids Soil Plant Water Systems*, 217–232. Academic Press.

Singh, D., Pandey, H., Thakur, K. et al. (2022f). Aquaporins and their functions in water transportation in different plant species. *Environmental Sustainability* 5 (4): 443–456. https://doi.org/10.1007/s42398-022-00251-8.

Singh, D., Sharma, D., and Agarwal, V. (2021). Screening of anti-microbial, anti-biofilm activity, and cytotoxicity analysis of a designed polyherbal formulation against shigellosis. *Journal of Ayurveda and Integrative Medicine* 12 (4): 601–606. https://doi.org/10.1016/j.jaim.2021.06.007.

Singh, D., Singh, V., Bhushan Mishra, S. et al. (2022d). Evaluation of anti-biofilm, anti-quorum, anti-dysenteric potential of designed polyherbal formulation: in vitro and in vivo study. *Journal of Applied Biomedicine* 20 (1): 7–14. https://doi.org/10.32725/jab.2022.005.

Singh, S., Parihar, P., Singh, R. et al. (2016b). Heavy metal tolerance in plants: role of transcriptomics, proteomics, metabolomics, and ionomics. *Frontiers in Plant Science* 6: 1143.

Singh, U.M., Sareen, P., Sengar, R.S., and Kumar, A. (2013). Plant ionomics: a newer approach to study mineral transport and its regulation. *Acta Physiologiae Plantarum* 35: 2641–2653.

Sjödin, A., Street, N.R., Sandberg, G. et al. (2009 Jun). The Populus Genome Integrative Explorer (PopGenIE): a new resource for exploring the Populus genome. *New Phytologist* 182 (4): 1013–1025.

Sunya, H., Sharma, A., Manpoong, C. et al. (2022). Evaluation of improved varieties of rice (*Oryza sativa* L.) under environmental condition of Namsai, Arunachal Pradesh, India. *Ecology, Environment and Conservation* 907.

Talukdar, D. and Sinjushin, A. (2015). Cytogenomics and mutagenomics in plant functional biology and breeding. In: *PlantOmics: The Omics of Plant Science* (ed. D. Barh, M.S. Khan, and E. Davies), 113–156. Springer.

Thottathil, G.P., Jayasekaran, K., and Othman, A.S. (2016). Sequencing crop genomes: a gateway to improve tropical agriculture. *Tropical Life Sciences Research* 27 (1): 93.

Varshney, R.K., Nayak, S.N., May, G.D., and Jackson, S.A. (2009). Next-generation sequencing technologies and their implications for crop genetics and breeding. *Trends in Biotechnology* 27 (9): 522–530.

Visser, R.G., Bachem, C.W., de Boer, J.M. et al. (2009). Sequencing the potato genome: outline and first results to come from the elucidation of the sequence of the world's third most important food crop. *American Journal of Potato Research* 86: 417–429.

Wang, W., Mauleon, R., Hu, Z. et al. (2018). Genomic variation in 3,010 diverse accessions of Asian cultivated rice. *Nature* 557 (7703): 43–49.

Wang, Y., Pang, C., Li, X. et al. (2017). Identification of tRNA nucleoside modification genes critical for stress response and development in rice and Arabidopsis. *BMC Plant Biology* 17: 261. https://doi.org/10.1186/s12870-017-1206-0.

Ware, D., Jaiswal, P., Ni, J. et al. (2002). Gramene: a resource for comparative grass genomics. *Nucleic Acids Research* 30 (1): 103–105.

Werner, T. (2010). Next generation sequencing in functional genomics. *Briefings in Bioinformatics* 11 (5): 499–511.

Xu, Z.S., Tan, H.W., Wang, F. et al. (2014 Jan). CarrotDB: a genomic and transcriptomic database for carrot. *Database* 1: 2014.

Yadav, G., Prajapati, P., Singh, D. et al. (2023). Emerging trends in plant metabolomics and hormonomics to study abiotic stress tolerance associated with rhizospheric probiotics. https://doi.org/10.1016/B978-0-323-91886-2.00012-4. In: *Plant Hormones in Crop Improvement* (ed. M.I.R. Khan, A. Singh, and P. Poór), 283–306. Academic Press.

Yang, Y., Saand, M.A., Huang, L. et al. (2021). Applications of multi-omics technologies for crop improvement. *Frontiers in Plant Science* 12: 563953.

Yang, W., Feng, H., Zhang, X. et al. (2020). Crop phenomics and high-throughput phenotyping: past decades, current challenges, and future perspectives. *Molecular Plant* 13 (2): 187–214.

Yu, J., Jung, S., Cheng, C.H. et al. (2021). CottonGen: the community database for cotton genomics, genetics, and breeding research. *Plants* 10 (12): 2805.

Yu, X., Meng, X., Liu, Y. et al. (2018). The chromatin remodeler ZmCHB101 impacts expression of osmotic stress-responsive genes in maize. *Plant Molecular Biology* 97: 451–465.

Zhao, C., Zhang, Y., Du, J. et al. (2019). Crop phenomics: current status and perspectives. *Frontiers in Plant Science* 10: 714.

4

Genetic Mapping of Valued Genes with Significant Traits in Crop Plants: Basic Principles, Current Practices, and Future Perspectives

Prasanta Kumar Majhi[1], *Akansha Guru*[2], *Suma C. Mogali*[3], *Prachi Pattnaik*[4], *Ritik Digamber Bisane*[5], *Lopamudra Singha*[5], *Partha Pratim Behera*[6], and *Prateek Ranjan Behera*[7]

[1] *Department of Genetics and Plant Breeding, Odisha University of Agriculture and Technology (OUAT), Bhubaneswar, Odisha, India*
[2] *Department of Genetics and Plant Breeding, Janta Vedic College, Chaudhary Charan Singh University, Baraut, Uttar Pradesh, India*
[3] *Department of Genetics and Plant Breeding, University of Agricultural Sciences, Dharwad, Karnataka, India*
[4] *Department of Horticulture, Institute of Agricultural Sciences, Banaras Hindu University (BHU), Varanasi, Uttar Pradesh, India*
[5] *Department of Genetics and Plant Breeding, Institute of Agricultural Sciences, Banaras Hindu University (BHU), Varanasi, Uttar Pradesh, India*
[6] *Department of Plant Breeding and Genetics, Assam Agricultural University, Jorhat, Assam, India*
[7] *Department of Plant Pathology, School of Agriculture, Gandhi Institute of Engineering and Technology University, Rayagada, Odisha, India*

4.1 Introduction

The process of ascertaining the relative positions and distances between markers along chromosomes is referred to as genetic mapping, also known as meiotic mapping or linkage mapping. In order to estimate the genetic map distances between two markers, the use of the average number of recombination events involving a specific chromatid in that location per meiosis is employed. A genetic map, which can be likened to signposts or landmarks along a highway, represents the positions and relative genetic distances between markers on chromosomes (Paterson 1996; Collard et al. 2005). Several crucial agricultural characteristics, such as productivity and quality, environmental stress tolerance, and certain types of disease resistance, exhibit quantitative traits. The determination of genetic variation in a quantitative feature is attributed to the collective influence of multiple genes referred to as quantitative trait loci (QTLs). The identification and utilization of agronomic QTLs in the genomes of crop species through the application of genetic markers is a crucial undertaking in the field of crop development. The discovery of QTL through the use of genetically linked DNA markers is crucial for several applications such as marker-assisted selection (MAS), map-based cloning of targeted genes, and the study of complicated trait genetics (Asins 2002). The majority of economically or scientifically significant animals possess genetic maps that are nearly saturated. Gelderman (1975) employed the abbreviation "QTL" as a mnemonic device to refer to polygenes. QTLs refer to specific genomic areas that are linked to the expression of quantitative traits.

Bioinformatics for Plant Research and Crop Breeding, First Edition. Edited by Jen-Tsung Chen.

The utilization of linkage mapping and association or linkage disequilibrium (LD) mapping has been extensively employed for genetic mapping in plants (Sehgal et al. 2016; Xu et al. 2017). In the past 20 years, linkage mapping has been employed by researchers in the field of plant species to successfully clone or tag many QTLs (Price 2006). Linkage mapping has low allele richness, limited mapping resolution, and slow speed. Association mapping (AM), an auxiliary to linkage mapping, uses historical recombination events over hundreds of generations to increase resolution and allele counts (Zhu et al. 2008). AM has been conducted for *Arabidopsis thaliana* (Zhao et al. 2007) as well as several significant crops including rice (Huang et al. 2010), maize (Li et al. 2013), wheat (Breseghello and Sorrells 2006), soybean (Zhou et al. 2015), barley (Wang et al. 2012), and sorghum (Zhou et al. 2015; Sauvage et al. 2014). In order to construct genetic maps, it is crucial to utilize genetic markers that possess certain characteristics. These markers should exhibit high-throughput capabilities, ensuring efficient and rapid analysis. Additionally, they should demonstrate reproducibility, allowing for consistent results across several experiments. Furthermore, the markers should be codominant, meaning that both alleles are expressed equally. Lastly, it is desirable for these markers to be transferable, enabling their utilization in different genetic mapping studies. This emphasis on markers derived from expressed regions is highlighted by Semagn et al. (2006). In the year 1911, T. H. Morgan, alongside his student Alfred Sturtevant, successfully generated the initial genetic map of genes associated with sex on a chromosome of the fruit fly. Despite being more sophisticated, genetic mapping and linkage analysis continue to be employed. The rapid advancement of molecular markers, which have supplanted the very limited variability observed in morphological features, mutants, and isozymes, has resulted in a substantial increase in genetic mapping investigations across numerous species over the last 20 years. In the field of plant breeding, the utilization of molecular markers necessitates the use of accurate genetic maps that encompass a significant portion of the genome (Tanksley et al. 1989). These maps are important for five reasons:

a) Permit thorough genetic characterization of qualitative and quantitative attributes to pinpoint genes or QTLs.
b) Using MAS, makes it easier for desirable genes or QTLs to enter the population.
c) Allow comparative mapping between species to determine gene order and phenotype expression.
d) Provide a framework for physically safeguarding DNA sequence, chromosomal translocation, or other direct measurement maps.
e) Initiate the process of positional or map-based cloning of genes that hold economic significance.

The process of genetic mapping can be categorized into two main types: family-based mapping, which involves the analysis of offspring resulting from crosses between two or more parents; and population-based mapping, which focuses on studying populations found in natural environments where the genetic relationships are not well-defined. This chapter aims to present a comprehensive introduction to QTL mapping techniques. It will specifically focus on different mapping populations, the utilization of molecular markers for the construction of genetic maps, the analysis of marker–trait associations through various statistical techniques and software programs, as well as AM or LD mapping. Applying this knowledge will facilitate the acquisition of a foundational comprehension of the analysis and utilization of natural variability for agricultural advancements in the forthcoming period.

4.2 Quantitative Trait Loci (QTLs) and Genetic Mapping of Traits

According to its definition, a QTL refers to a specific part of the genome or a gene locus that exhibits a linkage to the expression of a quantitative trait. A QTL refers to a specific genomic location, also known as a locus or polymorphic site, on a chromosome. At this locus, various alleles exert distinct effects on the expression of a quantitative characteristic. The term "QTL" was introduced by Gelderman in 1975. A QTL can manifest as either a solitary gene or a group of genes that collectively exert an influence on a specific attribute. The range of a quantitative trait is determined by the number of alleles that exert an influence on it. Quantitative traits are alternatively referred to as polygenic traits due to their regulation by multiple genes. Polygenic personality traits exhibit heightened susceptibility to environmental influences. Due to the inherent challenge of attributing these characteristics to a specific gene, researchers commonly designate them as "minor gene features." Height, yield, protein content, oil content, and similar metrics are all examples of quantitative plant features and the transgressive Segregations emerge for quantitative characteristics (i.e. segregation outside the range of parental value). In plant breeding, pinpointing the location of individual genes responsible for mediating quantitative traits is of paramount importance. QTL mapping relies heavily on the creation and comprehension of the statistical foundation.

The genome of a cell is its complete DNA. Animals have two types of genome likely; nuclear genome and mitochondrion genome, whereas, plants have three types of genomes likely, nuclear, mitochondrion and chloroplast genome. While all three are significant, the nuclear genome is the biggest and gives an organism most of its traits. Genome mapping is a common method for finding genes that cause a trait in an organism. Higher plants have 25,000 genes, most of which are unknown, although rapid progress is being made toward their characterization in selected botanical models. Genome mapping can isolate genes based on their phenotypic effects without a prior understanding of their molecular functions. A "genome map" can be compared to a road map in that it shows how far apart certain points of interest are. The nuclear genome is organized and transmitted as linear units termed chromosomes in higher species, allowing for genome mapping. In the same way that mileposts help drivers find their way along a road, "DNA markers" help scientists locate specific regions of each chromosome. Genome mapping methods can be broken down into two basic groups, each of which provides a new and improved level of resolution for studying DNA.

a) **Genetic mapping:** Recombination, or the "breaking and rejoining" of chromosomes that occur naturally, is the basis of genetic mapping, which uses the frequency with which DNA markers appear together on the same chromosome segment to infer the distance between them. Because recombination is a biological process, it is important to keep in mind that the rate of it varies a lot between genomic areas, populations, and environments. But it gives a direct way to measure a key part of crop improvement: the transfer of genes (or clusters of genes) from parent to child.

b) **Physical mapping:** The process of physical mapping entails directly measuring the physical amount of DNA that exists between known landmarks in a genome. This is analogous to genetic mapping and involves artificially breaking DNA molecules with radiation, DNA-modifying enzymes, or shear forces and determining the frequency of DNA landmarks on the same chromosome region.

4.2.1 Objectives of QTL Mapping

Mapping of a quantitative trait is done for several purposes but the major objectives are as follows:

1) Localize the genomic region or gene responsible for the observed quantitative variation.
2) Examining how the QTL influences the characteristic.
3) How much of the observed variation in the analyzed trait can be attributed to a given locus?
4) Determine whether a QTL is linked with an additive or dominant effect on a trait.
5) Which alleles are associated with beneficial outcome measures?

4.3 The Fundamentals of the QTL Mapping Approach

The process of identifying a QTL or gene inside a plant genome might be likened to the challenging task of locating a needle within a haystack. However, the application of QTL analysis can be employed to categorize the haystack into smaller clusters, hence enhancing the efficiency of the sifting process. QTL analysis, in its fundamental essence, revolves around the concept of identifying a correlation between the phenotype and the marker genotype of an organism. The ease of mapping crop plant populations is facilitated by the abundance of segregating markers that are readily accessible. The majority of these markers, conversely, are situated within regions of the genomes that are inactive and noncoding. Consequently, their impact on the trait under investigation may be negligible. Nonetheless, a subset of these markers may exhibit associations with genomic regions that do exert influence on the trait of interest. The occurrence of co-segregation between two quantitative traits and a marker locus is observed when the latter is genetically linked to the former (Prasanna 2003). In order to ascertain, the association between a QTL and a marker, it is necessary to partition the mapping population into distinct genotypic groups depending on the genotypes seen at the marker locus. Subsequently, correlative statistical analyses can be employed to determine if there is a substantial variation in the trait of interest among the different genotypes. LD is observed when there is a lack of segregation among genes. QTL analysis thus requires the presence of LD. Detecting a continuous association between QTL and marker genotypes in natural populations poses challenges, particularly when the marker is not in perfect LD with the QTL. QTL research often employs various types of populations, such as mapping populations, F_2-derived populations, recombinant inbred lines (RILs), near-isogenic lines (NILs), doubled haploids (DHs), and backcross populations (BCs), to facilitate the segregation process (Prasanna 2003).

4.3.1 Salient Requirements of QTL Mapping

The basic requirements of QTL mapping are as follows:

1) A suitable mapping population is created by selecting parents that exhibit contrasting traits.
2) Need a species-appropriate marker system and a rich marker linkage map.
3) The capability of phenotypically reliably screening the mapping population for the desired traits.
4) QTL information can be identified with the help of the right statistical package (software).
5) To make a genetic map, use these populations to figure out the pairwise recombination frequencies, set up linkage groups and estimate map distances, and use a statistical program to find out the order of the maps.

4.3.2 Basic Method for QTL Analysis

Biparental populations are commonly employed for QTL mapping. Alternatively, the utilization of wild populations, germplasm collections, and breeding lines can be employed for the purpose of mapping QTLs by leveraging the phenomenon of LD. Figure 4.1 presents a simplified depiction of the sequential procedures involved in the mapping of QTLs.

The following is a brief overview of the standard process for QTL linkage mapping.

1) To initiate the experimental process, it is essential to carefully choose two homozygous lines that exhibit contrasting phenotypes for the specific trait(s) under investigation.

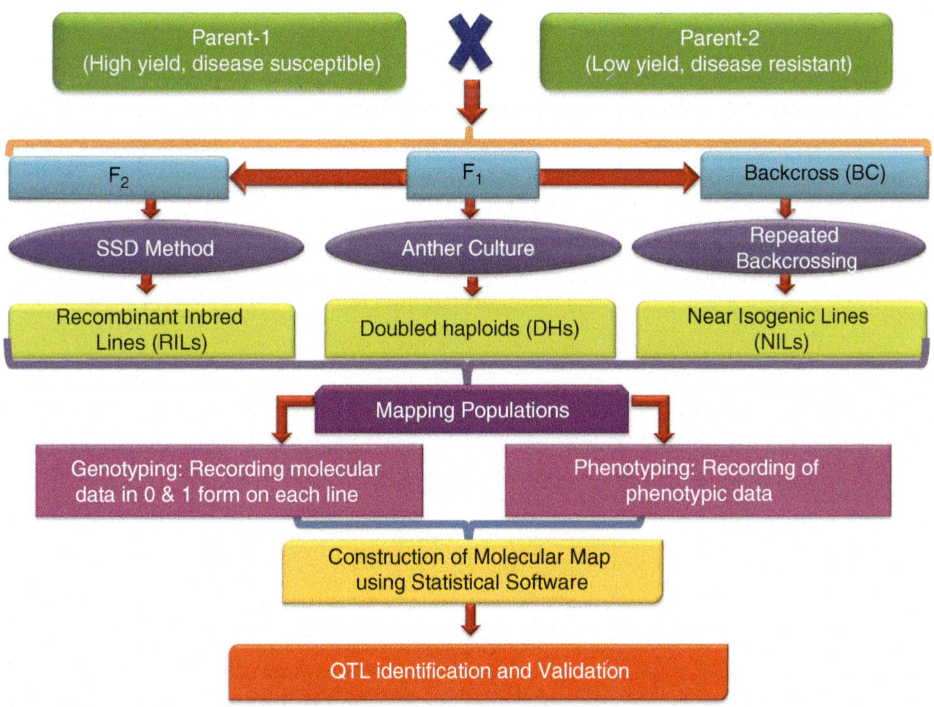

Figure 4.1 A schematic representation of genetic mapping of QTLs in plants.

These selected lines are then subjected to a controlled breeding process, resulting in the creation of a mapping population. Ideally, this population should consist of doubled haploid (DH) or RIL populations, which are commonly employed in such studies. The selection of parental lines for breeding should prioritize the presence of a significant disparity in the target trait(s) between the two homozygous lines. Additionally, it is desirable for these lines to have been generated through divergent selection specifically for the trait(s) in question.

2) Phenotyping encompasses the evaluation of the mapping population for the desired characteristic across many geographical regions and seasons.
3) Polymorphic markers are found by testing the mapping population's parents with many genome-wide markers. Polymorphic markers must be densely distributed across the genome.
4) Genotyping analyses all mapping population individuals/lines utilizing these polymorphic markers.
5) A linkage map framework for the population is established by utilizing the genotype data of the markers, which allows for the visualization of the markers' relative positions and the genetic distances separating them.
6) Finally, the genotypes of markers and the phenotypic data of traits are assessed for association. The plants are classified into marker genotype groups, and the mean and variation of the estimated trait phenotypes are utilized to compare these groups. If there is a significant difference in the genotype groups for a marker, it is probable that there is a connection between the marker and a QTL that influences the phenotype of the trait.

4.3.3 Conventional QTL Mapping (Linkage Mapping)

Below are the typical procedures of a classic QTL mapping study (Sehgal et al. 2016):

- **Step 1:** Determine the genes that influence variance in a trait and choose two parental strains that differ in these alleles. Because multiple allelic combinations can result in the same phenotypic mean, there is no requirement that the parents have distinct values for the characteristic on average.
- **Step 2:** Establish a suitable population for mapping by mating the desired parents.
- **Step 3:** Phenotyping the mapping population will allow you to examine the desired attributes in a safe setting, such as a greenhouse, a screen house, or a field.
- **Step 4:** Produce population molecular data with a sufficient number of evenly distributed polymorphic markers.
- **Step 5:** Generate a molecular map.
- **Step 6:** Look for analytical or statistical programs that can assist you in identifying molecular markers associated with the trait or traits of significance.

4.4 Mapping Populations Used in QTL Mapping Experiments

When referring to a population for the purposes of gene mapping, the term "mapping population" is widely used. Controlled hybridization between the ideal parents is the standard method for generating mapping populations. The goals of the research, the availability of markers, and the molecular map inform choices about which parents to utilize in developing the mapping population and which markers to employ. The genetic and phenotypic variety in the parents of mapping populations should be high enough to ensure accurate and reliable results. DNA polymorphism is crucial for recombination event tracking. Since there is greater variation in DNA sequences, it is simpler to identify polymorphic informative makers. Trait genetic diversity between parents is useful when trying to identify the genes responsible for a specific characteristic. There is a good possibility that parental phenotypic variability exists if there are big observable differences between the parents for a characteristic, yet large phenotypic variation may also result from uncontrolled environmental factors. Yet, the lack of phenotypic variance between parents does not rule out genetic variety; different sets of genes may have contributed to different phenotypes.

4.4.1 Parent Selection for Mapping Population Development

Selecting two genetically divergent parents with highly differing target characteristic phenotypes is the first stage in creating a mapping population (e.g. highly resistant and susceptible lines). One or more features of interest must be genetically different between parents. Parents should be genetically different enough to produce offspring with noticeable polymorphism without being so far apart that: (i) offspring are rendered sterile, or (ii) linkage analysis reveals extreme segregation distortion (Semagn et al. 2006). Map populations from highly homozygous parents using self-pollinating species (inbred). Complicating matters is the inability of the majority of cross-pollinating (outcrossing) species to tolerate inbreeding. To map outcrossing species, researchers employ two-way pseudo-test crosses, half-sib families, and full-sib families derived from controlled crossings. The process of constructing a map is contingent upon the selection of a parent. The economic value of a map is determined by the marker–trait connection. Therefore, it is advisable to use genetic pools that possess a wide range of qualitatively inherited morphological traits when generating mapping populations. Mapping population must consider parentage (adapted versus exotic). The introduction of wide crossings has been observed to reduce linkage lengths through the inhibition of chromosomal pairing and recombination (Albine and Jones 1987; Zamir and Tadmor 1986). In order to achieve a comprehensive and equally distributed collection of polymorphism markers throughout the genome, it is imperative to provide genetic diversity among the paternal lines. To fulfill the second requirement, it may be necessary to carry out a molecular polymorphism assessment on a variety of potentially beneficial lines to determine which ones are optimal for mapping population generation (Prasanna 2003). Broad crossings provide populations with more polymorphism than narrow crosses (adapted × adapted). A wide cross map must be collinear with a map created from adapted parents to be useful in crop improvement (Staub et al. 1996).

4.4.2 Family-based Mapping Populations

Various forms of mapping populations, such as nested association mapping (NAM) (Guo et al. 2010; Song et al. 2017) and multi-parent-based mapping populations multiparent advanced generation inter-cross (MAGIC) (Rakshit et al. 2012), can be generated for linkage mapping. Figure 4.2 illustrates the procedure employed in generating biparental mapping populations for the purpose of conducting QTL mapping. The MAGIC populations

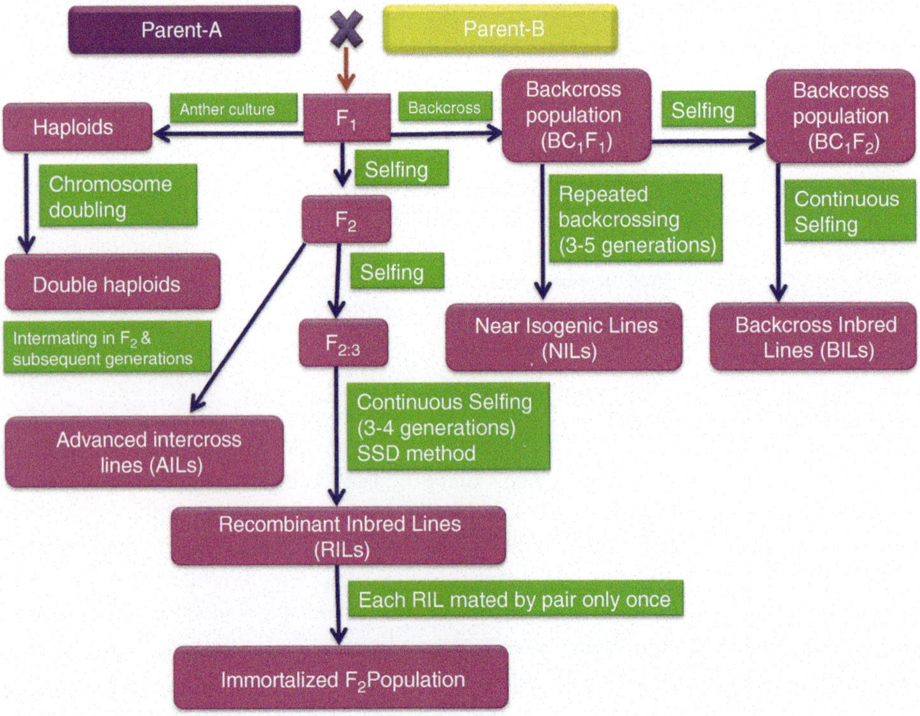

Figure 4.2 An overview of various biparental mapping populations that have been employed in plant quantitative trait loci (QTL) mapping studies.

consist of RILs derived from intricate crosses or a series of crosses involving more than two parents (Figure 4.3) (Bandillo et al. 2010, 2013; Scott et al. 2020). The proposition made by Yu et al. (2008) on the NAM population has the potential to contribute to both QTL linkage mapping and AM methodologies. The NAM system for maize is characterized by a higher level of precision compared to QTL linkage mapping, as it employs RILs derived from a diverse range of parental lines. Additionally, it should be noted that this method utilizes a smaller number of markers compared to genome-wide association studies (GWAS) in association panels based on population samples (Figure 4.4). NAM and related techniques face a significant statistical obstacle when endeavoring to forecast the likelihood of alleles from different loci, which share the same state, being similar during descent. Considering the imminent availability of almost entire genome sequences for the majority of species of interest, it is expected that this matter will become less pertinent. Nevertheless, several inquiries pertaining to the NAM scheme have yet to be resolved. These include determining the optimal number of parental lines required for generating NAM populations, establishing the appropriate criteria for selecting parental lines, ascertaining the optimal number of reference lines, devising suitable adjustments to effectively address population structure concerns, and elucidating the genetic architecture underlying the traits of interest (Myles et al. 2009). Several alleles at the target locus can be found and used for linkage and AM in such populations where parental differences exist for multiple variables. Table 4.1 summarizes, for quick reference, the positive and negative qualities shared by these various

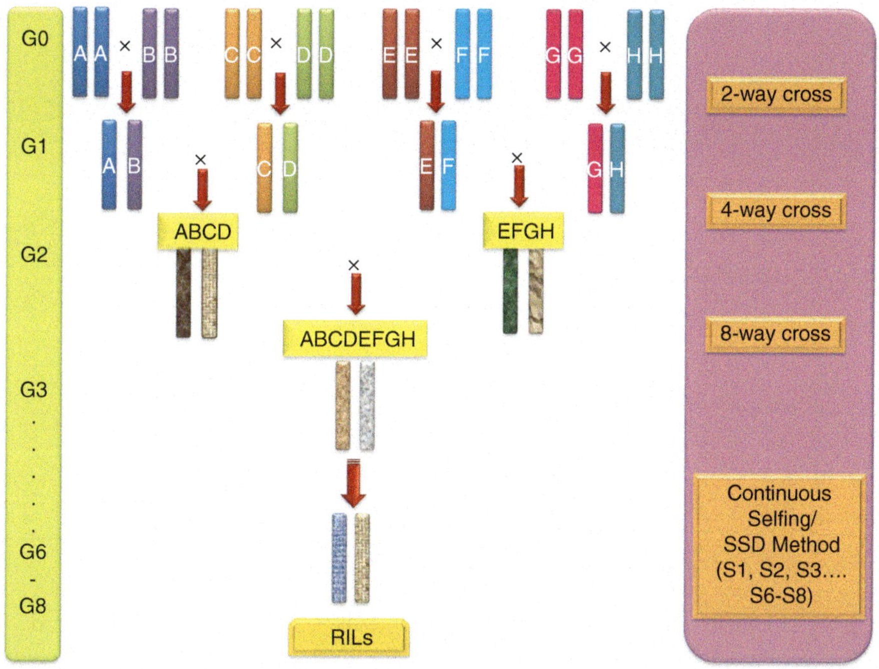

Figure 4.3 The present study employs a schematic depiction of multi-parent advanced generation inter-cross (MAGIC) populations as a means to conduct quantitative trait locus (QTL) mapping in plant species.

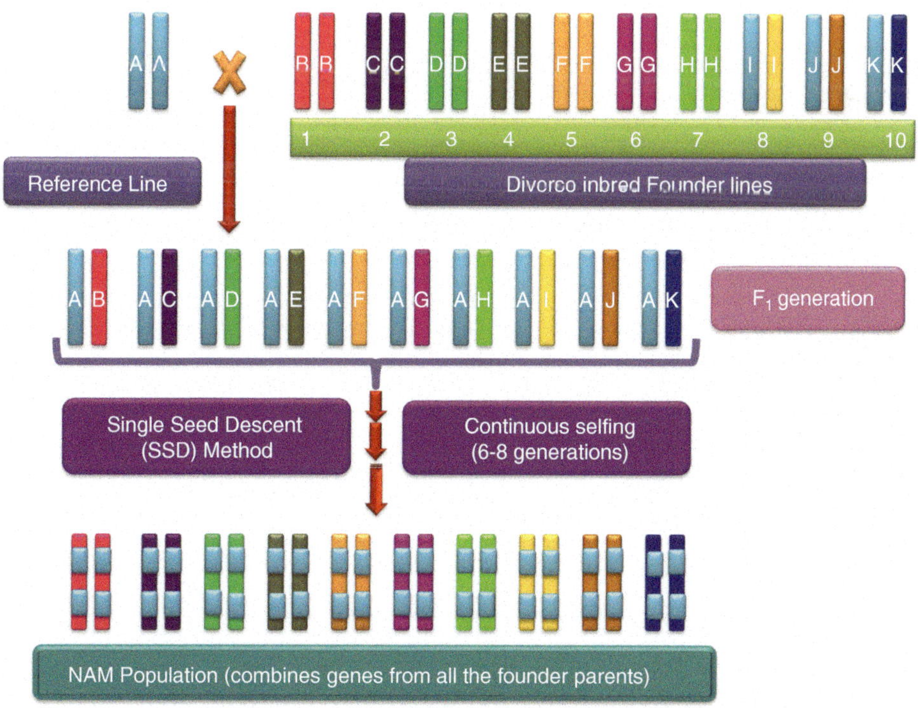

Figure 4.4 An investigation of the use of nested association mapping (NAM) populations in the context of plant quantitative trait locus (QTL) mapping.

Table 4.1 Different types of mapping populations used for Gene mapping in plants.

Mapping population	Development methods	Merits	Demerits
		Biparental mapping population	
F_2 population	Developed by sib mating or selfing the population that resulted from crossing the preferred parents. F_2 refers to the offspring of a single meiotic cycle. Predicted segregation ratios for dominant and codominant markers are 3:1 and 1:2:1, respectively	• The most suitable sample for initial map construction • Time spent on development is reduced • When compared to other populations, they require less work to cultivate	• Establishing a linkage in an F_2 population relies on data from just one meiotic division • When it comes to precise mapping, F_2 populations do not help all that much • Due to the presence of genetic diversity and the inherent limitations in conducting repeated tests across several places and time periods, F_2 populations are deemed unsuitable for accurately mapping quantitative traits. Therefore, it is not feasible to obtain precise estimations of the influence of the gene–environment interaction on the manifestation of quantitative characteristics • No permanent (mortal) population; no way to create a duplicate or multiply the seed stock
F_2 derived F_3 ($F_{2:3}$) population	Selfing the F_2 individuals for a single generation produces an $F_{2:3}$ population. Intended use includes pinpointing quantitative traits and recessive genes	$F_{2:3}$ populations are very suitable for the mapping of oligogenic traits that are governed by recessive genes and quantitative trait loci (QTLs) due to the ability to gather data from multiple plants within each $F_{2:3}$ family, thereby accounting for potential sampling error. The phenotype of an $F_{2:3}$ family can be approximated by calculating the mean values of many plants within the same generation	Like the F_2 population, it is not permanent. (i) These populations take an additional season to build than F_2 populations. (ii) Most F_3 families exhibit genetic diversity as a result of the segregation of many genes. Because of this, it is impossible to use numerous genetically identical plants from an $F_{2:3}$ family as replicates. (iii) A third round of segregation, recombination, and inbreeding has altered the genotype and, in particular, the phenotype of the F_3 population from that of the F_2 generation. For the same reasons stated above, the average phenotype of an F_3 plant is similar to but not identical to that of the parent F_2 plant. Last but not least (iv), the higher incidence of inbreeding in F_3 means that the data are more likely to understate dominance, overdominance, and some epistatic components of gene activity

Immortalized F_2 (IF_2) population	By mating pairs of randomly selected RILs from a cross in all feasible combinations (save for reciprocals), immortalized F_2 populations can be created. All potential genotype combinations (including the heterozygotes) found in the F_2 of the cross from which the RILs were created are accurately represented in the set of RILs used for crossing and the F_1s produced	Through the iterative selfing process, it is possible to maintain the recombinant inbred lines (RILs) and yield a desired quantity of F_1 seed via hybridization. The potential exists to identify heterotic quantitative trait loci (QTLs) and assess their interaction effects by utilizing data obtained from numerous sites within this particular population. The expected range of genotypes in the F_2 generation from the cross from which the RILs were produced is accurately represented in an IF_2 population	Every F_1 offspring in an IF_2 population is produced through the mating of two recombinant inbred lines (RILs). Consequently, by conducting a new hybridization between the parental RILs, a sufficient amount of F1 seeds can be generated. As a result, the utilization of IF_2 populations facilitates the discernment, delineation, and measurement of quantitative trait loci (QTLs), encompassing heterosis QTLs, as well as other epistatic effects. This is achieved by systematically analyzing F_2 genotypes across many sites, ensuring replicability in the investigation
Backcross mapping population	To create backcross populations, the F_1 is crossed with one of the parents. Often employed in genetics, the backcross with recessive parent (testcross) is a method of crossing off a trait of interest. Molecular markers would segregate in a 1:0 ratio for dominant markers and a 1:1 ratio for codominant markers in a backcross with the dominant parent (B_1)	Nonetheless, regardless of the type of marker, a backcross with the recessive parent (B2) or a testcross will segregate in a ratio of 1:1. Backcross populations can be established in a fraction of the time it takes to establish an F_2 population. The ability to use backcross populations in subsequent marker-assisted backcross breeding is a distinct benefit	• In the event of backcrosses, only a single parent's genetic material is used to determine recombination probabilities (F_1) • This is not "immortal"

(Continued)

Table 4.1 (Continued)

Mapping population	Development methods	Merits	Demerits
Doubled haploids (DHs)	The haploid pollen of F_1 plants is used to create DHLs, which are then treated to bring back a diploid condition in which all of the loci are homozygous. Because meiosis produces the pollen population, the DHLs are a direct representation of the gametes that are being segregated. Because both DHs and F_2s result from a single meiotic cycle, the expected segregation ratio for the marker is 1:1, regardless of its genetic makeup (whether dominant or codominant)	• Helpful for mapping qualitative and quantitative features, DHs are a permanent mapping population that can be repeated and assessed across places and years without any genotypic change • Time-efficient, instantaneous generation of homozygous lines	• Because it utilizes *in vitro* procedures, the development of these mapping populations requires a higher level of technical expertise than the creation of other mapping populations due to the inclusion of only male-to-male recombination • Since the tissue culture response of various plants varies greatly, it is not always possible to cultivate a number of crops using haploid production techniques. In addition, the variability brought on by another culture needs to be addressed
Recombinant inbred lines (RILs)	• RILs are generated through the repeated generational selfing or sib mating of offspring from an F2 population until complete homozygosity is reached • The ideal approach for creating RILs is the SSD method. It is also possible to use approaches like the bulk method and the pedigree method without any sort of selection • RILs balance out different kinds of markers like DHs so that the genetic segregation ratio for a dominant marker is the same as that of a codominant marker • A longer period of time is needed to create RILs through brother–sister mating than through selfing. If inbred lines are created through brother–sister mating instead of selfing, this is especially true for situations where the connection is not very tight, which doubles the number of inbred lines needed	• After homozygosity has been attained, RILs can be continuously bred without any further segregation • RILs are a valuable tool for mapping QTLs because they are immortal populations that can be duplicated over long periods of time and in different geographic regions • RILs, which are acquired after multiple meiotic cycles, are extremely helpful for pinpointing closely related producers • Recombination between extremely tightly connected markers is twice as common in RIL populations acquired through selfing as it is in populations derived from a single cycle of meiosis • A group of RILs can be thought of as a "permanent" or "immortal" mapping population	Producing RILs can be challenging in crops with severe inbreeding depression because of the long time (seasons/generations) needed to create the RILs

Near-isogenic lines (NILs)	• Non-inbreeding-loop (NIL) plants can be created through selfing or by backcrossing F_1 plants to their recurrent parents • Regardless of the type of marker, a 1:1 segregation ratio is to be expected	• NILs are an "immortal mapping population," just like DHs and RILs • When a suitable population for tagging the characteristic is available, it should be used • Functional genomics is one field where NILs shine	• Progress takes a lot of time, maybe even several generations • The sole direct application is molecular tagging of the relevant gene; linkage mapping cannot benefit from this • One issue that must be addressed when building NILs is linkage drag
		Multiparent-based mapping population	
MAGIC population	Breeders can take lines of interest from their breeding stock or germplasm collections and use them to create MAGIC populations. These populations can then be employed for a variety of improvements in one way or another. Furthermore, they provide openings for investigating how different genome segments interact with one another through introgressions and recombination of chromosomes. QTL mapping and variety development in rice have been aided by the creation of MAGIC populations in both wheat and rice (from indica and japonica lines, respectively, for the MAGIC Plus and World MAGIC populations)	• Through multiple rounds of recombination, the accuracy and resolution of QTL mapping are improved, and new genetic combinations for essential traits have been developed • Help people find, name, and control novel types of variability	• It is a laborious and time-consuming procedure • The need for extensive phenotyping is essential • There is clear evidence of parental incompatibility • Identification of QTL will necessitate a more refined marker system

(*Continued*)

Table 4.1 (Continued)

Mapping population	Development methods	Merits	Demerits
NAM populations	Using proper mating methods such as diallel mating, North Carolina design II, eight-way cross, single/double round robin, etc., NAM populations can be established in a variety of crop species to yield sets of RILs. Each male in a set of inbred lines is matched to a specific number of females from other inbred lines, and each female is mated to a certain number of males from other inbred lines, in a round-robin mating strategy. Each inbred is paired off with another inbred of the same sex and orientation in a single round-robin scheme. For instance, a single round-robin mating scheme would result in the following five crosses: (1×2, 2×3, 3×4, 4×5, and 5×1)	Due to this rationale, it may be argued that NAM (nested association mapping) is more efficacious compared to AM (association mapping) due to the implementation of regulated crossings in NAM populations. These regulated crossings serve to decrease the intricacy of the population structure and minimize the level of genetic relatedness within families. Furthermore, the North American Mixed (NAM) population predominantly consists of biparental families, leading to a higher occurrence of rare alleles	The recombinant inbred lines (RILs) inside a NAM population can also serve as valuable resources for conducting linkage mapping studies. The utilization of multiple recombinant inbred lines (RILs) enhances the precision of linkage mapping. The NAM approach facilitates the dissemination of the NAM panel, hence enabling researchers to conduct cost-effective genome-wide scans

Table 4.2 Determinants of segregation ratio between codominant and dominant markers in various mapping populations.

	Segregation ratio					
Marker type	**F_2**	**DHs**	**RILs**	**NILs**	**BC[a]**	**BC[b]**
Dominant	3:1	1:1	1:1	1:1	1:0	1:1
Codominant	1:2:1	1:1	1:1	1:1	1:1	1:1
Perpetuation	Ephemeral	Immortal	Immortal	Immortal	Ephemeral	Ephemeral

Dominant markers: RAPDs, AFLPs, most SCARs, ISSRs, SNPs, DArT, SFPs, RAD markers, and Codominant markers: RFLPs, SSRs, CAPSs.
[a] Backcross involving the parent which carries the marker's or trait's dominant allele.
[b] Crossbreeding with the parent carrying the marker's or trait's recessive allele.

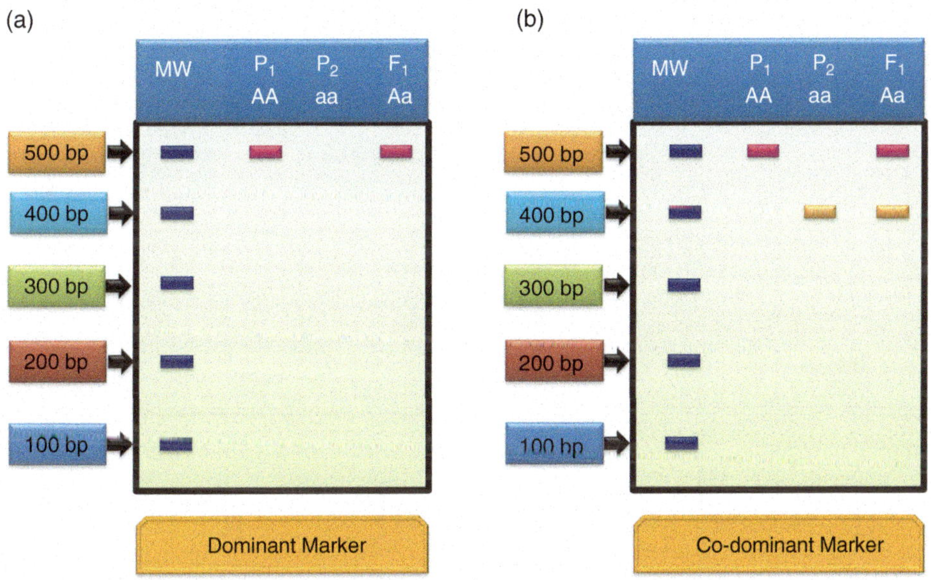

Figure 4.5 Comparison between (a) dominant and (b) codominant molecular markers. Dominant markers cannot distinguish homozygotes from heterozygotes, but codominant markers can. MW- Molecular weight of ladder, P1- Parent-1 (AA), P2-Parent-2 (aa), F_1-Hybrid (Aa).

populations. Table 4.2 and Figures 4.5 and 4.6 display the segregation ratio for several biparental mapping populations using codominant and dominant markers, respectively.

4.4.3 Natural Population-based Mapping

The utilization of natural population-based mapping has demonstrated considerable effectiveness in detecting natural variations associated with complex traits across a range of crops since 2001. This technique is characterized by its ability to provide high resolution

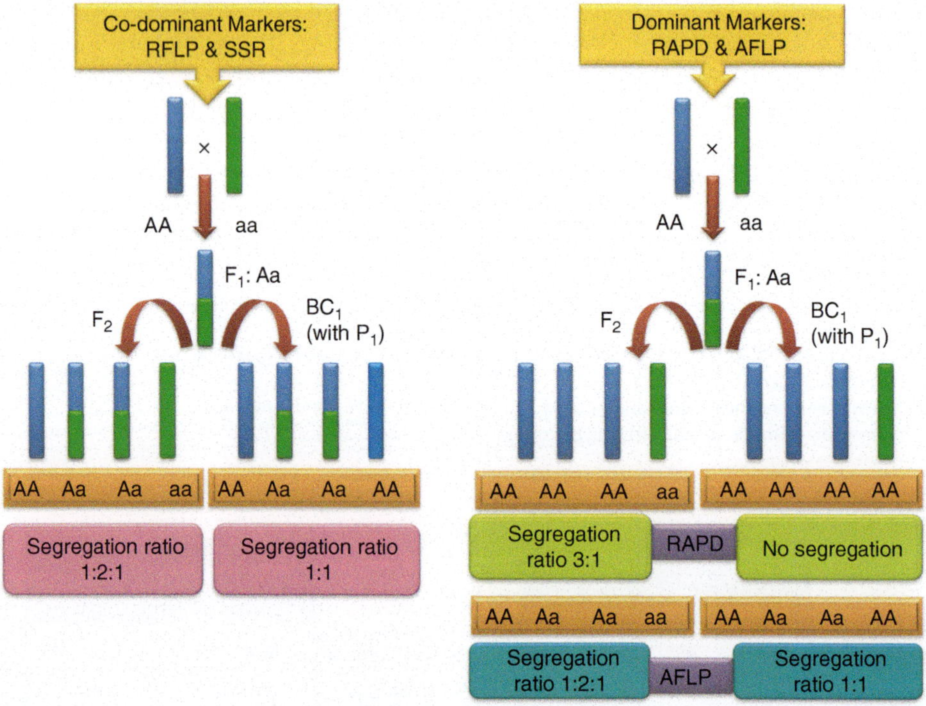

Figure 4.6 A schematic representation of the inheritance pattern of F_2 and backcross progenies by using molecular markers. The codominant markers easily differentiate among the homo- & heterozygotes, whereas by using dominant markers detection of heterozygosity is not possible or possible with rare exceptions as in AFLP (dominant) marker.

and allelic richness, while also eliminating the need for laborious development of a mapping population. The essential procedures involved in population-based mapping are depicted in Figure 4.7. This chapter presents an examination of the use of NAM populations in the context of plant QTL mapping. The proposed methodology involves several key steps: (i) Constructing a population sample that consists of elite cultivars, landraces, wild relatives, and exotic accessions. (ii) Conducting phenotypic assessments to evaluate specific traits of interest, estimating the broad-sense heritability of these traits, and identifying the genotypes of the entries. This can be done either by focusing on candidate genes or by conducting a genome-wide analysis. (iii) Quantifying the extent of LD within the selected population. (iv) Investigating the influence of population structure and kinship on the observed traits and genetic variations. According to Xu et al. (2017), the statistical power of natural population-based mapping is significantly influenced by factors such as sample size, minor allele frequency (MAF), level of LD, and population structure. The degree of LD between genotyped markers and QTL is a crucial factor in determining the utility of a natural population-based map. The measurement of the nonrandom association of alleles within a population is quantified by the concept of LD. Recombination, mutation, mating system, admixture, genetic drift, and selection represent a subset of the numerous factors that contribute to the phenomenon of LD. The estimation of LD can be conducted using

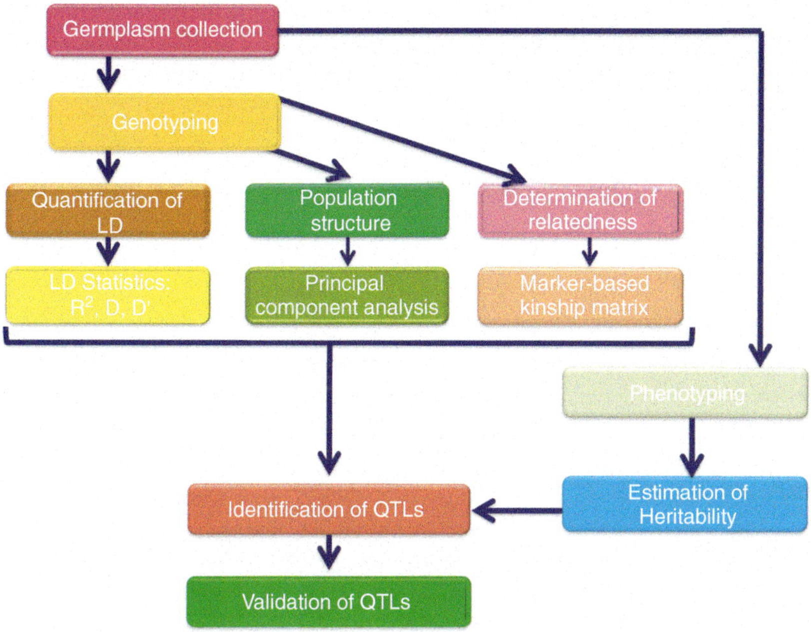

Figure 4.7 A diagram of the population used for QTL mapping that is based on natural populations. *Source:* Modified from Xu et al. (2017).

many statistical measures, with the most widely used ones being D′ and r2. The graphical depiction of LD degradation is commonly illustrated by utilizing a r2 value ranging from 0.1 to 0.2. The decline of the population's LD rate with physical distance has implications for the necessary marker density and achievable precision. The utilization of a limited number of markers for genome scanning may lead to a diminished mapping resolution, particularly in cases where LD exhibits a delayed decay within a specific region (Xu et al. 2017). Nevertheless, if the decay rate of LD is rapid, the enhancement in mapping resolution comes at the cost of a larger collection of markers. The extent of LD varies significantly across species due to their diverse mating systems. Self-fertilizing organisms have a greater extent of LD when compared to species that do not engage in self-fertilization. In their study, Garris et al. (2003) identified the presence of LD in the vicinity of locus Xa5 in rice, which was shown to span a distance of 100 kb. The strength of the LD, as measured by the r2 statistic, was determined to be 0.1. The rate of LD degradation is more rapid in species that exhibit outcrossing behavior, such as maize. The level of learning disabilities may exhibit significant variation, even across individuals within a single species. In the case of maize, it has been observed that LD can extend up to a length of 100 kb in commercially developed elite inbred lines. However, in landraces, LD rapidly diminishes to a range of 1–2 kb (Remington et al. 2001; Tenaillon et al. 2001). Maize, owing to its extensive genetic variety and rapid rate of LD degradation, serves as a robust model for conducting population-based mapping studies. While the overall amount of genome-wide LD throughout a population may not provide a realistic representation of LD at a specific genomic location, it is more pertinent to measure the decline of LD within a chromosome (Yan et al. 2009).

The presence of phenotypic differences within subpopulations, which are often linked to neutral markers, poses a substantial obstacle to the practical application of natural population-based mapping in crop studies. The most effective approach for addressing population structure (Q matrix) in natural population-based mapping is the mixed linear model (MLM), as demonstrated by Yu et al. (2006). The MLM considers population structure as a fixed effect.

4.4.4 Bulked Segregant Analysis (BSA)

In recent years, the technique of bulked segregant analysis (BSA) has been developed as a means of rapidly finding linkage, as discussed by Michelmore et al. (1991). In the process of BSA, two DNA samples that have been pooled together are utilized. These DNA samples originate from a segregating population resulting from a single-cross. These groupings consist of individuals who share the same feature, such as disease resistance or susceptibility, or a specific genetic area, but are otherwise unrelated to each other, meaning they are heterozygous. DNA polymorphism testing is performed on the masses, and the results are compared to a genetic background of unrelated regions to see if there are any significant differences. Inconsistencies between the two masses point to features connected with markers (like bands), which would explain the differences (Figure 4.8). Unlike virtually isogenic lines (NILs), which necessitate extensive backcrossing to generate, BSA does not have as many drawbacks. In contrast to NILs, which assert that only a subset of polymorphic loci (e.g. BC_5 at a rate of 50%) are genetically connected to a specific target region, BSA posits that genomic areas unlinked to the target region will remain unchanged throughout several bulked samples of individuals. The utilization of BSA can effectively mitigate the issue of linkage drag, as all loci will undergo segregation and subsequent mapping. This process involves the insertion of genes into lines through backcrossing, which are then flanked by DNA segments introduced from the donor parent. Traditional genetic mapping, in which all individuals in sample populations are assessed for features of interest, is more time-consuming and expensive than selective genotyping and bulked-segregant analysis. This is especially true for important gene-controlled traits and large-effect QTL. Bulk-segregant analysis can now be applied to both naturally occurring populations and biparentally segregating populations because of advancements in NGS technologies. Bulk sample analysis (BSA), according to Zou et al. (2016), is a sampling-bulking technique that chooses extreme or representative samples from all populations and then analyses them all at once. The strength of a bulked sample analysis is highly sensitive to the sample size and the selection of the samples within that sample. A bidirectional sample can be drawn from a distribution in which both tails are examined, while a unidirectional sample can be drawn from a distribution in which only one tail is (Sun et al. 2010). As bidirectional selection removes the impact of segregation distortion, it is frequently employed and generally more effective. On the other hand, single selection is better suited to features that are under heavy fatal or deleterious selection pressure. QTL for agronomically significant traits such as rust resistance in wheat (Forrest et al. 2014), kernel row number in maize (Yang et al. 2015), and salt tolerance and blast disease in rice have been successfully identified using chip-based BSA (Yang et al. 2015). BSA's widespread adoption in genetic mapping is assured by its superior detection power and low genotyping costs.

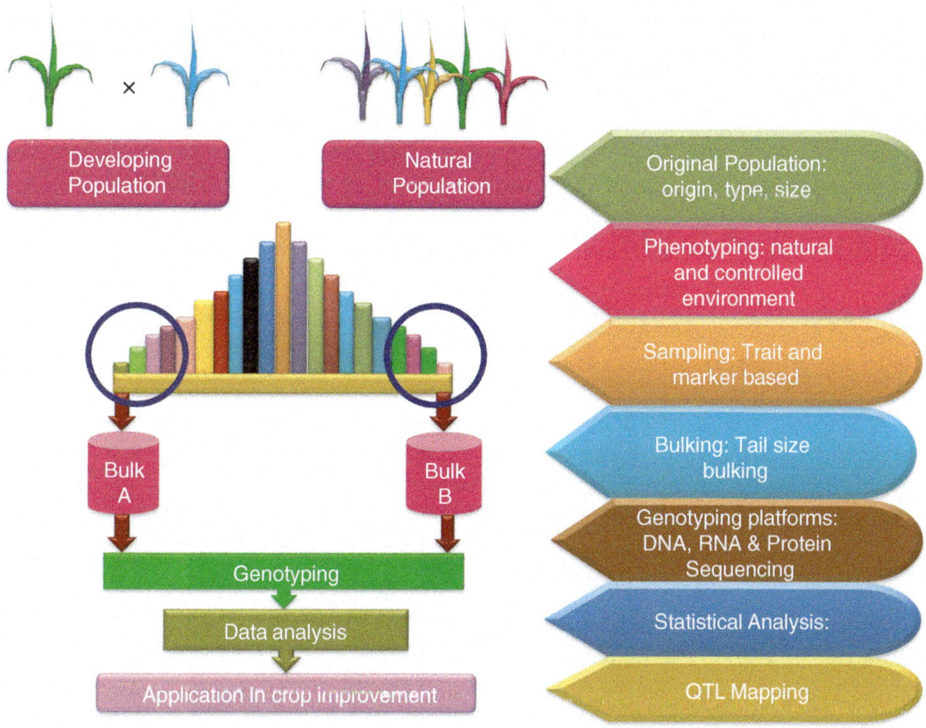

Figure 4.8 A general pipeline for bulked segregant analysis (BSA) in plants for QTL mapping. Using artificial populations and natural variants as examples, we see how the selection of individuals with extreme phenotypes can lead to the formation of two bulks (Bulk A and Bulk B). Genetically engineered, inbred groups are created by having children from two or more sets of parents. Natural variations: a population with a variety of naturally occurring variants chosen from a single or numerous sources.

4.4.5 Choice of Appropriate Mapping Population: Types and Size

The goals of the experiment, the timing, and the resources available for doing QTL analysis all have an impact on the mapping population used. There are benefits and drawbacks to using each of the aforementioned mapping populations. So, various aspects, such as the plant species, marker system, and mapped trait, influence the decision as to which type of mapping population to use. Because they are the most elementary forms of mapping populations to develop in self-fertile species, F_2 and F_2-derived populations, as well as BCs, have a greater capacity for detecting QTLs and containing information than other populations. Additive effect QTLs are easier to spot using F_2, and its estimated dominance can be employed for the same purpose. Its primary benefits lie in its low production costs and quick assembly time. These populations, however, cannot be improved upon because of their inherently heterozygous genetic makeup. Their usefulness in QTL study is consequently constrained. One of the greatest advantages of $F_{2:3}$ families is the ability to measure the effects of additive and dominant gene actions at specific loci. The most significant drawback is the time commitment involved in creating a RIL population; typically, six to eight generations are needed. To account for the diversity of human populations, the genetic model must be adapted (Prasanna 2003). As RILs are nearly homozygous, the combined effects of their genes can only be evaluated. Yet, the RILs have the benefit of conducting larger trials in a variety of settings and over a longer period of time. Producing enough seed for a multisite experiment with a population of $F_{2:3}$ families is not possible for many crops. Only species that are amenable to tissue culture can yield DH populations, which can be developed in a fraction of the time it takes to develop RIL populations (e.g. cereal species such as wheat, rice, and barley). Creating homozygous or "true-breeding" lines that can be replicated without genetic variation is one of the primary benefits of RIL and DH populations. This paves the way for repeated experiments at other times and places.

The kind of markers used to dictate the mapping population used (Tanksley et al. 1988). Several ways of mapping populations' genetic data on dominant versus codominant markers were evaluated by Staub et al. (1996). MAS can be used to cross progeny based on marker–trait map associations in early generations (F_2, and F_3) when linkage groups have not been completely disassociated by recombination events (i.e. maximal disequilibrium) (Staub et al. 1996). A F_2 population makes the most of the data provided by codominant markers, whereas RILs and DHLs make the most of the data offered by dominant marker systems (Burr et al. 1988). Double haploids, F_2 or F_3 families, or RILs are advantageous if the characteristic to be mapped cannot be adequately assessed on a single-plant basis and must be analyzed in replicated field experiments. Because LD is necessary to find QTLs using markers, QTL mapping in populations with random mating is more difficult (Mather 1938). In some cases, RILs, NILs, and DHs can discover QTLs, but they do not tell you anything about dominance relationships (Haley and Andersson 1997). If F_3 or BCF_2 progeny tests are performed to detect heterozygous F_2 individuals, information from a dominant marker system can be comparable to a thoroughly categorized F_2 population. This method is frequently impractical because progeny testing is expensive and time-consuming. Dominant markers are just as informative as codominant markers in BCs in the coupling phase, DHs, or RILs (i.e. a group of closely related lines; typically, $>F_8$) (Burr et al. 1988). Because all loci are homozygous, or nearly so, in RILs or DHs, the information gleaned from dominant markers is maximized. Dominant and codominant markers analyzed in RIL populations provide more information per individual than either marker type in BCs under conditions of tight linkage (i.e. roughly 10% recombination) (Reiter et al. 1992). When compared to codominant markers, however, the information received per individual in RIL populations drastically diminishes as the distance between markers increases (i.e. loci become more independent). BCs can be utilized to map dominant markers as long as the recurrent parent is homozygous for all loci and the donor and recurrent parent have distinct alleles for the polymorphic marker (Reiter et al. 1992). Since only one recombinant gamete is sampled per plant in BCs instead of two in F_2 populations, the information gained is limited. Yet, as the distance between linked loci rises in RILs (i.e. around >15% recombination), BCs become more informative (at low marker saturation) in comparison to RILs. Although increased recombination can aid in the resolution of highly intermingled links, it may be counterproductive when trying to build maps with low marker saturation. BC populations can be useful for mapping dominant markers if the donor and recurrent parent have different polymorphic marker alleles and the recurrent parent is homozygous for all loci (Reiter et al. 1992). As only one recombinant gamete is sampled per plant in BC populations, the information obtained from them using codominant or dominant markers is limited in comparison to that received from F_2 populations.

4.5 Molecular Markers for QTL Mapping

The success of artificial selection for the genetic improvement of a species relies on isolating and exploiting genetic influences that are distinct from environmental ones. The breeding population can undergo substantial, economically significant changes due to phenotypic selection based on features conditioned by additive allelic effects. Molecular markers, also called DNA markers, are heritable genetic differences between people that may be simply identified and tracked. Plant breeders can employ genetic markers, which are heritable entities linked to economically significant features, to make selections (Darvasi and Soller 1994). Molecular markers rely on polymorphism, or variation, in DNA banding patterns. Among the fortunate characteristics of molecular markers are their high polymorphism, codominant inheritance, even and frequent distribution across the genome, resistance to the effects of environmental factors, rapid and inexpensive detection, ease of assay, high reproducibility, and the ability to share data easily between laboratories. Given the variety of molecular marker approaches, such as hybridization-based and polymerase chain reaction-based methods, available for detecting polymorphism in DNA, making a choice might be challenging. There are pros and cons to using different marker systems, so it is important to weigh those before deciding which to implement. In Table 4.3, Kumar et al. (2009) review the benefits and drawbacks of a few frequently employed molecular markers. By enabling early selection and decreasing the size of the plant population employed during selection, MAS has the potential to increase selection efficiency. Despite the fact that marker loci can identify phenotypic variation, genetic linkage, pleiotropy, and environmental variables typically render this variation nonadditive (Lark et al. 1995). Hence, predicting responses to selection (R) or genetic gain is frequently difficult, and the success of using marker loci to predict phenotypic variance depends on a variety of factors. When a genetic marker is used in MAS, its inherent repeatability (Weeden et al. 1992), map position, and linkage with economically significant variables all increase its predictive value (quantitative or qualitative). It would be excellent if there was a substantial connection (10 cM) between a qualitative characteristic and a genetic marker for MAS to be effective (Kennard et al. 1994; Timmerman et al. 1994). Finding genes that affect quantitative traits using molecular markers is called mapping QTL. A high relationship between a quantitative trait and markers also increases the efficacy of selection for multiple loci or QTL using genetic markers (Edwards and Page 1994). Numerous studies have identified several yield-attributing QTLs in many crops including biotic and abiotic stress tolerance. The introgression of valuable mega QTLs through marker-assisted backcross breeding has changed the yield scenario of many crops due to the climate-resilient nature of the improved varieties. Several examples of QTL introgression include drought tolerance in rice (Vikram et al. 2011; Majhi et al. 2022a,b; Singh et al. 2022), submergence tolerance in rice (Voesenek and Bailey-Serres 2015; Sandhya et al. 2019), salt tolerance in rice (Reddy et al. 2017; Waziri et al. 2016), grain quality improvement in rice (Patra et al. 2022), bacterial blight resistance in rice (Fiyaz et al. 2022), bruchid tolerance in mung bean (Majhi et al. 2022a,b), drought tolerance in chickpea (Bharadwaj et al. 2021), stripe rust resistance in wheat (Elkot et al. 2015; Hu et al. 2020).

Table 4.3 Comparison of some important molecular markers used in QTL mapping.

Marker type	Variability	Cost	Speed to screen	Expression	Advantages	Disadvantages
Restriction fragment length polymorphism (RFLP)	Medium-high	Medium	Medium	Codominant	Very repeatable, solid, and reliable, and site-specific	Needs a lot of expensive, high-quality DNA, takes a long time, and is not very polymorphic
Amplified fragment length polymorphism (AFLP)	High	Low	Medium-fast	Dominant	Extremely repeatable, highly polymorphic, applicable to all organisms for whom sequencing data is unavailable	DNA of both high quality and quantity needs a laborious approach to analysis
Randomly amplified polymorphic DNA (RAPD)	High	Low	Medium	Dominant	Very little DNA is needed, and the procedure is fast and cheap	Not able to be duplicated, and usually not transferrable either
Simple sequence repeats (SSR) or (Microsatellite)	Highest	Medium	Medium	Codominant	Very repeatable, scalable to a population's size, stable and dependable, locus-specific, mechanized, and straightforward to implement	Costs are high because creating a primer takes a lot of time and effort
Single nucleotide polymorphism (SNP) or (Minisatellite)	Medium-high	Medium	Medium	Codominant	These mutations are helpful for mapping genes and doing positional cloning, and they can be found in both naturally occurring populations and those created through artificial means	Three times as many SNP markers are required as SSR in order to accurately map genes, but they provide far less information than SSR
Express sequence tag (EST)	High	High	Fast	Codominant	Significantly repeatable, stable, and trustworthy	While creating markers, scientists can only work with species for whom a sequencing database already exists

4.6 Statistical Approaches for Detection of QTLs

The main objective of QTL mapping investigations is to detect QTL while mitigating the occurrence of false positives, also known as Type I errors, which refers to the identification of a relationship between a marker and QTL when no such relationship actually exists. In order to examine the association between QTLs and traits, researchers frequently employ the following methodologies (Table 4.4).

4.6.1 Single-marker Analysis (Point Analysis)

The results of a univariate analysis demonstrated a significant association between the genotype of a specific marker and the observed trait value. Regression analysis is an essential component within the broader analytical framework. The single marker technique, alternatively referred to as single factor analysis of variance (SF-ANOVA) or single point analysis, has been effectively employed in numerous investigations, particularly in the context of isozymes (Edwards et al. 1987). Each marker locus is analyzed with SF-ANOVA separately from the others. F-tests reveal whether or not there are statistically significant distinctions between genotype classes at a marker locus. Commonly, this technique is used to rapidly screen the entire genome for the most useful QTLs. Marker loci that do not share any other genetic material with other places are given one. The location of QTLs cannot be determined using this method in most cases. The F-test is used to identify statistically significant differences amongst sets of genotypes. This technique exhibits several significant drawbacks, which encompass the following aspects: The strategy exhibits lower levels of accuracy in comparison to alternative methods due to the following factors: (i) The ability to determine whether the markers are associated with one or multiple QTLs is limited. (ii) The probability of detecting QTL decreases as the distance between markers increases. (iii) The estimated effects of QTL are inflated due to confounding with recombination frequencies. (iv) The method lacks the capability to determine whether the markers are linked to one or multiple QTLs (Sehgal et al. 2016). Examining individual genetic markers is the standard approach to finding QTLs near a marker. This study aims to compare the average phenotypes of offspring belonging to

Table 4.4 Comparison of different methods of QTL mapping.

Particulars	Simple interval mapping	Composite interval mapping	Multiple interval mapping	Bayesian interval mapping
Markers used	Two markers	Markers used as cofactors	Multiple markers	Two markers
Information obtained about	Position and number of QTLs	Position and number of QTLs and interaction of QTLs	Position and number of QTLs	Position and number of QTLs and their effects
Designated as	SIM	CIM	MIM	BIM
Precision	High	Very high	Very high	Very high

several marker classes, such as AA, Aa, and aa. The estimation of the phenotypic effect resulting from the substitution of a "A" allele with a "a" allele at the QTL can be determined by calculating the difference between two means. Basic statistical tests, such as the t-test or F-test, are utilized to ascertain whether the anticipated phenotypic impact significantly deviates from zero. If the value is high enough, a QTL must be close to the marker. An exhaustive molecular linkage map is unnecessary for single-point analysis. The probability of statistically detecting a QTL generally diminishes as the QTL's distance from the marker increases due to the occurrence of crossover events between the marker and the gene, as noted by Sehgal et al. (2016). Although this particular approach is very simple in terms of computer implementation, it is accompanied by many significant limitations. According to Prasanna (2003), the method employed is unable to differentiate between markers that are linked to a single QTL or several QTLs. Furthermore, the possibility of successfully identifying QTLs diminishes significantly as the distance between the marker and QTL increases. Additionally, it is probable that the impacts of QTLs are exaggerated due to recombination rates.

4.6.2 Interval Mapping by Maximum Likelihood

QTL interval mapping is widely regarded as the predominant method employed in QTL analysis. The underlying concept of interval mapping is to evaluate the presence of QTL in a model by examining numerous positions within the genomic interval delimited by two mapped marker loci. In order to assess the soundness of this model, the maximum likelihood method can be employed. According to Sehgal et al. (2016), when a QTL is hypothesized to be located between two markers, the genotypes of the markers at two loci (AABB, AABb, aaBB, and aabb for DH progeny) exhibit combinations of QTL genotypes. In order to obtain the most accurate estimation of the quantitative trait distributions associated with each marker class, it is advisable to identify QTL parameters that exhibit the highest likelihood. The evaluation of the models involves a comparison of the probability of the observed distributions under two scenarios: one with the inclusion of a QTL effect and one without. The investigation conducted by Sehgal et al. (2016) involves the analysis of the distribution of likelihood values, specifically log of odds ratio (LOD) scores. These scores represent the ratio between the likelihood that a particular impact is attributed to linkage and the likelihood that it is due to chance. The primary objective of this analysis is to identify the most probable position for a QTL.

4.6.3 Interval Mapping by Regression

Haley and Knott (1992) created the technique of interval mapping, which employs regression to enhance the efficiency of the maximum likelihood approach. Phenotypes are subjected to regression analysis using QTL genotypes, following a similar approach to traditional QTL analysis including regression with encoded marker genotypes. Probabilities based on the flanking markers are used in place of the actual QTL genotypes, which are unknown. Bayesian interval mapping, developed by Satagopan et al. (1996), provides data on the number and distribution of QTLs.

4.6.4 Simple Interval Mapping (SIM)

In their initial work, Lander and Botstein (1989) introduced SIM, which totally relies on a linkage map. The present study examines the correlation between trait values and the genotype of a hypothetical QTL, specifically referred to as the target QTL. This investigation is conducted at multiple analysis points inside the target interval, which is defined as the region between adjacent marker loci. The existence of a putative QTL is presumed if the LOD exceeds a certain threshold. Lander and Botstein (1989) derived mathematical equations to determine suitable significance levels for interval mapping, taking into account the known factors of genome size, number of chromosomes, number of marker intervals, and overall false positive rate. SIM has garnered significant attention due to the ease of utilization provided by statistical tools such as MAPMAKER/QTL. The acquisition of data from markers that are closely related can mitigate the influence of recombination between the markers and the QTL. This, in turn, enhances the statistical probability of identifying the QTL and facilitates an unbiased evaluation of its effects. However, the Simulation Model (SIM) fails to consider the genetic diversity resulting from other QTLs when numerous QTLs are present in a genetic cross. The utilization of SIM exhibits analogous limits to those encountered in a single marker analysis.

4.6.5 Composite Interval Mapping (CIM)

The technique of interval mapping is employed in conjunction with multiple regression analysis to examine markers that are connected to other QTLs in composite interval mapping (CIM) (Zeng 1994). Additionally, Jansen and Stam (1994) utilized the multiple-QTL model, also referred to as marker-QTL-marker analysis, known as MQM. Every study takes into account a marker interval and a select few individual markers, resulting in "n-1" tests for interval-QTL links on a chromosome containing "n" markers. This approach is used to reduce the impact of a cluster of connected QTLs. It relies on one QTL and other variables, such as different markers. Results are more accurate and bias from adjacent QTLs (nontarget QTLs) may be ruled out with this approach. In the past, only one QTL parameter could be fitted in a given time window. The utilization of the partial regression coefficient is employed in the estimation of the fraction of genetic variance that can be ascribed to QTLs that are not directly addressed. The number of tests for interval-QTL connections on a chromosome with n markers is determined by considering a marker interval and a few other single markers in each QTL research, resulting in "n-1" potential tests. There are several advantages associated with CIM. According to Prasanna (2003), the utilization of linked markers as covariates in the process of mapping multiple QTLs leads to an enhancement in the precision of QTL mapping. This improvement is attributed to a reduction in the number of additional QTLs that contribute to the genetic variation, hence increasing the efficacy of QTL determination. Despite the superior effectiveness of CIM compared to SIM, its utilization for QTL mapping remains limited. The inclusion of a single location as a variable in the model poses a challenge to the accuracy and reliability of interval mapping for identifying QTLs. The approach exhibits two shortcomings: firstly, the inclusion of additional

QTL effects would contribute to an increase in sample variation, and secondly, if two QTLs are linked, their collective effects would result in biased results. In order to tackle the matter at hand, researchers devised the CIM method, as documented by Utz and Melchinger (1996) and Zeng (1994). The analysis will be conducted using the customary approach of CIM. However, in order to accommodate the variability arising from other QTLs, partial regression coefficients derived from markers (referred to as cofactors) located in different genomic areas will be utilized. The efficacy and accuracy of standard interval mapping are inferior to those of CIM due to the inability to account for other QTL effects that do not present as residual variation (SIM). The utilization of CIM has the potential to mitigate the influence of QTLs associated with the testing site.

4.6.6 Multiple Interval Mapping (MIM)

The QTL mapping technique employed is at the forefront of scientific advancements. Many-interval mapping (MIM), a variant of interval mapping, incorporates several QTLs directly into the QTL mapping model, without the need for an intermediary marker. Just as multiple regression extends the analysis of variance, MIM enables the application of interval mapping to multiple QTLs. The utilization of this tool enables the mapping of a significant multitude of QTLs. According to Prasanna (2003), QTL–QTL interactions can be identified using this approach.

4.7 Software Used for QTL Mapping

There are over 100 distinct software available for the purpose of doing genetic analysis, encompassing various methodologies such as linkage analysis and QTL mapping. These devices were designed with the purpose of facilitating the analysis of DNA samples in the context of the mapping procedure. The populations that exhibit segregation patterns offer the necessary data for these programs to calculate the recombination frequency. This information, in turn, aids in determining the linear arrangement of genetic markers. By using data obtained from the segregation of populations, these applications make estimations of recombination frequencies, which are then employed to ascertain the linear arrangement of genetic markers by reducing the occurrence of recombination events. Several widely used software programs for conducting QTL analysis are available in the field. These include MAPMAKER/QTL (Lander et al. 1987), GMendel (Echt et al. 1992), Linkage1 (Suiter et al. 1983), MapManager, JoinMap (Stam 1993), QTL Cartographer, PLABQTL, QTL Mapper, and Qgene (Sehgal et al. 2016). Although there may be modest variations in the confidence intervals derived by different statistical techniques, the QTL locations and gene effects obtained from a certain data set tend to exhibit a high degree of consistency. Recent advancements in QTL mapping have seen the adoption of association-based methods and nonparametric statistics. These approaches have been effective in identifying QTLs and determining empirical critical values for the detection of significant QTLs (Prasanna 2003). Consequently, the use of these methods holds promise for enhancing QTL mapping techniques.

4.8 QTLs and the Signature of Selection

If one can show that there is a polarity to allelic substitution, then QTL data can be used as proof of the existence of directional selection. Sunflowers and Lake Malawi cichlids are two examples of animals that have benefited from this method, which has been used to better quantify the primary selection factors behind the evolution of these two very different groups.

1) **QTL detection and localization:** Figure 4.1 illustrates the process of developing and utilizing an NIL to identify QTLs with a significant likelihood. The initial step involves crossing a donor parent with a recurrent parent, followed by numerous backcrosses to the recurrent parent, resulting in a gradual decrease in the proportion of the donor genome contribution. MAS enables the construction of a genome-tiled panel of NILs. The panel members that are obtained can undergo testing for various phenotypic features in order to identify and locate potential QTL in numerous crop species.
2) **Introgressed lines development:** In their study, Schauer et al. (2006) conducted a comprehensive analysis to identify and map QTLs in tomato plants for the purpose of enhancing crop quality. This was achieved through the creation of introgression lines, which involved multiple rounds of self- and back-crossing between an elite cultivar, *Solanum lycopersicum* var. Roma, and a wild tomato plant, *Solanum pennellii*. The resulting 76 independent introgression lines contained chromosome segments derived from the wild relative. The researchers successfully mapped the QTLs associated with metabolic and fruit-quality traits in these introgression lines. The process of choosing particular homozygous, single, overlapping chromosome introgressions within this population serves to streamline the localization of QTL and establish associated DNA markers that can be utilized in the enhancement of crops.

Applications of QTL mapping:

i) The use of markers to aid in the introduction of advantageous genes or QTLs.
ii) Permit cross-species comparative mapping to assess genetic similarities in phenotype expression and gene order.
iii) This framework should be established and supported by the utilization of physical maps derived from DNA sequencing, chromosome translocations, or other direct measurement techniques.
iv) The initial step in the process of positional or map-based gene cloning involves identifying the primary stage responsible for economically significant features.

4.9 Factors Affecting the Power of QTL Mapping

The data from an experiment is used to make a statistical inference about the presence of QTLs. Because of the many test issues inherent in QTL mapping, however, statistical influence is not always indicative of biological importance (Liu 2002). The existing literature gives great descriptions of these difficulties as well as proposed ways to improve QTL mapping effectiveness (Kearsey and Farquhar 1998; Liu 2002). Methods for refining experimental designs have been proposed to increase the accuracy of QTL identification and QTL effect estimation. Several strategies that can be employed include repeat progeny analyses, selective genotyping, sample pooling, and sequential sampling.

The effectiveness of QTL mapping is influenced by the following parameters:

1) **Features of QTLs that are determined by their genetic makeup:** The genetic characteristics of QTLs encompass various factors, including their effect magnitude, potential linkage to other QTLs governing the same trait, interactions with other QTLs or the genetic background, and sensitivity to environmental influences. QTLs with substantial effects, accounting for 10% or more of the phenotypic variance, are more readily detected compared to QTLs with lesser effects. When two QTLs exist in close proximity (within a genetic distance of less than 20 cM), they are likely to be detected as a single QTL, unless the size of the mapping population exceeds 500 individuals or lines. QTLs that are influenced by environmental factors may exhibit detectability in certain contexts while remaining undetectable in others, especially when their magnitude of impact is relatively modest.
2) **Number and location of regulating genes for desirable qualities:** The success of QTL mapping is influenced by the genomic location of the gene on the chromosome. The proximity of genes to a specific genetic marker increases the likelihood of detecting target traits or target genes. The classification is determined by the distinctive pattern of markers employed. If genes are located at a distance from a specific genetic marker, there is a higher likelihood of crossing over occurring. It has an impact on the banding pattern of the markers employed. In this particular scenario, ascertaining the precise location of target genes is a challenging task.
3) **Background of genes:** The outcomes of QTL mapping are heavily influenced by the genetic background in which a QTL is situated. This matter will be a challenge, particularly when comparing results obtained from various mapping populations. QTL mapping enables the examination of interactions between QTLs, necessitating the presence of identical QTLs in many genetic backgrounds. However, this requirement incurs substantial expenses and demands a significant investment of time.
4) **The heritability of the segregating genes in a mapping population:** Characteristics controlled by single genes (or oligogenes) tend to be more heritable than those controlled by several genes.
5) **The QTL mapping population type:** QTL mapping requires a population in which mating is not completely random. It is the work of random chance, natural selection, mutation, and the like. The nature of the mapping population will dictate the types of gene effects uncovered by QTL analysis. For instance, only additive and additive × additive effects would be estimable by analyzing RIL and DH populations. Yet, in a backcross population, both additive and dominant effects are muddled.
6) **Mapping population size:** While studying a population, it is easier to spot QTLs with a small impact size if the population is large enough. The capability to discover QTLs, the accuracy of pinpointing their locations, and the reliability of estimating their effect sizes are all improved by a larger sample size.

7) **Number and types of markers:** Increasing the number of markers utilized will enhance the precision of estimating both the position and effect of QTLs. The codominant marker exhibits three distinct types of genetic variation, whereas the dominant marker demonstrates two types of genetic variation. The codominant marker offers a greater amount of information compared to the dominant marker with respect to recombination within marker intervals.
8) **Sample size and mapping population for phenotyping:** The quantitative qualities of interest are measured with a high degree of precision, and a small quantity of missing data is acceptable within acceptable limits. The ability to determine the location of QTL is primarily limited by two factors: the size of the sample population and the extent of genetic marker coverage across the genome. In many cases, the apparent size of a sample may be substantial, but the presence of missing data or imbalanced allele frequencies within the population can lead to a reduction in the effective sample size. Consequently, this reduction compromises the statistical power of the analysis. Occasionally, it becomes necessary to prioritize data quality over population size, resulting in a trade-off where only significant QTL with substantial effects may be identified. In order to derive a unified quantitative trait for a particular line, it is customary to aggregate QTL data across multiple sites and replications. It is advisable to assess the desired characteristic(s) in tests conducted at different (and suitable) sites in order to gain a more comprehensive comprehension of any potential interaction between QTL and the environment.
9) **QTL expression affected by environmental influence:** The manifestation of quantitative traits in terms of phenotype is significantly influenced by environmental circumstances, with the degree of influence being essentially determined by the specific trait under consideration. Typically, QTLs with substantial effects tend to display greater stability across many settings compared to QTLs with smaller effects, which account for less than 10% of the phenotypic variance. Hence, the outcomes of QTL analysis pertaining to characteristics exhibiting substantial heritability are considerably more dependable compared to those associated with traits demonstrating low heritability. Hence, it is advantageous to perform phenotypic assessment across several habitats, typically denoted by specific locations such as field, glasshouse, or greenhouse, as well as different years. This approach enables the evaluation of the consistency in the expression of QTLs and facilitates the identification of QTLs that exhibit relatively consistent expression across diverse environments.
10) **Experimental error:** The assessment of phenotypic traits in the mapping population and the genotyping of markers are the main sources of experimental error. The mapping population was subjected to phenotypic evaluation. Considering the impact of environmental factors on the expression of quantitative traits, it can be argued that measurements conducted on individual plants lack sufficient reliability. Moreover, it is imperative that phenotypic evaluation be grounded in meticulously executed repeatable trials. The presence of errors in genotyping and the absence of marker data can have an impact on the arrangement of markers in the linkage map as well as the distances between these mapped markers. Consequently, these factors may also influence the estimated locations of QTLs.

4.10 Merits of QTL Mapping

1) Each QTL that controls the phenotype of interest can be identified and mapped with high accuracy using linkage mapping.
2) It is possible to employ these markers for MAS, including recombinant selection when studying a QTL that has been mapped.
3) Third, it gives a rough idea of how much of an influence a given QTL has on the phenotype of a given trait. This gives breeders an approximate sense of the QTL's potential value in future breeding efforts.
4) In addition, the utilization of joint QTL analysis for several correlated traits enables the differentiation between the underlying mechanisms of trait correlations, namely close linkage and pleiotropy. Implementing this approach would enable us to assess the potential effectiveness of endeavors aimed at eliminating unfavorable characteristics.
5) High-resolution QTL mapping substantially facilitates the cloning of genes inside the QTL region due to the ability to detect a QTL with a narrow confidence interval of only 1 cM.
6) In order to reduce the number of putative candidate genes inside the genomic region containing the QTL associated with the target trait, a combination of selective DNA pooling and transcriptome analysis can be employed.
7) Suitable experimental plans and QTL analysis procedures can be used to detect and estimate QTL × QTL and QTL × environment interactions.
8) There are clear advantages associated with biparental populations in the context of QTL study as compared to AM. In contrast, linkage mapping efficiently facilitates the identification and mapping of even the most cryptic functional variations in genes and QTLs. This phenomenon occurs due to the inclusion of a specific line in the creation of the mapping population, which harbors the presence of the uncommon allele. Consequently, the QTL detection capacity of the mapping population will be enhanced, thereby facilitating the identification of the seldom allele.

4.11 Demerits of QTL Mapping

1) To commence, the quantitative traits' genetic variety inside the population can alone be elucidated by the discrepancies seen between the two parental individuals employed in the construction of the mapping population.
2) In mapping populations, researchers commonly limit their observations to the impact of only two alleles of the relevant genes or QTLs.
3) The resolving capability of QTL mapping is limited due to the minimal number of meiotic divisions that take place between parent hybridization and the subsequent usage of the resulting populations for mapping. Therefore, the location of a QTL can range from a few centimorgans to tens of centimorgans, often falling between the range of 5–20 cM. The spatial extent of this region often encompasses multiple megabases, ranging from 1.2 to 4.8 Mb, and has the potential to harbor a considerable number of genes, potentially numbering in the hundreds.

4) Fourth, locating QTLs at sub-centimorgan intervals and identifying candidate genes typically require high-resolution mapping efforts. This calls for extra work and increases the overall price.
5) It is exceedingly challenging to determine the gene responsible for a QTL impact, even when a QTL with a large effect has been established. Nevertheless, QTL mapping does not reveal information about the quantity, composition, or role of the genes associated with the observed QTLs.
6) In reality, multiple QTLs with minor or even antagonistic effects on the phenotype being studied can be found in close proximity to a major effect QTL.
7) It takes time and effort to create biparental mapping populations, which may not be possible in some species, such as tree species.
8) We cannot assume that a QTL found in a population with only two parents will be as effective in more complex families. This requires verifying the identified QTLs in genetic material with no genetic relationship to the tested material.
9) In the same vein, it is important to evaluate the transferability of the markers associated with QTLs/genes discovered in a biparental population to populations with unrelated genotypes. To do so necessitates the expenditure of more time and energy.
10) It is possible that conventional breeding methods have already introduced a new QTL into the breeding populations by the time it is discovered. This reduces the potential impact of MAS on breeding programs involving this QTL.
11) When mapping between populations, different QTLs for the same trait would be found. Therefore, it is unable to investigate possible interactions between the QTLs found in various populations.
12) Only QTLs that have a significant impact will consistently be identified in the immediate vicinity of a marker locus. In addition, the absence of appropriate experimental and analytical methodologies can provide challenges in identifying QTLs that exhibit significant epistatic effects and are influenced by environmental factors.

4.12 Conclusion and Way Forward

QTL analysis is a useful approach for determining the number of potential loci, their chromosomal location, the similarity of their effects, and the mechanism by which they function. QTL maps, which are constructed through linkage studies and marker–trait association, have the potential to enhance the screening of germplasm for resistance against biotic problems like as diseases and insects, as well as abiotic challenges including salinity, cold, salt, and drought. The identification and manipulation of genes that exert influence on quantitative features are of utmost importance in the field of plant breeding. DNA markers are commonly employed for the identification and quantification of QTLs due to their polymorphism nature, high abundance, and tendency to exhibit codominant inheritance patterns. In order to facilitate QTL analysis, it is necessary to construct high-resolution linkage maps utilizing a variety of molecular markers. QTL mapping requires careful planning as well as familiarity with the necessary statistical context. The utilization of MAS and QTL mapping should be extensively adopted for all significant crop species. Although there is less knowledge on the precise genetic characteristics of the QTLs, it is plausible that various crop species such as maize, tomato, and rice might potentially utilize MAS techniques to enhance breeding efforts targeting these QTLs. In the event that the QTL exhibits a substantial effect and can be successfully incorporated into a breeding program by marker-assisted backcrossing, the necessity for precise QTL mapping by plant breeders may be obviated. These technologies enable the identification of relevant QTLs that may have been overlooked by standard phenotypic selection methods. As an additional perk, the markers facilitate QTL introgression via backcrossing with less linkage drag. Although QTL analysis methods can shed light on the likely locations and effects of QTLs influencing a quantitative feature, they can reveal nothing about the molecular nature of the QTL. To rephrase, do these QTLs operate as gene regulators, code for specific enzymes in a specific pathway, or are they noncoding regions that modify the expression of the trait of interest? As QTLs and their associated DNA markers have been mapped, map-based cloning has become a viable method for isolating and characterizing QTLs at the molecular level. The strength and precision of QTL identification and QTL information exploitation for crop improvement are expected to increase with the development of improved theoretical models and high-throughput approaches, as well as the recent developments in molecular marker technology. Many plants' genetic maps were once constructed using SNP array-based genotyping techniques. As sequencing costs continue to drop, scientists are developing novel genotyping procedures using NGS technologies. A quick and affordable method for producing low-representation marker-discovery and genotyping libraries is genotyping by sequencing (GBS). Because it is quick, dependable, and affordable, GBS is already being used to genotype a variety of crop species with complicated genomes. Biparental mapping in conjunction with genome-wide association studies is proving to be a potent tool for analyzing complex plant characteristics. Because of advances in our understanding of genotypes over the last few decades, it is now possible to swiftly and cheaply sequence the complete genome and determine genotypes down to individual nucleotides.

References

Albine, S.M. and Jones, G.H. (1987). Synaptonemal complex spreading in *Allium cepa* and *A. fistulosum*. I. The initiation and sequence pairing. *Chromosoma* 95: 324–338.

Asins, M. (2002). Present and future of quantitative trait locus analysis in plant breeding. *Plant Breeding* 121: 281–291.

Bandillo, N., Muyco, P.A., Caspillo, C. et al. (2010). Development of multiparent advanced generation intercross (MAGIC) populations for gene discovery in rice (*Oryza sativa* L.). Philipp. *Journal of Crop Science* 35: 96.

Bandillo, N., Raghavan, C., Muyco, P.A. et al. (2013). Multi-parent advanced generation inter-cross (MAGIC) populations in rice: progress and potential for genetics research and breeding. *Rice* 6: 1–15.

Bharadwaj, C., Tripathi, S., Soren, K.R. et al. (2021). Introgression of "QTL-hotspot" region enhances drought tolerance and grain yield in three elite chickpea cultivars. *The Plant Genome* 14 (1): e20076.

Breseghello, F. and Sorrells, M.E. (2006). Association mapping of kernel size and milling quality in wheat (*Triticum aestivum* L.) cultivars. *Genetics* 172: 1165–1177.

Burr, B., Burr, F.A., Thompson, K.H. et al. (1988). Gene mapping with recombinant inbreds in maize. *Genetics* 118: 519–526.

Collard, B.C.Y., Jahufer, M.Z.Z., Brouwer, J.B., and Pang, E.C.K. (2005). An introduction to markers, quantitative trait loci (QTL) mapping and marker-assisted selection for crop improvement: the basic concepts. *Euphytica* 142: 169–196.

Darvasi, A. and Soller, M. (1994). Optimum spacing of genetic markers for determining linkage between marker loci and quantitative trait loci. *Theoretical and Applied Genetics* 89: 351–357.

Echt, C., Knapp, S., and Liu, B.H. (1992). Genome mapping with non-inbred crosses using GMendel 2.0. *Maize Genetics Cooperation Newsletter* 66: 27–29.

Edwards, M.D. and Page, N.J. (1994). Evaluation of marker-assisted selection through computer simulation. *Theoretical and Applied Genetics* 88: 376–382.

Edwards, M.D., Stuber, C.W., and Wendel, J.F. (1987). Molecular facilitated investigations of quantitative trait loci in maize. I. Number, genomic distribution and types of gene action. *Genetics* 116: 113–125.

Elkot, A.F.A., Chhuneja, P., Kaur, S. et al. (2015). Marker assisted transfer of two powdery mildew resistance genes PmTb7A. 1 and PmTb7A. 2 from *Triticum boeoticum* (Boiss.) to *Triticum aestivum* (L.). *PLoS One* 10 (6): e0128297.

Fiyaz, R.A., Shivani, D., Chaithanya, K. et al. (2022). Genetic improvement of rice for bacterial blight resistance: present status and future prospects. *Rice Science* 29 (2): 118–132.

Forrest, K., Pujol, V., Bulli, P. et al. (2014). Development of a SNP marker assay for the Lr67 gene of wheat using a genotyping by sequencing approach. *Molecular Breeding* 34: 2109–2118.

Garris, A.J., McCouch, S.R., and Kresovich, S. (2003). Population structure and its effect on haplotype diversity and linkage disequilibrium surrounding the *xa5* locus of rice (*Oryza sativa* L.). *Genetics* 165: 759–769.

Gelderman, H. (1975). Investigations on inheritance of quantitative characters in animals by gene markers. *Theoretical and Applied Genetics* 46: 319–330.

Guo, B., Sleper, D.A., and Beavis, W.D. (2010). Nested association mapping for identification of functional markers. *Genetics* 186: 373–383.

Haley, C.S. and Andersson, L. (1997). Linkage mapping of quantitative trait loci in plants and animals. In: *Genome Mapping: A Practical Approach* (ed. P.H. Dear), 49–71. New York: Oxford University Press.

Haley, C.S. and Knott, S.A. (1992). A simple regression method for mapping quantitative trait loci in line crosses using flanking markers. *Heredity* 69: 315–324.

Hu, T., Zhong, X., Yang, Q. et al. (2020). Introgression of two quantitative trait loci for stripe rust resistance into three Chinese wheat cultivars. *Agronomy* 10 (4): 483.

Huang, X.H., Wei, X.H., Sang, T. et al. (2010). Genome-wide association studies of 14 agronomic traits in rice landraces. *Nature Genetics* 42: 961–967.

Jansen, R.C. and Stam, P. (1994). High resolution of quantitative traits into multiple loci via interval mapping. *Genetics* 136: 1447–1455.

Kearsey, M.J. and Farquhar, A.G.L. (1998). QTL analysis in plants; where are we now? *Heredity* 80: 137–142.

Kennard, W.C., Poetter, K., Dijkhuizen, A. et al. (1994). Linkages among RFLP, RAPD, isozyme, disease resistance, and morphological markers in narrow and wide crosses of cucumber. *Theoretical and Applied Genetics* 89: 42–48.

Kumar, P., Gupta, V.K., Misra, A.K. et al. (2009). Potential of molecular markers in plant biotechnology. *Plant Omics Journal* 2: 141–162.

Lander, E.S. and Botstein, D. (1989). Mapping Mendelian factors underlying quantitative traits using RFLP linkage maps. *Genetics* 121: 185–199.

Lander, E.S., Green, P., Abrahamson, J. et al. (1987). MAPMAKER: an interactive computer package for constructing primary genetic linkage maps of experimental and natural populations. *Genomics* 1: 174–181.

Lark, K.G., Chase, K., Adler, F. et al. (1995). Interactions between quantitative trait loci in soybean in which trait variation at one locus is conditional upon a specific allele. *Proceedings of the National Academy of Sciences of the United States of America* 92: 4656–4660.

Li, H., Peng, Z.Y., Yang, X.H. et al. (2013). Genome-wide association study dissects the genetic architecture of oil biosynthesis in maize kernels. *Nature Genetics* 45: 43–50.

Liu, B.H. (2002). *Statistical Genomics: Linkage, Mapping and QTL Analysis*, 2e. Boca Raton, FL: CRC Press.

Majhi, P.K., Bhoi, T.K., Mogali, S.C. et al. (2022a). Advances in molecular breeding for bruchid (*Callosobruchus* spp.) resistance in mungbean [*Vigna radiata* (L.) Wilczek]: a review. *Legume Research-An International Journal* 45 (8): 933–941.

Majhi, P.K., Singh, S.K., Anandan, A. et al. (2022b). Parental polymorphism survey for evaluation and selection of contrasting parents for drought tolerance in rice (*Oryza sativa* L.) by using SSR markers. *International Journal of Environment and Climate Change* 12 (11): 2507–2519.

Mather, K. (1938). *Measurement of Linkage in Heredity*. London: Methuen and Co.

Michelmore, R.W., Paran, I., and Kesseli, R.V. (1991). Identification of markers linked to disease resistance genes by bulked segregant analysis: a rapid method to detect markers in specific genomic regions using segregating populations. *Proceedings of the National Academy of Sciences of the United States of America* 88: 9828–9832.

Myles, S., Peiffer, J., Brown, P.J. et al. (2009). Association mapping: critical considerations shift from genotyping to experimental design. *The Plant Cell* 21: 2194–2202.

Paterson, A.H. (1996). Making genetic maps. In: *Genome Mapping in Plants* (ed. A.H. Paterson), 23–39. San Diego, CA and Austin, TX: Academic Press.

Patra, B., Majhi, P.K., Tripathy, S.K. et al. (2022). Genomic-assisted breeding tools for grain and nutritional quality improvement in rice. *International Journal of Environment and Climate Change* 12 (1): 10–24.

Prasanna, B.M. (2003). *QTL Mapping and its Application in Crop Plants*. New Delhi: Indian Agricultural Research Institute.

Price, A.H. (2006). Believe it or not, QTLs are accurate. *Trends in Plant Science* 11: 213–216.

Rakshit, S., Rakshit, A., and Patil, J.V. (2012). Multiparent intercross populations in analysis of quantitative traits. *Journal of Genetics* 91: 111–117.

Reddy, I.N.B.L., Kim, B.K., Yoon, I.S. et al. (2017). Salt tolerance in rice: focus on mechanisms and approaches. *Rice Science* 24 (3): 123–144.

Reiter, R.S., Williams, J.G.K., Feldman, K.A. et al. (1992). Global and local genome mapping in *Arabidopsis thaliana* by using recombinant inbred lines and random amplified polymorphic DNAs. *Proceedings of the National Academy of Sciences of the United States of America* 89: 1477–1481.

Remington, D.L., Thornsberry, J.M., Matsuoka, Y. et al. (2001). Structure of linkage disequilibrium and phenotypic associations in the maize genome. *Proceedings of the National Academy of Sciences of the United States of America* 98: 11479–11484.

Sandhya, R.K., Kutubuddin, A.M., Krishnendu, C. et al. (2019). Introgression of *Sub1* (*SUB1*) QTL in mega rice cultivars increases ethylene production to the detriment of grain- filling under stagnant flooding. *Scientific Reports* 9: 18567.

Satagopan, J.M., Yandell, B.S., Newton, M.A., and Osborn, T.C. (1996). A Bayesian approach to detect quantitative trait loci using QTLs has not been fixed in advance, and is actually Markov Chain Monte Carlo. *Genetics* 144: 805–816.

Sauvage, C., Segura, V., Bauchet, G. et al. (2014). Genome-wide association in tomato reveals 44 candidate loci for fruit metabolic traits. *Plant Physiology* 165: 1120–1132.

Schauer, N., Semel, Y., Roessner, U. et al. (2006). Comprehensive metabolic profiling and phenotyping of interspecific introgression lines for tomato improvement. *Nature Biotechnology* 24: 447–454.

Scott, M.F., Ladejobi, O., Amer, S. et al. (2020). Multi-parent populations in crops: a toolbox integrating genomics and genetic mapping with breeding. *Heredity* 125: 396–416.

Sehgal, D., Singh, R., and Rajpal, V.R. (2016). Quantitative trait loci mapping in plants: concepts and approaches. In: *Molecular Breeding for Sustainable Crop Improvement*, Sustainable Development and Biodiversity 11 (ed. V.R. Rajpal, S.R. Rao, and S.N. Raina), 31–59. Switzerland: Springer International Publishing https://doi.org/10.1007/978-3-319-27090-6_2.

Semagn, K., Bjørnstad, A., and Ndjiondjop, M.N. (2006). Principles, requirements and prospects of genetic mapping in plants. *African Journal of Biotechnology* 5: 2569–2587.

Singh, S.K., Majhi, P.K., Anandan, A. et al. (2022). Microsatellites based parental polymorphism survey for moisture stress in rice (*Oryza sativa* L.) between the parental genotypes HUR-1309 and CR Dhan 801. *Emergent Life Sciences Research* 8: 248–258.

Song, Q., Yan, L., Quigley, C. et al. (2017). Genetic characterization of the soybean nested association mapping population. *The Plant Genome* 10 (2): 0109.

Stam, P. (1993). Construction of integrated genetic linkage maps by means of a new computer package: join map. *Plant Journal* 3: 739–744.

Staub, J.E., Serquen, F.C., and Gupta, M. (1996). Genetic markers, map construction, and their application in plant breeding. *Horticulture Science* 31: 729–741.

Suiter, K.A., Wendel, J.F., and Case, J.S. (1983). Linkage-1: a pascal computer program for the detection and analysis of genetic linkage. *Journal of Heredity* 74: 203–204.

Sun, Y.P., Wang, J.K., Crouch, J.H., and Xu, Y.B. (2010). Efficiency of selective genotyping for genetic analysis of complex traits and potential applications in crop improvement. *Molecular Breeding* 26: 493–511.

Tanksley, S.D., Miller, J., Paterson, A., and Bernatsky, R. (1988). Molecular mapping of plant chromosomes. In: *Chromosome structure and function: Impact of new concepts* (ed. J.P. Gustafson and R. Appels), 157–173. New York: Plenum Press.

Tanksley, S.D., Young, N.D., Patterson, A.H., and Bonierbale, M.W. (1989). RFLP mapping in plant breeding: new tools for an old science. *Nature Biotechnology* 7: 257–263.

Tenaillon, M.I., Sawkins, M.C., Long, A.D. et al. (2001). Patterns of DNA sequence polymorphism along chromosome 1 of maize (*Zea mays* ssp. *mays* L.). *Proceedings of the National Academy of Sciences of the United States of America* 98: 9161–9166.

Timmerman, G.M., Frew, T.J., Weeden, N.F. et al. (1994). Linkage analysis of er-1, a recessive *Pisum sativum* gene for resistance to powdery mildew fungus (*Erysiphe pisi* D.C.). *Theoretical and Applied Genetics* 88: 1050–1055.

Utz, H. and Melchinger, A. (1996). PLABQTL: a program for composite interval mapping of QTL. *Journal of Quantitative Trait Loci* 2: 1–5.

Vikram, P., Swamy, B.M., Dixit, S. et al. (2011). qDTY 1.1, a major QTL for rice grain yield under reproductive-stage drought stress with a consistent effect in multiple elite genetic backgrounds. *BMC Genetics* 12: 1–15.

Voesenek, L.A. and Bailey-Serres, J. (2015). Flood adaptive traits and processes: an overview. *New Phytologist* 206 (1): 57–73.

Wang, H., Smith, K.P., Combs, E. et al. (2012). Effect of population size and unbalanced data sets on QTL detection using genome-wide association mapping in barley breeding germplasm. *Theoretical and Applied Genetics* 124: 111–124.

Waziri, A., Kumar, P., and Purty, R.S. (2016). Saltol QTL and their role in salinity tolerance in rice. *Austin Journal of Biotechnology & Bioengineering* 3 (3): 1067.

Weeden, N.F., Timmerman, M., Hermmat, M. et al. (1992, 1992). Inheritance and reliability of RAPD markers. In: *Proceedings of Symposium Joint Plant Breeding Symposis Series. Applications of RAPD Technology to Plant Breeding* (ed. J. Nienhuis), 12–17. Minneapolis, MN.

Xu, Y., Li, P., Yang, Z., and Chenwu, X. (2017). Genetic mapping of quantitative trait loci in crops. *The Crop Journal* 5: 175–184.

Yan, J.B., Shah, T., Warburton, M.L. et al. (2009). Genetic characterization and linkage disequilibrium estimation of a global maize collection using SNP markers. *PLoS One* 4: e8451.

Yang, J.L., Jiang, H.Y., Yeh, C.T. et al. (2015). Extreme-phenotype genome-wide association study (XP-GWAS): a method for identifying trait-associated variants by sequencing pools of individuals selected from a diversity panel. *Plant Journal* 84: 587–596.

Yu, J.M., Pressoir, G., Briggs, W.H. et al. (2006). A unified mixed-model method for association mapping that accounts for multiple levels of relatedness. *Nature Genetics* 38: 203–208.

Yu, J., Holland, J.B., McMullen, M.D., and Buckler, E.S. (2008). Genetic design and statistical power of nested association mapping in maize. *Genetics* 178: 539–551.

Zamir, D. and Tadmor, Y. (1986). Unequal segregation of nuclear genes in plants. *Botanical Gazette* 147: 355–358.

Zeng, Z.B. (1994). Precision mapping of quantitative trait loci. *Genetics* 136: 1457–1468.

Zhao, K.Y., Aranzana, M.J., Kim, S. et al. (2007). An *Arabidopsis* example of association mapping in structured samples. *PLoS Genetics* 3: e4.

Zhou, Z.K., Jiang, Y., Wang, Z. et al. (2015). Resequencing 302 wild and cultivated accessions identifies genes related to domestication and improvement in soybean. *Nature Biotechnology* 33: 408–414.

Zhu, C., Gore, M., Buckler, E.S. et al. (2008). Status and prospects of association mapping in plants. *Plant Genome* 1: 5–20.

Zou, C., Wang, P., and Xu, Y. (2016). Bulked sample analysis in genetics, genomics and crop improvement. *Plant Biotechnology Journal* 14: 1941–1955.

5

Basic Bioinformatics for Identification and Analysis of Candidate Genes in Plants Toward Crop Improvement

Sadhana Singh

Department of Biochemistry, School of Sciences, U.P. Rajarshi Tandon Open University, Prayagraj, Uttar Pradesh, India

5.1 Introduction

Bioinformatics is an acronym derived from "biological informatics" and is proving to be a well-opted choice for obtaining biologically oriented data from nucleic acid sequences, protein sequences, structures, functions, pathways, and their interactions. Moreover, it involves assembling these databases, advancing approaches and tools to get meaningful knowledge from these repositories, and inventing methods to combine the associated data from diversified origins. Hence, it is also defined as the implementation of computational and investigative tools to represent and explain biological statistics. There are several crucial ways where bioinformatics can be applied to generate important information. Therefore, the objective of bioinformatics involves steps such as data acquisition, tools and database creation, data analysis, and data integration. Data acquisition is mainly focused on retrieving and storing data created directly from biological investigations. The data are arranged in various databases to get instant and current information and also put in new entries as and when they are generated. Protein Data Bank for 3D structures prediction data and Entrez Genome of NCBI for information on genome data are examples of such databases. Therefore, these datasets have no value until it is analyzed. Hence, the motive of bioinformatics expands much further. Also, a number of laboratories develop huge amounts of data for example- DNA sequences, high throughput screening, gene sequencing, 3D molecular structure, and gene expression data. Programs such as Fast Alignment (FASTA) and Basic Local Alignment Search Tool (BLAST) are some of the tools that quickly compare and give a biologically significant match. Now, the global analysis of the available data has to be conducted to get biologically meaningful results that highlight novel features. The user should be able to easily transform the data into a suitable format for the preferred analysis tools, once the data are generated from the database. The information thereby obtained has to be associated or integrated with the corresponding data from other databases. Furthermore, the scope of bioinformatics includes- genome and sequence

Bioinformatics for Plant Research and Crop Breeding, First Edition. Edited by Jen-Tsung Chen.

analysis, from sequence analysis to 3D structure prediction, genome-wide analysis and functional genomics, proteomics, drug discovery, mathematical modeling of life forms, systems biology, etc.

The major branches of bioinformatics involve- genomics, transcriptomics, proteomics, systems biology, functional genomics, and metabolomics. Genomics occupies a noteworthy aspect in modern biological research in which information regarding the nucleotide sequences throughout the chromosomes of an organism are depicted or aligned and the position of several distant genes and their sequences are calculated. Transcriptomics involves the study of transcriptome, which includes the whole set of transcripts, i.e. mRNAs of an individual/cell in a given environment conditions. This branch allows the researchers to depict the expression levels of the genes via next-generation sequencing (NGS) technologies using whole transcriptome sequencing, mRNA sequencing, targeted RNA sequencing, and smRNA (small RNA) sequencing methods. This involves total RNA, microRNA (miRNA), small interfering RNA (siRNA), long noncoding RNA (lncRNA), etc. On the other hand, proteomics is far more complicated than genomics, and it deals with proteins and their interconnections. It involves sequencing amino acids in a protein, predicting and analyzing its 3D structures, and correlating it with the protein function. Interestingly, proteome varies from cell to cell due to its changing biochemical associations with the genome and the environment while the genome is rather a regular structure. However, systems biology is the system-level conception of the biological system that mainly aims at the interconnections among the several different biological components, and how these associations bring about the function and attribute of that system. This includes understanding the gene regulatory networks, biochemical networks, and physical organization of the system and predicting the system dynamics. Moving ahead in the era of omics, functional genomics targets the functional identification of different genes. It further aims at the study of genes, their arising proteins, and the roles in which proteins take part. Therefore, this functional application to genomic clues or data is generated by a technique or method called functional genomics. The incorporation of cry genes obtained from *Bacillus thuringiensis* in several crop plant genomes is the result of the functional genomics approach. Another very important and modern-day branch of computational biology is metabolomics, which deals with the large-scale study of the unique chemical fingerprints that specific cellular process leaves behind, within cells, biofluids, tissues, or organisms. These chemical fingerprints are small molecules, commonly called metabolites. The collection of these tiny molecules and the way they interact with each other within a biological system is referred to as the metabolome.

Over time, extensive investigations have been conducted to reveal and describe the genes and molecular processes that regulate how both model and crop plants respond to abiotic stress such as extreme weather conditions (Singh et al. 2021a). Numerous transcription factors (TFs) and their specific DNA binding regions, known as cis-acting regulatory elements (CAREs), serve as molecular switches that control changes in gene expression in response to stress. This mechanism enables plants to enhance their ability to adapt when facing unfavorable environmental conditions, as discussed by Singh et al. (2021b). Discovering and altering the genes encoding novel TFs hold promise for creating genetically modified plants that can achieve higher crop yields and better withstand harsh environmental conditions like drought and extreme temperatures. In this context, exploring how crops can withstand and adapt to various environmental challenges is crucial due to the advantages they offer in terms of both nutrition and economic value. Therefore, it is extremely important to identify abiotic stress-associated genes in these grains.

5.2 Candidate Genes such as Transcription Factors and Gene Families

Candidate genes are particular genes that are examined as potential candidates for occupying a role in a specific biological process, trait, or disease built on several criteria like, their familiar functions, position within a genomic region related to the trait or disease, or differential expression patterns. These genes are adopted for further investigation because they are likely to be linked to the biological phenomenon or condition being investigated. Candidate genes are generally recognized via genetic, molecular, or bioinformatics analyses and are an important beginning point for research focused on interpreting the processes of various biological phenomena. Several key TFs play pivotal roles in regulating gene expression in plants. Among these, TFs are DREB/CBF, AP2/EREBP, MYB, MYC, bZIP, AREB/ABF, SCOF-1, WRKY, CBL-CIPK, DELLA, and NAC as highlighted in studies by Nakashima and Yamaguchi-Shinozaki (2010), Hussain et al. (2017), and Aslam et al. (2012). These TFs, often referred to as TFs, are believed to exert their regulatory influence by directly binding to specific cis-acting elements present in the promoter regions of their target genes. Through this direct interaction, they modulate the expression of these target genes, orchestrating a wide array of cellular processes and responses in plants.

5.2.1 Importance of NAC Transcription Factors

In the context of the above, the NAC (NAM – no apical meristem, ATAF – Arabidopsis transcription activation factor, and CUC – cup-shaped cotyledon) family of genes in plants represents one of the largest and most distinctive families of TFs within the plant kingdom. These genes are characterized by a remarkably conserved NAC domain situated at the N-terminus, spanning approximately 160 amino acid residues (Olsen et al. 2005a,b; Puranik et al. 2012; Singh et al. 2016; Singh et al. 2019). The versatile C-terminus of NAC proteins can function either as a transcriptional activator or a repressor, adding a layer of complexity to their regulatory roles in plant biology (Ooka et al. 2003; Grover et al. 2014). The discovery of the NAC TF family dates back more than 22 years, with its initial identification in Petunia by Souer et al. (1996). Since this groundbreaking discovery, a wealth of research has emerged, shedding light on the multifaceted roles played by NAC genes in a wide range of biological processes. Some key areas where NAC genes have been found to exert their influence include:

a) **Lateral root formation**: NAC genes have been implicated in the regulation of lateral root development, as evidenced by studies such as Xie et al. (2000).
b) **Seed development**: The role of NAC genes in seed development has been elucidated in studies like Park et al. (2011).

c) **Leaf senescence**: NAC genes have been linked to the regulation of leaf senescence, with notable contributions from research by Chen et al. (2016) and Kim et al. (2016).
d) **Stress-inducible flowering induction**: investigations such as Kim et al. (2007) have highlighted the involvement of NAC genes in the induction of flowering in response to stress.
e) **Regulation of secondary cell wall synthesis and cell division**: NAC genes have been found to play crucial roles in the regulation of secondary cell wall synthesis and cell division, as demonstrated by research by Zhong et al. (2007).
f) **Plant biotic and abiotic stress responses**: NAC genes have garnered significant attention for their involvement in both biotic and abiotic stress responses. Notably, they play pivotal roles in responses to drought stress, as emphasized in studies by Guo et al. (2015), Yu et al. (2016), and Zhao et al. (2016).

According to PlantTFDB (V5.0) (https://planttfdb.gao-lab.org), the largest number of NAC genes has been documented in various plant species, showcasing the remarkable diversity within this gene family. Notably, Arabidopsis boasts a total of 138 NAC genes, while rice stands out with 158. Poplar, scientifically known as *Populus trichocarpa*, amazes with a staggering 289 NAC genes. Maize, or *Zea mays*, follows closely with 189, while foxtail millet, identified as *Setaria italica* L., has 165. Soybean, recognized as *Glycine max*, showcases a substantial 269 NAC genes, demonstrating its genetic complexity. *Brassica napus*, with a remarkable 411 NAC genes, and Camelina sativa, with 350, further underscore the diversity in this gene family. *Eucalyptus grandis* rounds out the list with 200 NAC genes, highlighting the significant variations in NAC gene counts among different plant species. These findings illuminate the intricate genetic landscapes of various plants and offer valuable insights into the regulatory mechanisms governing their growth and development. In summary, the NAC family of TFs stands as a cornerstone in the intricate regulatory networks governing various aspects of plant growth, development, and stress responses. Their remarkable versatility and conservation across plant species underscore their gravity in plant biology.

5.2.2 Importance of DREBs/CBFs

The DREB/CBF TFs play a significant part in orchestrating the plant's response to various environmental cues, making them foremost players in the regulation of downstream genes linked with abiotic stress responses, (Gupta et al. 2013). When plants are subjected to low temperatures, a complex regulatory cascade is set in motion. This cascade involves the generation of particular transcriptional activators, namely CBF1, CBF2, and CBF3. These activators are useful in commencing the plant's response to cold stress. They achieve this by binding to a specific cis-acting regulatory element known as CRT/DRE (C-repeat/DNA regulatory element), which holds the extremely conserved CCGAC core sequence (Yamaguchi-Shinozaki and Shinozaki 1994). This binding event activates a sequence of downstream molecular mechanisms that eventually empower the plant to adjust and counter adequately to low-temperature stress. The finding of this complicated regulatory mechanism has significantly intensified our understanding of how plants cope with demanding environmental conditions.

In recent years, significant advancements have been made in understanding a central pathway known as the CBF (CRT/DRE binding factor) pathway within the model plant Arabidopsis. This pathway is recognized by the well-known part played by the TF CBF and a certain set of functionally essential cold-regulated (COR) genes. These crucial discoveries were precisely outlined in the seminal work of Chinnusamy et al. (2003), shedding light on the entangled mechanisms underlying cold stress responses in plants. These regulatory elements are found in the promoter regions of genes that respond to cold stress, including RD (responsive to dehydration), COR, LTI (low-temperature induced), and KIN (cold-inducible) genes. Remarkably, the expression of these genes increases quickly within just four hours of exposure to cold conditions, as noticed in Fowler and Thomashow (2002). The importance of CBF proteins in the process of cold adaptation is emphasized by the finding that the constitutive overexpression of CBF1, CBF2, or CBF3 in transgenic plants results in the initiation of several COR genes. This cluster of genes, often referred to as the CBF regulon becomes activated, and as a result, the plants display elevated freezing tolerance, even in the unavailability of prior exposure to cold temperatures, as exemplified in Medina et al. (2011). This innovative study sheds light on how plants can quickly acclimatize to and resist cold stress through the co-related operation of CBF TFs and their target genes.

While a considerable number of transgenic plants carry DREBs/CBFs and have shown promise in conferring abiotic stress tolerance in controlled experimental settings, there persist several challenges that must be addressed to unleash the full potential of DREB/CBF-transgenic plants in real-world stress environments. Consequently, it is all-important to delve into the finding of DREBs/CBFs TF genes, particularly those originating from crop plants adapted to the harsh conditions of high-altitude cold-arid regions. This exploration is crucial to enhance our understanding and harness the resilience of these plants in the face of natural stress conditions.

5.2.3 Availability of Draft Genome Sequences

The availability of draft genome sequences for important food crops such as *Cajanus cajan* (pigeon pea) (Varshney et al. 2012), *Cicer arietinum* (chickpea) (Varshney et al. 2013), and *Arachis hypogaea* (peanut) (Bertioli et al. 2019), etc. constitute an outstanding asset for biologists. These genomes unlock the powerful chance for performing comparative genome analysis, especially in the pursuit of predicting novel TFs. In a research endeavor, one can venture into a comprehensive genome-wide analysis directed at pinpointing NAC domain TFs in leguminous crops, e.g. chickpea, pigeon pea, and peanut, etc. In the study one can search deep into multiple facets of these NAC domain TFs, encompassing their genomic distribution, gene architecture, regulatory elements, protein–protein interactions (PPIs), conserved motifs, and expression profiles across varied developmental phases. The conclusion of extensive analyses can yield a promising roster of putative genes (NAC), that are likely to be related to drought stress. These genes hold remarkable promise for further investigation into their functions and lay the foundation for the development of superior crop genotypes that display improved resilience in these legume species against drought. Moreover, the functional characterization of these drought-responsive genes in food crops (legumes), e.g. chickpea, pigeon pea, and peanut, holds wider significance. This research will not only enrich the understanding of these food (leguminous) crops but will also create a beneficial resource that can be imposed in molecular breeding attempts across a spectrum of other crops. By applying biotechnology-related approaches or conventional breeding methods, we can surely pass on the understanding procured from these studies to boost crop varieties beyond the legume family, bestowing sustainability and resilience of agricultural systems across the globe. In essence, these types of studies set out a stepping stone toward advancing crop-boosting programs and harnessing the future of genomics for the welfare of global food security.

5.3 Methods

In this context, our attention will be directed solely toward the case study of the NAC gene family to facilitate a clear understanding of gene family identification and its subsequent analyses. Moreover, we propose that the identification of NAC family genes in food crops, e.g. legumes, can be achieved using the outlined methodology (Figure 5.1). Similarly, the gene identification of DREB, MYB, CBF, etc. related to various abiotic stresses can be conducted by adopting a similar methodology in any other food crops. In this context, a table summarizing the software and tools that can be utilized in this analysis and investigation is also presented (Table 5.1).

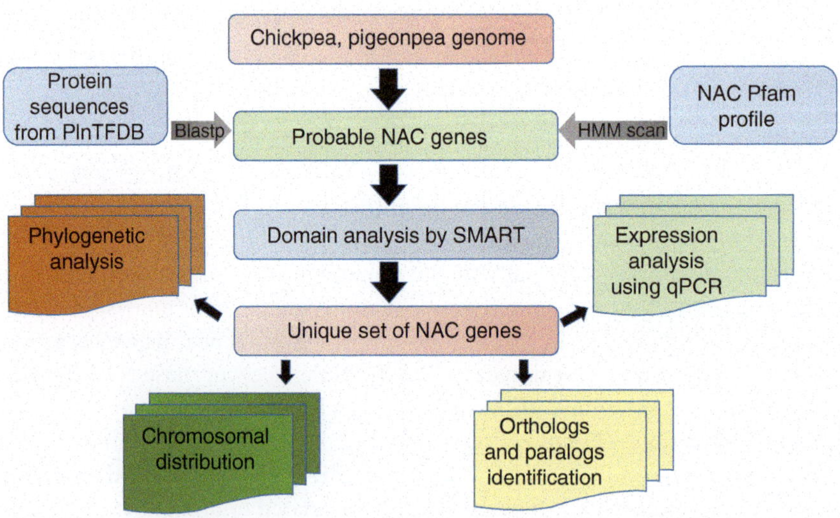

Figure 5.1 Outline for the identification and analysis of the candidate genes. NAC gene taken into consideration for representation and legume species has been chosen to better understand the methodology in this context.

Table 5.1 List of software and tools for the identification and analysis of candidate genes in plants toward crop improvement.

Software and tools	Link	Purpose	References (if available)
PrimerQuest tool	https://eu.idtdna.com/scitools/Applications/RealTimePCR/Default.aspx	Design primers or assays for PCR, qPCR, or sequencing (any species)	NA
MeV tool version 4.9.0	https://sourceforge.net/projects/mev-tm4/	Desktop application for the analysis, visualization, and data-mining of large-scale genomic data	Howe et al. (2011)
MEGA (V7.0) software	http://www.megasoftware.net/	Conducting statistical analysis of molecular evolution and for constructing phylogenetic trees	Kumar et al. (2016)
STRING (Search Tool for the Retrieval of Interacting Genes/Proteins) database version 11.0	https://string-db.org/	Database of known and predicted protein–protein interactions	Szklarczyk et al. (2011)
DeepTMHMM	https://dtu.biolib.com/DeepTMHMM	Detect and predict the topology of both alpha-helical and beta barrels proteins with unprecedented accuracy	Hallgren et al. (2022)
ExPASy	http://www.expasy.ch/tools/pi_tool.html	Known for its ability to impart accurate and exhaustive information concerning the proteins and gene features	NA
Softberry	http://www.softberry.com/berry.phtml?topic=index&group=programs&subgroup=proloc	Important resource that can be exploited for predicting subcellular localization	NA
MapChart 2.32 software	https://www.wur.nl/en/show/Mapchart.htm	Visualizing the chromosomal distribution of identified genes	Voorrips (2002)

(Continued)

Table 5.1 (Continued)

Software and tools	Link	Purpose	References (if available)
ProtComp – version 6: (Program for predicting protein sub-cellular location)	http://www.softberry.com/berry.phtml?topic=protcomp&group=help&subgroup=proloc	Predicting protein sub-cellular location	Klee and Ellis (2005)
TargetP 2.0	https://services.healthtech.dtu.dk/services/TargetP-2.0/	Subcellular location of proteins: mitochondrial, chloroplastic, secretory pathway, or other	Armenteros et al. (2019)
WoLF PSORT, CELLO, and DeepLoc	http://wolfpsort.org/; http://cello.life.nctu.edu.tw/; DeepLoc http://www.cbs.dtu.dk/services/DeepLoc/	Predicting the subcellular localization of proteins and often have web interfaces for users to input protein sequences and obtain predictions	Horton et al. (2007), Yu et al. (2006), and Thumuluri et al. (2022)
PlantCARE	https://bioinformatics.psb.ugent.be/webtools/plantcare/html/	Identification of cis-acting regulatory elements (CAREs)	Lescot et al. (2002)
HMMER v3.4	http://hmmer.org/	To identify homologous protein or nucleotide sequences, and to perform sequence alignments	Eddy (2011)
SMART (Simple Modular Architecture Research Tool)	http://smart.embl-heidelberg.de/	Identification and analysis of protein domains within protein sequences	Letunic and Bork (2018) and Letunic et al. (2021)
Pfam	http://pfam.xfam.org	Provides a complete and accurate classification of protein families and domains	Mistry et al. (2021)
MEME standalone V 5.5.3 (Multiple Em for Motif Elicitation)	https://meme-suite.org/meme/meme_5.5.3/tools/meme	Discover novel, *ungapped* motifs (recurring, fixed-length patterns) in your sequences (sample output from sequences)	Bailey et al. (2015)
GSDS 2.0 (Gene Structure Display Server)	http://gsds.gao-lab.org/	Designed for the visualization of gene features, such as the composition and position of exons, introns, conserved elements, etc.	Hu et al. (2015)

5.3.1 Identifying and Examining Genes in Crop Species to Understand their Characteristics through Data Analysis

Previously identified conserved domains (CDs) for the protein of interest (e.g. NAC protein sequences) from other model plant species such as *Arabidopsis*, *Lotus*, *Medicago*, *G. max*, etc. (especially in legume crops) can be used to search for matches within the predicted gene models of various food crops using BLAST program at an *E*-value threshold of 10^{-5}. BLAST-P (Basic Local Alignment Search Tool for Proteins) is a computer-generated program used by scientists to draw an analogy between several protein sequences and analyze protein sequences (Figure 5.2). Proteins are crucial molecules in living organisms, and knowing their similarities and differences can give perceptions into their functions and roles in biology. BLAST-P helps researchers find resemblance between different proteins, which can be valuable for assignments such as identifying genes, studying protein functions, and detecting evolutionary relationships.

Furthermore, one can acquire the hidden Markov model (HMM) profile for the NAC family from the Pfam database (Finn et al. 2016), and the NAC HMM profile would be scanned for potential matches within the anticipated gene models of the food crops being investigated, specifically targeting sequences that contain the NAC domain. This search can be performed using HMMER v3.4 (Eddy 2011). HMMER is a tool employed to explore sequence databases in search of sequences that share similarities, known as homologs, and to create alignments of these sequences. It achieves this by employing probabilistic models known as profile HMMs, often abbreviated as profile HMMs. HMMER is commonly employed in combination with a profile database like Pfam or various databases within the InterPro consortium (http://hmmer.org/). However, it is important to note that HMMER can also analyze query sequences directly, much like BLAST. Further, one can verify the

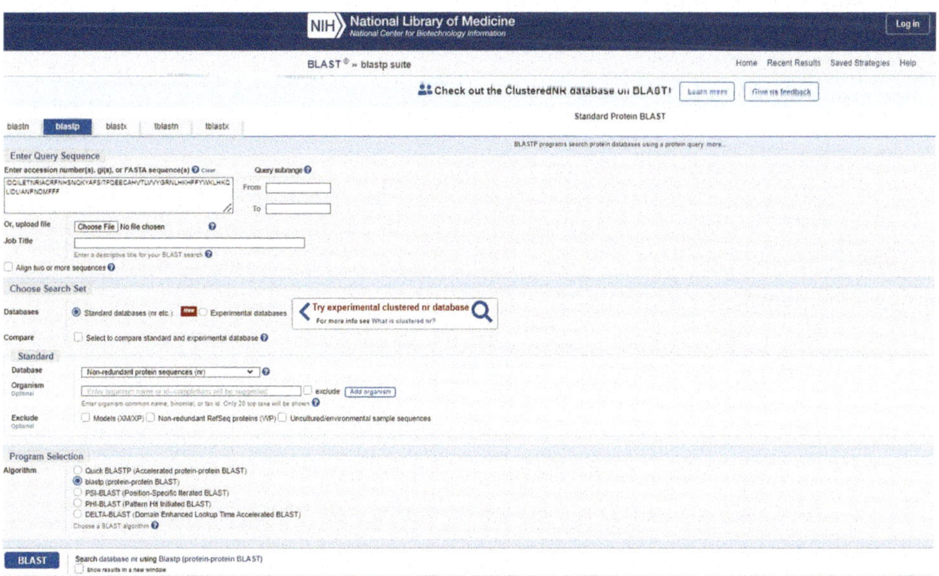

Figure 5.2 Screenshot displaying the BLAST-P suite for homology search.

existence of the NAC domain in the identified proteins/genes by conducting searches using SMART (Simple Modular Architecture Research Tool) and Pfam.

Conserved Domain Database (CDD) is generally based chiefly on the Pfam and SMART on the compilation of alignments that depict functional regions preserved throughout the course of evolution. One can find the CDs for a protein of interest from CD-search results such as Pfam and SMART (Figure 5.1). Pfam is a huge collection of multiple sequence alignments and HMM housing many prevalent protein domains and families while genomic SMART holds only the proteomes of fully sequenced genomes.

The NAC proteins characterization includes a detailed investigation of their physico-chemical properties. This investigation covers crucial features such as the length of the open reading frame (ORF), the isoelectric point (pI), the molecular weight (MW), and the overall gene length. To enable this analysis, one can make use of the ExPASy platform (http://www.expasy.ch/tools/pi_tool.html), a reliable resource in the bioinformatics domain, known for its ability to impart accurate and exhaustive information concerning the proteins and genes characteristics (Figure 5.3). Through ExPASy, researchers can acquire beneficial insights into the structural and chemical arrangement of the identified NAC proteins, assisting our understanding of their functional roles and potential implications within biological systems.

To predict the subcellular localization of the identified NAC family members, Softberry (http://www.softberry.com/berry.phtml?topic=index&group=programs&subgroup=proloc) is an important resource that can be exploited. It has established several bioinformatics tools, e.g. ProtComp, which is used for forecasting the subcellular localization of eukaryotic proteins. The most updated, ProtComp-Version 6 is a progressive application for deducing the sub-cellular localization of eukaryotic proteins in animals/fungi and plants as well (Figure 5.4a,b). It incorporates numerous approaches for protein localization speculation – prediction based on neural networks; direct matching with an upgraded base of homologous proteins of known localization; matching of pentamer distributions estimated for query and DB sequences; forecasting of definite functional peptide sequences, such as

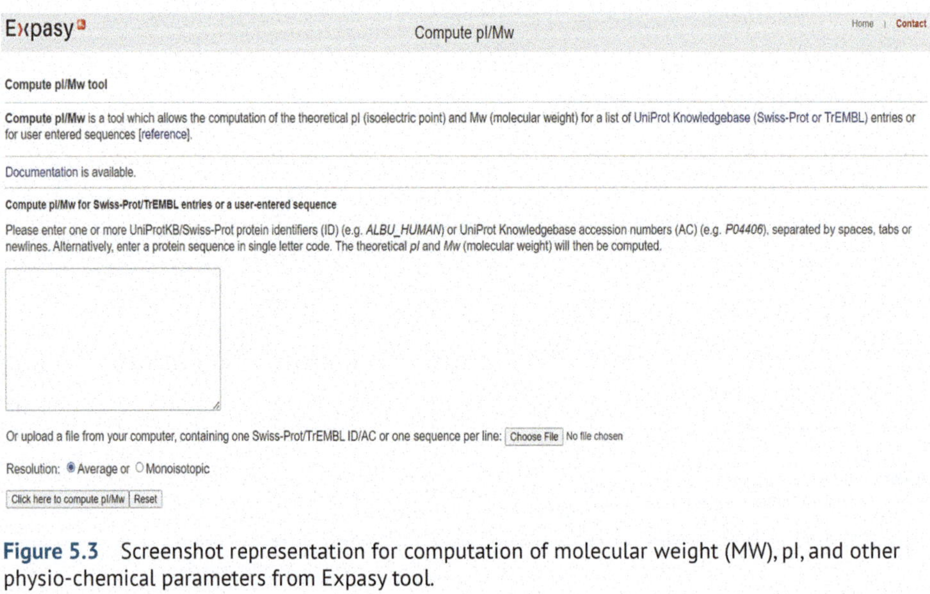

Figure 5.3 Screenshot representation for computation of molecular weight (MW), pI, and other physio-chemical parameters from Expasy tool.

(a)

```
ProtComp Version 6. Identifying sub-cellular location (Animals&Fungi)
Seq name: QUERY, Length=376
Significant similarity in Location DB -  Location:Cytoplasmic
Database sequence: AC=P08319 Location:Cytoplasmic  DE  Alcohol dehydrogenase class II pi chain precur
s
Score=14845, Sequence length=391, Alignment length=365
Predicted by Neural Nets - Extracellular (Secreted) with score    2.4
Integral Prediction of protein location: Cytoplasmic with score   14.7
Location weights:       LocDB / PotLocDB / Neural Nets / Pentamers / Integral
Nuclear                   0.0 /      0.0 /      0.71 /      0.00 /     0.71
Plasma membrane           0.0 /      0.0 /      0.73 /      0.00 /     0.73
Extracellular             0.0 /      0.0 /      2.42 /      0.00 /     2.42
Cytoplasmic           14845.0 /  18465.0 /      0.83 /      8.50 /    14.68
Mitochondrial             0.0 /      0.0 /      0.70 /      0.00 /     0.70
Endoplasm. retic.         0.0 /      0.0 /      0.70 /      0.50 /     1.21
Peroxisomal               0.0 /      0.0 /      0.49 /      0.00 /     0.49
Lysosomal                 0.0 /      0.0 /      0.33 /      0.00 /     0.33
Golgi                     0.0 /      0.0 /      0.40 /      0.00 /     0.40
```

(b)

Compartment	Percent predicted correctly		
	ver. 4	ver. 5	ver. 6
Nucleus	80	88	91
Plasma Membrane	80	87	100
Extracellular	69	83	86
Cytoplasm	46	63	88
Mitochondria	76	82	89
Endoplasmic Reticulum	67	83	89
Peroxisome	95	97	91
Lysosome	69	91	100
Golgi	57	77	91

Figure 5.4 (a) Screenshot of the ProtComp Version 6 for identification of sub-cellular location. (b) Accuracy comparison from previous versions of ProtComp i.e. version 4, 5, 6.

signal peptides, signal-anchors, glycosylphosphatidylinositol (GPI)-anchors, transit peptides of mitochondria and chloroplasts and transmembrane segments; and hunt for particular localization-specific motifs. It alludes that the application manages correctly full sequences only, carrying signal sequences, anchors, and further functional peptides, if any. The application encompasses independently trained recognizers for animal/fungal and plant proteins, which greatly enhances recognition precision. When interpreting the output results of ProtComp, it is required to examine certain key criteria:

a) **Interpreting scores**: ProtComp's scores, as weights created by composite neural networks, should not be misinterpreted as direct possibilities designating the protein's location within a particular cellular compartment. They constitute relative measures of likelihood and need careful evaluation.
b) **Homology significance**: The strong indicator of the query protein's probable site is when there is significant homology among the query protein and the proteins with established subcellular locations/positions. These matches, with familiar proteins, bestow strong corroboration for subcellular localization prediction.
c) **Relative neural network scores:** The significance of neural network scores lies more in their relative values across different compartments rather than their absolute values. If the second-best score notably lags beyond the best one, it intensifies the reliability of the prediction, regardless of the absolute numerical values.
d) **Convergence of predictions**: The reliability of the prediction is intensified or increased when both the neural network-based and homology-based indications point in the direction of a similar cellular compartment. The confidence in the localization prediction would be greatly increased when these two information sources are coherent.

In addition to the ProtComp – version 6, TargetP-2.0 (https://services.healthtech.dtu.dk/services/TargetP-2.0/), WoLF PSORT, https://wolfpsort.hgc.jp, CELLO (http://cello.life.nctu.edu.tw/) or DeepLoc (http://www.cbs.dtu.dk/services/DeepLoc/) are among other widely used tools developed for speculating the protein's subcellular localization, and usually have web interfaces for users to input protein sequences and get predictions.

Genome sequencing and annotation tools are generally used by researchers to position and annotate particular genes on chromosomes. These software tools/applications

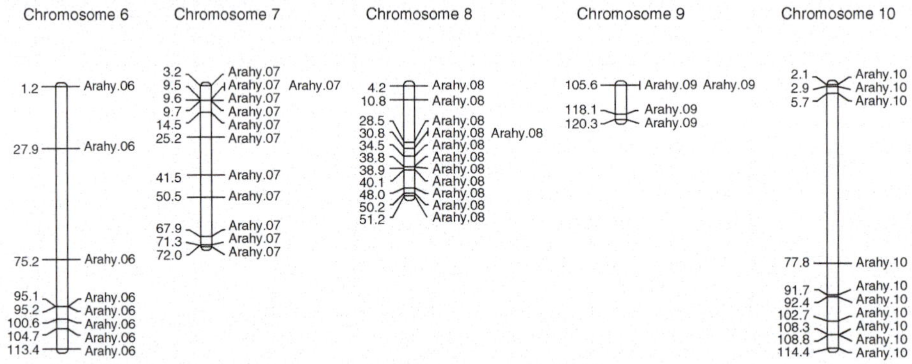

Figure 5.5 Representation of genes (NAC) distributed on peanut respective chromosomes.

typically use data from DNA sequencing to predict the gene locations, determine coding regions, and further impart knowledge on functional annotations. For this purpose, Genome browsers, such as the University of California, Santa Cruz (UCSC) Genome Browser or Ensembl Genome Browser, are commonly used.

MapChart 2.32, developed by Voorrips (2002), is a software widely utilized for visualizing the chromosomal distribution of identified (NAC) genes (Figure 5.5). It is particularly beneficial in developing or creating genetic maps. It is a computer package for the MS-Windows® platform that generates illustrations of genetic linkage maps and quantitative trait loci (QTL) information. These graphs constitute a sequence of vertical bars depicting the chromosomes or linkage groups. On these bars, the positions of loci/genes are demonstrated, and next to the bars gene intervals and gene names can be displayed. Therefore, MapChart is a valuable tool for visualizing genetic linkage maps; however, it does not directly locate or annotate genes on chromosomes. The positions of genes are typically established through genome sequencing and annotation processes.

5.3.2 Transmembrane Domains Prediction and Orthologs Distribution

DeepTMHMM (https://dtu.biolib.com/DeepTMHMM), a Deep Learning Model for Transmembrane Topology Prediction and Classification will be used to predict the transmembrane helices (https://www.biorxiv.org/content/10.1101/2022.04.08.487609v1). There have been breakthroughs in protein structure predictions using modern deep-learning techniques in recent years. DeepTMHMM is one of the exceptional tools among them due to its great comprehensiveness and high-performance approach for predicting the arrangement of the alpha-helical and beta-barrel transmembrane proteins (Figure 5.6). The primary amino acid sequence is utilized by this model, which further encodes it employing a pre-trained language model. To decode and create the predictions regarding the topology and the kind of proteins, it then utilizes a state space model with an unmatched accuracy level. Additionally, it has the exceptional ability to examine the whole proteomes, providing the scientific community to determine both categories of transmembrane proteins. This model holds enormous importance to the research community toward stimulating prospects for forecasting the protein structure and advancing our interpretation of these necessary biological molecules.

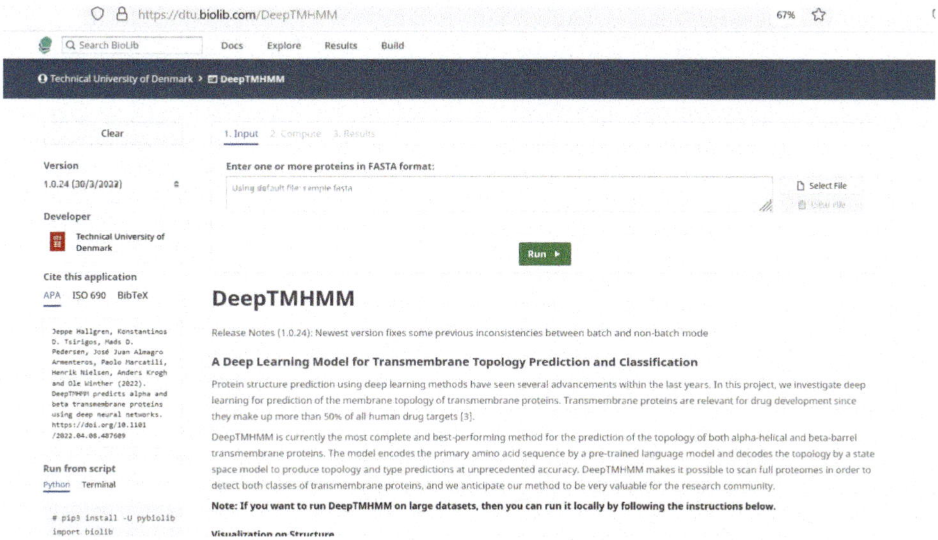

Figure 5.6 Screenshot image to detect and predict the topology of transmembrane proteins with unprecedented accuracy using DeepTHMM.

To determine orthologs, the NAC proteins of food legumes (such as chickpea, pigeon pea, and peanut) would undergo a search against the complete set of proteins in *Medicago* and soybean (as model legumes), utilizing the BLAST program with specified criteria: an *E*-value threshold of 10^{-10}, 80% similarity, and 80% query coverage. Additionally, the information graphics of these orthologous relationships can be achieved using circos, as outlined by Krzywinski et al. (2009).

5.3.3 Phylogenetic Analysis and Identification of Putative Stress-responsive NAC Genes

The analysis will utilize MEGA (Molecular Evolutionary Genetics Analysis) software (V7.0) available at http://www.megasoftware.net/ for demonstrating phylogenetic relationships, (Kumar et al. 2016). MEGA 7 is a bioinformatics software broadly employed for the analysis of molecular evolutionary genetics. Developed by the Kumar Lab, MEGA is designed for executing several analyses associated with molecular biology and evolutionary studies. The neighbor-joining (NJ) method will be employed to build unrooted phylogenetic trees, with a criterion of bootstrap values exceeding 30% (Figure 5.7a). To determine potential stress-responsive NAC genes, the study will integrate stress-related NAC protein sequences from *Arabidopsis*, *Oryza*, *Medicago*, and *G. max*, alongside NAC protein sequences from legume crops. Sequence alignments will be performed as part of this analysis (Figure 5.7b).

5.3.4 Conserved Motifs, and Gene Structure Analysis of Putative Stress-responsive Genes (NAC)

Conserved motifs are short, repetitive patterns/markings in biological sequences, such as DNA, RNA, or protein sequences, that are evolutionarily conserved over distinct species or within a set of linked genes. These motifs conservation suggested that they play vital

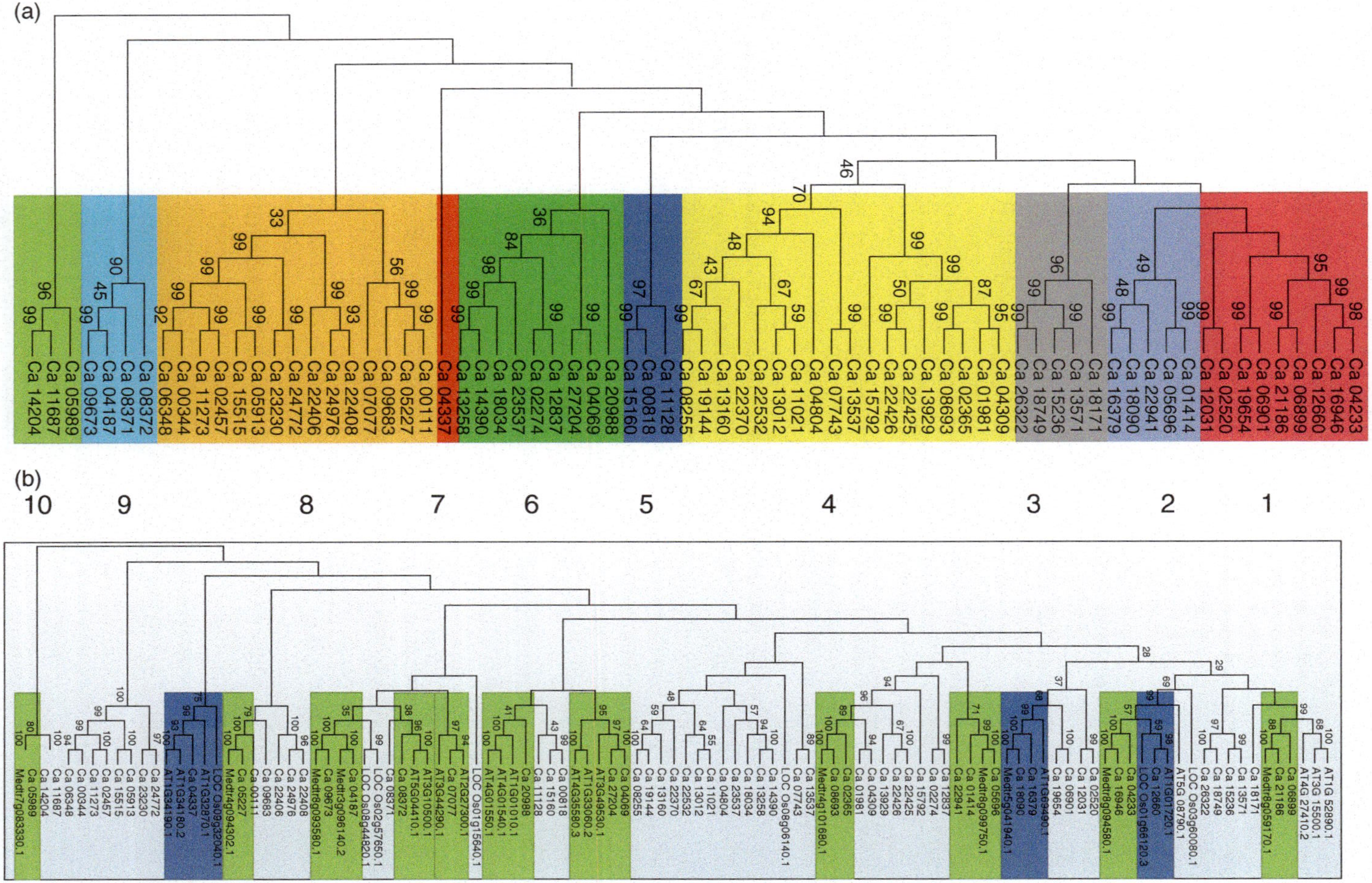

Figure 5.7 (a) The evolutionary history is inferred using the neighbor-joining method. The percentage of replicate trees in which the associated taxa clustered together in the bootstrap test (1000 replicates) are shown next to the branches. The evolutionary distances were computed using the p-distance method and are in the units of the number of amino acid differences per site. 10 Clades were identified from 72 NAC proteins from chickpea. Evolutionary analyses were conducted in MEGA7. (b) Putative stress-related NAC gene identification in chickpea using well-known abiotic stress-responsive genes from model crop species.

functional roles and are maintained all over the evolutionary processes. Conserved motif identification is a regular exercise in bioinformatics and computational biology. Tools like MEME (Multiple Em for Motif Elicitation) are utilized to analyze sequences and identify statistically over-represented motifs, indicating potential functional significance. These motifs frequently relate to TF-binding sites, protein structural elements, or other biologically relevant features. Inspecting conserved motifs aids researchers in recognizing the regulatory elements, functional genes, and protein characteristics. Conservation across species indicates a fundamental aspect of biological processes, and alterations in these motifs may contribute to phenotypic changes and evolutionary adjustments.

The MEME standalone version deals with the software tool that works separately and does not need an internet connection or external reliance to operate. In this context, it particularly relates to version 5.5.3 of MEME, a bioinformatics application used for the detection and interpretation of sequence motifs in biological data, specifically DNA, RNA, and protein sequences. Researchers and bioinformaticians use the MEME standalone version to determine conserved motifs in sets of sequences. It permits users to specify several parameters, such as the motif numbers to detect, the range of motif widths, the number of motif sites, and the *E*-value threshold. This standalone version contributes a flexible and customizable environment for motif examination, helping in the understanding of regulatory elements and functional patterns within biological sequences.

Motif prediction for legume stress-related genes (e.g. NAC) can be performed across various motif widths through the exploitation of MEME standalone version 5.5.3 (Bailey et al. 2015). The MEME investigation for predicting conserved motifs will require defined parameters such as 20 motifs to be identified, motif widths ranging from 10 to 50, motif sites spanning from 2 to 72 (Figure 5.8), and an *E*-value threshold set at 0.05. Motif sites spanned from 2 to 96 in the case of pigeon pea while 2–166 for groundnut crop.

GSDS 2.0 (Gene Structure Display Server) may use to visualize the exon/intron organization of the putatively identified stress-related NAC genes (Hu et al. 2015; Figure 5.9). It is a bioinformatics database intended to picture, scan, and examine the gene organization of eukaryotic genomes. Further, it allows users to input transcript or genomic sequences and prepare a graphical explanation of the structure of the gene, including exon–intron organization. Some of the Key features of GSDS may include:

a) **Gene structure visualization**: GSDS shows the gene structure graphically, displaying the exons, introns, and untranslated regions (UTRs) positions.

Figure 5.8 Conserved motif identification using MEME. Conserved motif-WYFF[SC]P[RK]D[RK]KYP[NT]GSR[TP]NRATX[SA]G[YF]WKATGKD[KR] for NAC gene(chickpea) subdomain-C in general.

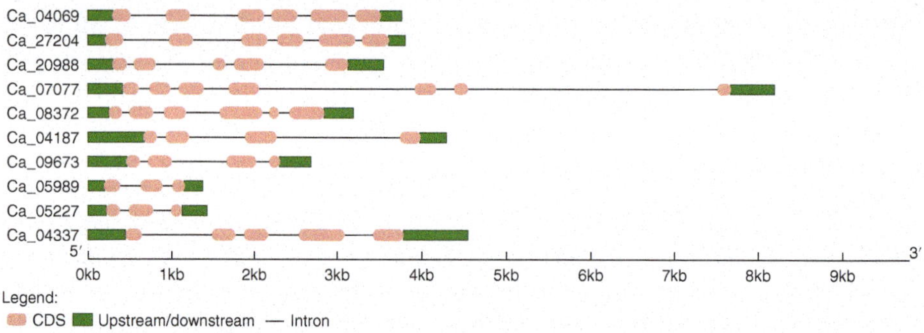

Figure 5.9 Gene structure display server GSDS analysis for a few representative NAC genes.

b) **Exon–intron boundaries**: It assists in visualizing the exon–intron margins and the arrangement of several structural elements within genes.
c) **Conserved motifs**: GSDS may also include features to show conserved motifs or domains within the gene structure.
d) **Comparative genomics**: Certain versions of GSDS assist in differentiating the structures of genes within the same species or among different species.
e) **Customization**: The facility to customize the display settings so that users may be able to focus on particular regions of interest along with the option to highlight specific/certain features.

5.3.5 Cis-acting Regulatory Elements in Stress-responsive Genes Identification

For researchers and scientists working in the field of plant molecular biology, the PlantCARE database is a valuable resource. It supplies an understanding of the CAREs in the promoter region of genes that are associated with plant gene expression regulation. Specific DNA sequences called CAREs play a crucial part in the control of gene expression. These elements are acknowledged and bound by TFs, which are proteins that control gene transcription. Understanding the existence and role of these regulatory elements is important for untangling the complicated regulatory networks that control plant growth, maturation, development, and response to environmental stimuli. PlantCARE contains information on a wide range of cis-acting elements, such as promoters, enhancers, silencers, and other regulatory elements. The database provides details about the sequences, positions, and functional roles of these elements in various plant species. Researchers can use PlantCARE to identify and analyze potential regulatory elements in the promoter regions of genes of interest. Access to the PlantCARE database typically involves visiting the official website or using bioinformatics tools that integrate its data. Keep in mind that databases may be updated over time, so it is a good idea to check for the latest information and features on the official PlantCARE website or related resources.

Researchers can analyze the 1500-bp upstream promoter sequences of stress-responsive genes (NAC) to determine probable CAREs. This investigation is assisted through the utilization of the PlantCARE databank (Lescot et al. 2002; Figure 5.10). Particularly, CAREs with a matrix score greater than five are presumed notable in this examination of regulatory elements related to stress-responsive genes (NAC).

5.3.6 STRING Evaluation for Protein–Protein Network Analysis

STRING, which refers to "Search Tool for the Retrieval of Interacting Genes/Proteins," is a bioinformatics web resource and tool/application/database that gives an understanding of PPIs. STRING is intended to aid researchers in analyzing the functional interconnections among proteins, providing awareness of cellular processes, pathways, and networks. Here are important characteristics and components of STRING:

a) **Protein interaction networks**: STRING incorporates experimental and putative protein–protein association data from several sources. It displays these interactions as networks, allowing researchers to visualize relationships among proteins.
b) **Species coverage**: STRING envelops a broad range of organisms, from bacteria to humans. Researchers can choose certain organisms of interest for their exploration.
c) **Data sources**: STRING collects information from varied sources, involving experimental data, co-expression surveys, text mining, and computational forecasting. This multi-source technique/tool/method aids in contributing an extensive view of protein interactions or networks.
d) **Functional enrichment survey**: In addition to PPI information, STRING provides functional enrichment study/inspection. Users can investigate the plant biological processes, plant molecular functions, and cellular components related to a set of proteins.
e) **Network visualization**: The database gives interconnected/interactions network visualization tools that permit users to explore and investigate protein interaction networks. The visualization can include data such as confidence scores for interactions.
f) **Customization and download**: Users can customize and download interaction networks for further study. This is particularly beneficial for researchers who want to incorporate STRING data with other bioinformatics tools or databases.
g) **Integration with other databases**: STRING can be used in combination with other bioinformatics databases and tools to accelerate the interpretation of protein interactions/associations in terms of numerous biological processes.

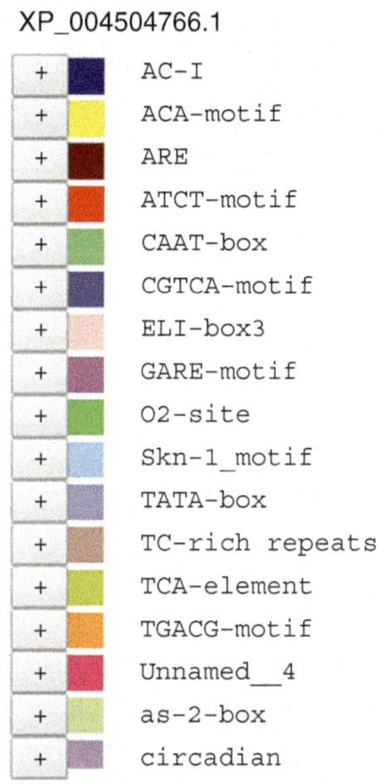

Figure 5.10 Pictorial representation of cis-elements identification using PlantCARE database.

Researchers in fields such as plant molecular biology, plant systems biology, and computational biology or bioinformatics often utilize STRING to investigate protein–protein interaction networks that further determine probable partners for uncharacterized proteins, and develop an understanding of the functional arrangements of cellular systems. To use the STRING application, scientists can look at the official website (https://string-db.org/) and

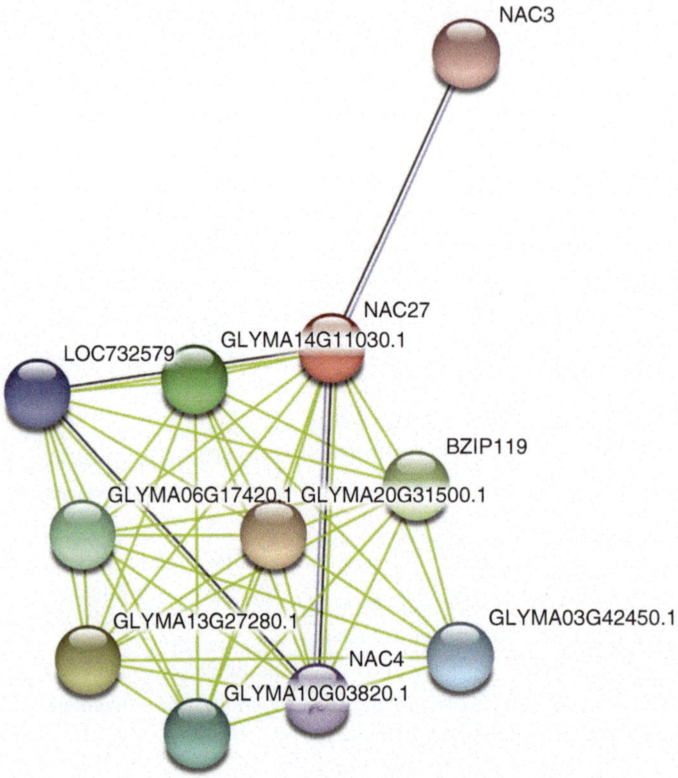

Figure 5.11 Pictorial representation of STRING analysis for protein–protein interaction of genes.

input the list of proteins or genes biologists are interested in, and the database will create a network of known and possible protein interconnections.

Building upon the aforestated approach, the STRING database (Szklarczyk et al. 2011), could be employed for performing protein–protein network analysis of legume *NAC* gene studies. Particularly, stress-related NAC protein sequences from chosen/selected crops (e.g. chickpea, pigeon pea, and peanut) could be queried against *Arabidopsis thaliana* proteins to detect the most pertinent matches, together with their annotations (Figure 5.11). Eventually, these recognized hits may be the foundation for performing an exhaustive analysis of protein–protein interaction in these food crops.

5.3.7 *In silico* Expression Analysis of NAC Genes in Food Legumes

Expression data for potential stress-responsive genes could be gathered from various gene expression atlases, including those specific to *C. arietinum* (Kudapa et al. 2018, https://doi.org/10.1111/pce.13210), *C. cajan* (Pazhamala et al. 2017, https://doi.org/10.1093/jxb/erx010), and *A. hypogaea* (Sinha et al. 2020, https://doi.org/10.1111/pbi.13374). The data encompass various tissues and developmental stages such as germination, seedling, vegetative, reproductive, and senescence. Analysis of the expression patterns of these genes (NAC) could be done, and the results can be visually represented as a heatmap (Figure 5.12).

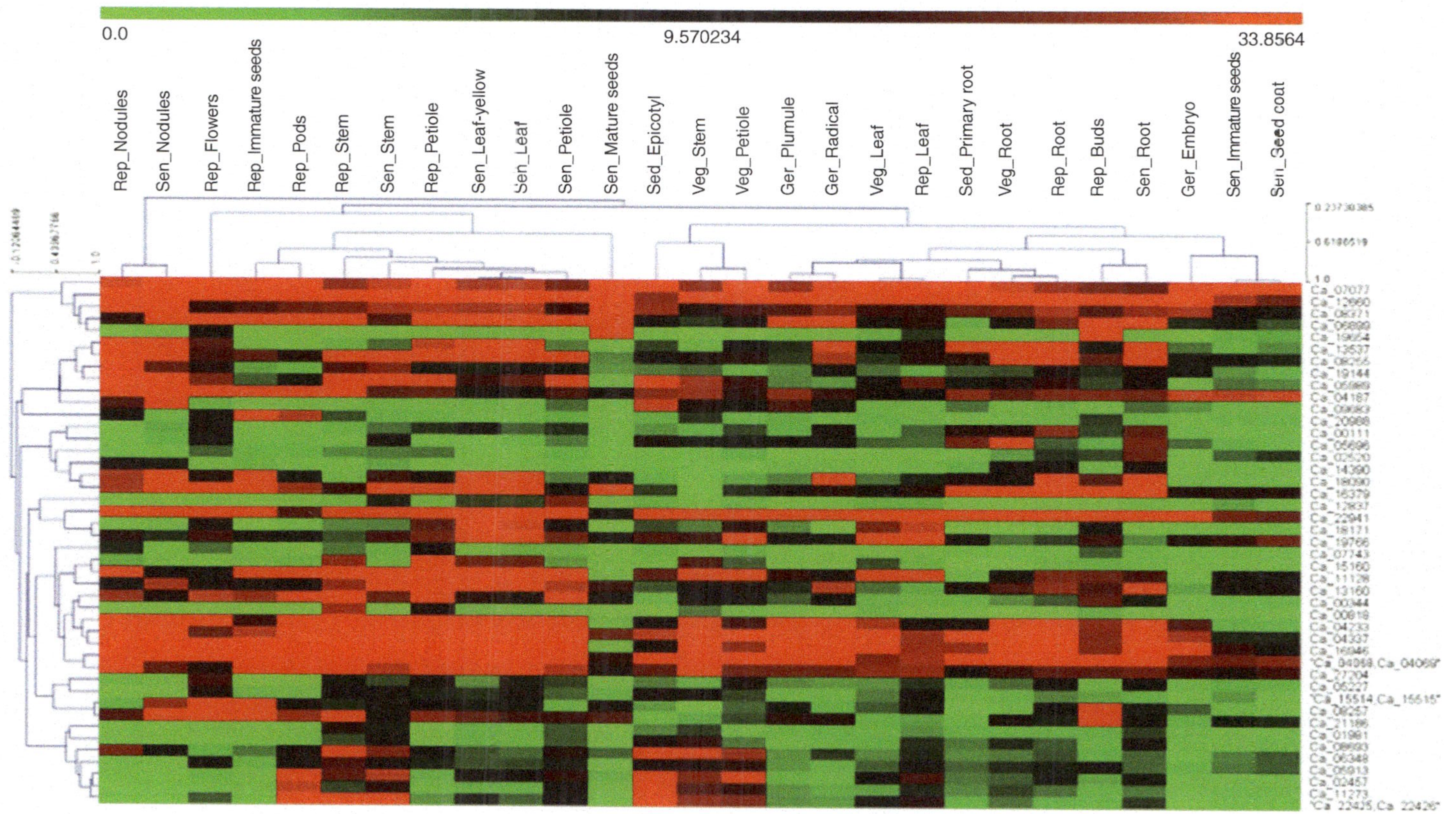

Figure 5.12 Heatmap representation for expression patterns of chickpea NAC genes across different tissues using MEV tool.

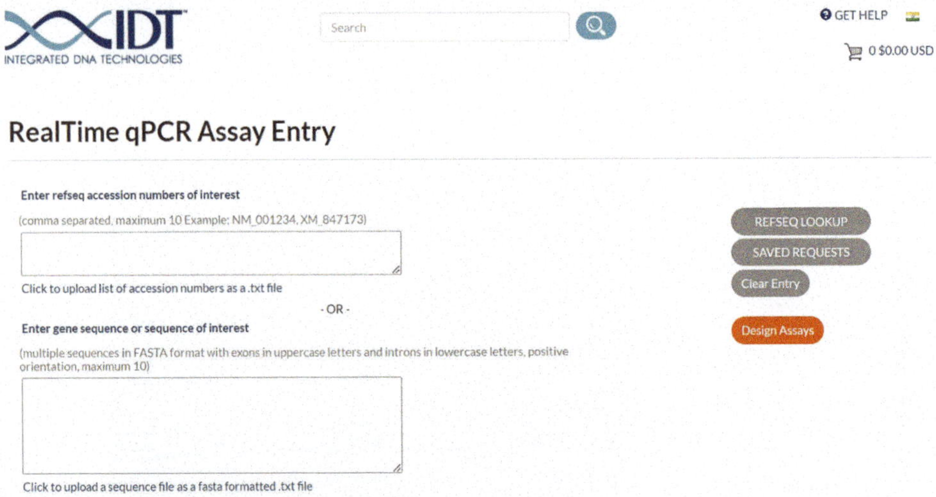

Figure 5.13 Screenshot of PrimerQuest tool/application for Real-time Primer designing.

The generated heatmap can be conveniently viewed and further explored using the MeV tool, accessible at https://webmev.tm4.org/about.

5.3.8 Quantitative PCR (Real Time) Analysis

The validation process for genes identified through *in silico* transcript mining involves employing real-time polymerase chain reaction (PCR) analysis. Relative expression levels will be calculated using the $2^{-\Delta\Delta CT}$ method, a method introduced by Livak and Schmittgen (2001).

For the design of specific primers essential for quantitative PCR (qPCR), the PrimerQuest tool will be utilized. This online tool, accessible at https://eu.idtdna.com/scitools/Applications/RealTimePCR/Default.aspx, facilitates the generation of primers tailored to the identified genes (Figure 5.13).

In summary, the steps include primer design using PrimerQuest and subsequent real-time PCR analysis using the $2^{-\Delta\Delta CT}$ method to validate the expression levels of the identified genes.

5.4 Conclusion

In conclusion, bioinformatics plays a central role in transforming the field of crop amelioration by assisting in the candidate plant gene identification and analysis. Through the seamless incorporation of computational tools and biological data, bioinformatics expedites an extensive understanding of these gene functions, pathways, and other characteristics. The advantages widen across several phases of crop enhancement, from early genome sequencing and annotation to the genetic variation's identification and the investigation of complicated regulatory networks.

By exploiting bioinformatics, researchers and scientists can effectively examine massive genomic datasets, discovering potential candidate genes related to beneficial traits such as tolerance against stress, disease resistance, and yield improvement. Functional annotation tools assist in unraveling the biological roles of these genes, while pathway interpretation imparts an understanding of the entangled networks controlling crucial physiological processes.

Furthermore, bioinformatics empowers the investigation of gene expression patterns across multiple conditions and tissues and lays the foundation for finding key regulators in response to environmental stimuli. Protein–protein interlinkage mesh and phylogenetic analyses provide valuable details about the evolutionary preservation and associations among candidate genes, supporting the prioritization of targets for crop improvement endeavors.

Population genetics investigations, aided by bioinformatics, offer a wider perspective on the genetic diversity within the plant community, leading breeders to choose genes that are not only advantageous but also applicable to varied agroecological circumstances. Machine learning approaches in bioinformatics boost predictive modeling, providing a significant technique for prioritizing candidate genes based on multidimensional data. This synergy of computational techniques and biological insights empowers researchers and breeders in making informed decisions for crop improvement schemes.

In essence, the marriage of bioinformatics and plant genomics opens new frontiers for crop improvement, accelerating the development of resilient, high-yielding, and sustainable crops that can address the challenges of a growing global population and changing environmental conditions. As technology continues to advance, the role of bioinformatics in crop improvement is poised to expand further, ensuring that agriculture remains at the forefront of innovation and productivity.

References

Armenteros, J.J.A., Salvatore, M., Emanuelsson, O. et al. (2019). Detecting sequence signals in targeting peptides using deep learning. *Life Science Alliance* 2 (5): e201900429.

Aslam, M., Grover, A., Sinha, V.B. et al. (2012). Isolation and characterization of cold responsive *NAC* gene from *Lepidium latifolium*. *Molecular Biology Reports* 39 (10): 9629–9638.

Bailey, T.L., Johnson, J., Grant, C.E., and Noble, W.S. (2015). The MEME suite. *Nucleic Acids Research* 43 (W1): W39–W49.

Bertioli, D.J., Jenkins, J., Clevenger, J. et al. (2019). The genome sequence of segmental allotetraploid peanut *Arachis hypogaea*. *Nature Genetics* 51 (5): 877–884.

Chen, S.P., Lin, I.W., Chen, X. et al. (2016). Sweet potato NAC transcription factor, IbNAC1, up-regulates sporamin gene expression by binding the SWRE motif against mechanical wounding and herbivore attack. *The Plant Journal* 86 (3): 234–248.

Chinnusamy, V., Ohta, M., Kanrar, S. et al. (2003). *ICE1*: a regulator of cold-induced transcriptome and freezing tolerance in *Arabidopsis*. *Genes & Development* 17 (8): 1043–1054.

Eddy, S.R. (2011). Accelerated profile HMM searches. *PLoS Computational Biology* 7 (10): e1002195.

Finn, R.D., Coggill, P., Eberhardt, R.Y. et al. (2016). The Pfam protein families database: towards a more sustainable future. *Nucleic Acids Research* 44 (D1): D279–D285.

Fowler, S. and Thomashow, M.F. (2002). Arabidopsis transcriptome profiling indicates that multiple regulatory pathways are activated during cold acclimatization in addition to the *CBF* cold response pathway. *The Plant Cell* 14: 1675–1690.

Grover, A., Singh, S., Pandey, P. et al. (2014). Over-expression of *NAC* gene from *Lepidium latifolium* L. enhances biomass, shortens life cycle and induces cold stress tolerance in tobacco: potential for engineering fourth generation biofuel crops. *Molecular Biology Reports* 41: 7479–7489.

Guo, W.L., Wang, S.B., Chen, R.G. et al. (2015). Characterization and expression profile of *CaNAC2* pepper gene. *Frontiers in Plant Science* 6: 755.

Gupta, S.M., Pandey, P., Negi, P.S. et al. (2013). DRE-binding transcription factor gene (*LlaDREB1b*) is regulated by various abiotic stresses in *Lepidium latifolium* L. *Molecular Biology Reports* 40: 2573–2580.

Hallgren, J., Tsirigos, K.D., Pedersen, M.D. et al. (2022). DeepTMHMM predicts alpha and beta transmembrane proteins using deep neural networks. bioRxiv preprint. https://doi.org/10.1101/2022.04.08.487609

Horton, P., Park, K.J., Obayashi, T. et al. (2007). WoLF PSORT: protein localization predictor. *Nucleic Acids Research* 35 (suppl_2): W585–W587.

Howe, E.A., Sinha, R., Schlauch, D., and Quackenbush, J. (2011). RNA-seq analysis in MeV. *Bioinformatics* 27 (22): 3209–3210.

Hu, B., Jin, J.P., Guo, A.Y. et al. (2015). GSDS 2.0: an upgraded gene features visualization server. *Bioinformatics* 31: 1296–1297.

Hussain, R.M., Ali, M., Feng, X., and Li, X. (2017). The essence of *NAC* gene family to the cultivation of drought-resistant soybean (*Glycine max* L. Merr.) cultivars. *BMC Plant Biology* 17 (1): 55.

Kim, S.G., Kim, S.Y., and Park, C.M. (2007). A membrane-associated NAC transcription factor regulates salt-responsive flowering via *FLOWERING LOCUS T* in *Arabidopsis*. *Planta* 226 (3): 647–654.

Kim, H.J., Nam, H.G., and Lim, P.O. (2016). Regulatory network of NAC transcription factors in leaf senescence. *Current Opinion in Plant Biology* 33: 48–56.

Klee, E.W. and Ellis, L.B. (2005). Evaluating eukaryotic secreted protein prediction. *BMC Bioinformatics* 6: 256.

Krzywinski, M., Schein, J., Birol, I. et al. (2009). Circos: an information aesthetic for comparative genomics. *Genome Research* 19 (9): 1639–1645.

Kudapa, H., Garg, V., Chitikineni, A., and Varshney, R.K. (2018). The RNA-seq-based high resolution gene expression atlas of chickpea (*Cicer arietinum* L.) reveals dynamic spatio-temporal changes associated with growth and development. *Plant, Cell & Environment* 41: 2209–2225.

Kumar, S., Stecher, G., and Tamura, K. (2016). MEGA7: molecular evolutionary genetics analysis version 7.0 for bigger datasets. *Molecular Biology and Evolution* 33 (7): 1870–1874.

Lescot, M., Dehais, P., Thijs, G. et al. (2002). PlantCARE, a database of plant cis-acting regulatory elements and a portal to tools for *in silico* analysis of promoter sequences. *Nucleic Acids Research* 30 (1): 325–327.

Letunic, I. and Bork, P. (2018). 20 years of the SMART protein domain annotation resource. *Nucleic Acids Research* 46 (D1): D493–D496.

Letunic, I., Khedkar, S., and Bork, P. (2021). SMART: recent updates, new developments and status in 2020. *Nucleic Acids Research* 49 (D1): D458–D460.

Livak, K.J. and Schmittgen, T.D. (2001). Analysis of relative gene expression data using a real-time quantitative PCR and the $2{-}\Delta\Delta^{CT}$ method. *Methods* 25 (4): 402–408.

Medina, J., Catalá, R., and Salinas, J. (2011). The CBFs: three arabidopsis transcription factors to cold acclimate. *Plant Science* 180 (1): 3–11.

Mistry, J., Chuguransky, S., Williams, L. et al. (2021). Pfam: the protein families database in 2021. *Nucleic Acids Research* 49 (D1): D412–D419.

Nakashima, K. and Yamaguchi-Shinozaki, K. (2010). Promoters and transcription factors in abiotic stress-responsive gene expression. In: *Abiotic Stress Adaptation in Plants, Physiological, Molecular and Genomic Foundation* (ed. A. Pareek, S.K. Sopory, H.J. Bohnert, and Govindjee), 199–216. Springer-Verlag.

Olsen, A.N., Ernst, H.A., Leggio, L.L., and Skriver, K. (2005a). DNA-binding specificity and molecular functions of NAC transcription factors. *Plant Science* 169: 785–797.

Olsen, A.N., Ernst, H.A., Leggio, L.L., and Skriver, K. (2005b). NAC transcription factors: structurally distinct, functionally diverse. *Trends in Plant Science* 10: 79–87.

Ooka, H., Satoh, K., Dosssi, K. et al. (2003). Comprehensive analysis of NAC family genes in *Oryza sativa* and *Arabidopsis thaliana*. *DNA Research* 10: 239–247.

Park, J., Kim, Y.S., Kim, S.G. et al. (2011). Integration of auxin and salt signals by the NAC transcription factor *NTM2* during seed germination in *Arabidopsis*. *Plant Physiology* 156 (2): 537–549.

Pazhamala, L.T., Purohit, S., Saxena, R.K. et al. (2017). Gene expression atlas of pigeonpea and its application to gain insights into genes associated with pollen fertility implicated in seed formation. *Journal of Experimental Botany* 68 (8): 2037–2054.

Puranik, S., Sahu, P.P., Srivastava, P.S., and Prasad, M. (2012). NAC proteins: regulation and role in stress tolerance. *Trends in Plant Science* 17 (6): 369–381.

Singh, S., Grover, A., and Nasim, M. (2016). Biofuel potential of plants transformed genetically with NAC family genes. *Frontiers in Plant Science* 7: 22.

Singh, S., Khalid, H., Grover, A. et al. (2019). Altered physiological responses of *LlaNAC* gene of *Lepidium latifolium* overexpressing tobacco plants. *Acta Physiologiae Plantarum* 41: 139.

Singh, S., Koyama, H., Bhati, K.K., and Alok, A. (2021a). Biotechnological importance of NAC family genes in improvement of food crop. *Journal of Plant Research* 134 (3): 475–495.

Singh, S., Kudapa, H., Garg, V., and Varshney, R.K. (2021b). Comprehensive analysis and identification of drought-responsive candidate NAC genes in three semi-arid tropics (SAT) legume crops. *BMC Genomics* 22: 289.

Sinha, P., Bajaj, P., Pazhamala, L.T. et al. (2020). *Arachis hypogea* gene expression atlas (*AhGEA*) for *fastigiata* subspecies of cultivated groundnut to accelerate functional and translational genomics applications. *Plant Biotechnology Journal* 18 (11): 2187–2200.

Souer, E., van Houwelingen, A., Kloos, D. et al. (1996). The no apical meristem gene of *Petunia* is required for pattern formation in embryos and flowers and is expressed at meristem and primordial boundaries. *Cell* 85 (2): 159–170.

Szklarczyk, D., Franceschini, A., Kuhn, M. et al. (2011). The STRING database in 2011: functional interaction networks of proteins, globally integrated and scored. *Nucleic Acids Research* 39 (suppl_1): D561–D568.

Thumuluri, V., Almagro Armenteros, J.J., Johansen, A.R. et al. (2022). DeepLoc 2.0: multi-label subcellular localization prediction using protein language models. *Nucleic Acids Research* 50 (W1): W228–W234.

Varshney, R.K., Chen, W., Li, Y. et al. (2012). Draft genome sequence of pigeonpea (*Cajanus cajan*), an orphan legume crop of resource-poor farmers. *Nature Biotechnology* 30 (1): 83–89.

Varshney, R.K., Song, C., Saxena, R.K. et al. (2013). Draft genome sequence of chickpea (*Cicer arietinum*) provides a resource for trait improvement. *Nature Biotechnology* 31 (3): 240–246.

Voorrips, R.E. (2002). MapChart: software for the graphical presentation of linkage maps and QTLs. *The Journal of Heredity* 93 (1): 77–78.

Xie, Q., Frugis, G., Colgan, D., and Chua, N.H. (2000). Arabidopsis *NAC1* transduces auxin signal downstream of *TIR1* to promote lateral root development. *Genes & Development* 14 (23): 3024–3036.

Yamaguchi-Shinozaki, K. and Shinozaki, K. (1994). A novel cis-acting element in an *Arabidopsis* gene is involved in responsiveness to drought, low temperature, or high-salt stress. *The Plant Cell* 6: 251–264.

Yu, C.S., Chen, Y.C., Lu, C.H., and Hwang, J.K. (2006). Prediction of protein subcellular localization. *Proteins: Structure, Function, and Bioinformatics* 64: 643–651.

Yu, X., Liu, Y., Wang, S. et al. (2016). CarNAC4, a NAC-type chickpea transcription factor conferring enhanced drought and salt stress tolerances in *Arabidopsis*. *Plant Cell Reports* 35: 613–627.

Zhao, X., Yang, X., Pei, S. et al. (2016). The miscanthus NAC transcription factor MlNAC9 enhances abiotic stress tolerance in transgenic *Arabidopsis*. *Gene* 586 (1): 158–169.

Zhong, R., Richardson, E.A., and Ye, Z.H. (2007). Two NAC domain transcription factors, SND1 and NST1, function redundantly in regulation of secondary wall synthesis in fibers of Arabidopsis. *Planta* 225 (6): 1603–1611.

6

Exploring Machine Learning Algorithms for Gene Function Prediction in Crops

Ruchi Jakhmola-Mani[1], *Sonali*[1], *Aniket Pandey*[2], *Dhananjay Raturi*[1], *Rishita Singh*[1], *Kusala Vanam*[3], *Manish D*[4], *Ritu Chauhan*[5], *Deepshikha Pande Katare*[1], *Potshangbam Nongdam*[6], *and Angamba Meetei Potshangbam*[6]

[1] *Proteomics and Translational Research Lab, Centre for Medical Biotechnology, Amity Institute of Biotechnology, Amity University Uttar Pradesh, Noida, Uttar Pradesh, India*
[2] *Department of Computer Science, Nguyen Engineering Building, Fairfax, VA, USA*
[3] *Department of Pharmaceutics, CMR College of Pharmacy, JNTU, Hyderabad, Telangana, India*
[4] *Department of Biotechnology, Dayanand Sagar College of Engineering, Bangalore, Karnataka, India*
[5] *Amity Institute of Biotechnology, Amity University Uttar Pradesh, Noida, Uttar Pradesh, India*
[6] *Department of Biotechnology, Manipur University, Imphal, Manipur, India*

6.1 Introduction

6.1.1 Why is the Function of Genes Important?

In molecular terms, a gene constitutes the complete nucleic acid sequence required for synthesizing a functional polypeptide or RNA (Lodish et al. 2000). The overarching objective is to comprehend how genes function alongside the proteins they encode within a living organism. This undertaking holds significance as it is ultimately the functional attributes of a gene that determine an organism's observable traits or phenotype. In recent years, plant biologists have sought to decipher the developmental pathways governing agriculturally essential traits. It has been observed that the regulatory pathways involving TCP and GRAS transcription factors (TFs) play pivotal roles in this context. Among these factors, members of the TCP group assume diverse functions in plant growth and development, including the regulation of apical dominance. The GRAS family derives its name from its initial three characterized members: GIBBERELLIC ACID INSENSITIVE (GAI), REPRESSOR of GAI (RGA), and SCARECROW (SCR). This family exerts significant control in various regulatory processes, such as signal transduction and meristem maintenance (Vaschetto 2017; Cubas et al. 1999). Phylogenetic analysis has categorized the GRAS family into at least 13 subfamilies. One of these, DELLA, has garnered considerable attention due to its evolutionarily conserved DELLA domain, which imposes a higher selective constraint and its ability to repress gibberellin (GA) activity within the nucleus, thereby restraining plant growth (Davière et al. 2014; Lewis et al. 2008).

Bioinformatics for Plant Research and Crop Breeding, First Edition. Edited by Jen-Tsung Chen.

Another notable TF family, the MADS-box family, has recently captured the interest of molecular breeders for its involvement in crucial plant developmental processes. The name MADS is derived from its founding members: MCM1 from *Saccharomyces cerevisiae* (baker's yeast), AGAMOUS (AG) from *Arabidopsis thaliana*, DEFICIENS (DEF) from *Antirrhinum majus* (snapdragon), and SRF from *Homo sapiens* (Schilling et al. 2018). MADS-box genes hold particular importance in controlling flowering timing, inflorescence arrangement, floral organ identity, embryo sac formation, and seed development. Considering protein–protein interactions (PPIs), these MADS-box TFs play substantial roles in early seed development, offering a potential route to enhance crop yield by regulating their expression. Instead of directly binding to DNA, various types of TFs can influence target genes via PPIs, a well-established mechanism seen in plants and other organisms (Schwarz-Sommer et al. 1990; Causier et al. 2010). Protein–protein interaction (PPI) is a critical mechanism through which TCP, DELLA, and MADS-box factors regulate gene expression. TCP factors possess a characteristic basic helix–loop–helix (bHLH) DNA-binding domain with amphipathic helices facilitating protein interactions via hydrophobic surfaces. An Arabidopsis study by Davière et al. in 2014 revealed that DELLA proteins regulate certain aspects of GA responses through PPIs with TCP factors, indicating that GA-regulated DELLA–TCP interactions in the inflorescence shoot apex offer a novel means to control plant height in response to changing environmental conditions. TCP transcriptional regulators oversee complex gene networks through PPIs, including those involving the microRNA (miRNA) pathway, a significant epigenetic regulatory mechanism employing noncoding RNAs to modulate gene expression levels. Similarly, DELLA and MADS-box factors contribute significantly to gene expression regulation and plant development by orchestrating protein interactions (Singh et al. 2021; Gayen et al. 2019).

6.1.2 Genetics and Crop Improvement in the Past

Since ancient times, there has been a change in the genetic make-up of different crops due to natural (polyploidization, recombination, translocation) as well as anthropogenic actions such as breeding practices to obtain superior varieties with higher crop yield as well as increased nutritional content and other medicinal aspects (Mbanjo et al. 2019; Martin et al. 2020; Hinnisdaels et al. 1992).

Food and nutrition have always played a central role in the development of human society. From nomadic hunters and food gatherers to a civilized society there is always a demand for increasing nutrition and food security. Since the beginning of the cultivation and sedentary lifestyle, cultivation of various crops has played a major role in the development of human society. Hence, there is a need for modification in the genetic make-up of the important food crops.

Traditional crop improvement practices mostly rely on visible traits such as increased yield, and better stress tolerance. Since the domestication of crops, the most primitive practice for crop improvement relies on the Agro-morphological marker system to identify the genotypes with desired traits. In this method, the genotypes with desired characters are selected and then used in the breeding programs. Different local cultivars, and landraces which are locally adapted and best suited for a given environment have been developed similarly. Since genetic variability is the basis of plant breeding and provides a spectrum of genotypes that can be selected for developing new varieties (Akinola et al. 2019; Roy and Shil 2020). Variability in terms of genetic divergence for traits of interest is the major component of breeding programs to broaden the crop gene pool.

6.1.3 The Impact of Precise Gene Location/Function Prediction on Crop Improvement

A trait or characteristic is a functional component of an organism's genetic makeup. With the discovery of DNA, the utilization of molecular markers to enhance crop improvement has become increasingly widespread. These molecular markers primarily rely on DNA and are based on whether specific genetic locations are present or absent in different varieties of a given crop. This presence or absence of particular genetic locations forms the basis for creating a molecular profile, which is then employed to analyze genetic diversity (Bisht et al. 2015; Roy and Shil 2020; Islam et al. 2018; Mathure et al. 2010; Salunkhe et al. 2013; Yadav and Chand 2018). Additionally, molecular markers are useful in marker-assisted selection through biparental crosses, a method commonly used to identify quantitative trait loci (QTL) in crop plants. QTL mapping is a widely used technique for selecting crop plants with desirable traits. However, it is a relatively laborious and time-consuming process and has limitations, such as a limited chance of recombination occurring within families and pedigrees with known ancestry, resulting in a relatively low mapping resolution.

Recently, another strategy gaining popularity in crop breeding is genome-wide association studies (GWAS). GWAS involves extensive genetic surveys of the entire genome to detect variants associated with a trait within natural populations. This approach is powerful for dissecting complex traits. GWAS employs statistical methods to identify associations between sequence polymorphisms in the genome and functional variations in traits. It offers several advantages over conventional QTL mapping using bi-parental populations. GWAS populations encompass a much broader genetic diversity compared to the two parental lines used in segregation populations, leading to a relatively higher mapping resolution due to the presence of numerous historical recombination events. Thanks to the development of efficient statistical algorithms suitable for plant populations, the GWAS platform has been extensively applied to dissect the genetic basis of various complex traits in different crop plants. GWAS proves to be an effective approach for breeding. The sequence variants detected can be further utilized to associate them with important agronomic traits, particularly those related to disease resistance, higher yield, and the accumulation of heavy-metal elements. These variants can be designed and used directly as molecular markers in crop breeding (Wang et al. 2014, 2020; Takeda and Matsuoka 2008; Listgarten et al. 2012; Zhou and Stephens 2012).

6.1.4 Importance of Crop Improvement in the 21st Century

As a result of rising population, climatic change, and dwindling agricultural resources, guaranteeing adequate food production becomes a crucial concern. A significant increase in agricultural productivity has become difficult due to the detrimental effects of environmental variability and the competitive demands of the socioeconomic on the limited resources.

Food distribution and accessibility are adversely influenced by the concentration of humans in big cities. Changes related to age such as the older population or fluctuating birth rates can impact dietary choices. Rising affluence in certain areas or nations may result in changes to diets, an increase in particular kinds of food consumption, and higher expectations for the quality and variety of food available.

The level of income, economic development, poverty, and inequality are some of the features that may be affected by socioeconomic variables. Demographic trends such as population growth, urbanization, and changes in age distribution have an impact on both the number of inhabitants and their composition. The need to have food increases with population growth.

Rising affluence in certain areas or nations may result in changes to diets, an increase in particular kinds of food consumption, and higher expectations for the quality and variety of food available. The need for agricultural supplies may increase as a result of economic growth and globalization, not just because of the increased need for food but also for animal feed or the generation of bioenergy. For some groups of the population, poverty and inequality might limit their access to enough nutritious and good-for-health food.

Crop production systems are being seriously impacted by climate change and its accompanying variability. The problems caused by climate change in the twentieth century are complex and have a significant influence on crop availability. The increase in sea levels, as well as the increased frequency of extreme weather events such as floods and droughts, are expected to put more strain on crop supplies. Crop yields will probably decline first. Temperature changes, patterns of precipitation, and severe weather can have a negative impact on plant development, resulting in decreased effectiveness and reduced yields. The length of the growing seasons is also projected to shorten. Large areas of peripheral farming may cease to exist as a consequence, shorter growing seasons may not provide crops enough time to mature and produce a useful harvest. Crop revenue might decline as a result of the aforementioned factors. Ranchers and rural economies vigorously depend on crop deals for money, and decreased yields can unfavorably affect their occupations and generally speaking agrarian benefit. Another major issue is water stress. The distribution of rainfall may vary due to climate change, and there may be more occurrences of drought. Due to the potential for water shortages, crops that need enough water for growth and development may be put under additional strain. Crops may no longer be economically viable in areas close to dry margins as a result of climate change. Particularly, temperate fruits and nuts may suffer since their development and reproduction processes are affected by diminished winter chill, potentially resulting in lower yields. Climate change may also make pests and illnesses more likely to spread. It is possible for horticultural pests, like the Queensland fruit fly, to spread into previously thought of as fruit fly-free regions. This will cause the health and productivity of crops to suffer. As a response to climate change, agricultural systems could also become more sensitive to environmental factors. Even small variations in temperatures or rainfall patterns can have a significant impact on crop growth and productivity. Explicit harvest yields are additionally liable to be impacted. Climate change can alter the yields of crops like cotton, soybeans, and barley, which can affect their profitability and market availability. Certain illnesses, for example, Fusarium head scourge in wheat, are supposed to represent a higher gamble. A favorable environment for the spread and proliferation of such diseases is provided by the favorable conditions created by climate change, which include increased humidity and altered precipitation patterns.

Finally, environmental change can work with the spread of intrusive species. Outsider microorganisms, growths, plants, and creatures might track down additional ideal circumstances in new regions, possibly disturbing environments and undermining local plant species, further affecting harvest accessibility. To offset the possible negative effects on crop output and assure food security, addressing these concerns needs proactive efforts, such as climate-resilient agricultural practices, the creation of stress-tolerant crop varieties, and efficient pest and disease control systems (Mba et al. 2012; Miflin 2000). These problems can be dealt with through effective use of the existing agricultural lands and other resources that are crucial to crop breeding. Now the way humans are consuming agricultural resources is not good for society, the planet, or humans themselves. Education around nutrition and eating habits should be promoted that encourage people to consciously adopt a lifestyle of eating products that require minimum agricultural resources to be produced. Well, all these solutions offered are temporary and impractical in this fast-paced society. The use of cutting-edge technologies is the best way possible to deal with these problems.

6.1.5 Global Food Demand, Changing Dietary Habits Challenges

Various economists, philosophers, and scientists have predicted the trends in increasing population over the years and the negative impacts that flowed from it, food insecurity concerning this study. Determined efforts and the use of innovation in science can prove to be the ultimate solution. The demand for animal-based products is continuously increasing and this expansion is ongoing which ultimately resulted in requiring more land on which animal feed grows. This shift in dietary habits is mostly seen in people who have relatively high incomes in low-/medium-income areas. Also, there is the transportation of these products to account for. Thus, socioeconomic conflict plays a great role in affecting food security (Khush 2001). Another change of dietary habits has thrown light on the population intaking enough calories but not enough nutrition. This is very popularly known as the "Hidden hunger" leading to skyrocketing cases of obesity and deficiency of essential micronutrients in the human system. Food portions and prices affect the quality and quantity of food intake. Two issues that have to be treated are identified here, with the use of technology either decreasing the prices of crops or increasing its yield (Fróna et al. 2019).

One can assume that all these problems can be solved if people stopped urbanization and increased the amount of land that can be used for crop cultivation and testing new ways of crop improvement but again it just does not work that way as there are problems associated with that too.

- More land will lead to more greenhouse gas emissions leading to consequences associated with this phenomena.
- More land means more water for irrigation and fresh water supply is already limited and not even reaching areas for enough human consumption. This will even more increase water scarcity (Webb and Buratini 2018).

6.1.6 Need for Efficient and Faster Crop Improvement Methods

Science and technology have now penetrated the deepest reaches of knowledge. Most of the time to study the impact of genes, coding mRNA sequences were paid attention to but with the development of methods such as deep sequencing, scientists are now able to find out crucial roles played by noncoding RNA sequences in the functioning of cells. IncRNAs are long noncoding RNAs that are generally neglected for understanding the mechanism of protein expression but recent advances have concluded the profound impact of these molecules, especially in regulation and effect produced by their interaction with other molecules. For example, lncRNA1459 of *Solanum lycopersicum* (tomato) plays an important role in ripening and lycopene production. Methods to edit the sequence of this IncRNA or change its structure which will impair its function can aid in increasing ripening time and therefore the shelf life of such tomatoes (Li et al. 2018).

When there is an understanding of such deep levels of gene function and regulations, it should be utilized for developing efficient and faster crop improvement methods that cater to the issues of the present world as discussed above.

6.1.7 Multidisciplinary Approach Combining Genetics, Plant Breeding, Biotechnology, and Data Science for New Crop Improvement Strategies

There is more than enough reason to integrate various sciences to come up with a solution that is effective yet sustainable. Sophisticated areas of genetics, molecular biology, evolution and taxonomy, agronomy and food sciences, ecology, computer science, and system biology can be clubbed together in various combinations to serve the purpose of innovating much-needed crop improvement methods. Long-term health impact studies of the new and improved crops have to be taken into consideration as well. Following this chapter, the focus is on the genetics behind the improvement of crops. Approaches that can be made with respect to enhancing the quality and quantity of crop includes various parameters, most of which can be modified at the genetic level. They include an increase in Yield, Resistance to diseases, Resistance to pests, Tolerance to heat, Tolerance to drought, Increased nutritional content, Increased shelf life, Enhanced nutrient utilization efficiency, Decreased susceptibility to climatic change, and management of weeds. Conventional approaches are time-consuming and extensive so there is a need to promote the use of methods that provide faster results. Immediate decisions in every field are taken using computational methods and when there is an understanding of such deep levels of gene function and regulations, it should be utilized for developing efficient and faster crop improvement methods by computational methods that cater to the issues of the present world as discussed above.

6.2 Computational Methods for Gene Function Prediction

6.2.1 Bioinformatics

Gene function prediction in crop plants using bioinformatics databases and tools involves the systematic analysis of genomic and transcriptomic data to define the role of specific genes in plant growth, development, and stress responses. By employing tools like BLAST for homology searches, gene ontology (GO) enrichment analysis, pathway mapping through KEGG, and PPI networks via STRING, researchers can uncover potential functions of candidate genes. This knowledge is crucial for informed crop breeding strategies, allowing for the selection and modification of genes associated with desired traits, such as yield, disease resistance, and nutritional content, ultimately accelerating the development of improved crop varieties.

Major bioinformatics databases and tools used for predicting the function of the gene include:

1) 1. Plant proteome databases, 2. PPI databases, 3. GO databases, and 4. Sequence alignment tools

Plant proteome databases: A key factor to consider here is the proteome. The proteome is closely related to gene function, as the expression of genes ultimately determines the composition of the proteome and the role of individual proteins. Analysis of the proteome will also provide insights into protein–protein interactions and the biological processes they are involved in. This can be done with the use of Plant proteome databases like the Rice Genome Annotation Project (RGAP) (http://rice.plantbiology.msu.edu/cgi-bin/ORF_infopage.cgi) and The Soybean Proteome Database (http://proteomics.iit.edu/soybean/) These are large-scale repositories of protein sequence and functional information that can be used to identify proteins with similar sequences or functions. By analyzing the proteome of a plant under different conditions, researchers can gain insights into the functions of unknown or poorly characterized genes. For example, if a protein is found to be upregulated in response to a particular stress, it may be involved in the plant's response to that stress and could be a target for further study. Therefore, proteomics can complement genomics and transcriptomics in providing a more complete understanding of plant biology.

PPI databases: These databases like the Rice Interactions Viewer (http://www.riceinteractions.info/) and the LegumeIP (http://plantgrn.noble.org/LegumeIP/) database provide information on the interactions between proteins, which can help in predicting the functions of unknown proteins based on their interactions with known proteins

Gene ontology (GO) databases: These databases provide a standardized vocabulary for describing gene function and can be used to annotate proteins based on their molecular function, biological process, and cellular component. The Gene Ontology Consortium (http://geneontology.org/), The Plant GO Slims database (http://www.geneontology.org/page/plant-go-slims), and The AgBase Gene Ontology database (https://www.agbase.msstate.edu/) for agriculture are some well-known examples of GO databases.

Sequence alignment tools: These tools can be used to compare gene and protein sequences and identify conserved domains or motifs that are associated with specific functions. Frequently used sequence alignment tools include the Basic Local Alignment Search Tool (BLAST) (https://blast.ncbi.nlm.nih.gov/Blast.cgi), Clustal Omega multiple sequence alignment tool (https://www.ebi.ac.uk/Tools/msa/clustalo/) and Hidden Markov Model (HMM) profiles. (https://www.ebi.ac.uk/Tools/hmmer/search/hmmsearch) (Subba et al. 2019).

6.2.1.1 Bioinformatics Tools in Plant Disease Management

With the capacity to predict plant genome function, a new level of prediction incorporating pathogen genomes and their connection can considerably contribute to crop development.

Bioinformatic tools that can be used to manage plant disease by predicting the function of certain genes include PhytoPath, EffectorP, PathoPlant, and InterSPPI. These tools are valuable for dissecting the mechanisms behind pathogen pathogenicity and pinpointing plant resistance proteins. Moreover, genome databases dedicated to plant pathogens have emerged to establish connections between the pathogen and host genomes, facilitating investigations into the connections between genetic diversity and disease. These databases serve as valuable resources for examining plant gene functionality and interactions between plants and pathogens. Uncovering the role of genes responsible for encoding plant resistance proteins offers a means to enhance crop breeding by shedding light on the molecular intricacies of plant-pathogen interactions and the plant immune system. This knowledge can be harnessed to develop novel crop varieties equipped with resistance against specific plant diseases. By understanding the function of plant resistance genes, researchers can identify the specific proteins that are involved in plant immunity and develop new strategies for disease management, such as the use of gene editing technologies to introduce resistance genes into crops. Additionally, understanding the function of plant resistance genes can help in the development of new treatments for plant diseases, such as the use of biopesticides or biological control agents that target specific plant-pathogen interactions. Overall, identifying the function of the gene that codes for plant-resistance proteins is an important step in developing new strategies for crop improvement and disease management (Dong et al. 2021).

6.2.2 Machine Learning

Machine learning (ML), a subset of artificial intelligence (AI), has emerged as a critical computational methodology in the field of functional genomics. This approach is built on the concept of training algorithms to recognize subtle and intricate patterns in massive biological datasets, eliminating the need for explicit rule-based programming. ML algorithms have become indispensable tools for the analysis of diverse data sources encompassing gene expression profiles, protein–protein interactions, and other relevant genomic attributes. The systematic application of these algorithms facilitates the prediction of gene functions for previously uncharacterized entities, predicated upon already discovered patterns within their genomic features. These algorithms can also be used to predict protein function based on sequence or structural features. Specific types of ML approaches are actively employed within the domain of gene function prediction to adapt to diverse targets and challenges.

6.2.2.1 Supervised Learning

A fundamental subset of ML is commonly used in situations requiring labeled training data. Such algorithms are well suited for tasks combining binary or multi-class classification and numerical trait value prediction, commonly referred to as regression. Random forests (RF), support vector machines (SVM), and k-nearest neighbors (kNN) are prominent examples in this category. For example, a SVM may be precisely tailored to predict the participation of a gene in biological pathways such as photosynthesis based on an examination of its expression levels and other relevant genetic characteristics.

6.2.2.2 Unsupervised Learning

Methods intended for tasks where labeled training data is either absent or insignificant. These techniques are mostly concerned with clustering and feature extraction. Examples of this category include principal components analysis, k-means clustering, and self-organizing maps. Unsupervised learning algorithms excel in delineating groups of genes characterized by similar genomic features that may be suggestive of specific biological activities in the context of gene function prediction. For example, k-means clustering may expertly classify genes based on their expression patterns, revealing commonality among genes within each cluster and providing insight into their functional associations.

Furthermore, gene function prediction is augmented through the utilization of feature-based methods, which operate on explicitly defined genomic features, frequently complementing supervised learning paradigms. Additionally, neural network-based approaches, inspired by the architectural and functional paradigms of biological neural networks, are increasingly adopted for complex analytical tasks in genomics. For gene function prediction, ANN-based methodologies offer a unique ability to decipher intricate patterns embedded within large and intricate datasets that may defy conventional analytical techniques (Mahood et al. 2020).

6.3 Machine Learning and Crop Improvement

There is a specific step-by-step process that is typically always followed when building ML models. This process has three main steps. The first one is called data preprocessing, where the data is imported, cleaned, and then split into training and test sets. In real-world situations, cleaning the data is quite an important step. The next step is to model, where the model is built, then trained, and later used to make the prediction (Figure 6.1).

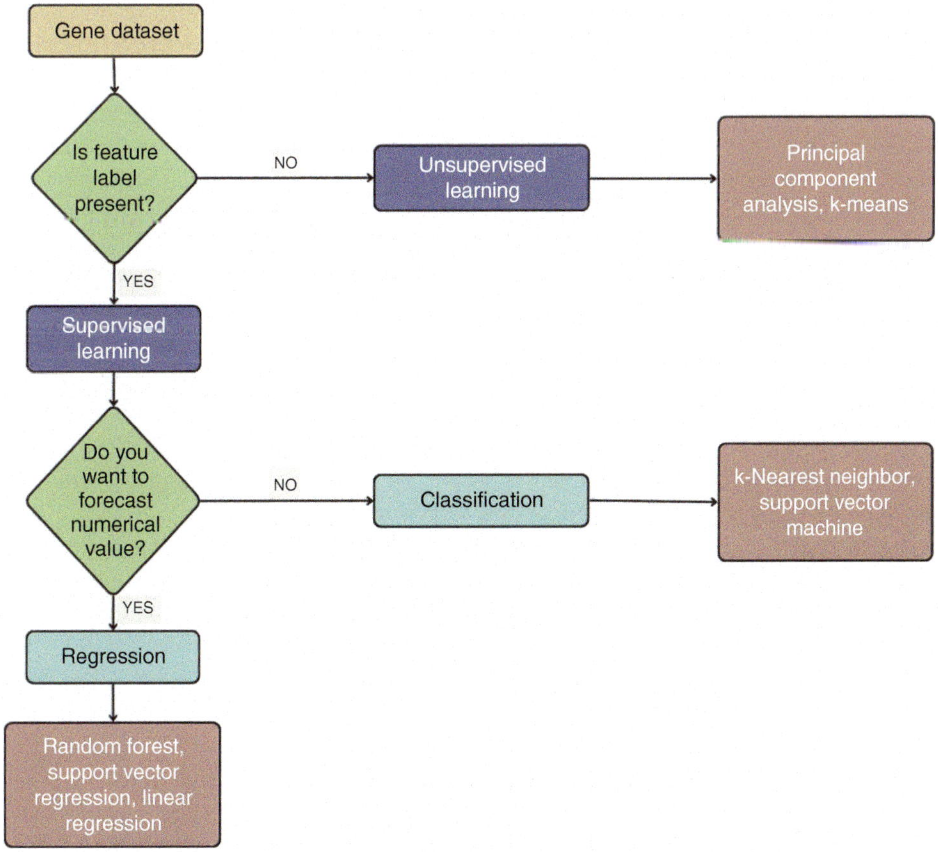

Figure 6.1 Schematic for choosing a machine learning (ML) workflow. A simplified decision tree for choosing the type of ML algorithm for a given task.

The last step is evaluations. Here performance metrics are calculated and a verdict about the model is made, whether it's a good-fitting model and if it works for the given data or not. This is a very important step to make sure that the models built serve the purpose that they're designed for.

6.3.1 Dataset

6.3.1.1 Data Integration

Strategies for Data Integration in gene prediction encompass techniques for amalgamating information from diverse sources, to enhance our understanding of the genetic and environmental factors influencing the success of plant genetics. These strategies involve the amalgamation of data derived from various trials, genotype, and phenotype data, or data collected through diverse omics methods. For researchers seeking to unravel the intricate interactions among different molecules, pathways, and networks underlying fundamental biological processes linked to specific traits, the effective integration of multiple datasets proves invaluable (Fabres et al. 2017). Furthermore, the identification of novel biomarkers and sources of data bias can enhance the precision of analytical outcomes through the application of data integration approaches. While substantial data have been amassed in various plant omics domains, their integration into gene studies remains largely unexplored, necessitating a range of approaches to harmonize these datasets. In the realm of Plant Genetics, common data integration strategies employ concepts, statistics, and models (Yoosefzadeh-Najafabadi et al. 2021). Conceptual integration involves synthesizing data from diverse origins to attain a more holistic grasp of the biological principles governing a particular trait. Subsequently, the data undergoes analysis to uncover patterns, relationships, and correlations, facilitating meaningful biological inferences (Cavill et al. 2016).

Statistical integration, on the other hand, entails the amalgamation of data from multiple sources into a unified perspective using statistical methodologies. This encompasses the application of statistical techniques to discern patterns and correlations across data from various sources, as well as the assessment of their impact on the overall dataset (Xu et al. 2022). Statistical approaches encompass the use of clustering algorithms to group similar datasets and the employment of regression models to establish associations among variables within diverse datasets (Acerbi et al. 2022). Therefore, the handling and development of high-performance computational strategies for managing vast gene datasets, given their scale and intricacy, pose substantial challenges. To address these challenges effectively, a thorough examination and evaluation of different data integration approaches should be undertaken, tailored to the specific characteristics of the dataset in question.

6.3.1.2 Splitting the Dataset

Data splitting involves the separation of a portion of the dataset before employing it to train the model. Approximately 20% of the data is set aside for this purpose. For instance, if there were 20 sequences of genes, about four gene sequences should be isolated for testing. The remaining 80% constitutes the training set. Following model construction using the training set, the gene sequences from the test set are utilized to evaluate the model's performance.

The model does not have any prior information concerning these test gene sequences, as they were not part of the model creation process. Subsequently, the model is applied to predict specific values and characteristics of the crop. Since the test set was preselected from the given data, the actual values for these sequences are known, enabling a comparison between the predicted and actual values. This facilitates model evaluation and helps determine its efficacy, identifying potential areas for improvement (Figure 6.2).

6.3.2 Overview of Machine Learning Algorithms

6.3.2.1 k-Nearest Neighbors

k-NN is a supervised learning algorithm that predicts the class of new instances based on the classes of the most similar instances. The kNN algorithm assumes that similar things exist in proximity (Mahood et al. 2020).

A step-by-step rule guide for the KNN algorithm will be constructed, and the algorithm's functionality will be manually observed. The algorithm is characterized by its simplicity.

The following procedure will be employed:

- A value for K, representing the number of neighbors in the algorithm, will be chosen. Typically, K will be set to 1, 2, 3, 5, or another specific number.
- The distance metric to be utilized will be determined. The most common choice is Euclidean distance, although other options such as Manhattan distance are also possible.
- The kNN of a new data point will be selected based on their Euclidean distance (or the chosen distance metric).
- The number of data points in each category among the K neighbors will be counted. This will be performed to ascertain how many data points belong to each category, especially in cases where the dataset contains more than two categories.
- The new data point will be assigned to the category that exhibits the highest count among the K neighbors.

The nomenclature "K-nearest neighbors" is derived from the essence of this process. It involves a straightforward algorithm that determines the category of a data point based on the majority of its nearest neighbors.

6.3.2.2 Random Forest

Random forest is an ensemble learning algorithm that improves prediction accuracy by combining multiple decision trees. The working process of the algorithm is as follows: Firstly, the starting point is the entire dataset. Subsequently, a subset of K data points is randomly selected from the training set. Rather than constructing a decision tree using the entire dataset, a decision tree is built solely based on this selected subset of data points. The construction of multiple decision trees is accomplished through a repetitive process. The number of trees to be built is determined, and for each tree, steps 2 and 3 are iterated. This results in the creation of a forest of decision trees. When making predictions, the random forest utilizes all the trees. Each tree predicts the value of y for the respective data point. The final prediction is determined by averaging the predicted y values from all the trees. By default, random forests often comprise around 500 trees, leading to 500 predictions for the value of y. The act of averaging these predictions enhances the accuracy of the overall prediction by incorporating the

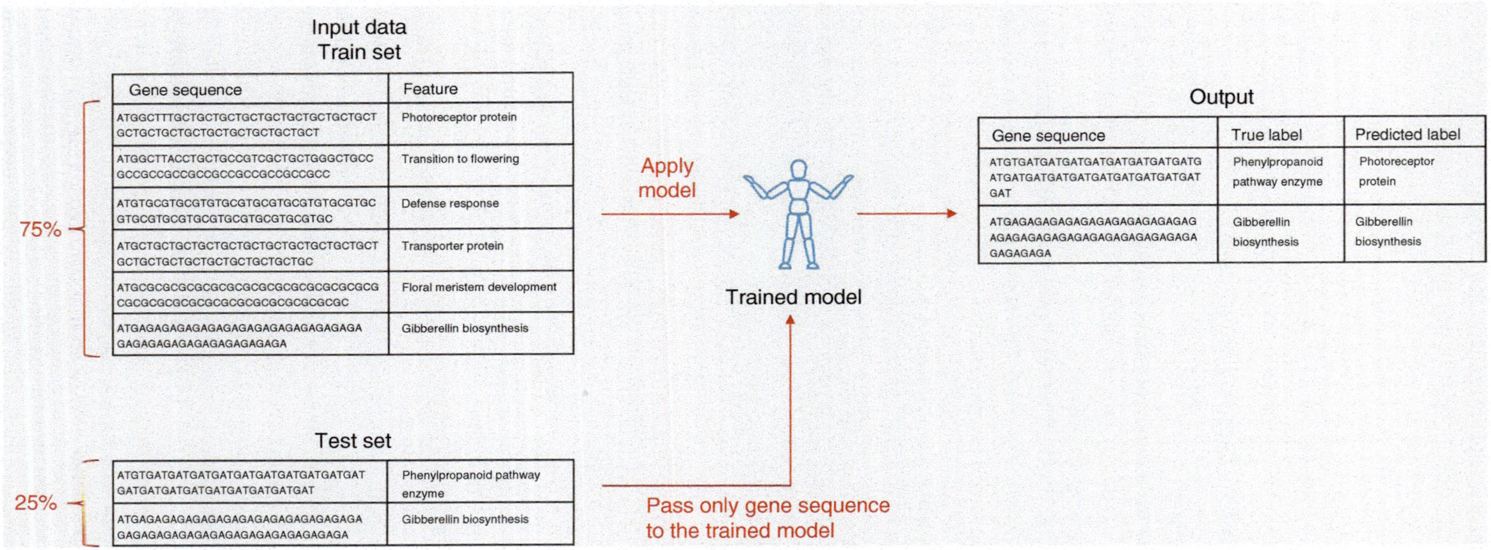

Figure 6.2 Schematic for implementing a machine learning (ML) workflow. A typical workflow for supervised, feature-based learning. An ML algorithm predicts the value of the response variable for each input data point (instance) based on its particular feature (dependent variable) values. We present an example with fictional data.

perspectives of multiple trees (Libbrecht and Noble 2015). This approach mitigates the impact of imperfections or biases in individual trees that may arise from how data points were selected.

An additional advantage of random forests lies in their stability. Compared to a single decision tree, ensemble algorithms like random forests exhibit greater robustness to variations or fluctuations in the dataset. Consequently, any changes in the dataset are less likely to significantly affect the collective predictions of the forest of trees. In conclusion, random forests harness the collective power of multiple decision trees to deliver accurate predictions. Through the averaging of predictions from a diverse set of trees, they offer improved accuracy and stability in predicting the value of *y* for new data points.

6.3.2.3 Principal Component Analysis

Principal component analysis (PCA) is a widely adopted unsupervised algorithm renowned for its efficacy in dimensionality reduction. Its utility spans a range of applications, encompassing tasks such as data visualization, feature extraction, and noise filtration. PCA plays pivotal roles in various domains, including stock market forecasting and gene analysis. Its fundamental objective revolves around identifying correlations among variables. When robust correlations are detected, PCA streamlines dimensionality by pinpointing the directions of maximum variance within high-dimensional data. It then projects this data into a lower-dimensional subspace, all while preserving the utmost vital information.

In PCA, the primary focus lies in the reduction of dimensions in a D-dimensional dataset by projecting it onto a K-dimensional subspace, where K is smaller than D. To encapsulate the essential steps of PCA, the initial procedure often involves standardizing the data. Subsequently, eigenvectors and eigenvalues are computed, followed by their arrangement in descending order. A projection matrix, denoted as W, is then meticulously crafted from the chosen K eigenvectors. Finally, the original dataset is transformed using this projection matrix. It's imperative to distinguish PCA from linear regression, despite superficial similarities. While linear regression is primarily employed for predicting values, PCA delves into comprehending the relationship between X and Y values, a relationship quantified through the determination of a set of principal axes.

6.3.2.4 Deep Learning Algorithms

Deep learning constitutes a subset of ML techniques, centering on the training of artificial neural networks endowed with multiple layers. These networks are inspired by the structure and operations of the human brain, featuring interconnected nodes, or neurons, that engage in the processing and transmission of information through weighted connections. The hallmark of deep learning models lies in their proficiency at learning and extracting intricate characteristics from high-dimensional datasets, encompassing various data types like images, audio, and text.

These models possess the capacity to autonomously acquire hierarchical representations of data by stacking multiple layers of neurons. Each layer progressively discerns more abstract features from the initial input data. This inherent capability empowers deep learning models to capture intricate relationships and patterns within the data, enabling them to make precise predictions or classifications. An eminent advantage of deep learning lies in its adeptness at tackling big data challenges, where the data volume or complexity exceeds the capacity of traditional ML algorithms. Furthermore, deep learning models exhibit strong generalization to novel, unseen data, rendering them valuable across a wide array of applications, including the prediction of gene and protein functions in crop plants (Bonetta and Valentino 2020).

6.3.3 Performance Assessment

Assessing the efficacy of an ML algorithm when applied to a specific dataset stands as a pivotal step in its utilization for data analysis. Numerous error evaluation techniques are at one's disposal to gauge the overall performance of ML algorithms in predicting accuracy. The choice of an appropriate error evaluation method hinges on the nature of the algorithm's outputs, whether they pertain to regression or classification tasks.

In the realm of regression, the most prevalent error evaluation methods encompass the mean squared error (MSE), the root mean squared error (RMSE) (Chai and Draxler 2014; Zhang et al. 2021), and the coefficient of determination (R2). MSE serves as a metric for quantifying the disparities between predicted values and actual values. It involves computing the average of the squared discrepancies between the predicted estimates and the true values (Das et al. 2004; Recchia 2010). A lower MSE signifies a superior alignment between the model and the dataset. In the formula below, Y' denotes the predicted value, Y represents the observed value, and n signifies the total number of observations.

$$\text{MSE} = \frac{\sum_{i=1}^{n}\left(Y' - Y\right)^2}{n}$$

RMSE assesses the goodness of fit of a regression model to the dataset. It is computed by averaging the squared differences between the predicted values and the true values.

$$\text{RMSE} = \sqrt{\frac{\sum_{i=1}^{n}\left(Y' - Y\right)^2}{n}}$$

R^2 quantifies the proportion of data variance accounted for by the model, ranging between zero and one, where higher values signify a superior fit. In the equation, SST stands for the total sum of squares, while SSE represents the sum of squares for the error.

$$R^2 = \frac{\text{SST} - \text{SSE}}{\text{SST}}$$

To assess the model's classification performance on the provided dataset, we compute evaluation metrics. We compare the predicted class of each instance to its actual class, where L denotes the positive class. TP stands for true positive, FP represents false positive, TN signifies true negative, and FN denotes false negative.

- **F1 Score**: This metric represents the harmonic mean of precision and recall, offering a more insightful measure of incorrectly classified cases compared to the accuracy matrix.

$$\text{F1 Score} = \left(\frac{\text{Recall}^{-1} + \text{Precision}^{-1}}{2}\right)^{-1}$$

- **Precision**: This metric represents the proportion of correctly identified positive cases out of all the predicted positive cases. It is particularly valuable when the costs associated with False Positives are significant.

$$\text{Precision} = \left(\frac{\text{True Positive}}{\text{True Positive} + \text{False Positive}} \right)$$

- **Recall**: This metric indicates the fraction of correctly identified positive cases out of all the actual positive cases. It holds particular significance when there is a substantial cost associated with False Negatives.

$$\text{Recall} = \left(\frac{\text{True Positive}}{\text{True Positive} + \text{False Negative}} \right)$$

- **Accuracy**: This metric provides a straightforward measure of correctly identified cases across all classes. It is particularly applicable when all classes carry equal importance.

$$\text{Accuracy} = \left(\frac{\text{True Positive} + \text{True Negative}}{\text{True Positive} + \text{False Positive} + \text{True Negative} + \text{False Negative}} \right)$$

6.4 Experiment

The primary objective of this study is to forecast the functions of a specific set of proteins. This endeavor is closely tied to predicting gene function because genes serve as the blueprints for proteins, and a protein's function is intricately linked to the sequence of amino acids comprising it. Consequently, by forecasting the function of a gene, we can extrapolate the function of the corresponding protein it encodes.

Our approach involved the development of a model, which underwent training using the amino acid sequences of these proteins and additional data. The outcomes of this research endeavor hold the potential to enhance researchers' comprehension of protein functionalities, a crucial aspect in unraveling the intricacies of genetic mechanisms. Furthermore, this knowledge could pave the way for advancements in the creation of new hybrid plants and crops, ultimately benefiting agricultural innovation.

Proteins are primarily involved in a wide variety of activities and play important roles in their structure and function. The protein is a huge molecule consisting of 20 kinds of building blocks, called amino acids. The crops create tens of thousands of proteins and each protein is composed of dozens or even hundreds of amino acids that are arranged in a sequential order. That amino acid sequence determines the protein's 3D structures and conformational dynamics, determining its biological function in turn. It is vital to understand life at the molecular level so that the biological function be accurately assigned to a protein. But the number of functions a protein has, as well as its ability to respond to different partners, may make it hard to assign a function to any specific protein. The cultivation of new hybrid crops in sectors such as medicine and agriculture could be facilitated by more knowledge about the functions associated with proteins.

6.4.1 Dataset Description

GO is a concept hierarchy that describes the biological function of genes and gene products at different levels of abstraction (Ashburner et al. 2000). It is an excellent model for describing the multifaceted nature of protein function. GO is a directed acyclic graph (https://geneontology.org/docs/ontology-documentation/; https://geneontology.org/docs/guide-go-evidence-codes/). The nodes in this graph are functional descriptors (terms or classes) connected by relational ties between them (is_a, part_of, etc.). For example, an is_a relationship is associated with the terms "protein binding activity" and "protein binding activity," but the edge of the graph is often reversed from binding to protein binding. This graph contains three subgraphs (subontologies): molecular function (MF), biological process (BP), and cellular component (CC), defined by their root nodes. Biologically, each subgraph represents a different aspect of the protein's function: what it does on a molecular level (MF), which biological processes it participates in (BP), and where in the cell it is located (CC).

A subset of the protein's function is thus represented by one or several subontologies. These annotations are supported by evidence codes, which can be broadly divided into experimental (e.g. as documented in a paper published by a research team of biologists) and nonexperimental. A computational method is usually the way nonexperimental terms can be inferred. The dataset used in the experiment is obtained from Kaggle (https://www.kaggle.com/competitions/cafa-5-protein-function-prediction/data) provided by the team at Iowa State University.

6.4.1.1 Training and Test Set

All proteins in the training set with a defined term that has been verified by experiments or high-throughput evidence shall be included. The test superset is a set of protein sequences on which the GO terms were predicted. It contains protein sequences and their functions.

6.4.1.2 File Description

Dataset file description.

6.4.1.3 Protein Embeddings for Train and Test Set

To train the model the alphabetical protein sequences in train_sequences.fasta cannot be used directly. They have to be converted into a vector format. Embeddings of the protein sequences are used to train the model. Consider protein embeddings to be similar to word embeddings used to train NLP models.

Protein embeddings are a machine-friendly method of capturing the protein's structural and functional characteristics, mainly through its sequence. One approach is to train a custom ML model to learn the protein embeddings of the protein sequences in the dataset being used. Since this dataset represents proteins using amino-acid sequences which is a standard approach, publicly available pretrained protein embedding models to generate the embeddings can be used. To make data preparation easier, the precalculated protein embeddings found at (https://www.kaggle.com/datasets/sergeifironov/t5embeds) created by Sergei Fironov using the Rost Lab's T5 protein language model are used (https://www.kaggle.com/datasets/sergeifironov/t5embeds).

6.4.2 Training

6.4.2.1 Model Description

Using Tensorflow a five-layer Deep Neural Net model was created. The model was trained on the training dataset using the T5 protein embeddings (Table 6.1).

6.4.2.2 Hyperparameters

The model was trained with hyperparameter tuning using Adam optimizer with a learning rate of 0.001 and the loss function was calculated using Binary Crossentropy.

6.4.3 Result

After training the model on given features (Table 6.2) for five epochs the model produced a binary accuracy of 0.98 and an accuracy of 0.9 on the training dataset. The final prediction data frame consists of Protein Id, GO Term Id, and Prediction (Table 6.3). This result with a probability of finding the GO Term in the Protein Id will describe the biological function of genes and gene products. These results will demonstrate the function a protein performs at the molecular level (MF), which biological processes it participates in (BP), and where in the cell it is located (CC). Having the answers to these questions will help researchers gain insight and perform better experiments (Figure 6.3).

Table 6.1 Sample output prediction result file.

Layer	Unit	Activation function
Batch normalization	1024	—
Dense	512	ReLU
Dense	512	ReLU
Dense	512	ReLU
Dense	1500	Sigmoid

Table 6.2 Description of each file in the dataset.

File	Description
train_sequences.fasta	Amino acid sequences for proteins in the training set
train_terms.tsv	The training set of proteins and corresponding annotated GO terms
train_taxonomy.tsv	Taxon ID for proteins in the training set
go-basic.obo	Ontology graph structure
testsuperset.fasta	Amino acid sequences for proteins on which the predictions should be made
testsuperset-taxon-list.tsv	Taxon ID for proteins in test superset
IA.txt	Information accretion for each term. This is used to weigh precision and recall

Table 6.3 Prediction results.

Seq. no	Protein Id	GO term Id	Predictions
0	Q9CQV8	GO:0005575	0.530618
1	QPCQV8	GO:0008150	0.524857
2	A0A3G2FQK2	GO:0006874	0.311140
3	A0A3G2FQK2	GO:0016887	0.267606

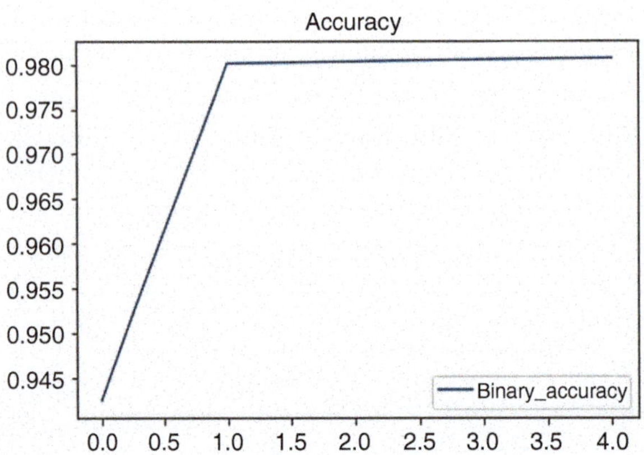

Figure 6.3 Graph demonstrating improvement in accuracy as the epoch increases.

6.5 Case Studies and Success Stories

6.5.1 Highlighting Successful Applications of Machine Learning in Crop Improvement

Classical methods of statistics involve very few parameters and simpler statistics to derive biological feature-based results. ML models on the other hand are processing large amounts of data for HTS and obtaining results that consider many factors at the same time like biological, physiochemical, phytochemical, and other pathways that are involved in vital processes. Let us understand with an example, for improving drought resistance in crops researchers might just look at two things or factors like Yp and Ys but in reality, there are numerous other factors to consider like how the cell functions, what kind of chemicals it has or produces and what is its response to the external environment. These are just to name a few.

ML is assisting in looking at a plethora of information, all at once, for successfully predicting the genetic sequence and associated function that will confer drought resistance for that particular plant species in that particular environment (Niazian and Niedbała 2020). Revisiting the study of IncRNA's role in the trait of plants, Because lncRNAs play a variety of different and context-dependent functions in cellular processes, predicting their activity can be difficult. To develop predictions of possible functions,

ML algorithms can examine trends in lncRNA sequences, structures, and interactions. ML may be applied in this situation as shown here:

Function prediction: ML algorithms may analyze a lncRNA's sequence and compare it to known sequences of other molecules (such as proteins or RNAs) to look for resemblances or patterns that could indicate possible activities. The algorithm may foretell a function in gene regulation, for instance, if the sequence of a lncRNA mimics a recognized regulatory element.

Structure prediction: An lncRNA's secondary structure can be predicted using ML methods based on its sequence. ML helps to reveal patterns where certain structural traits are linked to particular functions. The link between structure and function is not always clear-cut, and it is difficult to anticipate the secondary structure of RNA with accuracy.

Expression patterns: ML has the ability to examine the expression patterns of lncRNAs in diverse tissues or environments. Based on the genes it co-expresses with and whether or not a lncRNA is consistently expressed in a given environment, ML systems may be able to deduce its probable functional functions.

Network analysis: ML is able to examine how lncRNAs interact with other molecules (such as proteins, RNAs, and other compounds) in cellular networks. By examining these relationships, ML systems may be able to foretell whether a lncRNA functions as a network's regulator, scaffold, or another type of functional element.

6.5.2 Examples of Improved Crop Traits Through Gene Function Prediction

In a study that focused on breeding *Triticum aestivum* (wheat), the researcher's objective was to predict the phenotype of the gene using ML. Primitive genome selection models such as "RR-BLUP & GBLUP" are disadvantageous due to their dependence on statistical assumptions that are predefined. Moreover, handling the complexity of genotype-phenotype data and the relation between them was also a barrier to these approaches. Catering to these problems, (Ma et al. 2018), introduced an approach called DeepGS which is a deep CNN-based strategy. This type of approach does not require rules that are predetermined and can work on data that has many details put together in a complicated manner. The study made use of a dataset that included 2000 wheat accessions genotyped with 33,709 DArT markers and measured eight phenotypic traits. The most accurate potential prediction for phenotype was exhibited by the DeepGS model and this was concluded with cross-validation experiments. However, there is a need to deeply understand the biology behind these predictions and use the relevant design of the network to get better results.

With limited resources, it is crucial to make wise decisions about breeding, and deep learning has provided just the way to predict the features of wheat, even before planting. Its ability to deal with multi-dimensional genomic data aids in selecting superior genotypic characters like the height of wheat, content of protein, and size of the grain, to name a few. The wheat varieties bred with this method will meet the dynamic demands of present-time agriculture by reducing lengthy trials in the field and potentially improving global food security with higher yields (Ma et al. 2018).

Another example is a random forest model, a ML method, which was used to forecast the gene for phenotype (disease resistance) in potato plants. The goal of the study was to ascertain how cultivar, a genetic variable, affected the precision of disease resistance prediction. One model that has a cultivar as a predictor and another model without one was designed to calculate the impact of the mentioned cultivar on classification accuracy. The number of predictors randomly chosen at each split (mtry) and the number of trees created to build an ensemble (*n*-tree) were the two key parameters that the study improved for the performance of the random forest model. To estimate out-of-the-box accuracy, this optimization procedure used an 80:20 training/testing data split and tenfold cross-validation *kappa* values, ranging between 0 and 1 which indicates how the model has performed. The higher the *kappa* value, the greater the accuracy of the model applied.

The findings showed that although the potato plants showed some tolerance to late blight, they were not entirely resistant to the disease. According to the study, the variations between the potato cultivars "Katahdin" and "SP951" were particularly noticeable in the early stages of illness, this indicated that different behavior in reaction to disease was due to genetic variability. In the initial phase of infection, the RB gene, which reacts to certain effectors, affects plant responses. Potato breeding programs can greatly benefit from the use of ML techniques, such as the random forest, in identifying gene phenotypes relevant to disease resistance in potatoes. Breeders may choose which cultivars to cross and propagate in order to create new potato varieties with improved resistance to diseases like late blight by precisely measuring the disease resistance of various cultivars (Gold et al. 2020).

6.6 Challenges and Future Directions

6.6.1 Addressing Challenges and Limitations of Machine Learning in Crop Improvement

Although the potential benefits of using ML in research are great, several problems need to be solved before this type of innovation can reach widespread use in production. Massive, high-quality data sets and therefore substantial computing resources are required by ML algorithms. To a certain extent, ML algorithms are difficult to comprehend and interpret, making it hard for researchers who are not experts in the area of ML or large data analysis techniques to rely on results and adapt their strategies accordingly. Moreover, for a significant number of researchers who do not wish to use and incorporate the ML algorithm as part of their study, it remains controversial and demanding to choose the best possible ML algorithmics for that program and its characteristics. There are, in many cases, various types of ML algorithms, and their strengths and weaknesses vary depending on the study (Yoosefzadeh Najafabadi et al. 2023).

6.6.1.1 Limited Availability of Data

There are multiple studies performed on crops however there are limited data that are well structured and ready for performing the training. The datasets available publicly are smaller in size which is not sufficient to train the ML models. To obtain a good result the model needs millions of data to train itself and therefore availability of these datasets plays a very important role. Additionally, the limited data available is not structured and has significant noise in it which is an extra cost incurred to the team in cleansing the noisy data. The data may have missing values or even wrong data which may adversely train the model, the model may learn irrelevant features and even after good accuracy on the train set, the model may drop accuracy on test and real data. The effort from the crop specialist is required to label the missing data which is time-consuming. The specialist is also needed to validate whether the data is authentic and portrays correct values (Yoosefzadeh Najafabadi et al. 2023).

6.6.1.2 Learning Algorithm

It is very important to know what kind of ML models should be applied to the dataset. This choice of deciding the model is not straightforward. Multiple tests are performed on various algorithms which require resources and then finally an algorithm whose accuracy is reasonable is chosen to fine-tune and improve the result. The dataset available is of various structures and characteristics, so deciding which algorithm best suits the prediction is very important. The result of the algorithm depends highly on what algorithm is used. Selecting the wrong algorithm will result in failure and the model being incapable of giving good results (Hesami et al. 2022).

6.6.1.3 Multidisciplinary Expertise Gap

A thorough understanding of plant science, hereditary traits, farming, and ecological science is necessary for crop development. This knowledge is essential for identifying beneficial traits, comprehending the challenges crops confront, and developing effective rearing methods. On the other hand, AI expertise is crucial to combat the pressure of data analysis, prescient demonstration/modeling, and recognition of patterns to further advance crop reproducing procedures. The two areas often have few specialists who are experts in both, which results in a skill gap. This shortage may impede the use of ML in agricultural development initiatives. Strong communication and collaboration between these two groups are crucial, but differences in terminology, strategies, and goals can create barriers (Sumangala and Kini 2022).

6.6.2 Prospects and Potential Advancements in the Field

The application of ML in plant research has a bright future and is expected to increase continuously. Numerous phenotyping data sets have previously been analyzed, interpreted, and linked at various levels between genotypes and phenotypes owing in large part to ML. Technology's significance in plant research is anticipated to grow as it becomes more widely available, with the availability of high-throughput measurement data and developments in ML and AI. As it is clear by now, increased use of ML models will ultimately improve its accuracy and its ability to generate unprecedented results in predicting the function of genes (van Dijk et al. 2021). Certain ways ML should be applied in the future are:

6.6.2.1 ML for Gene Function Prediction in Large Multi-crop Datasets

Traditionally, it was tedious to obtain a sequence of crops therefore, small data sets were available to researchers with that the prediction of gene function was not so accurate. Now, with a publically available large dataset of genetic sequences as well as their associated function. ML models can be trained on these large datasets with data from one species as well as others i.e. multi-crop datasets. The larger the size of the dataset, the greater the diversification and accuracy of the results.

6.6.2.2 Revealing Potential Target for Crop Improvement

Strategic modification of genetic sequences to achieve the desired function in crops that will meet evolving agricultural demands. The predicting power of ML can be used to anticipate specific changes in genetic sequences to draw out better-performing new variety crops. Making sure that these enhanced functions are consistent across generations with minimum to no side effects enables efficient use of resources that ultimately contribute to meeting the food requirement.

6.6.2.3 Novel Gene Function Prediction

Leveraging extensive datasets for training purposes has the potential to revolutionize the way genetic sequences are predicted, marking a significant advancement in this field (Libbrecht and Noble 2015). Furthermore, these robust models offer the capability to anticipate the presence and function of conserved sequences shared among diverse species. To foster collaboration within the scientific plant breeding community, it is imperative to establish a consensus on essential principles encompassing data type, format, and content. This harmonization ensures that data generated by various breeders and institutions are not only compatible but also readily accessible to a wide array of machine-learning algorithms. The overarching objective is to facilitate the creation of substantial datasets conducive to meta-analysis, a crucial step in the progression of ML-driven plant breeding initiatives.

Moreover, we advocate for a collaborative endeavor aimed at seamlessly integrating existing databases, fostering a holistic approach to multi-omics data analysis, research, and advancement within the sphere of plant biology. This strategic initiative is poised to effectively confront the manifold challenges that confront plant breeding programs, charting a course toward groundbreaking solutions.

This innovative approach expedites the breeding process by diminishing the necessity for extensive field trials and laborious manual assessments. Furthermore, the increased utilization of well-trained ML models significantly curbs the margin of human error, which is inevitable when handling extensive multimodal datasets. In addition, it empowers breeders to pinpoint promising candidates for further breeding endeavors, ultimately culminating in the creation of more resilient crop varieties. These developments, in turn, hold the potential to enhance crop yields while reducing reliance on chemical pesticides, thereby promoting sustainable agriculture and bolstering food security (Yoosefzadeh Najafabadi et al. 2023).

References

Acerbi, F., Sassanelli, C., and Taisch, M. (2022). A conceptual data model promoting data-driven circular manufacturing. *Operations Management Research* 15 (3–4): 838–857.

Akinola, T.F., Odiyi, A.C., and Fayeun, L.S. (2019). Genetic variability and genetic diversity of 13 upland rice genotypes for agronomic traits and nutritional qualities. *Journal of Agricultural Science and Botany* 3 (1): 6–11.

Ashburner, M., Ball, C.A., Blake, J.A. et al. (2000). Gene ontology: tool for the unification of biology. *Nature Genetics* 25 (1): 25–29.

Bisht, I.S., Pandey, A., Yadav, M.C., and Rana, J.C. (2015). Population structure of upland red rice (*Oryza sativa* L.) landraces from North-western Indian Himalayas. *IJBT* 14: 42–48.

Bonetta, R. and Valentino, G. (2020). Machine learning techniques for protein function prediction. *Proteins: Structure, Function, and Bioinformatics* 88 (3): 397–413.

Causier, B., Schwarz-Sommer, Z., and Davies, B. (2010). Floral organ identity: 20 years of ABCs. *Seminars in Cell & Developmental Biology* 21 (1): 73–79.

Cavill, R., Jennen, D., Kleinjans, J., and Briedé, J.J. (2016). Transcriptomic and metabolomic data integration. *Briefings in Bioinformatics* 17 (5): 891–901.

Chai, T. and Draxler, R.R. (2014). Root mean square error (RMSE) or mean absolute error (MAE). *Geoscientific Model Development Discussions* 7 (1): 1525–1534.

Cubas, P., Lauter, N., Doebley, J., and Coen, E. (1999). The TCP domain: a motif found in proteins regulating plant growth and development. *The Plant Journal* 18 (2): 215–222.

Das, K., Jiang, J., and Rao, J.N.K. (2004). Mean squared error of empirical predictor. *Annals of Statistics* 32 (2): 818–840. https://doi.org/10.1214/009053604000000201.

Davière, J.M., Wild, M., Regnault, T. et al. (2014). Class I TCP-DELLA interactions in inflorescence shoot apex determine plant height. *Current Biology* 24 (16): 1923–1928.

van Dijk, A.D.J., Kootstra, G., Kruijer, W., and de Ridder, D. (2021). Machine learning in plant science and plant breeding. *Iscience* 24 (1).

Dong, A.Y., Wang, Z., Huang, J.J. et al. (2021). Bioinformatic tools support decision-making in plant disease management. *Trends in Plant Science* 26 (9): 953–967.

Fabres, P.J., Collins, C., Cavagnaro, T.R., and López, C.M.R. (2017). A concise review on multi-omics data integration for terroir analysis in *Vitis vinifera*. *Frontiers in Plant Science* 8: 01065.

Fróna, D., Szenderák, J., and Harangi-Rákos, M. (2019). The challenge of feeding the world *Sustainability* 11 (20): 5816.

Gayen, D., Gayali, S., Barua, P. et al. (2019). Dehydration-induced proteomic landscape of mitochondria in chickpea reveals large-scale coordination of key biological processes. *Journal of Proteomics* 192: 267–279.

Gold, K.M., Townsend, P.A., Herrmann, I., and Gevens, A.J. (2020). Investigating potato late blight physiological differences across potato cultivars with spectroscopy and machine learning. *Plant Science* 295: 110316.

Hesami, M., Alizadeh, M., Jones, A.M.P., and Torkamaneh, D. (2022). Machine learning: its challenges and opportunities in plant system biology. *Applied Microbiology and Biotechnology* 106 (9–10): 3507–3530.

Hinnisdaels, S., Mouras, A., Salesses, G. et al. (1992). Translocation events demonstrated by molecular, *in situ* hybridization and chromosome pairing analyses in highly asymmetric somatic hybrid plants. *Transgenic Research* 1: 170–176.

Islam, M.Z., Khalequzzaman, M., Prince, M.F.R.K. et al. (2018). Diversity and population structure of red rice germplasm in Bangladesh. *PLoS One* 13 (5): e0196096.

Khush, G.S. (2001). Challenges for meeting the global food and nutrient needs in the new millennium. *Proceedings of the Nutrition Society* 60 (1): 15–26.

Lewis, J.M., Mackintosh, C.A., Shin, S. et al. (2008). Overexpression of the maize *Teosinte Branched1* gene in wheat suppresses tiller development. *Plant Cell Reports* 27: 1217–1225.

Li, R., Fu, D., Zhu, B. et al. (2018). CRISPR/Cas9-mediated mutagenesis of lncRNA1459 alters tomato fruit ripening. *The Plant Journal* 94 (3): 513–524.

Libbrecht, M.W. and Noble, W.S. (2015). Machine learning applications in genetics and genomics. *Nature Reviews Genetics* 16 (6): 321–332.

Listgarten, J., Lippert, C., Kadie, C.M. et al. (2012). Improved linear mixed models for genome-wide association studies. *Nature Methods* 9 (6): 525–526.

Lodish, H., Berk, A., Zipursky, S.L. et al. (2000). *Molecular Cell Biology*, 4e. New York: WH Freeman.

Ma, W., Qiu, Z., Song, J. et al. (2018). A deep convolutional neural network approach for predicting phenotypes from genotypes. *Planta* 248: 1307–1318.

Mahood, E.H., Kruse, L.H., and Moghe, G.D. (2020). Machine learning: a powerful tool for gene function prediction in plants. *Applications in Plant Sciences* 8 (7): e11376.

Martin, G., Baurens, F.C., Hervouet, C. et al. (2020). Chromosome reciprocal translocations have accompanied subspecies evolution in bananas. *The Plant Journal* 104 (6): 1698–1711.

Mathure, S., Jawali, N., and Nadaf, A. (2010). Diversity analysis in selected non-basmati scented rice collection. *Rice Science* 17 (1): 35–42.

Mba, C., Guimaraes, E.P., and Ghosh, K. (2012). Re-orienting crop improvement for the changing climatic conditions of the 21st century. *Agriculture & Food Security* 1: 1–17.

Mbanjo, E.G.N., Jones, H., Caguiat, X.G.I. et al. (2019). Exploring the genetic diversity within traditional Philippine pigmented rice. *Rice* 12 (1): 1–18.

Miflin, B. (2000). Crop improvement in the 21st century. *Journal of Experimental Botany* 51 (342): 1–8.

Niazian, M. and Niedbała, G. (2020). Machine learning for plant breeding and biotechnology. *Agriculture* 10 (10): 436.

Recchia, A. (2010). R-squared measures for two-level hierarchical linear models using SAS. *Journal of Statistical Software* 32: 1–9.

Roy, S.C. and Shil, P. (2020). Assessment of genetic heritability in rice breeding lines based on morphological traits and caryopsis ultrastructure. *Scientific Reports* 10 (1): 7830.

Salunkhe, A., Tamhankar, S., Tetali, S. et al. (2013). Molecular genetic diversity analysis in emmer wheat (*Triticum dicoccon* Schrank) from India. *Genetic Resources and Crop Evolution* 60: 165–174.

Schilling, S., Pan, S., Kennedy, A., and Melzer, R. (2018). MADS-box genes and crop domestication: the jack of all traits. *Journal of Experimental Botany* 69 (7): 1447–1469.

Schwarz-Sommer, Z., Huijser, P., Nacken, W. et al. (1990). Genetic control of flower development by homeotic genes in *Antirrhinum majus*. *Science* 250 (4983): 931–936.

Singh, R.K., Sood, P., Prasad, A., and Prasad, M. (2021). Advances in omics technology for improving crop yield and stress resilience. *Plant Breeding* 140 (5): 719–731.

Subba, P., Narayana Kotimoole, C., and Prasad, T.S.K. (2019). Plant proteome databases and bioinformatic tools: an expert review and comparative insights. *OMICS: A Journal of Integrative Biology* 23 (4): 190–206.

Sumangala, N. and Kini, S. (2022). Impact of adopting machine learning methods on Indian Agriculture Industry-a case study. *International Journal of Case Studies in Business, IT and Education (IJCSBE)* 6 (2): 446–458.

Takeda, S. and Matsuoka, M. (2008). Genetic approaches to crop improvement: responding to environmental and population changes. *Nature Reviews Genetics* 9 (6): 444–457.

Vaschetto, L.M. (2017). Understanding the role of protein interaction motifs in transcriptional regulators: implications for crop improvement. *Briefings in Functional Genomics* 16 (3): 152–155.

Wang, Q., Tian, F., Pan, Y. et al. (2014). A SUPER powerful method for genome wide association study. *PLoS One* 9 (9): e107684.

Wang, Q., Tang, J., Han, B., and Huang, X. (2020). Advances in genome-wide association studies of complex traits in rice. *Theoretical and Applied Genetics* 133: 1415–1425.

Webb, R. and Buratini, J. (2018). Global challenges for the 21st century: the role and strategy of the agri-food sector. *Animal Reproduction (AR)* 13 (3): 133–142.

Xu, Y., Zhang, X., Li, H. et al. (2022). Smart breeding driven by big data, artificial intelligence, and integrated genomic-enviromic prediction. *Molecular Plant* 15 (11): 1664–1695.

Yadav, M.K. and Chand, P. (2018). Assessment of genetic diversity among twenty Indian wheat (*Triticum aestivum* L.) cultivars using simple sequence repeat (SSR) markers. *International Journal of Current Microbiology and Applied Sciences* 7 (3): 1708–1717.

Yoosefzadeh Najafabadi, M., Hesami, M., and Eskandari, M. (2023). Machine learning-assisted approaches in modernized plant breeding programs. *Genes* 14 (4): 777.

Yoosefzadeh-Najafabadi, M., Torabi, S., Tulpan, D. et al. (2021). Genome-wide association studies of soybean yield-related hyperspectral reflectance bands using machine learning-mediated data integration methods. *Frontiers in Plant Science* 12: 777028.

Zhang, N., Gupta, A., Chen, Z., and Ong, Y.S. (2021). Evolutionary machine learning with minions: a case study in feature selection. *IEEE Transactions on Evolutionary Computation* 26 (1): 130–144.

Zhou, X. and Stephens, M. (2012). Genome-wide efficient mixed-model analysis for association studies. *Nature Genetics* 44 (7): 821–824.

7

Omics and Bioinformatics Approaches for Abiotic Stress Tolerance in Plants

Santanu Samanta[1] *and Aryadeep Roychoudhury*[2]

[1] *Post Graduate Department of Biotechnology, St. Xavier's College (Autonomous), Kolkata, West Bengal, India*
[2] *Discipline of Life Sciences, School of Sciences, Indira Gandhi National Open University, New Delhi, India*

7.1 Introduction

As sessile organisms, plants are affected by several environmental constraints throughout their life cycle. These environmental factors are broadly categorized into two main groups: biotic and abiotic stress. Biotic stress encompasses mainly pests and pathogens such as fungi, bacteria, viruses, nematodes, and insects. However, in the present scenario, abiotic stress factors are the foremost limiting factors for declining agricultural productivity and crop yield (Qin et al. 2011; Grayson 2013). The most abundant abiotic stresses comprise salinity, limitation in water, temperature fluctuations (high and low), excess metal/metalloid, ultraviolet (UV), and flooding (hypoxia), responsible for the large gap existing between the potential yield and final harvest yield worldwide. As much as 80% of the crop harvest can be destroyed by these factors (Minhas et al. 2017; Oshunsanya et al. 2019). It is expected that both the intensity and duration of these abiotic stress factors will become the prime reason for delimiting the productivity of standing crops in the near future with the rapid climate change as evident from the reports by the Intergovernmental Panel on Climate Change (IPCC 2021). Therefore, understanding plant resistance is critical which is considered as a complex phenomenon involving a convoluted network of physiological, cellular, and molecular synchronization and is also difficult to untangle (Pascual et al. 2022).

For this purpose, it is imperative to gain knowledge of how plants perceive environmental signals and modulate adaptive responses under hostile environment. This has long been the focal point of extensive and rigorous studies by researchers. Using conventional methods in this regard is less effective due to limited success and requires effective supplements to feed the growing food demands worldwide. In recent times, a rapid rise in "omics"-based technologies has acquired significant attention in the area of agricultural research (Figure 7.1) (Roychoudhury et al. 2011; Samanta and Roychoudhury 2023). These high-throughput (HT) methodologies have proved persuasive for elucidating the alteration at the genome, transcriptome, proteome, metabolome, and various other levels required for gaining insights into

Bioinformatics for Plant Research and Crop Breeding, First Edition. Edited by Jen-Tsung Chen.

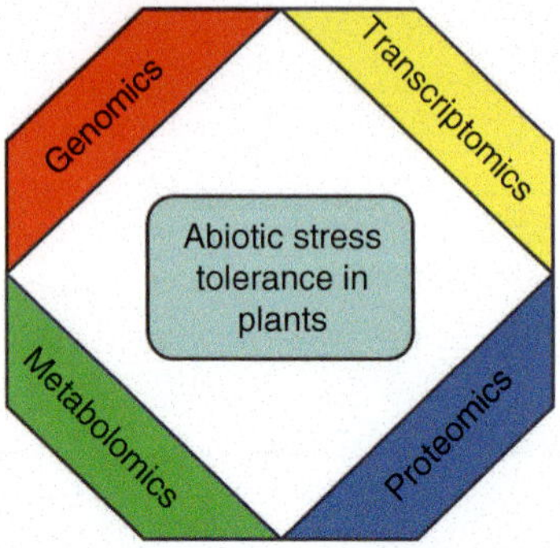

Figure 7.1 Different branches of omics to understand abiotic stress responses in plants.

the plant systems for real-time monitoring of the cellular processes (Soni et al. 2015; Bagati et al. 2018). Finally, integration of all the data obtained by different omics platforms into systems biology through bioinformatics approaches will be helpful in identifying novel genes or pathways. In this chapter, we summarize the recent progress of various omics approaches that have been discussed to provide a system-level understanding of how abiotic stress modulation plays a promising role in designing high-yielding, stress-resilient crop cultivars, and sustainable agricultural practices in the face of changing environmental conditions.

7.2 Genomic Approaches

Genomics is a branch of "omics" that deals with the study of the complex biological functions of the genetic material of a cell (genome) by using HT technologies. The advancement of genomic research, upscaled via the innovation of DNA sequencing technologies, rapidly produces huge amounts of sequence information in a short period of time. These advancements have escalated a paradigm shift from a single gene to entire genomes, which will provide a valuable resource for the identification of key stress-responsive gene pools for understanding the complexity of stress-responsive mechanisms in plants.

7.2.1 Structural Genomics

Structural genomics generally aims to delineate the structural repertoire of the genomes, aiming to identify, locate, and order genomic features (Skolnick et al. 2000). Several technological developments have been made in the structural genomics field that serves as the foundation for HT biological implications in genomics-based superior crop cultivation for enduring harsh environmental conditions.

7.2.1.1 Genome Sequencing

Genome sequencing involves detecting the arrangement of bases entirely along the genome of an organism. The progression of sequencing technologies can be divided into three generations, with Sanger's enzymatic dideoxy DNA sequencing method (sequencing by synthesis) (Zimmermann et al. 1989) and Maxam–Gilbert's chemical degradation chain termination method (sequencing by cleavage) (Maxam and Gilbert 1977) in 1977 constituting the "first-generation method." These two methods provided the foundation for the development of other sequencing platforms. To overcome the drawbacks of first-generation DNA sequencing, the first HT next-generation sequencing (NGS) was introduced by 454 Life Sciences (Roche) into the market in 2005 (Ansorge 2009). Later on, a number of NGS technologies are available, such as SOLiD sequencing (by Applied Biosystems) and Illumina sequencing (Solexa) platforms, which are collectively considered as "second-generation HT-NGS technology" methods (Bennett 2004; Bennett et al. 2005; Margulies et al. 2005). These HT-NGS platforms are able to sequence whole plant genomes in a short period of time (Bubnoff 2008; Mardis 2008). Although second-generation techniques have revolutionized genomics research and made possible the production of massively parallel data, it includes some limitations too. Therefore, overcoming those issues has given birth to "third-generation HT-NGS technology" based on sequencing from a single DNA molecule. More recently, Heliscope single molecule sequencer (by Helicos), nanopore DNA sequencer, PacBio RS (Pacific Biosciences) for single molecular sequencing, and Ion Torrent (Life Technologies) sequencing technology based upon semiconductor chip, revolutionized the genomics research to the next level (Milos 2008; Eid et al. 2009; Rothberg et al. 2011). The advantages including faster turnaround time, longer read length, small amounts of starting materials, lower cost, and higher consensus accuracy to enable rare variety detection made this technology highly popular among genomic researchers (Verma et al. 2017).

7.2.1.2 Molecular Mapping

For breeding efforts to improve abiotic stress tolerance in commercial cash crops, marker-assisted selection (MAS) is a great asset for plant breeders to attain this goal (Wani et al. 2018). Therefore, quantitative trait loci (QTL)-mapping studies are obvious for gene introgression or finding new or rare alleles with a trait of interest. These QTLs will not only contribute to the understanding of plant-environment interaction, but also to hasten the stable crop production that is more tolerant to multiple abiotic stresses (Gyenis et al. 2007; Rao et al. 2016). Mapping refers to putting markers (genes of interest) in order, which indicates the relative distances among them, and assigning them to their linkage groups on the basis of their recombination values from all pairwise combinations. Moreover, molecular mapping also allows researchers to construct a physical map of the whole genome, which is a vital tool for locating genomic regions controlling complex traits. On the basis of the mapping population used, molecular mapping is of two types: linkage mapping and association mapping. In linkage mapping, bi-parental segregating populations are utilized and considered to be useful so far for identifying major genes and mapping QTLs (Komatsuda et al. 2007).

7.2.1.3 Molecular Markers

The extensive use of molecular markers has been employed in plant breeding and inherited programs. A molecular marker is known as a fragment of DNA that is closely related to biological traits determined by allelic loci of genes (Younis et al. 2020). Molecular markers can be of two different types: classical markers and DNA-based markers. The classical markers consist of cytological markers, biochemical markers as well as morphological markers and were mainly used for traditional breeding programs in plants (Wani et al. 2018). Recently, efforts were made to develop different DNA-based markers including simple sequence repeats (SSRs), inter-simple sequence repeats (ISSRs), sequence characterized amplified regions (SCAR), sequence-tagged site (STS), cleaved amplified polymorphic sequence (CAPS), single nucleotide polymorphism (SNP), etc. Due to possessing numerous potential advantages over classical markers, the rapid utilization of DNA-based molecular markers speeds up the molecular-assisted breeding (MAB) programs (Oliveira et al. 2008; Kage et al. 2016). The DNA-based molecular markers can be applied to MAS in molecular breeding programs that exploit the genetic correlation between target gene(s) and respective phenotypes (Wei et al. 2009; Mantri et al. 2010). Of many kinds of markers, SSRs or microsatellites are the most familiar markers for genetic mapping studies (Varshney et al. 2005a). Molecular markers, specifically DNA-based molecular markers have been deployed to screen molecular diversity in abiotic stress tolerance (Roy et al. 2011). Such strategic marker-assisted breeding can quicken the development of the tolerant cultivars to withstand abiotic challenges. Identifying the molecular markers linked with desirable traits and enhancing MAS now made it possible to examine their usefulness in crop improvement with a special focus on stress endurance (Kage et al. 2016; Hasan et al. 2021).

7.2.2 Functional Genomics

Investigating the functional activity of genes by exploiting information gathered from structural genomics is known as functional genomics. The major aim of this genome-wide effort is to study gene functions in an HT manner for linking the genome with its corresponding phenome (Fields et al. 1999). Functional genomics is now a highly preferable tool for dissecting abiotic stress responses in cereal crops to unravel complex networks of stress perception and signal transduction to decipher improved/novel varieties for future outcomes in agricultural yield and productivity (Bouchez and Höfte 1998). The technologies related to functional genomics mostly utilize either sequence-based approaches or hybridization-based approaches which are discussed below.

7.2.2.1 Expressed Sequence Tags (ESTs)

Currently, plant researchers are utilizing expressed sequence tags (ESTs) that represent the most abundant nucleotide commodity from plant genomes (Jongeneel 2000; Dong et al. 2005). Since they represent the expressed portion of genomes, ESTs have proven to be a sequence resource that can be exploited for novel genes (Nagaraj et al. 2006). In the initial days, EST sequencing favored the 5′ end due to the presence of more protein-coding regions than the 3′ end. Later on, with significant improvements in the cDNA preparations along with the evolution of HT sequencing, 3′ end of a cDNA clone is often preferred owing to large numbers of untranslated regions (UTRs) which would help discriminate between gene paralogues. HT-EST sequencing can also be generated from both the 5′ or 3′ end of a cDNA clone (Rudd 2003; Parkinson and Blaxter 2009). The use of large-scale EST sequencing can also be used to circumvent complete plant genome sequencing. They therefore provide a suitable, low-cost alternative to full-genome sequencing, particularly for eukaryotic organisms that tend to have larger genome sizes (Mayer and Mewes 2002). ESTs are particularly attractive for novel sources of markers since they are physically associated with coding regions of the genome (Kantety et al. 2002; Varshney et al. 2005b). At present, there are >1.5 million ESTs available for dicotyledonous plants and >2 million ESTs for monocotyledonous plants available. The robust applications of the EST database (dbEST) prepared from complete genome sequence information of *Arabidopsis* and rice along with various other commercially grown crops provide a valuable resource for identifying or profiling genes expressed in various tissues, cell types, or developmental stages for developing multiple abiotic stress tolerance (Rudd 2003; Wani et al. 2010).

7.2.2.2 Serial Analysis of Gene Expression (SAGE)

Simultaneous quantification of a large number of transcripts can be accomplished by SAGE. SAGE is an HT sequencing-based genomic technology that aids highly accurate, quantitative analysis of short sequence tags (10–11 bp) derived from a defined location within each transcript (Velculescu et al. 1995). It is an extremely competent technology that provides a global gene expression profile. SAGE can detect even slight differences in expression levels between the samples. The major advantage of SAGE over microarray technology is that it does not require prior knowledge of gene sequences and therefore this technique can be useful for the identification of unknown transcripts (thus not cloned), whose genomes are largely uncharacterized. SAGE requires a less sophisticated device, which is another advantage of SAGE (Matsumura et al. 1999). Moreover, the portability of data generated through SAGE is an added advantage of SAGE over microarray. Despite having some advantages, it cannot be used for the analysis of small tissue samples, which is considered a disadvantage of SAGE. However, with recent modifications of this technique, in MicroSAGE, 500–5000-fold less material can be used than the original method (Datson et al. 1999). Another variant of SAGE, known as CAGE (cap analysis of gene expression) is based on sequencing tags from the 5′ end of an mRNA transcript only (Shiraki et al. 2003). Therefore, the technique is suitable for the identification of transcription start site which can be useful for promoter analysis.

7.2.2.3 Massively Parallel Signature Sequencing (MPSS)

Massively parallel signature sequencing (MPSS) is a refinement of SAGE that enables the simultaneous in-depth quantification of millions of transcripts on a genome-wide scale. MPSS, such as ESTs and SAGE, also utilizes tag-based arrays and can identify genes with higher efficiency and greater sensitivity due to ultra-HT analysis and longer tags (Brenner et al. 2000; Pollock 2002). In addition to identifying genes efficiently, MPSS provides quantitative expression information and allows the identification of novel transcripts with high sensitivity, particularly in non-model plants or crop species that lack whole genome sequences (Reinartz et al. 2002). MPSS has also been used to detect small RNA expression analysis (Vaughn and Martienssen 2005). Unlike microarray, MPSS has overcome the issues related to array design, cross-hybridization, and reproducibility, which provides this assay with wide coverage and dynamic range (Reinartz et al. 2002). Moreover, it provides a more robust means to assess transcript profiling in an mRNA population with no bias toward the transcriptome of certain tissues or cells. Expression data for several genotypes of commercially important crops such as rice, corn, and soybean is publicly available in the plant MPSS database (Nakano et al. 2006).

7.3 Transcriptomics Approaches

Transcriptomics is the complete set of RNA transcripts (mRNA, rRNA, tRNA, long ncRNA, miRNA, siRNA, circular RNA) in a specific cell type or a tissue at a certain time point. Transcriptomics mainly focuses on detailed and/or comparative analysis of quantitative genome-wide information responsible for the alteration of phenotypes of cells or populations of cells or entire organisms (Dong and Chen 2013). In other words, the transcriptomic study represents the transcriptional modifications at a particular developmental stage of plants and their ability to respond to environmental cues. With the introduction of next-generation HT sequencing technology, transcriptomic studies have made a widespread advancement in unraveling the RNA-based gene regulatory network.

7.3.1 Microarray and Gene Chip

Initial transcriptomic studies are largely based on hybridization-based microarray technologies, which have been one of the breakthrough technologies, since their first use in global gene expression profiling. It provides a systematic analysis of expression profiles on a genome-wide scale while processing a large number of genes at a time (Vij and Tyagi 2007). It has surpassed any other technologies in terms of its speed, accuracy, and comprehensiveness. At present, two types of microarrays are available: fragment-based cDNA microarrays and oligonucleotide-based chips. cDNA microarray allows simultaneous hybridization of two fluorescently labeled probes to an array of immobilized chemicallybound DNA fragments that form a duplex (Gul et al. 2016). Probes are the stretches of gene-specific DNA attached to a solid surface (slide or chip) designed for hybridization with target DNA or RNA strands in solution. Cyanidin dyes, such as Cy3 and Cy5, are commonly exploited to label the target. Measurement of intensities of fluorescence signals from two dyes that appear after laser scanning reflects the abundance and differential expression of the corresponding mRNA in the sample. On the contrary, Affymetrix chips (GeneChip) is a specialized microarray that utilizes *in silico* synthesized oligonucleotides (usually 20–25 bp) as probes complementary to the 3′ end of an expressed sequence in a genome (Lipshultz et al. 1999). Oligonucleotides are generally synthesized *in situ* on a glass surface using photolithographic masks to measure the correct sequence (Öktem et al. 2008; Seki et al. 2010). Significant improvement in the oligonucleotide/GeneChip microarray has analyzed whole-genome expression possible. Because of their broad gene coverage, oligonucleotide microarray is massively used to study the gene expression at genomic and transcriptomic levels in almost all model plants (*Arabidopsis*, tobacco) and commercially important cash crops (rice, wheat, barley, soybean, maize, potato, tomato) (Zhu and Wang 2000; Hunter et al. 2002; Zhu et al. 2003). Both cDNA and oligonucleotide microarray have given rise to rapid progress in identifying different genes, signaling determinants and associated pathways recruited to plant stress-related response under various abiotic stresses (Janská et al. 2010; Takehisa et al. 2013; Chen et al. 2014; Kim et al. 2014; Zhang et al. 2014; Wu et al. 2015) which will expedite our understanding toward complex signaling mechanisms in plants.

7.3.2 RNA Sequencing (RNA-seq)

RNA-seq technology is a combination of HT sequencing with sophisticated computational software to quantify whole transcriptome present in an RNA extract with high resolution (Wang et al. 2009). These technological advancements eliminated several difficulties posed by hybridization-based microarrays that were previously exploited for gene expression analysis. The starting input amount of RNA is much lower in RNA-seq with respect to microarray (Ozsolak and Milos 2011). RNA-seq has adapted to reveal the key aspects of RNA configuration, RNA–RNA, and RNA–protein interaction, which could illuminate regulatory networks and molecular insights that regulate cell fate, development, and modifications under varying environment (Ozsolak and Milos 2011; Stark et al. 2019). The RNA-seq technology uses mainly two methods: short-read cDNA sequencing, long-read cDNA sequencing, and long-read direct RNA sequencing. Nearly all RNA-seq methods (>95%) have been developed for short-read cDNA sequencing due to its cheaper and easier implementation along with its potential to generate comprehensive, high-quality data (currently 100–1000 times more reads per run as compared to long-read platforms) across the transcriptome (Su et al. 2014). Long-read cDNA sequencing enables single-molecule sequencing of complete individual RNA molecules after conversion to cDNA. This technique is particularly suited for isoform discovery, *de novo*, and other complex transcriptome analyses. The advantage of direct cDNA sequencing removes PCR bias, leading to high-quality results. In contrast, long-read direct RNA sequencing technology is able to sequence RNA directly and bypass reverse transcription steps (Garalde et al. 2018). Despite having some distinct advantages of long-read technology over short-read sequencing for evaluating transcriptome, they have some serious limitations as well. The long-read technologies can read only 10^6–10^7 reads per run currently (low throughput), while short-read sequencing can generate 10^9–10^{10} short reads (Stark et al. 2019). The dramatic growth of RNA-seq-based transcriptome profiling allows us to investigate the molecular basis of adaptation under various abiotic stress conditions (Shinde et al. 2020; Kumar et al. 2021; Hosseini et al. 2023; Sahoo et al. 2023). The additional benefit of the RNA-seq platform is that it can be employed for species without sequence information or with poor genome annotations.

7.4 Proteomics Approaches

In the post-genomic era, proteomics research has flourished with huge advances in accurate, sensitive, reliable, and reproducible analytical techniques (Monteoliva and Albar 2004). Recently, proteomics approaches have been utilized to investigate the complexity of plant–environment interaction along with the candidate genes responsible for genetic modifications to adapt under suboptimal conditions (Gong et al. 2015). In contrast to a static genome, proteome is more complex and dynamic in nature. For this purpose, researchers have developed various techniques to acquire a coordinated perspective of whole proteome coverage for the identification of protein biomarkers underlying phenotypic variation associated with plant abiotic stress tolerance.

7.4.1 Gel-based Techniques

Gel-based proteomics is the most widely used and versatile method of global protein separation and quantification. Despite having some limitations, this technique has been traditionally used in plant proteomics due to its easy-to-operate and lower cost as compared to gel-free proteomic approaches.

7.4.1.1 Two-dimensional Gel Electrophoresis (2-D GE)

2-D GE of proteins is one of the oldest and most widely utilized methods for protein separation tools in proteomics (Rabilloud et al. 2010). It is an orthogonal separation technique in which proteins are separated by two different physicochemical properties: (i) isoelectric point (pI) and (ii) molecular weight (MW). In the first dimension, proteins undergo isoelectric focusing (IEF) based on their net charge at different pH values along a pH gradient. Later on, the gel is transferred to the second dimension, where denatured proteins are separated according to their MW using sodium dodecyl sulfate (SDS)–polyacrylamide gel electrophoresis (PAGE). For the visualization of protein spots, colloidal Coomassie Blue or fluorescent dyes are generally used for good compatibility with mass spectrometry (MS) for protein identification (Garfin 2003). It can separate up to 10,000 protein spots on one gel (Ong and Pandey 2001). The major advantage of 2-D GE is the robustness of the technique (Rabilloud and Lelong 2011). This technique is inefficient in the analysis of proteins of (i) low abundance, (ii) extreme pI (<4 or >9) and masses (<15 or >200 kDa), and (iii) membrane or modified proteins due to poor solubility in the buffer solution and limited detection range (Gygi et al. 2000; Rabilloud et al. 2010). Despite having some difficulties, it is presently unparallel in its ability to separate and array complex mixtures of proteins by the scientific community.

7.4.1.2 Two-dimensional Difference in Gel Electrophoresis (2-D DIGE)

The introduction of 2-D DIGE has brought substantial improvement in the field of proteomics. It is an advancement of the classical 2-D GE technique with the ability to multiplex fluorescently labeled proteins on the same gel, thereby minimizing gel-to-gel variation (Alban et al. 2003; May et al. 2012). There are two types of fluorescent dyes available: Cy Dye minimal dyes (e.g. Cy2, Cy3, Cy5) and Refraction-2D G-Dyes (e.g. G-Dye100, G-Dye 200, G-Dye300). Both of these dyes react with the ε-amino group of lysine residues of samples (Beckett 2012). The fluorescent dyes used for the methodology possess similar molecular masses and are positively charged to match the charge for the replacement of lysine residues but with different excitation and emission spectra (Pasquali et al. 2017). It provides higher sensitivity and large-scale screening coverage with respect to normal protein staining methods, in association with a higher linear range of quantification. Recently, 2D-DIGE is extensively used to detect subtle changes in proteome analysis under abiotic stress responses like salinity in wheat (Gao et al. 2011), drought stress in wheat (Faghani et al. 2015), thermal stress in soybean (Das et al. 2016), chilling and freezing stress in *Pisum sativum* (Grimaud et al. 2013) and cadmium stress in *Medicago sativa* (Gutsch et al. 2018).

7.4.2 Gel-free Techniques

To fill the gap left by gel-based proteomic techniques, developmental endeavors have been focused on alternative approaches which resulted in the emergence of shotgun proteomic approaches. Gel-free proteomics offers various advantages over gel-based approaches, in terms of wide coverage, HT analyses of the proteome, and better resolution of protein separation of an organelle or cell types.

7.4.2.1 Multidimensional Protein Identification Technology (MudPIT)

MudPIT is a shotgun proteomic approach to analyze highly complex samples based on its robust sensitivity for large-scale proteome analysis. It can detect medium to low-abundance proteins by using a combination of multiple separation steps. The method relies on digestion in a solution of the protein mixture to be analyzed and the final separation of complex protein mixture via a bi- or tri-phasic microcapillary column, followed by mass spectrometric analysis. The first separation is generally aided by a cation exchanger based on charge distribution or molar mass. Afterward, the second separation is mediated by reverse-phase chromatography based on hydrophobicity (Yates 2004; Nägele et al. 2004; Mauri and Scigelova 2009). Using this technique, whole proteome, differential protein expression under variable growth conditions is made easy for the researchers (Washburn et al. 2003; Pan et al. 2004). Moreover, membrane proteins and large multi-protein complexes can also be quantified (Sato et al. 2004). The rapid, unbiased, and enhanced separation capability and sensitivity made this technology suitable for wide proteome coverage. The shorter requirement of the initial sample and lesser operational time for protein separation are also the added advantages of this method. The inability to provide information on protein isoforms and posttranslational modifications (PTMs) are some potential limitations of this technique.

7.4.2.2 Isotope-coded Affinity Tags (ICAT)

ICAT is the label-based (chemical labeling), gel-free proteomics approach widely utilized in quantitative proteomics to evaluate the abundance of expressed proteins with multiplexing capability (Shiio and Aebersold 2006). Three functional elements are necessary for each ICAT reagent: a specific reactive group, an isotopically coded light or heavy linker, and a biotin tag. For the experiment, protein samples are labeled with two different isotope tags generated by using linkers comprised of either eight deuterium atoms (heavy reagent) or eight hydrogen atoms (light isotope). Thereafter, light and heavy isotope-coded proteins are pooled and proteolytically cleaved into peptide fragments. The complex mixtures are separated through a multistep chromatographic separation step and the resulting peptides are identified with mass spectrometric (MS) analysis (Smolka et al. 2002; Tureček 2002). The difference in eight mass units is detected in MS analysis and relative quantification of peptides is inferred from ratios between intensities of peak (Sethuraman et al. 2004; Fu et al. 2008). This technique is suitable for large-scale screening for the detection of low-abundant proteins. ICAT generally possesses less protein sequence coverage than classical 2-D GE, which is a major disadvantage of this method. Since ICAT targets cysteine containing peptides, this technique can be exploited for quantitative analysis of redox proteome (Kozarova et al. 2007; Thamsen and Jakob 2011) and organelle proteomes using cellular organelle fractionation along with investigating the localization of proteins (Dunkley et al. 2004).

7.4.2.3 Isobaric Tags for Relative and Absolute Quantification (iTRAQ)

iTRAQ is also a label-based (chemical labeling) proteomic approach. The tags used in iTRAQ are a set of isobaric reagents that are amine-specific and primarily designed for the labeling of peptides rather than proteins (Ross et al. 2004). When iTRAQ-tagged peptides are subjected to MS/MS fragmentation, the unique reagents used for four reporter groups give rise to distinct isotope-encoded fragments, which can be utilized for the quantification of up to four samples (Wiese et al. 2007). Due to four different iTRAQ reagents being available, comparative analysis of up to four samples is feasible within a single MS run (Aggarwal et al. 2006; Evans et al. 2012). This technique can be useful in studying quantitative proteomics including post translational modifications (PTMs), such as phosphorylation, acetylation, methylation, fucosylation, and o-nitrosylation (Rosenzweig et al. 2008; Chiappetta et al. 2009; Wu et al. 2010; Hoffert et al. 2011). This platform represents a shotgun proteomic approach and can be applicable for a wide range of proteins including extreme pI and MW, even for low abundance proteins, which made this technique highly popular in the proteomics field. In addition, iTRAQ can be applicable for both cysteine and non-cysteine-containing peptides (Zieske 2006). Nevertheless, the technique involves multiple steps, expensive labeling reagents and very strict guidelines that need to be followed for a successful experiment. The potency of iTRAQ was widely employed for better understanding mechanisms of plant tolerance to study salinity in soybean (Ji et al. 2016), drought in maize (Ren et al. 2022), high temperature in *Brassica juncea* (Cheng et al. 2020), cold stress in watermelon (Shi et al. 2019) and waterlogging in wheat (Yang et al. 2022).

7.4.2.4 Stable Isotopic Labeling with Amino Acids in Cell Culture (SILAC)

SILAC is a metabolic labeling strategy, which is based on the direct addition of selected stable isotopically labeled amino acids into the entire proteome to distinguish two different proteomes by the MW of light and heavy amino acids during protein metabolism (Mann 2006). Lysine and arginine are established as preferred amino acids for isotope labeling (Ong and Mann 2006). After a sufficient number of cell divisions, two populations of cells are mixed and their proteomes are extracted and analyzed by LC-MS/MS (Sury et al. 2010; Chen et al. 2015). As the label is introduced in the amino acid level, it can be used to characterize the alteration of proteins between different samples, dynamics of protein PTMs in response to particular stimuli, protein–protein interaction as well as to quantify cellular protein turnover in the proteome-wide scale (Ong 2012). One of the major limitations of this technique is that it cannot be used on cell types that are unable to incorporate amino acids, such as plant systems. Being autotrophic in nature, they are able to synthesize all sets of amino acids via several pathways and the assimilation of inorganic nitrogen. However, two reports are available where SILAC labeling in *Arabidopsis thaliana* has been achieved partially (Gruhler et al. 2005; Schützet al. 2011).

7.5 Metabolomics Approaches

The metabolic perturbation during exposure to abiotic stress is an obvious phenomenon. The metabolic network is reconfigured by plants under hostile environment in order to maintain metabolic homeostasis as well as for the generation of essential metabolites to alleviate stress-induced injury (Obata and Fernie 2012). Metabolites represent a diverse group of low-MW molecules including lipids, carbohydrates, peptides, amino acids, organic acids, vitamins, thiols, polyphenols, and alkaloids that are responsible for the functional phenotype in a cell, tissue, or organism (Arakaki et al. 2008) since they are viewed as the phenotypic end product of cellular regulatory processes and are also considered as the ultimate response to fluctuating environmental conditions. Therefore, the aim of metabolomics study is the systematic identification and understanding of complex interconnected interactions among endogenous metabolites from a biological sample and their regulatory roles toward genes, transcripts, and proteins (Brown et al. 2005). For this purpose, there is a combination of different analytical techniques available in the quest for extended metabolome coverage as discussed below.

7.5.1 Gas Chromatography/MS (GC-MS)

GC-MS is a frequently used analytical tool for metabolomic analyses due to its robustness, excellent separation capacity, and high selectivity provided by MS (Beale et al. 2018). The combination of GC together with MS improves the efficacy. Before injecting into the GC-MS analysis, the analyte must be either volatile or can be converted into volatile form and thermally stable compounds (Hall 2006). One inherent disadvantage of GC-MS is that it can be useful for low-MW (50–600 Da) and volatile metabolites. Therefore, the identification and characterization of high-MW metabolites are difficult to study using this method (Beale et al. 2018). Owing to its low-cost technology and short running time, GC-MS has been extensively used in metabolic profiling in plants, especially to decipher the central primary metabolic pathways to better understand the complexity of plant metabolism (Gullberg et al. 2004; Kusano et al. 2007; Fukushima et al. 2009).

7.5.2 Liquid Chromatography-MS (LC-MS)

LC-MS is more practicable for the understanding of the overall richness of plant metabolome by widely covering a large group of secondary metabolites (Dixon et al. 2006; Hall 2006; De Vos et al. 2007), while GC-MS is mainly restricted to primary metabolites in plants for nontargeted metabolomic analysis (Vorst et al. 2005; Moco et al. 2006). The advantage of LC-MS over GC-MS is that it does not require sample pretreatment which is an essential step for GC-MS analysis. Recent advancements in techniques for improving resolution in LC, ion exchange, hydrophobic interaction, or reverse-phase columns demonstrate a high potential for analyzing a wide diversity of metabolites in plants. The high-MW and less thermostable metabolites can also be detected by LC-MS and it allows a collection of both quantitative and structural information and also can achieve pg/mL sensitivities (Theodoridis et al. 2012). Recently, the preparative LC-MS and LC-solid phase extraction-nuclear magnetic resonance (NMR)-MS systems have made it possible for HT identification of metabolite structures (Sturm and Seger 2012; Yang et al. 2014).

7.5.3 Capillary Electrophoresis-MS (CE-MS)

Both GC-MS and LC-MS are powerful techniques, but still have limitations. CE-MS are more powerful than GC-MS and LC-MS, in terms of separation efficiency. It is an efficient and promising method for the separation of polar (e.g. organic acids, amino acids) and charged metabolites (e.g. nucleotides, phosphates) with superior analyte resolution and sensitivity in a short period of time (Monton and Soga 2007; Levandi et al. 2008). CE separates the analytes based on their mass by charge (*m/z*) ratio, and then selectively detected by MS using monitoring ions over a large range of *m/z* values. One of the distinctive advantages of CE is that a very small quantity of starting material ranging from 1 to 20 nL is required for the analysis (Ramautar et al. 2006). The capillary tube used in CE is cheaper than the columns used in LC-MS. Despite having several advantages, CE is not yet applied widely in global metabolite profiling due to having some disadvantages too. In addition, the analytical system stability in CE is not as high as in GC-MS or LC-MS (Monton and Soga 2007; Britz-McKibbin 2011).

7.5.4 Fourier Transform Ion Cyclotron Resonance-MS (FT-ICR-MS)

FT-ICR-MS is a sophisticated analytical tool mainly used for nontargeted metabolic profiling. It offers HT analyses of small molecules with ultra-high resolving power and sensitivity, which can be capable of resolving >10,000 distinct species in a complex biological sample (Hughey et al. 2002). Owing to its extremely high mass accuracy (Hendrickson and Emmett 1999), this technique is useful for molecular identities of metabolites as well as the prediction of chemical formulae in integrated metabolomics research. The FT-based mass analyzers chiefly exploit cyclotron frequency in the fixed magnetic field for the measurement of ions based on the *m/z* ratio (Comisarow and Marshall 1996; Marshall et al. 1998). The only drawback of this technique is their relatively slow acquisition rates (Ghaste et al. 2016).

7.6 Bioinformatics Approaches

In recent years, bioinformatics has emerged as a comparatively newer, interdisciplinary field of science to analyze and interpret biological data using mathematical and statistical modeling of information technology (Singh et al. 2011). The availability of several bioinformatics tools such as databases, algorithms, software, and web resources has played a significant role in stress-related research. Integration of collected "multi-omics" (genomics, transcriptomics, proteomics, and metabolomics) outcome is firmly associated with bioinformatics and computational tools since it is fundamental to construct integrated models for detecting value-added biological information providing novel insights into the organization of biological systems (Iquebal et al. 2015; Esposito et al. 2016). This holistic approach has necessitated the importance of several databases and bioinformatics tools in analyzing and screening huge data sets in a short period of time, with nominal costs and without compromising the specificity of analysis.

7.6.1 Biological Databases

Biological databases are online web resources that refer to a vast collection of information about biological molecules. The primary aim of these databases and algorithms in biology is to allow biological data analysis in an accurate, fast, error free, and easily reproducible manner. These databases enable the researchers for easy retrieval of biological information based on their query (Thalor et al. 2019). Moreover, according to the annotation, information and expression profile databases can be categorized into two major types: primary or archived databases composed of information and annotation that are made accessible to the users without any change. These databases provide accession numbers for both data input and retrieval of those data from the database (Baxevanis and Bateman 2015). Some examples of primary databases include GenBank, DNA Data Bank of Japan (DDBJ) for nucleic acid databases and Protein Data Bank (PDB), Swiss-Prot, TrEMBL, and Metacyc for protein databases. On the contrary, secondary or derived databases are dependent on the result of analysis from the primary resources, which provide better and more valuable knowledge with respect to the primary databases. InterPro and UniProt Knowledge base are examples of secondary databases. The online resources, archives, and repositories allow for performing novel genome-wide analysis of plant stress responses as well as tolerance mechanisms for crop improvement.

7.6.2 Bioinformatics Tools and Software

Apart from biological databases, the recent development of different tools and techniques to resolve value-added biological information provides a novel insight into various problems including abiotic stress avoidance in plants. The alignment of two nucleotide or protein sequences enables the identification of homologs between two sequences for the prediction of structures and functions with great accuracy (Mehmood et al. 2014). Sequence alignment can be of two general types: (i) global alignment utilized to find out the best possible end-to-end alignment for aligning nucleotide or protein sequences when two or more closely related sequences are of the same length, and (ii) local alignment, which can be used to align when sequences are suspected to have similarity or even dissimilar sequences. ClustalW, MAFFT, MUSCLE, T-COFFEE, DIALIGN 2, and PROBCONS are commonly used sequence alignment tools widely utilized by biological researchers (Edgar and Batzoglou 2006). For predicting the three-dimensional (3D) structure of proteins, homology modeling is an excellent tool and also essential for resolving intriguing biological questions. This is particularly important for detecting abiotic stress-induced alterations in protein structures which ultimately affect their functions and cause deleterious effects in plants. IntFOLD, Biskit, RaptorX, ESyPred3D, and HHpred are widely utilized software for homology modeling purposes. Protein threading for the fold recognition method and *de novo* (*ab initio*) protein structure prediction is also used for the prediction of protein structure when no suitable template structure is available (Schwede et al. 2003; Klepeis et al. 2005).

7.7 Concluding Remarks

Agricultural productivity is highly dependent on the climate and the changes in climate result in various abiotic stresses causing significant yield losses each year. Since plants are the primary producers on earth, restraining crop productivity due to abiotic stresses is a global concern for the scientific community. Owing to polygenic traits, numerous genes are involved in abiotic stress tolerance at the molecular level. Therefore, the identification of various strategies to counteract the damages and crop loss imposed by abiotic stresses is a preliminary requirement in the current scenario. To date, the rigorous advent of "omics"-based technologies has gained momentum in plant science. The enormous HT multi-dimensional data obtained from various omics approaches including genomics, transcriptomics, proteomics, and metabolomics made it possible to use the systems biology approach to better understand plant responses under abiotic stress conditions (Figure 7.2). In an era of accelerated global climate change, the integrated multi-omics analysis is imperative for widening our knowledge to predict the agronomically important traits for crop improvement on the multitude of diverse abiotic stress conditions. Constructing regulatory mechanism hubs by integrating omics technologies will be helpful in deciphering

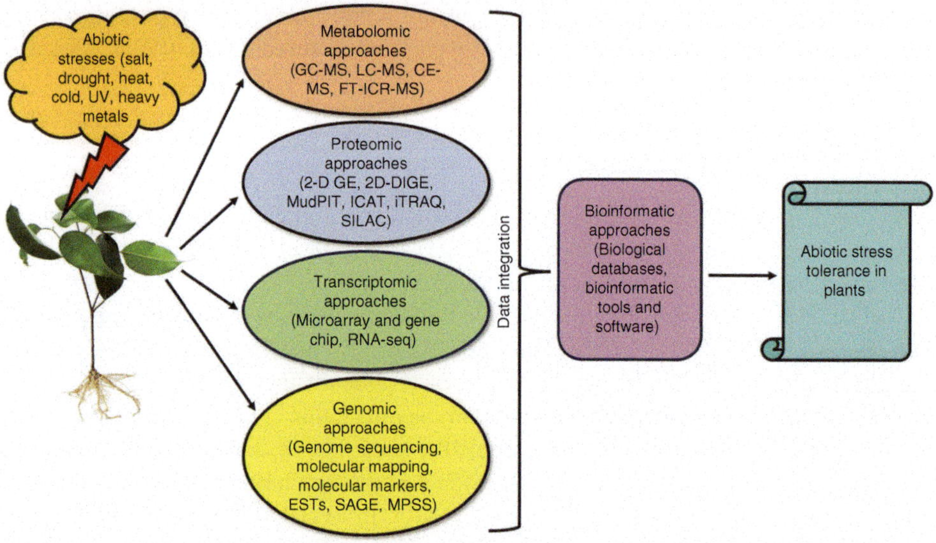

Figure 7.2 Schematic representation of the relationship among different omics technologies and bioinformatics for understanding abiotic stress tolerance in plants. *Source:* Acik/Adobe Stock.

systematic behavior and adaptations of crops under various abiotic stresses for reducing crop losses due to environmental vagaries with transformative advances in improving agronomic yield. Moreover, the collaborative integration of data derived from multi-omics platforms with the *in silico* bioinformatics tools has provided a holistic approach which made them cost-effective and easily accessible for different studies. This will speed up the future efforts for the successful management of abiotic stress resilience in plants which is ultimately helpful in crop improvement programs for generating high-yielding, stress-tolerant germplasm and climate-smart crop cultivars to meet the increasing global food demand for rapidly expanding populations in the next decade.

Acknowledgments

Financial assistance from the Science and Engineering Research Board, Government of India through the grant [EMR/2016/004799] and the Department of Higher Education, Science and Technology and Biotechnology, Government of West Bengal, through the grant [264(Sanc.)/ST/P/S&T/1G-80/2017] to Prof. Aryadeep Roychoudhury is gratefully acknowledged.

References

Aggarwal, K., Choe, L.H., and Lee, K.H. (2006). Shotgun proteomics using the iTRAQ isobaric tags. *Briefings in Functional Genomics & Proteomics* 5: 112–120.

Alban, A., David, S.O., Bjorkesten, L. et al. (2003). A novel experimental design for comparative two-dimensional gel analysis: two-dimensional difference gel electrophoresis incorporating a pooled internal standard. *Proteomics* 3: 36–44.

Ansorge, W.J. (2009). Next-generation sequencing techniques. *New Biotechnology* 25: 195–203.

Arakaki, A.K., Skolnick, J., and McDonald, J.F. (2008). Marker metabolites can be therapeutic targets as well. *Nature* 456: 443.

Bagati, S., Mahajan, R., Nazir, M. et al. (2018). "Omics": a gateway towards abiotic stress tolerance. In: *Abiotic Stress-mediated Sensing and Signaling in Plants: An Omics Perspective* (ed. S.M. Zargar and M.Y. Zargar), 1–45. Singapore: Springer Nature.

Baxevanis, A.D. and Bateman, A. (2015). The importance of biological databases in biological discovery. *Current Protocols in Bioinformatics* 50: 1.1.1–1.1.8.

Beale, D.J., Pinu, F.R., Kouremenos, K.A. et al. (2018). Review of recent developments in GC–MS approaches to metabolomics-based research. *Metabolomics* 14: 152.

Beckett, P. (2012). The basics of 2D DIGE. In: *Difference Gel Electrophoresis (DIGE)*, Methods in Molecular Biology (ed. R. Cramer and R. Westermeier), 9–19. Totowa, NJ: Humana Press.

Bennett, S. (2004). Solexa Ltd. *Pharmacogenomics* 5: 433–438.

Bennett, S.T., Barnes, C., Cox, A. et al. (2005). Toward the 1000 dollars human genome. *Pharmacogenomics* 6: 373–382.

Bouchez, D. and Höfte, H. (1998). Functional genomics in plants. *Plant Physiology* 118:725–732.

Brenner, S., Johnson, M., Bridgham, J. et al. (2000). Gene expression analysis by massively parallel signature sequencing (MPSS) on microbead arrays. *Nature Biotechnology* 18: 630–634.

Britz-McKibbin, P. (2011). Capillary electrophoresis–electrospray ionization-mass spectrometry (CE–ESI-MS)-based metabolomics. In: *Metabolic Profiling*, Methods in Molecular Biology (ed. T. Metz), 229–246. Totowa, NJ: Humana Press.

Brown, M., Dunn, W.B., Ellis, D.I. et al. (2005). A metabolome pipeline: from concept to data to knowledge. *Metabolomics* 1: 39–51.

Bubnoff, A.V. (2008). Next-generation sequencing: the race is on. *Cell* 132: 721–723.

Chen, J.-S., Lin, S.-C., Chen, C.-Y. et al. (2014). Development of microarray for two rice subspecies: characterization and validation of gene expression in rice. *BMC Research Notes* 7: 15.

Chen, X., Wei, S., Ji, Y. et al. (2015). Quantitative proteomics using SILAC: principles, applications, and development. *Proteomics* 15: 3175–3192.

Cheng, C., Liu, Y., Fang, W. et al. (2020). iTRAQ-based proteomic and physiological analyses of mustard sprouts in response to heat stress. *RSC Advances* 10: 6052–6062.

Chiappetta, G., Corbo, C., Palmese, A. et al. (2009). Quantitative identification of protein nitration sites. *Proteomics* 9: 1524–1537.

Comisarow, M.B. and Marshall, A.G. (1996). The early development of Fourier transform ion cyclotron resonance (FT-ICR) spectroscopy. *Journal of Mass Spectrometry* 31: 581–585.

Das, A., Eldakak, M., Paudel, B. et al. (2016). Leaf proteome analysis reveals prospective drought and heat stress response mechanisms in soybean. *BioMed Research International* 2016: 6021047.

Datson, N.A., van der Perk-de Jong, J., van den Berg, M.P. et al. (1999). MicroSAGE: a modified procedure for serial analysis of gene expression in limited amounts of tissue. *Nucleic Acids Research* 27: 1300–1307.

De Vos, R.C.H., Moco, S., Lommen, A. et al. (2007). Untargeted large-scale plant metabolomics using liquid chromatography coupled to mass spectrometry. *Nature Protocols* 2: 778–791.

Dixon, R.A., Gang, D.R., Charlton, A.J. et al. (2006). Applications of metabolomics in agriculture. *Journal of Agricultural and Food Chemistry* 54: 8984–8994.

Dong, Z. and Chen, Y. (2013). Transcriptomics: advances and approaches. *Science China Life Sciences* 56: 960–967.

Dong, Q., Kroiss, L., Oakley, F.D. et al. (2005). Comparative EST analyses in plant systems. *Methods in Enzymology* 395: 400–418.

Dunkley, T.P.J., Dupree, P., Watson, R.B., and Lilley, K.S. (2004). The use of isotope-coded affinity tags (ICAT) to study organelle proteomes in *Arabidopsis thaliana*. *Biochemical Society Transactions* 32: 520–523.

Edgar, R.C. and Batzoglou, S. (2006). Multiple sequence alignment. *Current Opinion in Structural Biology* 16: 368–373.

Eid, J., Fehr, A., Gray, J. et al. (2009). Real-time DNA sequencing from single polymerase molecules. *Science* 323: 133–138.

Esposito, A., Colantuono, C., Ruggieri, V., and Chiusano, M.L. (2016). Bioinformatics for agriculture in the next-generation sequencing era. *Chemical and Biological Technologies in Agriculture* 3: 9.

Evans, C., Noirel, J., Ow, S.Y. et al. (2012). An insight into iTRAQ: where do we stand now? *Analytical and Bioanalytical Chemistry* 404: 1011–1027.

Faghani, E., Gharechahi, J., Komatsu, S. et al. (2015). Comparative physiology and proteomic analysis of two wheat genotypes contrasting in drought tolerance. *Journal of Proteomics* 114: 1–15.

Fields, S., Kohara, Y., and Lockhart, D.J. (1999). Functional genomics. *Proceedings of the National Academy of Sciences* 96: 8825–8826.

Fu, C., Hu, J., Liu, T. et al. (2008). Quantitative analysis of redox-sensitive proteome with DIGE and ICAT. *Journal of Proteome Research* 7: 3789–3802.

Fukushima, A., Kusano, M., Redestig, H. et al. (2009). Integrated omics approaches in plant systems biology. *Current Opinion in Chemical Biology* 13: 532–538.

Gao, L., Yan, X., Li, X. et al. (2011). Proteome analysis of wheat leaf under salt stress by two-dimensional difference gel electrophoresis (2D-DIGE). *Phytochemistry* 72: 1180–1191.

Garalde, D.R., Snell, E.A., Jachimowicz, D. et al. (2018). Highly parallel direct RNA sequencing on an array of nanopores. *Nature Methods* 15: 201–206.

Garfin, D.E. (2003). Two-dimensional gel electrophoresis: an overview. *Trends in Analytical Chemistry* 22: 263–272.

Ghaste, M., Mistrik, R., and Shulaev, V. (2016). Applications of Fourier transform ion cyclotron resonance (FT-ICR) and orbitrap based high resolution mass spectrometry in metabolomics and lipidomics. *International Journal of Molecular Sciences* 17: 816.

Gong, F., Hu, X., and Wang, W. (2015). Proteomic analysis of crop plants under abiotic stress conditions: where to focus our research? *Frontiers in Plant Science* 6: 418.

Grayson, M. (2013). Agriculture and drought. *Nature* 501: S1.

Grimaud, F., Renaut, J., Dumont, E. et al. (2013). Exploring chloroplastic changes related to chilling and freezing tolerance during cold acclimation of pea (*Pisum sativum* L.). *Journal of Proteomics* 80: 145–159.

Gruhler, A., Schulze, W.X., Matthiesen, R. et al. (2005). Stable isotope labeling of *Arabidopsis thaliana* cells and quantitative proteomics by mass spectrometry. *Molecular & Cellular Proteomics* 4: 1697–1702.

Gul, A., Ahad, A., Akhtar, S. et al. (2016). Microarray: gateway to unravel the mystery of abiotic stresses in plants. *Biotechnology Letters* 38: 527–543.

Gullberg, J., Jonsson, P., Nordström, A. et al. (2004). Design of experiments: an efficient strategy to identify factors influencing extraction and derivatization of *Arabidopsis thaliana* samples in metabolomic studies with gas chromatography/mass spectrometry. *Analytical Biochemistry* 331: 283–295.

Gutsch, A., Zouaghi, S., Renaut, J. et al. (2018). Changes in the proteome of *Medicago sativa* leaves in response to long-term cadmium exposure using a cell-wall targeted approach. *International Journal of Molecular Sciences* 19: 2498.

Gyenis, L., Yun, S.J., Smith, K.P. et al. (2007). Genetic architecture of quantitative trait loci associated with morphological and agronomic trait differences in a wild by cultivated barley cross. *Genome* 50: 714–723.

Gygi, S.P., Corthals, G.L., Zhang, Y. et al. (2000). Evaluation of two-dimensional gel electrophoresis-based proteome analysis technology. *Proceedings of the National Academy of Sciences* 97: 9390–9395.

Hall, R.D. (2006). Plant metabolomics: from holistic hope, to hype, to hot topic. *The New Phytologist* 169: 453–468.

Hasan, N., Choudhary, S., Naaz, N. et al. (2021). Recent advancements in molecular marker-assisted selection and applications in plant breeding programmes. *Journal of Genetic Engineering and Biotechnology* 19: 128.

Hendrickson, C.L. and Emmett, M.R. (1999). Electrospray ionization Fourier transform ion cyclotron resonance mass spectrometry. *Annual Review of Physical Chemistry* 50: 517–536.

Hoffert, J.D., Pisitkun, T., Saeed, F. et al. (2011). Dynamics of the G protein-coupled vasopressin V2 receptor signaling network revealed by quantitative phosphoproteomics. *Molecular & Cellular Proteomics* 11: M111.014613.

Hosseini, S.S., Ramezanpour, S.S., Soltanloo, H., and Seifati, S.E. (2023). RNA-seq analysis and reconstruction of gene networks involved in response to salinity stress in quinoa (*cv.* Titicaca). *Scientific Reports* 13: 7308.

Hughey, C.A., Rodgers, R.P., and Marshall, A.G. (2002). Resolution of 11,000 compositionally distinct components in a single electrospray ionization Fourier transform ion cyclotron resonance mass spectrum of crude oil. *Analytical Chemistry* 74: 4145–4149.

Hunter, B.G., Beatty, M.K., Singletary, G.W. et al. (2002). Maize opaque endosperm mutations create extensive changes in patterns of gene expression. *Plant Cell* 14: 2591–2612.

Intergovernmental Pannel on Climate Change (2021) Climate change 2021: the physical science basis. *The Working Group I contribution to the Sixth Assessment*. https://www.ipcc.ch/report/sixth-assessment-report-working-group-i/ (accessed 6 August 2021).

Iquebal, M.A., Jaiswal, S., Mukhopadhyay, C.S. et al. (2015). Applications of bioinformatics in plant and agriculture. In: *PlantOmics: The Omics of Plant Science* (ed. D. Barh, M. Khan, and E. Davies), 755–789. New Delhi: Springer.

Janská, A., Maršík, P., and Zelenková, S. (2010). Cold stress and acclimation–what is important for metabolic adjustment? *Plant Biology* 12: 395–405.

Ji, W., Cong, R., Li, S. et al. (2016). Comparative proteomic analysis of soybean leaves and roots by iTRAQ provides insights into response mechanisms to short-term salt stress. *Frontiers in Plant Science* 7: 573.

Jongeneel, C.V. (2000). Searching the expressed sequence tag (EST) databases: panning for genes. *Briefings in Bioinformatics* 1: 76–92.

Kage, U., Kumar, A., Dhokane, D. et al. (2016). Functional molecular markers for crop improvement. *Critical Reviews in Biotechnology* 36: 917–930.

Kantety, R.V., Rota, M.L., Matthews, D.E., and Sorrells, M.S. (2002). Data mining for simple sequence repeats in expressed sequence tags from barley, maize, rice, sorghum and wheat. *Plant Molecular Biology* 48: 501–510.

Kim, C.-K., Lim, H.-M., Na, J.-K. et al. (2014). A multistep screening method to identify genes using evolutionary transcriptome of plants. *Evolutionary Bioinformatics Online* 10: 69.

Klepeis, J.L., Wei, Y., Hecht, M.H., and Floudas, C.A. (2005). Ab initio prediction of the three-dimensional structure of a de novo designed protein: a double-blind case study. *Proteins: Structure, Function, and Bioinformatics* 58: 560–570.

Komatsuda, T., Pourkeirandish, M., He, C. et al. (2007). Six-rowed barley originated from a mutation in a homeodomain-leucine zipper I–class homeobox gene. *Proceedings of the National Academy of Sciences of the United States of America* 104: 1424–1429.

Kozarova, A., Sliskovic, I., Mutus, B. et al. (2007). Identification of redox sensitive thiols of protein disulfide isomerase using isotope coded affinity technology and mass spectrometry. *Journal of the American Society for Mass Spectrometry* 18: 260–269.

Kumar, J., Sen Gupta, D., Kesari, R. et al. (2021). Comprehensive RNAseq analysis for identification of genes expressed under heat stress in lentil. *Physiologia Plantarum* 173: 1785–1807.

Kusano, T., Yamaguchi, K., Berberich, T., and Takahashi, Y. (2007). Advances in polyamine research in 2007. *Journal of Plant Research* 120: 345–350.

Levandi, T., Leon, C., Kaljurand, M. et al. (2008). Capillary electrophoresis time-of-flight mass spectrometry for comparative metabolomics of transgenic versus conventional maize. *Analytical Chemistry* 80: 6329–6335.

Lipshultz, R.J., Fodor, S.P.A., Gingeras, T.R., and Lockhart, D.J. (1999). High density synthetic oligonucleotide arrays. *Nature Genetics* 21: 20–24.

Mann, M. (2006). Functional and quantitative proteomics using SILAC. *Nature Reviews Molecular Cell Biology* 7: 953.

Mantri, N., Pang, E.C.K., and Ford, R. (2010). Molecular biology for stress management. In: *Climate Change and Management of Cool Season Grain Legume Crops* (ed. S. Yadav, D. McNeil, R. Redden, and S. Patil), 377–408. New York, USA: Springer.

Mardis, E.A. (2008). Next-generation DNA sequencing methods. *Annual Review of Genomics and Human Genetics* 9: 387–402.

Margulies, M., Egholm, M., Altman, W.E. et al. (2005). Genome sequencing in microfabricated high-density picolitre reactors. *Nature* 437: 376–380.

Marshall, A.G., Hendrickson, C.L., and Jackson, G.S. (1998). Fourier transform ion cyclotron resonance mass spectrometry: a primer. *Mass Spectrometry Reviews* 17: 1–35.

Matsumura, H., Nirasawa, S., and Terauchi, R. (1999). Transcript profiling in rice (*Oryza sativa* L.) seedlings using serial analysis of gene expression (SAGE). *The Plant Journal* 20: 719–726.

Mauri, P. and Scigelova, M. (2009). Multidimensional protein identification technology for clinical proteomic analysis. *Clinical Chemistry and Laboratory Medicine* 47: 636–646.

Maxam, M.A. and Gilbert, W. (1977). A new method for sequencing DNA. *Proceedings of the National Academy of Sciences* 74: 560–564.

May, C., Brosseron, F., Chartowski, P. et al. (2012). Differential proteome analysis using 2D-DIGE. In: *Quantitative Methods in Proteomics*, Methods in Molecular Biology (ed. K. Marcus, M. Eisenacher, and B. Sitek), 75–82. New York: Humana Press.

Mayer, K. and Mewes, H.-W. (2002). How can we deliver the large plant genomes? Strategies and perspectives. *Plant Biotechnology* 5: 173–177.

Mehmood, M.A., Sehar, U., and Ahmad, N. (2014). Use of bioinformatics tools in different spheres of life sciences. *Journal of Data Mining in Genomics & Proteomics* 5: 158.

Milos, P. (2008). Helicos BioSciences. *Pharmacogenomics* 9: 477–480.

Minhas, P.S., Rane, J., and Pasala, R.K. (2017). Abiotic stresses in agriculture: an overview. In: *Abiotic Stress Management for Resilient Agriculture* (ed. P. Minhas, J. Rane, and R. Pasala), 3–8. Singapore: Springer.

Moco, S., Bino, R.J., Vorst, O. et al. (2006). A liquid chromatography-mass spectrometry-based metabolome database for tomato. *Plant Physiology* 141: 1205–1218.

Monteoliva, L. and Albar, J.P. (2004). Differential proteomics: an overview of gel and non-gel based approaches. *Briefings in Functional Genomics & Proteomics* 3: 220–239.

Monton, M.R. and Soga, T. (2007). Metabolome analysis by capillary electrophoresis–mass spectrometry. *Journal of Chromatography A* 1168: 237–246.

Nagaraj, S.H., Gasser, R.B., and Ranganathan, S. (2006). A hitchhiker's guide to expressed sequence tag (EST) analysis. *Briefings in Bioinformatics* 8: 6–21.

Nägele, E., Vollmer, M., Hörth, P., and Vad, C. (2004). 2D-LC/MS techniques for the identification of proteins in highly complex mixtures. *Expert Review of Proteomics* 1: 37–46.

Nakano, M., Nobuta, K., Vemaraju, K. et al. (2006). Plant MPSS databases: signature-based transcriptional resources for analyses of mRNA and small RNA. *Nucleic Acids Research* 34: D731–D735.

Obata, T. and Fernie, A.R. (2012). The use of metabolomics to dissect plant responses to abiotic stresses. *Cellular and Molecular Life Sciences* 69: 3225–3243.

Öktem, H.A., Eyidoğan, F., Selçuk, F. et al. (2008). Revealing response of plants to biotic and abiotic stresses with microarray technology. *Genes Genomes Genomics* 2: 14–48.

Oliveira, L.K., Melo, L.C., Brondani, C. et al. (2008). Backcross assisted by microsatellite markers in common bean. *Genetics and Molecular Research* 7: 1000–1010.

Ong, S.-E. (2012). The expanding field of SILAC. *Analytical and Bioanalytical Chemistry* 404: 967–976.

Ong, S.-E. and Mann, M. (2006). A practical recipe for stable isotope labeling by amino acids in cell culture (SILAC). *Nature Protocols* 1: 2650–2660.

Ong, S.-E. and Pandey, A. (2001). An evaluation of the use of two-dimensional gel electrophoresis in proteomics. *Biomolecular Engineering* 18: 195–205.

Oshunsanya, S.O., Nwosu, N.J., and Li, Y. (2019). Abiotic stress in agricultural crops under climatic conditions. In: *Sustainable Agriculture, Forest and Environmental Management* (ed. M. Jhariya, A. Banerjee, R. Meena, and D. Yadav), 71–100. Singapore: Springer.

Ozsolak, F. and Milos, P.M. (2011). RNA sequencing: advances, challenges and opportunities. *Nature Reviews Genetics* 12: 87–98.

Pan, Y., Kislinger, T., Gramolini, A.O. et al. (2004). Identification of biochemical adaptations in hyper- or hypocontractile hearts from phospholamban mutant mice by expression proteomics. *Proceedings of the National Academy of Sciences* 101: 2241–2246.

Parkinson, J. and Blaxter, M. (2009). Expressed sequence tags: an overview. In: *Expressed Sequence Tags (ESTs)*, Methods in Molecular Biology, vol. 533 (ed. J. Parkinson), 1–12. Totowa, NJ: Humana Press.

Pascual, L.S., Segarra-Medina, C., Gómez-Cadenas, A. et al. (2022). Climate change-associated multifactorial stress combination: a present challenge for our ecosystems. *Journal of Plant Physiology* 276: 153764.

Pasquali, M., Serchi, T., Planchon, S., and Renaut, J. (2017). 2D-DIGE in proteomics. In: *Functional Genomics*, Methods in Molecular Biology (ed. M. Kaufmann, C. Klinger, and A. Savelsbergh), 245–254. New York: Humana Press.

Pollock, J.D. (2002). Gene expression profiling: methodological challenges, results, and prospects for addiction research. *Chemistry and Physics of Lipids* 121: 241–256.

Qin, F., Shinozaki, K., and Yamaguchi-Shinozaki, K. (2011). Achievements and challenges in understanding plant abiotic stress responses and tolerance. *Plant & Cell Physiology* 52: 1569–1582.

Rabilloud, T. and Lelong, C. (2011). Two-dimensional gel electrophoresis in proteomics: a tutorial. *Journal of Proteomics* 74: 1829–1841.

Rabilloud, T., Chevallet, M., Luche, S., and Lelong, C. (2010). Two-dimensional gel electrophoresis in proteomics: past, present and future. *Journal of Proteomics* 73: 2064–2077.

Ramautar, R., Demirci, A., and de Jong, G.J. (2006). Capillary electrophoresis in metabolomics. *Trends in Analytical Chemistry* 25: 455–466.

Rao, G.J.N., Reddy, J.N., Variar, M., and Mahender, A. (2016). Molecular breeding to improve plant resistance to abiotic stresses. In: *Advances in Plant Breeding Strategies: Agronomic, Abiotic and Biotic Stress Traits* (ed. J. Al-Khayri, S. Jain, and D. Johnson), 283–326. Cham: Springer.

Reinartz, J., Bruyns, E., Lin, J.-Z. et al. (2002). Massively parallel signature sequencing (MPSS) as a tool for in-depth quantitative gene expression profiling in all organisms. *Briefings in Functional Genomics & Proteomics* 1: 95–104.

Ren, W., Shi, Z., Zhou, M. et al. (2022). iTRAQ-based quantitative proteomic analysis provides insight into the drought-stress response in maize seedlings. *Scientific Reports* 12: 9520.

Rosenzweig, D., Smith, D., Myler, P.J. et al. (2008). Post-translational modification of cellular proteins during *Leishmania donovani* differentiation. *Proteomics* 8: 1843–1850.

Ross, P.L., Huang, Y.N., Marchese, J.N. et al. (2004). Multiplexed protein quantitation in *Saccharomyces cerevisiae* using amine-reactive isobaric tagging reagents. *Molecular & Cellular Proteomics* 3: 1154–1169.

Rothberg, J.M., Hinz, W., Rearick, T.M. et al. (2011). An integrated semiconductor device enabling non-optical genome sequencing. *Nature* 475: 348–352.

Roy, S.J., Tucker, E.J., and Tester, M. (2011). Genetic analysis of abiotic stress tolerance in crops. *Current Opinion in Plant Biology* 14: 232–239.

Roychoudhury, A., Datta, K., and Datta, S.K. (2011). Abiotic stress in plants: from genomics to metabolomics. In: *Omics and Plant Abiotic Stress Tolerance* (ed. N. Tuteja, S.S. Gill, and R. Tuteja), 91–120. Bentham Science Publishers.

Rudd, S. (2003). Expressed sequence tags: alternative or complement to whole genome sequences? *Trends in Plant Science* 8: 321–329.

Sahoo, S., Kusunoki, K., Goswami, K. et al. (2023). Differential transcriptional regulation of drought stress revealed by comparative RNA-seq analysis of contrasting *indica* rice from North East India. *Journal of Plant Growth Regulation* 42: 5780–5795.

Samanta, S. and Roychoudhury, A. (2023). Genomic and proteomic approaches to enhance stress tolerance in plants for creating stress-resilient species. In: *Frontiers in Plant-environment Interaction*, Environmental Science and Engineering (ed. T. Aftab). Switzerland AG: Springer Nature https://doi.org/10.1007/978-3-031-43729-8_12.

Sato, S., Tomomori-Sato, C., Parmely, T.J. et al. (2004). A set of consensus mammalian mediator subunits identified by multidimensional protein identification technology. *Molecular Cell* 14: 685–691.

Schütz, W., Hausmann, N., Krug, K. et al. (2011). Extending SILAC to proteomics of plant cell lines. *Plant Cell* 23: 1701–1705.

Schwede, T., Kopp, J., Guex, N., and Peitsch, M.C. (2003). SWISS-MODEL: an automated protein homology-modeling server. *Nucleic Acids Research* 31: 3381–3385.

Seki, M., Okamoto, M., Matsui, A. et al. (2010). Microarray analysis for studying the abiotic stress responses in plants. In: *Molecular Techniques in Crop Improvement* (ed. S. Jain and D. Brar), 333–355. Dordrecht: Springer.

Sethuraman, M., McComb, M.E., Huang, H. et al. (2004). Isotope-coded affinity tag (ICAT) approach to redox proteomics: identification and quantitation of oxidant-sensitive cysteine thiols in complex protein mixtures. *Journal of Proteome Research* 3: 1228–1233.

Shi, X., Wang, X., Cheng, F. et al. (2019). iTRAQ-based quantitative proteomics analysis of cold stress-induced mechanisms in grafted watermelon seedlings. *Journal of Proteomics* 192: 311–320.

Shiio, Y. and Aebersold, R. (2006). Quantitative proteome analysis using isotope-coded affinity tags and mass spectrometry. *Nature Protocols* 1: 139–145.

Shinde, H., Dudhate, A., Anand, L. et al. (2020). Small RNA sequencing reveals the role of pearl millet miRNAs and their targets in salinity stress responses. *South African Journal of Botany* 132: 395–402.

Shiraki, T., Kondo, S., Katayama, S. et al. (2003). Cap analysis gene expression for high-throughput analysis of transcriptional starting point and identification of promoter usage. *Proceedings of the National Academy of Sciences* 100: 15776–15781.

Singh, V.K., Singh, A.K., Chand, R., and Kushwaha, V. (2011). Role of bioinformatics in agriculture and sustainable development. *International Journal of Bioinformatics Research* 3: 221–226.

Skolnick, J., Fetrow, J.S., and Kolinski, A. (2000). Structural genomics and its importance for gene function analysis. *Nature Biotechnology* 18: 283–287.

Smolka, M., Zhou, H., and Aebersold, R. (2002). Quantitative protein profiling using two-dimensional gel electrophoresis, isotope-coded affinity tag labeling, and mass spectrometry. *Molecular & Cellular Proteomics* 1: 19–29.

Soni, P., Nutan, K.K., Soda, N. et al. (2015). Towards understanding abiotic stress signaling in plants: convergence of genomic, transcriptomic, proteomic, and metabolomic approaches. In: *Elucidation of Abiotic Stress Signaling in Plants* (ed. G.K. Pandey), 3–40. New York: Springer.

Stark, R., Grzelak, M., and Hadfield, J. (2019). RNA sequencing: the teenage years. *Nature Reviews Genetics* 20: 631–656.

Sturm, S. and Seger, C. (2012). Liquid chromatography-nuclear magnetic resonance coupling as alternative to liquid chromatography-mass spectrometry hyphenations: curious option or powerful and complementary routine tool? *Journal of Chromatography A* 1259: 50–61.

Su, Z. et al. (2014). A comprehensive assessment of RNA-seq accuracy, reproducibility and information content by the sequencing quality control consortium. *Nature Biotechnology* 32: 903–914.

Sury, M.D., Chen, J.-X., and Selbach, M. (2010). The SILAC fly allows for accurate protein quantification *in vivo*. *Molecular & Cellular Proteomics* 9: 2173–2183.

Takehisa, H., Sato, Y., Antonio, B.A., and Nagamura, Y. (2013). Global transcriptome profile of rice root in response to essential macronutrient deficiency. *Plant Signaling & Behavior* 8: e24409.

Thalor, A., Mehta, S., Singh, B. et al. (2019). Databases: a weapon from the arsenal of bioinformatics for plant abiotic stress research. In: *Recent Approaches in Omics for Plant Resilience to Climate Change* (ed. S. Wani), 135–169. Cham: Springer.

Thamsen, M. and Jakob, U. (2011). The redoxome: proteomic analysis of cellular redox networks. *Current Opinion in Chemical Biology* 15: 113–119.

Theodoridis, G.A., Gika, H.G., Want, E.J., and Wilson, I.D. (2012). Liquid chromatography–mass spectrometry based global metabolite profiling: a review. *Analytica Chimica Acta* 711: 7–16.

Tureček, F. (2002). Mass spectrometry in coupling with affinity capture–release and isotope-coded affinity tags for quantitative protein analysis. *Journal of Mass Spectrometry* 37: 1–14.

Varshney, R.K., Graner, A., and Sorrells, M.E. (2005a). Genic microsatellite markers in plants: features and applications. *Trends in Biotechnology* 23: 48–55.

Varshney, R.K., Sigmund, R., Börner, A. et al. (2005b). Interspecific transferability and comparative mapping of barley EST-SSR markers in wheat, rye and rice. *Plant Science* 168: 195–202.

Vaughn, M.W. and Martienssen, R. (2005). It's a small RNA world, after all. *Science* 309: 1525–1526.

Velculescu, V.E., Zhang, L., Vogelstein, B., and Kinzler, K.W. (1995). Serial analysis of gene expression. *Science* 270: 484–487.

Verma, M., Kulshrestha, S., and Puri, A. (2017). Genome sequencing. In: *Bioinformatics*, Methods in Molecular Biology, vol. 1525 (ed. J. Keith), 3–33. New York: Humana Press.

Vij, S. and Tyagi, A.K. (2007). Emerging trends in the functional genomics of the abiotic stress response in crop plants. *Plant Biotechnology Journal* 5: 361–380.

Vorst, O., de Vos, C.H.R., Lommen, A. et al. (2005). A non-directed approach to the differential analysis of multiple LC–MS-derived metabolic profiles. *Metabolomics* 1: 169–180.

Wang, Z., Gerstein, M., and Snyder, M. (2009). RNA-Seq: a revolutionary tool for transcriptomics. *Nature Reviews Genetics* 10: 57–63.

Wani, S.H., Singh, N.B., Saini, H.K. et al. (2010). Expressed sequence tags (ESTs) – a functional genomic approach for gene discovery. *International Journal of Current Research* 5: 74–79.

Wani, S.H., Choudhary, M., Kumar, P. et al. (2018). Marker-assisted breeding for abiotic stress tolerance in crop plants. In: *Biotechnologies of Crop Improvement*, vol. 3 (ed. S.S. Gosal and S.H. Wani), 1–23. Cham: Springer.

Washburn, M.P., Koller, A., Oshiro, G. et al. (2003). Protein pathway and complex clustering of correlated mRNA and protein expression analyses in *Saccharomyces cerevisiae*. *Proceedings of the National Academy of Sciences* 100: 3107–3112.

Wei, B., Jing, R., Wang, C. et al. (2009). *Dreb1* genes in wheat (*Triticum aestivum* L.): development of functional markers and gene mapping based on SNPs. *Molecular Breeding* 23: 13–22.

Wiese, S., Reidegeld, K.A., Meyer, H.E., and Warscheid, B. (2007). Protein labeling by iTRAQ: a new tool for quantitative mass spectrometry in proteome research. *Proteomics* 7: 340–350.

Wu, J., Warren, P., Shakey, Q. et al. (2010). Integrating titania enrichment, iTRAQ labeling, and orbitrap CID-HCD for global identification and quantitative analysis of phosphopeptides. *Proteomics* 10: 2224–2234.

Wu, M., Chen, A., Wang, Z. et al. (2015). Plant microarray for gene expression profiling and their application. *Journal of Agriculture Technology* 11: 93–105.

Yang, Z., Nakabayashi, R., Okazaki, Y. et al. (2014). Toward better annotation in plant metabolomics: isolation and structure elucidation of 36 specialized metabolites from *Oryza sativa* (rice) by using MS/MS and NMR analyses. *Metabolomics* 10: 543–555.

Yang, R., Li, M., Harrison, M.T. et al. (2022). iTRAQ proteomic analysis of wheat (*Triticum aestivum* L.) genotypes differing in waterlogging tolerance. *Frontiers in Plant Science* 13: 890083.

Yates, J.R. (2004). Mass spectral analysis in proteomics. *Annual Review of Biophysics and Biomolecular Structure* 33: 297–316.

Younis, A., Ramzan, F., Ramzan, Y. et al. (2020). Molecular markers improve abiotic stress tolerance in crops: a review. *Plants* 9: 1374.

Zhang, D., Jiang, S., Pan, J. et al. (2014). The overexpression of a maize mitogen-activated protein kinase gene (*ZmMPK5*) confers salt stress tolerance and induces defence responses in tobacco. *Plant Biology* 16: 558–570.

Zhu, T. and Wang, X. (2000). Large-scale profiling of the *Arabidopsis* transcriptome. *Plant Physiology* 124: 1472–1476.

Zhu, T., Budworth, P., Chen, W. et al. (2003). Transcriptional control of nutrient partitioning during rice grain filling. *Plant Biotechnology Journal* 1: 59–70.

Zieske, L.R. (2006). A perspective on the use of iTRAQ™ reagent technology for protein complex and profiling studies. *Journal of Experimental Botany* 57: 1501–1508.

Zimmermann, J., Voss, H., Schwager, C. et al. (1989). Automated Sanger dideoxy sequencing reaction protocol. *FEBS Letters* 223: 432–436.

8

Bioinformatics Approaches for Unraveling the Complexities of Plant Stress Physiology

Sneha Murmu[1], Himanshushekhar Chaurasia[2], Ipsita Samal[3], Tanmaya Kumar Bhoi[4], and Asit Kumar Pradhan[5]

[1] *Division of Agrciultural Bioinformatics, ICAR-Indian Agricultural Statistics Research Institute, New Delhi, India*
[2] *Mechanical Processing Division, ICAR-Central Institute for Research on Cotton Technology, Mumbai, Maharashtra, India*
[3] *Department of Entomology, Sri Sri University, Cuttack, Odisha, India*
[4] *Forest Protection Division, ICFRE-Arid Forest Research Institute, Jodhpur, Rajasthan, India*
[5] *Social Science Division, ICAR-National Rice Research Institute, Cuttack, Odisha, India*

8.1 Introduction

The understanding of stress in plants has undergone significant advancements over the course of the last six decades. Stress, as defined by Lichtenthaler (1998), refers to any adverse condition or component that hinders or disrupts a plant's metabolic processes, impeding its growth and development.

Plants are exposed to many environmental stressors that have detrimental effects on the productivity of agricultural crops, leading to reductions and limitations in their overall yield. Plants encounter two broad classifications of external stressors, specifically abiotic stress and biotic stress. Various abiotic stressors, including radiation, salinity, flooding, drought, temperature extremes, and heavy metal exposure, have been recognized as substantial factors contributing to the worldwide deterioration of crucial agricultural crops. On the other hand, biotic stressors refer to the adverse effects resulting from various pathogens, such as fungi, bacteria, oomycetes, nematodes, and herbivorous organisms. Based on the findings of several studies, it has been noted that abiotic stresses can lead to a drop in production surpassing 50%, whereas the decline in yield caused by biotic stresses has been estimated to be approximately 35% (Rahman et al. 2022). Given the projected growth of the global population and the consequent requirement for an adequate food supply, it is imperative to acquire a more comprehensive understanding of the intricate reactions exhibited by plants when subjected to individual and combined stressors. This knowledge is crucial in order to ultimately improve the ability of crops to withstand adverse conditions (Baillo et al. 2019). Due to their sessile nature, plants lack the ability to evade or avoid certain external stimuli. Plants have evolved a range of methods to mitigate the challenges posed by

Bioinformatics for Plant Research and Crop Breeding, First Edition. Edited by Jen-Tsung Chen.

biotic and abiotic stressors. According to Gull et al. (2019), organisms perceive the presence of external stressors, which then elicit a physiological reaction at the cellular level.

8.2 Understanding Plant Stress Response Mechanisms

8.2.1 Overview of Plant Stress Signaling Pathways

The immobile characteristics of plants have necessitated the development of intricate defense mechanisms in order to endure various environmental adversities. One of the strategies that enables organisms to tolerate stress is their capacity to perceive and interpret intricate stressors, and subsequently react in a suitable manner. The activation of intricate signaling networks facilitates the attainment of this goal in plants. Gaining a deeper comprehension of plant signaling networks would facilitate the manipulation of plants to enhance their adaptability. Nevertheless, the task at hand necessitates the simplification of the intricacies linked to signaling pathways (Sewelam et al. 2016). In addition, it should be noted that phytohormones, including salicylic acid (SA), jasmonic acid (JA), ethylene (ET), and abscisic acid (ABA), are endogenous compounds with low molecular weight, that play a crucial role in regulating the defensive mechanisms of plants against various biotic and abiotic stressors. Moreover, a wide range of hormone signaling pathways, kinases, and transcription factors (TFs) have been documented to play a role in both abiotic and biotic stress signaling (Fujita et al. 2006).

8.2.2 Role of Stress-responsive Genes and Proteins

The physiological and morphological reactions of plants to abiotic stressors frequently involve a diverse range of adaptation strategies. Maiti and Satya (2014) outline several tactics that can be employed, including stomatal closure, decreased photosynthetic activity, heightened reactive oxygen scavenging activity, limited leaf growth, and augmented root length. The phenomenon of stomatal closure in plants has been extensively reported as a physiological response to biotic stressors, particularly in the presence of pathogenic organisms. According to Bilgin et al. (2010), the aforementioned response results in a reduction in photosynthetic productivity. Plants exhibit several reactions to infections, which encompass the synthesis of detrimental chemicals, such as phytoalexins and reactive oxygen species, alongside the initiation of localized cell death. The work conducted by Nguyen et al. (2016) demonstrated the role of phytohormones in facilitating coordination among many responses. The control of tolerance toward abiotic stresses is primarily facilitated by two crucial regulators, specifically ABA and JA. The establishment of plant defense mechanisms against pathogens mostly depends on the signaling pathways mediated by SA, JA, and ethylene. The regulation of the abiotic stress response involves multiple families of TFs, which can be categorized into two distinct groups: ABA-dependent and ABA-independent. The group described above refers to a specific category of TFs known as basic leucine zipper (bZIP) TFs. These TFs have been found to be activated by a plant hormone called ABA (Banerjee and Roychoudhury (2017). As stated by Ghosh et al. (2016), the aforementioned TFs play a crucial role in the initiation of stomatal closure, the activation of genes associated with dehydration tolerance, and the facilitation of other adaptive physiological responses. However, previous studies have indicated that ABA has a tendency to increase the susceptibility of plants in biotic interactions (Peskan-Berghöfer et al. 2015). Additionally, it is well known to exhibit antagonistic effects when it interacts with SA (Ding et al. 2016).

8.2.3 Regulatory Networks Involved in Plant Stress Response

The investigation of the molecular mechanisms underlying plant responses to unfavorable stress has emerged as a significant area of academic interest, owing to advancements in molecular biology technology. The dynamic nature of the transcriptome is indicative of the varying levels of gene expression inside a certain cell, tissue, or organ of an individual organism at a given stage of growth and development. Transcriptomics has the capacity to reveal the comprehensive expression patterns of the entire genome in response to stress, hence facilitating the comprehension of the intricate regulatory network connected with the adaptation and tolerance of plants to stress (Wang et al. 2020). In response to these stressors, individuals have developed a diverse array of defense systems. The identification of a substantial quantity of noncoding RNA (ncRNA) species, such as microRNAs (miRNAs) and long noncoding RNAs (lncRNAs), has been made possible in tomato due to the swift advancement of genome sequencing technologies. In recent times, an increasing body of evidence suggests that several ncRNAs play a significant role in the plant's reaction to both biotic and abiotic challenges in the tomato species (Li et al. 2023).

Moreover, calcium ions (Ca^{2+}) serve as a ubiquitous second messenger that is promptly activated in signaling cascades. In the realm of plant signaling, Ca^{2+} plays a crucial role as a mediator in the response to various biotic and abiotic stimuli, as highlighted by Kudla et al. (2010). According to Dodd et al. (2010), the recognition and interpretation of both external and internal stimuli result in modifications to the patterns of cytosolic Ca^{2+} levels. These adjustments, in turn, give rise to variations in gene expression and cellular processes, ultimately leading to the activation of signaling pathways. Following the elicitor–receptor and effector–receptor interactions, the subsequent sequence of events that culminate in the activation of defense-related genes involves ion fluxes at the plasma membrane, specifically the influx of Ca^{2+} and the efflux of potassium ions (K^+) and chloride ions (Cl^-). Additionally, an oxidative burst occurs, accompanied by the activation of mitogen-activated protein kinases. Rapid fluctuations in intracellular Ca^{2+} levels are observed as one of the initial signaling events following the perception of microbe-associated molecular patterns (MAMPs) or pathogen-associated molecular patterns (PAMPs). These fluctuations have been identified as essential for subsequent responses in the detection of plant pathogens (Sanders et al. 2002; Ranf et al. 2011). It has been hypothesized that herbivore-associated molecular patterns (HAMPs) can be found in insect oral secretions (OS). These HAMP can be categorized as chemical elicitors that originate from insect OS and oviposition fluids, and plant-derived self-recognition factors that arise from a specific pattern of wounding (Heil 2009). The interplay between the signaling pathways of ABA, SA, JA, and ethylene (ET) has been well documented in the regulation of plant defensive responses. These pathways exhibit interactions at multiple points, including hormone-responsive TFs, thereby influencing the overall defense mechanisms of plants. Nevertheless, it is important to highlight that the ability of an entire plant to adapt and maintain growth is crucial for an effective defense response in the face of stressful conditions. Hence, the intercommunication between ABA, SA, JA, and ET with the primary growth-promoting hormones, namely auxins, gibberellins (GAs), and cytokinins (CKs), assumes a significant function in facilitating the stress response. According to Verma et al. (2016), the initiation of defensive mechanisms in plants in response to different stressors depends on the type of interaction, either positive or negative, between hormone signaling pathways. This interaction plays a crucial role in determining the overall response, rather than the individual impacts of each hormone.

The prevalence of signaling connections between several phytohormones in the regulation of diverse growth and developmental processes is evident. Plants have the ability to exert control over hormone function through many mechanisms, such as regulating the production of specific phytohormones, altering the pool of accessible hormone molecules, or employing intricate regulation of the signaling process (Verma et al. 2016).

The significance of omics in plant stress physiology lies in its comprehensive and high-throughput analysis of various molecular components, allowing for a holistic understanding of the complex biological processes underlying plant responses to stress. Omics approaches, including genomics, transcriptomics, proteomics, and metabolomics, enable the simultaneous examination of genes, transcripts, proteins, and metabolites on a genome-wide scale. These approaches provide valuable insights into the dynamic changes occurring at different levels of biological regulation in response to various environmental stressors, such as drought, salinity, and extreme temperatures.

8.3 Genome and Transcriptome Analysis for Plant Stress Physiology

8.3.1 Genomics

In recent years, the field of genome sequencing has experienced remarkable progress, with a significant shift from the laborious BAC-by-BAC (clone-by-clone shotgun strategy) approach to the more efficient and comprehensive whole-genome shotgun (WGS) methodologies. This transformation has been predominantly driven by the introduction of next-generation sequencing (NGS) technologies, which have revolutionized the process of sequencing by enabling the simultaneous analysis of billions of DNA fragments in a cost-effective and high-throughput manner.

The widespread adoption of NGS has brought about a fundamental change in the landscape of genome sequencing, making it not only feasible but also accessible to laboratories with varying levels of expertise. As a result, there has been a surge in genome-sequencing initiatives, expanding beyond the traditional focus on a limited number of reference model species to encompass the release of draft genome sequences from a diverse range of plant species and genotypes. These endeavors have been coupled with preliminary gene annotations, paving the way for a more comprehensive understanding of the molecular underpinnings of plant stress physiology and paving the way for advancements in plant breeding practices.

The proliferation of genome-sequencing initiatives has also led to the development of a multitude of web-accessible reference resources, which serve as repositories for processed data and value-added information for the wider scientific community. Apart from established reference centers like the EMBL-EBI sequence collection, the DNA Data Bank of Japan (DDBJ), and the National Center for Biotechnology Information (NCBI), which play a central role in collecting and curating data from global sequencing efforts, several specialized platforms tailored to the needs of the plant research community have also emerged.

For instance, resources such as Ensembl Plants, PlantGDB, and Phytozome offer comprehensive genomic data for numerous plant species, serving as valuable repositories for researchers exploring various aspects of plant stress physiology. Moreover, dedicated platforms, such as Plaza, TreeGenes, Gramene, and The Arabidopsis Information Resource (TAIR) cater to specific plant families or model species, offering a wealth of genomic data and analytical tools for in-depth investigation and analysis.

In the case of the well-studied plant family Solanaceae, a concerted effort has been made to provide specialized resources like The Sol Genomics Network (SGN) and Spud database, both of which have significantly contributed to enhancing our understanding of molecular mechanisms within the Solanaceae and closely related genomes. Similar initiatives have been undertaken for other prominent plant families, exemplified by resources like MaizeGDB for maize, the Legume Information System for leguminous plants, and Soybase for soybean, emphasizing the collective efforts to create comprehensive genomic databases tailored to the specific requirements of various plant species and families.

As activities in plant genome research continue to grow, collaborative efforts are being made to streamline various plant genome resources. Notably, issues with data release versions have emerged, even on prominent platforms like Ensembl Plants (Aken et al. 2017) and NCBI (O'Leary et al. 2016). This variability, linked to different genome assemblies and gene annotation versions, poses a significant challenge regarding the dependability and comparability of information across these resources. The diversity stems from rapid advancements in sequencing technologies, which have greatly increased the production of sequence data cost-effectively. However, the swift release of new genome assemblies complicates the establishment of high-quality reference standards (Chiusano 2015). Moreover, the continuous release of new or updated sequenced genomes faces hurdles in efficiently conducting subsequent analyses, including gene annotation, resource updating, and curation. This, in turn, affects the maintenance and accuracy of information in reference platforms. Therefore, the field of bioinformatics must evolve alongside the rapid development of new technologies, adapting to handle larger and more complex datasets. This adaptation is crucial for providing reliable information for both individual species analyses and comparative studies. Clear evaluations of assembly quality, reconciliation of annotation versions, and the use of appropriate methods for gene-based comparative analyses, including defining orthologs or paralog relationships, are imperative. This is especially crucial considering the limitations associated with protein-coding region predictions, particularly in initial gene annotations (Ambrosino et al. 2018).

While novel strategies are being introduced to address the challenges posed by big data in bioinformatics (Goodwin et al. 2015), the comprehensive assessment of information quality and the establishment of well-validated references require sustained long-term efforts to ensure their effective dissemination and subsequent utilization within the scientific community. This is critical as reliable and stable references are fundamental for the appropriate utilization of resources in associated endeavors, such as those in transcriptomics, epigenomics, proteomics, and metabolomics (Chiusano et al. 2008). Furthermore, potential challenges arising from limitations in functional data annotation should not be overlooked. Although significant efforts are dedicated to providing comprehensive gene function descriptions, such as those supported by the gene ontology consortium (GO) (Ashburner et al. 2000), and the integration of additional pertinent information (Nguyen and Wang 2020), continuous evolution is necessary to accommodate the expanding knowledge base specific to each species, as existing descriptions are inherently confined by the scope of current observations (Van Bel et al. 2019).

Various knowledge-driven repositories can introduce additional heterogeneity, given the availability of diverse resources often interconnected by related content. For instance, a query in the NCBI Gene database, which consolidates gene-specific information from multiple sources (Brown et al. 2015), using the general keyword "abiotic stress" yields a substantial collection of genes with functional annotations containing the specified term. In contrast, a similar query in the NCBI RefSeq database, renowned for its comprehensive and meticulously curated nonredundant sequences (O'Leary et al. 2016), produces different results. While acknowledging the potential bias introduced by nonspecific keywords and advocating for more stringent terms like "salt stress" or "drought stress," the divergent outcomes highlight that discrepancies can arise from a single reference source. Although experts understand the technical reasons behind these differences, the presence of ambiguous information across resources may confuse nonexperts and compromise subsequent analyses. Additionally, variations in the number of abiotic stress-related genes in different plant species, exemplified by differences between *Arabidopsis thaliana* and other species, emphasize the ongoing need for comparative initiatives facilitating the transfer of gene role assessments from reference models to other species.

8.3.2 Transcriptomics

The utilization of comparative transcriptome data plays a crucial role in unraveling the functionality of genes and understanding the intricate mechanisms that govern gene expression across various tissues and under different conditions (Roux et al. 2015). Transcriptomic data proves invaluable in depicting gene expression patterns and inferring the regulatory mechanisms that control cellular functions. Consequently, extensive investigations into how plants respond to abiotic stress have prominently employed microarray and RNA sequencing analyses, encompassing the exploration of expressed sequence tags (ESTs) and RNA-seq methods (Malde and Jonassen 2008).

Microarray technology has long stood as a pivotal method for investigating the transcriptional patterns within biological samples (Schena et al. 1995). The introduction of the GeneChip ATH1 Genome Array tailored for *A. thaliana*, encompassing probe sets for nearly 24,000 genes, has markedly propelled microarray analyses within the realm of plant sciences (Redman et al. 2004). This standardized array not only has facilitated data collection across various tissues and conditions but has also played a crucial role in constructing comprehensive gene expression atlases for Arabidopsis. Moreover, it has been instrumental in developing essential computational tools to support gene coexpression analyses (Di Salle et al. 2017).

The landscape of functional genomics data is primarily hosted in two vital public repositories, the ArrayExpress Archive (Kolesnikov et al. 2015) and the Gene Expression Omnibus

(GEO) (Edgar et al. 2002). A comprehensive search using the "abiotic stress" keywords in these platforms has yielded 157 and 436 array experiments in the ArrayExpress and GEO databases, respectively.

ESTs originating from randomly selected complementary DNA (cDNA) libraries (Adams et al. 1991) are stored in the specialized GenBank division called dbEST at the NCBI (Boguski et al. 1993). Currently, housing 221,400 ESTs derived from 77 distinct libraries within viridiplantae, these were identified using "abiotic stress" keywords (Figure 8.1). Although the sequencing of ESTs has been superseded by the adoption of cost-effective and robust NGS techniques such as RNA-seq and small RNA-seq, the preference for RNA-seq over microarrays has significantly increased due to its heightened sensitivity and cost-effectiveness (Iovieno et al. 2016). The versatility of RNA-seq analyses extends beyond model species, finding applicability in the study of transcriptomes in non-model species as well (Alagna et al. 2009; Zenoni et al. 2011).

The NCBI Sequence Read Archive (SRA) (100) stands out as a comprehensive resource, compiling publicly available NGS data from various platforms. Currently, the SRA records 330 accessions from 26 plant species, identified through querying with the "abiotic stress" keywords within the viridiplantae group.

Apart from these extensive repositories, significant efforts have been directed toward establishing species-specific or clade-specific databases. These databases have evolved from being predominantly EST-based to now incorporating NGS data. Such resources play a crucial role in aggregating and disseminating expression data and associated tools within the scientific community (Moretto et al. 2016).

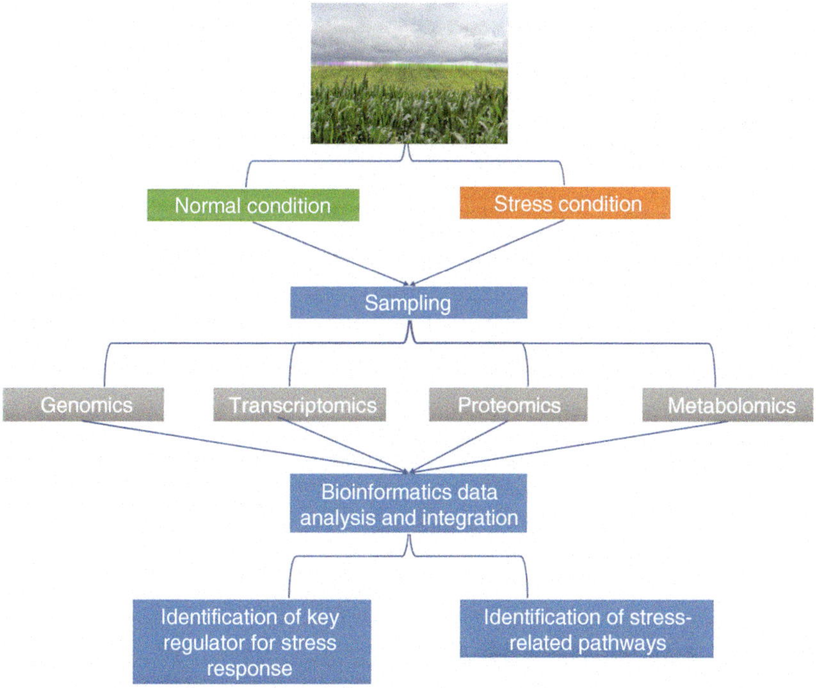

Figure 8.1 Workflow of omics approaches to study plant stress. *Source:* kyrychukvitaliy/Adobe Stock.

The active investigation into nonncRNAs, primarily facilitated by high-throughput sequencing technologies, has shed light on their crucial roles in plant physiology, pathology, and stress responses (Ramesh et al. 2019). Notably, small ncRNAs, such as microRNAs (miRNAs) with lengths ranging from 19 to 23 nucleotides, have demonstrated their responsiveness to abiotic stresses across various plant species (Liu et al. 2017). Several dedicated platforms and databases, such as RFAM, miRbase, plantDARIO, and PsRobot, have provided substantial support for the analyses of small RNAs and related research efforts within the field of plant sciences.

8.4 Proteomics and Metabolomics Approaches

8.4.1 Proteomics

Enzymes and proteins play vital roles in cellular functions, influencing how plants respond to abiotic stresses by shaping phenotypic traits and regulating cellular changes (Boguski et al. 1993). Proteomics has emerged as a robust investigative tool for exploring posttranslational modifications (PTMs), protein–protein interactions (PPI), and molecular impacts resulting from disturbances in plant metabolism (Meena et al. 2017). Through the scrutiny of protein accumulation, structure, and function under abiotic stresses, proteomic studies enhance our comprehension of plant cell functionality. These studies aid in the identification of protein effectors, targets, and variations within interactive networks (Zhang et al. 2020).

In its initial phase, proteomics heavily relied on two-dimensional gel electrophoresis (2D-gel) to segregate proteins based on their isoelectric point and molecular mass, coupled with mass spectrometry (MS) analysis to study differentially accumulated proteins (Rabilloud and Lelong 2011). Recent progress in MS platforms has given rise to the shotgun proteomics approach, involving the breakdown of protein mixtures into smaller peptides. Liquid chromatography (LC) and tandem mass spectrometry (MS/MS) are then employed for peptide separation and identification, supported by data analysis tools (Domon and Aebersold 2010). These peptides are subsequently pieced together to potentially reconstruct the entire proteome, leveraging bioinformatics tools such as IDPicker (Ma et al. 2009) and PAnalyzer (Prieto et al. 2012).

Proteomic explorations have been widely carried out across diverse plant species, including *A. thaliana*, *Oryza sativa*, *Glycine max*, *Triticum aestivum*, *Zea mays*, *Brassica napus*, *Solanum tuberosum*, and *Solanum lycopersicum* (Kosová et al. 2018). These investigations have unveiled dynamic changes in proteins and their isoforms, shedding light on signaling pathways, TFs, and PPI. Notably, factors such as genotype, environmental stress conditions, protein subcellular localization, and PTMs have emerged as crucial determinants. Furthermore, these studies have played a vital role in identifying proteins and enzymes potentially involved in stress-related processes and compound production (Kosová et al. 2015).

For instance, Chen et al. (2015) delved into the mechanisms of cold acclimation in *Medicago sativa* cultivars, revealing distinctions between freezing-tolerant and freezing-sensitive varieties. Ghabooli et al. (2013) unraveled the molecular underpinnings of water stress tolerance in barley induced by a fungus, pinpointing 45 differentially accumulated proteins associated with photosynthesis, signal transduction, and plant defense responses. Balbuena et al. (2011) probed into how sunflower adjusts its metabolism during cold treatment, unveiling varied proteome responses to cold acclimation across different sunflower lines. Other proteomic analyses have proposed mechanisms for chilling attenuation in plants and explored cross-tolerance processes. Recent studies have delved into proteome alterations in *S. lycopersicum* under high light stress (Parrine et al. 2018) and scrutinized plant responses to nitrogen deficiency in different *S. tuberosum* varieties using a shotgun proteomic approach (Jozefowicz et al. 2017).

Despite their invaluable contributions to unraveling biological mechanisms, the compilation and organization of extensive proteome data pose ongoing challenges. Public repositories for proteomics data storage and dissemination are relatively limited compared to genomics or transcriptomics resources. This scarcity is attributed to the diverse data types and experimental procedures involved, complicating the establishment of standardized collections. Furthermore, the intricate nature of data mining and interpretation in proteomics, influenced by various signals, analytical approaches, bioinformatics pipelines, and associated statistical analyses, adds an extra layer of complexity (den Ridder et al. 2020).

Addressing these limitations involves the establishment of the PAPPSO proteomic facility, playing a crucial role in advancing technology within the proteomics analysis pipeline. Various tools, including MassChroQ for LC-MS alignment and peak extraction, X!TandemPipeline for peptide filtering and identification, and PROTICdb for proteome data analysis and storage, have been developed to enhance the field of proteomics research. An impactful initiative, the Arabidopsis proteome draft, provides a comprehensive atlas covering protein accumulation, modification, regulation, and thermal stability, serving as a valuable resource for the scientific community (Mergner et al. 2020; Jarzab et al. 2020).

Prominent public databases for proteomics data encompass The ProteomicsDB (Samaras et al. 2020), The PRoteomics IDEntifications (PRIDE) database (Vizcaíno et al. 2012), PeptideAtlas (Farrah et al. 2014), and the Global Proteome Machine Database (GPMDB) (Craig et al. 2004). The Plant Proteomics Database (PPDB) focuses on Arabidopsis and maize studies, providing a wealth of proteomics experiments and tools for comparative and subcellular proteomics, biochemical pathways, and specific abiotic stress-related peptides. For the identification of stress-related integral membrane proteins in monocots and dicots, ARAMEMNON proves instrumental.

In addition, resources like Plprot, AT_CHLORO, and WallProtDB have been developed, specifically catering to plant subcellular proteome databases at the plastid and cell wall levels. P3DB facilitates the exploration of stress-induced phosphorylated proteins across various plant species. Simultaneously, NetPhos 3.1 and PhosPhat aid in predicting phosphorylation sites and retrieving information about kinase substrates. These collective resources significantly contribute to the investigation of PTM alterations, playing a crucial role in understanding plant responses to abiotic stress.

Bioinformatics approaches such as docking and molecular dynamics simulation are used in the study of plant stress physiology to unravel complex protein–ligand interactions and understand the dynamic behavior of biomolecules under stress conditions. These approaches aid in elucidating the molecular mechanisms underlying stress response pathways and contribute to the identification of potential targets for stress tolerance enhancement in plants. Docking studies enable the prediction of the binding modes and affinity between small molecules, such as potential stress-related compounds or ligands, and target proteins involved in stress response pathways. By employing computational algorithms and structural biology techniques, docking simulations provide insights into the key residues and binding pockets involved in the interaction between stress-related ligands and target proteins, thereby facilitating the design of novel compounds or agents for modulating stress responses in plants. Additionally, molecular dynamics simulations play a crucial role in understanding the conformational changes and dynamic behavior of plant stress-related proteins under different environmental conditions. By simulating the movement and interactions of biomolecules over time, molecular dynamics simulations provide a detailed understanding of the structural stability, flexibility, and functional dynamics of proteins involved in stress signaling and adaptation pathways. These simulations help in elucidating the impact of environmental stress factors on protein conformation and dynamics, offering valuable insights into the structural basis of stress tolerance mechanisms in plants.

Several studies have utilized these bioinformatics techniques to elucidate the interactions between stress-responsive plant proteins and their associated effectors (Murmu and Archak 2023). For example, studies have employed molecular docking to investigate the binding interactions between specific stress-related TFs and their target DNA sequences. Molecular dynamics simulations have been utilized to explore the conformational changes and stability of stress-related enzymes under different environmental stress conditions. Research focusing on the PPI between pathogen effectors and plant host proteins has shed light on the intricate molecular mechanisms underlying plant–pathogen interactions during stress conditions. Studies have employed a combination of experimental techniques and computational approaches to identify and characterize the key interacting domains and interfaces involved in the recognition and response of plant hosts to pathogen effectors. Understanding the dynamics of PPI networks can provide insights into the signaling pathways and regulatory mechanisms that govern plant stress responses, thereby facilitating the development of effective strategies for enhancing stress tolerance in crops. These bioinformatics approaches, coupled with experimental validation, contribute to a comprehensive understanding of the complexities of plant stress physiology and offer valuable insights into the molecular basis of plant stress responses and interactions between pathogens and their host plants.

Various dedicated web-based resources have been established, focusing exclusively on plant stress-related proteomics data, as detailed in the following section.

8.4.2 Metabolomics

Plant responses to abiotic stresses, such as extreme temperatures, drought, salinity, and nutrient limitations, lead to profound alterations not only at the transcriptome and proteome levels but also in the intricate network of plant metabolites. The plant metabolome, which encompasses the complete set of small molecules and organic compounds in a biological system, is directly influenced by the expression of stress-responsive genes and the accumulation of stress-induced proteins. This interplay between gene expression, protein dynamics, and metabolic shifts serves as a crucial indicator of how plants acclimate and adapt to adverse environmental conditions. Moreover, it is increasingly recognized that the metabolic changes in response to stress represent the most notable phenotypic effects, providing valuable insights into the molecular mechanisms underlying stress tolerance and resilience.

Metabolomics, as a powerful tool for comprehensive and high-throughput analysis of the plant metabolome, plays a pivotal role in deciphering the complex metabolic reprogramming that occurs in response to various abiotic stresses. By employing advanced analytical techniques, such as gas chromatography–mass spectrometry (GC-MS), liquid chromatography–mass spectrometry (LC-MS), and nuclear magnetic resonance (NMR) spectroscopy, metabolomics enables the detection, identification, and quantification of a wide array of primary and secondary metabolites in plant tissues.

The plant metabolome consists of two main components: the nonvolatile metabolome, which includes polar and nonpolar compounds soluble in hydroalcoholic or organic solvents, and the volatile metabolome, which encompasses semi-volatile and volatile compounds. Metabolomics employs both "untargeted" and "targeted" approaches. The "untargeted" approach involves a comprehensive analysis of mass chromatograms and subsequent bioinformatics analyses, enabling the identification of differentially accumulated metabolites without prior knowledge of their identities. On the other hand, the "targeted" approach focuses on the precise detection and quantification of known primary and secondary metabolites using established metabolomics databases and analytical platforms (Original Source).

Pioneering metabolomics studies in model plant species, such as Arabidopsis, have shed light on the dynamic metabolic changes occurring in response to diverse abiotic stresses. These studies have provided valuable insights into the metabolic pathways and key metabolites involved in stress acclimation and tolerance. Furthermore, investigations in various crop species, including maize, tomato, rice, and wheat, have elucidated the specific metabolic signatures associated with stress tolerance, enabling the identification of potential biomarkers and candidate genes involved in stress responses. Additionally, metabolomics has played a crucial role in uncovering the metabolic regulation of essential physiological processes, such as photosynthesis, respiration, and cell wall composition, under different stress conditions.

Although metabolomics offers a powerful approach to unraveling the intricate metabolic changes induced by abiotic stresses, it poses several challenges, including data heterogeneity, the complexity of metabolite identification, and the lack of comprehensive functional resources linking specific metabolites to their biological roles in stress responses. Despite these challenges, ongoing advancements in metabolomics technologies and bioinformatics tools continue to enhance our understanding of the intricate metabolic networks and adaptive strategies employed by plants to cope with environmental stresses. Various databases, such as MetaboLights, PlantMetabolomics.org, and Plant Metabolome Database (PMD) serve as valuable resources for the storage, annotation, and analysis of metabolomics data related to plant stress responses. Furthermore, the development of computational tools, including MZmine 2, MetAlign, MSClust, and MetFrag, has significantly contributed to the efficient processing, annotation, and interpretation of complex metabolomics data, facilitating the discovery of stress-responsive metabolites and metabolic pathways. Despite the challenges posed by the diversity and complexity of plant metabolites, the integration of diverse analytical approaches and the establishment of comprehensive metabolite annotation pipelines hold the potential to uncover novel insights into the intricate metabolic strategies employed by plants to withstand and adapt to challenging environmental conditions.

8.5 Data Integration and Systems Biology Approaches

Data integration stands as a cornerstone in the analysis of multifaceted -omics data, allowing researchers to merge diverse datasets from genomics, transcriptomics, proteomics, and metabolomics. This integration enables a holistic view of the complex regulatory networks and signaling pathways that govern plant stress responses. The growing magnitude of data generated through various -omics methodologies necessitates the continuous development and refinement of powerful and user-friendly bioinformatics tools. Key databases such as KEGG (Moriya et al. 2007), Reactome (Fabregat et al. 2016), and MetaCyc (Caspi et al. 2014) play pivotal roles in elucidating the intricate interactions of genes, proteins, and metabolites within specific pathways and biological processes. While KEGG provides insights into a limited number of *O. sativa* proteins involved in abiotic stress responses, Reactome offers a broader perspective, highlighting pathways related to cellular responses to heat and oxidative stress. Conversely, the absence of relevant results in MetaCyc emphasizes the need for comprehensive and curated data sources for plant stress-related studies.

In GO analysis, AmiGO (Carbon et al. 2009) emerges as a crucial platform for exploring and understanding the functional implications of genes and gene products. The search results in AmiGO underscore several GO terms associated with "abiotic stress," although the hierarchical nature of these terms can sometimes lead to redundancies, demanding careful interpretation of the enrichment results.

Blast2GO serves as a valuable tool for researchers dealing with datasets lacking curated annotations (Conesa and Götz 2008). Its ability to integrate data from various sources, including InterPro, enzyme codes, and GO direct acyclic graphs, contributes significantly to the annotation process and subsequent investigations of the biological information contained within the datasets. Furthermore, the introduction of AgriGO (Du et al. 2010) and its subsequent updated version, AgriGO2.0 (Tian et al. 2017), has revolutionized GO enrichment analysis, offering a comprehensive set of tools for various species and data types, thereby promoting a more efficient and user-friendly experience for researchers in the agricultural domain.

The versatility of MapMan software stands out in its capability to map transcripts, proteins, and metabolites onto cellular pathways and processes (Thimm et al. 2004). Its open and flexible nature allows for the creation of customized maps, thereby enabling a more holistic understanding of gene functions and their implications in stress responses. The integration of GOMapMan has further enhanced the annotation of gene functions, enabling researchers to delve deeper into the intricate networks and regulatory mechanisms associated with abiotic stress responses (Ramšak et al. 2014).

Moreover, the application of network theory approaches and correlation analysis tools, exemplified by Cytoscape, has significantly aided in the identification of coexpressed and accumulated elements within specific functional events. The visualization of complex

-omics datasets using network graphs and correlation matrices has provided valuable insights into the underlying molecular mechanisms governing plant stress responses.

8.6 Bioinformatics Resources for Plant Stress

8.6.1 Databases

Web-based databases and platforms that are publicly accessible play a crucial role in organizing and disseminating vital information related to plant stress responses, fostering collaborative engagement within the scientific community interested in abiotic stress in plants. An integral resource in this field is Plantstress, a comprehensive web repository that provides essential information and professional updates on various aspects of plant environmental abiotic stress. Within this resource, researchers can access valuable knowledge, including details about stress-responsive genes and their functional annotations. Another noteworthy database is the Plant Stress Gene Database, which contains a curated collection of 259 stress-related genes from diverse plant species (Prabha et al. 2011). Users can easily retrieve information by querying based on species, gene ID, or specific gene functions, as well as explore details about paralog or ortholog genes across multiple species. For a reference model in plant biology, the *A. thaliana* Stress Responsive Gene Database (ASRGD) stands out as a comprehensive collection of manually curated stress tolerance genes associated exclusively with *A. thaliana* (Borkotoky et al. 2013). This database encompasses 637 genes involved in approximately 50 different stress conditions, allowing users to access information through various search parameters, including keywords or stress types (e.g. osmotic, heat stress).

At the protein level, the Plant Stress Protein Database (PSPDB) provides extensive information on 2064 manually curated plant stress proteins from 134 plant species, highlighting their functional roles under the influence of 30 different biotic and abiotic stresses (Anil Kumar et al. 2014). Researchers can extract valuable insights from the database using diverse search options such as gene, species, keyword, citation, gene families, and taxonomic classifications. Additionally, the PlantPReS database (Mousavi et al. 2016) serves as a specialized resource in the field of proteomics, encompassing over 20,413 entries from 456 meticulously curated articles, providing a repository of more than 10,600 unique stress-responsive proteins. The user-friendly interface of PlantPReS offers various analytical tools, including a search engine, GO analysis, cross-referencing capabilities, and expression patterns of specific proteins involved in stress response pathways.

Additionally, there are specialized databases tailored to specific stress types, such as DroughtDB (Alter et al. 2015), which compiles a comprehensive gene list related to drought stress responses. This resource offers insights into computed ortholog genes across nine models and crop plants. In the domain of molecular and functional roles, the Stress Responsive Transcription Factor Database (STIFDB v.2) stands out as a valuable resource for targeted experimental and computational studies. It provides over 38,000 associations of stress signals, stress-responsive genes, and TF binding sites using the STIF algorithm based on an HMM model (Naika et al. 2013). Users can explore information by gene name, chromosome, TF, or stress signal. Similarly, RiceSRTFDB focuses on the rice–water stress community, offering a comprehensive database of rice TFs. It includes detailed expression patterns, cis-regulatory elements, and mutant information (Priya and Jain 2013). The website also provides an Expression Viewer for examining the differential expression of selected genes under salinity or drought stresses.

PASmiR is a database that consolidates information on miRNA molecular regulation in plant abiotic stress, incorporating data from around 200 published studies (Zhang et al. 2013). Documenting 1038 regulatory relationships between 682 miRNAs and 35 different types of abiotic stresses across 33 plant species, this resource allows users to search by miRNA name, species, and specific abiotic stress types, with the option to download the entire dataset. While specific metabolomics databases dedicated to abiotic stresses are not currently available, the Platform for RIKEN Metabolomics (PRIME) serves as a repository of data collected under various stress conditions in Arabidopsis, *Brachypodium distachyon*, and wheat. This resource proves valuable for researchers engaged in metabolomics and transcriptomics analyses. PhenoMeter (PM) takes an integrative approach to link different metabolomics resources and studies. It facilitates the connection between metabolic phenotypes and genetic materials, unraveling biological processes at both biochemical and molecular levels. PhenoMeter utilizes metabolite response patterns as queries, searching the MetaPhen database for statistically similar or inverse responses, thus highlighting functional associations.

8.6.2 Machine Learning Tools

The utilization of machine learning (ML) and associated tools has significantly progressed the comprehension of plant stress physiology by offering valuable insights into intricate biological processes and stress responses. Within the domain of bioinformatics strategies aiming to unravel the complexities of plant stress physiology, the integration of ML techniques has streamlined the analysis of extensive omics data, the identification of pivotal stress-responsive genes, and the anticipation of stress tolerance mechanisms in plants.

Numerous studies have underscored the effectiveness of ML-based approaches in decoding the molecular mechanisms at the core of plant stress physiology. Zandalinas et al. (2020) employed ML models to foresee the physiological and molecular responses of plants to elevated temperatures and oxidative stress.

ML-based tools prove instrumental in amalgamating multi-omics data to unveil the intricacies of plant stress responses, facilitating the identification of key biomarkers and signaling pathways linked to stress tolerance. Additionally, Chele et al.'s (2021) research accentuated the significance of ML-driven approaches in predicting crop performance under diverse environmental stress conditions, thereby aiding in the development of stress-resilient crop varieties.

The compilation of resources delineated in Table 8.1 underscores the variety of studies encompassing different plants and stress types. Nonetheless, the limited availability of integrated platforms exclusively dedicated to abiotic stresses underscores the imperative need for collective endeavors and comprehensive databases, ensuring robust data and information resources for subsequent investigations and data mining within this field.

Table 8.1 Bioinformatics resources for plant stress analysis.

Resources	Description	Website (URL, accessed on October 10, 2023)	References
Abiotic stress			
Plant Stress Gene Database	• Curates information on genes involved in plant stress responses • Provides data on various plant species and stress conditions	http://ccbb.jnu.ac.in/stressgenes/	Prabha et al. (2011)
RiceSRTFDB	• Focuses on stress-responsive transcription factors in rice • Provides comprehensive information on regulatory elements and networks	https://www.nipgr.ac.in/RiceSRTFDB.html	Priya and Jain (2013)
Stress Responsive Transcription Factor Database (version 2)	• A comprehensive resource for stress-responsive transcription factors in plants • Offers insights into the regulatory mechanisms governing plant stress responses	https://caps.ncbs.res.in/stifdb2 (inaccessible)	Naika et al. (2013)
Arabidopsis thaliana Stress Responsive Gene Database	• Specializes in stress-responsive genes in *Arabidopsis thaliana* • Provides curated information on gene expression and regulatory networks	http://srgdb.bicpu.edu.in/ (inaccessible)	Borkotoky et al. (2013)
PASmiR	• Facilitates the exploration of stress-responsive miRNAs in plants • Offers a comprehensive database of miRNA-mediated regulatory networks under stress	http://hi.ustc.edu.cn:8080/PASmiRhttp://pcsb.ahau.edu.cn:8080/PASmiR (inaccessible)	Zhang et al. (2013)
Plant Stress Protein Database	• Curates data on stress-responsive proteins in various plant species • Offers insights into the structural and functional aspects of stress-related proteins	https://bioclues.org/pspdb/	Kumar et al. (2014)

(Continued)

Table 8.1 (Continued)

Resources	Description	Website (URL, accessed on October 10, 2023)	References
DroughtDB	• Specializes in information related to plant drought response mechanisms • Provides data on genes, proteins, and regulatory pathways involved in drought tolerance	http://pgsb.helmholtz-muenchen.de/droughtdb/ (inaccessible)	Alter et al. (2015)
PlantPReS	• Focuses on plant proteins involved in stress responses • Provides comprehensive information on the functional roles of stress-responsive proteins	http://www.proteome.ir/	Mousavi et al. (2016)
Plantstress	• Serves as a comprehensive online platform for plant stress-related research • Offers resources, tools, and information on various aspects of plant stress physiology and molecular responses	https://plantstress.com/	2000
ASRpro	• ML-based computational tool designed for the identification of proteins associated with six major abiotic stresses, including heat, drought, cold, salinity, oxidative stress, and light stress	https://iasri-sg.icar.gov.in/asrpro/	Meher et al. (2022b)
ASLncR	• It employs the Support Vector Machine to identify potential abiotic stress-responsive lncRNAs	https://iasri-sg.icar.gov.in/aslncr/	Pradhan et al. (2023)
ASRmiRNA	• It specifically developed to predict miRNAs and pre-miRNAs that play a crucial role in regulating stress-responsive genes in plants under various abiotic stresses • The tool employs a support vector machine (SVM) for accurate prediction, utilizing the power of machine learning to identify potential abiotic stress-responsive miRNAs	http://cabgrid.res.in:8080/asrmirna/ (inaccessible)	Meher et al. (2022a)
DeepAProt	• Classifies proteins responsive to four abiotic stresses – cold, heat, salinity, and drought – from key crops within the Poaceae family	http://login1.cabgrid.res.in:5500/	Ahmed et al. (2023)

Biotic stress			
RiceMetaSysB	• It consolidates information on genes responsive to salt and drought stress, derived from various microarray experiments conducted in rice. The raw data from these experiments is accessible in NCBI GEO datasets • Additionally, the database includes microsatellite markers associated with salt and drought-responsive genes	http://14.139.229.201/RiceMetaSys/	Sandhu et al. (2017)
InterSPPI	• It predicts protein-protein interactions between Arabidopsis and pathogens based on random forest	http://systbio.cau.edu.cn/intersppi/AraPPI/index.php	Yang et al. (2019)
Abiotic and biotic stress both			
PncStress	• Curated repository containing experimentally confirmed microRNAs, lncRNAs, and circRNAs observed to respond to both abiotic and biotic stress conditions in plants	https://bis.zju.edu.cn/PncStress/	Wu et al. (2020)
Rice Stress-Resistant SNP (RSRS) Database	• Catalogs SNPs associated with the stress-resistant traits in rice, encompassing both biotic and abiotic stressors • Integrates information on SNP functionality, genome annotation, and phenotype details	http://bioinformatics.fafu.edu.cn/RSRS (inaccessible)	Tareke Woldegiorgis et al. (2019)

8.7 Conclusion

The convergence of diverse bioinformatics approaches, including data integration, systems biology analyses, and network-based modeling, has significantly enhanced our understanding of the molecular underpinnings and regulatory networks involved in plant stress adaptation and resilience. The comprehensive utilization of bioinformatics tools, coupled with the integration of multi-omics data, has provided a holistic perspective on the dynamic interplay between genes, proteins, and metabolites in plant stress physiology. By leveraging sophisticated computational algorithms and predictive models, researchers have successfully elucidated the complex regulatory networks that orchestrate plant responses to diverse environmental stressors, including drought, salinity, temperature extremes, and pathogen attacks.

Furthermore, the collaborative efforts of the scientific community in curating and expanding publicly available databases and resources have facilitated the systematic annotation and analysis of plant genomes, enabling researchers to unravel the functional significance of specific genes, pathways, and regulatory elements involved in stress response mechanisms. The continuous development of user-friendly bioinformatics platforms and the implementation of standardized protocols have streamlined data analysis and interpretation, fostering a more collaborative and integrative research environment in the field of plant stress physiology.

Moving forward, the continued exploration and refinement of bioinformatics tools, in conjunction with experimental validation and functional studies, hold the key to unlocking novel insights into the adaptive strategies employed by plants under stressful environmental conditions. By harnessing the power of bioinformatics, researchers are poised to make significant strides in the development of resilient crop varieties and the implementation of sustainable agricultural practices, thereby addressing global challenges related to food security and environmental sustainability.

References

Adams, M.D., Kelley, J.M., Gocayne, J.D. et al. (1991). Complementary DNA sequencing: expressed sequence tags and human genome project. *Science* 252 (5013): 1651–1656.

Ahmed, B., Haque, M.A., Iquebal, M.A. et al. (2023). DeepAProt: deep learning based abiotic stress protein sequence classification and identification tool in cereals. *Frontiers in Plant Science* 13: 1008756.

Aken, B.L., Achuthan, P., Akanni, W. et al. (2017). Ensembl 2017. *Nucleic Acids Research* 45 (D1): D635–D642.

Alagna, F., D'Agostino, N., Torchia, L. et al. (2009). Comparative 454 pyrosequencing of transcripts from two olive genotypes during fruit development. *BMC Genomics* 10x(1): 1–15.

Alter, S., Bader, K.C., Spannagl, M. et al. (2015). DroughtDB: an expert-curated compilation of plant drought stress genes and their homologs in nine species. *Database* 2015: bav046.

Ambrosino, L., Ruggieri, V., Bostan, H. et al. (2018). Multilevel comparative bioinformatics to investigate evolutionary relationships and specificities in gene annotations: an example for tomato and grapevine. *BMC Bioinformatics* 19: 85–101.

Anil Kumar, S., Hima Kumari, P., Sundararajan, V.S. et al. (2014). PSPDB: plant stress protein database. *Plant Molecular Biology Reporter* 32: 940–942.

Ashburner, M., Ball, C.A., Blake, J.A. et al. (2000). Gene ontology: tool for the unification of biology. *Nature Genetics* 25 (1): 25–29.

Baillo, E.H., Kimotho, R.N., Zhang, Z., and Xu, P. (2019). Transcription factors associated with abiotic and biotic stress tolerance and their potential for crops improvement. *Genes (Basel)* 10 (10): 771. https://doi.org/10.3390/genes10100771. PMID: 31575043; PMCID: PMC6827364.

Balbuena, T.S., Salas, J.J., Martinez-Force, E. et al. (2011). Proteome analysis of cold acclimation in sunflower. *Journal of Proteome Research* 10 (5): 2330–2346.

Banerjee, A. and Roychoudhury, A. (2017). Abscisic-acid-dependent basic leucine zipper (bZIP) transcription factors in plant abiotic stress. *Protoplasma* 254: 3–16.

Bilgin, D.D. et al. (2010). Biotic stress globally downregulates photosynthesis genes. *Plant, Cell & Environment* 33: 1597–1613.

Boguski, M.S., Lowe, T.M., and Tolstoshev, C.M. (1993). dbEST—database for "expressed sequence tags". *Nature Genetics* 4 (4): 332–333.

Borkotoky, S., Saravanan, V., Jaiswal, A. et al. (2013). The Arabidopsis stress responsive gene database. *International Journal of Plant Genomics* 2013: 949564.

Brown, G.R., Hem, V., Katz, K.S. et al. (2015). Gene: a gene-centered information resource at NCBI. *Nucleic Acids Research* 43 (D1): D36–D42.

Carbon, S., Ireland, A., Mungall, C.J. et al. (2009). AmiGO: online access to ontology and annotation data. *Bioinformatics* 25 (2): 288–289.

Caspi, R., Altman, T., Billington, R. et al. (2014). The MetaCyc database of metabolic pathways and enzymes and the BioCyc collection of Pathway/Genome Databases. *Nucleic Acids Research* 42 (D1): D459–D471.

Chele, K.H., Steenkamp, P., Piater, L.A. et al. (2021). A global metabolic map defines the effects of a Si-based biostimulant on tomato plants under normal and saline conditions. *Metabolites* 11 (12): 820.

Chen, J., Han, G., Shang, C. et al. (2015). Proteomic analyses reveal differences in cold acclimation mechanisms in freezing-tolerant and freezing-sensitive cultivars of alfalfa. *Frontiers in Plant Science* 6: 105.

Chiusano, M.L. (2015). On the multifaceted aspects of bioinformatics in the next generation era: the run that must keep the quality. *Next Generation Sequencing & Applications* 2: e106.

Chiusano, M.L., D'Agostino, N., Traini, A. et al. (2008). ISOL@: an Italian SOLAnaceae genomics resource. *BMC Bioinformatics* 9 (2): 1–11.

Conesa, A. and Götz, S. (2008). Blast2GO: a comprehensive suite for functional analysis in plant genomics. *International Journal of Plant Genomics* 2008: 619832.

Craig, R., Cortens, J.P., and Beavis, R.C. (2004). Open source system for analyzing, validating, and storing protein identification data. *Journal of Proteome Research* 3 (6): 1234–1242.

Di Salle, P., Incerti, G., Colantuono, C., and Chiusano, M.L. (2017). Gene co-expression analyses: an overview from microarray collections in *Arabidopsis thaliana*. *Briefings in Bioinformatics* 18 (2): 215–225.

Ding, Y., Dommel, M., and Mou, Z. (2016). Abscisic acid promotes proteasome-mediated degradation of the transcription coactivator NPR 1 in *Arabidopsis thaliana*. *The Plant Journal* 86: 20–34.

Dodd, A.N., Kudla, J., and Sanders, D. (2010). The language of calcium signaling. *Annual Review of Plant Biology* 61: 593–620.

Domon, B. and Aebersold, R. (2010). Options and considerations when selecting a quantitative proteomics strategy. *Nature Biotechnology* 28 (7): 710–721.

Du, Z., Zhou, X., Ling, Y. et al. (2010). agriGO: a GO analysis toolkit for the agricultural community. *Nucleic Acids Research* 38 (suppl_2): W64–W70.

Edgar, R., Domrachev, M., and Lash, A.E. (2002). Gene expression omnibus: NCBI gene expression and hybridization array data repository. *Nucleic Acids Research* 30 (1): 207–210.

Fabregat, A., Sidiropoulos, K., Garapati, P. et al. (2016). The reactome pathway knowledgebase. *Nucleic Acids Research* 44 (D1): D481–D487.

Farrah, T., Deutsch, E.W., Omenn, G.S. et al. (2014). State of the human proteome in 2013 as viewed through PeptideAtlas: comparing the kidney, urine, and plasma proteomes for the biology-and disease-driven Human Proteome Project. *Journal of Proteome Research* 13 (1): 60–75.

Fujita, M., Fujita, Y., Noutoshi, Y. et al. (2006). Crosstalk between abiotic and biotic stress responses: a current view from the points of convergence in the stress signaling networks. *Current Opinion in Plant Biology* 9 (4): 436–442.

Ghabooli, M., Khatabi, B., Ahmadi, F.S. et al. (2013). Proteomics study reveals the molecular mechanisms underlying water stress tolerance induced by *Piriformospora indica* in barley. *Journal of Proteomics* 94: 289–301.

Ghosh, T.K., Kaneko, M., Akter, K. et al. (2016). Abscisic acid-induced gene expression in the liverwort *Marchantia polymorpha* is mediated by evolutionarily conserved promoter elements. *Physiologia Plantarum* 156: 407–420.

Goodwin, S., Gurtowski, J., Ethe-Sayers, S. et al. (2015). Oxford nanopore sequencing, hybrid error correction, and *de novo* assembly of a eukaryotic genome. *Genome Research* 25 (11): 1750–1756.

Gull, A., Ahmad Lone, A., and Wani, N.U.I. (2019). *Biotic and Abiotic Stresses in Plants*. London: IntechOpen https://doi.org/10.5772/intechopen.85832.

Heil, M. (2009). Damaged-self recognition in plant herbivore defence. *Trends in Plant Science* 14 (7): 356–363.

Iovieno, P., Punzo, P., Guida, G. et al. (2016). Transcriptomic changes drive physiological responses to progressive drought stress and rehydration in tomato. *Frontiers in Plant Science* 7: 371.

Jarzab, A., Kurzawa, N., Hopf, T. et al. (2020). Meltome atlas—thermal proteome stability across the tree of life. *Nature Methods* 17 (5): 495–503.

Jozefowicz, A.M., Hartmann, A., Matros, A. et al. (2017). Nitrogen deficiency induced alterations in the root proteome of a pair of potato (*Solanum tuberosum* L.) varieties contrasting for their response to low N. *Proteomics* 17 (23–24): 1700231.

Kolesnikov, N., Hastings, E., Keays, M. et al. (2015). ArrayExpress update—simplifying data submissions. *Nucleic Acids Research* 43 (D1): D1113–D1116.

Kosová, K., Vítámvás, P., Urban, M.O. et al. (2015). Biological networks underlying abiotic stress tolerance in temperate crops—a proteomic perspective. *International Journal of Molecular Sciences* 16 (9): 20913–20942.

Kosová, K., Vítámvás, P., Urban, M.O. et al. (2018). Plant abiotic stress proteomics: the major factors determining alterations in cellular proteome. *Frontiers in Plant Science* 9: 122.

Kudla, J., Batistič, O., and Hashimoto, K. (2010). Calcium signals: the lead currency of plant information processing. *The Plant Cell* 22 (3): 541–563.

Kumar, S.A., Kumari, P.H., Sundararajan, V.S. et al. (2014). PSPDB: plant stress protein database. *Plant Molecular Biology Reporter* 32: 940–942.

Li, Q., Shen, H., Yuan, S. et al. (2023). miRNAs and lncRNAs in tomato: roles in biotic and abiotic stress responses. *Frontiers in Plant Science* 13: 1094459.

Lichtenthaler, H.K. (1998). The stress concept in plants: an introduction. *Annals of the New York Academy of Sciences* 851: 187–198. https://doi.org/10.1111/j.1749-6632.1998.tb08993.x. PMID: 9668620.

Liu, M., Yu, H., Zhao, G. et al. (2017). Profiling of drought-responsive microRNA and mRNA in tomato using high-throughput sequencing. *BMC Genomics* 18: 1–18.

Ma, Z.Q., Dasari, S., Chambers, M.C. et al. (2009). IDPicker 2.0: improved protein assembly with high discrimination peptide identification filtering. *Journal of Proteome Research* 8 (8): 3872–3881.

Maiti, R. and Satya, P. (2014). Research advances in major cereal crops for adaptation to abiotic stresses. *GM Crops & Food* 5: 259–279.

Malde, K. and Jonassen, I. (2008). Repeats and EST analysis for new organisms. *BMC Genomics* 9: 1–7.

Meena, K.K., Sorty, A.M., Bitla, U.M. et al. (2017). Abiotic stress responses and microbe-mediated mitigation in plants: the omics strategies. *Frontiers in Plant Science* 8: 172.

Meher, P.K., Begam, S., Sahu, T.K. et al. (2022a). ASRmiRNA: abiotic stress-responsive miRNA prediction in plants by using machine learning algorithms with pseudo K-tuple nucleotide compositional features. *International Journal of Molecular Sciences* 23 (3): 1612.

Meher, P.K., Sahu, T.K., Gupta, A. et al. (2022b). ASRpro: a machine-learning computational model for identifying proteins associated with multiple abiotic stress in plants. *The Plant Genome* e20259.

Mergner, J., Frejno, M., List, M. et al. (2020). Mass-spectrometry-based draft of the Arabidopsis proteome. *Nature* 579 (7799): 409–414.

Moretto, M., Sonego, P., Pilati, S. et al. (2016). VESPUCCI: exploring patterns of gene expression in grapevine. *Frontiers in Plant Science* 7: 633.

Moriya, Y., Itoh, M., Okuda, S. et al. (2007). KAAS: an automatic genome annotation and pathway reconstruction server. *Nucleic Acids Research* 35 (suppl_2): W182–W185.

Mousavi, S.A., Pouya, F.M., Ghaffari, M.R. et al. (2016). PlantPReS: a database for plant proteome response to stress. *Journal of Proteomics* 143: 69–72.

Murmu, S. and Archak, S. (2023). In-silico study of protein-protein interactions in wheat blast using docking and molecular dynamics simulation approach. *Journal of Biomolecular Structure and Dynamics* 1–11.

Naika, M., Shameer, K., Mathew, O.K. et al. (2013). STIFDB2: an updated version of plant stress-responsive transcription factor database with additional stress signals, stress-responsive transcription factor binding sites and stress-responsive genes in Arabidopsis and rice. *Plant and Cell Physiology* 54 (2): e8.

Nguyen, N.D. and Wang, D. (2020). Multiview learning for understanding functional multiomics. *PLoS Computational Biology* 16 (4): e1007677.

Nguyen, D., Rieu, I., Mariani, C., and van Dam, N.M. (2016). How plants handle multiple stresses: hormonal interactions underlying responses to abiotic stress and insect herbivory. *Plant Molecular Biology* 91: 727–740.

O'Leary, N.A., Wright, M.W., Brister, J.R. et al. (2016). Reference sequence (RefSeq) database at NCBI: current status, taxonomic expansion, and functional annotation. *Nucleic Acids Research* 44 (D1): D733–D745.

Parrine, D., Wu, B.S., Muhammad, B. et al. (2018). Proteome modifications on tomato under extreme high light induced-stress. *Proteome Science* 16: 1–15.

Peskan-Berghöfer, T., Vilches-Barro, A., Müller, T.M. et al. (2015). Sustained exposure to abscisic acid enhances the colonization potential of the mutualist fungus *Piriformospora indica* on *Arabidopsis thaliana* roots. *The New Phytologist* 208: 873–886.

Prabha, R., Ghosh, I., and Singh, D.P. (2011). Plant stress gene database: a collection of plant genes responding to stress condition. *ARPN Journal of Science and Technology* 1 (1): 28–31.

Pradhan, U.K., Meher, P.K., Naha, S. et al. (2023). ASLncR: a novel computational tool for prediction of abiotic stress-responsive long non-coding RNAs in plants. *Functional & Integrative Genomics* 23 (2): 113.

Prieto, G., Aloria, K., Osinalde, N. et al. (2012). PAnalyzer: a software tool for protein inference in shotgun proteomics. *BMC Bioinformatics* 13: 1–8.

Priya, P. and Jain, M. (2013). RiceSRTFDB: a database of rice transcription factors containing comprehensive expression, cis-regulatory element and mutant information to facilitate gene function analysis. *Database* 2013: bat027.

Rabilloud, T. and Lelong, C. (2011). Two-dimensional gel electrophoresis in proteomics: a tutorial. *Journal of Proteomics* 74 (10): 1829–1841.

Rahman, M.U., Zulfiqar, S., Raza, M.A. et al. (2022). Engineering abiotic stress tolerance in crop plants through CRISPR genome editing. *Cells* 11 (22): 3590. https://doi.org/10.3390/cells11223590. PMID: 36429019; PMCID: PMC9688763.

Ramesh, S.V., Govindasamy, V., Rajesh, M.K. et al. (2019). Stress-responsive miRNAome of *Glycine max* (L.) Merrill: molecular insights and way forward. *Planta* 249: 1267–1284.

Ramšak, Ž., Baebler, Š., Rotter, A. et al. (2014). GoMapMan: integration, consolidation and visualization of plant gene annotations within the MapMan ontology. *Nucleic Acids Research* 42 (D1): D1167–D1175.

Ranf, S., Eschen-Lippold, L., Pecher, P. et al. (2011). Interplay between calcium signalling and early signalling elements during defence responses to microbe-or damage-associated molecular patterns. *The Plant Journal* 68 (1): 100–113.

Redman, J.C., Haas, B.J., Tanimoto, G., and Town, C.D. (2004). Development and evaluation of an *Arabidopsis* whole genome Affymetrix probe array. *The Plant Journal* 38 (3): 545–561.

den Ridder, M., Daran-Lapujade, P., and Pabst, M. (2020). Shot-gun proteomics: why thousands of unidentified signals matter. *FEMS Yeast Research* 20 (1): foz088.

Roux, J., Rosikiewicz, M., and Robinson-Rechavi, M. (2015). What to compare and how: comparative transcriptomics for Evo-Devo. *Journal of Experimental Zoology Part B: Molecular and Developmental Evolution* 324 (4): 372–382.

Samaras, P., Schmidt, T., Frejno, M. et al. (2020). ProteomicsDB: a multi-omics and multi-organism resource for life science research. *Nucleic Acids Research* 48 (D1): D1153–D1163.

Sanders, D., Pelloux, J., Brownlee, C., and Harper, J.F. (2002). Calcium at the crossroads of signaling. *The Plant Cell* 14 (suppl_1): S401–S417.

Sandhu, M., Sureshkumar, V., Prakash, C. et al. (2017). RiceMetaSys for salt and drought stress responsive genes in rice: a web interface for crop improvement. *BMC Bioinformatics* 18: 1–11.

Schena, M., Shalon, D., Davis, R.W., and Brown, P.O. (1995). Quantitative monitoring of gene expression patterns with a complementary DNA microarray. *Science* 270 (5235): 467–470.

Sewelam, N., Kazan, K., and Schenk, P.M. (2016). Global plant stress signaling: reactive oxygen species at the cross-road. *Frontiers in Plant Science* 7: 187.

Tareke Woldegiorgis, S., Wang, S., He, Y. et al. (2019). Rice stress-resistant SNP database. *Rice* 12: 1–12.

Thimm, O., Bläsing, O., Gibon, Y. et al. (2004). MAPMAN: a user-driven tool to display genomics data sets onto diagrams of metabolic pathways and other biological processes. *The Plant Journal* 37 (6): 914–939.

Tian, T., Liu, Y., Yan, H. et al. (2017). agriGO v2.0: a GO analysis toolkit for the agricultural community, 2017 update. *Nucleic Acids Research* 45 (W1): W122–W129.

Van Bel, M., Bucchini, F., and Vandepoele, K. (2019). Gene space completeness in complex plant genomes. *Current Opinion in Plant Biology* 48: 9–17.

Verma, V., Ravindran, P., and Kumar, P.P. (2016). Plant hormone-mediated regulation of stress responses. *BMC Plant Biology* 16: 1–10.

Vizcaíno, J.A., Côté, R.G., Csordas, A. et al. (2012). The PRoteomics IDEntifications (PRIDE) database and associated tools: status in 2013. *Nucleic Acids Research* 41 (D1): D1063–D1069.

Wang, X., Li, N., Li, W. et al. (2020). Advances in transcriptomics in the response to stress in plants. *Global Medical Genetics* 7 (2): 30–34.

Wu, W., Wu, Y., Hu, D. et al. (2020). PncStress: a manually curated database of experimentally validated stress-responsive non-coding RNAs in plants. *Database* 2020: baaa001.

Yang, S., Li, H., He, H. et al. (2019). Critical assessment and performance improvement of plant–pathogen protein–protein interaction prediction methods. *Briefings in Bioinformatics* 20 (1): 274–287.

Zandalinas, S.I., Fritschi, F.B., and Mittler, R. (2020). Signal transduction networks during stress combination. *Journal of Experimental Botany* 71 (5): 1734–1741.

Zenoni, S., D'Agostino, N., Tornielli, G.B. et al. (2011). Revealing impaired pathways in the *an11* mutant by high-throughput characterization of *Petunia axillaris* and *Petunia inflata* transcriptomes. *The Plant Journal* 68 (1): 11–27.

Zhang, S., Yue, Y., Sheng, L. et al. (2013). PASmiR: a literature-curated database for miRNA molecular regulation in plant response to abiotic stress. *BMC Plant Biology* 13 (1): 1–8.

Zhang, F., Ge, W., Ruan, G. et al. (2020). Data-independent acquisition mass spectrometry-based proteomics and software tools: a glimpse in 2020. *Proteomics 20* (17–18): 1900276.

9

Bioinformatics Tools for Assessing Drought Stress Tolerance in Crops

Nageswara Rao Reddy Neelapu[1] *and Kolluru Viswanatha Chaitanya*[2]

[1] *Biochemistry and Bioinformatics Department, GITAM School of Science, GITAM University, Visakhapatnam, Andhra Pradesh, India*
[2] *Microbiology and Food Science Technology Department, GITAM School of Science, GITAM University, Visakhapatnam, Andhra Pradesh, India*

9.1 Introduction

Food security for all and the growing population in the future is the major problem (Challa et al. 2019). The future of food security for humans is based on identifying and arranging food for all, and one of the ways to provide food is with appropriate agricultural production. Crops are produced via appropriate agricultural practices and thus become the major food security resource. Disruption of food security leads to food insecurity and threatens humankind (Challa et al. 2020). Crop production is based on agricultural land and water availability, agricultural diseases, soil fertility, and environmental conditions while the crop is grown. All the above factors induce either biotic or abiotic stress directly or indirectly while the crop grows (Surekha et al. 2015; Neelapu et al. 2015). The different types of stresses are water, high and low temperatures, freezing stress, nutrient stress, wind stress, metal stress, and photooxidative or light stress. Advancements in molecular genetics and genomics technologies helped to capture data on the above stresses and resulted in the enormous growth of biological information. Rapid, continuous development and technological advancement have resulted in the generation of massive data, facilitating the growth and development of bioinformatics (Nageswara Rao Reddy et al. 2021). The expansion of omics technologies aids in generating the genome and proteome-related information of various plants for their efficient exploration as resources to humans and has further increased the applications of bioinformatics. The biological information or data must be structured (more organized) and available to manage and analyze to enable logical interrogation. Bioinformatics is an interdisciplinary field with tools for efficiently arranging and classifying biological information in databases. It applies biology, mathematics, and computer science to analyze and interpret biological data. This chapter discusses the utilization of bioinformatics resources and tools to determine abiotic stress tolerance, along with the challenges and limitations.

Bioinformatics for Plant Research and Crop Breeding, First Edition. Edited by Jen-Tsung Chen.

9.2 Bioinformatics for Plant Research and Crop Breeding

Crops are commonly exposed to challenging ecological abiotic stress conditions, unfavorably influencing their yield. One of the most damaging environmental stresses that reduces crop yield is drought stress. Drought stress inhibits plant growth and development and substantially reduces biomass accumulation. Some consequences of drought stress are a reduction in cell division and cell expansion, leaf size, root proliferation, stem elongation, disturbed plant water relations, plant nutrient relations, stomatal oscillations, decreased water use efficiency, and diminished crop productivity (Farooq and Abdul 2012). Drought stress triggers many molecular, biochemical, physiological, and morphological changes in plants, which governs their adaptation to drought stress conditions (Sallam et al. 2019).

Bioinformatics has become a significant tool of research in biology. The omics technologies expanded the utilization of bioinformatics and have increased the applications in various fields such as genomics, epigenomics, transcriptomics, proteomics, metabolomics, interactomics, and phenomics. These fields further witnessed the generation of omics data and analysis of data via genome-wide association studies (GWAS), transcriptome analysis, microRNAs (miRNAs), proteome analysis, metabolome analysis, and phenome analysis to understand drought stress and its tolerance.

9.3 Genomics and Drought Stress Tolerance

The advent of genomics facilities for precise and quick outcomes has witnessed five generations of sequencing technologies (Neelapu et al. 2015). "The next-generation sequencing (NGS) technologies available to generate NGS data are first, second, third, fourth, and fifth-generation sequencing technologies." The "first-generation sequencing technologies" include "Sanger sequencing and the Maxam–Gilbert sequencing method." "First-generation sequencing technologies," exclusively "Sanger sequencing," continued to be significant for nearly three decades. Nevertheless, the two biggest obstacles, cost and time, have resulted in the creation of "second-generation sequencing technologies." The "second-generation sequencing technologies" include "Roche/454 sequencing, Ion torrent sequencing, Illumina/Solexa sequencing, and ABI/SOLiD sequencing." The "third-generation sequencing methods comprise the single-molecule real-time (SMRT) and Oxford Nanopore Technology (ONT) sequencing approaches." "Third-generation sequencing technologies" run at low cost and are significantly faster when compared to "second-generation sequencing technologies." It is possible to prepare the sample without "Polymerase Chain Reaction (PCR) amplification." These sophisticated techniques solve the assembly issue and repetitive regions of complex genomes by producing long reads in kilobases. "The fourth-generation sequencing method includes nanopore-based sequencing by biological nanopores and solid-state nanopores." "Fourth-generation sequencing technologies" are less precise and unable to sequence "ultra-short DNA." Nanopore sequencing offers an inexpensive, portable substitute for sequencing-by-synthesis techniques. "The fifth-generation sequencing method includes high-fidelity nanopore sequencing of ultra-short DNA targets and cyclomics and ultra-sensitive nanopore sequencing of cell-free tumor DNA" (Satam et al. 2023). Despite various obstacles in the field of bioinformatics concerning gene

Table 9.1 List of plant genome databases.

S. No.	Server	Details
1.	Ensembl Plants	Genomic database of 35 plant species comprising genome sequence, gene models, and functional annotation
2.	Gene Expression Omnibus	Functional genomics database
3.	Phytozome	Comparative genomics database
4.	Plant Promoter Database	Plant promoter sequence database. Data regarding the transcription start site of different plants that are experimentally tested
5.	PLEXdb	A database featuring structural genomics and gene expression profiles
6.	DRAGO2	Annotation of genes resistant to abiotic and biotic stress
7.	The Bio-Analytic Resource	Platform for Genomics, gene expression, gene mapping, and molecular markers
8.	Array Express	Database for functional genomics derived from microarray and sequencing technologies

annotation, genome comparison, and sequence assembly, a deeper comprehension of plant genomes has created new opportunities for enhancing our utilization and preservation of plants. More data on plant transcriptomes, proteomes, and genomes has made it possible to assess, contrast, and develop the best product. The list of plant genome databases is provided in Table 9.1.

There are several databases on abiotic stress, such as the plant stress gene database (Prabha et al. 2011), Plant Stress Protein Database (PSPDB) (Kumar et al. 2014), DroughtDB (Alter et al. 2015), and PlantPReS (Mousavi et al. 2016). Two gene databases, DroughtDB and Plant stress gene database, will also be added to the list. Whole Genome Sequencing (WGS) generates the whole genome sequence data, which will be searched against the above databases to retrieve the genes on stress or drought stress. These studies will help the researchers understand a particular plant's genomic repertoire against abiotic stress. Plant genomes are used to build "metabolic pathways" using various methods. "Kyoto Encyclopedia of Genes and Genomes (KEGG) and MetaCyc" are used to build "metabolic pathways" (Aoki-Kinoshita and Kanehisa 2007). This type of genome analysis can help to predict and identify the "metabolic pathways" associated to "abiotic stress" or any other specific stress. Therefore, the whole genome sequence, in addition, can be used to predict "metabolic pathways" connected to "abiotic stress."

9.3.1 Genome-wide Association Studies (GWAS)

Tolerance to drought stress is a complicated characteristic that is regulated by numerous genes. Some of them have significant and some of them have minor effects (Gataa et al. 2022). Further, drought stress has low heritability due to its polygenic inheritance and significant genotype by environmental interaction. Present knowledge of the genes and molecular pathways underlying drought stress tolerance has been considerably improved due to high-throughput genotyping and phenotyping techniques and GWAS. Also referred to as whole genome association studies, these studies look into an extensive collection of genetic variants across the genome in various varieties to show how a variant is related to a particular trait. GWAS emphasizes the associations between SNP and a trait by comparing the genetic makeup of plant varieties that differ in a particular trait's phenotype (Challa and Nageswara 2018). GWAS relies on genotyping technology, design, and statistical analysis to interpret and monitor associations. This approach confers two advantages – i. minimizes the cost and time, and ii. effectively uses the crossover events in the association panel, providing a high resolution (Saini et al. 2022). GWAS studies have detected the regions in the wheat genome connected to the characteristics that increase resistance to "drought stress" at the "seedling and flowering stages"(Sallam et al. 2022).

A GWAS for "drought stress" resistance is designed by considering the stressed plant versus control for comparing DNA from different plants with different phenotypes for a specific trait related to tolerance. Moreover, the trait may be measured quantitatively. The variety without or with various phenotypes for a given characteristic can be considered for control. Assessment of SNPs in the genome of the targeted phenotype is the concept behind the GWAS. Two distinct methods are used to evaluate SNPs affecting phenotypes: direct association and indirect association. Genotyping of the phenotype influenced by the SNP and its statistical association with the trait is known as direct association. Non-genotyping of the SNP, but its statistical association with phenotype is indirect. The relationship between an allele and an SNP in a population is correlated for both "single-locus and multilocus" analyses. It is a "single locus" statistic test if the relationship between every SNP and phenotype is looked at separately. It is a "multilocus" analysis if each SNP's relationship to a phenotype or trait is looked at separately. GWAS analysis for control and abiotic stress traits is performed with an analysis of variance (ANOVA) and a generalized linear model. GWAS analysis has an edge over other techniques when linking SNPs to characteristics and phenotypes. It is extensively employed to identify novel QTLs in plants. As observed in the rice plant, its drought stress resistance is controlled by multiple QTLs. Identifying phenotypically associated QTLs and candidate genes is essential for drought stress tolerance genes and molecular breeding. Drought stress can cause a reduction in grain weight, grain size, and spikelet fertility. Additional consequences of drought stress include increased sterile grains in rice panicles, causing yield loss. GWAS studies had identified eight QTLs "*qDTY1.1*, *qDTY2.4*, *qDTY3.1*, *qDTY3.3*, *qDTY6.1*, *qDTY6.3*, *qDTY11.2*, and *qDTY12.1*," in the rice subjected to drought stress (Wang et al. 2023). Despite its advantages, GWAS also carries certain limitations. Due to population stratification, SNPs have a higher probability of being connected to the characteristics in the analysis. As genetic analysis measures a population sub-structure, these structures can be studied. Principal component analysis (PCA) can be used to reduce their impact on the data.

9.4 Transcriptome Analysis for the Drought Stress Tolerance

Crop plants are frequently exposed to drought stress, affecting various biological processes that include water and nutrient absorption, photosynthesis, and transport of the assimilate to other plant parts. The ongoing drought stress in plants will re-adjust their morphological, physiological, and metabolic processes by altering their gene expression, decreasing the yield and considerably reducing the production quality (Cao et al. 2021; Cai et al., 2019). Regulation of transcription factor expression during drought stress is involved in plant resurrection is regulated by ABA and ethylene hormonal signals, leading to the induced multiple stress-responsive genes (Jiang et al. 2021). A transcriptome is the entire collection of transcripts and the amount of each transcript in a cell under a particular physiological state. The functional elements in an organism's genomc can be interpreted by understanding its transcriptome. A transcriptome's primary goals are to catalog all of a cell's transcripts, including small RNAs, mRNAs, and ncRNAs; evaluate each transcript's continuously fluctuating expression levels; and ascertain the transcriptional structure of genes with regard to their 5′ and 3′ ends, splicing patterns, and posttranscriptional modifications (Wang et al. 2009).

Various techniques have been developed for quantifying either hybridization or sequence-based transcriptome. The foundation of hybridization-based techniques is the incubation of fluorescently labeled cDNA on high-density or specially made oligo microarrays. Specialized microarrays, such as genome tiling microarrays (map the transcribed region of a genome with high resolution. Microarrays with probe-spanning junctions can detect and quantify the splicing regions. The hybridization methods are relatively inexpensive but have a number of restrictions, including the dependence on prior knowledge about the targeted genome and elevated background levels as a result of hybridization. The targeted cDNA's sequence is ascertained using sequence-based techniques. Tag-based methods such as SAGE, MPSS, and CAGE give exact levels of digital gene expression. It is not possible to map these pricy techniques and tags to the reference genome. This method renders the isoforms identical and only allows for partial transcript analysis. These drawbacks were overcome by novel "high-throughput sequencing methods," such as RNA Seq, developed using deep sequencing technologies.

All "high-throughput sequencing technologies," including RNA-Seq, need the development of efficient methods for storing, retrieving, and processing enormous volumes of data with few errors and for eliminating poor-quality reads. After obtaining the high-quality reads, they must be mapped to the reference genome or assembled into contigs to develop a transcription structure. Several programs, such as ELAND and SOAP31, were developed for mapping the reads to the genome. Short transcriptome reads also contain reads spanning the exon junctions, and Poly A tail ends that cannot be analyzed. Similarly, the genomes with rare splicing events, such *as Saccharomyces cerevisiae* poly-A tails and exon–exon junctions, must be targeted. Mapping the reads spanning for splice junctions is difficult for complex transcriptomes due to alternative splicing and trans-splicing.

Using RNA-Seq, 3147 Ggnes that were expressed differently in "drought-tolerant W9706" and "drought-susceptible B73" maize varieties by exposing them to drought stress (Jiang et al. 2023). The upregulated genes belong to "ABA signaling, wax biosynthesis, CHO metabolism, signal transduction, and brassinosteroid biosynthesis-related processes." Transcriptome analysis of the maize reproductive and leaf meristem has shown decreased "transcript abundance of cell division and cell cycle-regulated genes" (Kakumanu et al. 2012). RNA seq analysis of drought-sensitive Liaodou soybean variety upon exposure to 7, 17, and 27 days of "drought stress" during the "flowering stage," followed by rehydration until harvest, have resulted in the differential expression of 2214, 3684, and 2985 genes, in that order (Li et al. 2022). WRKY, MYB, bZI, and other drought-responsive transcription factors that might have been upregulated during drought stress were identified by "weighted gene co-expression network analysis (WGCNA)."

9.4.1 MicroRNAs (miRNAs) and Drought Stress

Regulation of gene expression at the transcription level during drought stress is critical for altering the downstream expressed genes. Altering the expression of transcription factors will improve the resistance of plants to abiotic stress. MicroRNAs (miRNAs), through translational inhibition, posttranscriptional gene silencing, and heterochromatin modification, play a significant role in mechanisms of gene regulation. miRNAs are "single-stranded noncoding RNAs of 20–24 nucleotides in length" and serve as gene regulators by affecting biological processes. Constantly growing evidence confirms the role of miRNAs in plant response to "drought stress." MicroRNAs mediate the responses by modifying timing, location, and the amount of proteins synthesized from the expression of genes during drought. Most miRNA gene regulation occurs at the posttranscriptional level, as the miRNAs are transcribed but not translated into proteins. RNA polymerase II transcribes the single-stranded RNA precursors (pri-miRNA) from the miRNA genes to produce microRNAs. DCL1 and other protein factors process the transcript to produce a 20–24 nucleotide long miRNA (Wu et al. 2010). MiRNAs associated with drought stress are divided into three classes: I, II, and III. Class I microRNAs specifically target transcription factors that control genes involved in the stress response. "miR156, miR159, miR165, miR169, miR171, miR172, miR319, and miR396" belong to Class I miRNAs. Class II miRNAs are involved in the direct stress response. "miRNAs miR167, miR168, miR393, and miR394" are involved in this function. "miR167 targets auxin-responsive factors, and miR168 targets ARGONAUTE"(Wu and Poethig 2006). F-box proteins involved in abiotic stress tolerance are the targets of "miR393 and miR394"(Jain et al. 2007). Class III miRNAs, such as "miR397 and miR408, target hydrolase and oxidoreductase genes" while responding to abiotic stress conditions (Apel and Hirt 2004), and stress from drought changes miRNA expression. Drought stress-responsive miRNAs were reported in *Arabidopsis*, cowpea, rice, wheat, soybean, and tobacco (Jannatul et al. 2015). miRNAs of *Arabidopsis* "miR156, miR159, miR167, miR168, miR171, miR172, miR319, miR393, miR394a, miR395c, miR395e, miR396, and miR397" were upregulated; whereas "miR161, miR168a, miR168b, miR169, miR171a and miR319c" were down-regulated during drought stress (Liu et al. 2012b).

Bioinformatics tools were developed to understand miRNA biogenesis and annotation. A storage platform for miRNA data comprising the information related to miRNA sequence, secondary structure, and annotation information is necessary. miRBase is an efficient platform that serves as a repository for miRNA storage and contains all miRNA sequences and annotations. A system for RNA annotation is called Rfam, which consists of information regarding the miRNA family. miRIAD provides details about host genes and intragenic miRNA. mirtronPred is a tool for predicting the mirtrons and intron sequences, whereas mirSTP identifies the miRNA transcription start site. MetaMirClust contains information regarding miRNA clusters. ViennaRNA is a software comprising tools for predicting and comparing the RNA structure. It consists of algorithms for predicting and analyzing the

Table 9.2 Plant miRNA identification and prediction tools.

Tool Name	Platform	Link
Triplet-SVM	Stand-alone	http://bioinfo.au.tsinghua.edu.cn/mirnasvm/
UEA sRNA Workbench	Stand-alone	http://srna-workbench.cmp.uea.ac.uk/mircat2/
miRanalyzer	Web-based	http://bioinfo2.ugr.es/miRanalyzer/stand alone.html
MicroPC	Web-based	http://www3a.biotec.or.th/micropc/
MirDeep-P	Stand-alone	https://sourceforge.net/projects/mirdp
miRNAFold	Web-based	https://evryrna.ibisc.univ-evry.fr/evryrna/mirnafold/mirnafold_home
miRPlant	Stand-alone	https://sourceforge.net/projects/mirplant/
TarBase	Web-based	http://carolina.imis.athena-innovation.gr/diana_tools/web/index.php?r¼tarbasev8
miRTarBase	Web-based	http://mirtarbase.mbc.nctu.edu.tw

RNA secondary structure. miRNAFold is a rapid abinitio method that predicts the miRNAs in a genome at a larger scale. Tools such as "miRanda, TargetScan, PicTar, RNAhybrid, and PITA" are used for predicting the miRNA targets (Liang et al. 2019). The list of plant miRNA databases is depicted in Table 9.2.

9.5 Proteome and Drought Stress

All physiological processes in a cell are catalyzed by proteins, including signal processing and integration as well as the biochemical activities in intermediate metabolism. Proteins are synthesized as a translated product of an mRNA chain in ribosomes and function in large complexes. Bioinformatics plays a pivotal role in proteome analysis like data processing – preparation, data gathering, and its separation from raw data; its integration into a database; database search; and functional integration of the data set into a database. A proteomics analysis approach that is iterative and involves systemic difficulties and data interpretation will lead to developing and modifying a hypothesis and generating new knowledge.

Proteins from a sample of interest are extracted, then one or more proteases are used to breakdown the proteins to produce peptides for proteome analysis. Next, liquid chromatography–mass spectrometry (LC-MS) is used to examine the peptides. To achieve deep cover of the proteome, two methods are used: targeted MS and shotgun MS. In an MS, the peptides are broken apart, producing tandem mass spectra that reveal the peptide's sequence – a crucial detail for identification. Prior knowledge of the peptides in the sample is not necessary when using the shotgun technique. The identification of the samples that emerge from the column depends on the data. Along with fragmented spectra chosen during data-dependent cycles, the data is presented as a 3D map with mass-to-charge ratios (m/z), retention durations (RT), and intensities for the observed peptides as an axis.

To determine the relevant peptide's chromatographic peak, the intensity of a peptide is plotted alongside the RT. The related peptide is then quantified using the area under the curve (AUC). Peptide identification for the focused approach is accomplished by means of the sample's fragmented spectrum. Different versions of mass spectrometers produced a significant number of mass spectra, which means that search engines are needed to locate and measure the peptides under analysis. Using a predefined database, search algorithms infer a peptide sequence from a spectrum by producing a list of peptide sequences that have a probability score that indicates how well they fit the data. The protein databases, mRNA databases, and the ones containing protein data translated from a genome are the predefined databases. The identified peptides are assembled into proteins in the final step of the analysis. This step is tedious, especially while handling redundant peptides and alternatively spliced proteins (Nesvizhskii and Aebersold 2005). To reduce the false discovery, a few approaches comprising of shotgun or discovery were made, identifying a whole lysate containing hundreds of proteins with a range of 104–105. (Nagaraj et al. 2012). The two main shortcomings of this method are its sensitivity and repeatability.

These limitations were addressed by targeted proteomics, a "selected reaction monitoring (SRM)" method, where predefined peptides at scheduled RT are subjected to fragmentation in MS, and the status of 2–3 fragments is monitored. With an increased scan speed, SRM can be simultaneously performed, leading to a multiplexing of SRMs known as "multiple reaction monitoring (MRM)." Several hundred proteins with a wide range as well as proteins expressed at low copy numbers (50 copies per cell) are quantified using this multiplexing capability (Picotti et al. 2013). The majority of the stored proteome datasets are directly accessible through large repositories such as PRIDE, Proteome Commons, and Peptide Atlas, which are invaluable resources for bioinformatics data mining.

Adapting plants to drought stress involves activating integrated processes that illicit cellular and molecular responses. Proteomics is one of the directed approaches for relating the function of genes to the associated products, such as stress-related proteins (Yahoueian et al. 2021). The application of proteomics for the identification of drought stress-induced proteins has been reported in rice (Salekdeh et al. 2002), sunflower (Ghaffari et al. 2017), common bean (Gebeyehu et al. 2010), barley (Rollins et al. 2013), soybean (Yu et al. 2016). Proteins expressed during drought stress will alter or modify their expression. These alterations in a proteome composition will enable the plants to carry out their biological functions and adapt to the changing environmental conditions. Tandem mass tags (TMT)-Label LC-MS/MS was used to identify the protein alterations in mulberries in response to drought stress (Liu et al. 2012a).

Protein changes of mulberry in response to drought stress were determined using TMT-Label LC-MS/MS (Liu et al. 2012a). The levels of proteins involved in photosynthesis, energy and sugar metabolism, antioxidants, phytohormones, and cellular homeostasis were found to be changed. The levels of 123 differentially changed proteins were identified in the delicate roots of cotton plants subjected to 30 days of drought stress. The number was increased to 1273 upon exposure of plants to 45 days of drought stress (Xiao et al. 2020).

9.6 Metabolomics and Drought Stress Tolerance

The collection of an organism's many low molecular weight primary and secondary metabolites is known as metabolomics. It is one of the most rapidly evolving fields dealing with primary and secondary metabolites characterization and studying the metabolite composition in cells, tissues, and organs using high-throughput methods. Metabolomics allows the measurement of metabolites that are produced by an organism. Plants create primary metabolites that are derived into secondary metabolites in response to various stress-induced physiological changes, providing adaptation to the plants against stress. One of the most important abiotic stresses is drought, and plants have evolved to withstand its features by triggering secondary metabolites. The capacity of plants to sustain the proper concentrations of secondary metabolites for defense responses indicates how tolerant they are to drought stress. Targeted and nontargeted strategies are the two main methods used to comprehend the function of metabolites during drought stress. Targeted metabolomics can detect, estimate, and analyze known metabolites in plants during stress, whereas, during stress, nontargeted metabolomics will give a summary of the most prevalent compounds in plants (Van Meulebroek et al. 2016).

The biosynthesis of secondary metabolites during drought stress is regulated by environmental factors, which play a pivotal role in protecting the plants from drought. Alkaloids (glycosides and glucosinolates), phenolic compounds (tannins, flavonoids, and lignins), and terpenes (terpenoids and isoprenoids) are the three groups of secondary metabolites that the plants create. The plants synthesize secondary metabolites via "mevalonic acid, shikimic acid, phenyl propanoic acid, and methylerythritol phosphate pathways." Secondary metabolites are involved in actively scavenging ROS produced during drought stress. Secondary metabolites remove ROS and protect the plant cells from lipid peroxidation (Treml and Smejkal 2016). "Phenylalanine ammonia-lyase (PAL), 4-coumaroyl CoA ligase (4CL), and coumarate-4-hydroxylase (C4H)" are the key enzymes of phenyl propanoid pathway implicated under drought stress. Volatile compounds in the plants are used as signaling molecules for activating the drought stress mitigating functions (Jogawat et al. 2021).

Metabolic regulation is a vital mechanism in the plant cells to maintain the osmotic potential during drought stress. Lack of water or a high transpiration rate will trigger the secondary metabolite production. "MYB, NAC, bZIP, AP2/ERF, and AREB/ABF" are transcription factors that control stomatal mobility and expression of genes that respond to drought either upstream or downstream of a metabolic pathway. Microarrays and real-time PCR studies investigate genes related to drought stress and their transcript levels (Fabregas and Fernie 2019). Various genes were isolated and characterized from the plants for their role in enhancing drought stress tolerance. During drought, these genes provide tolerance by accumulating metabolites such as proline, glycine betaine, trehalose, flavonoids, and polyamines. These metabolites will build up and trigger cellular reactions that will increase resistance to drought stress. Upregulation of the "C1A cysteine protease family" during drought stress is important for reorganizing the metabolism, degrading the repairing harmed or superfluous proteins, and redistributing nutrients (Gomez-Sanchez et al. 2019). In comparison to wild-type plants, Arabidopsis has demonstrated improved drought stress tolerance through overexpression of the *Triticum aestivum* cysteine protease gene (Zhang et al. 2010). Flavonoids are the secondary metabolites in maintaining plant cells' cellular redox balance. The essential enzyme in the route leading to the manufacture of flavonoids, chalcone synthase, is elevated in response to drought stress. Overexpression of chalcone synthase gene NtCHS in tobacco has enhanced its drought stress tolerance and oxidative stress response during drought stress (Zhao et al. 2019). Overexpression of the "sweet potato cinnamate-4-hydroxylase gene" in tobacco has conferred drought stress tolerance by promoting the accumulation of phenolic compounds (Wang et al. 2017). The induction of metabolite biosynthesis in plants will provide protection and enhance their tolerance to drought stress.

Several bioinformatics tools must be used in order to analyze metabolomics data. Utilizing methods to separate and describe metabolites and their groupings is the initial stage in the metabolomics analysis. This is achieved by two different platforms: "MS and nuclear magnetic resonance (NMR) spectrometry." Both these platforms have their advantages and disadvantages. MS-based metabolomics consists of a separation step for reducing the biological sample's complexity, which makes it possible to analyze various substances whose data is acquired in the form of a spectrum based on the mass-to-charge ratio and the ionized compound's relative intensity. The basis of NMR as a spectroscopic technique is the idea that atomic nuclei absorb and re-emit energy in response to changes in an external magnetic field. The chemical structure and quantity of the substance are described and quantified using the NMR spectra. Metabolite isolation and identification is a laborious process that is not automatable. Careful data processing and statistical analysis are required. MS is ideal for detecting moderate to high polarity compounds, such as "fatty acids, alcohols, phenols, vitamins, organic acids, nucleic acids," etc. The second step is the preparation of spectra generated by MS and NMR to analyze the compounds. This is achieved by specific software such as XCMS, MAVEN, and MZmine. Noise reduction, retention time adjustment, detection of peaks, alignment, and their integration are the processes in this step. Data generated in step two and its quality control are processed in step three. The list of metabolites and their analytical methods were tabulated in Table 9.3.

Data obtained from the quality control samples is used to separate high or low-quality data, the variance of metabolite features, and correct the noise in the signal. Subsequent spectrum or peak data is used to annotate and identify the metabolite by comparing it with authentic standard compound data available in the databases. Metabolomics Standard Initiative (MSI) was established in 2005 to enable the effective application, sharing, and reuse of data. There are four stages in identifying metabolites as per MSI: (i) the first stage is identifying metabolites, (ii) the second stage is presumptively annotating compounds, (iii) the third stage is the characterization of annotated compounds into classes, and (iv) the fourth stage is the categorization of unknown compounds. The Tandem MS database (METLIN) is used to identify the metabolites from spectra (Chen et al. 2022).

9.7 Phenome and Drought Stress

One of the things that limits plant growth and reduces agricultural yield is drought stress. Plant respiration, photosynthesis, and stomatal movement are all impacted by drought stress, which has an impact on physiological metabolism and plant growth. This illicits

Table 9.3 Key metabolites involved in drought stress tolerance.

Biomolecule	Analytical Method	Metabolites
Carbohydrates	GC/MS, GC-MS	"Glucose, glycerol, mannobiose, maltose, sucrose, galactose, galactinol, mannitol, mannose, fructose, trehalose, maltose, D-ribose, D-xylopyronose, xylonic acid, α-D-glucopyranose, 2-deoxyribose, L-manopyranose, myo-inositol, galactosoxime, D-fructose, D-turanose, malic acid, succinic acid, 2 butenedoic acids, 2-deoxyribose"
Aminoacids	GC-MS-EI, ^{1}H-NMR	"Proline, glutamine, threonine, glycine, aspartate, serine, aromatic amino acids, norvaline, glutamate, arginine, serine, asparagine, threonine, alanine, L-arginine, L-histidine, L-isoleucine, tryptophan"
Lipids	GC	"Monoacylglycerol, diacylglycerol, triacyl glycerols free fatty acids, triacylglyceride"
Fatty acids	GC	"Palmitic acid, linolenic acid, stearic acid, pentadecanoic acid, 8,11-octadecadienoic acid, oleic acid, linoleic acid, linolenic acid, myristic acid, arachidic acid, palmitic acid"
Secondary metabolites	GC-TOF, SPME-GC-MS, HPLC-DAD-MSn	"Flavone glycosides, chlorogenic acids, caffeoyl-hexose, sinapoyl-hexoses, feruloyl-hexose, hydroxycinnamic acids, 5-methoxytryptamine, 4-hydroxycinnamic acid, ferulic acid, salicylic acid, quercetin, quercetin-3-*O*-glucoside, kaempferol-3-*O*-glucoside"
Other metabolites	GC-MS	"Isocitric acid, α-ketoglutaric acid, malic acid, fumaric acid, succinate, citrate, aspartate, gluconate, glutathione, tricarboxylic cycle metabolite, Succinate, acetate, isocitric acid, nicotinic acid, 2-methyl-butanal phenylacetaldehyde"

response mechanisms in plants, which include morphological and structural changes, producing osmoregulatory chemicals and hormones that reduce the stress brought on by drought. During drought, plants produce signaling molecules such as "ABA, Ca^{2+}, Inositol-1,4,5-triphosphate (IP3), and Cyclic adenosine-diphosphate ribose," leading to morphological and physiological changes. Plant response to drought stress will be reflected in its morphological form. Plants adapt themselves to drought stress through phenotype plasticity. Drought-resistant plants will adjust to arid environments through morphological and structural characteristics such as roots, stems, and leaves. Drought stress decreases plant height, reduces the number and area of the leaf, and causes wilting.

Cell enlargement and leaf senescence are linked to the decrease in plant height caused by drought stress. The decreased cell expansion, impaired mitosis, and increased leaf shredding might have reduced the plant height. Plant leaf area is another morphological feature subjected to a decrease during drought stress. As a direct response to drought, plant leaves adapt to small areas with large thicknesses and a high tissue density (Werner et al. 1999). Variations in the leaf area will immediately impact photosynthesis and, ultimately, productivity. The elements causing the change in plant leaf area are leaf turgor pressure, canopy temperature, and the availability of photoassimilates. During drought, leaf turgor pressure and photosynthesis rate in the plant leaves decrease, reducing the plant leaf area. Leaf rolling is another significant morphological change observed during drought stress due to water loss from the upper epidermis. Wheat flag leaf undergoes severe rolling under drought stress (Willick et al. 2018).

A major function of plant roots will also be played under drought stress. Fully developed roots will absorb the water to their maximum potency. Root system configurations, including root hair, root branches, and root density, will affect water absorption during low water availability. Drought stress promotes the thinning and elongation of roots, shortens their life, promotes root hair elongation, and reduces the angle of root branching, leading to a steep and deeper root system facilitating the adaptation to drought stress. Drought stress inhibited the distribution of roots in soybean, chickpea, and field pea plants (Du et al. 2020). Drought stress has completely destroyed the root system in sugarcane and increased the internode length by 39% in its vegetative stage (Benjamin and Nielsen 2006). Drought stress has significantly reduced the root-stem biomass of soybeans and plant height, stem girth, and leaf size in maize (Bismillah Khan et al. 2015).

Genomics has completely transformed the scope of experimental inquiry and plant breeding. The detailed genetic understanding of a plant's performance has been expanded by the high-resolution capability of genomics. The complex biological mechanisms that comprise the abiotic stress responses, such as drought, provide a quality analysis of a plant's performance, making a systemic functional analysis. This analysis generates amounts of data that are analyzed using specific software and stored in the database as a part of plant bioinformatics analysis. Genome initiatives are in progress for 60 important crop plants, and efforts are being made to use comparative genome methods.

With an increase in the genetic map resolution of major crops, the molecular basis for the specific traits and their phenotypic responses are elucidated with better clarity, facilitating the possibility of associating the candidate genes that were discovered in model organisms with corresponding loci in crop plants. Databases associated with the model crop plants further increased the association by providing accessibility for a gene sequence. Genetic markers are used to utilize computational models to formulate the predictive hypothesis and create the phenotypes of interest. This will enable linking the vast amount of plant breeding knowledge to basic plant biology for enhancing the ability to elucidate the genes responsible for complex phenotypic traits.

9.8 Future of the Omics technologies

Understanding biology at this point depends on realizing the components of the genome, epigenome, transcriptome, proteome, metabolome, interactome, and phenome. This kind of approach is known as the reductionist paradigm or approach. In reality, understanding biology requires a holistic approach or paradigm, i.e., understanding the interaction of the components of the genome, epigenome, transcriptome, proteome, metabolome, interactome, and phenome in a cell, tissue, or organism. To implement a holistic approach, research groups started employing the integrated genome-to-phenome and systems biology approaches.

The integrated genome-to-phenome approach uses the data generated by omics technologies. It integrates genome to phenome, and this integrated data is used to understand an organism's biology and abiotic stress (Figure 9.1). The phrase "genome to phenome"

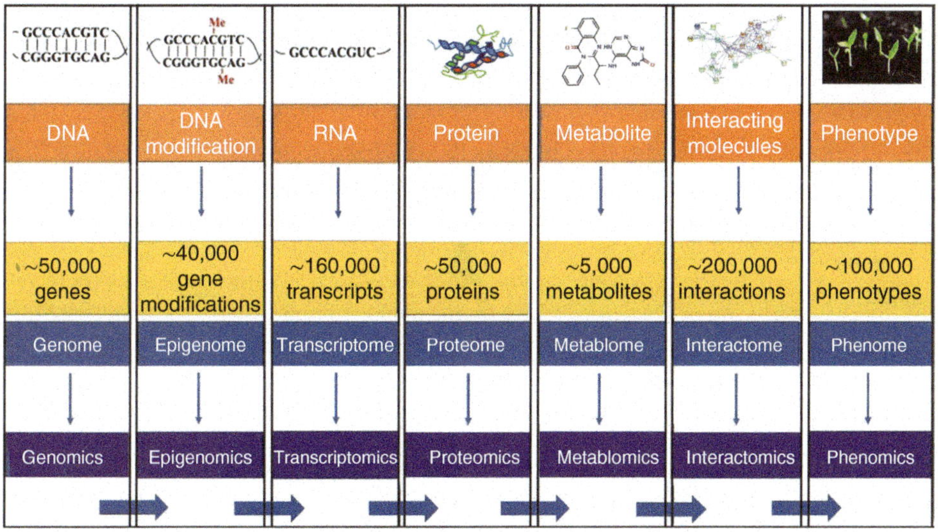

Figure 9.1 Integrated genome-to-phenome approach by Omics technologies for abiotic stress tolerance of crop plants. *Source:* tycoon101/Adobe Stock.

refers to the relationship and causality between a plant's genome, genetic composition, phenome, or all of its observable physical or physiological qualities or characteristics. Gaining more insight into the composition and functions of plant genomes and how they interact with nongenetic elements of production systems (such as environment and nutrition) will be necessary to increase plant productivity. This will allow management strategies to be tailored to maximize performance.

The data from the genome to phenome can be used as an input model as a system. What is a system? theoretically, cells, tissues, and organisms function as a system. Thus, the computational and mathematical study and modeling of intricate biological systems is known as systems biology. It is an interdisciplinary field based on biology that uses a holistic approach to biological study, focusing on complex interactions within biological systems. Therefore, holistic approaches like the integrated genome-to-phenome approach and systems biology approach can used to understand an organism's biology and abiotic stress.

9.9 Conclusions

Bioinformatics has become a major tool of research in biology. Rapid, continuous development and advancement in molecular genetics and genomics technologies have resulted in the enormous growth of biological information, which needs to be more organized and should be made available to manage and analyze. Bioinformatics is an interdisciplinary field with tools for efficiently sorting and organizing biological data into databases. It applies mathematics, biology, and computer science to analyze and interpret the genome and proteome data. Bioinformatics aids in generating the genome-related information of various plants for their efficient exploration as resources to humans. Bioinformatics is the structuring of biological information to enable logical interrogation. Advancements in technologies have resulted in the generation of huge data, facilitating the growth and development of bioinformatics. The expansion of omics technologies has further increased the applications of bioinformatics. The development of technology to annotate all genes, proteins, and metabolites that are expressed in crop plants and economically important plants subjected to abiotic stress, such as drought, has opened new doors for crop improvement and hybrid research to generate plants with improved yield under drought stress. The development of systemic analysis for functional annotation of the genes expressed during drought stress will further open new avenues in this field. Such studies will uncover novel concepts related to the performance of the plants during abiotic stress conditions, leading to improved crop yield.

References

Alter, S., Bader, K.C., Spannagl, M. et al. (2015). Drought DB: an expert-curated compilation of plant drought stress genes and their homologs in nine species. *Database* 2015: 1–7.

Aoki-Kinoshita, K.F. and Kanehisa, M. (2007). Gene annotation and pathway mapping in KEGG. *Methods in Molecular Biology* 396: 71–91.

Apel, K. and Hirt, H. (2004). Reactive oxygen species: metabolism, oxidative stress, and signal transduction. *Annual Review of Plant Biology* 55: 373–399.

Benjamin, J.G. and Nielsen, D.C. (2006). Water deficit effects on root distribution of soybean, field pea and chickpea. *Field Crops Research* 97: 248–253.

Bismillah Khan, M., Hussain, M., Raza, A. et al. (2015). Seed priming with $CaCl_2$ and ridge planting for improved drought resistance in maize. *Turkish Journal of Agriculture Forests* 39: 937–952.

Cai, Y.F., Wang, J.H., Zhang, L. et al. (2019). Physiological and transcriptomic analysis highlight key metabolic pathways in relation to drought tolerance in *Rhododendron delavayi*. *Physiology and Molecular Biology of Plants* 25: 991–1008.

Cao, L., Lu, X., Wang, G. et al. (2021). Transcriptional regulatory networks in response to drought stress and rewatering in maize (*Zea mays* L.). *Molecular Genetics and Genomics* 296: 1203–1219.

Challa, S. and Nageswara, R.R.N. (2018). Genome wide association studies (GWAS) for abiotic stress tolerance in plants. In: *Biochemical, Physiological and Molecular Avenues for Combating Abiotic Stress Tolerance in Plants* (ed. S. Wani), 135–150. Elsevier.

Challa, S., Dutta, T., and Neelapu, N.R.R. (2019). Fungal white biotechnology applications for food security: opportunities and challenges. In: *Recent Advancement in White Biotechnology Through Fungi, Fungal Biology*, vol. II (ed. A.N. Yadav, S. Singh, S. Mishra, and A. Gupta), 119–148. Cham: Springer.

Challa, S., Dutta, T., and Neelapu, N.R.R. (2020). Microbiomes associated with plant growing under the hypersaline habitats and mitigation of salt stress. In: *Advances in Plant Microbiome and Sustainable Agriculture. Microorganisms for Sustainability*, vol. 20 (ed. A. Yadav, A. Rastegari, N. Yadav, and D. Kour), 151–178. Singapore: Springer.

Chen, Y., Li, E.-M., and Xu, L.-Y. (2022). Guide to metabolomics analysis: a bioinformatics workflow. *Metabolites* 12: 357.

Du, Y., Zhao, Q., Chen, L. et al. (2020). Effect of drought stress on sugar metabolism in leaves and roots of soybean seedlings. *Plant Physiology and Biochemistry* 146: 1–12.

Fabregas, N. and Fernie, A.R. (2019). The metabolic response to drought. *Journal of Experimental Botany* 70: 1077–1085.

Farooq, M.H. and Abdul, W.S. (2012). Drought stress in plants: an overview. In: *Plant Responses to Drought Stress* (ed. R. Aroca), 1–33. Berlin Heidelberg: Springer-Verlag.

Gataa, E.L.Z., Samir, K., and Tadesse, W. (2022). Genetic dissection of drought tolerance of elite bread wheat (*Triticum aestivum* L.) genotypes using genome wide association study in morocco. *Plants* 11: 2705.

Gebeyehu, S., Wiese, H., and Schubert, S. (2010). Effects of drought stress on seed sink strength and leaf protein patterns of common bean genotypes. *African Crop Science Journal* 18: 75–88.

Ghaffari, M., Toorchi, M., Valizadeh, M., and Shakiba, M. (2017). Proteomic prospects for tolerance of sunflower (Helianthus annuus) to drought stress during the flowering stage. *Crop and Pasture Science* 68: 457–465.

Gomez-Sanchez, A., González-Melendi, P., Santamaria, M.E. et al. (2019). Repression of drought-induced cysteine-protease genes alters barley leaf structure and responses to abiotic and biotic stresses. *Journal of Experimental Botany* 70: 2143–2155.

Jain, M., Nijhawan, A., Arora, R. et al. (2007). F-box proteins in rice. genome-wide analysis, classification, temporal and spatial gene expression during panicle and seed development, and regulation by light and abiotic stress. *Plant Physiology* 143: 1467–1483.

Jannatul, F., Syed, S.H., and Bu-Jun, S. (2015). Role of microRNAs in plant drought tolerance. *Plant Biotechnology Journal* 13: 293–305.

Jiang, C., Li, X., Zou, J. et al. (2021). Comparative transcriptome analysis of genes involved in the drought stress response of two peanut (*Arachis hypogaea* L.) varieties. *BMC Plant Biology* 21: 64.

Jiang, Y., Su, S., Chen, H. et al. (2023). Transcriptome analysis of drought-responsive and drought-tolerant mechanisms in maize leaves under drought stress. *Physiologia Plantarum* 175 (2): e13875.

Jogawat, A., Yadav, B., Lakra, N. et al. (2021). Crosstalk between phytohormones and secondary metabolites in the drought stress tolerance of crop plants: a review. *Physiologia Plantarum* 172: 1106–1132.

Kakumanu, A., Ambavaram, M.M., Klumas, C. et al. (2012). Effects of drought on gene expression in maize reproductive and leaf meristem tissue revealed by RNA-seq. *Plant Physiology* 160: 846–867.

Kumar, S.A., Kumari, P.H., Sundararajan, V.S. et al. (2014). PSPDB: plant stress protein database. *Plant Molecular Biology Reporter* 32: 940–942.

Li, M., Li, H., Sun, A. et al. (2022). Transcriptome analysis reveals key drought-stress responsive genes in soybean. *Frontiers in Genetics* 13: 1060529.

Liang, C., Liisa, H., Changliang, W. et al. (2019). Trends in the development of miRNA bioinformatics tools. *Briefings in Bioinformatics* 20: 1836–1852.

Liu, N., Chen, N.Y., Cui, R.X. et al. (2012a). Prognostic value of a microRNA signature in nasopharyngeal carcinoma: a microRNA expression analysis. *The Lancet Oncology* 13: 633–641.

Liu, Z., Kumari, S., Zhang, L. et al. (2012b). Characterization of miRNAs in response to short-term waterlogging in three inbred lines of *Zea mays*. *PLoS One* 0039786.

Mousavi, S.A., Pouya, F.M., Ghaffari, M.R. et al. (2016). Plant PReS: a database for plant proteome response to stress. *Journal Proteome* 143: 69–72.

Nagaraj, N., Alexander Kulak, N., Cox, J. et al. (2012). System-wide perturbation analysis with nearly complete coverage of the yeast proteome by single-shot ultra HPLC runs on a bench top orbitrap. *Molecular and Cellular Proteomics* 11 (M111): 013722.

Neelapu, N.R.R., Deepak, K.G.K., and Surekha, C. (2015). Transgenic plants for higher antioxidant. In: *Managing Salt Tolerance in Plants: Molecular and Genomic Perspectives* (ed. S.H. Wani and M.A. Hossain), 391–406. Taylor & Francis.

Neelapu, N.R.R., Dutta, T., Shabir, H.W., and Challa, S. (2021). Oxidative stress tolerance in response to salinity in plants. In: *Organic Solutes, Oxidative Stress, and Antioxidant Enzymes Under Abiotic Stressors* (ed. A.H. Arafat and L. Abdel), 123–132. CRC Press.

Nesvizhskii, A.I. and Aebersold, R. (2005). Interpretation of shotgun proteomic data: the protein inference problem. *Molecular and Cellular Proteomics* 4: 1419–1440.

Picotti, P., Clement-Ziza, M., Lam, H. et al. (2013). A complete mass-spectrometric map of the yeast proteome applied to quantitative trait analysis. *Nature* 494: 266–270.

Prabha, R., Ghosh, I., and Singh, D.P. (2011). Plant stress gene database: a collection of plant genes responding to stress condition. *ARPN Journal of Science and Technology* 1: 28–31.

Rollins, J., Habte, E., Templer, S. et al. (2013). Leaf proteome alterations in the context of physiological and morphological responses to drought and heat stress in barley (*Hordeum vulgare* L.). *Journal of Experimental Botany* 64: 3201–3212.

Saini, D.K., Chopra, Y., Singh, J. et al. (2022). Comprehensive evaluation of mapping complex traits in wheat using genome-wide association studies. *Molecular Breeding* 42: 1.

Salekdeh, G.H., Siopongco, J., Wade, L. et al. (2002). A proteomic approach to analyzing drought- and salt- responsiveness in rice. *Field Crops Research* 76: 199–219.

Sallam, A., Alqudah, A.M., Dawood, K.H.M. et al. (2019). Drought stress tolerance in wheat and barley: advances in physiology, breeding and genetics research. *International Journal of Molecular Sciences* 20: 3137.

Sallam, A., Eltaher, S., Alqudah, A.M. et al. (2022). Genomics combined GWAS and QTL mapping revealed candidate genes and SNP network controlling recovery and tolerance traits associated with drought tolerance in seedling winter wheat. *Genomics* 114: 110358.

Satam, H., Joshi, K., Mangrolia, U. et al. (2023). Next-generation sequencing technology: current trends and advancements. *Biology* 12: 997.

Surekha, C., Aruna, L., Hossain, M.A. et al. (2015). Present status and future prospects of transgenic approaches for salt tolerance in plants/crop plants. In: *Managing Salt Tolerance in Plants, Molecular and Genomic Perspectives* (ed. S.H. Wani and M.A. Hossain), 329–352. Taylor & Francis.

Treml, J. and Smejkal, K. (2016). (2016) Flavonoids as potent scavengers of hydroxyl radicals. *Comprehensive Review on Food Science and Food Safety* 15: 720–738.

Van Meulebroek, L., Hanssens, J., Steppe, K., and Vanhaecke, L. (2016). Metabolic fingerprinting to assess the impact of salinity on carotenoid content in developing tomato fruits. *International Journal of Molecular Sciences* 17: 821.

Wang, Z., Gerstein, M., and Snyder, M. (2009). RNA-Seq: a revolutionary tool for transcriptomics. *Nature Reviews. Genetics* 10 (1): 57–63. https://doi.org/10.1038/nrg2484. PMID: 19015660; PMCID: PMC2949280.

Wang, A., Zhu, M., Luo, Y. et al. (2017). A sweet potato cinnamate 4-hydroxylase gene, IbC4H, increases phenolics content and enhances drought tolerance in tobacco. *Acta Physiolgia Plantarum* 39: 276.

Wang, N., Gao, Z., Zhang, W. et al. (2023). Genome-wide association analysis reveals the gene loci of yield traits under drought stress at the rice reproductive stage. *Agronomy* 13: 2096.

Werner, C., Correia, O., and Beyschlag, W. (1999). Two different strategies of Mediterranean macchia plants to avoid photoinhibitory damage by excessive radiation levels during summer drought. *Acta Oecologica* 20: 15–23.

Willick, I.R., Lahlali, R., Vijayan, P. et al. (2018). Wheat flag leaf epicuticular wax morphology and composition in response to moderate drought stress are revealed by SEM, FTIR-ATR and synchrotron X-ray spectroscopy. *Physiologia Plantarum* 162: 316–332.

Wu, G. and Poethig, R. (2006). Temporal regulation of shoot development in *Arabidopsis thaliana* by miR156 and its target SPL3. *Development* 133: 3539–3547.

Wu, L., Zhou, H., Zhang, Q. et al. (2010). DNA methylation mediated by a microRNA pathway. *Molecular Cell* 38: 465–475.

Xiao, S., Liu, L., Zhang, Y. et al. (2020). Tandem mass tag-based (TMT) quantitative proteomics analysis reveals the response of fine roots to drought stress in cotton (*Gossypium hirsutum* L.). *BMC Plant Biology* 20: 328.

Yahoueian, S.H., Bihamta, M.R., Babaei, H.R., and Bazargani, M.M. (2021). Proteomic analysis of drought stress response mechanism in soybean (*Glycine max* L.) leaves. *Food Science & Nutrition* 9: 2010–2020.

Yu, X., James, A.T., Yang, A. et al. (2016). A comparative proteomic study of drought-tolerant and drought-sensitive soybean seedlings under drought stress. *Crop and Pasture Science* 67: 528–540.

Zhang, Q.W., Wang, C.X., Li, X.Y. et al. (2010). Isolation and characterization of a gene encoding a polyethylene glycol-induced cysteine protease in common wheat. *Journal of Biosciences* 35: 379–338.

Zhao, M., Jin, L., Hu, B. et al. (2019). Overexpression of chalcone synthase gene improves flavonoid accumulation and drought tolerance in tobacco. *Research Square 2019*: 1–11.

10

Bioinformatics Tools and Resources for Plant Transcriptomics: Challenges and Opportunities

Sona Charles[1] *and Merlin Lopus*[2]

[1] *Bioinformatics and Integrative Genomics Facility, Division of Crop Improvement and Biotechnology, ICAR-Indian Institute of Spices Research, Kozhikode, Kerala, India*
[2] *Environmental Sciences and Sustainable Engineering Centre, Indian Institute of Technology, Palakkad, Kerala, India*

10.1 Introduction

Plants possess an extraordinary ability to adapt and thrive in diverse environments. Behind this remarkable resilience lies a multifaceted, intricate ensemble of gene expressions, in which the genes are first transcribed to RNA molecules and ultimately into proteins that orchestrate every aspect of plant growth, progress, and the reaction given to external stimuli (Anderson et al. 2011). The pattern of gene expression holds implications in the study of plants since they are an integral part of the exploration of functional gene discovery, establishing complex gene regulatory networks, understanding secondary metabolites for their role and mechanisms, identifying resistant genes, and assessing the diversity at the genetic level. Additionally, this contributes valuable perceptions to the development and differentiation of cells, tissues, and organs, shedding light on fundamental processes of the plant system, including photosynthesis, growth, response to environmental cues, and adaptation to stressors. Furthermore, research on plant-microbe interactions, cross-species comparisons, and data integration with metabolomics and proteomics are enhancing our holistic understanding of plant biology. Hence the current sequencing technology trends in plant studies are marked by a dynamic landscape of advancements and innovations.

The transcriptome is a dynamic system (Schwabe et al. 2020) that transmits information to the discrepancy of the expression level of genes and pathways of secondary metabolites at various times and spaces (Tyagi et al. 2022). This dynamicity is visible at the species level as various growth environments and periods are expressed by plants. Thus, it influences the expression levels of the tissue. This leads to the variation in the accretion of biologically active compounds in different plant tissues. Because of its unique features, the transcriptome is preferred over the genome for understanding genes associated with therapeutic plant components (Wang et al. 2009). Transcriptomics is important for studying the functional aspects of plant genes, understanding the gene regulatory domains, and analyzing genetic variance, and bioactive compounds (Tyagi and Ranjan 2023).

Bioinformatics for Plant Research and Crop Breeding, First Edition. Edited by Jen-Tsung Chen.

Next-generation sequencing (NGS) has piloted into a transformative era in transcriptomics, revolutionizing the way we approach gene expression within a cell (Lee-Liu et al. 2012). Transcriptomics delves into the dynamics of gene expression within plants, offering a comprehensive view of how genetic information is translated into functional molecules. The chain-termination technique pioneered by Sanger et al proved to be more pragmatic and it caused the initial wave of automated DNA sequencing technology (Sanger et al. 1977). Over the years, transcriptomics has emerged from the fundamental DNA microarray platform driven by ongoing developments in sequencing technology to the cutting-edge RNA-Seq technology (Mironova et al. 2015). DNA Microarray gained prominence during the early 2000s, allowing the simultaneous investigation of thousands of gene expressions in a genome. However, the limitations of microarrays, such as the need for a priori knowledge of the genes being studied, led to the development of RNA-Seq technology (Bumgarner 2013). RNA-Seq extends various advantages, including increased sensitivity, high throughput, and remarkable efficiency. The major benefit of NGS in plant transcriptomics is its ability to produce massive quantities of sequencing data in a relatively short time. This helps in capturing a detailed snapshot of the entire set of RNA transcripts produced by a plant under diverse conditions, developmental stages, or in response to various stressors. Markedly, it allows for the comprehensive analysis of the entire transcriptome even in the absence of a reference genome. Consequently, RNA-Seq has become a widely adopted sequencing technique in the fields of molecular biology, biotechnology, and bioinformatics.

Long-read sequencing technologies are enabling the comprehensive study of full-length transcripts and alternative splicing, shedding light on transcript isoform diversity (Amarasinghe et al. 2020). The functional annotation of noncoding RNAs, including miRNAs, lncRNAs, and circRNAs, is a growing focus, revealing their significant roles in post-transcriptional regulation (Vaucheret 2006). The integration of transcriptomics with epigenomic data and the emerging field of epitranscriptomics are unraveling the intricate regulation of gene expression in plants. Computational methods and machine learning are pivotal for managing big datasets. Bioinformatics plays a vital role in analyzing big complex datasets created by transcriptomics studies. Since the field of transcriptomics continues to evolve very fast, computational approaches are adapting to address emerging challenges and opportunities. This chapter focuses on the bioinformatic tools and resources aiding the analysis of NGS data.

10.2 Evolution of Transcriptomic Technologies

The genetic revolution accompanied by the advancements in sequencing chemistry has risen to unparalleled prominence transforming DNA sequencing into an affordable and accessible technology. The development of sequencing techniques has witnessed three distinct generations, each with its unique advantages and applications. The first-generation sequencing (FGS), pioneered by Sanger and Coulson, relied on the selective integration of chain-terminating dideoxynucleotides during DNA replication (Sanger et al. 1977). Sanger sequencing was limited by its low throughput and high cost. FGS could generate read lengths under one kilobase (kb). To produce longer DNA fragments, shotgun sequencing was employed in which overlapping DNA fragments were individually cloned and sequenced before being computationally assembled into a single contig (Anderson 1981).

Second-generation sequencing (SGS), also known as NGS, emerged in the mid-2000s and is defined by immensely parallel sequencing of short DNA fragments. Platforms like Illumina HiSeq could generate billions of reads, making NGS invaluable for various applications, including genome assembly, gene expression analysis, and mutation identification (Shendure and Ji 2008) resulted in high throughput and low cost; however, it had limitations in attaining long read lengths. It also introduced errors during the "wash and scan" cycles during polymerase chain reaction (PCR) amplification.

The rise of third-generation sequencing technology was supported by the advantage of sequencing single molecules. The SMRT technology by Pacific Biosciences (PacBio) sidesteps the PCR amplification process. When a labeled dNTP is incorporated into the DNA molecule, a laser activates fluorescent signals, whose color is recorded by a camera system. These signals are meticulously recorded in real-time within a specialized flow cell featuring zero-mode waveguides (ZMWs) (Eid et al. 2009). The presence of any DNA modifications is indicated by an extended impulse duration. Nanopore Sequencing Technology (ONT) ushered in a novel approach by integrating nanopores within an electrically resistant membrane. A potential difference is given across the membrane directing a current exclusively through the nanopore. As dsDNA traversed the nanopore, attached polymerase or helicase enzymes played a pivotal role in guiding the DNA through the pore. At the same time, sequencing disruptions in the electrical current can be observed as individual nucleotides pass through the nanopore, which allows their detection (Schatz 2017). Third-generation sequencers not only streamline the sequencing workflow but also reduce the risk of introducing errors or bias during amplification. The growth of sequencing technology is depicted in Figure 10.1.

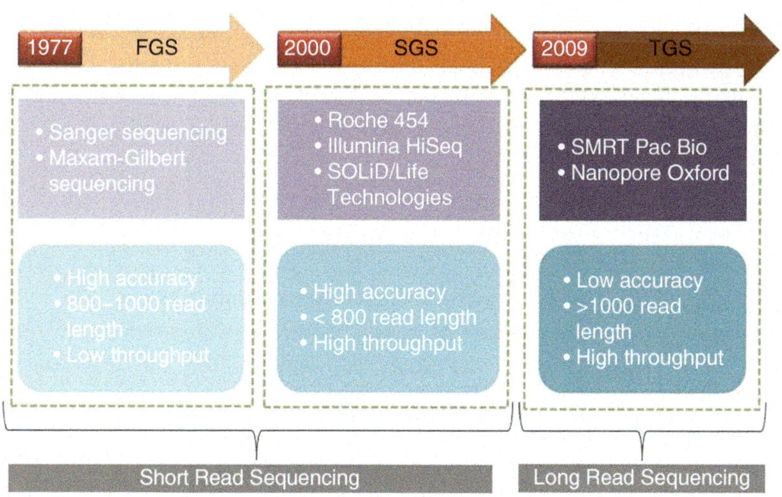

Figure 10.1 Generations of sequencing technologies and their features.

10.3 Steps in Transcriptomic Data Analysis

10.3.1 Raw Data Preprocessing

NGS has allowed higher sequencing throughput at lower sequencing costs. Illumina reads include normally 36–300 nucleotide bases and are formed by a reversible-terminator cyclic reaction linked to base-specific colorimetric signals within the sequencing machine. Reads can be differentiated as "single-end" or "paired-end," and they epitomize both extremities of the same nucleotide fragment. The calorimetric signals will be translated to base calls by the internal Illumina software known as CASAVA and are further denoted in the FASTQ format (Cock et al. 2010). Each of the nucleotides in the FASTQ format is linked to an ASCII-encoded quality number corresponding to a PHRED score (Q) (Ewing and Green 1998). Recently, Illumina runs it ranges from 0 to 41 and the error rate at each position ranges from 7.94e-5 to 1. If the product is a low-quality read or nucleotides with a higher probability of error such as air bubbles, spot-specific signal noise, malfunctioning laser or lens, etc. To avoid such mistakes encoding the Q value to the sequence information will be useful for the subsequent analysis (He et al. 2020). To achieve reliable conclusions from downstream analyses, quality control (QC) processes are vital and removal of low-quality reads through filtration is important. Moreover, the removal of the adapter is a crucial part of the downstream analysis as this contamination can lead to NGS alignment errors and an increased number of unaligned reads, specifically in the sequencing of small RNA (Liu et al. 2019). There are numerous tools have been developed with different functions to deal with FASTQ files generated by Illumina sequencing (Table 10.1).

10.3.2 Genome Assembly and Mapping

A genome assembly follows after procuring the quality-checked raw reads. This process is meant to assemble the resulting short DNA sequences from various sequencing techniques to reconstruct the actual chromosomes. Two categories are there in genome assembly technique, (i) De novo assembly, which needs no prior knowledge of the source of DNA sequence length, layout, or composition, and (ii) mapping to the reference genome. Due to the progress in sequencing technology, more and more genomes of various species are sequenced. Moreover, there is a chance that the genome of a different but related species is available which in turn gives a chance of a significant proportion of reads getting mapped. The reference-guided genome assembly retrieves additional information through the similarity between the target and reference species which in turn gives a more complete and enhanced genome (Pop et al. 2004; Schneeberger et al. 2011). Though the sequencing is done at a low coverage, genome assembly guided by a reference genome can provide more insights into genomic resources (Card et al. 2014; Vezzi et al. 2011). There are two reference-guided assembly strategies, and in the first method, the raw reads will be mapped to an available reference genome. Later, the mapped sequence will be used to generate an alternate consensus sequence (Vezzi et al. 2011). Whereas the second method takes a slightly different approach in that it first *de novo* assembles the raw reads and further the resulting scaffolds are aligned against the available reference genome. Through this genome will get annotated and the misassembled

Table 10.1 Tools for preprocessing of FASTQ files.

Software	Main functions	References
FastQC	Produces a quality analysis report on FASTQ files	Andrews (2010)
PIQA	Produces a quality analysis report on FASTQ files	Martínez-Alcántara et al. (2009)
FASTX-Toolkit	A collection of tools to filter low-quality reads and remove adapters	http://hannonlab.cshl.edu/fastx_toolkit/
Fqtools	A software suite for fast processing of FASTQ files and various file manipulations is supported	Droop (2016)
seqtk	It is a fast and lightweight tool for processing sequences in the FASTA or FASTQ format	https://github.com/lh3/seqtk.
SeqKit	A tool kit for common manipulations of FASTA/Q file includes converting, searching, filtering, deduplication, splitting, shuffling, and sampling	Shen et al. (2016)
PRINSEQ	It is a tool for quality control of metagenomic sequence data. The package also contains functions for trimming, filtering, and data reformatting	Schmieder and Edwards (2011)
Cutadapt	It identifies and eliminates adapter sequences, primers, poly-A tails and other types of undesirable sequences from high-throughput sequencing reads	Martin (2011)
FastP	It has functions for quality control, trimming of adapters, filtering by quality, and read pruning	Chen et al. (2018)
Trimmomatic	Tool to filter low-quality reads and remove adapters	Bolger et al. (2014)
Multiqc	A tool to create a single report visualizing output from multiple tools across many samples	Ewels et al. (2016)
NGS QC Toolkit	A collection of tools to filter low-quality reads and remove adapters	Patel and Jain (2012)
FaQCs	Filter low-quality reads and trim adapters	Lo and Chain (2014)
FastProNGS	Filter low-quality reads and trim adapters	Liu et al. (2019)

scaffolds will be identified (Vezzi et al. 2011). Sometimes, a combination of the two approaches is also applied (Wang et al. 2014). Genome assemblers commonly and their respective algorithms are given in Table 10.2.

10.3.3 Differential Expression Analysis

Identifying genes with distinct expression patterns across different conditions or samples is a fundamental step in unraveling the molecular mechanisms of any cellular process. In plant research, differentially expressed genes (DEGs) offer crucial clues about which genes are activated or suppressed in response to specific conditions or stimuli and they impose an important role in stress-related study (Soneson and Delorenzi 2013). After the genome assembly, it will be scrutinized for DEGs and the techniques for differential gene expression analysis of RNA-Seq can be broadly categorized into two main groups: parametric and

Table 10.2 List of genome assemblers.

Software	Algorithm used	References
NECAT	NECAT algorithm	Chen et al. (2021)
NextDenovo	String graph algorithm	https://github.com/Nextomics/NextDenovo
Flye	De-Bruijn Graph algorithms	Kolmogorov et al. (2019)
Euler		Pevzner et al. (2004)
Shasta toolkit		Shafin et al. (2020)
SPAdes		Bankevich and Pevzner (2016)
Arachne	Overlap-layout-consensus algorithms	Batzoglou et al. (2002)
AMOS		https://sourceforge.net/projects/amos/files/
Atlas		Havlak et al. (2004)
Celera Assembler		Myers et al. (2000)
CAP3		Huang and Madan (1999)
PCAP		Huang et al. (2003)
Phrap		De La Bastide and McCombie (2007)
RePS		Wang et al. (2002)
Phusion		Mullikin and Ning (2003)
Canu		Koren et al. (2017)
wtdbg2		Ruan and Li (2020)
Miniasm		Lu et al. (2016)
Smartdenovo		https://github.com/ruanjue/Smartdenovo

nonparametric approaches. The parametric method focuses on the distribution of gene expression values and it follows a normalization step in which each gene's expression value conforms to a specific distribution, such as Poisson or negative binomial. Conversely, non-parametric methods are valuable while working with non-normally distributed RNA-Seq data when dealing with small sample sizes (Li and Tibshirani 2013). Numerous algorithms are available for differential expression analysis out of which DeSeq2, edgeR, and limma have gained much popularity over the years.

DeSeq2 employs a generalized linear model (GLM) framework to model the relationship between gene expression counts and covariates, such as experimental conditions or biological factors. It takes into account sample-specific variability, estimates dispersions, and employs a negative binomial distribution to model count data (Love et al. 2014). EdgeR (Empirical Analysis of Digital Gene Expression data in R) is another tool used for the analysis of DEGs and it is well-suited for datasets with a limited number of replicates, where the difference in gene expression measures can be substantial. It employs a negative binomial distribution-based model to account for both biological and technical sources of variation, allowing for the identification of DEGs while controlling for false discovery rates (FDR). EdgeR uses count data from RNA-Seq experiments to estimate dispersion and applies GLMs to evaluate variations in gene expression upon various experimental conditions or groups. It also incorporates methods for library size normalization, which is crucial for accurate comparisons between samples (Robinson et al. 2010). Limma (Linear Models for Microarray and RNA-Seq Data) utilizes linear modeling, estimating the relationships between gene expression and experimental factors while effectively accounting for both systematic and random sources of variation. It utilizes empirical Bayes methods to moderate variance estimates, especially when dealing with limited sample sizes, thereby improving the accuracy of differential expression analysis (Ritchie et al. 2015). Online webservers like Differential Expression Gene Explorer (DrEdGE) support the visualizations of differential gene expression through graphs (Tintori et al. 2020).

10.3.4 Functional Enrichment Analysis

Once the DEGs are identified, the next step in the downstream analysis is the identification of function functional enrichment analysis which seeks to elucidate the biological significance of the observed expression changes by evaluating whether specific functional categories, gene ontology (GO) terms, or pathways are overrepresented among the DEGs. Identifying the functional roles and associations connected with genes delivers valuable insights into key pathways and processes that are relevant to the observed differential expression patterns.

Functional enrichment tools are typically categorized into two groups: (i) overrepresentation analysis (ORA) and (ii) functional class scoring (FCS), with their prime application being in the context of analyzing differential gene expression. In ORA, the analysis includes setting a cut-off based on fold change for DEGs and comparing them against curated pathways or gene sets followed by a statistical test to determine whether the number of DEGs related to a particular gene set is significantly higher than what would be expected by random chance, using a background gene list as a reference. FCS tools assign a differential expression score value for each gene and assess whether the scores match that expected by chance for a set of genes (Wijesooriya et al. 2022).

10.3.5 Pathway Analysis

Pathway analysis of the DEGs and proteins allows us to understand the underlying mechanisms of them more deeply. Many studies involve comparing two different phenotypes, such as disease versus control or the effects of various treatments or drugs. Various statistical methods like the t-test (Tian et al. 2005), Z-score (Kim and Volsky 2005), and ANOVA (Al-Shahrour et al. 2005) are employed to identify DEGs between these phenotypes. While DEG lists provide valuable insights into phenotypic differences, they cannot fully elucidate the complex mechanisms involved. With the availability of freely available pathway databases such as KEGG (Kanehisa and Goto 2000), Reactome (Jassal et al. 2020), Biocarta (Biocarta 2004), WikiPathways (Martens et al. 2021), MetaCyc (Karp et al. 2002), PathBank (Wishart et al. 2020), etc. and commercial databases associated with GeneGo (https://portal.genego.com/), Protein Lounge (http://www.proteinlounge.com/), andIngenuity Pathway Analysis, it is crucial to emphasize that the choice of an appropriate pathway database should be dependent upon the precise biological context under investigation. To address this, the knowledge found in pathway databases can be leveraged as a common technique for gaining deeper biological insights. Pathway analysis methods may be categorized into two: topology-based and non-topology-based tools. Topology-based methods consider all existing knowledge to attain a conclusion about the genes in the list. Non-topology-based approaches do not utilize the existing knowledge about gene positions, roles within pathways, the flow of signals between genes, or other pathway-specific details (Nguyen et al. 2019). The list of topology and non-topology based tools are given in Table 10.3.

Table 10.3 List of pathway analysis tools

Sl No	Pathway analysis tools	References	Type
1	Onto-Express	Khatri et al. (2002)	Non-topology-based
2	GeneMAPP	Dahlquist et al. (2002)	Non-topology-based
3.	GeneMerge	Castillo-Davis and Hartl (2003)	Non-topology-based
4.	EASE	Hosack et al. (2003)	Non-topology-based
5.	FuncAssociate	Berriz et al. (2003)	Non-topology-based
6.	FatiGO	Al-Shahrour et al. (2004)	Non-topology-based
7.	GOstats	Beissbarth and Speed (2004)	Non-topology-based
8.	GOToolBox	Martin et al. (2004)	Non-topology-based
9.	GoMiner	Zeeberg et al. (2003)	Non-topology-based
10.	DAVID	Huang et al. (2009)	Non-topology-based
11.	WebGestalt	Wang et al. (2013).	Non-topology-based
12.	Pathway-Express	Draghici et al. (2007)	Topology-based methods
13.	SPIA	Tarca et al. (2009)	Topology-based methods
14.	NetGSA	Shojaie and Michailidis (2009)	Topology-based methods
15.	TopoGSA	Glaab et al. (2010)	Topology-based methods
16.	TopologyGSA	Massa et al. (2010)	Topology-based methods
17.	PWEA	Hung et al. (2010)	Topology-based methods
18.	PathOlogist	Greenblum et al. (2011)	Topology-based methods
19.	GGEA	Geistlinger et al. (2011)	Topology-based methods. The method and accompanying visualization capabilities are bundled into an R package and tied to the Galaxy workflow environment.
20.	cepaORA	Gu et al. (2012)	Topology-based methods
21.	cepaGSA	Gu et al. (2012)	Topology-based methods
22.	PathNet	Dutta et al. (2012)	Topology-based methods
23.	ROntoTools	Voichiţa et al. (2023)	Topology-based methods
24.	BLMA	Nguyen and Draghici (2017)	Topology-based methods

10.3.6 GO Analysis

The GO is the most extensive repository of annotations representing precisely defined terms, each interconnected by well-established relationships. The connections between genes and ontology terms that describe their functions are substantiated by supporting evidence (Ashburner et al. 2000). GO terms describe three nonoverlapping ontologies such as molecular function, biological process, and cellular component. Biological processes encompass a wide range of activities, from broad and high-level descriptions like "cell cycle" or "cellular respiration" to more specific and lower-level terms such as "DNA replication" or "protein phosphorylation" which pertains to a specific biological goal that is influenced by the involvement of a gene or protein. Molecular function describes the molecular actions performed by the gene products, without delving into the specifics of where these events occur within a biological system e.g.: "kinase activity" or "DNA binding." The cellular component provides critical insights into the specific locations within a cell where gene products exert their functions. The cellular component includes terms such as "nuclear membrane" or "Golgi apparatus," elucidating the distinct cellular structures and compartments that play vital roles in various cellular processes.

10.4 R/Bioconductor Packages for Transcriptomic Analysis

R/Bioconductor (https://www.bioconductor.org/) packages have emerged as indispensable tools in the realm of omics studies, offering a diverse array of functionalities to analyze and interpret high-throughput biological data (Huber et al. 2015). These packages provide a robust and flexible environment for researchers to tackle a wide range of omics datasets, including genomics, transcriptomics, proteomics, and metabolomics. Bioconductors continually evolve to keep pace with emerging technologies and data types, ensuring that they remain at the forefront of omics research through bi-annual updations. The latest release of Bioconductor (v 3.17) accommodates 2230 software packages, 913 annotation data, 421 experimental data, and 30 workflows.

Each package is linked to a documentation/manual, which assists its usage. Bioconductor documentation is structured into three tiers:

- **Workflows**: These documents cover comprehensive analyses that encompass various tools and processes.
- **Package vignettes**: These narratives offer insights into the intended applications of specific packages, complete with detailed executable code examples.
- **Function manual pages**: These pages contain precise descriptions of inputs and outputs for functions, accompanied by practical working examples.

R seamlessly integrates data visualization with data analysis. The data visualization packages in R enable the depiction of informative and visually appealing plots and graphics tailored to the specific needs of genomics and biological data analysis. Bioconductor provides various extensions and integrations of visualization packages like ggplot2 specifically designed for genomic data visualization.

10.5 Galaxy Server for Transcriptome Analysis

The Galaxy platform (https://usegalaxy.org/) provides a user-friendly and accessible environment for performing transcriptomics analysis (Jalili et al. 2020) and it hosts many specific tools and workflows for transcriptomics research. This platform streamlines the execution of complex bioinformatics analyses by integrating the requisite tools within its web interface. Galaxy offers a wide range of support to experimental scientists starting from data uploading and extending to the visualization and functional enrichment analysis of genes exhibiting differential expression. It offers tools and workflows for diverse tasks, including sequence analysis, variant calling, RNA-Seq and ChIP-Seq data processing, structural biology, metagenomics, etc. By providing a preconfigured series of interconnected tools and parameters, Galaxy simplifies the complexity of transcriptomics analysis. Users can design custom workflows tailored to their research questions, incorporating steps such as data preprocessing, read alignment, transcript assembly, differential expression analysis, and downstream functional annotation. This streamlined approach not only saves time but also ensures reproducibility in transcriptomics studies. Galaxy democratizes access to sophisticated bioinformatics tools and empowers researchers to efficiently analyze complex biological data while promoting good practices in data analysis and sharing.

10.6 Stress Transcriptomics – A Case Study

Transcriptomics can reveal the expression at the whole genome level in response to various stress from the whole transcriptional level, which is advantageous for understanding the underlying mechanisms associated with the response of plants to stress (Wang et al. 2020). In a meta-analysis study, Balan et al. (2018) revealed the response of Malus × domestica Borkh or the domesticated apple which is the main fruit crop of the temperate regions of the world for various biotic stresses and observed the response of plants for Apple Stem Grooving Virus (ASGV). It was observed that there was an upregulation in the majority of GO-terms relating to transcription regulation, response to chitin, and phosphorylation. Similar upregulation was also identified with *Erwinia amylovora* where the upregulated GO-terms were related to ethylene and jasmonic acid defense responses. Moreover, it was recognized that the upregulation of above mentioned GO terms was higher with *E. amylovora* than with fungal pathogens. As a result of *E. amylovora* exposure a reduction in responses to xyloglucan metabolic process, actin, cellular responses to gravity, and lipid transport was observed. Attacks of virus significantly repressed jasmonic acid-related GO terms, response to bacterium and fungal pathogens. Another study using RNA-seq analysis carried out on two cotton cultivars, Xinluzao-36 (susceptible) and Zhongzhimian-2 (disease resistant) infected by *Verticillium dahliae* a fungal plant pathogen revealed that the infection elicited variation in gene expression patterns in these varieties. Moreover, the DEGs were more in the susceptible cultivar than in the resistant cultivars. This study also revealed numerous genes encompassed in disease resistance and defense mechanisms (Zhang et al. 2021).

Transcriptomics can be applied in various levels of intervention in biotic stress. One of the major problems in agriculture is the adaptive strategy developed for pests for insecticide resistance. A comparative transcriptomic study on both imidacloprid-resistant (IR) and imidacloprid-susceptible (IS) *Myzus persicae* identified 252 DEGs among the IR and IS *M. persicae* transcriptomes. These candidate genes included 160 and 92 genes that were down- and up-regulated, respectively, in the IR strain (Meng et al. 2019). This study provides a valuable basis for further investigation into the complicated mechanisms of imidacloprid resistance in *M. persicae*.

The genes that are expressed differently under abiotic stress can also be identified through a transcriptome approach. Identification of key influencing factors will be helpful in plant cultivation and breeding, as well as simplifying the selection of improved varieties (Rastogi et al. 2019). One of the major environmental stresses in the scenario of climate change is salinity and it affects plant growth and productivity which in turn affects about 20% of the cultivable and about 50% of the irrigated lands worldwide (Tuteja 2007). Mangrove plants exhibit exceptional salt tolerance. When they are exposed to salinity, the roots of the mangrove plants show adaptive mechanisms to tolerate salt stress. To understand the saline tolerance of mangrove trees, Krishnamurthy et al. (2017) conducted a study on NaCl-treated seedlings of *Avicennia officinalis* a mangrove tree species and generated the gene expression profiles through RNA sequencing. They have identified a total of 6547 DEGs in response to NaCl treatment, of which 1404 genes were up-regulated and 5213 genes were down-regulated significantly. By comparative genomics, this study identified 93 key salt tolerance genes.

Drought is another threat to the productivity and survival of crops worldwide. To tolerate drought stress different plant tissues such as leaves, roots, and sap exhibits specific response. Upon drought stress many protein-coding genes are regulated by the abscisic acid (ABA)-dependent and the ABA-independent signaling pathways (Yoshida et al. 2014). The transcriptomics study conducted by Gamboa-Tuz et al. (2018) with papaya plants is an example of recording plant response upon drought. The papaya plants were grown under control and drought-stressed conditions and transcriptomes were generated using Illumina RNA-Seq. They investigated a total of 18 cDNA libraries made from leaves, sap, and roots of papaya plants. From the reference and de novo transcriptomics analysis, a total of 8549 drought-responsive genes and 6089 unigenes were identified followed by GO enrichment analysis. Through this, it was discovered that there is an upregulation in the processes related to cell cycle and DNA repair in leaves and sap under the state of moderate drought stress. Similarly, upregulation in processes related to abiotic stress, hormone signaling, sucrose metabolism, and suberin biosynthesis was observed in roots. Whereas under severe drought stress, biological processes related to abiotic stress, hormone signaling, and oxidation-reduction will be up-regulated in all tissues (Gamboa-Tuz et al. 2018).

Submergence stress generates a hypoxic environment. Roots are the parts of the plant getting affected by these low oxygen conditions first. A transcriptomic analysis conducted on orchardgrass (*Dactylis glomerata* L.) roots for understanding the plant response to submergence revealed numerous DEGs. The stress was induced at different time points viz., 0, 8, and 24 h in two species viz., submergence-tolerant "Dianbei" and submergence-sensitive "Anba." The study revealed a total of 6663 and 7894 DEGs at 8 h post submergence stress in Dianbei and Anba respectively. Whereas a total of 9857 and 11215 DEGs were identified in Dianbei and Anba at 24 h post-submergence-stress (Qu et al. 2022). Furthermore, significant DEGs of plant hormones, mitogen-activated protein kinase (MAPK), and Ca^{2+} signals were only observed in Dianbei.

Multiple stress responses of plants were also analyzed through a single transcriptomic study. The transcriptomic study conducted in holy basil or *Ocimum tenuiflorum* a generally used medicinal plant is an example of this. They have adopted a de novo transcriptome sequencing of leaves and investigated the response of the plant against multiple stress conditions viz., cold, drought, flood, and salinity (Rastogi et al. 2019). The study observed that among the four stresses, the plant was more susceptible to cold stress. Moreover, severe treatments of all these abiotic stresses also decreased eugenol which is the main secondary metabolite present in the *O. tenuiflorum* plant.

Differential transcriptomics is essential for understanding how plants fine-tune their gene expression to cope with stressors such as drought, salinity, pathogens, and extreme temperatures. By comparing the transcriptomes of plants grown under control and stressed conditions, we can recognize stress-responsive genes, uncover regulatory networks, and elucidate the pathways included in stress adaptation. In the case of biotic stresses, the gene expression profiles of the host and pathogen under diverse conditions, often before and after the pathogen infection in plants may be studied to understand the insights into the underlying mechanisms of the host's response and the pathogen's way of adaptation for survival and proliferation. Analysis of DEGs also sheds light on the activation of immune-related genes, the modulation of cellular pathways, and the interplay between host defense mechanisms and pathogen virulence factors.

10.7 Conclusion and Way Forward

The future of plant transcriptomics, empowered by advanced bioinformatics tools, holds great promise for revolutionizing our understanding of various molecular mechanisms in the plant system and enhancing agricultural practices. As sequencing technology continues to advance, more precise and comprehensive insights into the transcriptome dynamics of plants across various developmental stages, environmental conditions, and genotypes may be anticipated. Integration of multi-omics data, including genomics, proteomics, and metabolomics, will enable holistic systems biology approaches, shedding light on complex gene regulatory networks and signaling pathways. Machine learning and artificial intelligence techniques will play a pivotal role in deciphering the "big datasets," leading to the identification of novel regulatory elements and potential targets for combating biotic and abiotic stresses leading to crop improvement. Moreover, as transcriptomics continues to evolve, the development of user-friendly bioinformatics tools and platforms will democratize access to these technologies. The future of plant transcriptomics using bioinformatics tools promises to be both exciting and transformative, with profound implications for crop yield improvement, stress resistance, and ecosystem management.

References

Al-Shahrour, F., Díaz-Uriarte, R., and Dopazo, J. (2004). FatiGO: a web tool for finding significant associations of Gene Ontology terms with groups of genes. *Bioinformatics* 20 (4): 578–580.

Al-Shahrour, F., Díaz-Uriarte, R., and Dopazo, J. (2005). Discovering molecular functions significantly related to phenotypes by combining gene expression data and biological information. *Bioinformatics* 21 (13): 2988–2993.

Amarasinghe, S.L., Su, S., Dong, X. et al. (2020). Opportunities and challenges in long-read sequencing data analysis. *Genome Biology* 21 (1): 30.

Anderson, S. (1981). Shotgun DNA sequencing using cloned DNase I-generated fragments. *Nucleic Acids Research* 9 (13): 3015–3027.

Anderson, J.T., Willis, J.H., and Mitchell-Olds, T. (2011). Evolutionary genetics of plant adaptation. *Trends in Genetics* 27 (7): 258–266.

Andrews, S. (2010). FastQC: a quality control tool for high throughput sequence data. http://www.bioinformatics.babraham.ac.uk/projects/fastqc/ (accessed 3 March 2024).

Ashburner, M., Ball, C.A., Blake, J.A. et al. (2000). Gene ontology: tool for the unification of biology. The Gene Ontology Consortium. *Nature Genetics* 25 (1): 25–29.

Balan, B., Marra, F.P., Caruso, T., and Martinelli, F. (2018). Transcriptomic responses to biotic stresses in Malus × domestica: a meta-analysis study. *Scientific Reports* 8 (1): 1970. https://doi.org/10.1038/s41598-018-19348-4.

Bankevich, A. and Pevzner, P.A. (2016). TruSPAdes: barcode assembly of TruSeq synthetic long reads. *Nature Methods* 13 (3): 248–250.

de la Bastide, M. and McCombie, W.R. (2007). Assembling genomic DNA sequences with PHRAP. *Current Protocols in Bioinformatics* 17 (1): 11.4.1–11.4.15. https://doi.org/10.1002/0471250953.bi1104s17.

Batzoglou, S., Jaffe, D.B., Stanley, K. et al. (2002). ARACHNE: a whole-genome shotgun assembler. *Genome Research* 12 (1): 177–189.

Beissbarth, T. and Speed, T.P. (2004). GOstat: find statistically overrepresented Gene Ontologies within a group of genes. *Bioinformatics* 20 (9): 1464–1465.

Berriz, G.F., King, O.D., Bryant, B. et al. (2003). Characterizing gene sets with FuncAssociate. *Bioinformatics* 19 (18): 2502–2504.

BioCarta (2004). BioCarta - Charting Pathways of Life. Technical report, *BioCarta*.

Bolger, A.M., Lohse, M., and Usadel, B. (2014). Trimmomatic: a flexible trimmer for Illumina sequence data. *Bioinformatics* 30 (15): 2114–2120.

Bumgarner, R. (2013). Overview of DNA microarrays: types, applications, and their future. *Current Protocols in Molecular Biology* 101: 22.1.1–22.1.11.

Card, D.C., Schield, D.R., Reyes-Velasco, J. et al. (2014). Two low coverage bird genomes and a comparison of reference-guided versus *de novo* genome assemblies. *PLoS One* 9 (9): e106649.

Castillo-Davis, C.I. and Hartl, D.L. (2003). GeneMerge-post-genomic analysis, data mining, and hypothesis testing. *Bioinformatics* 19 (7): 891–892.

Chen, S., Zhou, Y., Chen, Y., and Gu, J. (2018). fastp: an ultra-fast all-in-one FASTQ preprocessor. *Bioinformatics* 34 (17): i884–i890.

Chen, Y., Nie, F., Xie, S.-Q. et al. (2021). Efficient assembly of nanopore reads via highly accurate and intact error correction. *Nature Communications*. State Key Laboratory of Ophthalmology, Zhongshan Ophthalmic Center, Sun Yat-sen University, #7 Jinsui Road, Tianhe District, Guangzhou, People's Republic of China.; 12: 60. http://europepmc.org/abstract/MED/33397900 (accessed 20 August 2023).

Cock, P.J.A., Fields, C.J., Goto, N. et al. (2010). The Sanger FASTQ file format for sequences with quality scores, and the Solexa/Illumina FASTQ variants. *Nucleic Acids Research* 38 (6): 1767–1771.

Dahlquist, K.D., Salomonis, N., Vranizan, K. et al. (2002). GenMAPP, a new tool for viewing and analyzing microarray data on biological pathways. *Nature Genetics United States* 31: 19–20.

Draghici, S., Khatri, P., Tarca, A.L. et al. (2007). A systems biology approach for pathway level analysis. *Genome Research* 17: 1537–1545. Accessed 20 September 2023.

Droop, A.P. (2016). fqtools: an efficient software suite for modern FASTQ file manipulation. *Bioinformatics* 32 (12): 1883–1884.

Dutta, B., Wallqvist, A., and Reifman, J. (2012). PathNet: a tool for pathway analysis using topological information. *Source Code for Biology and Medicine* 7 (1): 10.

Eid, J., Fehr, A., Gray, J. et al. (2009). Real-time DNA sequencing from single polymerase molecules. *Science* 323 (5910): 133–138.

Ewels, P., Magnusson, M., Lundin, S., and Käller, M. (2016). MultiQC: summarize analysis results for multiple tools and samples in a single report. *Bioinformatics* 32 (19): 3047–3048.

Ewing, B. and Green, P. (1998). Base-calling of automated sequencer traces using phred. II. Error probabilities. *Genome Research* 8 (3): 186–194.

Gamboa-Tuz, S.D., Pereira-Santana, A., Zamora-Briseño, J.A. et al. (2018). Transcriptomics and co-expression networks reveal tissue-specific responses and regulatory hubs under mild and severe drought in papaya (*Carica papaya* L.). *Scientific Reports*. Biotechnology Unit, Yucatan Center for Scientific Research (CICY), 97205, Merida, Yucatan, Mexico.; 8: 14539. http://europepmc.org/abstract/MED/30267030 (accessed 5 August 2023).

Geistlinger, L., Csaba, G., Küffner, R. et al. (2011). From sets to graphs: towards a realistic enrichment analysis of transcriptomic systems. *Bioinformatics* 27 (13): i366–i373.

Glaab, E., Baudot, A., Krasnogor, N., and Valencia, A. (2010). TopoGSA: network topological gene set analysis. *Bioinformatics* 26 (9): 1271–1272.

Greenblum, S.I., Efroni, S., Schaefer, C.F., and Buetow, K.H. (2011). The PathOlogist: an automated tool for pathway-centric analysis. *BMC Bioinformatics* 12: 133.

Gu, Z., Liu, J., Cao, K. et al. (2012). Centrality-based pathway enrichment: a systematic approach for finding significant pathways dominated by key genes. *BMC Systems Biology* 6: 56.

Havlak, P., Chen, R., Durbin, K.J. et al. (2004). The Atlas genome assembly system. *Genome Research* 14 (4): 721–732.

He, B., Zhu, R., Yang, H. et al. (2020). Assessing the impact of data preprocessing on analyzing next generation sequencing data. *Frontiers in Bioengineering and Biotechnology* 8: 817.

Hosack, D.A., Dennis, G., Sherman, B.T. et al. (2003). Identifying biological themes within lists of genes with EASE. *Genome Biology* 4 (10): R70. https://doi.org/10.1186/gb-2003-4-10-r70.

Huang, X. and Madan, A. (1999). CAP3: a DNA sequence assembly program. *Genome Research* 9 (9): 868–877.

Huang, X., Wang, J., Aluru, S. et al. (2003). PCAP: a whole-genome assembly program. *Genome Research* 13 (9): 2164–2170.

Huang, D.W., Sherman, B.T., and Lempicki, R.A. (2009). Systematic and integrative analysis of large gene lists using DAVID bioinformatics resources. *Nature Protocols* 4 (1): 44–57. https://doi.org/10.1038/nprot.2008.211.

Huber, W., Carey, V.J., Gentleman, R. et al. (2015). Orchestrating high-throughput genomic analysis with Bioconductor. *Nature Methods* 12 (2): 115–121.

Hung, J.-H., Whitfield, T.W., Yang, T.-H. et al. (2010). Identification of functional modules that correlate with phenotypic difference: the influence of network topology. *Genome Biology* 11 (2): R23.

Jalili, V., Afgan, E., Gu, Q. et al. (2020). The Galaxy platform for accessible, reproducible and collaborative biomedical analyses: 2020 update. *Nucleic Acids Research* 48 (W1): W395–W402. https://doi.org/10.1093/nar/gkaa434.

Jassal, B., Matthews, L., Viteri, G. et al. (2020). The reactome pathway knowledgebase. *Nucleic Acids Research* 48 (D1): D498–D503.

Kanehisa, M. and Goto, S. (2000). KEGG: kyoto encyclopedia of genes and genomes. *Nucleic Acids Research* 28 (1): 27–30.

Karp, P.D., Riley, M., Paley, S.M., and Pellegrini-Toole, A. (2002). The MetaCyc Database. *Nucleic Acids Research* 30 (1): 59–61.

Khatri, P., Draghici, S., Ostermeier, G.C., and Krawetz, S.A. (2002). Profiling gene expression using onto-express. *Genomics* 79 (2): 266–270.

Kim, S.-Y. and Volsky, D.J. (2005). PAGE: parametric analysis of gene set enrichment. *BMC Bioinformatics* 6 (1): 144. https://doi.org/10.1186/1471-2105-6-144.

Kolmogorov, M., Yuan, J., Lin, Y., and Pevzner, P.A. (2019). Assembly of long, error-prone reads using repeat graphs. *Nature Biotechnology* 37 (5): 540–546. https://doi.org/10.1038/s41587-019-0072-8.

Koren, S., Walenz, B.P., Berlin, K. et al. (2017). Canu: scalable and accurate long-read assembly via adaptive k-mer weighting and repeat separation. *Genome Research* 27 (5): 722–736.

Krishnamurthy, P., Mohanty, B., Wijaya, E. et al. (2017). Transcriptomics analysis of salt stress tolerance in the roots of the mangrove Avicennia officinalis. *Scientific Reports* 7 (1): 10031. https://doi.org/10.1038/s41598-017-10730-2.

Lee-Liu, D., Almonacid, L.I., Faunes, F. et al. (2012). Transcriptomics using next generation sequencing technologies. *Methods in Molecular Biology* 917: 293–317.

Li, J. and Tibshirani, R. (2013). Finding consistent patterns: a nonparametric approach for identifying differential expression in RNA-Seq data. *Statistical Methods in Medical Research* 22 (5): 519–536.

Liu, X., Yan, Z., Wu, C. et al. (2019). FastProNGS: fast preprocessing of next-generation sequencing reads. *BMC Bioinformatics* 20 (1): 345.

Lo, C.-C. and Chain, P.S.G. (2014). Rapid evaluation and quality control of next generation sequencing data with FaQCs. *BMC Bioinformatics* 15 (1): 366.

Love, M.I., Huber, W., and Anders, S. (2014). Moderated estimation of fold change and dispersion for RNA-seq data with DESeq2. *Genome Biology* 15 (12): 550. https://doi.org/10.1186/s13059-014-0550-8.

Lu, H., Giordano, F., and Ning, Z. (2016). Oxford nanopore MinION sequencing and genome assembly. *Genomics, Proteomics & Bioinformatics* 14 (5): 265–279.

Martens, M., Ammar, A., Riutta, A. et al. (2021). WikiPathways: connecting communities. *Nucleic Acids Research* 49 (D1): D613–D621.

Martin, M. (2011). Cutadapt removes adapter sequences from high-throughput sequencing reads. *EMBnet.journal* 17 (1): 10–12. https://doi.org/10.14806/ej.17.1.200. https://journal.embnet.org/index.php/embnetjournal/article/view/200 (accessed 3 March 2024).

Martin, D., Brun, C., Remy, E. et al. (2004). GOToolBox: functional analysis of gene datasets based on Gene Ontology. *Genome Biology* 5 (12): R101.

Martínez-Alcántara, A., Ballesteros, E., Feng, C. et al. (2009). PIQA: pipeline for Illumina G1 genome analyzer data quality assessment. *Bioinformatics* 25 (18): 2438–2439.

Massa, M.S., Chiogna, M., and Romualdi, C. (2010). Gene set analysis exploiting the topology of a pathway. *BMC Systems Biology* 4: 121.

Meng, J., Chen, X., and Zhang, C. (2019). Transcriptome-based identification and characterization of genes responding to imidacloprid in *Myzus persicae*. *Scientific Reports* 9 (1): 13285. https://doi.org/10.1038/s41598-019-49922-3.

Mironova, V.V., Weinholdt, C., and Grosse, I. (2015). RNA-seq data analysis for studying abiotic stress in horticultural plants. In: *Abiotic Stress Biology in Horticultural Plants* (ed. Y. Kanayama and A. Kochetov), 197–220. Tokyo: Springer Japan https://doi.org/10.1007/978-4-431-55251-2_14.

Mullikin, J.C. and Ning, Z. (2003). The phusion assembler. *Genome Research* 13 (1): 81–90.

Myers, E.W., Sutton, G.G., Delcher, A.L. et al. (2000). A whole-genome assembly of Drosophila. *Science* 287 (5461): 2196–2204.

Nguyen T, Draghici S. BLMA: a package for bi-level meta-analysis. 2017. R package

Nguyen, T.-M., Shafi, A., Nguyen, T., and Draghici, S. (2019). Identifying significantly impacted pathways: a comprehensive review and assessment. *Genome Biology* 20 (1): 203. https://doi.org/10.1186/s13059-019-1790-4.

Patel, R.K. and Jain, M. (2012). NGS QC toolkit: a toolkit for quality control of next generation sequencing data. *PLoS One* 7 (2): e30619.

Pevzner, P.A., Tang, H., and Tesler, G. (2004). *De novo* repeat classification and fragment assembly. *Genome Research* 14 (9): 1786–1796.

Pop, M., Phillippy, A., Delcher, A.L., and Salzberg, S.L. (2004). Comparative genome assembly. *Briefings in Bioinformatics* 5 (3): 237–248. https://doi.org/10.1093/bib/5.3.237.

Qu, M., Zheng, Y., Bi, L. et al. (2022). Comparative transcriptomic analysis of the gene expression and underlying molecular mechanism of submergence stress response in orchardgrass roots. *Frontiers in Plant Science* 13: 1104755.

Rastogi, S., Shah, S., Kumar, R. et al. (2019). Ocimum metabolomics in response to abiotic stresses: cold, flood, drought and salinity. *PLoS One* 14 (2): e0210903.

Ritchie, M.E., Phipson, B., Wu, D. et al. (2015). limma powers differential expression analyses for RNA-sequencing and microarray studies. *Nucleic Acids Research* 43 (7): e47.

Robinson, M.D., McCarthy, D.J., and Smyth, G.K. (2010). edgeR: a Bioconductor package for differential expression analysis of digital gene expression data. *Bioinformatics* 26 (1): 139–140.

Ruan, J. and Li, H. (2020). Fast and accurate long-read assembly with wtdbg2. *Nature Methods* 17 (2): 155–158.

Sanger, F., Nicklen, S., and Coulson, A.R. (1977). DNA sequencing with chain-terminating inhibitors. *Proceedings of the National Academy of Sciences of the United States of America* 74 (12): 5463–5467.

Schatz, M.C. (2017). Nanopore sequencing meets epigenetics. *Nature Methods* 14 (4): 347–348. https://doi.org/10.1038/nmeth.4240.

Schmieder, R. and Edwards, R. (2011). Quality control and preprocessing of metagenomic datasets. *Bioinformatics* 27 (6): 863–864.

Schneeberger, K., Ossowski, S., Ott, F. et al. (2011). Reference-guided assembly of four diverse *Arabidopsis thaliana* genomes. *Proceedings of the National Academy of Sciences of the United States of America* 108 (25): 10249–10254.

Schwabe, D., Formichetti, S., Junker, J.P. et al. (2020). The transcriptome dynamics of single cells during the cell cycle. *Molecular Systems Biology* 16 (11): e9946.

Shafin, K., Pesout, T., Lorig-Roach, R. et al. (2020). Nanopore sequencing and the Shasta toolkit enable efficient de novo assembly of eleven human genomes. *Nature Biotechnology* 38 (9): 1044–1053. https://doi.org/10.1038/s41587-020-0503-6.

Shen, W., Le, S., Li, Y., and Hu, F. (2016). SeqKit: a cross-platform and ultrafast toolkit for FASTA/Q file manipulation. *PLoS One* 11 (10): e0163962.

Shendure, J. and Ji, H. (2008). Next-generation DNA sequencing. *Nature Biotechnology* 26 (10): 1135–1145.

Shojaie, A. and Michailidis, G. (2009). Analysis of gene sets based on the underlying regulatory network. *Journal of Computational Biology* 16 (3): 407–426.

Soneson, C. and Delorenzi, M. (2013). A comparison of methods for differential expression analysis of RNA-seq data. *BMC Bioinformatics* 14: 91.

Tarca, A.L., Draghici, S., Khatri, P. et al. (2009). A novel signaling pathway impact analysis. *Bioinformatics* 25 (1): 75–82.

Tian, L., Greenberg, S.A., Kong, S.W. et al. (2005). Discovering statistically significant pathways in expression profiling studies. *Proceedings of the National Academy of Sciences of the United States of America* 102 (38): 13544–13549.

Tintori, S.C., Golden, P., and Goldstein, B. (2020). Differential expression gene explorer (DrEdGE): a tool for generating interactive online visualizations of gene expression datasets. *Bioinformatics* 36 (8): 2581–2583.

Tuteja, N. (2007). Mechanisms of high salinity tolerance in plants. *Methods in Enzymology* 428: 419–438.

Tyagi, P. and Ranjan, R. (2023). Comparative study of the pharmacological, phytochemical and biotechnological aspects of *Tribulus terrestris* Linn. and *Pedalium murex* Linn: an overview. *Acta Ecologica Sinica* 43 (2): 223–233. https://www.sciencedirect.com/science/article/pii/S1872203221000834 (accessed 10 September 2023).

Tyagi, P., Singh, D., Mathur, S. et al. (2022). Upcoming progress of transcriptomics studies on plants: an overview. *Frontiers in Plant Science* 13: 1030890.

Vaucheret, H. (2006). Post-transcriptional small RNA pathways in plants: mechanisms and regulations. *Genes & Development* 20 (7): 759–771.

Vezzi, F., Cattonaro, F., and Policriti, A. (2011). e-RGA: enhanced reference guided assembly of complex genomes. *EMBnet.journal* 17 (1) Next Gener Seq Data Anal [Internet].: https://journal.embnet.org/index.php/embnetjournal/article/view/208 (accessed 10 September 2023).

Voichita, C., Ansari, S., and Draghici, S. (2023). *ROntoTools: R Onto-Tools suite.* doi:10.18129/B9.bioc.ROntoTools, R package version 2.30.0, https://bioconductor.org/packages/ROntoTools (accessed 20 September).

Wang, J., Wong, G.K.-S., Ni, P. et al. (2002). RePS: a sequence assembler that masks exact repeats identified from the shotgun data. *Genome Research* 12 (5): 824–831.

Wang, Z., Gerstein, M., and Snyder, M. (2009). RNA-Seq: a revolutionary tool for transcriptomics. *Nature Reviews Genetics* 10 (1): 57–63.

Wang, J., Duncan, D., Shi, Z., and Zhang, B. (2013). WEB-based GEne SeT AnaLysis Toolkit (WebGestalt): update 2013. *Nucleic Acids Research* 41 (Web Server issue): W77–W83.

Wang, B., Ekblom, R., Bunikis, I. et al. (2014). Whole genome sequencing of the black grouse (*Tetrao tetrix*): reference guided assembly suggests faster-Z and MHC evolution. *BMC Genomics* 15 (1): 180. https://doi.org/10.1186/1471-2164-15-180.

Wang, X., Li, N., Li, W. et al. (2020). Advances in transcriptomics in the response to stress in plants. *Global Medical Genetics* 7 (2): 30–34.

Wijesooriya, K., Jadaan, S.A., Perera, K.L. et al. (2022). Urgent need for consistent standards in functional enrichment analysis. *PLoS Computational Biology* 18 (3): e1009935.

Wishart, D.S., Li, C., Marcu, A. et al. (2020). PathBank: a comprehensive pathway database for model organisms. *Nucleic Acids Research* 48 (D1): D470–D478.

Yoshida, T., Mogami, J., and Yamaguchi-Shinozaki, K. (2014). ABA-dependent and ABA-independent signaling in response to osmotic stress in plants. *Current Opinion in Plant Biology* 21: 133–139.

Zeeberg, B.R., Feng, W., Wang, G. et al. (2003). GoMiner: a resource for biological interpretation of genomic and proteomic data. *Genome Biology* 4 (9): R28. https://doi.org/10.1186/gb-2003-4-4-r28.

Zhang, G., Zhao, Z., Ma, P. et al. (2021). Integrative transcriptomic and gene co-expression network analysis of host responses upon *Verticillium dahliae* infection in *Gossypium hirsutum*. *Scientific Reports* 11 (1): 20586. doi: 10.1038/s41598-021-99063-9.

11

Development of a Core Set from Large Germplasm Collections in Genebank

Pradeep Ruperao

Center of Excellence in Genomics and Systems Biology, International Crops Research Institute for the Semi-Arid Tropics, Patancheru, Telangana, India

11.1 Introduction

The genebank resources are a rich source of genetic diversity and for plant breeding and genetic research. On the other hand, it can be challenging to manage the huge collection of germplasm, especially to preserve and use it appropriately. Therefore, developing the core collection is optimal for identifying and utilizing genetic variations (Zhang et al. 2011). The core collection of crop species represents limited accessions with maximum diversity and minimum duplication (Arber et al. 1984; Brown 1989a). Evaluating such core collection from the germplasm includes certain methods to capture the maximum genetic diversity. Such methods include the physiological, morphologic, agronomic traits, and genetic characteristics. According to the Brown sampling theory of selection (Brown 1989a), the core subset should be nearly 10–30% of the total collection, retaining more than 70% of the alleles of the entire collection. The discriminating data in germplasm is a good classifier for forming a group of accessions (Zeuli and Qualset 1993). The cluster size depends on the sampling strategy used and the resulting core set should capture maximum diversity with low redundancy.

It is possible to assess a population's genetic diversity using genomic resources, such as DNA-based markers. Evaluating the genetic diversity will provide information about germplasm to fit in a breeding program to accelerate crop improvement. Single nucleotide polymorphism (SNPs), simple sequence repeats (SSRs), insertions and deletions (indels), heterozygosity, and k-mers sequences are the most useful methods of investigating the genetic diversity of a crop plant (Ruperao et al. 2023a). The developed core collection that can capture the maximum genetic diversity of germplasm using such genomic resources efficiently reduces the number of accessions for core collection. This chapter discusses the methods to assess genetic information in different dimensions that should be considered to develop the representative core sample collection.

Bioinformatics for Plant Research and Crop Breeding, First Edition. Edited by Jen-Tsung Chen.

11.2 Developing a Core Collection

The core collection aims to represent all material of a crop in genebank collection and have a major role in conserving genetic resources and using them in plant improvement. Core collection refers to selecting a subset that represents the entire germplasm diversity with minimum redundancy (Frankel 1984; Brown 1989b). A core collection is expected to have representativeness, data availability, reusability, and low redundancy (Odong et al. 2013). The core collection has been developed in plant species including the cereals [maize (Sa et al. 2021), sorghum (Grenier et al. 2001), rice (Zhang et al. 2011), wheat (Mourad et al. 2020), peanut (Upadhyaya et al. 2002)], pulses [chickpea (Upadhyaya et al. 2001), pigeon pea (Reddy et al. 2005), soybean (Wang et al. 2006) core collections with agronomic traits and SSR markers], [common bean (Rivera et al. 2018), mung bean (Schafleitner et al. 2015)], and oilseeds [safflower (Kumar et al. 2016), sesame (Liu et al. 2017)].

11.3 Constructing a Core Collection

The development of core collection has been implemented in various perspectives for different crops (as listed in the above session). Due to the wide variation of available resources of the respective germplasm, there is no such thumb rule to develop a core collection construction method. Generally, the construction of the core collection mainly includes the identification of diverse accessions from the big population. However, it is recommended to consider a few parameters to minimize the genetic similarity between its entire and could be considered as conserved genetic resources (Brown and Schoen 1994). Different approaches identify the representative samples from a group of samples. Such approaches discussed in this chapter mainly focus on assessing the genetic variations and identifying the representative samples (Figure 11.1). Such diverse accessions can be determined with the genomic parameters evaluated in recent studies (Ruperao et al. 2023a).

11.3.1 Single Nucleotide Polymorphisms (SNPs)

SNPs play a crucial role in finding diverse accessions in the context of genetic diversity analysis in populations, particularly in plant breeding, and evolutionary studies.

SNPs are the most common type of genetic variation in the plant genome and are single base-pair differences in DNA sequences. By identifying SNPs within a population or group of accessions, researchers can pinpoint specific genetic variations that contribute to diversity. Such SNPs can be used to assess the genetic structure of populations. Researchers can determine the relatedness and genetic distance between individuals by analyzing the presence or absence of specific SNPs across a set of accessions. This information helps in identifying distinct subpopulations or clusters within the larger population.

Researchers can use SNP data to calculate various diversity metrics, such as genetic diversity, heterozygosity, and allele frequency distribution. These metrics provide insights into the overall genetic diversity within a population or group of accessions. SNPs associated with traits of interest can be used in marker-assisted selection programs. Accessions with specific SNP profiles linked to desirable traits can be selected for breeding purposes,

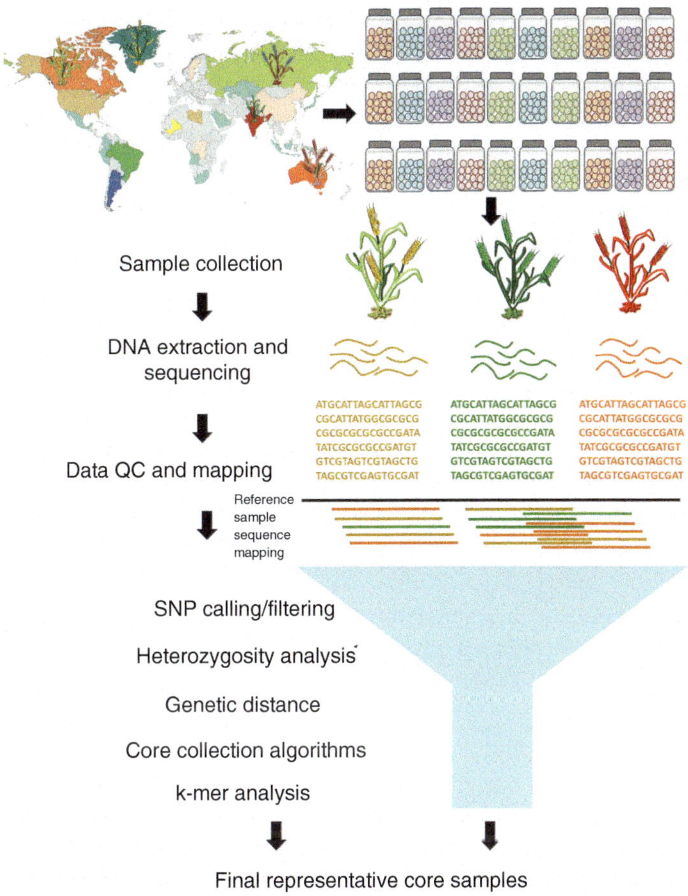

Figure 11.1 A schematic representation and identification of core samples from a global collection of accessions.

enhancing the efficiency of breeding programs. In the context of identifying representative accessions and preserving biodiversity, SNPs can help identify and prioritize accessions with unique genetic profiles. This information aids in the selection of individuals or populations for conservation efforts, ensuring the preservation of genetic diversity within a species.

The SNPs are powerful genetic markers that assist in identifying diverse accessions within populations and they enable researchers to assess genetic diversity, population structure, and the genetic basis of traits of interest, ultimately aiding in breeding, and conservation studies across various species.

11.3.2 k-mer Analysis

k-mer sequences, short contiguous DNA sequences of a fixed length (k), can be useful in finding diverse accessions within populations or datasets. While k-mers are not as specific as SNPs in pinpointing genetic variation, they offer several advantages and can complement SNP-based approaches for assessing diversity:

Apart from k-mer sequences used commonly in *de novo* genome assembly and estimating the genome size, the variations in genome size may reflect structural differences such as duplications, deletions, or transpositions, which contribute to diversity. k-mers can be used to identify regions of homology or orthology between genomes. This information helps us understand conserved genetic elements across diverse accessions and regions undergoing divergence or rearrangement. By comparing the k-mer profiles of different accessions, researchers can infer evolutionary relationships and identify the diversity of species or subspecies. As WGRS reads, k-mer can be used to identify genetic variants, including single nucleotide variants (SNVs), insertions, deletions, and structural variations can help find genetic differences between accessions. Assessing the diversity of functional elements (such as promoter sequences, enhancers, or protein-binding sites) can provide insights into the potential for phenotypic diversity among accessions. k-mers can be the basis for developing genetic markers, such as microsatellites or SSRs and SNP (Ruperao et al. 2023b). These markers can be used for diversity analysis and population genetics studies.

11.3.3 Heterozygosity

Heterozygosity is a measure of genetic diversity within an individual organism, and it can be useful in finding diverse plant accessions in several ways.

Heterozygosity reflects the presence of different alleles at the same genetic locus. Higher heterozygosity indicates greater genetic variation within an individual. When assessing a population of plant accessions, individuals with high levels of heterozygosity are more likely to carry a diverse set of alleles, making them valuable for preserving genetic diversity.

With the average heterozygosity within a population of plant accessions, it is possible to estimate the overall genetic diversity of that population. A population with high heterozygosity typically contains a wider array of genetic variants, which is desirable for maintaining genetic diversity. With the wide range of heterozygosity, accessions with exceptionally high heterozygosity may harbor unique or rare genetic variants. Identifying and preserving these individuals can be crucial for conserving rare or potentially valuable alleles that may be lost without conservation efforts.

Accessions with higher heterozygosity can serve as valuable genetic resources in plant breeding. They may have greater potential for producing genetically diverse offspring, which can be important for developing new varieties with improved traits, disease resistance, or adaptability.

When prioritizing which plant accessions to conserve, those with higher heterozygosity may be given preference. Preserving individuals with high genetic diversity can help maintain the adaptive potential of the population and ensure that a broad range of genetic traits is safeguarded. Greater heterozygosity can also lead to increased phenotypic diversity within a population. This phenotypic diversity may be valuable for breeding programs seeking to develop plants with a wide range of characteristics. Accessions with higher heterozygosity may exhibit greater adaptability and resilience to changing environmental conditions.

In summary, heterozygosity is an important measure of genetic diversity within the crop germplasm. It serves as an indicator of genetic variation, adaptability, and resilience, making it a valuable tool in identifying and conservating diverse plant accessions.

11.3.4 Genetic Distance

Genetic distance measures the genetic divergence between two or more individuals or populations. It plays a significant role in finding diverse plant accessions by quantifying the degree of genetic variation within a collection of accessions. Genetic distance measures help assess the overall genetic diversity within a set of plant accessions. Researchers can identify the range of genetic differences present by calculating genetic distances between all pairs of accessions in a collection. Genetic distance analysis can also provide insights into the adaptive potential of plant accessions. Accessions that are genetically distant from each other may possess a broader range of adaptive traits, making them valuable resources for addressing changing environmental conditions.

Genetic distance is often used as input for cluster analysis techniques such as hierarchical clustering or multidimensional scaling. These methods group accessions that are genetically similar into clusters or groups, making identifying distinct subpopulations within the collection easier. Genetic distance metrics can assist in prioritizing diverse accessions for conservation efforts. Accessions with high genetic distances from others in the collection may represent unique or rare genetic variants, and preserving them can help maintain genetic diversity. Genetic distance information is valuable for designing effective conservation strategies. Accessions that are genetically distant from each other may be prioritized to preserve a wide range of genetic diversity. Genetic distance can guide plant breeding programs by identifying parents for crosses. Crosses between genetically distant accessions often result in offspring with increased heterosis or hybrid vigor. Breeders can select parents that maximize genetic diversity to produce plants with desirable traits.

The genetic distance also helps characterize the genetic diversity within collections, assists in population structure analysis, guides breeding efforts, and informs conservation strategies to ensure the preservation of valuable plant genetic resources.

11.3.5 Available Tools

Core collections are subsets of a larger germplasm or genetic resource collection that aim to capture the genetic diversity of the entire collection in a more manageable and accessible form. Core collection tools and methods are used to select and maintain these representative subsets. Several core collection tools and strategies are employed in plant genetic resources management to select and maintain a representative subset of accessions from a larger germplasm collection. These tools aim to capture the genetic diversity of the entire collection while reducing redundancy and resource-intensive maintenance. Here are some pros and cons of core collection tools:

Core collections help preserve the genetic diversity of the entire germplasm collection, ensuring that valuable genetic resources are conserved. Maintaining a core collection is more efficient in terms of space and resources than managing the entire collection. It reduces the cost and labor required for maintenance. A smaller core collection is more accessible and manageable for researchers, breeders, and conservators, making it easier to study, utilize, and maintain. Core collections can be used for focused research and breeding efforts, as they represent a diverse set of accessions with potential traits of interest.

Core collections facilitate the analysis of genetic diversity, population structure, and trait distribution within a manageable subset of accessions. Rare or unique accessions that might otherwise be overlooked in a large collection can be included in the core collection to ensure their preservation.

On the other hand, selecting a core collection involves subjective decisions, and if not done carefully, it can introduce sampling bias. Certain genetic groups or traits may be underrepresented. Rare alleles or traits present in a small number of accessions may be lost when creating a core collection, particularly if those accessions are not included. The genetic makeup of accessions can change over time due to environmental factors, genetic drift, or adaptation. A static core collection may not fully represent evolving genetic diversity. Determining which accessions to include in the core collection can be challenging, especially when dealing with large and diverse germplasm collections. The initial creation of a core collection requires a substantial investment of time and resources for genetic and phenotypic characterization of accessions. Maintaining the genetic integrity of the core collection over the long term can be challenging, as it may require periodic updates and validation.

Core collection tools offer many advantages in terms of efficient genetic resource management, but they also come with challenges and potential drawbacks. Careful planning, continuous monitoring, and regular updates are essential to ensure that core collections accurately represent the genetic diversity and adapt to changing circumstances over time.

The few algorithms employed in defining the core collection are the (i) **Random sampling:** This method involves randomly selecting accessions from the larger collection without any specific criteria. However, it may not guarantee an optimal representation of genetic diversity. (ii) **Stratified sampling:** Accessions are grouped into strata based on specific criteria such as geographic origin, traits, or genetic markers. A representative number of accessions are then selected from each stratum to ensure diversity across predefined categories. (iii) **Climatic or ecogeographic data:** Accessions can be selected based on their geographic origin and environmental data, such as climate. This approach helps ensure that the core collection captures accessions adapted to various environmental conditions. (iv) **Morphological or phenotypic data:** Accessions can be selected based on specific morphological or phenotypic traits of interest. This approach is useful when conserving or studying accessions with particular agronomic or adaptive characteristics. (v) **Genotypic data**: Genetic markers, such as SNPs or microsatellites, can be used to assess genetic diversity. Accessions are selected to maximize genetic diversity or to capture specific allelic variation. (vi) **Dissimilarity or diversity indices**: Mathematical algorithms can be employed to calculate the genetic diversity between accessions, and a representative set is chosen to maximize diversity or minimize redundancy. (vii) **Maximal-length substring (M-STR)**: In this method, DNA sequences of accessions are compared, and those that share a high degree of similarity are grouped. A representative from each group is selected for the core collection. (viii) **Cluster analysis**: Hierarchical or nonhierarchical clustering techniques group accessions based on genetic, phenotypic, or geographic similarity. Accessions from different clusters are then selected to form the core collection. (ix) **Agronomic traits**: Accessions can be chosen based on their agronomic performance or breeding value, especially in crop improvement programs. (x) **Phylogenetic diversity**: Accessions can be selected to maximize phylogenetic diversity, ensuring that the core collection captures a broad range of evolutionary lineages.

Table 11.1 Most used core collection defining tools.

Tool	Availability	References
GenoCore	https://github.com/lovemun/Genocore	(Jeong et al. 2017)
PowerCore	http://genebank.rda.go.kr/powercore	(Kim et al. 2007)
Core Hunter	http://www.corehunter.org	(Thachuk et al. 2009)
		(De Beukelaer et al. 2012)
		(De Beukelaer et al. 2018)
MSTRAT	https://www.agap-ge2pop.org/eu-genres-088/download-mstrat4-1/	(Gouesnard et al. 2001)
Roger distance	https://doi.org/10.3389/fpls.2020.612106	(Kao et al. 2021)

It's important to note that core collection creation is not a one-time but an ongoing process. The composition of a core collection may need to be updated periodically as new accessions are acquired or as research priorities change. Careful planning and documentation of the selection criteria and methods are essential to ensure the effectiveness and representativeness of the core collection (Table 11.1).

11.4 Assessing the Core Collections

Assessing the quality and effectiveness of a core collection set is a crucial step in plant genetic resources management. The assessment helps ensure that the selected subset accurately represents the genetic diversity and characteristics of the entire germplasm collection. Here are a few generalized factors for assessing a core collection set:

11.4.1 Background Information

Gather comprehensive information about the original germplasm collection, including passport data (origin, collection date, etc.), phenotypic traits, genotypic data (e.g. molecular markers, DNA sequencing), and any other relevant information. Document the criteria and methods for selecting the core collection, including the rationale for choosing specific accessions. Seek feedback from researchers, breeders, and other stakeholders who use the core collection can provide insights into the usability and representativeness of the core set. Core collections should not be static but periodically reviewed and updated. Incorporate new accessions or data as they become available to maintain the core collection's relevance.

11.4.2 Diversity Analysis

Calculate diversity indices, such as genetic diversity (e.g. Nei's genetic distance), allelic richness, and Shannon's diversity index, for both the entire collection and the core collection. Assess the coverage of genetic diversity in the core collection compared to the whole collection. The core collection should capture a substantial portion of the diversity present in the entire collection.

11.4.3 Phenotypic Characterization

If the core collection was selected based on phenotypic traits, assess whether the chosen accessions exhibit the expected traits and whether they represent the full range of variation. Perform field trials or phenotypic evaluations to confirm that the core accessions display the desired traits accurately.

11.4.4 Genetic Marker Analysis

If genotypic data were used to select the core collection, analyze the genetic structure, population differentiation, and admixture within both the core and whole collections. Use principal component analysis (PCA), hierarchical clustering, or other methods to visualize genetic relationships. Confirm that the core collection adequately captures the genetic diversity in the entire collection based on marker data.

11.4.5 Phylogenetic Analysis

Construct a phylogenetic tree using molecular data (e.g. DNA sequences) for both the core and whole collections. Examine the branching patterns to determine if the core accessions represent various evolutionary lineages or groups in the whole collection.

11.4.6 Trait Performance Assessment

If the core collection was selected for agronomic traits, conduct field trials or greenhouse experiments to evaluate the performance of the core accessions and compare it to the whole collection. Assess whether the core accessions exhibit the desired traits and if there is significant variation in trait expression within the core set.

11.4.7 Evaluate Representativeness

Compare the geographic distribution of accessions in the core collection with that of the whole collection. Ensure that accessions from diverse geographic regions are adequately represented. Consider the representation of different ecotypes or subspecies, if relevant.

11.4.8 Allele Coverage

Assessing a core collection with allele coverage involves evaluating whether the selected subset of accessions adequately represents the genetic diversity regarding allele variation in the entire germplasm collection. This includes examining the genetic data for allelic variation at each marker locus. For each locus, determining the allelic difference present in the whole collection and calculating the frequency distribution of each allele at every marker locus in the entire germplasm collection, will give an idea of each locus's most common and rare alleles. Examine the allele frequency distribution in the core collection and compare it to the distribution in the whole collection, ensuring that the core collection contains accessions with rare alleles not captured by the most common alleles (Presence of Rare Alleles) and confirm that each allele observed in the whole collection is represented in the core collection to some extent (Representation of All Alleles) are the good sign of better core collection. Additionally, assessing whether rare alleles in the whole collection are adequately represented in the core collection, may be valuable for specific traits or adaptive traits (Allele Rarity). Assessing core collections for allele coverage is crucial for ensuring they effectively capture the genetic diversity in the entire germplasm collection. This assessment process requires a thorough understanding of genetic data analysis and diversity metrics.

11.4.9 Heterozygosity

Heterozygosity involves evaluating whether the selected subset of accessions adequately represents the genetic diversity and level of heterozygosity present in the entire germplasm collection. Here are the few steps and considerations for assessing core collections with respect to heterozygosity:

The genetic data for the entire germplasm and core collection includes SNP, microsatellite, or other molecular marker information. Measuring the heterozygosity levels for each accession in both the whole collection and the core collection indicates the proportion of heterozygous loci out of the total number of loci examined, i.e. examine the heterozygosity levels within the core collection and compare them to the levels observed in the whole collection. Maintaining various levels of heterozygosity (highly and less heterozygous accessions) is good for representing the diversity within the whole collection. Based on your defined criteria, calculate the proportion of heterozygosity levels captured by the core collection. This gives a quantitative measure of heterozygosity coverage.

Assessing core collections for heterozygosity coverage is important for ensuring that they effectively capture the genetic diversity present in the entire germplasm collection, with respect to levels of heterozygosity. This assessment process requires a thorough understanding of genetic data analysis and genetic diversity metrics.

11.4.10 Shannon Diversity

Assessing a core collection with Shannon diversity involves evaluating whether the selected subset of accessions adequately represents the genetic diversity present in the entire germplasm collection using Shannon's diversity index (also known as Shannon–Weaver index or Shannon entropy). Shannon diversity quantifies the diversity of alleles or traits in a population. Assessing core collections with Shannon diversity helps ensure that they effectively capture the genetic diversity in the entire germplasm collection, using a quantitative measure that accounts for diversity and evenness of distribution. This assessment process requires a good understanding of genetic data analysis and diversity metrics.

11.4.11 Genetic Distance

Assessing a core collection with genetic distance involves evaluating whether the selected subset of accessions adequately represents the genetic diversity and divergence present in the entire germplasm collection. Genetic distance measures the degree of genetic dissimilarity or divergence between individuals or populations. Assessing core collections with genetic distance helps ensure they effectively capture the genetic diversity and divergence in the entire germplasm collection. This assessment process requires understanding genetic data analysis and clustering techniques.

11.4.12 Agro-climatic Conditions

Assessing a core collection with respect to agro-climatic conditions involves evaluating whether the selected subset of accessions adequately represents the genetic diversity and adaptability of crops to different environmental conditions. Agro-climatic conditions encompass factors such as temperature, precipitation, soil type, and other environmental variables that influence crop growth and performance.

Obtain comprehensive agro-climatic data for the geographic regions where the germplasm collection is located. This data may include temperature, rainfall, soil properties, altitude, and other relevant environmental factors. Divide the geographic regions covered by the germplasm collection into agro-climatic zones or regions based on similar environmental conditions. Agro-climatic zones may vary in terms of temperature, rainfall patterns, or soil types. Integrate the agro-climatic data with the genetic and phenotypic data of the germplasm collection. This allows you to link each accession to specific agro-climatic conditions. Assessing core collections with agro-climatic conditions is essential for ensuring that they are representative of the range of environmental conditions where the crop is grown. This assessment process requires a multidisciplinary approach, combining genetic, climatic, and agronomic knowledge. It helps breeders and researchers access germplasm resources suitable for different agro-climatic regions and adapt to changing climate conditions.

Assessing a core collection set is an ongoing process that requires careful consideration of genetic, phenotypic, and geographic diversity and the evolving needs of researchers and breeders. Regular updates and revisions ensure that the core collection remains a valuable genetic resource for research and breeding programs.

11.5 Conclusion and Future Considerations

As plant genetic resources management continues to evolve and adapt to the changing needs of agriculture, research, and conservation, several future considerations for core collections become apparent. Here are some key factors to consider when working with core collections in the future:

11.5.1 Dynamic Updating

Core collections should not be static but rather dynamic and adaptable. As new accessions or genetic data become available, core collections should be periodically updated to reflect the latest knowledge and diversity.

11.5.2 Integration of Multi-omics Data

Future core collections may incorporate genetic data and other omics data, such as transcriptomics, proteomics, and metabolomics, to provide a more comprehensive understanding of plant diversity and function.

11.5.3 Climate Resilience

With climate change posing new challenges to agriculture, core collections should consider the inclusion of accessions with traits related to climate resilience, such as drought tolerance, heat resistance, or disease resistance.

11.5.4 Genomic Selection and Breeding Applications

Core collections can be designed to facilitate genomic selection and breeding programs, with a focus on capturing the genetic diversity and traits of interest needed for crop improvement.

11.5.5 Precision Conservation

Advances in genomics and data analytics can enable precision conservation efforts, ensuring that unique or endangered genetic resources are preserved in the most effective and targeted manner.

11.5.6 Community Engagement and Indigenous Knowledge

Recognize the importance of involving local communities and indigenous knowledge in the management of core collections, respecting traditional agricultural practices, and preserving culturally significant plant varieties.

11.5.7 Digital Repositories and Data Sharing

Core collections will likely be integrated into digital repositories and databases, making genetic and phenotypic data easily accessible to researchers, breeders, and the public.

11.5.8 Crop Wild Relatives

Consider including crop wild relatives (CWRs) in core collections to harness their potential for crop improvement and conservation.

11.5.9 Machine Learning and AI

Utilize machine learning and artificial intelligence (AI) techniques to analyze and interpret the vast amounts of genetic and phenotypic data associated with core collections.

11.5.10 International Collaboration

Promote international collaboration and cooperation in developing and maintaining core collections to ensure a global perspective on conservation and utilization of genetic resources.

11.5.11 Education and Awareness

Raise awareness among policymakers, researchers, and the public about the importance of core collections in preserving genetic diversity and enhancing food security.

11.5.12 Food Security and Nutrition

Address the nutritional needs of a growing global population by selecting accessions that contribute to improved food security and nutrition.

11.5.13 Crop Disease and Pest Resistance

Given the increasing challenges posed by crop diseases and pests, prioritize accessions with resistance genes or traits that can help combat these threats.

In conclusion, core collections will continue to play a vital role in ensuring plant genetic resources conservation and sustainable use. Future considerations should incorporate advances in genomics, climate resilience, precision conservation, and ethical and legal frameworks to meet the evolving needs of agriculture and research.

References

Arber, W., Illmensee, K., Peacock, W.J., and Starlinger, P. (ed.) (1984). *Genetic Manipulation: Impact on Man and Society*. Cambridge: Cambridge University Press.

Brown, A.H.D. (1989a). Core collections: a practical approach to genetic resources management. *Genome* 31: 818–824. https://doi.org/10.1139/g89-144.

Brown, A.H.D. (1989b). The case for core collections. In: *The Use of Plant Genetic Resources* (ed. A.H.D. Brown, O.H. Frankel, D.R. Marshall, and J.T. Williams), 136–156. Cambridge: Cambridge University Press.

Brown, A.H.D. and Schoen, D.J. (1994). Optimal sampling strategies for core collections of plant genetic resources. In: *Conservation Genetics* (ed. V. Loeschcke, S.K. Jain, and J. Tomiuk), 357–370. Basel: Birkhäuser https://doi.org/10.1007/978-3-0348-8510-2_28.

De Beukelaer, H., Smýkal, P., Davenport, G.F., and Fack, V. (2012). Core Hunter II: fast core subset selection based on multiple genetic diversity measures using Mixed Replica search. *BMC Bioinformatics* 13: 312. https://doi.org/10.1186/1471-2105-13-312.

De Beukelaer, H., Davenport, G.F., and Fack, V. (2018). Core Hunter 3: flexible core subset selection. *BMC Bioinformatics* 19: 203. https://doi.org/10.1186/s12859-018-2209-z.

Frankel, O. (1984). Genetic perspective of germplasm collection. In: *Genetic Manipulations: Impact on Man and Society* (ed. W. Arber, K. Limensee, W.J. Peacock, and P. Stralinger), 1.Cambridge: Cambridge University Press.

Gouesnard, B., Bataillon, T.M., Decoux, G. et al. (2001). MSTRAT: an algorithm for building germ plasm core collections by maximizing allelic or phenotypic richness. *The Journal of Heredity* 92: 93–94. https://doi.org/10.1093/jhered/92.1.93.

Grenier, C., Bramel-Cox, P.J., and Hamon, P. (2001). Core collection of sorghum: I. Stratification based on eco-geographical data. *Crop Science* 41: 234–240. https://doi.org/10.2135/cropsci2001.411234x.

Jeong, S., Kim, J.Y., Jeong, S.C. et al. (2017). GenoCore: a simple and fast algorithm for core subset selection from large genotype datasets. *PLoS One* 12: e0181420. https://doi.org/10.1371/journal.pone.0181420.

Kao, C.F., He, S.S., Wang, C.S. et al. (2021). A modified Roger's distance algorithm for mixed quantitative–qualitative phenotypes to establish a core collection for Taiwanese vegetable soybeans. *Frontiers in Plant Science* 11: 612106. https://doi.org/10.3389/fpls.2020.612106.

Kim, K.W., Chung, H.K., Cho, G.T. et al. (2007). PowerCore: a program applying the advanced M strategy with a heuristic search for establishing core sets. *Bioinformatics* 23: 2155–2162. https://doi.org/10.1093/bioinformatics/btm313.

Kumar, S., Ambreen, H., Variath, M.T. et al. (2016). Utilization of molecular, phenotypic, and geographical diversity to develop compact composite core collection in the oilseed crop, safflower (*Carthamus tinctorius* L.) through maximization strategy. *Frontiers in Plant Science* 7: 01554. https://doi.org/10.3389/fpls.2016.01554.

Liu, Y., Mei, H., Du, Z. et al. (2017). Construction of core collection of sesame based on phenotype and molecular markers. *Scientia Agricultura Sinica* 50: 2433–2441. https://doi.org/10.3864/j.issn.0578-1752.2017.13.003.

Mourad, A.M.I., Belamkar, V., and Baenziger, P.S. (2020). Molecular genetic analysis of spring wheat core collection using genetic diversity, population structure, and linkage disequilibrium. *BMC Genomics* 21: 434. https://doi.org/10.1186/s12864-020-06835-0.

Odong, T.L., Jansen, J., van Eeuwijk, F.A., and van Hintum, T.J.L. (2013). Quality of core collections for effective utilisation of genetic resources review, discussion and interpretation. *Theoretical and Applied Genetics* 126: 289–305. https://doi.org/10.1007/s00122-012-1971-y.

Reddy, L.J., Upadhyaya, H.D., Gowda, C.L.L., and Singh, S. (2005). Development of core collection in pigeonpea [*Cajanus cajan* (L.) Millspaugh] using geographic and qualitative morphological descriptors. *Genetic Resources and Crop Evolution* 52: 1049–1056. https://doi.org/10.1007/s10722-004-6152-7.

Rivera, A., Plans, M., Sabaté, J. et al. (2018). The spanish core collection of common beans (*Phaseolus vulgaris* L.): an important source of variability for breeding chemical composition. *Frontiers in Plant Science* 871: 01642. https://doi.org/10.3389/fpls.2018.01642.

Ruperao, P., Bajaj, P., Subramani, R. et al. (2023a). A pilot-scale comparison between single and double-digest RAD markers generated using GBS strategy in sesame (*Sesamum indicum* L.). *PLoS One* 18: e0286599. https://doi.org/10.1371/journal.pone.0286599.

Ruperao, P., Gandham, P., Odeny, D.A. et al. (2023b). Exploring the sorghum race level diversity utilizing 272 sorghum accessions genomic resources. *Frontiers in Plant Science* 14: 1143512. https://doi.org/10.3389/fpls.2023.1143512.

Sa, K.J., Kim, D.M., Oh, J.S. et al. (2021). Construction of a core collection of native *Perilla germplasm* collected from South Korea based on SSR markers and morphological characteristics. *Scientific Reports* 11: 23891. https://doi.org/10.1038/s41598-021-03362-0.

Schafleitner, R., Nair, R.M., Rathore, A. et al. (2015). The AVRDC - The World Vegetable Center mungbean (*Vigna radiata*) core and mini core collections. *BMC Genomics* 16: 344. https://doi.org/10.1186/s12864-015-1556-7.

Thachuk, C., Crossa, J., Franco, J. et al. (2009). Core Hunter: an algorithm for sampling genetic resources based on multiple genetic measures. *BMC Bioinformatics* 10: 243. https://doi.org/10.1186/1471-2105-10-243.

Upadhyaya, H.D., Bramel, P.J., and Singh, S. (2001). Plant genetic resources: development of a chickpea core subset using geographic distribution and quantitative traits. *Crop Science* 41: 206–210. https://doi.org/10.2135/cropsci2001.411206x.

Upadhyaya, H.D., Bramel, P.J., Ortiz, R., and Singh, S. (2002). Developing a mini core of peanut for utilization of genetic resources. *Crop Science* 42: 2150–2156. https://doi.org/10.2135/cropsci2002.2150.

Wang, L., Guan, Y., Guan, R. et al. (2006). Establishment of Chinese soybean (*Glycine max*) core collections with agronomic traits and SSR markers. *Euphytica* 151: 215–223. https://doi.org/10.1007/s10681-006-9142-3.

Zeuli, P.L.S. and Qualset, C.O. (1993). Evaluation of five strategies for obtaining a core subset from a large genetic resource collection of durum wheat. *Theoretical and Applied Genetics International Journal of Plant Breeding Research* 87: 295–304. https://doi.org/10.1007/BF01184915.

Zhang, H., Zhang, D., Wang, M. et al. (2011). A core collection and mini core collection of *Oryza sativa* L. in China. *Theoretical and Applied Genetics* 122: 49–69. https://doi.org/10.1007/s00122-010-1421-7.

12

Bioinformatics Approaches to Determine Plant microRNA Targets

Shree Prakash Pandey

College of Pastoral Agriculture Science and Technology, Lanzhou University, Lanzhou, China

12.1 Introduction

microRNAs (miRNAs) are typically 20–24 nucleotide (nt) long and form an inevitable layer of regulators of gene expression. They are involved in regulating nearly all aspects of the life of a plant, including its growth, reproduction, and adapting to abiotic and biotic challenges during its interactions with the environment (Li et al. 2017; Sunkar et al. 2012). miRNAs are transcribed from bona-fide miRNA genes (miR- or *MIR* genes), the majority of which are located in the non-coding intergenic regions of the genome (Xie et al. 2005; Yang et al. 2012). A fraction of these miR-genes may also be found in the intronic regions (Rajagopalan et al. 2006). Like protein-coding genes, the transcription of miR-genes is known to be regulated by transcription factors in the context of their biological and regulatory roles in imparting phenotypic plasticity. Their promoters contain motifs that facilitate binding of general (e.g. NOT2, CDC5), as well as specific transcription factors, such as ARFs, ABI3, FUS3, MYBs, MYC2, SPLs, and WRKYs (Yu et al. 2017; Zhan and Meyers 2023; Pandey and Somssich 2009). The miR-genes are transcribed in an RNA Polymerase II-dependent manner (Xie et al. 2005) to generate polyadenylated and single-stranded RNA molecules. Due to their self-complementary nature, these single-stranded RNA molecules imperfectly fold like a hairpin to form a stem-look structure; these are referred to as primary miRNAs or pri-miRNAs (Xie et al. 2005). Compared to animal miRNAs, plant miRNA hairpins are more heterogeneous in nature (Reinhart et al. 2002).

The pri-miRNAs are processed into precursor miRNAs (pre-miRNAs) and would involve Dicer-like (DCL) RNase-III endonucleases, HYPONASTIC LEAVES 1 (HYL1) double-stranded RNA-binding proteins, and the SERRATE (SE) zinc finger protein (Kurihara and Watanabe 2004; Han et al. 2004; Lobbes et al. 2006). DCL1 is shown to further process the pre-miRNAs into imperfect duplexes of miRNA/miRNA*, which comprise a 2-nucleotide (nt) overhang at the 3′-end. Then, HUA ENHANCER 1 (HEN1) methyltransferase enzyme 2′-*O*-methylates the

Bioinformatics for Plant Research and Crop Breeding, First Edition. Edited by Jen-Tsung Chen.

3′-end so that miRNA is protected from degeneration (Yu et al. 2005; Baranauske et al. 2015). Finally, the miRNA* (star or the passenger strand of the miRNA/miRNA* duplex) is degraded while the miRNA strand acts as a guide. This preferential selection of miRNAs over miRNA* strands is facilitated by the dsRNA-binding DRB proteins (Eamens et al. 2009). A miRNA is loaded on the Argonaute (AGO) protein, and together these form an RNA-induced silencing complex (RISC). miRNA-loaded RISC is also often referred to as miRISC (Zhan and Meyers 2023). The loading of the guide strand and the assembly of a RISC is a complex process that involves ATP (Iki et al. 2010).

It is debated that RISC could be assembled in the nucleus as well as in the cytoplasm. Several pieces of evidence, which include the involvement of heat shock protein 90 (HSP90) in RISC favor a model where RISC-loading happens in the cytoplasm (Iki et al. 2010; Yu et al. 2017; Zhan and Meyers 2023). Nevertheless, the export of miRNAs from the nucleus to the cytoplasm is an important factor in its regulatory actions (Kohler and Hurt 2007; Rogers and Chen 2013). AGOs and other components of the non-coding RNA pathways may exert an auto-regulatory or feedback system, in a context-dependent manner (development stage or stress-response), on the biogenesis and accumulation levels of miRNAs (Pandey et al. 2008; Pradhan et al. 2017, 2020, 2023; Dalmadi et al. 2021). This could directly impact the outcomes of the miRNA-regulated pathway and the expression of targets.

In general, miRNAs regulate the expression of their targets in a highly sequence-dependent manner at a posttranscriptional level in the process that could be referred to as posttranscriptional gene silencing (PTGS). Posttranscriptional action of miRNAs on the expression of genes involve two important mechanisms or modes of action – mRNA/transcript cleavage and translation inhibition/repression. Further, miRNAs also participate in the execution of DNA methylation and the generation of secondary, phased small-interfering RNAs (Chellappan et al. 2010; Wu et al. 2010). For executing the silencing of the expression of genes, it was often believed that the degree or level of complementarity between the miRNA and the mRNA transcript could dictate how miRNAs and targets might engage. It was earlier believed that high levels of sequence complementarity, often in the coding region of the genes, might dictate mRNA cleavage, whereas comparatively limited complementarity, possibly in the untranslated regions (UTRs), would lead to translational inhibition. In earlier investigations, plant miRNAs had displayed nearly perfect levels of sequence complementarity to their targets (Rhoades et al. 2002). Therefore, it was believed that miRNA cleavage would be a dominant mode of action of plant miRNAs (Rhoades et al. 2002; Jones-Rhoades and Bartel 2004). Clearly, this has been proven to be a misconception. One recent study has claimed Arabidopsis would have considerably fewer than 300 miRNA-cleavage-driven interactions (Wong and Millar 2022). Further, several miRNAs display a comparatively limited degree of complementarity, especially in species other than Arabidopsis (Srivastava et al. 2014; Liu et al. 2014). While mRNA cleavage is an important mode of action in plant miRNAs, it is now also clearly evident that translational inhibition could be a more widespread mode of action of plant miRNAs than earlier thought (Axtell 2013; Yu et al. 2017). mRNA degradation and translational repression could even co-occur for many plant targets (Axtell 2013).

12.2 Characteristic Features and Principles of miRNA-targeting in Plants

Plant miRNAs often have target sites in the coding region of the genes (open-reading frames), while those in the 5′- or in the 3′- UTRs, or the noncoding regions of a gene have also been reported (Addo-Quaye et al. 2008; Allen et al. 2005; German et al. 2008). miRNA–target interactions in plants display interesting characteristics, such as sequence complementarity, and evolutionary conservation. Such characteristics have been utilized as principles for computationally exploring and identifying plant miRNA targets at individual and genomic scales in across plant species, and even in a trans-kingdom manner. Some of the important principles and characteristics are summarized below.

12.2.1 Sequence Complementarity

One of the most fundamental principles of miRNA target recognition is the base-pairing ability of miRNAs with a potential site on the mRNA. On the basis of the patterns of complementarity between miRNAs and the mRNAs, the target sites could broadly be classified into canonical, noncanonical, and mimics. Traditionally, plant miRNAs and their canonical targets have displayed high sequence complementarity (encompassing 5 or less number of mismatches). This high level of sequence complementarity (base-pairing), typically at positions 2–13 of 5′-end (of miRNAs), has been regarded as a critical factor for miRNA-mediated repression of a canonical target in plants. This region includes the slicing sites at positions 9–11 that are recognized by the AGOs (Mallory et al. 2004; Schwab et al. 2005). In contrast, in animals, a smaller seed region of 7-nt (positions 2–8 of 5′-end of miRNAs) could be sufficient for miRNA–target interaction (Lewis et al. 2005). The 3′-regions of plant miRNA are believed to have a higher tolerance for mismatches to their targets. Canonical animal seed-pairing that spans only 6–7 nt has not been experimentally validated in plants, although conceptual analogies could be drawn where longer regions of 5′-end of miRNAs (~10–12 nt) could be used as "seeds" for predicting miRNA-targets in plants (Pandey et al. 2008, 2018, 2019; Pradhan et al. 2017, 2020). Results of such predictions have been experimentally validated [e.g. in gain-of-function studies by miRNA over-expressions (Pradhan et al. 2023)]. Over-expression studies also show that some miRNA-target pairs might acquire new/novel biological functions due to their selection constraints (Pradhan et al. 2021).

Noncanonical miRNA target sites have also been identified in plants. These are not conventional sites as the pairing between the 5′-region of a miRNA and a target site, which is thought to be critical for canonical sites, is strongly disrupted to form a bulge (Brousse et al. 2014). Yet, such interactions are able to execute mRNA-cleavage with the help of AGO1 (like a conventional or canonical target) between the 10th and the 11th nt opposite of miRNA. The interaction between miR398 and the blue copper-binding protein (BCBP) target is an example of noncanonical targeting (Brousse et al. 2014). Furthermore, plants also harbor a noncanonical miRNA pathway where RNA-silencing is executed in a cleavage-independent manner. Here, DRB2, DRB3, and DRB5 are deployed with an alternate AGO protein (instead of AGO1) for miRNA-mediated translational repression of targets (Eamens et al. 2012).

Although rare, plant miRNA targets could also be "mimics," where nonprotein coding RNA competes with a "typical" target for associating with a miRNA. Target mimics display base-pairing patterns different than the canonical sites in the sense that they harbor bulges or mismatches in the central region. This prevents the slicing activity of the AGOs. miR399-targeted IPS1/At4 family of noncoding RNA is one of the first examples of natural target mimics in plants (Franco-Zorrilla et al. 2007). Artificial target-mimicking sites are now easy to design and are deployed as tools for determining miRNA functions (Todesco et al. 2010).

The traditional belief that a high degree of sequence complimentarily would lead to target cleavage whereas low complementarity would funnel translational repression is not necessarily universally correct. Translationally repressed targets are equally composed of high complementarity-driven canonical sites (Chen 2004; Aukerman and Sakai 2003; Gandikota et al. 2007; Brodersen et al. 2008; Li et al. 2013; Liu et al. 2014). For instance, complementarity requirements of sites for actions of miRNAs were investigated for their efficacies, position-specific mismatches, location (open reading frames [ORF] or UTRs), and pairing patterns (Liu et al. 2014). From such types of investigations, it was evident that the location of a perfectly paired site in the ORF or the 3′-UTR could determine its efficacy or behavior. Perfectly complemented sites had higher efficacies in the ORFs than in the 3′-UTRs. The performance/function of perfect sites decreased when they were placed in the 3′-UTRs. On the other hand, natural miRNA target sites with 2–3 mismatches were more effective in the 3′-UTRs than the perfect sites (Liu et al. 2014; Iwakawa and Tomari 2013). It was found that 5′-mismatches have a strong impact on the reduction in efficiency of target sites. The 3′-region of the miRNAs is more tolerant to consecutive mismatches than the 5′-miRNA-regions. The 3′-miRNA region could tolerate up to 3 consecutive mismatches without affecting the efficacy when compared to a perfect site. An increase of mismatches to 4 or 5 in the target site at the 3′-end of the miRNAs significantly impacted its efficiency. Furthermore, 6 or more mismatches at 3′-end had no efficacy. A single mismatch in the target site at the 5′-end of the miRNA could reduce the efficacy by 2–4-fold, and it supported a canonical viewpoint as discussed above. It was concluded that maximally effective sites need not be essentially perfectly complementary to miRNAs, and up to 3 mismatches in the 3′-end of the miRNAs are equally or more effective. So, introducing penalties for these mismatches in scoring matrices for computational prediction may not necessarily be effective.

Independent computational investigations of experimentally validated target sites across Arabidopsis and non-Arabidopsis species revealed the relevance of following properties: continuous match in the region of 2–12 nt (of 5′-end of miRNA), 5′- and 3′-mismatches, G:U wobble and gaps, as well as training bias for data from Arabidopsis and non-Arabidopsis species (Srivastava et al. 2014). For target prediction purposes, 2–12 nt of 5′- end of miRNA could be loosely described as a "seed" region for plant miRNAs. Bias for complementarity could be introduced in the prediction algorithms if the datasets are excessively trained or tested for only Arabidopsis. Clearly, additional factors beyond sequence complementarity are relevant for the efficient identification and functioning of miRNA targets.

12.2.2 Target Accessibility and Flanking Sequences

Another important criterion for miRNA–mRNA interaction is the accessibility of the target site. Factors that could impact target site accessibility include the G:C content around a target site, RNA structures, and thermodynamic stability, often evaluated in terms of the free energy of the miRNA–mRNA-duplex. The regions flanking the miRNA-binding sites in the targets tend to prefer the G:C poor codons (Gu et al. 2012). It is believed that synonymous codons are selected near the miRNA-targeting sites so that the binding of miRNAs is efficient and miRNA functions are appropriately executed (Gu et al. 2012). Furthermore, for miRNAs to get access, their binding sites are relatively less structured as compared to the neighboring regions so it could facilitate an efficient miRNA–mRNA interaction (Li et al. 2012). Poor G:C codons would automatically reduce the structural complexity of a miRNA–mRNA interaction site.

The structural accessibility of miRNA to their target sites and the efficiency of their binding are often evaluated in terms of the thermodynamic properties of the complex. Here, the free energy (ΔG) of the predicted structures was often used as an important parameter by target identification tools (Srivastava et al. 2014). However, when ΔG values of miRNA-binding sites were correlated to the mRNA transcript lengths, it was evident that transcripts that were longer had lower ΔG (Srivastava et al. 2014). As computational tools do not take into account such correlations of transcript lengths to ΔG in their algorithms, it could potentially introduce a bias in prediction (Srivastava et al. 2014).

12.2.3 Conservation

Evolutionary conservation of miRNA-mediated regulation of targets across species has been another important feature for the development of bioinformatics tools. It has been widely believed that miRNAs as well as their binding sites are usually conserved during an evolutionary landscape, and therefore could be exploited to identify miRNA-targets (Jones-Rhoades and Bartel 2004; Chorostecki et al. 2012; Cuperus et al. 2011). miRNAs form an evolutionary dynamic group of regulators. Their numbers have substantially expanded during the course of the evolution of plants, including in the early land plants, such as Physcomitrella (Fattash et al. 2007), and changed subsequently in a lineage-specific manner after the divergence of monocots and eudicots (Nozawa et al. 2012). Differences in the origin of miR-genes could reflect the differences in their binding to their targets (Nozawa et al. 2012). Divergence in miR-genes could enhance the divergence in the expression of their targets (Shi et al. 2017). The evolution of miRNA genes has been extremely dynamic in the sense that nearly half of their gene families in the ancestors of flowering plants have been lost in due course (Nozawa et al. 2012). At the same time, novel miRNA genes and families are found in species in the context of unique selection pressures of a species in their native habitats (Pandey et al. 2018; Pradhan et al. 2017, 2020, 2023). Consequently, their functions have also expanded and diversified even if the binding sites of miRNAs are conserved (Pradhan et al. 2021).

12.2.4 Expression Patterns

An important feature of miRNA-mediated regulation of gene expression is its impact on the levels of accumulation of target mRNAs. miRNA-mediated regulations often result in a reduction in the accumulation of mRNAs, and levels of accumulation of miRNAs and mRNAs often show an inverse relationship (Voinnet 2009). This characteristic has long been extensively utilized to study miRNA-mediated regulations (Mallory et al. 2004; Schwab et al. 2005; Pandey et al. 2008; Liu et al. 2014). Consequently, gene expression- and miRNA-accumulation patterns have been integrated and utilized heavily for the purpose of target prediction (Zhang et al. 2016; Pradhan et al. 2017, 2020, 2023; Pandey et al. 2018, 2019). Such studies have revealed that miRNAs and mRNAs show a one-many::many-one relationships, i.e. a miRNA could target more than one gene, and a gene could be a target of more than one miRNA. Moreover, such complex relationships of a network of miRNAs are very prominently seen during stress-response where specific pathways are tailored in a highly spatiotemporal and stimulus-dependent context.

12.3 Tools for miRNA Target Prediction in Plants

The first generation of tools for miRNA target prediction were mostly developed for Arabidopsis and were mainly based on a high degree of sequence complementarity between the miRNAs and their corresponding mRNA targets (Rhoades et al. 2002; Jones-Rhoades and Bartel 2004; Fahlgren and Carrington 2010). They utilized initial observations in *Arabidopsis thaliana* that in many target-miRNA interactions, positions 2–13 (in the context of miRNA's 5′-end) were without any mismatches and contained very limited numbers of G:U pairs. Most of the mismatches were either at the 1st position or after the 13th position. These observations were utilized for deploying the technique of PSM (position-based scoring matrix) for scoring potential targets with the help of FASTA alignments and Patscan tool (Rhoades et al. 2002; Jones-Rhoades and Bartel 2004; Fahlgren and Carrington 2010). Since then, a large number of prediction tools have been developed for computational miRNA–target identification. Most of these tools supplement the complementarity feature with other characteristics, such as conservation, minimum free energy, site accessibility, site location (coding sequence [CDS] or UTR), interaction outcomes, and expression patterns (these characteristics are discussed above). More sophisticated approaches beyond PSMs, including machine learning (ML) or artificial neural networks (ANN) have also been implemented. Many of the target prediction tools have focused on datasets from Arabidopsis or other handful of species. When some of these tools were comprehensively evaluated for their performance (in terms of their "precision" and "recall"), a distinct bias toward Arabidopsis was evident (Srivastava et al. 2014).

For Arabidopsis, the maximum number of true positive miRNA–target sites were predicted by Targetfinder (Fahlgren et al. 2007). Similarly, for non-Arabidopsis/model species, a combination of results from psRNATarget (Dai and Zhao 2011) and Tapirhybrid (Bonnet et al. 2010) produced the best outcomes (Srivastava et al. 2014). As the training of many of those tools on the dataset from an extremely limited number of (model) species is a caveat of computational target identification, their use on datasets of non-models produces a large number of false positive (FP) predictions. This challenge of FPs could be circumvented by applying more than one tool. Yet again, an arithmetic union of two tools would strongly increase FPs. In contrast, an intersection of results significantly reduces the FP numbers (i.e. increases specificity), but it may also decrease the sensitivity of the prediction. Therefore, a balancing act of optimization of cut-off values for the criteria or scores of prediction, followed by a statistical combination of results is recommended (Srivastava et al. 2014).

As Targetfinder, psRNATarget, and Tapirhybrid were the best-performing tools in a comparative evaluation, we shall briefly discuss these below. We shall also review some other tools that have been developed since the evaluation of Srivastava et al. (2014) was published. Many of these tools specialize in particular niches, like conserved targets. Further, deploying datasets from degradome sequencing in bioinformatic identification of miRNA targets is becoming popular. Another area that has gained significant traction relates to noncanonical targets as well as target-mimics. Therefore, we shall discuss a few recent tools as examples of a variety of techniques that are being currently deployed for predicting targets and for addressing specific roles of miRNAs.

Targetfinder: In Targetfinder (Fahlgren et al. 2007), a 'FASTA' algorithm is implemented, which incorporates a scheme for scoring penalties for bulges, mismatches, and gaps during the alignment of miRNA and target sequences. This tool was developed on Arabidopsis datasets.

Tapirhybrid: TAPIR webserver (Bonnet et al. 2010) is imbedded with two tools/options for target prediction, Tapirhybrid and Tapirfasta. Tapirhybrid implements RNA hybrid in its search engine to increase the precision of the results. On the other hand, Tapirfasta implemented a FASTA search option and was amongst the fastest tools during an evaluation but had lower precision and recall (Srivastava et al. 2014). TAPIR is also able to predict target mimics.

psRNATarget: psRNATarget (Dai and Zhao 2011) uses an approach of dynamic programming, and utilizes alignment and target site accessibility features. It uses a modified Smith–Waterman-based predefined scoring scheme for the alignment, and RNAup algorithm of ViennaRNA package (Lorenz et al. 2011) for accessibility analysis of targets. psRNATarget, originally an upgrade of the miRU tool (Zhang 2005), has now again been updated by developing a new scoring scheme (Dai et al. 2018). This scoring procedure now allows for customization and detection of noncanonical sites along with canonical sites. The tool was also updated with resources, including its user web-based interface, so that it allows analysis of big data.

comTar: comTar (Chorostecki and Palatnik 2014) was developed as a web tool to predict miRNA targets on the basis of the principle of their conservation across plant species. The authors claim that the tool could be used to predict new candidates on the basis of their evolutionary conservation. A MySQL database of miRNA targets of 33 angiosperm species was claimed to have been developed so that the information related to miRNA target pairs, and their functions and evolutionary conservation could be displayed.

MTide: MTide (Zhang et al. 2015) was developed as a package by combining several tools/scripts for exploring miRNA–target interactions in plants. The authors used a modified version of miRDeep2, a miRNA sequence identification pipeline for animals (Friedlander et al. 2012) that now incorporated plant-specific parameters for predicting miRNAs. For target identification, they used CleaveLand, a tool that was developed for analyzing degradome data (Addo-Quaye et al. 2009). The authors used TAPIR's 'precise search' option for target prediction and the targets were scored on a GO similarity to reduce FPs. The MTide tool was made available as an integrated command-line Perl script (Zhang et al. 2015).

isomiR2Function: isomiR2Function (Yang et al. 2017) is claimed to be a comprehensive profiling tool that performs the discovery as well as characterization of functional roles of isomiRs. It is now well known that individual miRNAs could display heterogeneity in terms of their lengths and sequences, which are referred to as isomiRs (Neilsen et al. 2012).

Specific isomiRs could be recruited in a highly stimulus-dependent manner for regulating specific targets so that plants could achieve phenotypic plasticity in the context of its environmental signal (Neilsen et al. 2012; Pandey et al. 2018; Pradhan et al. 2023). isomiR2Function could perform the following tasks on a miR-Seq dataset: identify both, templated as well as non-templated isomiRs, provide information on their differential accumulations, predict their targets from transcript or PARE-Seq data by deploying Targetfinger or Cleaveland tools, and perform functional enrichment with the help of the TopGO package (Yang et al. 2017). Understandably, its authors have described isomiR2Function as a single, standalone, comprehensive tool for investigating and characterizing isomiRs.

PAREsnip2: PAREsnip2 (Thody et al. 2018) is a software tool for the analysis of degradome datasets for small RNAs (smRNAs). It can overcome the constraints of the previous degradome analysis tools in terms of time and resource requirements, scalability, and user-defined configuration for targeting rules. This tool was later embedded into another tool, PAREameters, as described below (Thody et al. 2020). It is recommended that the users also review PAREameters while considering PAREsnip2.

TarHunter: TarHunter (Ma et al. 2018) was developed to study miRNA targets that are conserved. It implements a strict filter for cross-species conservation, which increases the precision of its prediction. TarHunter is also capable of detecting endogenous target mimics. In combination with degradome analysis, TarHunter could also identify some noncanonical targets.

PAREameters: PAREameters (Thody et al. 2020) infer targeting rules for miRNA–mRNA interactions from sequencing datasets of smRNAs and degradomes. It echoes the inference (Srivastava et al. 2014) that most of the earlier criteria for miRNA–target predictions were inferred and adapted largely from a limited number of validated targets from Arabidopsis and were not optimal across several other species or treatment conditions. It could also quantitatively identify the differences in miRNA–target interactions in model and non-model species. PAREsnip2 is centrally deployed in PAREameters for miRNA–target analysis.

TRUEE: Targets ranked using experimental evidence, TRUEE (Wong and Millar 2022), is a bioinformatics pipeline that aims to identify functional miRNA–target interactions. Here, the authors used the degradome data and publicly available prediction tools in their pipeline. They developed a ranking system (category scores) to determine the extent to which a target could be subjected to mRNA cleavage. As a result, an Arabidopsis miRNA-targetome of high stringency was mapped. TRUEE proposes to rank the miRNA targets in a manner independent of complementarity. The authors acknowledge that this pipeline suffers from some limitations (Wong and Millar 2022): First, it cannot identify targets that are modulated by translational repression. Next, TRUEE relies on degradome signals, which would dictate the outcomes of the analysis. Further, it focuses on miRNA–target interactions of widespread occurrences only. So, it fails to identify interactions that have specific/narrow windows in terms of time and space.

DIGITAL: DIGITAL (Bu et al. 2023) is a deep-learning-based prediction program that aims at identifying miRNA-mediated generation of phasiRNAs. In terms of its architecture, DIGITAL is comprised of "a multi-scale residual network" and a "bi-directional long-short term memory" network. The authors also claim that the program could be extended for miRNA target prediction. This tool is available as a source code (Bu et al. 2023).

P-TarPmiR: P-TarPmiR (Ajila et al. 2023) was developed by modifying TarPmiR, which was originally trained on the human dataset and used a random-forest-based approach for target prediction. The tool claims to work for all miRNAs, including noncanonicals. It was used to predict targets of all the miRNAs in the genome of soybeans (Ajila et al. 2023). The tool is available as a web-based application.

12.4 Bioinformatics Identification of miRNA and mRNA at a Genome-scale

miRNAs profoundly impact the physiology of a plant by modulating functions of biological networks, including signaling pathways, in a highly sophisticated manner. As discussed above, a pathway could be regulated by a number of miRNAs, at the same time, several biological pathways may recruit common miRNA candidates. Such regulation at the pathway/network level exert an impact that fine-tunes the expression of the participating genes. Studying these molecular scenarios now often requires the identification of miRNA-targets at a whole genome/transcriptome scale. Moreover, miRNA–target interactions could be highly context-dependent (e.g. a particular type of stress or occurring in a particular time frame) that warrants kinetic or time-course investigations of multiple treatments. For such scenarios, two main principles could be utilized to computationally identify miRNA targets: (i) the miRNA–mRNA complementarity principle, and (ii) the inverse or negative correlation of accumulation levels of the mRNAs and miRNA. Tools of network analysis significantly assist such genome-wide investigations of miRNA–target identification. As an example, we shall now briefly discuss the broad steps for a genome-wide bioinformatics investigation of canonical miRNA target sites.

Recent advances in deep sequencing have enabled simultaneous profiling of the mRNA- and the miRNA portions of the transcriptome. Such technological advances in sequencing are now allowing the investigators to ask questions related to spatio-temporal reprogramming of gene expression in complex networks and the role of miRNAs at a scale that was not possible in the past. At the same time, it has introduced the aspects of 'big data' to biological investigations in ways not seen earlier in traditional genomics. Therefore, finding miRNA targets in the whole genome has become computationally demanding. The following approach has proven to be effective in multiple studies (Pandey et al. 2018, 2019; Pradhan et al. 2017, 2020, 2023) for identifying miRNA-mediated networks of regulated targets, at a genomic scale, in large datasets from experiments that tries to resolve spatial and temporal reprogramming under specific contexts: (i) identification of canonical sites in the CDS region of the genes in the transcriptomes, (ii) investigating the patterns of correlation between the (changes in) accumulation of miRNAs and the expression of genes (here we use the term expression in a broad way that incorporates the estimations of mRNAs and/or proteins). Practically, all the miRNA-modulated regulations are expected to display a negative correlation between miRNAs and mRNAs due to the repressive nature of miRNA action on direct targets. (iii) Then, the outcomes of (i) and (ii) are combined, and network topologies are established and displayed for the interaction of miRNAs and mRNAs at a genome scale.

For determining the presence of any canonical sites for miRNA binding, the first step would be to assemble the datasets. Here, complete CDSs (gene's coding sequences), or the transcriptome, is assembled. Similarly, all the miRNA sequences of the miRNome are organized (in a separate file). Then, the 3′-end sequence of the mRNAs would be compared to the 5′-end sequences of the miRNA. Here, "seeds" are defined as sub-sequences of 8–13 nt of the miRNAs, which are extracted from positions 1 and 2 of the miRNA's 5′-end. These seeds are searched in the genes (CDSs) to define a perfect Watson–Crick base-pairing scenario. To simplify this step for computation, a reverse complement of the gene sequences could be generated and the miR-seeds could be searched as a step of 'pattern matching'. Once a pattern matches, it is extended with the help of a 1 nt sliding window. The longest seed could be regarded as a putative hit. The confidence in this putative miRNA–mRNA interaction due to the presence of a canonical targeting site is increased by determining the correlation coefficient of levels between the miRNA and mRNA expression. A significant inverse correlation of expression along with an above-described positive hit is indicative of miRNA-mediated canonical targeting. An interaction network is further inferred, plotted, and visualized. The use of Cytoscape for this purpose has proven helpful.

Of the several popular tools, Cytoscape (http://www.cytoscape.org/) is computationally platform-independent (so, it is available for most popular operating systems); it is capable of integrating with other databases and highly useful for visualizing complex networks, like of miRNA–mRNAs. It is equipped with functions that offer various layouts for networks such as force-directed, grid, hierarchical, organic, and circular, each of which offers a unique way of representing a graph. For instance, the 'force-directed' layout is useful for unraveling networks that are highly connected as is the case with gene networks (Pandey et al. 2019). Force-directed graphs view the entities as a system of particles with forces acting between them to generate minimal local energy.

The next step in the process requires analysis of the topologies of the networks so that hubs can be identified. This involves determining the statistical frameworks, such as degrees of nodes (defined as the number of edges) and 'betweenness', which estimates the centrality of a node.

Further, insights into the biological relevance of the information in the miRNA–mRNA regulatory network could be gained by conducting over-representation analysis. This is achieved with the help of deploying other databases such as GO (gene ontology). BiNGO (Maere et al. 2005) is a popular App/plugin for such over-representation analysis of GO terms. It uses a hypergeometric test and could be applied over only a part, or the complete network.

12.5 Conclusion

The bioinformatics analysis of plant miRNA targets is now at the center of miRNA-mediated investigations. With the rapid advances in deep sequencing technologies and a reduction in sequencing costs, the profiling of miRNAs and exploring its reprogramming during several environmental, genetic, and complex genomic contexts has become increasingly feasible. The field of computational tool development for various applications during miRNA–mRNA interactions has also seen changes in its landscape. Yet, there is still scope for developing optimum tools that could be applied across the experimental contexts, especially in non-model organisms, and in scenarios where novel miRNAs are often found. It is expected that the field of computational/bioinformatics-driven miRNA target interactions is going to further expand and evolve and that integrated platforms are available for this purpose.

References

Addo-Quaye, C., Eshoo, T.W., Bartel, D.P., and Axtell, M.J. (2008). Endogenous siRNA and miRNA targets identified by sequencing of the Arabidopsis degradome. *Current Biology* 18: 758–762.

Addo-Quaye, C., Miller, W., and Axtell, M.J. (2009). CleaveLand: a pipeline for using degradome data to find cleaved small RNA targets. *Bioinformatics* 25: 130–131.

Ajila, V., Colley, L., Ste-Croix, D.T. et al. (2023). P-TarPmiR accurately predicts plant-specific miRNA targets. *Scientific Reports* 13: 332.

Allen, E., Xie, Z., Gustafson, A.M., and Carrington, J.C. (2005). microRNA-directed phasing during trans-acting siRNA biogenesis in plants. *Cell* 121: 207–221.

Aukerman, M.J. and Sakai, H. (2003). Regulation of flowering time and floral organ identity by a MicroRNA and its APETALA2-like target genes. *Plant Cell* 15: 2730–2741.

Axtell, M.J. (2013). Classification and comparison of small RNAs from plants. *Annual Review of Plant Biology* 64: 137–159.

Baranauske, S., Mickute, M., Plotnikova, A. et al. (2015). Functional mapping of the plant small RNA methyltransferase: HEN1 physically interacts with HYL1 and DICER-LIKE 1 proteins. *Nucleic Acids Research* 43: 2802–2812.

Bonnet, E., He, Y., Billiau, K., and Van de Peer, Y. (2010). TAPIR, a web server for the prediction of plant microRNA targets, including target mimics. *Bioinformatics* 26: 1566–1568.

Brodersen, P., Sakvarelidze-Achard, L., Bruun-Rasmussen, M. et al. (2008). Widespread translational inhibition by plant miRNAs and siRNAs. *Science* 320: 1185–1190.

Brousse, C., Liu, Q., Beauclair, L. et al. (2014). A non-canonical plant microRNA target site. *Nucleic Acids Research* 42: 5270–5279.

Bu, Y., Zheng, J., and Jia, C. (2023). An efficient deep learning based predictor for identifying miRNA-triggered phasiRNA loci in plant. *Mathematical Biosciences and Engineering* 20: 6853–6865.

Chellappan, P., Xia, J., Zhou, X. et al. (2010). siRNAs from miRNA sites mediate DNA methylation of target genes. *Nucleic Acids Research* 38: 6883–6894.

Chen, X. (2004). A microRNA as a translational repressor of APETALA2 in Arabidopsis flower development. *Science* 303: 2022–2025.

Chorostecki, U. and Palatnik, J.F. (2014). comTAR: a web tool for the prediction and characterization of conserved microRNA targets in plants. *Bioinformatics* 30: 2066–2067.

Chorostecki, U., Crosa, V.A., Lodeyro, A.F. et al. (2012). Identification of new microRNA-regulated genes by conserved targeting in plant species. *Nucleic Acids Research* 40: 8893–8904.

Cuperus, J.T., Fahlgren, N., and Carrington, J.C. (2011). Evolution and functional diversification of MIRNA genes. *Plant Cell* 23: 431–442.

Dai, X. and Zhao, P.X. (2011). psRNATarget: a plant small RNA target analysis server. *Nucleic Acids Research* 39: W155–W159.

Dai, X., Zhuang, Z., and Zhao, P.X. (2018). psRNATarget: a plant small RNA target analysis server (2017 release). *Nucleic Acids Research* 46: W49–W54.

Dalmadi, A., Miloro, F., Balint, J. et al. (2021). Controlled RISC loading efficiency of miR168 defined by miRNA duplex structure adjusts ARGONAUTE1 homeostasis. *Nucleic Acids Research* 49: 12912–12928.

Eamens, A.L., Smith, N.A., Curtin, S.J. et al. (2009). The *Arabidopsis thaliana* double-stranded RNA binding protein DRB1 directs guide strand selection from microRNA duplexes. *RNA* 15: 2219–2235.

Eamens, A.L., Wook Kim, K., and Waterhouse, P.M. (2012). DRB2, DRB3 and DRB5 function in a non-canonical microRNA pathway in *Arabidopsis thaliana*. *Plant Signaling & Behavior* 7: 1224–1229.

Fahlgren, N. and Carrington, J.C. (2010). miRNA target prediction in plants. *Methods in Molecular Biology* 592: 51–57.

Fahlgren, N., Howell, M.D., Kasschau, K.D. et al. (2007). High-throughput sequencing of Arabidopsis microRNAs: evidence for frequent birth and death of MIRNA genes. *PLoS One* 2: e219.

Fattash, I., Voss, B., Reski, R. et al. (2007). Evidence for the rapid expansion of microRNA-mediated regulation in early land plant evolution. *BMC Plant Biology* 7: 13.

Franco-Zorrilla, J.M., Valli, A., Todesco, M. et al. (2007). Target mimicry provides a new mechanism for regulation of microRNA activity. *Nature Genetics* 39: 1033–1037.

Friedlander, M.R., Mackowiak, S.D., Li, N. et al. (2012). miRDeep2 accurately identifies known and hundreds of novel microRNA genes in seven animal clades. *Nucleic Acids Research* 40: 37–52.

Gandikota, M., Birkenbihl, R.P., Hohmann, S. et al. (2007). The miRNA156/157 recognition element in the 3′ UTR of the Arabidopsis SBP box gene SPL3 prevents early flowering by translational inhibition in seedlings. *The Plant Journal* 49: 683–693.

German, M.A., Pillay, M., Jeong, D.H. et al. (2008). Global identification of microRNA-target RNA pairs by parallel analysis of RNA ends. *Nature Biotechnology* 26: 941–946.

Gu, W., Wang, X., Zhai, C. et al. (2012). Selection on synonymous sites for increased accessibility around miRNA binding sites in plants. *Molecular Biology and Evolution* 29: 3037–3044.

Han, M.H., Goud, S., Song, L., and Fedoroff, N. (2004). The Arabidopsis double-stranded RNA-binding protein HYL1 plays a role in microRNA-mediated gene regulation. *Proceedings of the National Academy of Sciences of the United States of America* 101: 1093–1098.

Iki, T., Yoshikawa, M., Nishikiori, M. et al. (2010). In vitro assembly of plant RNA-induced silencing complexes facilitated by molecular chaperone HSP90. *Molecular Cell* 39: 282–291.

Iwakawa, H.O. and Tomari, Y. (2013). Molecular insights into microRNA-mediated translational repression in plants. *Molecular Cell* 52: 591–601.

Jones-Rhoades, M.W. and Bartel, D.P. (2004). Computational identification of plant microRNAs and their targets, including a stress-induced miRNA. *Molecular Cell* 14: 787–799.

Kohler, A. and Hurt, E. (2007). Exporting RNA from the nucleus to the cytoplasm. *Nature Reviews Molecular Cell Biology* 8: 761–773.

Kurihara, Y. and Watanabe, Y. (2004). Arabidopsis micro-RNA biogenesis through Dicer-like 1 protein functions. *Proceedings of the National Academy of Sciences of the United States of America* 101: 12753–12758.

Lewis, B.P., Burge, C.B., and Bartel, D.P. (2005). Conserved seed pairing, often flanked by adenosines, indicates that thousands of human genes are microRNA targets. *Cell* 120: 15–20.

Li, F., Zheng, Q., Vandivier, L.E. et al. (2012). Regulatory impact of RNA secondary structure across the Arabidopsis transcriptome. *Plant Cell* 24: 4346–4359.

Li, J.F., Chung, H.S., Niu, Y. et al. (2013). Comprehensive protein-based artificial microRNA screens for effective gene silencing in plants. *Plant Cell* 25: 1507–1522.

Li, S., Castillo-Gonzalez, C., Yu, B., and Zhang, X. (2017). The functions of plant small RNAs in development and in stress responses. *The Plant Journal* 90: 654–670.

Liu, Q., Wang, F., and Axtell, M.J. (2014). Analysis of complementarity requirements for plant microRNA targeting using a *Nicotiana benthamiana* quantitative transient assay. *Plant Cell* 26: 741–753.

Lobbes, D., Rallapalli, G., Schmidt, D.D. et al. (2006). SERRATE: a new player on the plant microRNA scene. *EMBO Reports* 7: 1052–1058.

Lorenz, R., Bernhart, S.H., Siederdissen, C.H.Z. et al. (2011). ViennaRNA Package 2.0. *Algorithms for Molecular Biology* 6: 26.

Ma, X., Liu, C., Gu, L. et al. (2018). TarHunter, a tool for predicting conserved microRNA targets and target mimics in plants. *Bioinformatics* 34: 1574–1576.

Maere, S., Heymans, K., and Kuiper, M. (2005). BiNGO: a Cytoscape plugin to assess overrepresentation of gene ontology categories in biological networks. *Bioinformatics* 21: 3448–3449.

Mallory, A.C., Reinhart, B.J., Jones-Rhoades, M.W. et al. (2004). MicroRNA control of PHABULOSA in leaf development: importance of pairing to the microRNA 5′ region. *The EMBO Journal* 23: 3356–3364.

Neilsen, C.T., Goodall, G.J., and Bracken, C.P. (2012). IsomiRs-the overlooked repertoire in the dynamic microRNAome. *Trends in Genetics* 28: 544–549.

Nozawa, M., Miura, S., and Nei, M. (2012). Origins and evolution of microRNA genes in plant species. *Genome Biology and Evolution* 4: 230–239.

Pandey, S.P. and Somssich, I.E. (2009). The role of WRKY transcription factors in plant immunity. *Plant Physiology* 150: 1648–1655.

Pandey, S.P., Shahi, P., Gase, K., and Baldwin, I.T. (2008). Herbivory-induced changes in the small-RNA transcriptome and phytohormone signaling in *Nicotiana attenuata*. *Proceedings of the National Academy of Sciences of the United States of America* 105: 4559–4564.

Pandey, P., Wang, M., Baldwin, I.T. et al. (2018). Complex regulation of microRNAs in roots of competitively-grown isogenic *Nicotiana attenuata* plants with different capacities to interact with arbuscular mycorrhizal fungi. *BMC Genomics* 19: 937.

Pandey, P., Srivastava, P.K., and Pandey, S.P. (2019). Prediction of plant miRNA targets. *Methods in Molecular Biology* 1932: 99–107.

Pradhan, M., Pandey, P., Gase, K. et al. (2017). Argonaute 8 (AGO8) mediates the elicitation of direct defenses against herbivory. *Plant Physiology* 175: 927–946.

Pradhan, M., Pandey, P., Baldwin, I.T., and Pandey, S.P. (2020). Argonaute4 modulates resistance to *Fusarium brachygibbosum* infection by regulating jasmonic acid signaling. *Plant Physiology* 184: 1128–1152.

Pradhan, M., Rocha, C., Halitschke, R. et al. (2021). microRNA390 modulates *Nicotiana attenuata*'s tolerance response to *Manduca sexta* herbivory. *Plant Direct* 5: e350.

Pradhan, M., Baldwin, I.T., and Pandey, S.P. (2023). Argonaute7 (AGO7) optimizes arbuscular mycorrhizal fungal associations and enhances competitive growth in *Nicotiana attenuata*. *The New Phytologist* 240: 382–398.

Rajagopalan, R., Vaucheret, H., Trejo, J., and Bartel, D.P. (2006). A diverse and evolutionarily fluid set of microRNAs in *Arabidopsis thaliana*. *Genes & Development* 20: 3407–3425.

Reinhart, B.J., Weinstein, E.G., Rhoades, M.W. et al. (2002). MicroRNAs in plants. *Genes & Development* 16: 1616–1626.

Rhoades, M.W., Reinhart, B.J., Lim, L.P. et al. (2002). Prediction of plant microRNA targets. *Cell* 110: 513–520.

Rogers, K. and Chen, X. (2013). Biogenesis, turnover, and mode of action of plant microRNAs. *Plant Cell* 25: 2383–2399.

Schwab, R., Palatnik, J.F., Riester, M. et al. (2005). Specific effects of microRNAs on the plant transcriptome. *Developmental Cell* 8: 517–527.

Shi, T., Wang, K., and Yang, P. (2017). The evolution of plant microRNAs: insights from a basal eudicot sacred lotus. *The Plant Journal* 89: 442–457.

Srivastava, P.K., Moturu, T.R., Pandey, P. et al. (2014). A comparison of performance of plant miRNA target prediction tools and the characterization of features for genome-wide target prediction. *BMC Genomics* 15: 348.

Sunkar, R., Li, Y.F., and Jagadeeswaran, G. (2012). Functions of microRNAs in plant stress responses. *Trends in Plant Science* 17: 196–203.

Thody, J., Folkes, L., Medina-Calzada, Z. et al. (2018). PAREsnip2: a tool for high-throughput prediction of small RNA targets from degradome sequencing data using configurable targeting rules. *Nucleic Acids Research* 46: 8730–8739.

Thody, J., Moulton, V., and Mohorianu, I. (2020). PAREameters: a tool for computational inference of plant miRNA-mRNA targeting rules using small RNA and degradome sequencing data. *Nucleic Acids Research* 48: 2258–2270.

Todesco, M., Rubio-Somoza, I., Paz-Ares, J., and Weigel, D. (2010). A collection of target mimics for comprehensive analysis of microRNA function in *Arabidopsis thaliana*. *PLoS Genetics* 6: e1001031.

Voinnet, O. (2009). Origin, biogenesis, and activity of plant microRNAs. *Cell* 136: 669–687.

Wong, G.Y. and Millar, A.A. (2022). TRUEE; a bioinformatic pipeline to define the functional microRNA targetome of Arabidopsis. *The Plant Journal* 110: 1476–1492.

Wu, L., Zhou, H., Zhang, Q. et al. (2010). DNA methylation mediated by a microRNA pathway. *Molecular Cell* 38: 465–475.

Xie, Z., Allen, E., Fahlgren, N. et al. (2005). Expression of Arabidopsis MIRNA genes. *Plant Physiology* 138: 2145–2154.

Yang, G.D., Yan, K., Wu, B.J. et al. (2012). Genomewide analysis of intronic microRNAs in rice and Arabidopsis. *Journal of Genetics* 91: 313–324.

Yang, K., Sablok, G., Qiao, G. et al. (2017). isomiR2Function: an integrated workflow for identifying microRNA variants in plants. *Frontiers in Plant Science* 8: 322.

Yu, B., Yang, Z., Li, J. et al. (2005). Methylation as a crucial step in plant microRNA biogenesis. *Science* 307: 932–935.

Yu, Y., Jia, T., and Chen, X. (2017). The 'how' and 'where' of plant microRNAs. *The New Phytologist* 216: 1002–1017.

Zhan, J. and Meyers, B.C. (2023). Plant small RNAs: their biogenesis, regulatory roles, and functions. *Annual Review of Plant Biology* 74: 21–51.

Zhang, Y. (2005). miRU: an automated plant miRNA target prediction server. *Nucleic Acids Research* 33: W701–W704.

Zhang, Z., Jiang, L., Wang, J. et al. (2015). MTide: an integrated tool for the identification of miRNA-target interaction in plants. *Bioinformatics* 31: 290–291.

Zhang, W., Le, T.D., Liu, L. et al. (2016). Predicting miRNA targets by integrating gene regulatory knowledge with expression profiles. *PLoS One* 11: e0152860.

13

Machine Learning for the Discovery of DNA-binding Proteins in Plants

Upendra Kumar Pradhan[1], *Prabina Kumar Meher*[1], *and Pushpendra Kumar Gupta*[2,3,4]

[1] *Division of Statistical Genetics, ICAR-Indian Agricultural Statistics Research Institute, New Delhi, India*
[2] *Department of Genetics and Plant Breeding, Chaudhary Charan Singh University, Meerut, Uttar Pradesh, India*
[3] *Murdoch's Centre for Crop and Food Innovation, Murdoch University, Perth, Australia*
[4] *Borlaug Institute for South Asia (BISA), Ludhiana, Punjab, India*

13.1 Introduction

DNA-binding proteins (DBPs) are sequence-specific proteins that contain functional groups for the identification of specific DNA sequences to allow interaction with major grooves of B-DNA; however, DBPs interacting with minor grooves are also known (Zimmer and Wähnert 1986). These DBPs account for a wide range of proteins with a variety of sequences and structures. Structurally, these proteins have been placed in 8 groups and 54 families that are available in published literature (Lin et al. 2013; Luscombe et al. 2000). The prediction of DBPs has major implications for future research in the fields of proteomics and genomics (Luscombe et al. 2000; Siggers and Gordân 2014; Ferraz et al. 2021).

Based on their functions, eukaryotic DBPs are classified into the following four major classes (Zheng and Wang 2011): (i) transcription factors (TFs) involved in transcriptional regulation, such as transcription activators; (ii) DNA replications factors (DRFs), which carry out *in vivo* and *in vitro* synthesis of DNA at all levels; (iii) repair factors, having a role in removing from the DNA segment either single base pairs or specific oligonucleotides, followed by filling these gaps with suitable nucleotides, and (iv) histones, which are involved in chromosome packaging and transcription. DBPs in bacteria are mostly restriction enzymes, which detect and cleave DNA at specific sites with nucleotide sequences to provide host (bacteria) defense against invasion by viruses, etc.

During the last two decades, due to the availability of newer and high-throughput (HTP) genome sequencing tools/techniques (Wheeler et al. 2008; Harris et al. 2008), whole-genome sequences of >1000 eukaryotes are now available. In prokaryote genomes, the proportion of these whole-genome sequences encoding DBPs is 2–3%, whereas in eukaryotic genomes, it is 6–7% (Luscombe et al. 2000; Walter et al. 2009). A plethora of protein

Bioinformatics for Plant Research and Crop Breeding, First Edition. Edited by Jen-Tsung Chen.

sequences have also become available (Wu et al. 2006), which need to be annotated to elucidate their functions. Identification of DBPs is a part of this exercise.

Identification of DBPs was initially carried out by a variety of wet lab techniques including the following: filter binding assays, electrophoretic mobility shift assay (EMSA), cross-linking, *in vitro* analysis involving yeast one-hybrid (Y1H), and protoplast transactivation assays (PTAs); these methods are listed in Table 13.1. Despite being accurate, these experimental procedures for assigning functions to DBPs are cost-ineffective and time-consuming, making it difficult for this exercise to keep up with the volume of sequenced data generated (Zaitzeff et al. 2021). Therefore, there has been a demand for developing methods involving computational predictions of DBPs and downstream protein analysis.

The task of recognizing DBPs within the proteomes is challenging due to the presence of numerous DBP families with a variety of characteristic features from a wide range of eukaryotes and prokaryotes. Using these features, several models have been developed; however, these models tend to lack the level of accuracy needed to annotate specific DBPs. For example, TFs are highly specific for the plant and animal kingdoms, with up to 47% of DNA binding TFs belonging to lineage-specific families (Riechmann et al. 2000; Motion

Table 13.1 Summary of experimental approaches for identification of DNA binding proteins (DBPs).

Techniques	Pros	Cons	References
1. Filter binding assay	Simple and inexpensive	The interaction may not hold up to the filtering process; the resulting products are irreversible	Yarus and Berg (1967); Riggs et al. (1970); Stockley (2009)
2. Electrophoretic mobility shift assay (EMSA)	Quick and simple, strong and sensitive, semi-quantitative	—	Fried and Crothers (1981); Garner and Revzin (1981); Hellman and Fried (2007)
3. Cross-linking (UV and Laser)	UV crosslinking is noninvasive with no effect on molecular structures; laser cross-linking is easy and quick, and reduces the risk of disrupting molecular structures	The conventional cross-linking procedures of formaldehyde, glutaraldehyde, and UV are nonspecific	Guille and Kneale (1997); Welsh and Cantor (1984)
4. EMSA combined with Western Blotting techniques	Rapid and simple; when coupled with immune-blotting, just one diffusion blotting is performed; the electrophoretic' supershift' assay requires only one gel and no diffusion blotting	Need purification of antibody required for electrophoretic' supershift'; Shift-Western Blotting needs two blotting membranes	Ofran et al. (2007); Hellman and Fried (2007); Chen and Chang (2001)
5. *In vivo* analysis: Y1H and PTA	Y1H and PTA produce consistent results that are both direct and sensitive	PTA is labor-intensive and requires an array of processes	Cai and Huang (2012); Ueda et al. (2011)

et al. 2015; Pradhan et al. 2023a,b). Although several of these models have been evaluated using *Arabidopsis thaliana*, these approaches have shown low accuracy during the prediction of DBPs for other plant species. Significant advances in prediction of DBPs in humans have also been made in recent years; the development of plant-based models is rather slow, thus necessitating the development of efficient computational models to predict DBPs with higher accuracy in plant systems.

The approaches for computational prediction of DBPs can be placed in two major categories, namely template-based and machine learning-based. The template-based approaches involve the detection of significant structural or sequence similarity between the query and the template DNA, while machine learning algorithms (MLAs) provide prediction models for the identification of a pattern in the input feature space. In this chapter, we describe the available computational models for the prediction of DBPs. We believe that this chapter provides valuable information for biologists, who wish to learn and use methods for computational prediction of plant-specific DBPs.

13.2 Steps Involved in Identification of DBPs Using Machine Learning

The identification of DBPs using machine learning involves the following three major steps: (i) collection and processing of sequence data; (ii) extraction of numeric features from sequence data; and (iii) prediction using a machine learning model. The steps involved in the prediction of DBPs using a machine-learning algorithm are shown in Figure 13.1.

13.2.1 Collection and Processing of Sequence Data

The construction of a benchmark dataset is a key step in DBP recognition. It determines the accuracy of predictive models including MLAs (Liang et al. 2017; Su et al. 2021). For the prediction of DBPs using MLAs, the main sources include the databases including the PDB database (Rose et al. 2017), Uniprot (UniProt Consortium 2021), and other protein–DNA interaction databases. Specifically, plant-specific proteins can be collected from the UniProt database (UniProt Consortium 2021) using the Taxonomy id: 3398. Further, the protein sequences that are annotated as "DNA-binding and their child terms" by GO term (GO: 0003677) are treated as DBP sequences, and the proteins with no association with the GO term "DNA-binding" are considered non-DBP sequences. The UniProt Knowledgebase consists of two parts, (i) the part containing reviewed and manually-annotated sequences, and (ii) the part having un-reviewed protein sequences with computationally generated annotations. In particular, the UniProtKB/Swiss-Prot (reviewed) option can be used to obtain experimentally validated and nonredundant (NR) protein sequences. In doing this, the protein sequences with <50 amino acids are removed to avoid protein fragments. The sequence dataset must be subjected to homology reduction, which can help to improve the accuracy of sequence analysis by eliminating biases that may arise due to the inclusion of highly similar sequences. This can also help reduce the computational cost of sequence

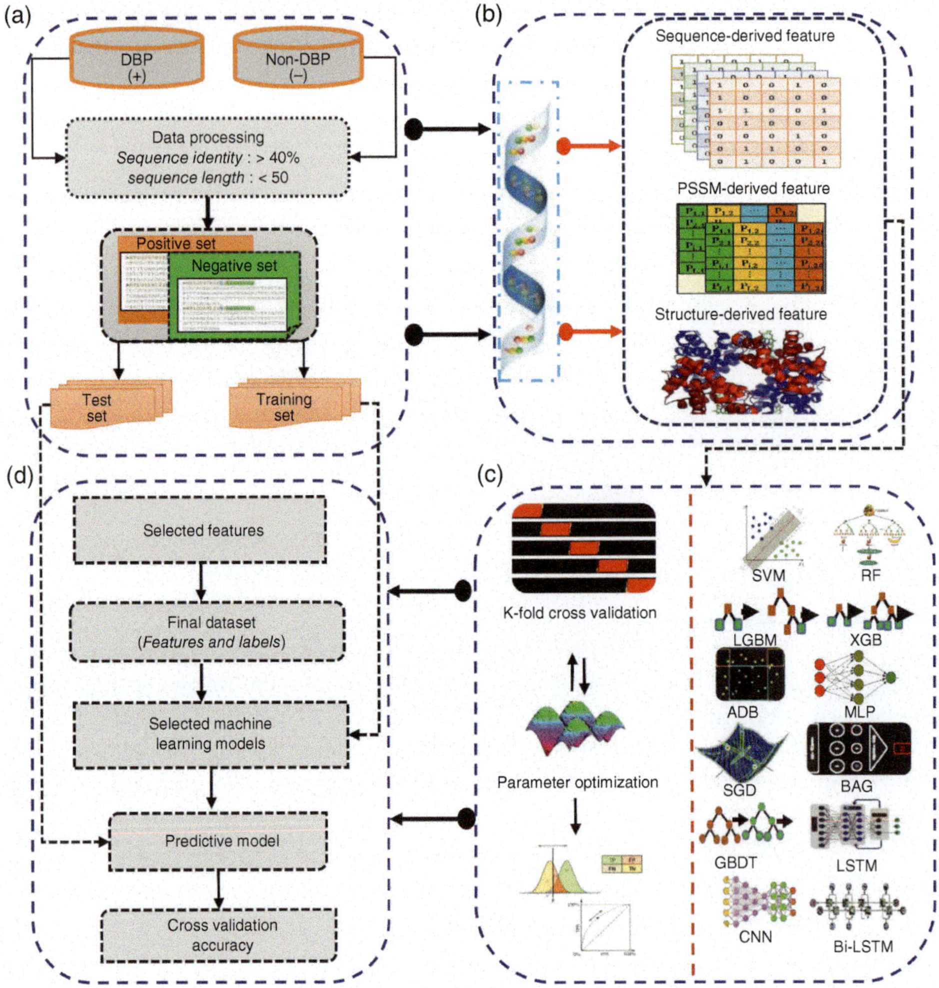

Figure 13.1 General outline for machine learning-based computational model for DBP prediction. (a) Retrieval of DBP and non-DBP sequences from different databases and processing of sequence data; (b) Feature generation from sequence data using different encoding schemes. (c) Selection of machine learning algorithm for model building using training dataset based on their prediction accuracy. (d) Model building using selected machine learning techniques and assessment of cross-validation accuracy.

analysis. The CD-HIT algorithm (Huang et al. 2010) can be used to eliminate similar sequences using a suitable threshold.

13.2.2 Extraction of Numeric Features from Sequence Data

Sequence data cannot be used directly as input in MLAs for prediction purposes. Therefore, transformation of sequence data to numeric features is a necessary step for prediction using this approach. Different schemes for the feature encoding have been put forward to extract meaningful patterns from protein sequences. The following three different types of features are used for this purpose: (i) features derived from sequence (both compositional and evolutionary), (ii) features based on protein structure, and (iii) features based on physico-chemical properties. The following sections briefly describe some of the important characteristics of these three types of features that have been successfully used for the identification of DBPs.

13.2.2.1 Sequence-derived Features

13.2.2.1.1 Compositional Features

***k*-mer feature:** The k-mer is the most common method to convert the protein sequence into a vector (Bhasin and Raghava 2004), in which k-mer frequency refers to all possible combinations of k amino acids in a given sequence. For the prediction of DBPs, the k-mer features obtained for $k = 1$ to 3 have been mostly employed, where the features are respectively called amino acid composition (AAC), di-peptide composition (DPC), and tri-peptide composition (TPC). The formulae for computing AAC, DPC, and TPC are as follows.

$$\text{AAC}: f(r) = \frac{N_r}{N} \cdots \tag{13.1}$$

$$\text{DPC}: f(r,s) = \frac{N_{rs}}{N-1} \cdots \tag{13.2}$$

$$\text{TPC}: f(r,s,t) = \frac{N_{rst}}{N-2} \cdots \tag{13.3}$$

In the above equations, N_r is the number of amino acid r, N_{rs} is the number of dipeptides represented by amino acids r and s, N_{rst} is the number of tri-peptides represented by amino acids r, s, and t, N is the length of the sequence, and $r, s, t = 1, 2, \ldots, 20$ (for 20 different amino acids).

Pseudo amino acid composition (PAAC): The PAAC feature is used to encode amino acid sequences for classification and prediction purposes (Chou 2001). It considers both local and global sequence ordering effects. PAAC for a protein is a numeric feature vector of dimension $20 + \lambda$ for λ-tier correlation structure. The λ factor represents the rank of correlation and is a nonnegative integer, where the value of λ should not exceed the sequence length (Shen and Chou 2008).

Evolutionary features: Position-specific scoring matrices (PSSM) have been widely used in prior DBP prediction methods due to their ability to predict the likelihood of a given residue substitution based on evolutionary evidence. The PSSM recognizes the conservation pattern in multiple alignments and records it as a scoring matrix for every position in the alignment. The PSSM profile of each protein sequence can be generated by running PSI-BLAST on the NCBI NR database (ftp://ftp.ncbi.nih.gov/blast/db/nr), which is an $L \times 20$-dimensional matrix, as shown below.

$$\text{PSSM}_{L\times 20} = \begin{bmatrix} p_{1,1} & p_{1,2} & \cdots & p_{1,20} \\ p_{2,1} & p_{2,2} & \cdots & p_{2,20} \\ \vdots & \vdots & \cdots & \vdots \\ p_{L,1} & p_{L,2} & \cdots & p_{L,20} \end{bmatrix}$$

In the matrix $PSSM_{L\times 20}$, L is the protein length, $p_{k,j}$ denotes the probability of amino acid j occurring at position k in the protein sequence. Important PSSM-based features used for DBP prediction are briefly explained below.

13.2.2.1.2 AAC_PSSM The 20-dimensional feature vector AAC_PSSM (y_j) is similar to the amino acid compositional features derived from the PSSM profiles, and can be obtained as $y_j = \frac{1}{L}\sum_{k=1}^{L} p_{k,j}; \left(1 \le j \le 20\right)$, where L is the protein sequence length.

13.2.2.1.3 DPC_PSSM The DPC_PSSM is a 400-dimensional feature vector derived from the PSSM matrix (Liu et al. 2010) which can be computed as $y_{i,j} = \frac{1}{L-1}\sum_{k=1}^{L-1} p_{k,i}p_{k+1,j}; \left(1 \le i, j \le 20\right)$, where L is the protein sequence length.

13.2.2.1.4 PSSM400 The positions in the protein sequence are first obtained for each amino acid, and the appropriate rows from the PSSM matrix are extracted to generate a new sub-matrix. As a result, there will be 20 such sub-matrices, each corresponding to 20 standard amino acids. The average of each submatrix's columns is then computed, yielding a vector of length 20. Finally, by concatenating these 20 vectors, a feature vector of length 400 for each protein is generated, which is called the PSSM400 feature vector (Nanni et al. 2014).

13.2.2.1.5 EDP_PSSM and EEDP_PSSM The evolutionary score between adjacent amino acids is first computed as $\text{Aver}_1 = {\left(p_{k-1,i} + p_{k,j}\right)}/{2}$ and $\text{Aver}_2 = {\left(p_{k,j} + p_{k+1,l}\right)}/{2}$, $k = 1, 2, ..., L$ and $l, s, t = 1, 2, ..., 20$. Here, Aver_1 is the mean score of positions k and $k-1$, and Aver_2 is the mean score of positions k and $k+1$. Then, compute the evolutionary difference (EDF) as $x_{k-1,k+1} = (\text{Aver}_1 - \text{Aver}_2)^2$, where $x_{k-1,k+1}$ indicates the average evolutionary divergence between the residues of a particular protein sequence. The 400-dimensional ED-PSSM feature vector can be computed as $e_{j,l} = \sum_{k=2}^{L-1} x_{k-1,k+1} / L-2; j,l = 1,2,\ldots,20$. Based on the EDF matrix, the 20-dimensional EEDP_PSSM feature vector is defined as $\left[\psi_1,\psi_2,\ldots,\psi_{20}\right]^T$ where $\psi_l = \sum_{j=1}^{20} e_{j,l} / 20; l = 1,2,\ldots,20$.

13.2.2.1.6 PSE_PSSM This feature vector's length depends on the lag value. The mean of the first 20 PSSM columns made up the elements of this feature vector. The mean squares of the differences between the elements of row k and $k+\text{lag}$ in this column represent the subsequent values for each column. For each lag value, there will be another 20-dimensional feature vector. For instance, for lag $= 1$, there will be a 40-dimensional feature vector, and so on. The following equation can be used for computing the feature vector for each lag value.

$$p(m) = \frac{1}{L-\text{lag}} \sum_{k=1}^{L-\text{lag}} \left(p_{k,j} - p_{k+\text{lag},j}\right)^2 \quad j = 1,2,\ldots,20 \text{ and } m = 20 + 20 * \text{lag}$$

13.2.2.1.7 kbigram-PSSM This is a 400-dimensional feature vector, the elements of which can be computed by using the formula $T_{i,j}(m)=\sum_{k=1}^{L-m} p_{k,i}p_{k+m,j};\ (1\le i,j\le 20)$, where L is the protein sequence length and m denotes *kbigram* value.

13.2.2.2 Structure-based Features

Different features of the secondary structure of proteins are used as input in MLAs to predict DBPs. When the protein structure is not known, a method like PSIPRED can be used to predict the secondary structure (McGuffin et al. 2000) and when it is known, the DSSPcont (Carter et al. 2003) can be used for the same purpose. The accessible solvent area (ASA) is also one of the important features computed from a given protein structure, and can be calculated using NACCESS (Yu et al. 2006); it cannot be calculated without the availability of the DNA molecule.

13.2.2.3 Physico-chemical Features

13.2.2.3.1 Residue-wise Contact Energy Matrix (RCEM) Feature DBPs often contain intrinsically disordered proteins (IDPs) (Babu et al. 2011), especially in disordered tails (Dosztányi et al. 2005; Vuzman et al. 2012). Thus, anticipated residue-wise contact energies can be used as features to automatically incorporate crucial information about the amino acid interactions and intrinsically disordered regions (IDRs). According to Dosztányi et al. (2005), these contact energies are generated by least square fitting the contact energies acquired from the tertiary structure of proteins with the protein primary sequences. The rows and columns of the 20×20-dimensional matrix known as the RCEM correspond to the 20 standard amino acids.

13.2.2.3.2 Physicochemical Distance Transformation (PDT) Liu et al. (2012) introduced the concept of physicochemical distance transformation (PDT), which takes into account all 531 significant amino acid indices that are accessible through the AAIndex1 database (Kawashima et al. 2008). A protein sequence S having L amino acids can be represented as a 531-dimensional vector $X_{\mathrm{PDT}}=[X_1, X_2, \ldots, X_{531}]^T$, where $X_j=\dfrac{\sum_{i=1}^{L-1} D_j(A_i, A_{i+1})}{L-1}$; j represents the jth amino acid index in AAIndex1; A_i is the amino acid at ith position of the protein sequence. The value of $D_j(A_i, A_{i+1})$ can be calculated as $(I_j(A_i)-I_j(A_{i+1}))^2$, where $I_j(A_i)$ and $I_j(A_{i+1})$ represent the normalized physicochemical property values of amino acid A_i and A_{i+1} in property j.

13.2.3 Prediction Using Machine Learning Algorithms

From the above discussion, it is apparent that a number of ML models are available for the prediction of DBPs. These models can be placed in two categories: (i) shallow learning models, which include support vector machines (SVMs) (Kumar et al. 2007; Zhang et al. 2014; Liu et al. 2014; Mishra et al. 2019; Pradhan et al. 2023a), artificial neural networks (Ofran et al. 2007), random forest (RF) (Ma et al. 2016; Rahman, et al. 2018; Wang et al. 2022; Jia et al. 2021; Barukab et al. 2022), extreme gradient boosting (XGB) (Zhao et al. 2022) and light gradient boosting machine (LGBM) (Ali et al. 2022), and (ii) deep learning models, which include deep neural network (Shadab et al. 2020), convolution neural network (CNN) (Zhang et al. 2020, 2021) and Bi-LSTM (Li et al. 2021). For prediction of plant-specific DBPs, algorithms such as SVM, RF, XGB, and adaptive boosting (ADB) were successfully implemented (Motion et al. 2015; Pradhan et al. 2023a). The four major shallow learning models (SVM, RF, XGB, ADB) and two deep learning models (CNN, LSTM), which are relatively more important, are briefly described below.

13.2.3.1 Shallow Learning Models

13.2.3.1.1 Support Vector Machine (SVM) One widely used machine learning technique in binary and multi-class classification is the SVM (Vapnik 1999). The binary SVM searches for the best separation hyperplane that maximizes the margin between the observations of the two classes. In other words, the hyperplane serves as the decision boundary which divides the feature space into two regions corresponding to two classes. SVM can be used to separate data that is both linearly and nonlinearly separable. In the case of nonlinear separable data, the kernel approach is employed to migrate the input dataset to a high dimensional feature space with linearly separable observations. Polynomials, radial basis functions, and sigmoid functions are examples of appropriate kernel functions. Because of its capacity to handle big and noisy datasets and high generalized predictive accuracy, SVM has proven advantageous in a variety of fields of study, including bioinformatics. A brief mathematical description is as follows.

Consider a binary classification problem of separating the set of training vectors belonging to two separate classes, $(y_1, x_1), (y_2, x_2), \ldots (y_n, x_n), x \in R^m, y \in (-1, +1)$. The objective is to construct the hyperplane $w^T x + b = 0$ such that $w^T x + b \geq 1$ if $y_i = 1$ and $w^T x + b \leq 1$ if $y_i = -1$, where w and b are the weight vector and bias, respectively. In this formulation, $2/\|w\|$ is the width of the margin which needs to be maximized. In the case of nonlinear separable data points, the objective function becomes

$$\text{minimize}: \frac{1}{2}\|w\|^2 - C\sum_{i=}^{N} \alpha_i$$

$$\text{subjected to}: y_i\left(w^T x + b\right) \geq 1 - \alpha_i \alpha_j \geq 0,$$

where $\alpha_i \geq 0$, $(i = 1, 2, \ldots, N)$ is the slack variable which allows for the introduction of soft margins. Furthermore, the trade-off between maximizing the margin and decreasing the misclassification is managed by the regularization parameter C. While a higher value of C puts more focus on correctly classifying observations, a lower value of C emphasizes a wider margin and can allow more misclassifications. The training of SVM involves finding the optimal values of the weight vector w and the bias term b that satisfy the constraints and minimize the objective function. Once the SVM model is trained, the decision rule for classifying new data points involves evaluating the $f_{w,b}(x) = \text{sgn}(w^T x + b)$. If $f_{w,b}(x) > 0$, the point belongs to one class, otherwise, it belongs to the other class.

13.2.3.1.2 Random Forest (RF) RF is an ensemble learning method widely used for binary classification(Breiman 2001). Based on the principle of bagging (bootstrap aggregating), RF constructs an ensemble of diverse decision tree classifiers in which each classifier is trained based on a random subset of the training dataset. Each decision tree is built based on different features and splitting criteria, where the splitting is done in a way that optimally separates the data points of the two classes. When it comes to making predictions, the RF aggregates the predictions of all individual decision trees. In particular, the mode of all the tree predictions is taken as the final classification task. There are mainly two phases in classification using the RF algorithm (Training phase and Prediction phase) briefly described as follows.

Training phase: (i) Consider a training dataset with feature vector $X = (x_1, x_2, ..., x_p)$ and corresponding observed labels $y \in \{-1, 1\}$.

Let T be the total number of decision trees constructed. Then, for each decision tree $t = 1$ to T:

a) Randomly select a subset of the training data X_t, by random sampling without replacement
b) Randomly select a subset of the feature set F_t, from the available features
c) Build a decision tree using the dataset (X_t, y) and the selected features F_t

Prediction phase: To predict a new feature vector x_n;

- For each decision tree $t = 1$ to T, let h_t be the prediction label assigned by the decision tree for input x_n
- Then the final prediction label is the majority vote or mode of the prediction from all the decision trees, i.e. $\hat{y} = \text{mode}(h_1(x), h_2(x), ..., h_T(x))$

13.2.3.1.3 XGBoost XGBoost (eXtreme Gradient Boosting) is a well-known ensemble-supervised machine learning technique (Chen and Guestrin 2016). Boosting is the process of merging multiple weak classifiers (decision trees) to produce a stronger and more accurate model. The strategy adds decision trees to the ensemble repeatedly, with each succeeding tree aiming to correct the errors produced by the preceding trees. The trees are added in a sequential order, with each new tree trained to minimize the ensemble's residual errors. Regularization, gradient-based optimization, tree pruning, and handling of missing values are some of the important features of XGBoost that contribute to its success. A brief description of XGBoost is as follows.

The objective function of XGBoost classification represents the error to be minimized during the training. The objective function is defined as $L(y, \hat{y}) + \varnothing(f)$, where $L(y, \hat{y})$ is the softmax loss function that measures the difference between the observed class labels (y) and the predicted probabilities ($\hat{y}$) for each class, $\varnothing(f)$ is the regularization term that penalizes the overfitting, and f represents the ensemble of weak classification trees. The softmax loss function is used to convert the output of the ensemble model into probabilities of each class and then to calculate the error or loss. It is defined as $L(y, \hat{y}) = -\sum_i y_i \log(\hat{y}_i)$, where y_i and $\hat{y}_i$ are the true probability and predicted probability of ith class. XGBoost for classification builds the ensemble model in an additive manner. Each weak model is trained to minimize the gradient of the objective function, and the predictions of all weak models are summed to obtain the final predicted probabilities. The predicted probability at tth iteration is given by $\hat{y}_i^t = \text{softmax}\left(\hat{y}_i^{t-1}\right) + f_t\left(x_i\right)$, where $\hat{y}_i^t$ is the predicted probability for the ith instance at iteration t and $f_t(x_i)$ is the prediction of the tth weak model for the ith instance. The gradients of the softmax loss function with respect to the predicted probabilities are calculated to update the weak models, which are given by $g_i = \hat{y}_i - y_i$ with g_i as the gradient for the ith instance. By optimizing the objective function, updating the weak models based on gradients, and constructing decision trees, XGBoost for classification builds an ensemble model that can accurately predict class probabilities for different input instances.

13.2.3.1.4 AdaBoost AdaBoost (Adaptive Boosting) is a commonly used machine learning technique used for classification (Freund and Schapire 1997). It is an ensemble method that combines multiple weak classifiers (tree-based classifiers) to produce a strong classifier. Each instance in the training data is assigned an initial weight, typically equal for all instances. These are adjusted for each iteration based on the weak classifier's performance on the training data. The weak classifier is trained on the weighted training data in each iteration. The weak classifier's performance is evaluated by measuring the classification error, which is the weighted sum of misclassified instances. The weak classifier's performance feeds the updates to the training instances' weights. Higher weights are given to instances that the weak classifier incorrectly classified, whereas lower weights are given to instances that are correctly identified. This gives more emphasis to the misclassified instances. Besides weighting the misclassified instances, weights are also assigned to the weak classifiers based on their accuracy in classifying the training instances. The weak classifiers that are more accurate receive higher weights. AdaBoost uses a weighted voting scheme, in which the votes of each weak classifier are weighted by the classifier weight, to aggregate the predictions of all the weak classifiers in order to predict a test instance. The final class label is then determined by comparing the weighted sum of the predictions made by the weak classifiers to a threshold value. A brief mathematical description of AdaBoost for classification is as follows.

Given a training dataset with n instances and their corresponding class labels, where each instance is denoted by x_i and the class label by y_i, the goal is to learn a strong classifier that accurately predicts the class labels of unseen instances. Classification using the AdaBoost algorithm involves mainly three steps, i.e. initialization, iterative training, and final classification

Step 1: Initialization:

- Initialize the weights for each observation/instance: $w_i = \frac{1}{n}$, for $i = 1, 2, \ldots, n$

Step 2: Iterative training:

- For $t = 1$ to T (number of iterations)
- Train a weak classifier, $h_t(x)$, on the training data using the instance weights w_i. The weak classifier takes an instance x and outputs a binary prediction (e.g. +1 or −1)
- Calculate the weighted error of the weak classifier: $\varepsilon_t = \sum_{i=1}^{n}\left(w_i * \left(h_t\left(x_i\right) \neq y_i\right)\right)$
- Compute the classifier weight for the weak classifier: $\alpha_t = 0.5 * \ln((1 - \varepsilon_t)/\varepsilon_t)$
- Update the weights: $w_i = w_i * \exp(-\alpha_t * y_i * (h_t(x_i))$, for $i = 1, 2, \ldots, n$

- Normalize the instance weights: $w_i = w_i/Z_t$, where Z_t is a normalization factor that ensures that the sum of the weights equal to 1

Step 3: Final classification:

- Combine the predictions of weak classifiers using the corresponding classifier weights: $H(x) = sign\left(\sum_{t=1}^{T} \alpha_t * h_t(x)\right)$. The sign function assigns the class label based on the sign of the weighted sum of the weak classifiers' predictions.

13.2.3.2 Deep Learning Models

Deep learning models are another approach to predict DBPs (Zhang et al. 2020, 2021; Li et al. 2021) and include CNN and long short-term memory (LSTM), which are briefly described.

13.2.3.2.1 Convolutional Neural Network (CNN) CNN incorporates statistical concepts and operations to extract meaningful features from the input data (mostly image data; Kim 2014). In the field of bioinformatics, CNN has been successfully applied to various tasks, including sequence analysis and genomics. Here is a brief description of the CNN algorithm.

Convolution operation: Convolution is the central process of CNNs, and it includes applying a collection of filters (also known as kernels) to the input data. Mathematically, convolution can be expressed as:

$$S(i,j) = (I * K)(i,j) = \Sigma\Sigma I(m,n) * K(i-m, j-n)$$

where $S(i, j)$ represents the output value at position (i, j) in the feature map, I is the input image, K is the convolutional kernel, and (m, n) represents the coordinates within the kernel. For DBP prediction, the input data arc numeric features instead of images, and hence one-dimensional convolution is applicable.

Nonlinear activation function: Following the convolution operation, a nonlinear activation function is used to introduce nonlinearity into the model element by element. The rectified linear unit (ReLU) is the most often used activation function, which can be represented as:

$$f(x) = \max(0, x)$$

where x represents the input to the activation function.

Pooling operation: To minimize the spatial dimensions of the learned features, pooling layers are used. A typical pooling technique in CNNs is max pooling represented as:

$$P(i,j) = \max(I(i', j'))$$

where $P(i, j)$ represents the pooled value at the position (i, j) in the pooled feature map, $I(i', j')$ represents the input values within the pooling window.

Fully connected layers: Convolutional and pooling layers are followed by fully connected layers. Like a regular neural network, these layers link every neuron in one layer to each subsequent layer's neuron. The output of a completely connected layer can be represented mathematically as follows:

$$Y = W * X + b$$

where Y represents the output, W represents the weight matrix, X represents the input, and b represents the bias vector.

Loss function and optimization: CNNs are trained by minimizing a loss function that quantifies the difference between predicted and actual values using labeled datasets. Mean squared error (MSE) and cross-entropy loss are typical loss functions (A **loss function** is a function that compares the target and predicted output values). The network's parameters are iteratively updated and the loss function is minimized through the use of optimization techniques such as stochastic gradient descent or its variations.

13.2.3.2.2 *Long Short-term Memory (LSTM)*

A specific type of recurrent neural network architecture called the LSTM (Hochreiter and Schmidhuber 1997) technique is made especially to model and capture long-term dependencies in sequential input. A brief description of the different components of the LSTM algorithm follows:

Memory cell: The memory cell, which enables the network to store and retrieve data over extended periods of time, is a vital component of LSTM. Three primary gates comprise the memory cell: an input gate, an output gate, and a forget gate. The information that enters, exits, and is stored within the memory cell is managed by these gates.

Input gate: The input gate controls how much new information is stored in the memory cell. It takes the current input and the prior hidden state as inputs and uses a sigmoid activation function to calculate the input activation. Mathematically, the input gate can be expressed as:

$$i(t) = \sigma\left(W_i * \left[h(t-1), x(t)\right] + b_i\right)$$

where $i(t)$ represents the input gate activation at time t, W_i and b_i are the weight matrix and bias vector associated with the input gate, $h(t-1)$ is the previous hidden state, and $x(t)$ is the current input.

Forget gate: The forget gate regulates how much information is removed from the memory cell. It uses a sigmoid activation function to calculate the forget activation given the current input and the previous hidden state. Mathematically, the forget gate can be expressed as:

$$f(t) = \sigma\left(W_f * \left[h(t-1), x(t)\right] + b_f\right)$$

where $f(t)$ represents the forget gate activation at time t, W_f, and b_f are the weight matrix and bias vector associated with the forget gate.

Memory update: The memory cell updates its content by combining the new input information and the existing memory content. The input activation is multiplied by a candidate activation, which is computed by using a hyperbolic tangent (tanh) activation function. The resulting information is added to the memory cell content using the input gate activation. Mathematically, the memory update can be expressed as:

$$g(t) = \tanh\left(W_g * \left[h(t-1), x(t)\right] + b_g\right)$$
$$c(t) = f(t) * c(t-1) + i(t) * g(t)$$

where $g(t)$ represents the candidate activation, W_g and b_g are the weight matrix and bias vector associated with the memory update, and $c(t)$ represents the updated memory cell content at time t.

Output gate: The output gate controls how much information is output from the memory cell. It takes the current input and the prior hidden state as inputs and uses a sigmoid activation function to determine the output activation. Mathematically, the output gate can be expressed as:

$$o(t) = \sigma\left(W_o * \left[h(t-1), x(t)\right] + b_o\right)$$

where $o(t)$ represents the output gate activation at time t, W_o and b_o are the weight matrix and bias vector associated with the output gate.

Hidden state and output: The hidden state of the LSTM is computed by applying the output gate activation to the memory cell content, which is passed through a hyperbolic tangent (tanh) activation function. Mathematically, the hidden state and output can be expressed as:

$$h(t) = o(t) * \tanh\left(c(t)\right)$$
$$y(t) = \text{softmax}\left(W_h * h(t) + b_h\right)$$

where $h(t)$ represents the hidden state at time t, $y(t)$ represents the output, W_h and b_h are the weight matrix and bias vector associated with the output layer, and softmax is the activation function used for classification tasks.

13.3 Assessment of Learning Algorithms for DBP Prediction Using Sequence- and PSSM-derived Features

The performance of four shallow learning algorithms (AdaBoost, RF, SVM, and XGBoost) and two deep learning algorithms (CNN, LSTM) was evaluated using the dataset used in one of our recent studies (Pradhan et al. 2023a). This dataset comprises 849 DBP sequences and 849 non-DBP sequences. The protein sequences were transformed into numeric feature vectors using both sequence-derived features (AAC, DPC, TPC, PAAC) and evolutionary features (PSSM400, AAC_DPC_PSSM, EDP_PSSM, *kbigram*_PSSM, and PSE_PSSM).

These feature vectors were subsequently employed as input in machine-learning algorithms for the classification of DBP and non-DBP sequences. Using a fivefold cross-validation approach, the MLAs' performance was assessed. The accuracy of the classification models was assessed using metrics including accuracy, area under the receiver operating characteristic curve (AU-ROC), and area under the precision-recall curve (AU-PRC) (Table 13.2). The prediction accuracy of the machine learning models was observed to be higher while PSSM-derived features were employed as compared to the other sequence-derived features. Among the shallow learning algorithms, XGBoost achieved higher prediction accuracy. The accuracy of CNN was achieved higher as compared to the LSTM deep

Table 13.2 Performance metrics of machine learning models with different feature sets.

Metrics	Feature	AdaBoost	RF	SVM	XGBoost	CNN	LSTM
Accuracy	AAC	75.58	78.87	49.97	78.28	76.46	76.04
	DPC	73.16	76.87	49.97	77.34	76.10	76.52
	TPC	71.75	76.93	49.97	74.57	75.28	74.40
	PAAC	73.50	77.56	49.97	76.73	75.95	75.65
	PSSM400	76.27	74.14	66.55	76.97	84.45	81.98
	AAC_DPC_PSSM	82.74	83.51	78.51	83.33	85.69	83.16
	EDP_PSSM	81.68	83.86	50.06	84.10	85.75	83.39
	*kbigram*_PSSM	64.69	69.70	68.11	67.93	73.58	69.99
	PSE_PSSM	81.55	83.90	82.61	83.79	80.97	63.23
AU-ROC	AAC	82.71	86.06	49.06	85.50	85.03	82.69
	DPC	81.39	83.73	23.20	84.22	85.05	86.12
	TPC	78.53	84.21	24.72	82.58	82.42	81.86
	PAAC	80.88	84.67	32.33	84.10	84.16	83.56
	PSSM400	82.47	81.95	74.34	84.07	91.10	88.23
	AAC_DPC_PSSM	89.92	90.75	85.10	91.84	93.30	90.34
	EDP_PSSM	89.15	91.03	15.32	92.03	93.46	91.06
	*kbigram*_PSSM	69.80	73.51	73.17	72.15	87.07	73.65
	PSE_PSSM	89.53	92.04	89.94	92.73	90.04	61.68
AU-PRC	AAC	79.78	82.82	54.99	83.58	83.47	80.44
	DPC	80.34	81.47	41.77	83.53	85.60	84.46
	TPC	78.29	82.12	42.13	82.84	82.52	80.98
	PAAC	79.47	82.14	46.30	83.32	83.86	81.96
	PSSM400	79.98	80.02	71.31	83.58	90.97	87.90
	AAC_DPC_PSSM	89.52	88.72	82.73	91.68	92.92	89.40
	EDP_PSSM	89.36	88.94	33.13	91.54	93.36	90.71
	*kbigram*_PSSM	70.47	72.04	72.80	72.68	87.36	76.36
	PSE_PSSM	89.48	90.98	88.66	92.61	89.94	58.38

learning models. Among shallow and deep learning algorithms, higher prediction accuracy was observed with the CNN model (Table 13.2). Among the machine learning models considered, the SVM performed the worst.

13.4 Evaluation of Existing Tools for DBP Prediction in Plants

Several methods for the prediction of DBPs, described above are generally trained with human or mouse datasets which may not be appropriate for the prediction of plant-specific DBPs. Some recently developed computational models for the prediction of DBPs are listed in Table 13.3. Of these models, PlDBPred is the plant-specific model. In the existing tools, both shallow and deep learning algorithms have been employed for prediction. Besides, both sequence-derived and evolutionary features were utilized for encoding sequence data into numeric features. Performance of these DBP prediction tools such as

Table 13.3 List of some of commonly used the DBP prediction tools.

Name of the tool	MLA used	Features employed	Availability	References
DNAbinder	SVM	AAC, DPC	http://crdd.osdd.net/raghava/dnabinder/index.html	Kumar et al. (2007)
DPP-PseAAC	SVM	PseAAC	http://77.68.43.135:8080/DPP-PseAAC/	Rahman et al. (2018)
StackDPPred	SVM	Evolutionary features	http://bmll.cs.uno.edu/add	Mishra et al. (2019)
iDRBP_MMC	Multi-label with CNN	Motif- based CNN	http://bliulab.net/iDRBP_MMC	Zhang et al. (2020)
iDRBP-PPCT	RF	Evolutionary information	http://bliulab.net/IDRBP-PPCT	Wang et al. (2022)
PDBP-fusion	LSTM	One-hot encoding; embedding encoding	http://119.45.144.26:8080/PDBP-Fusion/	Li et al. (2021)
DeepDRBP-2L	LSTM and CNN	Evolutionary information	http://bliulab.net/DeepDRBP-2L	Zhang et al. (2021)
PlDBPred	SVM	Evolutionary features	https://iasri-sg.icar.gov.in/pldbpred/	Pradhan et al. (2023a)

	Sensitivity	Specificity	Accuracy	Precision	F.score	MCC
DNAbinder	50.8	51.5	51	67.4	57.9	
DPP_PseAAC	56.1	54.9	55.4	49.8	52.8	10.9
iDRBP_MMC	72.7	67.7	69.9	63.8	67.9	40.1
iDRBP_PPCT	63.9	84.9	69.9	91.4	75.2	44.1
PDBP-fusion	86.4	70	75.8	61.2	71.7	54
DeepDRBP-2L	61.4	70.6	64.7	79	69.1	30.7
StackDPPred	65.8	61.1	63	54.2	59.4	26.4
PlDBPred	82.6	93.4	88	92.6	87.3	76.4

Figure 13.2 Performance of different DBP prediction tools on plant dataset.

PlDBPred, DNAbinder, DPP_PseAAC, iDRBP_MMC, IDRBP-PPCT, PDBP-fusion, DeepDRBP-2L, and StackDPPred was also evaluated using plant datasets. The test datasets used in PlDBPred were used for measuring the performance of the models. The performances were evaluated using different performance metrics such as sensitivity, specificity, accuracy, precision, F-score, and MCC values. PlDBPred outperformed the compared tools in all of the performance metrics (Figure 13.2). Some of these tools exhibited major fluctuation in the performance metrics, indicating the unreliable prediction for plant-specific DBPs.

13.5 Conclusion and Future Perspectives

Prediction of DBPs in plants is a critical area of research with broad implications for understanding plant biology, crop improvement, and addressing global challenges such as food security and environmental sustainability. In the past, this area has largely been ignored due to the nonavailability of suitable plant datasets, barring a few tools recently developed. The future of research for the prediction of DBPs in plants should involve the integration of -omics data, including genomics, transcriptomics, proteomics, and epigenomics. Combining these diverse datasets can provide a more comprehensive view of the regulatory networks in plants. Continued development of MLAs and deep learning techniques will lead to more accurate and robust predictions. The use of a variety of MLAs, pretrained models, and novel architectures tailored to plant-specific data will become common. Predictions should not only focus on the identification of DBPs but also on understanding their functional roles in plant growth, development, and stress responses. Integration of functional annotations and pathway analysis will be crucial. DBPs predictions can play a pivotal role in crop improvement by identifying key regulators of traits like yield, stress tolerance, and disease resistance. Future research may involve designing strategies to modify these proteins for enhanced agricultural performance. The prediction of DBPs in plants is a rapidly evolving field that holds great promise for advancing plant science, agriculture, and environmental sustainability. Future research will continue to refine prediction methods and leverage emerging technologies to unlock the full potential of plant DBP predictions for the benefit of society.

References

Ali, F., Kumar, H., Patil, S. et al. (2022). Target-DBPPred: an intelligent model for prediction of DNA-binding proteins using discrete wavelet transform based compression and light eXtreme gradient boosting. *Computers in Biology and Medicine* 145: 105533. https://doi.org/10.1016/j.compbiomed.2022.105533.

Babu, M.M., van der Lee, R., de Groot, N.S., and Gsponer, J. (2011). Intrinsically disordered proteins: regulation and disease. *Current Opinion in Structural Biology* 21: 432–440. https://doi.org/10.1016/j.sbi.2011.03.011.

Barukab, O., Khan, Y.D., Khan, S.A., and Chou, K.-C. (2022). DNAPred_Prot: identification of DNA-binding proteins using composition- and position-based features. *Applied Bionics and Biomechanics* 2022: 5483115. https://doi.org/10.1155/2022/5483115.

Bhasin, M. and Raghava, G.P.S. (2004). Classification of nuclear receptors based on amino acid composition and dipeptide composition. *The Journal of Biological Chemistry* 279: 23262–23266. https://doi.org/10.1074/jbc.M401932200.

Breiman, L. (2001). Random forests. *Machine Learning* 45: 5–32. https://doi.org/10.1023/A:1010933404324.

Cai, Y.-H. and Huang, H. (2012). Advances in the study of protein-DNA interaction. *Amino Acids* 43: 1141–1146. https://doi.org/10.1007/s00726-012-1377-9.

Carter, P., Andersen, C.A.F., and Rost, B. (2003). DSSPcont: continuous secondary structure assignments for proteins. *Nucleic Acids Research* 31: 3293–3295. https://doi.org/10.1093/nar/gkg626.

Chen, H. and Chang, G.D. (2001). Simultaneous immunoblotting analysis with activity gel electrophoresis in a single polyacrylamide gel. *Electrophoresis* 22: 1894–1899. https://doi.org/10.1002/1522-2683(200106)22:10<1894::AID-ELPS1894>3.0.CO;2-5.

Chen, T. and Guestrin, C. (2016). XGBoost: a scalable tree boosting system. *Proceedings of the 22nd ACM SIGKDD International Conference on Knowledge Discovery and Data Mining KDD'16: ACM*, San Francisco, CA, USA, 785–794.

Chou, K.C. (2001). Prediction of protein cellular attributes using pseudo-amino acid composition. *Proteins* 43: 246–255. https://doi.org/10.1002/prot.1035.

Dosztányi, Z., Csizmók, V., Tompa, P., and Simon, I. (2005). The pairwise energy content estimated from amino acid composition discriminates between folded and intrinsically unstructured proteins. *Journal of Molecular Biology* 347: 827–839. https://doi.org/10.1016/j.jmb.2005.01.071.

Ferraz, R.A.C., Lopes, A.L.G., da Silva, J.A.F. et al. (2021). DNA–protein interaction studies: a historical and comparative analysis. *Plant Methods* 17: 82. https://doi.org/10.1186/s13007-021-00780-z.

Freund, Y. and Schapire, R.E. (1997). A decision-theoretic generalization of on-line learning and an application to boosting. *Journal of Computer and System Sciences* 55: 119–139. https://doi.org/10.1006/jcss.1997.1504.

Fried, M. and Crothers, D.M. (1981). Equilibria and kinetics of lac repressor-operator interactions by polyacrylamide gel electrophoresis. *Nucleic Acids Research* 9: 6505–6525. https://doi.org/10.1093/nar/9.23.6505.

Garner, M.M. and Revzin, A. (1981). A gel electrophoresis method for quantifying the binding of proteins to specific DNA regions: application to components of the *Escherichia coli* lactose operon regulatory system. *Nucleic Acids Research* 9: 3047–3060. https://doi.org/10.1093/nar/9.13.3047.

Guille, M.J. and Kneale, G.G. (1997). Methods for the analysis of DNA-protein interactions. *Molecular Biotechnology* 8: 35–52. https://doi.org/10.1007/BF02762338.

Harris, T.D., Buzby, P.R., Babcock, H. et al. (2008). Single-molecule DNA sequencing of a viral genome. *Science (New York, N.Y.)* 320: 106–109. https://doi.org/10.1126/science.1150427.

Hellman, L.M. and Fried, M.G. (2007). Electrophoretic mobility shift assay (EMSA) for detecting protein-nucleic acid interactions. *Nature Protocols* 2: 1849–1861. https://doi.org/10.1038/nprot.2007.249.

Hochreiter, S. and Schmidhuber, J. (1997). Long short-term memory. *Neural Computation* 9: 1735–1780. https://doi.org/10.1162/neco.1997.9.8.1735.

Huang, Y., Niu, B., Gao, Y. et al. (2010). CD-HIT Suite: a web server for clustering and comparing biological sequences. *Bioinformatics* 26: 680–682. https://doi.org/10.1093/bioinformatics/btq003.

Jia, Y., Huang, S., and Zhang, T. (2021). KK-DBP: a multi-feature fusion method for DNA-binding protein identification based on random forest. *Frontiers in Genetics* 12: 811158. https://doi.org/10.3389/fgene.2021.811158.

Kawashima, S., Pokarowski, P., Pokarowska, M. et al. (2008). AAindex: amino acid index database, progress report 2008. *Nucleic Acids Research* 36: D202–D205.

Kim, Y. (2014). Convolutional neural networks for sentence classification. *Proceedings of the 2014 Conference on Empirical Methods in Natural Language Processing (EMNLP)*, EMNLP 2014, 1746–1751. Doha, Qatar: Association for Computational Linguistics.

Kumar, M., Gromiha, M.M., and Raghava, G.P. (2007). Identification of DNA-binding proteins using support vector machines and evolutionary profiles. *BMC Bioinformatics* 8: 463. https://doi.org/10.1186/1471-2105-8-463.

Li, G., Du, X., Li, X. et al. (2021). Prediction of DNA binding proteins using local features and long-term dependencies with primary sequences based on deep learning. *PeerJ* 9: e11262. https://doi.org/10.7717/peerj.11262.

Liang, Z.-Y., Lai, H.-Y., Yang, H. et al. (2017). Pro54DB: a database for experimentally verified sigma-54 promoters. *Bioinformatics (Oxford, England)* 33: 467–469. https://doi.org/10.1093/bioinformatics/btw630.

Lin, C., Zou, Y., Qin, J. et al. (2013). Hierarchical classification of protein folds using a novel ensemble classifier. *PLoS One* 8: e56499. https://doi.org/10.1371/journal.pone.0056499.

Liu, T., Zheng, X., and Wang, J. (2010). Prediction of protein structural class for low-similarity sequences using support vector machine and PSI-BLAST profile. *Biochimie* 92: 1330–1334. https://doi.org/10.1016/j.biochi.2010.06.013.

Liu, B., Wang, X., Chen, Q. et al. (2012). Using amino acid physicochemical distance transformation for fast protein remote homology detection. *PLoS One* 7: e46633. https://doi.org/10.1371/journal.pone.0046633.

Liu, B., Xu, J., Lan, X. et al. (2014). iDNA-Prot|dis: identifying DNA-binding proteins by incorporating amino acid distance-pairs and reduced alphabet profile into the general pseudo-amino acid composition. *PLoS One* 9: e106691. https://doi.org/10.1371/journal.pone.0106691.

Luscombe, N.M., Austin, S.E., Berman, H.M., and Thornton, J.M. (2000). An overview of the structures of protein-DNA complexes. *Genome Biology* 1: reviews001.1. https://doi.org/10.1186/gb-2000-1-1-reviews001.

Ma, X., Guo, J., and Sun, X. (2016). DNABP: identification of DNA-binding proteins based on feature selection using a random forest and predicting binding residues. *PLoS One* 11: e0167345. https://doi.org/10.1371/journal.pone.0167345.

McGuffin, L.J., Bryson, K., and Jones, D.T. (2000). The PSIPRED protein structure prediction server. *Bioinformatics* 16: 404–405. https://doi.org/10.1093/bioinformatics/16.4.404.

Mishra, A., Pokhrel, P., and Hoque, M.T. (2019). StackDPPred: a stacking based prediction of DNA-binding protein from sequence. *Bioinformatics* 35: 433–441. https://doi.org/10.1093/bioinformatics/bty653.

Motion, G.B., Howden, A.J.M., Huitema, E., and Jones, S. (2015). DNA-binding protein prediction using plant specific support vector machines: validation and application of a new genome annotation tool. *Nucleic Acids Research* 43: e158–e158. https://doi.org/10.1093/nar/gkv805.

Nanni, L., Lumini, A., and Brahnam, S. (2014). An empirical study of different approaches for protein classification. *The Scientific World Journal* 2014: e236717. https://doi.org/10.1155/2014/236717.

Ofran, Y., Mysore, V., and Rost, B. (2007). Prediction of DNA-binding residues from sequence. *Bioinformatics* 23: i347–i353. https://doi.org/10.1093/bioinformatics/btm174.

Pradhan, U.K., Meher, P.K., Naha, S. et al. (2023a). PlDBPred: a novel computational model for discovery of DNA binding proteins in plants. *Briefings in Bioinformatics* 24: bbac483. https://doi.org/10.1093/bib/bbac483.

Pradhan, U.K., Meher, P.K., Naha, S. et al. (2023b). DBPMod: a supervised learning model for computational recognition of DNA-binding proteins in model organisms. *Briefings in Functional Genomics* elad039. https://doi.org/10.1093/bfgp/elad039.

Rahman, M.S., Shatabda, S., Saha, S. et al. (2018). DPP-PseAAC: a DNA-binding protein prediction model using Chou's general PseAAC. *Journal of Theoretical Biology* 452: 22–34. https://doi.org/10.1016/j.jtbi.2018.05.006.

Riechmann, J.L., Heard, J., Martin, G. et al. (2000). Arabidopsis transcription factors: genome-wide comparative analysis among eukaryotes. *Science* 290: 2105–2110. https://doi.org/10.1126/science.290.5499.2105.

Riggs, A.D., Bourgeois, S., and Cohn, M. (1970). The *lac* represser-operator interaction: III. Kinetic studies. *Journal of Molecular Biology* 53: 401–417. https://doi.org/10.1016/0022-2836(70)90074-4.

Rose, P.W., Prlić, A., Altunkaya, A. et al. (2017). The RCSB protein data bank: integrative view of protein, gene and 3D structural information. *Nucleic Acids Research* 45: D271–D281. https://doi.org/10.1093/nar/gkw1000.

Shadab, S., Alam Khan, M.T., Neezi, N.A. et al. (2020). DeepDBP: deep neural networks for identification of DNA-binding proteins. *Informatics in Medicine Unlocked* 19: 100318. https://doi.org/10.1016/j.imu.2020.100318.

Shen, H.-B. and Chou, K.-C. (2008). PseAAC: a flexible web server for generating various kinds of protein pseudo amino acid composition. *Analytical Biochemistry* 373: 386–388. https://doi.org/10.1016/j.ab.2007.10.012.

Siggers, T. and Gordân, R. (2014). Protein-DNA binding: complexities and multi-protein codes. *Nucleic Acids Research* 42: 2099–2111. https://doi.org/10.1093/nar/gkt1112.

Stockley, P.G. (2009). Filter-binding assays. *Methods in Molecular Biology (Clifton, N.J.)* 543: 1–14. https://doi.org/10.1007/978-1-60327-015-1_1.

Su, W., Liu, M.-L., Yang, Y.-H. et al. (2021). PPD: a manually curated database for experimentally verified prokaryotic promoters. *Journal of Molecular Biology* 433: 166860. https://doi.org/10.1016/j.jmb.2021.166860.

The UniProt Consortium (2021). UniProt: the universal protein knowledgebase in 2021. *Nucleic Acids Research* 49: D480–D489. https://doi.org/10.1093/nar/gkaa1100.

Ueda, M., Zhang, Z., and Laux, T. (2011). Transcriptional activation of *Arabidopsis* axis patterning genes *WOX8/9* links zygote polarity to embryo development. *Developmental Cell* 20: 264–270. https://doi.org/10.1016/j.devcel.2011.01.009.

Vapnik, V.N. (1999). An overview of statistical learning theory. *IEEE Transactions on Neural Networks* 10: 988–999. https://doi.org/10.1109/72.788640.

Vuzman, D., Hoffman, Y., and Levy, Y. (2012). Modulating protein-DNA interactions by post-translational modifications at disordered regions. *Pacific Symposium on Biocomputing*, 188–199.

Walter, M.C., Rattei, T., Arnold, R. et al. (2009). PEDANT covers all complete RefSeq genomes. *Nucleic Acids Research* 37: D408–D411. https://doi.org/10.1093/nar/gkn749.

Wang, N., Zhang, J., and Liu, B. (2022). IDRBP-PPCT: identifying nucleic acid-binding proteins based on position-specific score matrix and position-specific frequency matrix cross transformation. *IEEE/ACM Transactions on Computational Biology and Bioinformatics* 19: 2284–2293. https://doi.org/10.1109/TCBB.2021.3069263.

Welsh, J. and Cantor, C.R. (1984). Protein—DNA cross-linking. *Trends in Biochemical Sciences* 9: 505–508. https://doi.org/10.1016/0968-0004(84)90271-8.

Wheeler, D.A., Srinivasan, M., Egholm, M. et al. (2008). The complete genome of an individual by massively parallel DNA sequencing. *Nature* 452: 872–876. https://doi.org/10.1038/nature06884.

Wu, C.H., Apweiler, R., Bairoch, A. et al. (2006). The Universal Protein Resource (UniProt): an expanding universe of protein information. *Nucleic Acids Research* 34: D187–D191. https://doi.org/10.1093/nar/gkj161.

Yarus, M. and Berg, P. (1967). Recognition of tRNA by aminoacyl tRNA synthetases. *Journal of Molecular Biology* 28: 479–490. https://doi.org/10.1016/s0022-2836(67)80098-6.

Yu, X., Cao, J., Cai, Y. et al. (2006). Predicting rRNA-, RNA-, and DNA-binding proteins from primary structure with support vector machines. *Journal of Theoretical Biology* 240: 175–184. https://doi.org/10.1016/j.jtbi.2005.09.018.

Zaitzeff, A., Leiby, N., Motta, F.C. et al. (2021). Improved datasets and evaluation methods for the automatic prediction of DNA-binding proteins. *Bioinformatics* 38: 44–51. https://doi.org/10.1093/bioinformatics/btab603.

Zhang, Y., Xu, J., Zheng, W. et al. (2014). newDNA-Prot: prediction of DNA-binding proteins by employing support vector machine and a comprehensive sequence representation. *Computational Biology and Chemistry* 52: 51–59. https://doi.org/10.1016/j.compbiolchem.2014.09.002.

Zhang, J., Chen, Q., and Liu, B. (2020). iDRBP_MMC: identifying DNA-binding proteins and RNA-binding proteins based on multi-label learning model and motif-based convolutional neural network. *Journal of Molecular Biology* 432: 5860–5875. https://doi.org/10.1016/j.jmb.2020.09.008.

Zhang, J., Chen, Q., and Liu, B. (2021). DeepDRBP-2L: a new genome annotation predictor for identifying DNA-binding proteins and RNA-binding proteins using convolutional neural network and long short-term memory. *IEEE/ACM Transactions on Computational Biology and Bioinformatics* 18: 1451–1463. https://doi.org/10.1109/TCBB.2019.2952338.

Zhao, Z., Yang, W., Zhai, Y. et al. (2022). Identify DNA-binding proteins through the extreme gradient boosting algorithm. *Frontiers in Genetics* 12: 821996.

Zheng, Z. and Wang, Y. (2011). DNA binding proteins: outline of functional classification. *Biomol Concepts* 2: 293–303. https://doi.org/10.1515/bmc.2011.023.

Zimmer, C. and Wähnert, U. (1986). Nonintercalating DNA-binding ligands: specificity of the interaction and their use as tools in biophysical, biochemical and biological investigations of the genetic material. *Progress in Biophysics and Molecular Biology* 47: 31–112. https://doi.org/10.1016/0079-6107(86)90005-2.

14

Bioinformatics for Gene Identification and Crop Improvement in Wheat

Pushpendra Kumar Gupta[1,2,3], *Jyoti Chaudhary*[1], *and Tinku Gautam*[1,4]

[1] *Department of Genetics and Plant Breeding, CCS University, Meerut, Uttar Pradesh, India*
[2] *Murdoch's Centre for Crop and Food Innovation, Murdoch University, Perth, Australia*
[3] *Borlaug Institute for South Asia (BISA), Ludhiana, Punjab, India*
[4] *Agriculture and Agri-Food Canada, Swift Current Research and Development Centre, Swift Current, Canada*

14.1 Introduction

Knowledge of the genetics of all important traits of agronomic value in a crop is a prerequisite for developing a result-oriented crop improvement program. In recent years, interval mapping and genome-wide association studies generated massive data on QTLs and associated markers for important traits in all major crops, such that markers can be used for marker-assisted selection (MAS) (Singh et al. 2021, 2022). This has already been successfully achieved through a large number of studies leading to crop improvement programs in all major crops including wheat (Song et al. 2023). However, specific genes associated with these QTLs are not known in most cases, so that robust functional markers cannot be developed for breeding programs. In this connection, *in silico* identification and characterization of novel useful genes is a relatively recent area of research, which has started generating useful information that has yet to be utilized for breeding programs.

The *in silico* identification of genes has been facilitated by the availability of whole-genome sequences (including transcriptome and ESTs) for almost all major crops including wheat, where a large number of useful genes have been identified and characterized. If individual genes are known, it facilitates further planning of the breeding programs, more so when genes are known for a variety of traits, so that functional markers may be developed and utilized with a high level of confidence. In this chapter, we provide the current status of *in silico* studies conducted for the identification and characterization of genes of agronomic value and encoded proteins in wheat.

14.2 Databases and Tools for Individual Genes and Proteins

Wheat carries a large genome (17Gb) with >80% repetitive DNA, which makes the identification of genes using bioinformatic tools a challenge. The exercise involving the

Bioinformatics for Plant Research and Crop Breeding, First Edition. Edited by Jen-Tsung Chen.

Table 14.1 Databases for identification of genes in wheat.

Database/Tool, Description, Link, Reference (if available)
1. NCBI. A resource for browsing genomes, searching sequences, and annotating genes for almost all species. https://www.ncbi.nlm.nih.gov/
2. EnsemblPlants. Genome annotations, gene models, and functional information for plants. https://plants.ensembl.org/Triticum_aestivum/Info/Index
3. Phytozome. Comparative genomics of plant species, including wheat; gene families; functional annotations. https://phytozome-next.jgi.doe.gov/info/Taestivumcv_ChineseSpring_v2_1
4. IWGSC RefSeq. High-quality reference sequences for wheat. https://www.wheatgenome.org/Projects/Reference-Genome-Project/RefSeq-v2.1
5. PlantTFDB. Database for transcription factors from 165 species. http://planttfdb.gao-lab.org/aboutus.php. Jin et al. (2016)
6. expVIP. Collection of gene expression profiles in wheat, encompassing various conditions and tissues. http://www.wheat-expression.com/. Borrill et al. (2016)
7. KEGG. A collection of databases dealing with genomes, biological pathways, diseases, drugs, and chemical substances, aiding in the understanding of the functions of genes & gene families. https://www.genome.jp/kegg/kegg2.html. Kanehisa and Goto (2000)
8. PlantGF. This is a database for the search and statistics of plant gene families. http://biodb.sdau.edu.cn/PGF/index.html; Li et al. (2022a)
9. PlantTribes2. Linux-based gene family analysis tool. https://github.com/dePamphilis/PlantTribes; Wafula et al. (2023)
10. PlantCARE. Database of plant cis-acting regulatory elements. https://bioinformatics.psb.ugent.be/webtools/plantcare/html/; Lescot et al. (2002)
11. miRBase. Database of published miRNA sequences and annotation. https://www.mirbase.org/; Kozomara et al. (2019)
12. The PlantGDB EST Database, https://www.plantgdb.org/. Dong et al. (2004)
13. NCBI Plant Transcript Assemblies Database-NCBI, (http://www.ncbi.nlm.nih.gov). Acland et al. (2013)
14. DFCI Gene Index. (http://compbio.dfci.harvard.edu/tgi/tgipage.html). Antonescu et al. (2010)
15. ConGenIE EST Database. (http://v22.popgenie.org/microdisection/). Sundell et al. (2015)
16. TIGR Plant Transcript Assemblies Database-(http://plantta.tigr.org). Childs et al. (2007)

identification of genes makes use of nucleic acid and protein databases and tools, some of them designed for wheat genomics. These resources help in gene discovery, annotation, and functional analysis of genes for important traits. In this Section (14.2), we summarize the key databases and bioinformatics tools utilized for the identification of genes in wheat. On the basis of their functionality, we can divide them into three major categories, namely genomics, transcriptomics (including ESTs), and proteomics.

The major resources for wheat genomics and transcriptomics research include a number of databases, which are listed in Table 14.1. Similarly, protein databases are listed in Table 14.2. The bioinformatics tools available for the study of genes and their proteins are similarly listed in Tables 14.3 and 14.4.

Table 14.2 Databases for identification/characterization of proteins in wheat.

1. InterProScan. InterPro provides functional analysis of proteins by classifying them into families and predicting domains and important sites. https://www.ebi.ac.uk/interpro/ (Quevillon et al. 2005)
2. Pfam. Domain organization of a protein sequence, find the domains on a PDB structure. http://pfam-legacy.xfam.org/ (Punta et al. 2012)
3. BatchCD. Identifies protein domains of more than one protein. https://www.ncbi.nlm.nih.gov/Structure/bwrpsb/bwrpsb.cgi (Lu et al. 2020)
4. CDD. It is a protein annotation resource that consists of a collection of well-annotated multiple sequence alignment models for ancient domains and full-length proteins. https://www.ncbi.nlm.nih.gov/Structure/cdd/cdd.shtml (Lu et al. 2020)
5. TMHMM. Used for predicting Transmembrane domains in proteins that have been recently found. https://services.healthtech.dtu.dk/services/TMHMM-2.0/ (Krogh et al. 2001)
6. SignalP. Predicts the presence and location of signal peptide cleavage sites in amino acid sequences from different organisms. https://services.healthtech.dtu.dk/services/SignalP-4.1/ (Teufel et al. 2022)
7. SWISS-MODEL. Server for automated comparative modeling of three-dimensional (3D) protein structures. https://swissmodel.expasy.org/ (Waterhouse et al. 2018)
8. MODELLER. Prediction of protein three-dimensional structures. https://salilab.org/modeller (Webb and Sali 2016)
9. I-TASSER. Predicting three-dimensional structure model of protein. https://zhanggroup.org/I-TASSER// (Zhou et al. 2022)
10. AMBER. Amber is a suite of biomolecular simulation programs. https://ambermd.org/ (Salomon-Ferrer et al. 2013)
11. STRING. Protein–Protein Interaction Networks Functional Enrichment Analysis. https://string-db.org/ (Szklarczyk et al. 2019)
12. Cytoscape. Platform for visualizing complex networks and integrating these with any type of attribute data. https://cytoscape.org/ (Shannon et al. 2003)

Table 14.3 Tools for identification/characterization of genes.

1. Plaza. Combines plant sequence data and comparative genomics methods, offering a powerful tool for conducting evolutionary analyses and data mining. https://bioinformatics.psb.ugent.be/plaza/versions/plaza_v5_monocots/; Van Bel et al. (2022)
2. BLAST. Used for sequence similarity searches against various databases to identify homologous genes. https://blast.ncbi.nlm.nih.gov/Blast.cgi
3. HMMER. Makes use of Hidden Markov models (HMMs) to identify protein families and conducts a comprehensive search of sequence databases to identify homologous proteins. http://hmmer.org/; Finn et al. (2011)
4. OrthoFinder. Tool for identification of orthologous gene families across multiple species for evolutionary relationships. https://github.com/davidemms/OrthoFinder Emms and Kelly (2019)
5. Gene Ontology (GO). Offers functional annotations to genes, facilitating the categorization and interpretation of gene function. http://geneontology.org/ Gene Ontology Consortium (2021)
6. GSDS. A tool for the visual representation of gene features, (arrangement and location of exons, introns). http://gsds.gao-lab.org/; Hu et al. (2015)
7. psRNATarget. miRNA/sRNA target prediction. https://www.zhaolab.org/psRNATarget/home; Dai et al. (2018)
8. TBtools. Incorporates over 130 functions with user-friendly interface. https://github.com/CJ-Chen/TBtools/releases; Chen et al. (2020)

Table 14.4 Tools for study of proteins encoded by genes (identified *in silico*).

Database/Tool, Description, Link, Reference (if available)
1. SMART. A simple modular architecture research tool allows the identification and annotation of genetically mobile domains and the analysis of domain architectures (http://SMART.embl-heidelberg.de) (Schultz et al. 2000)
2. CDART:Domain Architectures. Finds protein similarities across significant evolutionary distances using sensitive domain profiles rather than direct sequence similarity. Proteins similar to the query are grouped and scored by architecture. https://www.ncbi.nlm.nih.gov/Structure/lexington/lexington.cgi (Geer et al. 2002)
3. CDTree. CDTree works with Cn3D as its alignment viewer/editor, it is used in the CDD curation process and is a both classification and research tool for functional annotation and the study of protein and protein domain families. https://www.ncbi.nlm.nih.gov/Structure/cdtree/cdtree.shtml (Lu et al. 2020)
4. MEME suite. Tools for motif discovery and searching. https://meme-suite.org/meme/tools/meme (Bailey et al. 2006)
5. ClustalW/Clustal Omega. It is a new multiple sequence alignment program that uses seeded guide trees and HMM profile-profile techniques to generate alignments between three or more sequences. https://www.ebi.ac.uk/Tools/msa/clustalo/ (Thompson et al. 2003)
6. MUSCEL. For multiple sequence alignment. https://www.ebi.ac.uk/Tools/msa/muscle/(Edgar 2004)
7. MAFFT. For multiple sequence alignment. https://www.ebi.ac.uk/Tools/msa/mafft/ (Katoh et al. 2002)
8. T-Coffee. For multiple sequence alignment. https://www.ebi.ac.uk/Tools/msa/tcoffee/ (Di et al. 2011)
9. ProbCons. For multiple sequence alignment. http://probcons.stanford.edu/ (Do et al. 2005)
10. ProtParam. Tool which allows the computation of various physical and chemical parameters for a given protein stored in Swiss-Prot or TrEMBL or for a user-entered protein sequence. https://web.expasy.org/protparam/ (Garg et al. 2016)
11. CELLO. Subcellular localization predictor. http://cello.life.nctu.edu.tw/
12. CELLO2GO. Screening various properties of a targeted protein and its subcellular localization. http://cello.life.nctu.edu.tw/cello2go/ (Yu et al. 2014)
13. APSLAP. Prediction of apoptosis protein subcellular Localization. http://apslap.bicpu.edu.in/ (Saravanan and Lakshmi 2013)
14. WoLF PSORT. It is an advanced bioinformatics tool that can be used to predict the subcellular localization of your protein of interest in different organisms. https://www.genscript.com/wolf-psort.html (Horton et al. 2007)
15. PSI-BLAST. Uses position-specific scoring matrices (PSSMs) to score matches between query and database sequences. https://www.ebi.ac.uk/Tools/sss/psiblast/ (Margelevičius and Venclovas 2005)
16. PROCHECK. Checks the stereochemical quality of a protein structure, producing a number of PostScript plots analyzing its overall and residue-by-residue geometry. https://www.ebi.ac.uk/thornton-srv/software/PROCHECK/ (Laskowski et al. 1993)
17. VERIFY3D. Determines the compatibility of an atomic model (3D) with its own amino acid sequence (1D) by assigning a structural class based on its location and environment (alpha, beta, loop, polar, nonpolar, etc.). https://www.doe-mbi.ucla.edu/verify3d/ (Eisenberg et al. 1997)

Table 14.4 (Continued)

18. ERRAT. A program for verifying protein structures determined by crystallography. https://saves.mbi.ucla.edu/ (Colovos and Yeates 1993)
19. GROMACS. Molecular dynamics simulations. https://www.gromacs.org/ (Abraham et al. 2015)
20. Hex. Molecular docking analysis http://hex.loria.fr/ (Ritchie and Kemp 2000)
21. BindUP. Molecular docking analysis. https://bindup.technion.ac.il/ (Paz et al. 2016)
22. Molegro Molecular Viewer: Molecular docking analysis. http://molexus.io/molegro-molecular-viewer/ (Thomsen and Christensen 2006)
23. **FATCAT**. Superimposition of proteins. https://fatcat.godziklab.org/ (Ye and Godzik 2004)
24. NAMD. Molecular dynamics simulations. https://www.ks.uiuc.edu/Research/namd/ (Phillips et al. 2005)
25. VMD. For visualization of molecular dynamic simulation. https://www.ks.uiuc.edu/Research/vmd/ (Humphrey et al. 1996)
26. UCSF Chimera. It is a program for the interactive visualization and analysis of molecular structures and related data. https://www.cgl.ucsf.edu/chimera/
27. MEGA. The phylogenetic analysis can be performed using Molecular Evolutionary Genetics Analysis. https://www.megasoftware.net/ (Tamura et al. 2021)
28. PhyML. Phylogenetic estimation software using Maximum Likelihood. http://www.atgc-montpellier.fr/phyml/ (Guindon et al. 2005)
29. iTOL. Manage, annotate, and visualize phylogenetic tree. https://itol.embl.de/ (Letunic and Bork 2021)
30. Autodock vina. Open-source program for doing molecular docking. https://vina.scripps.edu/ (Eberhardt et al. 2021)
31. Pymol. Molecular visualization system. https://pymol.org/2/

14.3 Identification/Characterization of Genes/Gene Families at the DNA Level

14.3.1 Identification of Genes

In silico identification of specific genes can be achieved either by similarity-based searches or ab initio prediction. These two methods will be briefly described in Sections 14.3.1.1 and 14.3.1.2, although our major emphasis in this chapter will be on the identification of genes through similarity search.

14.3.1.1 Identification of Genes by Similarity-based Search and Sequence Retrieval

Step-wise identification and characterization of genes and gene families involves several steps, which are shown in Figure 14.1 and include the following: Sequences of the target genes are generally retrieved using available gene/cDNA/protein databases (e.g. EnsemblPlants) or using published literature and used as query against the wheat genome, transcriptome (including ESTs), or proteome in NCBI or EnsemblPlants to identify potential matches. BLAST/tBLAST/tBLASTx may be used for this purpose. Alternatively, the sequence of the protein encoded by the gene of interest may be retrieved from the Pfam

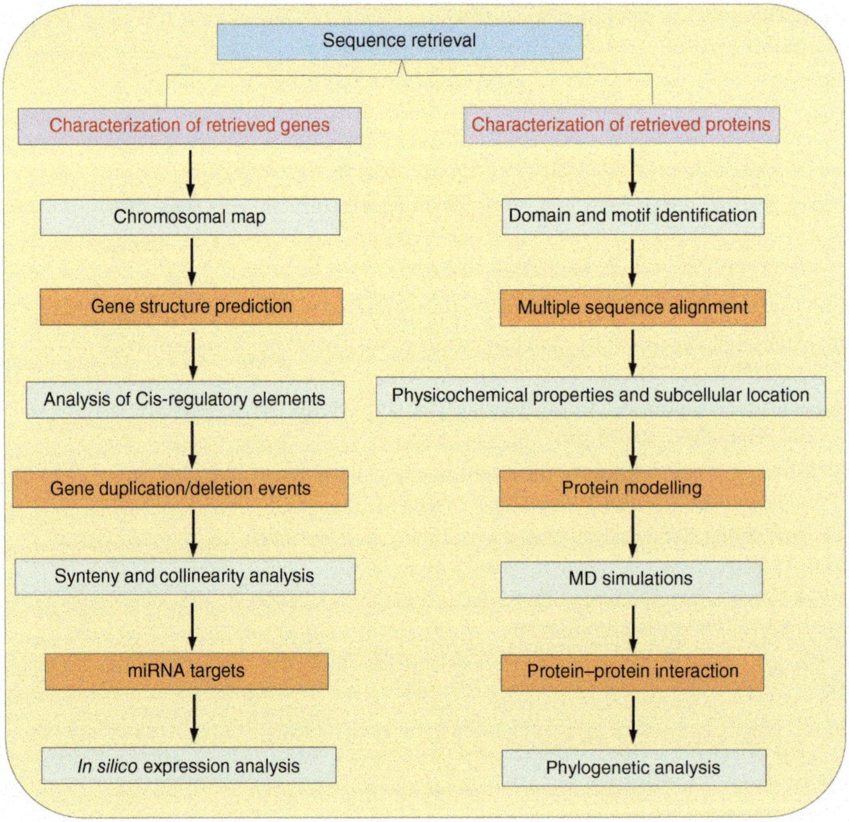

Figure 14.1 Steps involved in the identification and characterization of genes/gene families in wheat using bioinformatics.

database in the form of profile HMM of the domain and used as a query using HMMER 3.0. Transcriptome and EST databases may also be used for the retrieval of sequences to be used as queries. Sequences showing the highest similarity and lowest e-value with the query sequence will be retrieved for further use. Two other software that may be used for alignment include /GeneWise and CSTfinder, which allow pairwise genome comparison to identify genes.

14.3.1.2 Identification of Genes through Ab Initio Prediction

Ab initio prediction of genes is based on known general patterns of gene structure and signal-based searches. The known gene structure includes short signal sequences (e.g. start codons, stop codons, splicing sites, and branch points), unique patterns of codons, and other major distinct features of the gene of interest. This allows the identification of coding sequences in the split genes carrying other noncoding sequences using any of the available algorithms.

Ab initio identification of genes is routinely undertaken using assembled partial or complete sequences of a genome. In case of wheat, several attempts have been made for such ab initio identification of genes. For instance, early on 182 genes from partial wheat genome sequences and five low-molecular-weight (LMW) glutenin subunit genes from the D-genome were identified and tested for their prediction accuracy (Nasiri et al. 2013). According to the latest estimates, the total number of genes involving the pangenome in wheat is ~120,000 involving all three subgenomes (A, B, and D). The algorithms that are used for modeling gene structure include the following: linear discriminant analysis, dynamic programming, hidden Markov model (HMM), linguist methods, and neural networks. Using these models, the ab initio prediction programs that were developed included the following: FGENESH, GeneID, GeneParser, GENSCAN, GlimmerM, etc. (Nasiri et al. 2013).

Expressed sequence tags (ESTs) belonging to specific traits and KEGG database for metabolic pathways have also been used for *in silico* identification of genes for specific traits like tolerance against drought and salinity. For instance, the available contigs (obtained after assembly of raw sequences) are analyzed for transcription factors (TFs) using the tool PlantTFcat (http://plantgrn.noble.org/PlantTFcat). Similarly, the KEGG database (https://www.genome.jp/kegg/) is utilized for the identification of genes involved in metabolic pathways and enzymes. Some details of the steps involved and bioinformatics tools that were also used for this purpose were listed by Nahas et al. (2019).

14.3.1.3 Identification of Alien Genes in Wheat Cultivars

Alien genes that were introgressed into wheat in the past during wheat breeding have also been identified. Although in most cases, this exercise involved identification of alien chromosome segments transferred, in some cases specific genes were also identified (Gupta 2016; Rey and Prieto 2017; Keilwagen et al. 2022).

14.3.2 Characterization of Retrieved Gene Sequences

Characterization of retrieved gene sequences involves the following steps, which are also depicted in Figure 14.1: (i) **Visualization on chromosome map**. For this purpose, genomic coordinates of the identified genes are obtained from EnsemblPlants, and used for chromosome mapping using the gene location visualization (advanced) function of TBtools (Chen et al., 2020). (ii) **Prediction of gene structure**. Different parts of a gene including introns, exons, UTRs, and intron phases of individual genes are obtained by comparing the gene sequences with full-length cDNA sequences. A number of tools including the following are available for this purpose: GeneMark, Augustus, Glimmer, and GSDS. (iii) **Analysis of regulatory elements**: Cis-regulatory elements (CREs), including promoters, are identified using a 1.5–2.0 kb long genomic region upstream of the translation start site using PlantCARE or any other database (e.g. PlantPAN database). Only CREs showing a matrix score of ≥5 are selected. (iv) **Gene duplications/deletions** are also identified using information on genes from related species using CDS as a query for BLAST against each other and with other species (identity >90%). In case of wheat, three progenitors (*Aegilops tauschii*, *Triticum urartu*, *Triticum turgidum*) and related species like rice, maize, Brachypodium, and barley can also be used as references. (v) **Synteny and collinearity analysis.** This is achieved using diploid and tetraploid wheat progenitors and related species like barley and rice. For this purpose, sequence files (whole-genome files, genome annotation files) can be downloaded from the Phytozome database or NCBI database, or EnsemblPlants. For synteny analysis, MCScanX (http://chibba.pgml.uga.edu/ mcscan2/), based on Python can be used. Those with no experience with Linux-based analysis may like to use MCScanX and Dual Synteny Plot (for syntenic plots) available in the TBtools package (Chen et al. 2020). Similarly, information on collinearity can be obtained by comparing a block of 21 genes (10 genes on either side of each gene) of wheat with those of genes from other species in the URGI database (https://wheat-urgi.versailles.inra.fr/). (vi) ***In silico* expression analysis:** This is achieved using the wheat expression browser (http://www.wheat-expression.com/) or by retrieving transcriptome data manually from the NCBI-SRA database for comparing expression across different experimental conditions. Alternatively one may sample groups by using tools like DESeq2 or edgeR to identify differentially expressed genes. Heat maps of the differentially expressed genes can also be generated using online tools like ClustVis (http://biit.cs.ut.ee/clust vis/). (vii) **Identification of microRNA (miRNA) targets:** This is achieved by submitting wheat miRNAs available in miRBase and other published literature with cDNA sequences of the identified genes into the psRNAtarget server (Dai et al. 2018). Target hits with ≤2 e-value are selected for further analysis.

14.4 Characterization of Genes at the Protein Level

The protein sequences are obtained from genes with the help of genetic code and characterized using the following steps. (i) **Domain and motif identification**: This is achieved using tools like the following: InterProScan, Pfam, SMART, BatchCD, CDART, CDTree, and CDD; of these, Conserved Domain Database (CDD) is generally used for domains in individual proteins, one at a time (Marchler-Bauer et al. 2017); BatchCD is used for multiple sequence alignment (MSA). (ii) **Identification of conserved motifs**: these can be identified using MEME v3.5.454 (http://meme-suite.org/tools/meme). The sites for individual motifs can be set at 0 or 1 and the number of motifs to be identified can be set to anywhere between 10 and 20. The minimum length of a motif may range from 6 to 50 amino acids; these motifs can be annotated using **the** InterProScan database. (iii) **Multiple sequence alignment (MSA)**: this is needed for the identification of conserved regions; bioinformatics tools available for this purpose include the following: ClustalW/Clustal Omega, MUSCLE, MAFFT, T-Coffee, ProbCons. Each of these tools has its unique features, which can be adjusted based on the specific requirements. The most appropriate tool can be selected by comparing the results obtained using more than one tool. ClustalW is the most widely used tool for MSA and ClustalΩ (clustal Omega) is the latest and improved version that handles larger datasets more efficiently. Both tools make use of a progressive alignment approach, where sequences are aligned pair-wise based on similarity scores, and then combined to achieve the final multiple alignment.

14.4.1 Physico-chemical Properties and Subcellular Location

14.4.1.1 Physico-chemical Properties

The physicochemical properties of proteins can be obtained using the ProtParam tool of Expasy (http://web.expasy.org/protparam/). The physical and chemical properties computed by this software include molecular weight, isoelectric point (pI), amino acid composition, atomic composition, extinction coefficient, half-life, instability index, aliphatic index, and grand average of hydropathicity (GRAVY).

14.4.1.2 Subcellular Localization

Subcellular localization is also important for the study of the cellular function of a protein; this is obtained using one or more of the following tools, which differ in minor details about the information each provides: (i) CELLO2GO provides gene ontology (GO), subcellular localization(s), and other features like GO by combining the CELLO localization-predicting and BLAST homology-searching approaches. (ii) DeepLoc predicts whether a protein has only one or multiple localizations inside the cell; (iii) APSLAP predicts subcellular localization of apoptosis proteins; (iv) WoLF PSORT is an extension of the PSORT II program and converts amino acid sequences of a protein into numerical localization features, including sorting signals, amino acid composition, and functional motifs; (v) TMHMM predicts whether or not the protein carries transmembrane helixes; (vi) SignalP predicts if the protein has a signal peptide; and (vii) PSORTb predicts subcellular localization of transmembrane helices and signal peptide, if the protein carries both.

14.4.2 3D Structure

Three-dimensional (3D) structure of proteins is obtained through homology modeling based on the similarity of unknown protein to the structure of a known protein used as a template. This involves aligning the target protein sequence with the template protein and building a model of the target protein based on this alignment. The following steps are involved: (i) **Identification of template**. A suitable template for a protein is identified based on high sequence similarity to the target protein using sequence alignment tools such as PSI-BLAST or SWISS-MODEL. (ii) **Alignment of the target protein with the template**. The template is used for alignment with the target protein sequence to identify corresponding amino acid residues and specific regions, representing domains and motifs. (iii) **Generation of the 3D model**. The aligned target and template sequences are used to generate a 3D model of the target protein. Following servers/databases are available for this purpose: MODELLER (used for homology or comparative modeling of protein 3D structures); SWISS-MODEL (provides quality estimates at several stages of the modeling process) or I-TASSER (ranked as the best method and used for thousands of proteins). (iv) **Evaluation and verification of 3D models**. The 3D models of target proteins are evaluated and verified using tools such as PROCHECK, VERIFY3D, and ERRAT to get valuable insights into the quality and reliability of the predicted 3D models. Among these, PROCHECK provides information regarding the overall conformation of the protein by analyzing the relative proportion of amino acids in favored regions compared to other regions of the protein, thus helping in identification of potential structural anomalies, if any. VERIFY3D provides information regarding compatibility between the atomic model and the amino acid sequence; a measure of agreement of the model with experimental data is also obtained. ERRAT analyzes nonbonded interactions within the model to assess its overall quality. In addition, the Ramachandran plots are used for assessing the quality of protein structures. These plots represent the distribution of phi (φ) and psi (ϕ) torsion angles, which define the backbone conformation of amino acid residues. The analysis of Ramachandran plots helps in the identification of regions with favorable or disallowed backbone conformations, providing insights into the overall structural integrity of the protein model. By incorporating these tools, one can gain confidence in the accuracy and quality of the homology models.

14.4.3 Protein–Protein Interactions (PPI)

Protein–protein interactions (PPI) among several given proteins or between target proteins and other proteins using either the database called "Search Tool for Retrieval of Interacting Genes" (STRING) or the software Cytoscape. For STRING, accession numbers of the target proteins are submitted and the results with the highest score and lowest e-value are considered to be relevant for each protein involved in protein–protein interaction. An interface to the STRING protein–protein interactions is also available in the form of a R package called STRINGdb.

14.4.4 Molecular Docking and Molecular Dynamics (MD) Simulations

Molecular docking, which is undertaken to visualize interactions between a protein and its target molecule, is performed through one or more of the following software: Autodock Vina, Hex BindUP, and Molegro Molecular Viewer. Among these, Autodock vina allows significant improvement in the accuracy of predictions of binding between protein and ligand. The best structures of a docking complex are selected and analyzed using Pymol software.

MD simulation is used to study the behavior and dynamics of protein–protein or protein–nucleic acid complexes at the atomic level. It allows the study of the behavior of molecules in a dynamic environment by simulating the movement of atoms over time and by solving Newton's equations of motion. This involves the following steps: (i) the 3D structure of the protein of interest is prepared by adding missing atoms or residues, assigning partial charges, and solvating the molecule in an appropriate solvent model (e.g. water); (ii) field selection is forced and the system is further subjected to an energy minimization step to relieve any steric clashes or unfavorable interactions. (iii) the solvent and ions around the protein are equilibrated, which allows the system to adjust and reach a stable state. It involves NVT equilibration to control temperature and NPT equilibration to reach the desired pressure (N = number of molecules; V = volume; T = temperature; P = Pressure). Once the system is equilibrated, an MD simulation is performed. The simulation integrates the equations of motion to propagate the system forward in time. Typical simulations range from nanoseconds to microseconds or longer, depending on the objectives. In most cases, MD simulation is allowed for 1000 ns (Gautam et al. 2019; Gautam et al. 2022, Prasad et al. 2023). (iv) After MD simulation, the trajectories of simulated proteins can be analyzed to assess stability. Root mean square deviation (RMSD) and root mean square fluctuation

(RMSF) of Cα atoms and backbones are calculated to track structural changes over time. The total and potential energies of the final simulated structure are also worked out. These analyses provide valuable insights into the dynamic behavior and stability of the simulated protein structure. The following tools are available for analysis of functionalities for MD simulations: GROMACS, NAMD, VMD, UCSF Chimera, or AMBER.

14.5 Phylogenetic Analysis

Phylogenetic analysis can be carried out using the target proteins with proteins of the related species. In the case of wheat, generally, all the related species including the following diploid and tetraploid progenitors are used: *T. urartu*, *A. tauschii*, *T. turgidum*. Initially, all protein sequences are aligned by multiple sequence alignment (MSA) through ClustalW or MUSCLE; the aligned files are then used to generate a phylogenetic tree. The phylogenetic analysis can be performed using Molecular Evolutionary Genetics Analysis (MEGA) or PhyML. Various algorithms, such as neighbor-joining, maximum likelihood, or Bayesian inference, can be used for tree construction. For the neighbor-joining method, a bootstrap involving 1000 iterations (for robustness) and pairwise deletion are recommended to obtain an unrooted tree. The phylogenetic tree thus obtained can be visualized using the tool iTOL (Letunic and Bork 2021).

14.6 Present Status of Wheat Genes Identified *in silico*

A large number of wheat genes (numbering in thousands), belonging to hundreds of gene families and involving a wide range of traits have been identified *in silico*; most of these genes were identified during the last five years (Table 14.5). Some of these genes each control more than one trait. These genes are waiting for their direct or indirect use in wheat breeding. Efforts have also been made in this direction in several centers in India including the laboratory of the authors at CCS University Meerut (see Table 14.6 and the next paragraph).

Studies involving *in silico* identification of genes in wheat have also been an ongoing activity in the authors' laboratory in Meerut (India). These studies supplement the ongoing genetic studies being conducted using interval mapping and GWAS. Genes belonging to nine gene families have already been identified in the authors' laboratory and results published (Table 14.6). In these studies, the main focus has been on the identification of genes for the following traits: (i) tolerance to biotic stresses including disease resistance against spot blotch, leaf rust, and nematodes; (ii) tolerance to abiotic stresses, including drought and heat; (iii) biofortification mainly involving grain Fe and grain Zn; (iv) nitrogen use efficiency including uptake, translocation and utilization of nitrogen. The gene information, protein information, and their functions for individual gene families are included in Table 14.6.

Table 14.5 A summary of wheat genes identified using *in sillico* analysis.

Gene family/number of genes	Function	References
	(i) Tolerance against abiotic stresses	
Salt		
1. Sulfate transporter (SULTR) durum [22 (444–2097)]	Absorption and distribution of sulfur	Puresmaeli et al. (2023)
2. bHLH TF (225)	Stress/development/ metabolite biosynthesis	Guo and Wang (2017)
3. DUF966 (28)	Growth and development	Zhou et al. (2020)
4. Aquaporin (*T. turgidum*) (80)	Membrane water transport	Yaghobi and Heidari (2023)
5. Glutathione S-transferase (346)	Growth and development	(Hao et al. 2021)
6. RAV (26)	Development and stress	Luo et al. (2022)
Drought		
7. RPK1 [15 (1830–3372)]	Growth and development	Rahim et al. (2022)
8. LOG (45)	Cytokinin pathway	Chen et al. (2022)
9. CDPK (85)	Calcium signaling	Linghu et al. (2023)
10. PYL (38)	Development and stress	Lei et al. (2021)
11. NF-Y (37)	Ubiquitous promoter binding	Stephenson et al. (2007)
12. Aquaporins (AQPs) (113)	Transmembrane proteins, control the flow of water/ molecule	Madrid-Espinoza et al. (2018)
Heat		
13. AP2/ERF superfamily genes (517)	Plant growth and development under heat stress	Magar et al. (2022)
14. Heat shock protein family (753)	Protein folding	Kumar et al. (2020)
15. HSP101/CLPB (13)	Heat acclimation/ development	Erdayani et al. (2020)
16. HSP70s (84)	Growth/development/stress	Lu et al. (2022)
17. bZIP gene family (227)	Growth/Stress	Liang et al. (2022)
18. 20S proteasome [67 (1126–6043)]	Ubiquitination of cellular component	Sharma et al. (2022a)
Heat and drought		
19. CCHC-ZFP (50)	Growth, development, and adaptation	Sun et al. (2022a)
20. NAC (488)	Development/stresses	Guerin et al. (2019)
21. bZIP gene family [266 (422–3669)]	Stress response, growth / development	Sharma et al. (2022b)
22. Amino acid transporter (296)	Alleviating plant damage under stresses	Tian et al. (2020)

Table 14.5 (Continued)

Gene family/number of genes	Function	References
23. WRKY (160)	Growth/development/stresses	Gupta et al. (2019)
24. OPR (37)	Growth/development/stresses	Mou et al. (2019)
25. PAL (37)	Adoption/growth/development	Rasool et al. (2021)
26. AGPase [31 (432–7962)]	Starch biosynthesis	Batra et al. (2017)
27. RuvBL [9 (1647–21,970))	Helicase binding	Chaudhary et al. (2023)
Drought and salt		
28. Expansin (241)	Cell wall modification	Han et al. (2019)
29. SCPL (210)	Biotic and abiotic stress	Xu et al. (2021)
30. Dof (108)	Plant growth and stress response	Fang et al. (2020)
31. GATA TF (79)	Plant growth, development, morphogenesis, stress	Du et al. (2022b)
32. GSK (22)	Growth/abiotic stress	Zhang et al. (2022c)
33. NAC (*T. durum*) (168)	Development/stresses	Saidi et al. (2017)
34. WRKY (124)	Stress response	Ye et al. (2021)
35. AP2/ERF (*T. durum*) (271)	Development, biosynthesis of metabolites, stress	Faraji et al. (2020)
Heat and salt		
36. LEA (179)	Dehydration, salinity, high temperature/cold	Liu et al. (2019)
Heat, drought, and salt		
37. HD-Zip (46)	Development, growth, and stress	Yue et al. (2018)
38. LBD [90 (350–4097)]	Architecture and stress	Wang et al. (2021)
39. AP2/ERF gene family (322)	Signal transduction, growth, development, stresses	Riaz et al. (2021)
Heat, drought, and cold		
40. Aconitase [15 (1533–3024)]	Growth, developmental	Kesawat et al. (2022a)
Heat, drought, salt, and cold		
41. CAMTA [15 (2841–3988)]	Hormone signal/developmental/stress	Yang et al. (2020)
42. CaM (18) and CAML (230)	Growth and development	Liu et al. (2022b)
43. RRAs [20 (447–2013)]	Cytokinin signaling	Sun et al. (2022b)
Heat, drought, and salt		
44. SOD (26)	Growth and development	Jiang et al. (2019)
45. VQ motif (113)	Growth/development/stresses	Zhang et al. (2022b)

(Continued)

Table 14.5 (Continued)

Gene family/number of genes	Function	References
Drought, salt, and cold		
46. Dehydrin genes (55)	Respond to stressors	Hao et al. (2022)
Heat, salt, and cold		
47. CHYR (18)	Growth and development	Liu et al. (2022a)
Heat, drought, and cold		
48. GSTs (330)	Development, metabolism as well as abiotic and biotic stress	Wang et al. (2019)
49. TPP [31 (750–1680)]	Development and stress	Du et al. (2022a)
50.PERKs [37 (1206–3159)]	Cellular processes	Kesawat et al. (2022b)
Drought, cold, and salt		
51. Trihelix (94)	Growth/development/stresses	Xiao et al. (2019)
Salt and osmotic stress		
52. Auxin-responsive GH3 (36)	Growth and development	(Jiang et al. 2020b)
Heat, drought, salt, and heavy metal stress		
53. Dof TF [96 (717–4157)]	Growth and abiotic stress responses	Liu et al. (2020)
Heat, osmotic stress, oxidative stress		
54. GASA (Durum) (19)	Growth and development	Bouterra et al. (2023)
	(ii) Biotic stresses (disease resistance)	
Fusarium head blight		
55. MATE (44)	Removal of toxic substances	Debnath et al. (2022)
56. WAK (320)	Defense against FHB	Xia et al. (2022)
Fusarium head blight, fusarium crown rot, stripe rust, septoria tritici blotch		
57. Serpin (189)	Development/defense of pathogen	Benbow et al. (2019)
Leaf rust		
58. Trx (42)	Redox regulators	Bhurta et al. (2022)
Stem rust disease		
59. SWEET (59)	Sugar transporters	Gao et al. (2018)
Stripe rust and powdery mildew		
60. NAC (186)	Physiological processes, stress, stripe rust, powdery mildew	Lv et al. (2020)

Table 14.5 (Continued)

Gene family/number of genes	Function	References
Wheat sheath-blight resistance		
61. PKSs (81)	Secondary metabolites	Geng et al. (2022)
(iii) Tolerance to biotic + abiotic stresses		
Heat, drought, cold, disease resistance		
62. BZR (20)	Development/stresses	Kesawat et al. (2021)
Heat, drought, salt, Fusarium graminearum		
63. LIM gene family (28)	Biological processes	Li et al. (2022b)
Heat, drought, cold + FHB		
64. MADS-Box (300)	Growth/development	Raza et al. (2022)
Heat, drought + Septoria tritici blotch and powdery mildew		
65. PIN (44)	Development	Kumar et al. (2021)
Heat, drought + leaf rust		
66. SET Domain [166 (819–7516)]	Histone methylation	Batra et al. (2020)
Heat, drought + leaf rust		
67. SWEET [108 (1100–10 300)]	Development/stress	Gautam et al. (2019)
Heat + Fusarium graminearum, B. graminis, P. striiformis, P. striiformis		
68. PIP5K (75)	Inositol signal transduction	Liu et al. (2021)
Heat + Fusarium graminum		
69. TUBBY (40)	Abiotic and biotic stress	Altaf et al. (2022)
Drought + Fusarium graminearum		
70. RLK (64)	Abiotic and biotic stress	Yan et al. (2023)
Salt + Fusarium graminearum		
71. EIN3/EIL (21)	Ethylene signaling pathway	Yi-Quin et al. (2020)
(iv) Development-related traits		
Growth/development		
72. AP2 TF (62)	Different developing stages	Zhao et al. (2019)
73. BRI1 (106)	Embryogenesis	Sharma and Khurana (2022)
Circadian rhythm/flowering/light signaling/development/stress		
74. CCT genes (127)	Vernalization	Zhang et al. (2022a)
Cell proliferation and differentiation		
75. CLE (104)	Involved in stem development	Li et al. (2019)

(Continued)

Table 14.5 (Continued)

Gene family/number of genes	Function	References
Protein degradation		
76. F-Box Genes (1796)	Effect on different developmental stages	Hong et al. (2020)
Development/stress		
77. HSF (61)	Anther development	Ye et al. (2020)
Carbohydrate partitioning		
78. Invertase (130)	Pollen development	Ye et al. (2022)
Lignification, cell elongation, and stress		
79. Laccase (156)	Arbuscular mycorrhizal (AM) symbiosis	Zhong et al. (2023)
	(v) Nutrient-related traits	
Cadmium		
80. HSF (78)	Cellular protection/folding/ intracellular distribution/ degradation of proteins	Zhou et al. (2019)
Aluminum		
81. MATE (211)	Growth/Development	Duan et al. (2022)
Fe		
82. NRAMP (29)	Iron Transport	Kumar et al. (2022b)
Fe and Zn		
83. ZIP (58)	Uptake, transport, and accumulation of microelements	Li et al. (2021)
84. VMT [6 (4108–18126)]	Fe homeostasis	Sharma et al. (2023)
Nitrogen use efficiency (NUE)		
85. NPF/NRT2/CLC/SLAC1/SLAH (412)	Nitrate transportation	Kumar et al. (2022a)
86. RWP-RK 37 –d NLP (1064–5768), RKD (2865–7340)	Nitrogen starvation	Kumar et al. (2018)
Phosphorus-use efficiency PUE)		
87. CCD8 [3 (2196–2210)]	Growth/development	Batra et al. (2019)
	(vi) Abiotic/biotic stress + development/nutrient-related traits	
Heat/drought/crown rot/fusarium/ powdery mildew/stripe rust/septoria/ phosphorus stress		
88. CAT [19 (873–1479)]	Development/stress	Zhang et al. (2022d)
Heat/drought/cold + phosphate stress		
89. GATA (79)	Growth/development	Feng et al. (2022)
Heat/drought/cold + phosphorus stress + development stages + Septoria tritici blotch infection/powdery mildew/ stripe rust//Fusarium head blight/ F. pseudograminearum/Magnaporthe oryzae		

Table 14.5 (Continued)

Gene family/number of genes	Function	References
90. SRO (30)	Growth/development	Jiang et al. (2020a)
Heat/cold + phosphorus stress/ Zymoseptoria tritici, F. graminearum, stripe rust pathogen		
91. YTH domain (39)	m6A-containing RNA-binding protein	Sun et al. (2020)
Heat/cold/salt/mannitol and ABA treatment		
92. bZIP (191)	Development /stress	Agarwal et al. (2019)
Drought/phytohormone/Fusarium graminearum		
93. PRX (374)	Wall loosening/lignification/ and abiotic and biotic stress response	Yan et al. (2019)
Developmental stages/phosphate stress		
94. SPX [46 (654–3021)]	Ubiquitous	Kumar et al. (2019)
Salt, cold, PEG + pollen fertility + stamen development + SA- and MeJA phytohormones		
95. COI (18)	Growth/development	Bai et al. (2018)
Cold/nitrogen stress		
96. CLC [23 (1722–3246)]	Nutrient transport/stress	Mao et al. (2022)
Inflorescence and spike development + ABA hormone		
97. SPL (56)	Growth/development	Zhu et al. (2020)

14.7 Utility of Predicted Genes for Crop Improvement

We know that conventional plant breeding is time-consuming and generally ineffective in dealing with the improvement of many complex quantitative traits including traits like tolerance against abiotic stresses (e.g. drought, heat, salinity, etc.). In this connection, three effective recent approaches include the following: (i) MAS, (including marker-assisted recurrent selection [MARS]), (ii) gene editing, and (iii) transgenic crops. The progress made in this direction will be briefly discussed in this section (Section 14.7).

14.7.1 Molecular Breeding Involving Molecular Markers

As mentioned above, the genes identified *in silico* can be directly utilized in developing functional DNA markers, which in turn can be utilized for MAS and MARS in conventional breeding programs. These markers may sometimes also be utilized for selection of markers to be used in genomic selection (GS).

Table 14.6 A summary of details about genes identified in authors' lab.

Genes information	Protein information			
Gene/gene family name (gene length [kb])	**Protein length as number of aa**	**Mol wt (kDa)/Pi**	**Function**	**References**
(i) Heat/drought				
1. AGPase [432(1DL)-7725(1BL), 6719(7DS)-7962(7AS)]	(458–574 aa in LS and 472–586 aa in SS)	—	Starch biosynthesis	Batra et al. (2017)
2. SET [166 (819–7516)]	272–2264	(7.5–255.5)/ (4.6–9.5)	Histone methylation	Batra et al. (2020)
3. RuvBL [9 (1647–21,970)]	455–491	(31.8–53.8)/ (5.3–5.8)	Helicases	Chaudhary et al. (2023)
(ii) Heat				
4. 20 S Proteasome [67(1126–6043)]	199–298	(21.7–35.9)/ (4.6–9.6)	Ubiquitination	Sharma et al. (2022a)
(iii) Heat/leaf rust				
5. SWEET [108 (1110–10,300)]	145–340	(17.1–36.9)/ (5.7–10.2)	Development/ stress	Gautam et al. (2019)
(iv) Iron uptake				
6. NRAMP [29 (196–1023)]	163–1042	(17471–75347.7)/ (5.2–9.99)	Involved in iron transport.	Kumar et al. (2022b)
(vii) Iron/zinc uptake				
7. VMT [6 (4108–18,126)]	395–522	(49,034–64,209)/ (5.9–8.9)	Transportation/ utilization of Fe/Mg	Sharma et al. (2023)
(viii) Nitrogen use efficiency				
8. RWP-RK [37- NLP (1064–5768), RKD (2865–7340)]	RKD (204–927) NLP (559–938)	(23.7–103.7)/ (5.07–9.8)	Nitrogen starvation	Kumar et al. (2018)
P-use efficiency				
9. CCD8 [3(2196-2210)]	560	(47.6–63.9)/ (5.7–6.9)	Synthesis of strigolactone	Batra et al. (2019)
10. NAAT [24 (1891–12,942)]	403–514	(43.5–54.6)/ (5.15–7.1)	Convert nicotianamine to deoxymugineic acid	Gautam et al. (2023)

QTL interval mapping and GWAS are the two major approaches, which have been used effectively for the study of genetic architecture of almost all kinds of traits in all major crops. A database for wheat-QTLs for different traits has also been developed by the authors of this chapter, jointly with their colleagues (Singh et al. 2021, 2022). The markers associated with QTLs have also been used for MAS and MARS leading to the improvement of wheat germplasm. The work in this direction has been reviewed, providing a list of suitable robust markers that are already being used and will be increasingly used in the future (Gupta et al. 2010, 2022; Song et al. 2023).

In the past, detailed information about individual genes/QTLs could be obtained mainly through mapping of these genes followed by map-based cloning; both genetic maps and physical maps were needed for this purpose, which is time-consuming, and resource-demanding. Therefore, *in silico* identification and characterization of desirable genes is a short-cut to obtain this information, which can be used for a variety of purposes In other words, the knowledge generated by *in silico* identification and characterization of genes is equivalent to cloning of genes by recombinant DNA technology followed by study of the structure of these genes. As evident from the information and the data summarized in this chapter, *in silico* identification and characterization of thousands of genes belonging to ~100 gene families for a variety of traits of value has been possible (Tables 14.5 and 14.6). This is a good resource for wheat improvement, using molecular techniques, which include the development of functional markers, to be used for marker-assisted selection (MAS). These functional markers may be evaluated for their utility in wheat improvement.

14.7.2 Gene Editing

Gene editing, including the CRISPR/Cas approach, also depends on the availability of information about the gene of interest, so that *in silico* identification and characterization of genes should prove useful. The causal polymorphisms should also be known, so that one knows, which part of the gene or gene sequences needs to be subjected to gene editing. Moreover, in a crop like wheat, the structure of all three homoeologues for a gene should be known and unique consensus sites for all three homoeologues should be worked out so that while designing gRNA (guideRNA), which is a prerequisite for gene editing, these sites representing consensus sites among three homoeologues are taken into consideration so that the same gRNA should be sufficient for editing all the three homoeologues of a gene. In the case of wheat, there are at least half a dozen examples, where gene editing has been successfully achieved. Some details about these studies along with. appropriate vectors and transformation protocols (needed for successful gene editing) are listed in a recent study on gene editing in wheat, where gene editing for the leaf rust resistance gene *Lr67* and mildew resistance locus *MLO* were selected for gene editing (Biswal et al. 2023).

14.7.3 Transgenic Wheat

It is known that transgenic (GM = genetically modified) wheat can be produced, although it is plagued with regulatory issues thus making it difficult to utilize GM wheat. Despite this, some progress has been made. The latest example of the successful use of transgenics is HB4 wheat, which has been produced using a regulatory gene from *Helianthus annuus* and has been approved for cultivation in several countries including Argentina, Brazil, USA, Australia, Nigeria, etc. (Gupta, 2023). But in many other countries like India, there seems to be no future for transgenic wheat.

14.8 Conclusion and Prospects

The large number of useful wheat genes discovered through *in silico* approach demonstrates the power of this approach, although these genes are yet to be validated and utilized in an actual breeding programs. Perhaps, there is only one example, where a gene identified *in silico* has been utilized further for validation studies. This study involved the transcription factor gene *TaWRKY2* providing drought tolerance that was identified using *in silico* approach involving transcriptome data (Okay et al. 2014). This gene was later utilized for the identification of multiple regulatory *cis*-elements in the promoter region of this gene and utilized for developing transgenic wheat. The transgenic wheat seedlings exhibited significantly enhanced tolerance to drought stress.

There can be several other approaches, which can be used for the successful utilization of genes identified *in silico* for wheat breeding. The simplest approach is to develop gene-based DNA markers and use them for MAS for the transfer of desirable genes from a donor to a recipient genotype. Another approach is to identify regulatory sequences and utilize the same for gene editing or for developing transgenics, as done in the recent study described above (Biswal et al. 2023).

References

Abraham, M.J., Murtola, T., Schulz, R. et al. (2015). GROMACS: high performance molecular simulations through multi-level parallelism from laptops to supercomputers. *SoftwareX* 1: 9–25.

Acland, A., Agarwala, R., Barrett, T. et al. (2013). Database resources of the national center for biotechnology information. *Nucleic Acids Research* 41: D8–D20.

Agarwal, P., Baranwal, V.K., and Khurana, P. (2019). Genome-wide analysis of bZIP transcription factors in wheat and functional characterization of a TabZIP under abiotic stress. *Scientific Reports* 9 (1): 4608.

Altaf, A., Zada, A., Hussain, S. et al. (2022). Genome-wide identification, characterization, and expression analysis of Tubby gene family in wheat (*Triticum aestivum* L.) under biotic and abiotic stresses. *Agronomy* 12 (5): 1121.

Antonescu, C., Antonescu, V., Sultana, R., and Quackenbush, J. (2010). Using the DFCI gene index databases for biological discovery. *Current Protocols in Bioinformatics* 29: 1–6.

Bai, J.F., Wang, Y.K., Wang, P. et al. (2018). Genome-wide identification and analysis of the COI gene family in wheat (*Triticum aestivum* L.). *BMC Genomics* 19 (1): 1–17.

Bailey, T.L., Williams, N., Misleh, C., and Li, W.W. (2006). MEME: discovering and analyzing DNA and protein sequence motifs. *Nucleic Acids Research* 34: 369–373.

Batra, R., Saripalli, G., Mohan, A. et al. (2017). Comparative analysis of AGPase genes and encoded proteins in eight monocots and three dicots with emphasis on wheat. *Frontiers in Plant Science* 8: 19.

Batra, R., Agarwal, P., Tyagi, S. et al. (2019). A study of CCD8 genes/proteins in seven monocots and eight dicots. *PLoS One* 14 (3): e0213531.

Batra, R., Gautam, T., Pal, S. et al. (2020). Identification and characterization of SET domain family genes in bread wheat (*Triticum aestivum* L.). *Scientific Reports* 10 (1): 14624.

Benbow, H.R., Jermiin, L.S., and Doohan, F.M. (2019). Serpins: genome-wide characterisation and expression analysis of the serine protease inhibitor family in *Triticum aestivum*. *G3: Genes, Genomes, Genetics* 9 (8): 2709–2722.

Bhurta, R., Hurali, D.T., Tyagi, S. et al. (2022). Genome-wide identification and expression analysis of the thioredoxin (trx) gene family reveals its role in leaf rust resistance in wheat (*Triticum aestivum* L.). *Frontiers in Genetics* 13: 563.

Biswal, A.K., Hernandez, L.R.B., Castillo, A.I.R. et al. (2023). An efficient transformation method for genome editing of elite bread wheat cultivars. *Frontiers in Plant Science* 14: 1135047. https://doi.org/10.3389/fpls.2023.1135047.

Borrill, P., Ramirez-Gonzalez, R., and Uauy, C. (2016). expVIP: a customizable RNA-seq data analysis and visualization platform. *Plant Physiology* 170 (4): 2172–2186.

Bouteraa, M.T., Romdhane, W.B., Hsouna, A.B. et al. (2023). Genome-wide characterization and expression profiling of GASA gene family in *Triticum turgidum* ssp. *durum* (desf.) husn. (Durum wheat) unveils its involvement in environmental stress responses. *Phytochemistry* 206: 13544.

Chaudhary, J., Gautam, T., Gahlaut, V. et al. (2023). Identification and characterization of RuvBL DNA helicase genes for tolerance against abiotic stresses in bread wheat (*Triticum aestivum* L.) and related species. *Functional & Integrative Genomics* 23 (3): 255.

Chen, C., Chen, H., Zhang, Y. et al. (2020). TBtools: an integrative toolkit developed for interactive analyses of big biological data. *Molecular Plant* 13 (8): 1194–1202.

Chen, L., Jameson, G.B., Guo, Y. et al. (2022). The LONELY GUY gene family: from mosses to wheat, the key to the formation of active cytokinins in plants. *Plant Biotechnology Journal* 20 (4): 625–645.

Childs, K.L., Hamilton, J.P., Zhu, W. et al. (2007). The TIGR plant transcript assembly's database. *Nucleic Acids Research* 35: D846–D851.

Colovos, C. and Yeates, T.O. (1993). Verification of protein structures: patterns of nonbonded atomic interactions. *Protein Science* 2 (9): 1511–1519.

Dabab Nahas, L., Al-Husein, N., Lababidi, G., and Hamwieh, A. (2019). *In-silico* prediction of novel genes responsive to drought and salinity stress tolerance in bread wheat (*Triticum aestivum*). *PLoS One* 14 (10): e0223962.

Dai, X., Zhuang, Z., and Zhao, P.X. (2018). psRNATarget: a plant small RNA target analysis server (2017 release). *Nucleic Acids Research* 46: W49–W54.

Debnath, S., Mohanta, D., Perveen, K. et al. (2022). Structural and functional characterization at the molecular level of the MATE gene family in wheat *in silico*. *Contrast Media & Molecular Imaging* 2022: 9289007.

Di Tommaso, P., Moretti, S., Xenarios, I. (2011). T-Coffee: a web server for the multiple sequence alignment of protein and RNA sequences using structural information and homology extension. *Nucleic acids research* 39 (suppl_2): W13–W17.

Do, C.B., Mahabhashyam, M.S., Brudno, M., and Batzoglou, S. (2005). ProbCons: probabilistic consistency-based multiple sequence alignment. *Genome Research* 15 (2): 330–340.

Dong, Q., Schlueter, S.D., and Brendel, V. (2004). PlantGDB, plant genome database and analysis tools. *Nucleic Acids Research* 32: D354–D359.

Du, L., Li, S., Ding, L. et al. (2022a). Genome-wide analysis of trehalose-6-phosphate phosphatases (TPP) gene family in wheat indicates their roles in plant development and stress response. *BMC Plant Biology* 22 (1): 120.

Du, X., Lu, Y., Sun, H. et al. (2022b). Genome-wide analysis of wheat GATA transcription factor genes reveals their molecular evolutionary characteristics and involvement in salt and drought tolerance. *International Journal of Molecular Sciences* 24 (1): 27.

Duan, W., Lu, F., Cui, Y. et al. (2022). Genome-wide identification and characterisation of wheat MATE genes reveals their roles in aluminium tolerance. *International Journal of Molecular Sciences* 23 (8): 4418.

Eberhardt, J., Santos-Martins, D., Tillack, A.F., and Forli, S. (2021). AutoDock Vina 1.2.0: new docking methods, expanded force field, and python bindings. *Journal of Chemical Information and Modeling* 61 (8): 3891–3898.

Edgar, R.C. (2004). MUSCLE: a multiple sequence alignment method with reduced time and space complexity. *BMC Bioinformatics* 5 (1): 1–19.

Eisenberg, D., Lüthy, R., and Bowie, J.U. (1997). VERIFY3D: assessment of protein models with three-dimensional profiles. *Methods in Enzymology* 277: 396–404.

Emms, D.M. and Kelly, S. (2019). OrthoFinder: phylogenetic orthology inference for comparative genomics. *Genome Biology* 20: 1–14.

Erdayani, E., Nagarajan, R., Grant, N.P., and Gill, K.S. (2020). Genome-wide analysis of the HSP101/CLPB gene family for heat tolerance in hexaploid wheat. *Scientific Reports* 10 (1): 3948.

Fang, Z., Jiang, W., He, Y. et al. (2020). Genome-wide identification, structure characterization, and expression profiling of Dof transcription factor gene family in wheat (*Triticum aestivum* L.). *Agronomy* 10 (2): 294.

Faraji, S., Filiz, E., Kazemitabar, S.K. et al. (2020). The *AP2/ERF* gene family in *Triticum durum*: genome-wide identification and expression analysis under drought and salinity stresses. *Genes* 11 (12): 1464.

Feng, X., Yu, Q., Zeng, J. et al. (2022). Genome-wide identification and characterization of GATA family genes in wheat. *BMC Plant Biology* 22 (1): 1–15.

Finn, R.D., Clements, J., and Eddy, S.R. (2011). HMMER web server: interactive sequence similarity searching. *Nucleic Acids Research* 39: W29–W37.

Gao, Y., Wang, Z.Y., Kumar, V. et al. (2018). Genome-wide identification of the SWEET gene family in wheat. *Gene* 642: 284–292.

Garg, V.K., Avashthi, H., Tiwari, A. et al. (2016). MFPPI–multi FASTA ProtParam interface. *Journal of Bioinformatics* 12: 74–77.

Gautam, T., Saripalli, G., Gahlaut, V. et al. (2019). Further studies on sugar transporter (SWEET) genes in wheat (*Triticum aestivum* L.). *Molecular Biology Reports* 46: 2327–2353.

Gautam, T., Dutta, M., Jaiswal, V. et al. (2022). Emerging roles of SWEET sugar transporters in plant development and abiotic stress responses. *Cells* 11 (8): 1303.

Gautam, T., Jan, I., Batra, R. (2023). Further studies on nicotianamine aminotransferase (NAAT) genes involved in biofortification in bread wheat (*Triticum aestivum* L.). *Plant Gene* 33: 100389.

Geer, L.Y., Domrachev, M., Lipman, D.J., and Bryant, S.H. (2002). CDART: protein homology by domain architecture. *Genome Research* 12 (10): 1619–1623.

Gene Ontology Consortium (2021). The Gene Ontology resource: enriching a Goldgold mine. *Nucleic Acids Research* 49: D325–D334.

Geng, X., Chen, Y., Zhang, S. et al. (2022). Genome-wide analysis of Type-III polyketide synthases in wheat and possible roles in wheat sheath-blight resistance. *International Journal of Molecular Sciences* 23 (13): 7187.

Guerin, C., Roche, J., Allard, V. et al. (2019). Genome-wide analysis, expansion and expression of the NAC family under drought and heat stresses in bread wheat (*T. aestivum* L.). *PLoS One* 14 (3): e0213390.

Guindon, S., Lethiec, F., Duroux, P., and Gascuel, O. (2005). PHYML Online—a web server for fast maximum likelihood-based phylogenetic inference. *Nucleic Acids Research* 33 (suppl_2): W557–W559.

Guo, X.J. and Wang, J.R. (2017). Global identification, structural analysis and expression characterization of bHLH transcription factors in wheat. *BMC Plant Biology* 17 (1): 1–12.

Gupta, P.K. (2016). Use of alien genetic variation for wheat improvement. In: *Molecular Breeding for Sustainable Crop Improvement*, vol. 2 (ed. V.R. Rajpal et al.), 1–30. Springer.

Gupta, P.K., Langridge, P., and Mir, R.R. (2010). Marker-assisted wheat breeding: present status and future possibilities. *Molecular Breeding* 26: 145–161.

Gupta, P.K., Balyan, H.S., Chhuneja, P. et al. (2022). Pyramiding of genes for grain protein content, grain quality, and rust resistance in eleven Indian bread wheat cultivars: a multi-institutional effort. *Molecular Breeding* 42 (4): 21.

Gupta, S., Mishra, V.K., Kumari, S. (2019). Deciphering genome-wide WRKY gene family of *Triticum aestivum* L. and their functional role in response to abiotic stress. *Genes & genomics* 41: 79–94.

Han, Z., Liu, Y., Deng, X. et al. (2019). Genome-wide identification and expression analysis of expansin gene family in common wheat (*Triticum aestivum* L.). *BMC Genomics* 20 (1): 1–19.

Hao, Y., Xu, S., Lyu, Z. et al. (2021). Comparative analysis of the Glutathione S-Transferase gene family of four triticeae species and transcriptome analysis of GST genes in common wheat responding to salt stress. *International Journal of Genomics* 2021: 1–11.

Hao, Y., Hao, M., Cui, Y. et al. (2022). Genome-wide survey of the dehydrin genes in bread wheat (*Triticum aestivum* L.) and its relatives: identification, evolution and expression profiling under various abiotic stresses. *BMC Genomics* 23 (1): 1–18.

Hong, M.J., Kim, J.B., Seo, Y.W., and Kim, D.Y. (2020). F-box genes in the wheat genome and expression profiling in wheat at different developmental stages. *Genes* 11 (10): 1154.

Horton, P., Park, K.J., Obayashi, T. et al. (2007). WoLF PSORT: protein localization predictor. *Nucleic Acids Research* 35 (suppl_2): W585–W587.

Hu, B., Jin, J., Guo, A.Y. et al. (2015). GSDS 2.0: an upgraded gene featurefeatures visualization server. *Bioinformatics* 31: 1296–1297.

Humphrey, W., Dalke, A., and Schulten, K. (1996). VMD: visual molecular dynamics. *Journal of Molecular Graphics* 14 (1): 33–38.

Jiang, W., Yang, L., He, Y. et al. (2019). Genome-wide identification and transcriptional expression analysis of superoxide dismutase (SOD) family in wheat (*Triticum aestivum*). *PeerJ* 7: e8062.

Jiang, W., Geng, Y., Liu, Y. et al. (2020a). Genome-wide identification and characterization of SRO gene family in wheat: molecular evolution and expression profiles during different stresses. *Plant Physiology and Biochemistry* 154: 590–611.

Jiang, W., Yin, J., Zhang, H. et al. (2020b). Genome-wide identification, characterization analysis and expression profiling of auxin-responsive GH3 family genes in wheat (*Triticum aestivum* L.). *Molecular Biology Reports* 47: 3885–3907.

Jin, J., Tian, F., Yang, D.C. et al. (2016). PlantTFDB 4.0: toward a central hub for transcription factors and regulatory interactions in plants. *Nucleic Acids Research* 982: D1040–D1045.

Kanehisa, M. and Goto, S. (2000). KEGG: kyoto encyclopedia of genes and genomes. *Nucleic Acids Research* 28: 27–30.

Katoh, K., Misawa, K., Kuma, K.I., and Miyata, T. (2002). MAFFT: a novel method for rapid multiple sequence alignment based on fast Fourier transform. *Nucleic Acids Research* 30 (14): 3059–3066.

Keilwagen, J., Lehnert, H., Berner, T. et al. (2022). Detecting major introgressions in wheat and their putative origins using coverage analysis. *Scientific Reports* 12 (1): 1908.

Kesawat, M.S., Kherawat, B.S., Singh, A. et al. (2021). Genome-wide identification and characterization of the brassinazole-resistant (BZR) gene family and its expression in the various developmental stage and stress conditions in wheat (*Triticum aestivum* L.). *International Journal of Molecular Sciences* 22 (16): 8743.

Kesawat, M.S., Kherawat, B.S., Ram, C. et al. (2022a). Genome-wide identification and expression profiling of aconitase gene family members reveals their roles in plant development and adaptation to diverse stress in *Triticum aestivum* L. *Plants* 11 (24): 3475.

Kesawat, M.S., Kherawat, B.S., Singh, A. et al. (2022b). Genome-wide analysis and characterization of the proline-rich extensin-like receptor kinases (PERKs) gene family reveals their role in different developmental stages and stress conditions in wheat (*Triticum aestivum* L.). *Plants* 11 (4): 496.

Kozomara, A., Birgaoanu, M., and Griffiths-Jones, S. (2019). miRBase: from microRNA sequences to function. *Nucleic Acids Research* 47: D155–D162.

Krogh, A., Larsson, B., Von Heijne, G., and Sonnhammer, E.L. (2001). Predicting transmembrane protein topology with a hidden Markov model: application to complete genomes. *Journal of Molecular Biology* 305 (3): 567–580.

Kumar, A., Batra, R., Gahlaut, V. et al. (2018). Genome-wide identification and characterization of gene family for RWP-RK transcription factors in wheat (*Triticum aestivum* L.). *PLoS One* 13 (12): e0208409.

Kumar, A., Sharma, M., Gahlaut, V. et al. (2019). Genome-wide identification, characterization, and expression profiling of SPX gene family in wheat. *International Journal of Biological Macromolecules* 140: 17–32.

Kumar, A., Sharma, S., Chunduri, V. et al. (2020). Genome-wide identification and characterization of Heat Shock Protein Family revealsreveal's role in development and stress conditions in *Triticum aestivum* L. *Scientific Reports* 10 (1): 1–12.

Kumar, M., Kherawat, B.S., Dey, P. et al. (2021). Genome-wide identification and characterization of PIN-FORMED (PIN) gene family revealsreveal's role in developmental and various stress conditions in *Triticum aestivum* L. *International Journal of Molecular Sciences* 22 (14): 7396.

Kumar, A., Sandhu, N., Kumar, P. et al. (2022a). Genome-wide identification and *in silico* analysis of NPF, NRT2, CLC and SLAC1/SLAH nitrate transporters in hexaploid wheat (*Triticum aestivum*). *Scientific Reports* 12 (1): 11227.

Kumar, A., Singh, G., Prasad, B.K. et al. (2022b). Genome wide analysis and identification of nramp gene family in wheat (*Triticum aestivum* L.). *The Pharma Innovation Journal* 11 (1): 499–504.

Laskowski, R.A., MacArthur, M.W., Moss, D.S., and Thornton, J.M. (1993). PROCHECK - a program to check the stereochemical quality of protein structures. *Journal of Applied Crystallography* 26: 283–291.

Lei, P., Wei, X., Gao, R. et al. (2021). Genome-wide identification of PYL gene family in wheat: evolution, expression and 3D structure analysis. *Genomics* 113 (2): 854–866.

Lescot, M., Déhais, P., Thijs, G. et al. (2002). PlantCARE, a database of plant cis-acting regulatory elements and a portal to tools for *in silico* analysis of promoter sequences. *Nucleic Acids Research* 30 (1): 325–327.

Letunic, I. and Bork, P. (2021). Interactive tree of life (iTOL) v5: an online tool for phylogenetic tree display and annotation. *Nucleic Acids Research* 49 (W1): W293–W296.

Li, J., Yang, S., Yang, X. et al. (2022a). PlantGF: an analysis and annotation platform for plant gene families. *Database* 2022: baab088.

Li, S., Liu, Z., Guo, L. (2021). Genome-wide identification of wheat ZIP gene family and functional characterization of the TaZIP13-B in plants. *Frontiers in Plant Science* 12 (748146).

Li, Y., Liu, X., Xiao, Y. et al. (2022). Genome-wide characterization and function analysis uncovered roles of wheat LIMs in responding to adverse stresses and TaLIM8-4D function as a susceptible gene. *The Plant Genome* 15 (3): p.e20246.

Li, Z., Liu, D., Xia, Y. et al. (2019). Identification and functional analysis of the CLAVATA3/EMBRYO SURROUNDING REGION (CLE) gene family in wheat. *International Journal of Molecular Sciences* 20 (17): 4319.

Liang, Y., Xia, J., Jiang, Y. et al. (2022). Genome-wide identification and analysis of bZIP gene family and resistance of TaABI5 (TabZIP96) under freezing stress in wheat (*Triticum aestivum*). *International Journal of Molecular Sciences* 23 (4): 2351.

Linghu, B., Xu, Z., Chu, Y. et al. (2023). Genome-wide analysis of calcium-dependent protein kinase (CDPK) family and functional characterization of TaCDPK25-U in response to drought stress in wheat. *Environmental and Experimental Botany* 209: 105277.

Liu, H., Xing, M., Yang, W. et al. (2019). Genome-wide identification of and functional insights into the late embryogenesis abundant (LEA) gene family in bread wheat (*Triticum aestivum*). *Scientific Reports* 9 (1): 13375.

Liu, Y., Liu, N., Deng, X. et al. (2020). Genome-wide analysis of wheat DNA-binding with one finger (Dof) transcription factor genes: evolutionary characteristics and diverse abiotic stress responses. *BMC Genomics* 21 (1): 1–18.

Liu, H., Sun, Z., Hu, L., and Yue, Z. (2021). Genome-wide identification of PIP5K in wheat and its relationship with anther male sterility induced by high temperature. *BMC Plant Biology* 21 (1): 598.

Liu, H., Yang, W., Zhao, X. et al. (2022a). Genome-wide analysis and functional characterization of CHYR gene family associated with abiotic stress tolerance in bread wheat (*Triticum aestivum* L.). *BMC Plant Biology* 22 (1): 204.

Liu, Y., Chen, W., Liu, L. et al. (2022b). Genome-wide identification and expression analysis of calmodulin and calmodulin-like genes in wheat (*Triticum aestivum* L.). *Plant Signaling & Behavior* 17 (1): 2013646.

Lu, S., Wang, J., Chitsaz, F. et al. (2020). CDD/SPARCLE: the conserved domain database in. *Nucleic Acids Research* 48 (D1): D265–D268.

Lu, Y., Zhao, P., Zhang, A. et al. (2022). Genome-wide analysis of HSP70s in hexaploid wheat: tandem duplication, heat response, and regulation. *Cells* 11 (5): 818.

Luo, Y.X., Chen, S.K., Wang, P.D. et al. (2022). Genome-wide analysis of the RAV gene family in wheat and functional identification of TaRAV1 in salt stress. *International Journal of Molecular Sciences* 23 (16): 8834.

Lv, S., Guo, H., Zhang, M. et al. (2020). Large-scale cloning and comparative analysis of TaNAC genes in response to stripe rust and powdery mildew in wheat (*Triticum aestivum* L.). *Genes* 11 (9): 1073.

Madrid-Espinoza, J., Brunel-Saldias, N., Guerra, F.P. et al. (2018). Genome-wide identification and transcriptional regulation of aquaporin genes in bread wheat (*Triticum aestivum* L.) under water stress. *Genes* 9 (10): 497.

Magar, M.M., Liu, H., and Yan, G. (2022). Genome-wide analysis of AP2/ERF superfamily genes in contrasting wheat genotypes reveals heat stress-related candidate genes. *Frontiers in Plant Science* 13: 799.

Mao, P., Run, Y., Wang, H. et al. (2022). Genome-wide identification and functional characterization of the chloride channel TaCLC gene family in wheat (*Triticum aestivum* L.). *Frontiers in Genetics* 13: 846795.

Marchler-Bauer, A., Bo, Y., Han, L. et al. (2017). CDD/SPARCLE: functional classification of proteins via subfamily domain architectures. *Nucleic Acids Research* 45: D200–D203.

Margelevičius, M. and Venclovas, Č. (2005). PSI-BLAST-ISS: an intermediate sequence search tool for estimation of the position-specific alignment reliability. *BMC Bioinformatics* 6 (1): 1–10.

Mou, Y., Liu, Y., Tian, S. (2019). Genome-wide identification and characterization of the OPR gene family in wheat (*Triticum aestivum* L.). *International Journal of Molecular Sciences*, 20 (8): 1914.

Nasiri, J., Naghavi, M., Rad, S.N. et al. (2013). Gene identification programs in bread wheat: a comparison study. *Nucleosides, Nucleotides and Nucleic Acids* 32 (10): 529–554.

Okay, S., Derelli, E., and Unver, T. (2014). Transcriptome-wide identification of bread wheat WRKY transcription factors in response to drought stress. *Molecular Genetics and Genomics* 289: 765–781.

Paz, I., Kligun, E., Bengad, B., and Mandel-Gutfreund, Y. (2016). BindUP: a web server for non-homology-based prediction of DNA and RNA binding proteins. *Nucleic Acids Research* 44: 568–574.

Phillips, J.C., Braun, R., Wang, W. et al. (2005). Scalable molecular dynamics with NAMD. *Journal of Computational Chemistry* 26 (16): 1781–1802.

Prasad, P., Jain, N., Chaudhary, J. et al. (2023). Candidate effectors for leaf rust resistance gene *Lr28* identified through transcriptome and *in-silico* analysis. *Frontiers in Microbiology* 14: 1143703.

Punta, M., Coggill, P.C., Eberhardt, R.Y. et al. (2012). The Pfam protein families database. *Nucleic Acids Research* 40: D290–D301.

Puresmaeli, F., Heidari, P., and Lawson, S. (2023). Insights into the sulfate transporter gene family and its expression patterns in durum wheat seedlings under salinity. *Genes* 14 (2): 333.

Quevillon, E., Silventoinen, V., Pillai, S. et al. (2005). InterProScan: protein domains identifier. *Nucleic Acids Research* 33: 116–120.

Rahim, A.A., Uzair, M., Rehman, N. et al. (2022). Genome-wide identification and characterization of receptor-like protein kinase 1 (RPK1) gene family in *Triticum aestivum* under drought stress. *Frontiers in Genetics* 13: 912251.

Rasool, F., Uzair, M., Naeem, M.K. et al. (2021). Phenylalanine ammonia-lyase (PAL) genes family in wheat (*Triticum aestivum* L.): genome-wide characterization and expression profiling. *Agronomy* 11 (12): 2511.

Raza, Q., Riaz, A., Atif, R.M. et al. (2022). Genome-wide diversity of MADS-Box genes in bread wheat is associated with its rapid global adaptability. *Frontiers in Genetics* 12: 818880.

Rey, M.D. and Prieto, P. (2017). Detection of alien genetic introgressions in bread wheat using dot-blot genomic hybridisation. *Molecular Breeding* 37 (3): 32.

Riaz, M.W., Lu, J., Shah, L. et al. (2021). Expansion and molecular characterization of AP2/ERF gene family in wheat (*Triticum aestivum* L.). *Frontiers in Genetics* 12: 632155.

Ritchie, D.W. and Kemp, G.J. (2000). Protein docking using spherical polar Fourier correlations. *Proteins: Structure, Function, and Genetics* 39: 178–194.

Saidi, M.N., Mergby, D., and Brini, F. (2017). Identification and expression analysis of the NAC transcription factor family in durum wheat (*Triticum turgidum* L. ssp. durum). *Plant Physiology and Biochemistry* 112: 117–128.

Salomon-Ferrer, R., Case, D.A., and Walker, R.C. (2013). An overview of the Amber biomolecular simulation package. *Wiley Interdisciplinary Reviews: Computational Molecular Science* 3 (2): 198–210.

Saravanan, V. and Lakshmi, P.T.V. (2013). APSLAP: an adaptive boosting technique for predicting subcellular localization of apoptosis protein. *Acta Biotheoretica* 61: 481–497.

Schultz, J., Copley, R.R., Doerks, T. et al. (2000). SMART: a web-based tool for the study of genetically mobile domains. *Nucleic Acids Research* 28 (1): 231–234.

Shannon, P., Markiel, A., Ozier, O. et al. (2003). Cytoscape: a software environment for integrated models of biomolecular interaction networks. *Genome Research* 13 (11): 2498–2504.

Sharma, N. and Khurana, P. (2022). Genome-wide identification, characterization and expression analysis of the *BRI1* gene family in *Triticum aestivum* L. *Plant Biotechnology Reports* 16 (6): 777–791.

Sharma, H., Batra, R., Kumar, S. et al. (2022a). Identification and characterization of 20S proteasome genes and their relevance to heat/drought tolerance in bread wheat. *Gene Reports* 27: 101552.

Sharma, M.K., Kumar, S., and Sharma, M.K. (2022b). Genome-wide identification, characterization and expression profiling of bzip gene family in wheat using bioinformatics approach. *Bhartiya Krishi Anusandhan Patrika* 37 (1): 35–42.

Sharma, H., Shayaba, Kumar, R. et al. (2023). Comparative analysis of VMT genes/proteins in selected plant species with emphasis on bread wheat (*Triticum aestivum* L.). *Genes & Genomics* 45 (11): 1445–1461.

Singh, K., Batra, R., Sharma, S. et al. (2021). WheatQTLdb: a QTL database for wheat. *Molecular Genetics and Genomics* 296 (5): 1051–1056.

Singh, K., Saini, D.K., Saripalli, G. et al. (2022). WheatQTLdb V2. 0: a supplement to the database for wheat QTL. *Molecular Breeding* 42 (10): 56.

Song, L., Wang, R., Yang, X. et al. (2023). Molecular markers and their applications in marker-assisted selection (MAS) in bread wheat (*Triticum aestivum*). *Agriculture* 13: 642. https://doi.org/10.3390/agriculture13030642.

Stephenson, T.J., McIntyre, C.L., Collet, C., and Xue, G.P. (2007). Genome-wide identification and expression analysis of the NF-Y family of transcription factors in *Triticum aestivum*. *Plant Molecular Biology* 65: 77–92.

Sun, J., Bie, X.M., Wang, N. et al. (2020). Genome-wide identification and expression analysis of YTH domain-containing RNA-binding protein family in common wheat. *BMC Plant Biology* 20 (1): 1–14.

Sun, A., Li, Y., Zou, X. et al. (2022a). Comprehensive genome-wide identification, characterization, and expression analysis of CCHC zinc finger gene family in wheat (*Triticum aestivum* L.). *Frontiers in Plant Science* 13: 892105.

Sun, L., Lv, L., Zhao, J. et al. (2022b). Genome-wide identification and expression analysis of the TaRRA gene family in wheat (*Triticum aestivum* L.). *Frontiers in Plant Science* 13: 108.

Sundell, D., Mannapperuma, C., Netotea, S. et al. (2015). The plant genome integrative explorer resource: PlantGenIE.org. *New Phytologist* 208: 1149–1156.

Szklarczyk, D., Gable, A.L., Lyon, D. et al. (2019). STRING v11: protein–protein association networks with increased coverage, supporting functional discovery in genome-wide experimental datasets. *Nucleic Acids Research* 47: 607–613.

Tamura, K., Stecher, G., and Kumar, S. (2021). MEGA11: molecular evolutionary genetics analysis version 11. *Molecular Biology and Evolution* 38: 3022–3027.

Teufel, F., Almagro Armenteros, J.J. et al. (2022). SignalP 6.0 predicts all five types of signal peptides using protein language models. *Nature biotechnology* 40 (7): 1023–1025.

Thompson, J.D., Gibson, T.J., and Higgins, D.G. (2003). Multiple sequence alignment using ClustalW and ClustalX. *Current Protocols in Bioinformatics* 1: 2–3.

Thomsen, R. and Christensen, M.H. (2006). MolDock: a new technique for high-accuracy molecular docking. *Journal of Medicinal Chemistry* 49: 3315–3321.

Tian, R., Yang, Y., Chen, M. (2020). Genome-wide survey of the amino acid transporter gene family in wheat (*Triticum aestivum* L.): Identification, expression analysis and response to abiotic stress. *International Journal of Biological Macromolecules* 162: 1372–1387.

Van Bel, M., Silvestri, F., Weitz, E.M. et al. (2022). PLAZA 5.0: extending the scope and power of comparative and functional genomics in plants. *Nucleic Acids Research* 50 (D1): D1468–D1474.

Wafula, E.K., Zhang, H., Von Kuster, G. et al. (2023). PlantTribes2: tools for comparative gene family analysis in plant genomics. *Frontiers in Plant Science* 13: 1011199.

Wang, R., Ma, J., Zhang, Q. et al. (2019). Genome-wide identification and expression profiling of glutathione transferase gene family under multiple stresses and hormone treatments in wheat (*Triticum aestivum* L.). *BMC Genomics* 20 (1): 1–15.

Wang, Z., Zhang, R., Cheng, Y. et al. (2021). Genome-wide identification, evolution, and expression analysis of LBD transcription factor family in bread wheat (*Triticum aestivum* L.). *Frontiers in Plant Science* 12: 721253.

Waterhouse, A., Bertoni, M., Bienert, S. et al. (2018). SWISS-MODEL: homology modelling of protein structures and complexes. *Nucleic Acids Research* 46: 296–303.

Webb, B. and Sali, A. (2016). Comparative protein structure modeling using MODELLER. *Current Protocols in Bioinformatics* 54 (1): 5–6.

Xia, X., Zhang, X., Zhang, Y. et al. (2022). Characterization of the WAK gene family reveals genes for FHB resistance in bread wheat (*Triticum aestivum* L.). *International Journal of Molecular Sciences* 23 (13): 7157.

Xiao, J., Hu, R., Gu, T. et al. (2019). Genome-wide identification and expression profiling of trihelix gene family under abiotic stresses in wheat. *BMC Genomics* 20: 1–14.

Xu, X., Zhang, L., Zhao, W. et al. (2021). Genome-wide analysis of the serine carboxypeptidase-like protein family in *Triticum aestivum* reveals TaSCPL184-6D is involved in abiotic stress response. *BMC Genomics* 22 (1): 1–19.

Yaghobi, M. and Heidari, P. (2023). Genome-wide analysis of aquaporin gene family in *Triticum turgidum* and its expression profile in response to salt stress. *Genes* 14 (1): 202.

Yan, J., Su, P., Li, W. et al. (2019). Genome-wide and evolutionary analysis of the class III peroxidase gene family in wheat and *Aegilops tauschii* reveals that some members are involved in stress responses. *BMC Genomics* 20: 1–19.

Yang, F., Dong, F.S., Hu, F.H. et al. (2020). Genome-wide identification and expression analysis of the calmodulin-binding transcription activator (CAMTA) gene family in wheat (*Triticum aestivum* L.). *BMC Genetics* 21 (1): 1–10.

Yan, J., Su, P., Meng, X. et al. (2023). Phylogeny of the plant receptor-like kinase (RLK) gene family and expression analysis of wheat RLK genes in response to biotic and abiotic stresses. *BMC genomics* 24 (1): 224.

Ye, Y. and Godzik, A. (2004). FATCAT: a web server for flexible structure comparison and structure similarity searching. *Nucleic Acids Research* 32: W582–W585.

Ye, J., Yang, X., Hu, G. et al. (2020). Genome-wide investigation of heat shock transcription factor family in wheat (*Triticum aestivum* L.) and possible roles in anther development. *International Journal of Molecular Sciences* 21 (2): 608.

Ye, H., Qiao, L., Guo, H. et al. (2021). Genome-wide identification of wheat WRKY gene family reveals that TaWRKY75-A is referred to drought and salt resistances. *Frontiers in Plant Science* 12: 663118.

Ye, J., Bu, Y., He, M. et al. (2022). Genome-wide analysis of invertase gene family in wheat (*Triticum aestivum* L.) indicates involvement of TaCWINVs in pollen development. *Plant Growth Regulation* 98 (1): 77–89.

Yi-Qin, H., Wen-di, H., Lei, Y. et al. (2020). Genome-wide analysis of ethylene-insensitive3 (EIN3/EIL) in *Triticum aestivum*. *Crop Science* 60 (4): 2019–2037.

Yu, C.S., Cheng, C.W., Su, W.C. et al. (2014). CELLO2GO: a web server for protein subCELlular LOcalization prediction with functional gene ontology annotation. *PLoS One* 9 (6): e99368.

Yue, H., Shu, D., Wang, M. et al. (2018). Genome-wide identification and expression analysis of the HD-zip gene family in wheat (*Triticum aestivum* L.). *Genes* 9 (2): 70.

Zhang, H., Jiao, B., Dong, F. et al. (2022a). Genome-wide identification of CCT genes in wheat (*Triticum aestivum* L.) and their expression analysis during vernalization. *PLoS One* 17 (1): e0262147.

Zhang, L., Wang, K., Han, Y. et al. (2022b). Genome-wide analysis of the VQ motif-containing gene family and expression profiles during phytohormones and abiotic stresses in wheat (*Triticum aestivum* L.). *BMC Genomics* 23 (1): 1–20.

Zhang, P., Zhang, L., Chen, T. et al. (2022c). Genome-wide identification and expression analysis of the GSK gene family in wheat (*Triticum aestivum* L.). *Molecular Biology Reports* 49 (4): 2899–2913.

Zhang, Y., Zheng, L., Yun, L. et al. (2022d). Catalase (CAT) gene family in wheat (*Triticum aestivum* L.): evolution, expression pattern and function analysis. *International Journal of Molecular Sciences* 23 (1): 542.

Zhao, Y., Ma, R., Xu, D. et al. (2019). Genome-wide identification and analysis of the AP2 transcription factor gene family in wheat (*Triticum aestivum* L.). *Frontiers in Plant Science* 10: 1286.

Zhong, X., Li, M., Zhang, M. et al. (2023). Genome-wide analysis of the laccase gene family in wheat and relationship with arbuscular mycorrhizal colonization. *Planta* 257 (1): 15.

Zhou, M., Zheng, S., Liu, R. et al. (2019). Genome-wide identification, phylogenetic and expression analysis of the heat shock transcription factor family in bread wheat (*Triticum aestivum* L.). *BMC Genomics* 20 (1): 1–18.

Zhou, X., Zhu, X., Shao, W. et al. (2020). Genome-wide mining of wheat DUF966 gene family provides new insights into salt stress responses. *Frontiers in Plant Science* 11: 569838.

Zhou, X., Zheng, W., Li, Y. et al. (2022). I-TASSER-MTD: a deep-learning-based platform for multi-domain protein structure and function prediction. *Nature Protocols* 17 (10): 2326–2353.

Zhu, T., Liu, Y., Ma, L. et al. (2020). Genome-wide identification, phylogeny and expression analysis of the SPL gene family in wheat. *BMC Plant Biology* 20 (1): 1–14.

15

Bioinformatics for Analyzing the Role of Epigenetics in Plant Disease Resistance

Kalpana Singh[1,2]*, Harindra Singh Balyan*[1]*, and Pushpendra Kumar Gupta*[1,3,4]

[1] *Department of Genetics and Plant Breeding, CCS University, Meerut, Uttar Pradesh, India*
[2] *Department of Bioinformatics, College of Animal Biotechnology, Guru Angad Dev Veterinary and Animal Sciences University, Ludhiana, Punjab, India*
[3] *Murdoch's Centre for Crop and Food Innovation, Murdoch University, Perth, Australia*
[4] *Borlaug Institute for South Asia (BISA), Ludhiana, Punjab, India*

15.1 Introduction

Epigenetic modifications include chemical or physical changes in chromatin, which regulate the expression of gene(s) with no alteration in DNA sequence. These epigenetic modifications include histone modifications, chromatin accessibility, DNA methylation, DNA–protein interactions, synthesis of specific noncoding RNAs (ncRNAs), etc., which activate or suppress gene activity, leading to alteration in all kinds of traits including disease resistance/susceptibility in almost all plant species including all crops. Due to the availability of improved high-throughput next-generation sequencing (NGS) techniques, it has become possible now to capture the epigenetic modifications at the level of individual nucleotides. Identification of different epigenetic modifications requires DNA/RNA sequencing, followed by sequence analysis that needs bioinformatics tools. In this chapter, we provide a brief account of epigenetic modifications and bioinformatics tools/techniques that are used to study these epigenetic modifications. While doing so, we will cover only the following three major epigenetic modifications: (i) histone modifications; (ii) DNA methylation; and (iii) noncoding RNAs (miRNA, circRNA, and lncRNA).

15.2 Histone Modifications

15.2.1 Types of Histone Modifications (Acetylation, Methylation, Ubiquitylation)

Histone modifications represent an important epigenetic modification in chromatin associated with heritable changes in phenotype in plants. These modifications include acetylation, methylation, sulphonation, and ubiquitylation (also written as ubiquitination),

Bioinformatics for Plant Research and Crop Breeding, First Edition. Edited by Jen-Tsung Chen.

each regulating gene expression and chromatin structure. This subject has been adequately reviewed in recent literature (Kimura 2013; Martire and Banaszynski 2020; Zhi and Chang 2021; Millán-Zambrano et al. 2022). The subject will be briefly discussed in this section before we present an account of the involvement of bioinformatics in the study of the role of histone modifications in disease resistance. Only histone acetylation, methylation, and ubiquitylation will be discussed. An account of these three histone modifications will be followed by the techniques used for the analysis of these histone modifications. These techniques include the generation and analysis of data generated by ChIP seq and CUT&Tag techniques.

15.2.1.1 Histone Acetylation

Reversible histone acetylation involves the use of histone acetyltransferases (HATs) and histone deacetylases (HDACs), the former inducing gene transcription, and the latter causing repression (Kong et al. 2020). Exhaustive studies involving model plant Arabidopsis and other plant systems provided evidence for reversible histone acetylation in the plant–pathogen interactions (for reviews, see, Wang et al. 2017; Latrasse et al. 2017; Ramirez-Prado et al. 2018a,b). Reports are also available in wheat and rice, where histone acetylation involving gene activation has been shown to regulate defense responses negatively. Similarly, deacetylation often involves gene repression leading to enhanced resistance.

The importance of reversible histone acetylation in the regulation of gene expression associated with host-pathogen interactions has been emphasized in several studies. As an example, binding of the effector PsAvh23 secreted by *Phytophthora sojae* (a soybean oomycete pathogen) with the HAT complex disrupts the assembly of this complex causing a disturbance in H3K9 acetylation, leading to repression of defense genes (Kong et al. 2017). Another example involves the inhibition of HDAC in the host by HC-toxin produced by the fungal pathogen *Cochliobolus carbonum* (*Helminthosporium carbonum*) (Walley et al. 2018). Following are some other important examples, where plants have been shown to utilize histone acetylation and deacetylation in their interactions with the pathogens: (i) In tomato, the host genes are involved in transcriptional gene silencing (TGS) of disease-causing genes of tomato yellow leaf curl virus (TYLCV; a geminivirus). In turn, the V2 protein of the virus is responsible for causing the disease through suppression of TGS activity of the host. This is achieved through modification of the activity of HDA6 (a deacetylase) of the host by interfering in the recruitment of MET1 (a methyltransferase). (ii) In rice, the fungus *Magnaporthe oryzae* causes blast disease through autophagy caused due to several autophagy genes (*MoATG6*, *15*, *16*, *22*) of the pathogen encoding MoSNT2 proteins; this activity of the pathogen is regulated by H3 histone (de) acetylation due to *M. oryzae* protein MoSNT2 (*M. oryzae* homolog of SNT2 gene of yeast) at the promoter sites of the above autophagy genes of the pathogen (He et al. 2018). (iii) Sometimes, a bacterium belonging to the microbiome of the host interferes with the activity of the pathogen. An important example is the bacterium *Pseudomonas piscium*, which occurs in wheat microbiome and secretes phenazine-1-carboxamide, which interferes with the activity of the fungal HAT, named FgGcn5, leading to reduction in activity of the fungus *Fusarium graminearum*, causing fusarium head blight in wheat (Chen et al. 2018a). The above three examples suggest that histone acetylation and (de) acetylation could regulate processes like defense responses during pathogen growth and infection, crop-pathogen interactions, and control of crop diseases.

15.2.1.2 Histone Methylation

Reversible histone methylation has also been shown to be involved in gene activation/repression for the regulation of plant defense systems; this has been shown to involve histone methyltransferases and demethylases (Black et al. 2012). Among these events, methylations of H3K4 and H3K36 are known as transcription activation marks, whereas methylation of H3K9 and H3K27 are known as gene repression marks (Ding and Wang 2015; Ramirez-Prado et al. 2018a,b; Wang et al. 2019; Saripalli et al. 2020a). This will be illustrated using some examples.

In Arabidopsis, plant–pathogen interactions are regulated by several histone methyltransferases and demethylases (Alvarez-Venegas et al. 2007; Berr et al. 2010; Lee et al. 2016a; Dutta et al. 2017; Chan and Zimmerli 2019; Zhang et al. 2020a). In at least two crops, namely rice and potato, histone (de)methylation has been shown to regulate plant–pathogen interactions (Hou et al. 2015; Meller et al. 2018). In rice, it was first observed that di- and trimethylation of some lysine residues in H3 histone proteins was dramatically altered on infection by *Xanthomonas oryzae* pv oryzae (Xoo). This was associated with the upregulation of 15 Jumonji C (JmjC) domain-containing proteins, which were already known to be involved in the epigenetic regulation of a number of traits. This suggested the involvement of methylation of JmjC proteins in providing defense against Xoo. A mutant, named jmjC704 that was found to be susceptible to bacterial blast was also found to have increased histone methylation, thus suggesting the role of wild-type JmjC in providing defense through histone demethylation. This histone modification repressed transcription of genes like *NRR*, *Os-11N3*, and *OsWRKY62* by negative rice defense regulator genes through reduced H3K4me2/3 at promoter sites (Hou et al., 2015).

Histone methylation has also been artificially manipulated by chemicals like azadeoxycytidine (5AzaC, a DNA de-methylating agent) in rice and by β-aminobutyric acid (BABA) in potatoes. Spraying by BABA was shown to increase the level of H3K4me2 and H3K27me3 histone marks. These marks influence the expression of two genes (*NPR1* and *SNI1*) that are involved in systemic acquired resistance (SAR). Response of some PR genes, like *PR1* and *PR2*, associated with resistance of potato plants against *Phytophthora infestans*, also suggested that JMJ704 potentiates rice defense response against Xoo infection (Meller et al. 2018).

15.2.1.3 Histone Ubiquitination

Ubiquitination in histone proteins like H2A and H2B is also known to cause alterations in gene expression. Following are two important examples: (i) In Arabidopsis, HUB1 and HUB2 ubiquitin ligases have been shown to mediate histone H2B mono-ubiquitination (H2Bub1). This contributes to resistance against two pathogens, namely *Pseudomonas syringae* (a bacterial pathogen) and *Verticillium dahliae* (*Vd;* a fungal pathogen). In both cases, resistance is achieved through upward regulation of the transcription of resistance genes including *SNC1* and *RPP4* (Dhawan et al. 2009; Hu et al. 2014; Zou et al. 2014; Zhau and Zhang 2020). In addition to HUBs, binding proteins of mono-ubiquitinated histones were also reported to be involved in the regulation of the defense pathway. (ii) In rice, it is known that the *BRHIS1*-containing complex (carrying SWI/SNF2 ATPase genes) is involved in the repression of some genes like *OsPBZc* and *OsSIRK1* that are known to be involved in defense. These genes are implicated in rice blast resistance caused by *M. oryzae* race EL0917, through their specific interaction with monoubiquitinated H2B and H2A variants (Li et al. 2015a). Hopefully, in the future, many more cases of histone ubiquitinated binding proteins (HUBs) and their involvement in plant–pathogen interactions will be discovered.

15.2.2 Generation of ChIP-seq/CUT&Tag Data for the Study of Histone Modification

ChIP-seq and CUT&Tag are two routinely used techniques for the study of genome-wide histone modifications. In the case of ChIP, the first step in this exercise is the precipitation of chromatin (with DNA) associated with modified histones (in the form of nucleosomes) using specific antibodies (Figure 15.1a). The DNA associated with immunoprecipitated chromatin is then purified, and used for adapter ligation, polymerase chain reaction (PCR) amplification, and sequencing. This provides ChIP-seq libraries for further bioinformatic analysis. In most cases, ChIP-seq libraries are prepared from replicated samples of resistant and susceptible plants along with the control.

In the case of CUT&Tag (Cleavage Under Targets and Tagmentation), the cells or nuclei are incubated with an antibody that is specific for the pA-Tn5 protein (fusion protein involving Tn5 transposase and **protein** A). No immunoprecipitation is involved in CUT&Tag (unlike ChIP-seq), as explained in Figure 15.1b. It involves antibody incubation

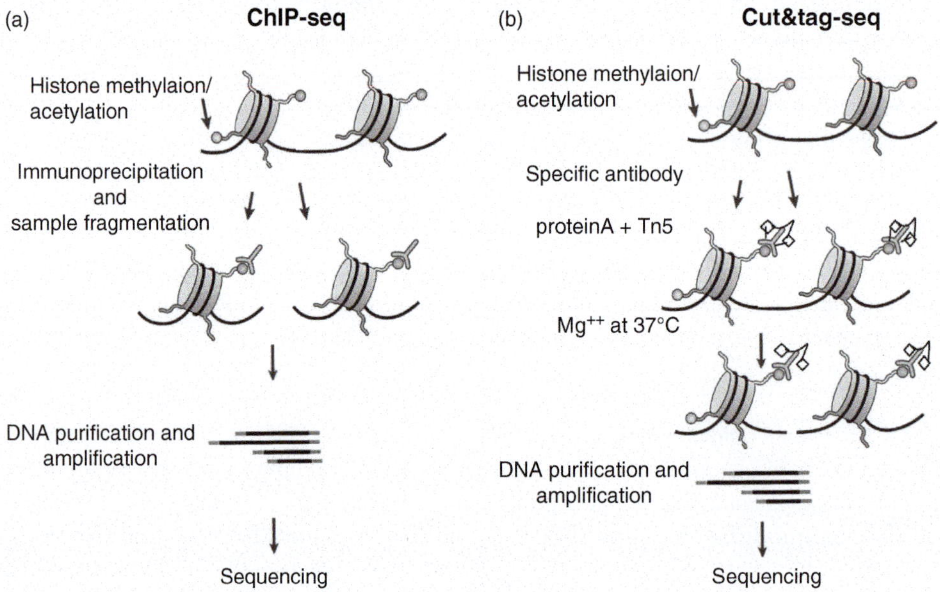

Figure 15.1 Steps involved in two different techniques involved in epigenetic analysis; both techniques involve the use of antibodies for specific proteins associated with the chromatin, (a) ChIP-seq (chromatin immunoprecipitation sequencing) involves immunoprecipitation (b) CUT&Tag-seq (cleavage under targets and segmentation, sequencing) needs relatively little DNA and involves no immunoprecipitation.

followed by chromatin shearing and library preparation. It also involves the use of the transposase enzyme Tn5, which is fused with a specific protein to direct the enzyme Tn5 to the site, where the antibody is associated with this protein.

In both cases (ChIP-seq and CUT&Tag), the control library is prepared from susceptible plants used for the normalization of data (Liang and Keleş 2012a). The CUT&Tag-seq approach has certain advantages over ChIP, including the requirement of a relatively small quantity of DNA and rapid profiling of histone marks (Ouyang et al. 2021).

15.2.3 Bioinformatic Analysis of ChIP-seq and CUT&Tag Data

Several steps are involved in bioinformatics analysis of ChIP-seq and CUT&Tag data; these will be briefly described.

15.2.3.1 Checking Quality of DNA Sequences

Libraries of raw ChIP-seq and CUT&Tag data are first passed through quality check using one or more of the following tools that are available for this purpose: (i) NGS QC (Patel and Jain 2012), (ii) NGS QCbox (Katta et al. 2012), (iii) KAT (Mapleson et al. 2017), (iv) ClinQC (Pandey et al. 2016), (v) HTQC (Yang et al. 2013), (vi) BIGpre (Zhang et al. 2011), (vii) fastp (Chen et al. 2018b), (viii) FastQC (http://www.bioinformatics.babraham.ac.uk/projects/fastqc), (ix) FASTX-toolkit (http://hannonlab.cshl.edu/fastx_toolkit/), (x) QC-Chain (Zhou et al. 2013), (xi) Quack (Thrash et al. 2018), and (xii) HTSQualC (Bedre et al. 2021) etc. A summary of a few quality assessment software is provided in Table 15.1.

Table 15.1 Important features of some quality assessment software.

Features	FastQC	NGS QC	fastp	FASTX-Toolkit	QC-Chain	NGS QCbox	HTSQualC
Automatic identification of quality variant	×	√	×	×	×	×	√
Removal of Uncalled bases	×	√	×	×	×	×	√
Parallel computing	√	√	√	×	√	√	√
Removal of min. length reads	×	×	√	×	×	√	√
Cutting low-quality parts of reads	×	√	√	×	√	√	√
Paired-end reads	×	√	√	×	√	√	√
Removal of low-quality reads	√	√	√	√	√		√
Availability on GitHub	√	×	√	√	×	√	√
Output in FASTA format	×	×	×	×	×	×	√
Input in compressed form	×	√	√	×	×	√	√
Analysis in batch	√	√	×	×	×	√	√
Adapter removal	√	√	√	√	√	×	√

15.2.3.2 Elimination of Low-quality Reads and Bases

After quality check and removal of low-quality read, removal of adapters and low-quality bases is undertaken. The tools available for adapter removal include the following six tools, besides others: (i) Cutadapt (Martin 2012), (ii) Trimmomatic (Bolger et al. 2014), (iii) TrimGalore (https://github.com/FelixKrueger/TrimGalore), (iv) SeqPurge (Sturm et al. 2016), (v) Trimmomatic (http://www.usadellab.org/cms/?page=trimmomatic), and (vi) Cutadapt/TrimGalore 3 (http://cutadapt.readthedocs.io/en/stable/index.html), etc.

15.2.3.3 Correlation Studies

Before starting the mapping exercise for the identification of histone marks, correlation analysis is undertaken for the comparison of samples based on read distribution to identify and eliminate other samples of doubtful merit (Nakato and Sakata 2021). Pearson correlation coefficient (PCC) is calculated using deepTools (Ramírez et al. 2014). Filtered reads are mapped with the reference genome. The different available tools for sequence alignment include the following: (i) MAQ (Li et al. 2008), (ii) Bowtie (Langmead et al. 2009), (iii) SOAP2 (Li et al. 2009), (iv) BWA (Li and Durbin 2009), (v) Novoalign (http://novocraft.com/), (vi) Stampy (Lunter and Goodson 2011), and (vii) Bowtie2 (Langmead and Salzberg 2012). A higher mapping rate reflects a high quality of reads (Nakato and Sakata 2021).

15.2.3.4 Retrieval of Histone Modification Sites

To retrieve the histone modification sites, the first step is peak calling. The peaks obtained thus depict the binding sites of modified histone proteins. Two types of peaks are generally available, broad peaks for H3K27me3 and narrow peaks for H3K4me3. These peaks are used for selection depending on the objective of the study. The available tools include the following: (i) MACS (Zhang et al. 2008), (ii) F-Seq (Boyle et al. 2008a), (iii) PePr and SISSRs (Jothi et al. 2008), (iv) CisGenome (Ji et al. 2008), (v) MACS2; (vi) SICER (Zang et al. 2009), (vii) BayesPeak (Spyrou et al. 2009), (viii) SIPeS (Site Identification from Paired-end Sequencing) (Wang et al. 2010a), (ix) Genrich (https://gensoft.pasteur.fr/docs/Genrich/0.6.1/), (x) MOSAiCS (Kuan et al. 2011), (xi) ZINBA (Rashid et al. 2011), (xii) GEM (Guo et al. 2012), (xiii) MACS2 (Feng et al. 2012), (xiv) Peakzilla (Bardet et al. 2013), (xv) BCP (Xing et al. 2012), (xvi) MUSIC (Harmanci et al. 2014), (xvii) PePr (Zhang et al. 2014), (xviii) Q (Hansen et al. 2015), (xix) JAMM (Ibrahim et al. 2015), (xx) GoPeaks (Yashar et al. 2022), and (xxi) ChIP-R (Newell et al., 2021). Among these tools, the most frequently used tools include MACS and MACS2; these are designed for both single and paired-end read sequencing. GoPeaks is specially designed for the study of histone modification in CUT&Tag data.

15.2.3.5 Reproducibility Analysis

Extracted peaks are then passed through a reproducibility test with a user-defined threshold (e.g. IDR $=0.05$) of irreproducible discovery rate (IDR), which can also be utilized for the assessment of peak callers (Li et al. 2011; Chen et al. 2012) and the quality of experiments (Landt et al. 2012). The peaks assessed on the reproducibility metric are more comparable across experiments rather than based on enrichment metric (e.g. FDR or *p*-value) (Landt et al. 2012; Bailey et al. 2013). IDR (Li et al. 2011) is an R-package to calculate the IDR of ChIP-seq data, and SeqBox (Beccuti et al. 2018) can perform reproducibility analysis for both RNA-seq and ChIP-seq data.

15.2.3.6 Retrieval of Histone Binding Sites

Differential binding sites (DBS) for modified histones are extracted from the peaks retrieved from libraries of susceptible and resistant plants. The tools for extraction of DBS include the following: (i) ChIPDiff (Xu et al. 2008), (ii) DIME (Taslim et al. 2011), (iii) RSEG (Song and Smith 2011), (iv) DBChIP (Liang and Keles 2012b), (v) MAnorm (Shao et al. 2012), (vi) POLYPHEMUS (Mendoza-Parra et al. 2012), (vii) DiffBind (Ross-Innes et al. 2012), (viii) diffReps (Shen et al. 2013), (ix) MultiGPS (Mahony et al. 2014), (x) QChIPat (Liu et al. 2013), (xi) PePr (Zhang et al. 2014), (xii) ChIPComp (Chen et al. 2015), (xiii) histoneHMM (Heinig et al. 2015), and (xiv) ODIN (Allhoff et al. 2014).

An account of algorithms used and important features of various peak calling and DBS finding approaches are listed in Table 15.2. According to the cut-off value of log2Fold change, as defined by the user, *p*-value and/or FDR, high and low-affinity binding sites are determined. These sites are involved in positive/negative effects on gene expression; the place of localization in the genome and the type of enrichment mark of these sites determine heterochromatization or euchromatization of the gene.

15.2.3.7 Genomic and Functional Annotation of Histone binding Sites

These genomic marks can be intergenic or intragenic [exon, intron, promoter (−1 kb to +100 bp), TSS, and TTS (−100 bp to +1 kb)]. This is achieved using an annotation file of reference genome usually in .gff and .gtf formats. The tools for genomic localization include the following: (i) HOMER (Heinz et al. 2010), (ii) CEAS (Shin et al. 2009), and (iii) ChIPseeker (Wang et al. 2022a). Among these, CEAS is used for the annotation of cis-regulatory elements, and ChIPseeker (an R package) is used for ChIP peak annotation, comparison, and visualization.

The functional annotation of DBS is also achieved to know the differential binding genes (DBG) using an annotation file of reference genes. Further, functional annotation of DBGs is conducted by gene ontology analysis using ChipPeakAnno (Zhu et al. 2010) and Blast2GO (Conesa et al. 2005) along with pathway analysis using GREAT (McLean et al. 2010), GSEA (Subramanian et al. 2005) and KEGG pathway analysis. One of the web-server for automated annotation and pathway identification is KOBAS (Wu et al. 2006).

15.3 Chromatin Accessibility

Chromatin accessibility allows the determination of binding sites for RNA polymerases, transcription factors, and other related features (e.g. nucleosome positioning, identification of enhancer sites) involved in the regulation of gene expression. Bioinformatics tools are also available to identify disease-relevant and cell-type-specific regulatory mechanisms, cell-type-specific regulation, etc. Following are the four important techniques that are available for this purpose (Figure 15.2): (i) DNase-seq (DNase I hypersensitive sites sequencing) involves cleavage with DNase I followed by sequencing to identify the regions of regulatory sequences (Boyle et al. 2008b) (Figure 15.2a); (ii) FAIRE-seq (formaldehyde-assisted isolation of regulatory elements sequencing) involves formaldehyde cross-linking, chromatin shearing, phenol extraction, and DNA sequencing, which allows determination of regulatory sequences (Giresi et al. 2007) (Figure 15.2b); (iii) MNase-seq is a DNA

Table 15.2 Some features of peak calling and DBS finding software.

Features	Replicates	Peak calling	Narrow peaks	Broad peaks	Normalization	Statistics	Significance test
MultiGPS	√	Needs	√	×	edgeR	edgeR	*p*-value
ODIN	×	doesn't need	√	√	SES/library size	HMM/Binomial/Poisson distribution	*p*-value
RSEG	×	doesn't need	√	×	–	HMM	–
MAnorm	×	Needs	√	√	MA plot + LOWESS regression	Bayesian model	*p*-value
MMDiff	√	Needs	√	×	DESeq	Nonparametric test	*p*-value
PePr	√	Needs	√	√	edgeR	Binomial distribution	*p*-value
SICER	×	Needs	×	√	Library size	Poisson distribution	FDR
HOMER	×	Needs	√	√	Library size	Poisson distribution	*p*-value/FDR
diffReps	√	Needs	×	√	Linear	Negative binomial test	*p*-value
DBChip	√	Needs	√	×	DESeq	Negative binomial test	FDR
DiffBind	√	Needs	√	√	DESeq2/DESeq/edgeR	DESeq2/DESeq/edgeR	FDR/*p*-value
MASC2	×	doesn't need	√	√	Library size	Predefined cut-off	Likelihood ratio
QChIPat	×	Needs	√	√	Linear/Quartile/Bayes correction	Wilcoxon test	*p*-value
ChIPComp	√	Needs	√	×	Poisson distribution	Wald's test	Posterior probability

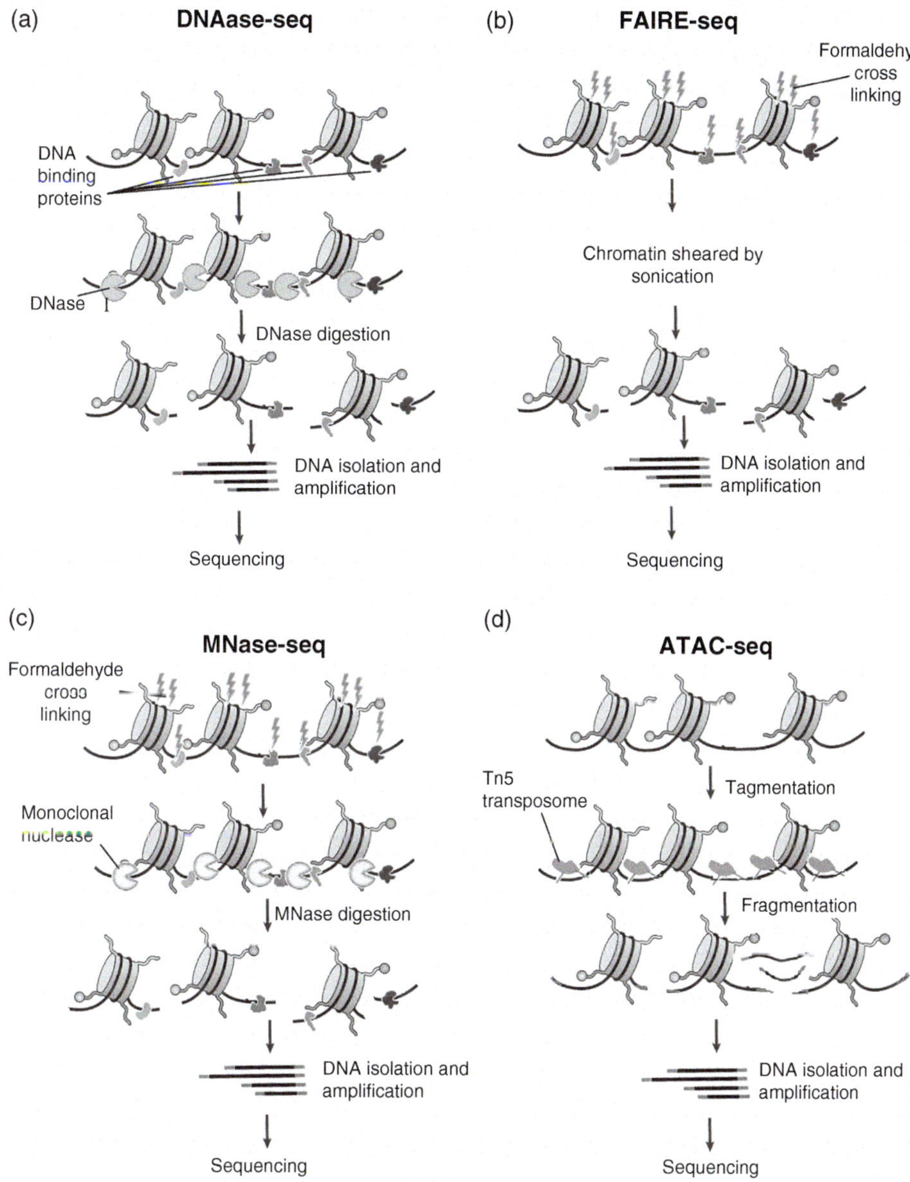

Figure 15.2 Steps involved in four different techniques involved in the mapping of accessible chromatin associated with epigenetic modifications (for details see text).

footprinting method, which allows study of nucleosome positioning and occupancy throughout the genome; it involves digestion of DNA with Micrococcal nuclease (MNase) to examine DNA accessibility (Figure 15.2c);. (iv) ATAC-seq (assay for transposase-accessible chromatin sequencing) involves the use of Tn5 transposase and sequencing adapters (as in CUT&Tag) to identify regions of open chromatin accessible to Tn5 and other important regions (Buenrostro et al. 2015) (Figure 15.2d). CUT&Tag-seq mentioned in the previous section is also utilized to study chromatin accessibility such as transcription factors binding sites etc.

Of the above four approaches, ATAC-seq is a particularly popular method to sequence the regions of open chromatin associated with gene expression, without any prior knowledge of regulatory elements. Further data analysis in ATAC-seq is the same, as used for analysis of ChIP-seq and CUT&Tag data described above. Two specially designed algorithms to extract differential RNA pol III binding sites for prediction of protein binding events include GPS (Guo et al. 2010) and PICS (Zhang et al. 2011). Another tool is POLYPHEMUS (Mendoza-Parra et al. 2012), which is specifically used to extract differential RNA pol III binding sites.

15.4 DNA Methylation

15.4.1 DNA Methylation Sites in the Genome

Among plants, the most important epigenetic modification of wide occurrence is the methylation of carbon-5 of cytosine (5-mC) in DNA. This methylation may involve methylated cytosine either as symmetrical CG or CHG or as asymmetric CHH (where H=A, C, or T) (Henderson and Jacobsen 2007). In plants, on average, CG is highly methylated (80–100%), CHG is moderately methylated (20–100%) and CHH is least methylated (=/<10%) (see Matzke and Mosher 2014; Deleris et al. 2016). DNA methylation can occur in any of the following genomic regions: intergenic regions or intragenic regions including promoters, introns, exons, transcription termination sites (TTSs), and repetitive sequences including transposable elements (TEs) (Saripalli et al. 2020b; Jain et al. 2021).

15.4.2 Methylation/Demethylation Enzymes

DNA methylation is a reversible process, where methylation is carried out with the help of DNA (cytosine-5)-methyltransferase and other enzymes involving a number of different pathways (Cuerda-Gil and Slotkin 2016; Huettel et al. 2006; Liu et al. 2014; Zhang et al. 2018; Huang and Jin 2022). Similarly, demethylation is either a passive process, where the newly synthesized DNA remains unmethylated (this does not require demethylase), or is an active process, where direct removal of methyl group from DNA, is mediated by any of the following four different DNA demethylases: DEMETER (DME), REPRESSOR OF SILENCING 1 (ROS1), DEMETER-LIKE 2 (DML2), and DML3 (Kumar et al. 2018). Mostly, methylation in promoters and TTSs leads to downregulation of gene expression, whereas methylation in exons, introns, and intergenic regions promotes gene expression; although these relationships are not absolute (Mette et al. 2000; Curradi et al. 2002; Dowen et al. 2012; Saripalli et al. 2020b).

15.4.3 Role of DNA Methylation in Disease Resistance

Reversible DNA methylation is involved in differential gene expression in host-pathogen systems, although available results are sometimes contradictory (Curradi et al. 2002; Mette et al. 2000; Dowen et al. 2012; Singh et al. 2022; Deleris et al. 2016; Wang et al. 2018; Huang and Jin 2022; Thakur et al. 2022; Xie and Duan 2023). Some examples of the role of DNA methylation in disease resistance/susceptibility including our studies in *Triticum aestivum-Puccinia triticina* Erikss. & E. Henn system (wheat-leaf rust pathosystem) will be described first. Later we describe the bioinformatics pipelines for facilitating the analysis of the large body of sequencing data for DNA methylation.

DNA hypomethylation is associated with improved resistance in several studies including the following: (i) In Arabidopsis, green peach infestation is associated with DNA hypomethylation in hundreds of loci (Annacondia et al. 2021); regulation of gene activity following differential methylation of CHH context has also been reported in *Arabidopsis-P. syringae* (Dowen et al. 2012). (ii) In tomato, CHH hypomethylation was shown to be associated with nematode pattern-triggered immunity (Atighi et al. 2020); following flg22 treatment, transcriptional activation of some defense-related genes and TEs was also shown to be associated with CHH demethylation (Yu et al. 2013). (iii) In wheat, infestation with *Blumeria graminis* f. sp. *tritici* was found to be associated with hypomethylation of CHH (Geng et al. 2019). Our recent study involving wheat-leaf rust pathosystem also revealed that, although the CHH methylation represents only ~2% of the total genome-wide DNA methylation, its differential methylation in a pair of NILs (with and without the resistance gene *Lr28*) represented >50% of the methylated sites (Singh et al. 2022). Differential methylation of DNA has also been reported in *Ae. tauschii*-powdery mildew (*B. graminis*) (Geng et al. 2019) and wheat-leaf rust pathosystems (Singh et al. 2022).

15.4.4 Techniques and Data Analysis for the Study of DNA Methylation

At least three different approaches for the study of DNA methylation are available, which include MSAP (methylation-sensitive amplified polymorphism), MeDIP (methylated DNA immunoprecipitation) and BS-seq (bisulfite sequencing); all these three methods have been utilized in our laboratory for the study of wheat-leaf rust pathosystems. Omony et al. (2020) described various methods for DNA methylation analysis in plants. These three approaches are briefly described.

15.4.4.1 MSAP and MSAP-seq

MSAP is a modification of the AFLP (amplified fragment length polymorphism) approach that is routinely used for the development of DNA-based molecular markers. In this approach, a pair of isoscizomers (*Hpa*II/*Eco*RI) are used, one of these enzymes (*Hpa*II) being sensitive to methylation and the other not sensitive to methylation (*Eco*RI). Therefore, the patterns of bands obtained due to electrophoresis following digestion by these two enzymes individually will differ. This pattern of electrophoretic bands is utilized to infer the extent of methylation, although actual methylation sites can not be identified by this approach (Figure 15.3).

MSAP-seq involves sequencing of MSAP bands, which differ due to a pair of isoschizomers. This allows the identification of differentially methylated CCGG sites in plants with large genomes, like the wheat genome (Figure 15.4; Chwialkowska et al. 2017). The raw MSAP-seq data will first pass through quality check and adapter removal, followed by mapping with reference genome as mentioned in the section dealing with ChIP-seq data analysis. Mapped reads are then normalized using RPKM (Reads Per Kilobase Million) values and used for extraction of differentially methylated regions (DMR) using DESeq (Anders and Huber, 2010) and DESeq2 (Love et al. 2014). Genomic localization and functional annotation of DMRs are obtained as described earlier in the ChIP-seq data analysis section.

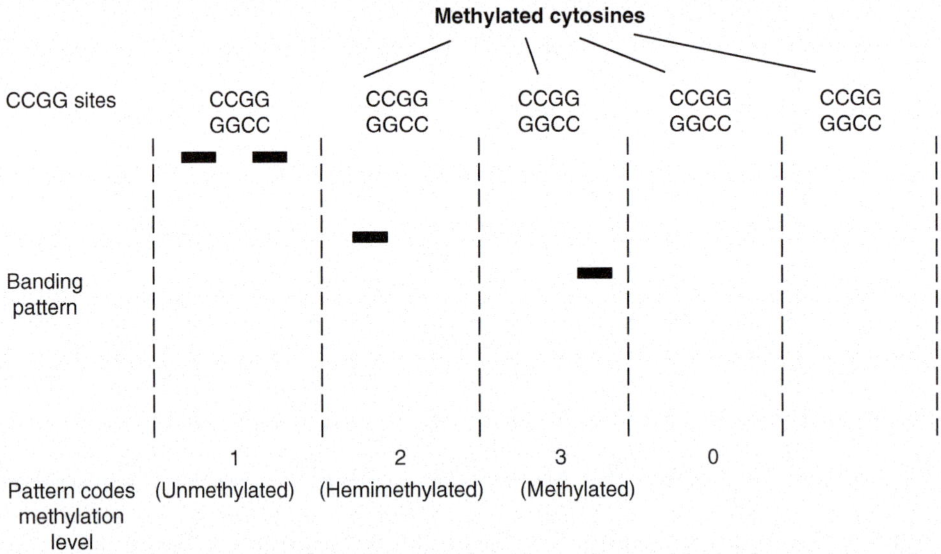

Figure 15.3 A banding pattern commonly observed in methylation-sensitive amplified polymorphism (MSAP) due to cleavage of different methylated CCGG sites by a pair of isoschizomers (*Hpa*II/*Msp*I). Black squares indicate methylated cytosines.

15.4.4.2 MeDIP-seq

MeDIP-seq (methylated DNA immunoprecipitation sequencing) (Figure 15.4b) is a relatively superior technique (relative to MSAP-seq described above) for whole-genome DNA methylation with a resolution of up to 100 bp. First, chromatin is precipitated with 5-methylcytosine (5mC) monoclonal antibodies to extract methylated regions of the genome, and DNA is isolated and purified, which is subjected to sequencing to obtain MeDIP-seq libraries (Neary and Carless 2020). The raw MeDIP-seq data is first subjected to quality check and adapter removal, followed by mapping with reference genome. This is followed by Peak calling, DMR extraction, and annotation. Following MeDIP-seq-specific tools are available for analysis: (i) Batman (Down et al. 2008), (ii) MEDME (Pelizzola et al. 2008), (iii) MeQA (Huang et al. 2012), (iv) MeDUSA (Wilson et al. 2012), (v) MethylCRF (Stevens et al. 2013), (vi) MEDIPS (Lienhard et al. 2014), (vii) BayMeth (Riebler et al. 2014), (viii) coMET (Martin et al. 2015), (ix) DISMISS (Niazi et al. 2016), (x) MethRaFo (Ding and Bar-Joseph 2017), (xi) QSEA (Lienhard et al. 2017), (xii) SIMD (Zhou et al. 2018), and (xii) MeDEStrand (Xu et al. 2018).

15.4.4.3 BS-seq (WGBS/BS-seq and RRBS-seq)

The techniques of whole-genome bisulfite sequencing (WGBS-seq) and reduced representation bisulfite sequencing (RRBS) facilitate genome-wide identification of 5-methyl cytosine (5-mC). The technique is routinely utilized for the study of DNA methylation associated with different traits including plant disease response (Gardiner et al. 2015; Geng et al. 2019; Singh et al. 2022). In both cases, libraries are prepared from sodium bisulfite-treated DNA

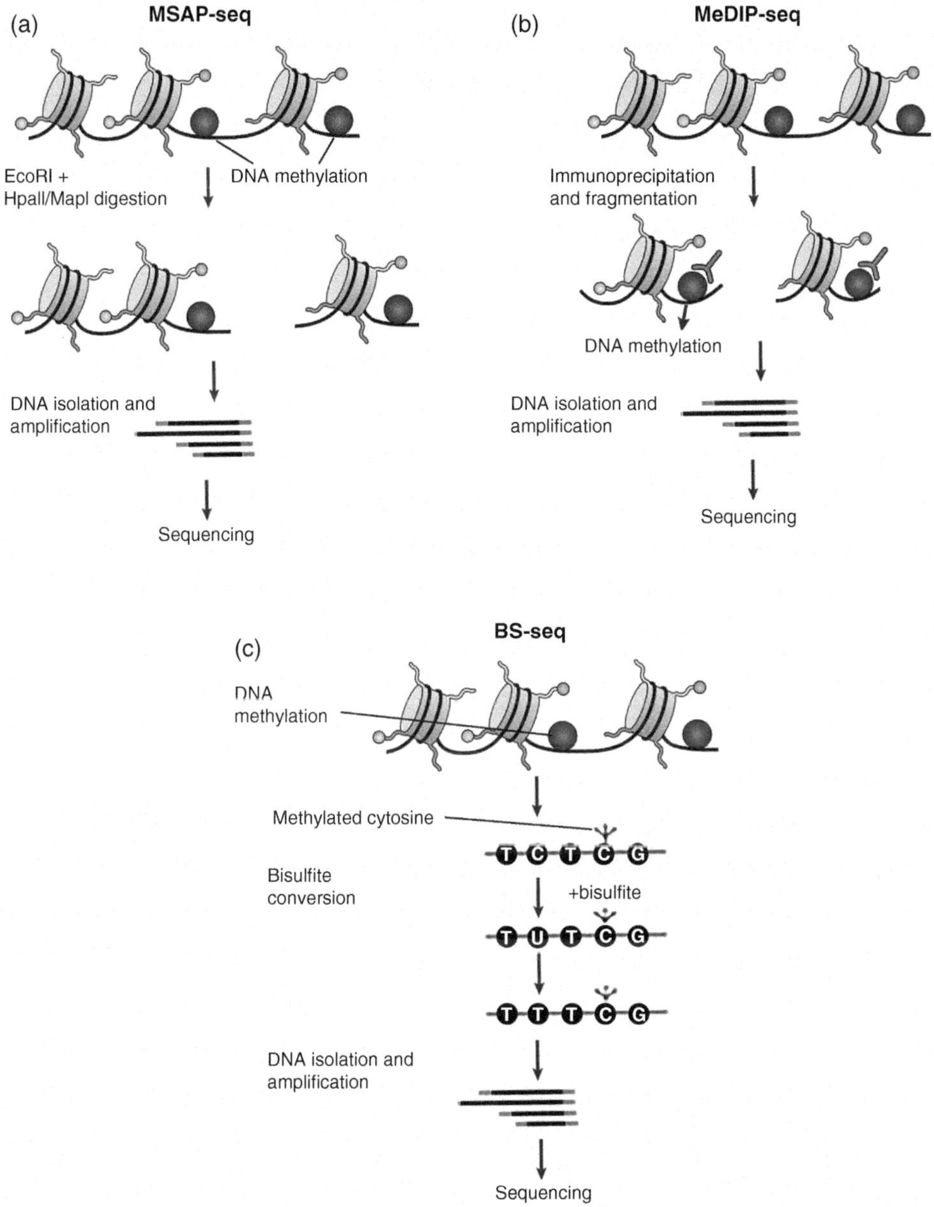

Figure 15.4 Steps involved in three different techniques involved in epigenetic analysis (a) MSAP-seq (methylation sensitive amplification polymorphism) for identification of DNA methylation and (c) BS-seq (Bisulphite sequencing), (b) MeDIP-seq (methylated DNA immunoprecipitation sequencing).

to convert non-methylated cytosines to uracils (Figure 15.4c). These uracils are then read as thymine, while methylated ^{m5}Cs and ^{hm5}Cs are read as cytosines during PCR amplification. WGBS-seq is capable of capturing whole-genome methylation; RRBS-seq is relatively cost-effective but reduces the depth of WGBS-seq by reducing the level of nucleotides to 1% of the genome (Meissner et al. 2005). Despite the reduced depth in RRBS-seq, the sample still includes the majority of promoters and repeat sequences that are generally missing in WGBS-seq (Gu et al. 2011).

BS-seq provides single base-pair resolution and therefore is considered to be the best method to extract information on DNA methylation at the whole-genome level (Kurdyukov and Bullock 2016). It provides resolution for all three contexts (CG, CHG, and CHH). As usual, raw BS-seq data is passed through quality check and adapter removal. The analysis is conducted using tools like BiQ Analyzer (Bock et al. 2005) and BSeQC (Lin et al. 2013), which are specific to BS-seq data for quality control and visualization. Due to the conversion of non-methylated cytosines to uracils, mapping to reference genome and methylation is achieved through tools having the wild card (allowing either Cs or Ts to map to Cs) or those aligning three letters (all Cs converted to Ts). The following tools are available for this purpose: (i) BSMAP (Xi and Wei 2009), (ii) Bismark (Krueger and Andrews 2011), (iii) methylKit (Akalin et al. 2012), (iv) BS-Seeker (Chen et al. 2010), (v) BS-Seeker2 (Guo et al. 2013), (vi) MOABS (Sun et al. 2014), (vii) MethGO (Liao et al. 2015), (viii) BRAT-nova (Harris et al. 2016), (ix) WALT (Chen et al. 2016a), (x) gemBS (Merkel et al. 2018), (xi) BS-Seeker3 (Huang et al. 2018), and (xii) MethylIT (Hafner and Mackenzie 2023). Of these tools, BS-Seeker3 is an ultrafast pipeline for BS-seq mapping and methylation extraction. Later, differentially methylated regions are extracted using one or more of the following tools: (i) bsseqpackage from Bioconductor (Hansen et al. 2012), (ii) DMRfinder (Gaspar and Hart 2017), (iii) DMAP (Stockwell et al. 2014), (iv) MOABS (Sun et al. 2014), (v) DSS-general (Park and Wu 2016), (vi) metilene packages (Jühling et al. 2016), (vii) ChAMP (Tian et al. 2017), and (viii) MethCP (Gong and Purdom 2019). The DMRs obtained thus are subjected to genomic localization into intergenic or intragenic [exon, intron, promoter (−1 kb to +100 bp), TSS, and TTS (−100 bp to +1 kb)] regions using HOMER (Heinz et al. 2010). Further, functional annotation of DMRs is performed with MethylMix 2.0 (Cedoz et al. 2018), gene ontology analysis using Blast2GO (Conesa et al. 2005) along with pathway analysis using GREAT (McLean et al. 2010), GSEA (Subramanian et al. 2005) and KEGG pathway analysis. WBSA (Liang et al. 2014) is a web-based server for BS-seq data analysis. A summary of the bioinformatics tools and methods used at various stages of BS-seq data analysis is presented in Table 15.3.

15.4.4.4 Software for Methylated Transposable Elements (TEs)

Usually, TEs cause plasticity, genetic variation, and evolution due to their ability to move and accumulate in genomes (Del Angel et al. 2018). DNA methylation in TEs provides stability to the genome since hypermethylation of TEs renders them incapable of moving. The stationary position of TEs due to DNA methylation indirectly affects the expression, structure, and function of associated genes (Lisch 2013) through epigenetic mechanisms (Slotkin and Martienssen 2007). A widely used tool for the detection of TEs in large eukaryotic genomes is the REPET package (https://urgi.versailles.inra.fr/Tools/REPET) using TEdenovo; the annotation is achieved using TEannot. RepeatMasker (https://www.repeatmasker.org/cgi-bin/WEBRepeatMasker) is used for the identification of TEs from the Dfam (Hubley et al. 2016) and Repbase (Bao et al. 2015) databases. Some other tools for the identification and classification of TEs include the following: (i) Recon (Bao and Eddy 2002), (ii) RepeatScout (Price et al. 2005), (iii) MITE-hunter (Han and Wessler 2010), (iv) MITE

Table 15.3 A list of different software for bisulfite-sequencing data analyses for mapping, statistical analysis, and complete pipeline.

(a) Mapping Software	Mapping using	(b) Statistical analysis Software	Feature	(c) Complete pipeline Software	Feature
BS Seeker	bowtie2	Bumphunter	Finds DMRs	BiQ Analyzer HT	Quality assessment
Segemehl	Segemehl	MethylMapper	Maps by MethylMapper		
Bismark	Bowtie	BISMA	Uses ClustalW	gemBS	Finds DMRs post-mapping
BRAT-nova	FM-index	RnBeads	Uses bed files		
RRBSMap	RRBSMAP	DMAP	Finds DMRs	SAAP-RRBS	Finds DMRs using hashing/ bitwise masking reads mapping
BS Seeker2	Bowtie2	BSPAT	Uses Bowtie		
BSmooth	bowtie/ bowtie2/ Merman	EpiGRAPH	Uses machine learning		
WALT	WALT	MethGo	Scales at global and gene level	BiQ Analyzer HiMod	An updated version of BiQ Analyzer HT
BS Seeker3	SNAP-aligner				
Metilene	DMR finder	MethylMix	Uses BIC		
BSMap	SOAP	CyMATE	Maps non/ CpG methylation	QUMA	EMBOSS using needleman wunsch
BRAT-BW	FM-index				

Digger (Yang 2013), (v) detectMITE (Ye et al. 2016), (vi) MITE Tracker (Crescente et al. 2018), and (vii) RepeatModeler2 (Flynn et al. 2020). DMRs extracted from DNA methylation analysis (discussed earlier) have to pass through TEs identification, to find how many and which class of TEs are methylated.

15.5 Noncoding RNAs (miRNAs, lncRNA, circRNA)

A variety of noncoding RNAs (ncRNAs) have also been shown to be involved in disease resistance; these ncRNAs include miRNA (microRNA), lncRNA (long ncRNA), and circRNA (circular RNA), which are transcribed from genomic sequences through intermediate long primary transcripts, which are obtained with the help of DNA dependent RNA Polymerase II. Their origin, cellular localization, and functions are shown in Figure 15.5. Among linear ncRNAs, miRNAs are typically 22–24 nucleotide long and lncRNAs are >200bp long (Cech and Steitz 2014). The circRNAs are characterized by a closed-loop structure generated by a covalent bond linking the 5′ and 3′ ends; thus these are more stable than the linear RNAs and are not easily degraded by Ribonuclease R (RNase R); the majority of these circRNAs are transcribed from the protein-coding genes and contain one or more exons (for more details, see Zhang et al. 2020b).

Different types of ncRNA have also been implicated in the response of the host to different diseases (Liu et al. 2017; Xiang et al. 2018; Sun et al. 2018; Yu et al. 2020; Jain et al. 2020;

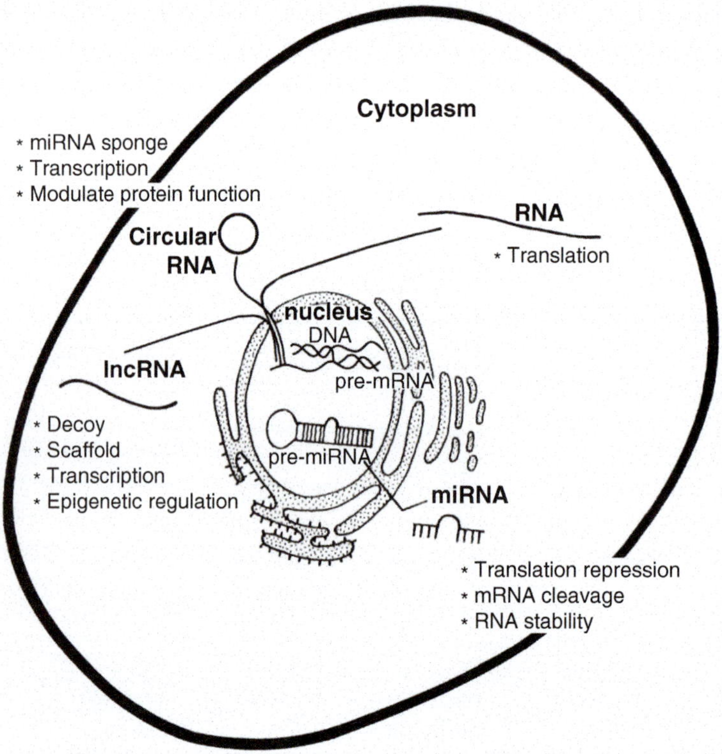

Figure 15.5 Origin, cellular localization, and functions of three noncoding RNAs (miRNA, lncRNA, ciecRNA).

Yang et al. 2021; Song et al. 2021; Chidambara et al. 2022; Zhang et al. 2023; Jain et al. 2023). A few examples of the involvement of ncRNAs in disease response including our studies in wheat-leaf rust pathosystem will be described here along with the bioinformatics approaches used for the analysis of different types of ncRNAs from the available sequencing data.

15.5.1 Small RNA Sequencing and miRNAs

The miRNAs play a role as secondary regulatory molecules by targeting mRNA and lncRNA at post transcription level and cleaving or binding with them to inactivate them. These miRNAs control a variety of traits, but in the present context, these are also known to be associated with defense responses in plant–pathogen interactions (Yang et al. 2015). The miRNAs are involved in both, PTI and ETI. The miR393 was the first miRNA that was identified to be involved in resistance to bacteria in plants by suppressing the auxin response (Navarro et al. 2006). Following are some examples: (i) In cotton, the miR166 and miR159 have been shown to inhibit the expression of virulence genes of the pathogen *Verticillium dahlia* (Zhang et al. 2016). (ii) In potato, miRNAs (e.g. miR482) have also been reported to suppress the NBS-LRR defense genes during *Vd* infection (Yang et al. 2021); (iii) In poplar, miRNAs have been shown to work against *Colletotrichum gloeosporioides* and *Cytospora chrysosperma* infection (Su et al. 2018). Another miRNA, namely miR482/2118, targets NBS-LRR genes in several eudicots (Zhang et al. 2022).

In our studies involving the wheat-leaf rust pathosystem, eight miRNAs showing differential expression between resistant and susceptible NILs were found to target several important disease-responsive genes (Jain et al. 2020). Later, using the same pathosystem, we identified eight miRNAs (miR156, miR158, miR159, miR168, miR169, miR172, miR319, miR396) that were induced during downstream disease resistance pathways (Jain et al. 2023).

For the identification of miRNAs involved in disease resistance, small RNA-seq (sRNA-seq) data libraries are used. The following steps are involved: (i) As a usual step, sRNA-seq raw reads are passed through quality check and adapter trimming. (ii) The rRNAs and TEs are removed; (iii) the filtered reads are mapped with Rfam (Kalvari et al. 2020), and TREP (Wicker et al. 2007) databases for rRNAs, and TEs, respectively. (iv) Remaining unmapped reads are potential miRNAs and are mapped with miRbase (Kozomara and Griffiths-Jones 2010) database to extract conserved miRNAs by BLAST (Altschul et al. 1990) or miReader (Jha and Shankar 2013), which involves imperfect DICER processing and permits mismatches of six nucleotides at 3′end and three nucleotides at 5′end. (v) The unmapped reads are mapped with miRNA precursor sequences, which allow mismatches, treating them as point mutations or RNA editing. Thus, the remaining unmapped reads should contain novel miRNAs, which can be identified by aligning them with reference genome using one or more of the following tools: (i) RSEM (Merino et al. 2019), (ii) Alexa-seq (Griffith et al. 2010), (iii) Cufflinks (Trapnell et al. 2010), (iv) StringTie (Pertea et al. 2015), (v) FluxCapacitor (Montgomery et al. 2010), and (vi) MISO (Katz et al. 2010).

The mapped reads along with ~100 nt flanking sequences are passed through RNA folding and/or star strand expression. The following tools are used for RNA folding with cut-off free energy <25 kcal/mol: (i) RNAmicro (Hertel and Stadler 2006), (ii) miRNAminer (Artzi et al. 2008), (iii) miRNAFold (Tempel and Tahi 2012), and (iv) deepMiRGene (Park et al. 2016). After, ensuring stability, the post-trimming and refolding novel precursor miRNAs are extracted using miRdeep (Yang and Li 2011) based on the star strand expression method to identify novel miRNAs based on higher expression of mature miRNA in comparison to the star strand and loop sequences. The abundance of miRNA expression is calculated after normalizing the read count for each identified miRNA with a total number of mapped miRNAs. In case of sRNA-seq data, libraries prepared from susceptible and resistant plants to study disease resistance, abundance of miRNAs is calculated using one of the following tools, which are specific for small RNAs: (i) miRDeep2 (Friedlander et al. 2012), (ii) sRNAbench (Barturen et al. 2014), (iii) miRge (Baras et al. 2015), (iv) ShortStack (Johnson et al. 2016), and (v) Manatee (Handzlik et al. 2020). Some other tools, which could be used for extraction of differentially expressed miRNAs, included the following: (i) DEGseq (Wang et al. 2010b), (ii) edgeR (Robinson et al. 2010), (iii) PoissonSeq (Li et al. 2012), (iv) DEseq2 (Love et al. 2014), (v) Cuffdiff (Trapnell et al. 2013), (vi) Ballgown (Frazee et al. 2015), (vii) Sleuth (Pimentel et al. 2017), and (viii) baySeq (Hardcastle and Kelly, 2010).

Most *in silico* tools are based on the target binding site of miRNAs, which are conserved. These tools include the following: (i) EMBL (Stark et al. 2005), (ii) PicTar (Krek et al. 2005). (iii) PITA Top (Kertesz et al. 2007), (iv) EIMMo (Gaidatzis et al. 2007), (v) mirWIP

(Hammell et al. 2008), (vi) miRBase Targets (Griffiths-Jones et al. 2008), (vii) miRanda (Betel et al. 2010), (viii) DIANA-microT (Reczko et al. 2012), (ix) TargetScan (Nam et al. 2014), and (x) DeepTarget (Lee et al. 2016b). Few tools such as RNA22 (Miranda et al. 2006), PITA All (Kertesz et al. 2007), and TargetScan All (Nam et al. 2014), etc. are independent of conservation status.

Degradome-seq and PARE (parallel analysis of RNA ends)-seq are other high-throughput sequencing approaches to identify miRNA targets and miRNA cleavage sites. For the identification of mRNA targets (both novel and conserved) of miRNAs and to identify miRNA cleavage sites, degradome sequencing (Degradome-Seq) or parallel analysis of RNA ends (PARE) is used. A database termed starBase (Yang et al., 2011) is also available for the identification of miRNAs through their cleavage sites that are obtained from Degradome-seq.

15.5.2 Long ncRNA (lncRNAs)

Hundreds of differentially expressed lncRNA are now known, which are associated with response to fungal diseases in a variety of plant systems (for a review see Zhang et al. 2020c). For instance, tomato lncRNA16397 is known to induce the expression of ROS accumulation, which contributes to resistance against *P. infestans* (Cui et al. 2017). The upregulation of a lncRNA (LINC-AP2) has also been shown to be negatively correlated with the expression of the AP2 gene, following the infection of Arabidopsis by the Turnip crinkle virus (Gao et al. 2016). Several lncRNAs have also been shown to be co-regulated with TFs (NAC17L, NAC68L, WRKY55L, C2H2, and WRKY64/70) involved in plant response to pathogen stress suggesting the possible role of lncRNAs in differential expression of downstream defense-related genes (Hong et al. 2018). In rice, 39 jasmonic acid (JA)-related genes were identified as the targets of 73 lncRNAs following infection by *Xoo*. Overexpression of a specific lncRNA (ALEX1) also suggested its involvement in the activation of the JA pathway and resistance to bacterial blight (Yu et al. 2020). In our studies involving the wheat-leaf rust pathosystem, 22 differentially expressed lncRNAs were identified in a pair of near-isogenic lines (NILs), which differed for the gene *Lr28* (Jain et al. 2020). Three of these lncRNAs were shown to function as target mimics (TMs), which compete with the mRNA targets for miRNA, thus regulating the expression of target genes. lncRNAs are also extracted from RNA-seq data libraries using the following steps. (i) RNA-seq data libraries are subjected to routine quality check and adapter removal (as also done in ChIP-seq data analysis); (ii) the filtered reads are mapped with the reference genome, followed by transcript assembly using splice aligner such as HISAT (Kim et al., 2015), HISAT2 (Kim et al. 2019), and HISAT-3N (Zhang et al. 2021); (iii) the transcripts with ≥200 nt are subjected to evaluation of coding potential using one of the following tools: CPC (Kong et al. 2007), CNCI (Sun et al. 2013), CPC2 (Kang et al. 2017), and CNIT (Guo et al. 2019); (iv) this followed by PhyloCSF (Phylogenetic Codon Substitution Frequencies) (Lin et al. 2011), and PFAM protein domain analyses (Mistry et al. 2021).

Long ncRNAs have also been shown to function as transcription factors for regulating the expression of mRNA, miRNA, and regulatory proteins. The following tools are used to find mRNA targets of lncRNAs: lncTar (Li et al. 2015b), lncRNATargets (Hu and Sun 2016), LncRRIsearch (Fukunaga et al. 2019), and WAFNRLTG (Li et al. 2022), etc. Similarly, miRNA targets of lncRNAs could be identified using LncRRIseach (Fukunaga et al. 2019), GCLMI (Huang et al. 2019), etc. lncRNA and protein interactions could be predicted using RPI-SAN

(Yi et al. 2018), RPITER (Peng et al. 2019), etc. Differential expression of lncRNAs is routinely done as in other cases. The databases that are specific to lncRNAs include LncAtlas (Mas-Ponte et al. 2017), DIANA-LncBase (Karagkouni et al. 2019), lncRNAdb (Quek et al. 2015), and NONCODE (Fang et al. 2017). Complete details about lncRNA and their analysis are available in a recent review (Mattick et al. 2023). Some of the databases are available for lncRNA such as PLncDB V2.0 (Jin et al. 2021) and GreeNC 2.0 (Di Marsico et al. 2022).

15.5.3 Circular RNAs

The circRNAs in plants are another class of ncRNA, which is involved in the regulation of the expression of a variety of genes including those involved in disease resistance. Following are some examples: (i) In Arabidopsis, co-expression of circR194 and circR4022 with their parent genes was observed in response to infection by *P. syringae*. Similarly, co-expression of circR11208 with its parent gene was observed in response to *Botrytis cinerea* infection (Wang et al. 2022b); (ii) In tomato, during response to infection by tomato leaf curl Bengaluru virus (ToLCBaV), a correlation was observed between the relative expression of three circRNAs and the expression of their respective parent genes (Chidambara et al. 2022).

To extract circular RNAs, RNA-seq data libraries are first prepared from total RNA, which is first digested by RNase R to remove linear RNAs containing G-quadruplexes or structured 3′ ends (Xiao and Wilusz 2019). Raw reads of RNA-seq data are passed through routine quality checks and adapter removal. This is followed by mapping with reference genome using mapping tools like the following: (i) Browniel (Langmead et al. 2009), (ii) Bowtie2 (Langmead and Salzberg 2012), (iii) BWA (Li and Durbin 2009), (iv) STAR (Dobin et al. 2013), (v) BLAT (Kent 2002), and (vi) TopHat-Fusion (Kim and Salzberg 2011) etc.

CircRNA is identified from the mapped reads using tools based on the candidate approach such as [KNIFE (Szabo et al. 2015), PTESFinder (Izuogu et al. 2016), NCLScan (Chuang et al. 2016), CIRCexplorer (Dong et al. 2019), and CircRNAwrap (Li et al. 2019)] or based on segmented read approach involving the following tools: (i) circRNA_finder (Westholm et al. 2014), (ii) CIRI (Gao et al. 2015), (iii) find_circ (Memczak et al. 2013), (iv) CIRCexplorer (Dong et al. 2019), (v) CircPro (Meng et al. 2017), and (vi) circtools (Jakobi et al. 2018), (vii) DCC (Cheng et al. 2016), (viii) MapSplice (Wang et al. 2010c), (ix) UROBORUS (Song et al. 2016), and (x) Segemehl (Hoffmann et al. 2014). Further, genomic localization of lncRNA genes is performed using an annotation file of the reference genome.

A few databases such as circBase (Glazar et al. 2014), CIRCpedia (Dong et al. 2018), and PlantcircBase (Chu et al. 2017) are available for circRNAs along with their annotation. There are only a few tools available to identify the translational potential of circRNAs such as CirCode (Sun and Li 2019); however, CPC (Kong et al. 2007), CNCI (Sun et al. 2013), CPC2 (Kang et al. 2017), and CNIT (Guo et al. 2019), etc. are the tools used for all ncRNAs. Differential expression analysis of circRNA is similar to miRNA. Ferrero et al. (2021) described the details of computational analysis of circRNA expression data. Some of the databases such as circRNADb (Chen et al. 2016b), circBase (Glazar et al. 2014), and circNet (Liu et al. 2016) are available.

15.6 Conclusions and Future Perspectives

Epigenetics is a relatively recent area of research, which involves the identification of heritable changes in chromatin and DNA with no change in the sequences of DNA. Epigenetic changes influence almost all traits including disease resistance, the subject of this chapter. Epigenetic modifications involving resistance against leaf rust in wheat have also been the area of recent research for a multi-institutional team involving the authors of this chapter.

The major epigenetic modifications discussed in this chapter include histone modifications, DNA methylation, and ncRNAs (including miRNAs, lncRNA, and circRNA), each involved in the regulation of expression of a variety of genes involved in disease resistance. For analysis of each of these epigenetic modifications and their involvement in disease resistance, the available techniques have been described in this chapter; most of these techniques make use of high-throughput NGS techniques at the whole-genome level. The emphasis in this chapter has, however, been laid on bioinformatics analysis and the tools utilized at different steps for the analysis of data for each epigenetic modification.

The development of methods and bioinformatics tools for the analysis of data is an ongoing activity in different parts of the world, such that newer methods and tools are regularly developed and used in epigenetics research. In this connection, the development of Alphafold and its modified versions developed by DeepMind Technologies based in the UK is a good example. This software has already been used for determination of the 3D structure of >200 million proteins (https://alphafold.ebi.ac.uk/). Another important development in recent years is the use of third-generation long single molecule SMRT PacBio and nanopore techniques for analysis of epigenetic modifications (including methylation of 5-methyl cytosine and histone modifications) through direct DNA sequencing (Yue et al. 2022; Ni et al. 2023; Wang et al. 2023). These long single-molecule sequencing technologies have been suitably modified for the detection of epigenetic modifications. PacBio's "ccsmeth" and "ccsmethphase" are two such recently modified robust tools for detecting DNA 5-methylcytosines (Ni et al. 2023).

Epigenetic modifications may also be artificially manipulated through the use of chemicals, which function as "writers" and "erasers." This would require complete knowledge of the effect of different epigenetic modifications. At least three classes of receptors including cell surface-localized PRR (pattern-recognition receptors) and intracellular receptors like leucine-rich NLRs [nucleotide-binding and oligomerization domain (NOD) like receptors], and RLRs [retinoid acid-inducible gene-I (RIG-I)-like receptors] are known to be involved in PTI and ETI. Pathogenesis-related (PR) genes are also involved in defense pathways, involving classical plant hormones including salicylic acid (SA), jasmonic acid (JA), and ethylene (ET), which are known to play key roles in the plant immune response.

Some chemical agents have been identified that enhance plant resistance through an epigenetic mechanism. Following are some examples (mentioned earlier also): (i) In rice, application of 5-azadeoxycytidine (5AzaC, a DNA de-methylating agent) enhances resistance against *X. oryzae* (a bacterial pathogen causing bacterial blight) (Akimoto et al. 2007). (ii) In several plant species, BABA has been shown to induce histone modifications (like H3K4me3 or H3K36me3) leading to SAR that imparts improved resistance against a number of pathogens like *P. syringae* (a hemibiotrophic bacterium causing a number of diseases in fruit trees including grapes), *Botrytis cinerea* (a necrotrophic fungus causing diseases like grey mold disease), *Phytophthora parasitica* (oomycetes causing diseases like root rot and crown rot), *Pectobacterium carotovorum* (a necrotrophic bacterium causing soft rot disease). It is thus obvious that BABA can be used to improve disease resistance involving epigenetic modifications in a variety of crops (for a review, see Xie and Duan 2023).

Epialleles for individual R genes and many other individual genes, including those mentioned above, are involved in defense pathways associated with epigenetics. Similarly, epiQTLs using epiRILs and marker-trait associations (MTAs) using EWAS (epigenome-wide association studies) or MWAS (methylation-wide association studies (like GWAS) have also been identified in some plant species (Dubin et al. 2015). This subject is sometimes described as quantitative epigenetics, which is another important area of research that needs attention. Similarly, epiRILs have been developed and used for the identification of epiQTLs for disease resistance in Arabidopsis. DNA-based markers associated with these epialleles and epiQTLs may also be designed, validated, and used for MAS (marker assisted selection), MARS (marker-assisted recurrent selection), and GS (genomic selection).

References

Akalin, A., Kormaksson, M., Li, S. et al. (2012). MethylKit: a comprehensive R package for the analysis of genome-wide DNA methylation profiles. *Genome Biology* 13: R87. https://doi.org/10.1186/gb-2012-13-10-r87.

Akimoto, K., Katakami, H., Kim, H.J. et al. (2007). Epigenetic inheritance in rice plants. *Annals of Botany* 100: 205–217. https://doi.org/10.1093/aob/mcm110.

Allhoff, M., Seré, K., Chauvistré, H. et al. (2014). Detecting differential peaks in ChIP-seq signals with ODIN. *Bioinformatics* 30: 3467–3475. https://doi.org/10.1093/bioinformatics/btu722.

Altschul, S.F., Gish, W., Miller, W. et al. (1990). Basic local alignment search tool. *Journal of Molecular Biology* 215: 403–410. https://doi.org/10.1016/S0022-2836(05)80360-2.

Alvarez-Venegas, R., Abdallat, A.A., Guo, M. et al. (2007). Epigenetic control of a transcription factor at the cross section of two antagonistic pathways. *Epigenetics* 2: 106–113. https://doi.org/10.4161/epi.2.2.4404.

Anders, S. and Huber, W. (2010). Differential expression analysis for sequence count data. *Genome Biology* 11: R106. https://doi.org/10.1186/gb-2010-11-10-r106.

Annacondia, M.L., Markovic, D., Reig-Valiente, J.L. et al. (2021). Aphid feeding induces the relaxation of epigenetic control and the associated regulation of the defense response in Arabidopsis. *New Phytologist* 230: 1185–1200. https://doi.org/10.1111/nph.17226.

Artzi, S., Kiezun, A., and Shomron, N. (2008). miRNAminer: a tool for homologous microRNA gene search. *BMC Bioinformatics* 9: 39. https://doi.org/10.1186/1471-2105-9-39.

Atighi, M.R., Verstraeten, B., De Meyer, T. et al. (2020). Genome wide DNA hypomethylation shapes nematode pattern-triggered immunity in plants. *New Phytologist* 227: 545–558. https://doi.org/10.1111/nph.16532.

Bailey, T., Krajewski, P., Ladunga, I. et al. (2013). Practical guidelines for the comprehensive analysis of ChIP-seq data. *PLoS Computational Biology* 9: e1003326. https://doi.org/10.1371/journal.pcbi.1003326.

Bao, Z. and Eddy, S.R. (2002). Automated *de novo* identification of repeat sequence families in sequenced genomes. *Genome Research* 12: 1269–1276. https://doi.org/10.1101/gr.88502.

Bao, W., Kojima, K.K., and Kohany, O. (2015). Repbase update, a database of repetitive elements in eukaryotic genomes. *Mobile DNA* 6: 11. https://doi.org/10.1186/s13100-015-0041-9.

Baras, A.S., Mitchell, C.J., Myers, J.R. et al. (2015). miRge – a multiplexed method of processing small RNA-Seq data to determine microRNA entropy. *PLoS One* 10: e0143066. https://doi.org/10.1371/journal.pone.0143066.

Bardet, A.F., Steinmann, J., Bafna, S. et al. (2013). Identification of transcription factor binding sites from ChIP-seq data at high resolution. *Bioinformatics* 29: 2705–2713. https://doi.org/10.1093/bioinformatics/btt470.

Barturen, G., Rueda, A., Hamberg, M. et al. (2014). sRNAbench: profiling of small RNAs and its sequence variants in single or multi-species high-throughput experiments. *Methods in Next Generation Sequencing* 1: 21–31. https://doi.org/10.2478/mngs-2014-0001.

Beccuti, M., Cordero, F., Arigoni, M. et al. (2018). SeqBox: RNAseq/ChIPseq reproducible analysis on a consumer game computer. *Bioinformatics* 34: 871–872. https://doi.org/10.1093/bioinformatics/btx674.

Bedre, R., Avila, C., and Mandadi, K. (2021). HTSQualC is a flexible and one-step quality control software for high-throughput sequencing data analysis. *Scientific Reports* 11: 18725. https://doi.org/10.1038/s41598-021-98124-3.

Berr, A., McCallum, E.J., Alioua, A. et al. (2010). Arabidopsis histone methyltransferase SET DOMAIN GROUP8 mediates induction of the jasmonate/ethylene pathway genes in plant defense response to necrotrophic fungi. *Plant Physiology* 154: 1403–1414. https://doi.org/10.1104/pp.110.161497.

Betel, D., Koppal, A., Agius, P. et al. (2010). Comprehensive modeling of microRNA targets predicts functional non-conserved and non-canonical sites. *Genome Biology* 11: R90. https://doi.org/10.1186/gb-2010-11-8-r90.

Black, J.C., Van Rechem, C., and Whetstine, J.R. (2012). Histone lysine methylation dynamics: establishment, regulation, and biological impact. *Molecular Cell* 48: 491–507. https://doi.org/10.1016/j.molcel.2012.11.006.

Bock, C., Reither, S., Mikeska, T. et al. (2005). BiQ analyzer: visualization and quality control for DNA methylation data from bisulfite sequencing. *Bioinformatics* 21: 4067–4068. https://doi.org/10.1093/nar/gkr312.

Bolger, A.M., Lohse, M., and Usadel, B. (2014). Trimmomatic: a flexible trimmer for Illumina sequence data. *Bioinformatics* 30: 2114–2120. https://doi.org/10.1093/bioinformatics/btu170.

Boyle, A.P., Guinney, J., Crawford, G.E. et al. (2008a). F-Seq: a feature density estimator for high-throughput sequence tags. *Bioinformatics* 24: 2537–2538. https://doi.org/10.1093/bioinformatics/btn480.

Boyle, A.P., Davis, S., Shulha, H.P. et al. (2008b). High-resolution mapping and characterization of open chromatin across the genome. *Cell* 132: 311–322. https://doi.org/10.1016/j.cell.2007.12.014.

Buenrostro, J., Wu, B., Chang, H. et al. (2015). ATAC-seq: a method for assaying chromatin accessibility genome-wide. *Current Protocols in Molecular Biology* 109: 21.29.1–21.29.9. https://doi.org/10.1002/0471142727.mb2129s109.

Cech, T.R. and Steitz, J. (2014). The noncoding RNA revolution-trashing old rules to forge new ones. *Cell* 157: 77–94. https://doi.org/10.1016/j.cell.2014.03.008.

Cedoz, P.L., Prunello, M., Brennan, K. et al. (2018). MethylMix 2.0: an R package for identifying DNA methylation genes. *Bioinformatics* 34: 3044–3046. https://doi.org/10.1093/bioinformatics/bty156.

Chan, C. and Zimmerli, L. (2019). The histone demethylase IBM1 positively regulates Arabidopsis immunity by control of defense gene expression. *Frontiers in Plant Science* 10: 1587. https://doi.org/10.3389/fpls.2019.01587.

Chen, P.-Y., Cokus, S.J., and Pellegrini, M. (2010). BS seeker: precise mapping for bisulfite sequencing. *BMC Genomics* 11: 203. https://doi.org/10.1186/1471-2105-11-203.

Chen, Y., Negre, N., Li, Q. et al. (2012). Systematic evaluation of factors influencing ChIP-seq fidelity. *Nature Methods* 9: 609–614. https://doi.org/10.1038/nmeth.1985.

Chen, L., Wang, C., Qin, Z.S. et al. (2015). A novel statistical method for quantitative comparison of multiple ChIP-seq datasets. *Bioinformatics* 31: 1889–1896. https://doi.org/10.1093/bioinformatics/btv094.

Chen, H., Smith, A.D., and Chen, T. (2016a). WALT: fast and accurate read mapping for bisulfite sequencing. *Bioinformatics* 32: 3507–3509. https://doi.org/10.1093/bioinformatics/btw490.

Chen, X., Han, P., Zhou, T. et al. (2016b). circRNADb: a comprehensive database for human circular RNAs with protein-coding annotations. *Scientific Reports* 6: 34985. https://doi.org/10.1038/srep34985.

Chen, Y., Wang, J., Yang, N. et al. (2018a). Wheat microbiome bacteria can reduce virulence of a plant pathogenic fungus by altering histone acetylation. *Nature Communications* 9: 3429. https://doi.org/10.1038/s41467-018-05683-7.

Chen, S., Zhou, Y., Chen, Y. et al. (2018b). fastp: an ultra-fast all-in-one FASTQ preprocessor. *Bioinformatics* 34: i884–i890. https://doi.org/10.1093/bioinformatics/bty560.

Cheng, J., Metge, F., and Dieterich, C. (2016). Specific identification and quantification of circular RNAs from sequencing data. *Bioinformatics* 32: 1094–1096. https://doi.org/10.1093/bioinformatics/btv656.

Chidambara, B., Elangovan, D., Avverahally, S. et al. (2022). Identification of circular RNAs in resistant tomato genotype in response to ToLCBaV infection. *Journal of Horticultural Sciences* 17: 496–504. https://doi.org/10.24154/jhs.v17i2.1520.

Chu, Q., Zhang, X., Zhu, X. et al. (2017). PlantcircBase: a database for plant circular RNAs. *Molecular Plant* 10: 1126–1128. https://doi.org/10.1016/j.molp.2017.03.003.

Chuang, T.J., Wu, C.S., Chen, C.Y. et al. (2016). NCLscan: accurate identification of non-co-linear transcripts (fusion, trans-splicing and circular RNA) with a good balance between sensitivity and precision. *Nucleic Acids Research* 44: e29. https://doi.org/10.1093/nar/gkv1013.

Chwialkowska, K., Korotko, U., Kosinska, J. et al. (2017). Methylation sensitive amplification polymorphism sequencing (MSAP-Seq)—a method for high-throughput analysis of differentially methylated CCGG sites in plants with large genomes. *Frontiers in Plant Science* 8: 2056. https://doi.org/10.3389/fpls.2017.02056.

Conesa, A., Götz, S., García-Gómez, J.M. et al. (2005). Blast2GO: a universal tool for annotation, visualization and analysis in functional genomics research. *Bioinformatics* 21: 3674–3676. https://doi.org/10.1093/bioinformatics/bti610.

Crescente, J., Zavallo, D., Helguera, M. et al. (2018). MITE Tracker: an accurate approach to identify miniature inverted-repeat transposable elements in large genomes. *BMC Bioinformatics* 19: 348. https://doi.org/10.1186/s12859-018-2376-y.

Cuerda-Gil, D. and Slotkin, R.K. (2016). Non-canonical RNA-directed DNA methylation. *Nature Plants* 2: 16163. https://doi.org/10.1038/nplants.2016.163.

Cui, J., Luan, Y., Jiang, N. et al. (2017). Comparative transcriptome analysis between resistant and susceptible tomato allows the identification of lncRNA16397 conferring resistance to *Phytophthora infestans* by co-expressing glutaredoxin. *The Plant Journal* 89: 577–589. https://doi.org/10.1111/tpj.13408.

Curradi, M., Izzo, A., Badaracco, G. et al. (2002). Molecular mechanisms of gene silencing mediated by DNA methylation. *Molecular and Cellular Biology* 22: 3157–3173. https://doi.org/10.1128/MCB.22.9.3157-3173.2002.

Del Angel, V.D., Hjerde, E., Sterck, L. et al. (2018). Ten steps to get started in genome assembly and annotation. *F1000Research* 7: ELIXIR-148. https://doi.org/10.12688/f1000research.13598.1.

Deleris, A., Halter, T., and Navarro, L.D. (2016). DNA methylation and demethylation in plant immunity. *Annual Review of Phytopathology* 54: 579–603. https://doi.org/10.1146/annurev-phyto-080615-100308.

Dhawan, R., Luo, H., Foerster, A.M. et al. (2009). HISTONE MONOUBIQUITINATION1 interacts with a subunit of the mediator complex and regulates defense against necrotrophic fungal pathogens in Arabidopsis. *Plant Cell* 21: 1000–1019. https://doi.org/10.1105/tpc.108.062364.

Di Marsico, M., Gallart, A.P., Sanseverino, W., and Cigliano, R.A. (2022). GreeNC 2.0: a comprehensive database of plant long non-coding RNAs. *Nucleic Acids Research* 50: D1442–D1447. https://doi.org/10.1093/nar/gkab1014.

Ding, J. and Bar-Joseph, Z. (2017). MethRaFo: MeDIP-seq methylation estimate using a Random Forest Regressor. *Bioinformatics* 33: 3477–3479. https://doi.org/10.1093/bioinformatics/btx449.

Ding, B. and Wang, G.L. (2015). Chromatin versus pathogens: the function of epigenetics in plant immunity. *Frontiers in Plant Science* 6: 675. https://doi.org/10.3389/fpls.2015.00675.

Dobin, A., Davis, C.A., Schlesinger, F. et al. (2013). STAR: ultrafast universal RNA-seq aligner. *Bioinformatics* 29: 15–21. https://doi.org/10.1093/bioinformatics/bts635.

Dong, R., Ma, X.K., Li, G.W. et al. (2018). CIRCpedia v2: an updated database for comprehensive circular RNA annotation and expression comparison. *Genomics, Proteomics & Bioinformatics* 16: 226–233. https://doi.org/10.1016/j.gpb.2018.08.001.

Dong, R., Ma, X.K., Chen, L.L. et al. (2019). Genome-wide annotation of circRNAs and their alternative back-splicing/splicing with CIRCexplorer Pipeline. In: *Epitranscriptomics* (ed. N. Wajapeyee and R. Gupta), 137–149. New York: Springer https://doi.org/10.1007/978-1-4939-8808-2_10.

Dowen, R.H., Pelizzola, M., Schmitz, R.J. et al. (2012). Widespread dynamic DNA methylation in response to biotic stress. *Proceedings of the National Academy of Sciences of the United States of America* 109: E2183–E2189. https://doi.org/10.1073/pnas.1209329109.

Down, T.A., Rakyan, V.K., Turner, D.J. et al. (2008). A Bayesian deconvolution strategy for immunoprecipitation-based DNA methylome analysis. *Nature Biotechnology* 26: 779–785. https://doi.org/10.1038/nbt1414.

Dubin, M.J., Zhang, P., Meng, D. et al. (2015). DNA methylation in Arabidopsis has a genetic basis and shows evidence of local adaptation. *eLife* 4: e05255. https://doi.org/10.7554/eLife.05255.

Dutta, A., Choudhary, P., Caruana, J. et al. (2017). JMJ27, an Arabidopsis H3K9 histone demethylase, modulates defense against *Pseudomonas syringae* and flowering time. *Plant Journal* 91: 1015–1028. https://doi.org/10.1111/tpj.13623.

Fang, S., Zhang, L., Guo, J. et al. (2017). NONCODEV5: a comprehensive annotation database for long non-coding RNAs. *Nucleic Acids Research* 46: D308–D314. https://doi.org/10.1093/nar/gkx1107.

Feng, J., Liu, T., Qin, B. et al. (2012). Identifying ChIP-seq enrichment using MACS. *Nature Protocols* 7: 1728–1740. https://doi.org/10.1038/nprot.2012.101.

Ferrero, G., Licheri, N., De Bortoli, M. et al. (2021). Computational analysis of circRNA expression data. *Methods in Molecular Biology* 2284: 181–192. https://doi.org/10.1007/978-1-0716-1307-8_10.

Flynn, J.M., Hubley, R., Goubert, C. et al. (2020). RepeatModeler2 for automated genomic discovery of transposable element families. *Proceedings of the National Academy of Sciences of the United States of America* 117: 9451–9457. https://doi.org/10.1073/pnas.1921046117.

Frazee, A.C., Pertea, G., Jaffe, A.E. et al. (2015). Ballgown bridges the gap between transcriptome assembly and expression analysis. *Nature Biotechnology* 33: 243–246. https://doi.org/10.1038/nbt.3172.

Friedlander, M.R., Mackowiak, S.D., Li, N. et al. (2012). miRDeep2 accurately identifies known and hundreds of novel microRNA genes in seven animal clades. *Nucleic Acids Research* 40: 37–52. https://doi.org/10.1093/nar/gkr688.

Fukunaga, T., Iwakiri, J., Ono, Y. et al. (2019). LncRRIsearch: a web server for lncRNA-RNA interaction prediction integrated with tissue-specific expression and subcellular localization data. *Frontiers in Genetics* 10: 462. https://doi.org/10.3389/fgene.2019.00462.

Gaidatzis, D., van Nimwegen, E., Hausser, J. et al. (2007). Inference of miRNA targets using evolutionary conservation and pathway analysis. *BMC Bioinformatics* 8: 69. https://doi.org/10.1186/1471-2105-8-69.

Gao, Y., Wang, J., and Zhao, F. (2015). CIRI: an efficient and unbiased algorithm for *de novo* circular RNA identification. *Genome Biology* 16: 4. https://doi.org/10.1186/s13059-014-0571-3.

Gao, R., Liu, P., Irwanto, N. et al. (2016). Upregulation of LINC-AP2 is negatively correlated with *AP2* gene expression with Turnip crinkle virus infection in *Arabidopsis thaliana*. *Plant Cell Reports* 35: 2257–2267. https://doi.org/10.1007/s00299-016-2032-9.

Gardiner, L.J., Quinton-Tulloch, M., Olohan, L. et al. (2015). A genome-wide survey of DNA methylation in hexaploid wheat. *Genome Biololgy* 16: 273. https://doi.org/10.1186/s13059-015-0838-3.

Gaspar, J.M. and Hart, R.P. (2017). DMRfinder: efficiently identifying differentially methylated regions from MethylC-seq data. *BMC Bioinformatics* 18: 528. https://doi.org/10.1186/s12859-017-1909-0.

Geng, S., Kong, X., Song, G. et al. (2019). DNA methylation dynamics during the interaction of wheat progenitor *Aegilops tauschii* with the obligate biotrophic fungus *Blumeria graminis* f. sp. *tritici*. *New Phytologist* 221: 1023–1035. https://doi.org/10.1111/nph.15432.

Giresi, P.G., Kim, J., McDaniell, R.M. et al. (2007). FAIRE (formaldehyde-assisted isolation of regulatory elements) isolates active regulatory elements from human chromatin. *Genome Research* 17: 877–885. https://doi.org/10.1101/gr.5533506.

Glazar, P., Papavasileiou, P., and Rajewsky, N. (2014). circBase: a database for circular RNAs. *RNA* 20: 1666–1670. https://doi.org/10.1261/rna.043687.113.

Gong, B. and Purdom, E. (2019). MethCP: differentially methylated region detection with change point models. In: *Research in Computational Molecular Biology. RECOMB 2019*, Lecture Notes in Computer Science, vol. 11467 (ed. L. Cowen). Cham: Springer https://doi.org/10.1007/978-3-030-17083-7_5.

Griffith, M., Griffith, O.L., Mwenifumbo, J. et al. (2010). Alternative expression analysis by RNA sequencing. *Nature Methods* 7: 843–847. https://doi.org/10.1038/nmeth.1503.

Griffiths-Jones, S., Saini, H.K., Dongen, S.V. et al. (2008). miRBase: tools for microRNA genomics. *Nucleic Acids Research* 36: D154–D158. https://doi.org/10.1093/nar/gkm952.

Gu, H., Smith, Z.D., Bock, C. et al. (2011). Preparation of reduced representation bisulfite sequencing libraries for genome-scale DNA methylation profiling. *Nature Protocols* 6: 468–481. https://doi.org/10.1038/nprot.2010.190.

Guo, Y., Papachristoudis, G., Altshuler, R.C. et al. (2010). Discovering homotypic binding events at high spatial resolution. *Bioinformatics* 26: 3028–3034. https://doi.org/10.1093/bioinformatics/btq590.

Guo, Y., Mahony, S., and Gifford, D.K. (2012). High resolution genome wide binding event finding and motif discovery reveals transcription factor spatial binding constraints. *PLoS Computational Biology* 8: e1002638. https://doi.org/10.1371/journal.pcbi.1002638.

Guo, W., Fiziev, P., Yan, W. et al. (2013). BS-Seeker2: a versatile aligning pipeline for bisulfite sequencing data. *BMC Genomics* 14: 774. https://doi.org/10.1186/1471-2164-14-774.

Guo, J.C., Fang, S.S., Wu, Y. et al. (2019). CNIT: a fast and accurate web tool for identifying protein-coding and long non-coding transcripts based on intrinsic sequence composition. *Nucleic Acids Research* 47: W516–W522. https://doi.org/10.1093/nar/gkz400.

Hafner, A. and Mackenzie, S. (2023). Re-analysis of publicly available methylomes using signal detection yields new information. *Scientific Reports* 13: 3307. https://doi.org/10.1038/s41598-023-30422-4.

Hammell, M., Long, D., Zhang, L. et al. (2008). mirWIP: microRNA target prediction based on microRNA-containing ribonucleoprotein-enriched transcripts. *Nature Methods* 9: 813–819. https://doi.org/10.1038/nmeth.1247.

Han, Y. and Wessler, S.R. (2010). Mite-hunter: a program for discovering miniature inverted-repeat transposable elements from genomic sequences. *Nucleic Acids Research* 38: 199. https://doi.org/10.1093/nar/gkq862.

Handzlik, J.E., Tastsoglou, S., Vlachos, I.S. et al. (2020). Manatee: detection and quantification of small non-coding RNAs from next-generation sequencing data. *Scientific Reports* 10: 705. https://doi.org/10.1038/s41598-020-57495-9.

Hansen, K.D., Langmead, B., and Irizarry, R.A. (2012). BSmooth: from whole genome bisulfite sequencing reads to differentially methylated regions. *Genome Biology* 13: R83. https://doi.org/10.1186/gb-2012-13-10-r83.

Hansen, P., Hecht, J., Ibrahim, D.M. et al. (2015). Saturation analysis of ChIP-seq data for reproducible identification of binding peaks. *Genome Research* 25: 1391–1400. https://doi.org/10.1101/gr.189894.115.

Hardcastle, T.J. and Kelly, K.A. (2010). baySeq: empirical Bayesian methods for identifying differential expression in sequence count data. *BMC Bioinformatics* 11: 422. https://doi.org/10.1186/1471-2105-11-422.

Harmanci, A., Rozowsky, J., and Gerstein, M. (2014). MUSIC: identification of enriched regions in ChIP-Seq experiments using a mappability-corrected multiscale signal processing framework. *Genome Biology* 15: 474. https://doi.org/10.1186/s13059-014-0474-3.

Harris, E.Y., Ounit, R., and Lonardi, S. (2016). BRAT-nova: fast and accurate mapping of bisulfite-treated reads. *Bioinformatics* 32: 2696–2698. https://doi.org/10.1093/bioinformatics/btw226.

He, M., Xu, Y., Chen, J. et al. (2018). MoSnt2-dependent deacetylation of histone H3 mediates MoTor-dependent autophagy and plant infection by the rice blast fungus *Magnaporthe oryzae*. *Autophagy* 14: 1543–1561. https://doi.org/10.1080/15548627.2018.1458171.

Heinig, M., Colomé-Tatché, M., Taudt, A. et al. (2015). histoneHMM: differential analysis of histone modifications with broad genomic footprints. *BMC Bioinformatics* 16: 60. https://doi.org/10.1186/s12859-015-0491-6.

Heinz, S., Benner, C., Spann, N. et al. (2010). Simple combinations of lineage-determining transcription factors prime cis-regulatory elements required for macrophage and B cell identities. *Molecular Cell* 38: 576–589. https://doi.org/10.1161/ATVBAHA.114.304051.

Henderson, I.R. and Jacobsen, S.E. (2007). Epigenetic inheritance in plants. *Nature* 447: 418–424. https://doi.org/10.1038/nature05917.

Hertel, J. and Stadler, P.F. (2006). Hairpins in a Haystack: recognizing microRNA precursors in comparative genomics data. *Bioinformatics* 22: e197–e202. https://doi.org/10.1093/bioinformatics/btl257.

Hoffmann, S., Otto, C., Doose, G. et al. (2014). A multi-split mapping algorithm for circular RNA, splicing, trans-splicing and fusion detection. *Genome Biology* 15: R34. https://doi.org/10.1186/gb-2014-15-2-r34.

Hong, Z., Shikai, L., Changyou, W. et al. (2018). The role of transcription factor in wheat defense against pathogen and its prospect in breeding. *Journal of Plant Biology: Crop Research* 1: 1005. https://doi.org/10.33582/2637-7721/1005.

Hou, Y., Wang, L., Wang, L. et al. (2015). JMJ704 positively regulates rice defense response against *Xanthomonas oryzae* pv. oryzae infection via reducing H3K4me2/3 associated with negative disease resistance regulators. *BMC Plant Biology* 15: 286. https://doi.org/10.1186/s12870-015-0674-3.

Hu, R. and Sun, X. (2016). lncRNATargets: a platform for lncRNA target prediction based on nucleic acid thermodynamics. *Journal of Bioinformatics and Computational Biology* 14: 1650016. https://doi.org/10.1142/S0219720016500165.

Hu, M., Pei, B.L., Zhang, L.F. et al. (2014). Histone H2B monoubiquitination is involved in regulating the dynamics of microtubules during the defense response to *Verticillium dahliae* toxins in Arabidopsis. *Plant Physiology* 164: 1857–1865. https://doi.org/10.1104/pp.113.234567.

Huang, C.Y. and Jin, H. (2022). Coordinated epigenetic regulation in plants: a potent managerial tool to conquer biotic stress. *Frontiers in Plant Science* 12: 3047. https://doi.org/10.3389/fpls.2021.795274.

Huang, J., Renault, V., Sengenes, J. et al. (2012). MeQA: a pipeline for MeDIP-seq data quality assessment and analysis. *Bioinformatics* 28: 587–588. https://doi.org/10.1093/bioinformatics/btr699.

Huang, K.Y., Huang, Y.J., and Chen, P.Y. (2018). BS-Seeker3: ultrafast pipeline for bisulfite sequencing. *BMC Bioinformatics* 19: 111. https://doi.org/10.1186/s12859-018-2120-7.

Huang, Y.A., Huang, Z.A. et al. (2019). Predicting lncRNA-miRNA interaction via graph convolution auto-encoder. *Frontiers in Genetics* 10: 758. https://doi.org/10.3389/fgene.2019.00758.

Hubley, R., Finn, R.D., Clements, J. et al. (2016). The Dfam database of repetitive DNA families. *Nucleic Acids Research* 44: D81–D89. https://doi.org/10.1093/nar/gkv1272.

Huettel, B., Kanno, T., Daxinger, L. et al. (2006). Endogenous targets of RNA-directed DNA methylation and Pol IV in Arabidopsis. *The EMBO Journal* 25: 2828–2836. https://doi.org/10.1038/sj.emboj.7601150.

Ibrahim, M.M., Lacadie, S.A., and Ohler, U. (2015). JAMM: a peak finder for joint analysis of NGS replicates. *Bioinformatics* 31: 48–55. https://doi.org/10.1093/bioinformatics/btu568.

Izuogu, O.G., Alhasan, A.A., Alafghani, H.M. et al. (2016). PTESFinder: a computational method to identify post-transcriptional exon shuffling (PTES) events. *BMC Bioinformatics* 17: 31. https://doi.org/10.1186/s12859-016-0881-4.

Jain, N., Sinha, N., and Krishna, H. (2020). A study of miRNAs and lncRNAs during *Lr28*-mediated resistance against leaf rust in wheat (*Triticum aestivum* L.). *Physiological and Molecular Plant Pathology* 112: 101552. https://doi.org/10.1016/j.pmpp.2020.101552.

Jain, N., Batra, R., Saripalli, G. et al. (2021). Methylome changes during *Lr48*-mediated APR for leaf rust in wheat (*Triticum aestivum* L.). *Physiological and Molecular Plant Pathology* 116: 101726. https://doi.org/10.1016/j.pmpp.2021.101726.

Jain, N., Shiv, A., Sinha, N. et al. (2023). Leaf rust responsive miRNA and their target genes in wheat. *Functional & Integrative Genomics* 23: 14. https://doi.org/10.1007/s12042-019-09235-1.

Jakobi, T., Uvarovskii, A., and Dieterich, C. (2018). Circtools—a one-stop software solution for circular RNA research. *Bioinformatics* 35: 2326–2328. https://doi.org/10.1093/bioinformatics/bty948.

Jha, A. and Shankar, R. (2013). miReader: discovering novel miRNAs in species without sequenced genome. *PLoS One* 8: e66857. https://doi.org/10.1371/journal.pone.0066857.

Ji, H., Jiang, H., Ma, W. et al. (2008). An integrated software system for analyzing ChIP-chip and ChIP-seq data. *Nature Biotechnology* 26: 1293–1300. https://doi.org/10.1038/nbt.1505.

Jin, J., Lu, P., Xu, Y. et al. (2021). PLncDB V2.0: a comprehensive encyclopedia of plant long noncoding RNAs. *Nucleic Acids Research* 49: D1489–D1495.

Johnson, N.R., Yeoh, J.M., Coruh, C. et al. (2016). Improved placement of multi-mapping small RNAs. *G3: Genes, Genomes, Genetics* 6: 2103–2111. https://doi.org/10.1534/g3.116.030452.

Jothi, R., Cuddapah, S., Barski, A. et al. (2008). Genome-wide identification of *in vivo* protein-DNA binding sites from ChIP-Seq data. *Nucleic Acids Research* 36: 5221–5231. https://doi.org/10.1186/1471-2105-11-81.

Jühling, F., Kretzmer, H., Bernhart, S.H. et al. (2016). Metilene: fast and sensitive calling of differentially methylated regions from bisulfite sequencing data. *Genome Research* 26: 256–262. https://doi.org/10.1101/gr.196394.115.

Kalvari, I., Nawrocki, E.P., Ontiveros-Palacios, N. et al. (2020). Rfam 14: expanded coverage of metagenomic, viral and microRNA families. *Nucleic Acids Research* 49: D192–D200. https://doi.org/10.1093/nar/gkaa1047.

Kang, Y.J., Yang, D.C., Kong, L. et al. (2017). CPC2: a fast and accurate coding potential calculator based on sequence intrinsic features. *Nucleic Acids Research* 45: W12–W16. https://doi.org/10.1093/nar/gkx428.

Karagkouni, D., Paraskevopoulou, M.D., Tastsoglou, S. et al. (2019). DIANA-LncBase v3: indexing experimentally supported miRNA targets on non-coding transcripts. *Nucleic Acids Research* 48: D101–D110. https://doi.org/10.1093/nar/gkz1036.

Katta, M.A., Khan, A.W., Doddamani, D. et al. (2012). NGS-QCbox and raspberry for parallel, automated and rapid quality control analysis of large-scale next generation sequencing (Illumina) data. *PLoS One* 10: e0139868. https://doi.org/10.1371/journal.pone.0030619.

Katz, Y., Wang, E.T., Airoldi, E.M. et al. (2010). Analysis and design of RNA sequencing experiments for identifying isoform regulation. *Nature Methods* 7: 1009–1015. https://doi.org/10.1038/nmeth.1528.

Kent, W.J. (2002). BLAT—the BLAST-like alignment tool. *Genome Research* 12: 656–664. https://doi.org/10.1101/gr.229202.

Kertesz, M., Iovino, N., Unnerstall, U. et al. (2007). The role of site accessibility in microRNA target recognition. *Nature Genetics* 39: 1278–1284. https://doi.org/10.1038/ng2135.

Kim, D. and Salzberg, S.L. (2011). TopHat-Fusion: an algorithm for discovery of novel fusion transcripts. *Genome Biology* 12: R72. https://doi.org/10.1186/gb-2011-12-8-r72.

Kim, D., Langmead, B., and Salzberg, S.L. (2015). HISAT: a fast spliced aligner with low memory requirements. *Nature Methods* 12: 357–360. https://doi.org/10.1038/nmeth.3317.

Kim, D., Paggi, J.M., Park, C. et al. (2019). Graph-based genome alignment and genotyping with HISAT2 and HISAT-genotype. *Nature Biotechnology* 37: 907–915. https://doi.org/10.1038/s41587-019-0201-4.

Kimura, H. (2013). Histone modifications for human epigenome analysis. *Journal of Human Genetics* 58: 439–445. https://doi.org/10.1038/jhg.2013.66.

Kong, L., Zhang, Y., Ye, Z.Q. et al. (2007). CPC: assess the protein-coding potential of transcripts using sequence features and support vector machine. *Nucleic Acids Research* 35: W345–W349. https://doi.org/10.1093/nar/gkm391.

Kong, L., Qiu, X., Kang, J. et al. (2017). A phytophthora effector manipulates host histone acetylation and reprograms defense gene expression to promote infection. *Current Biology* 27: 981–991. https://doi.org/10.1016/j.cub.2017.02.044.

Kong, L., Liu, Y., Wang, X., and Chang, C. (2020). Insight into the role of epigenetic processes in abiotic and biotic stress response in wheat and barley. *International Journal of Molecular Sciences* 21: 1480. https://doi.org/10.3390/ijms21041480.

Kozomara, A. and Griffiths-Jones, S. (2010). miRBase: integrating microRNA annotation and deep-sequencing data. *Nucleic Acids Research* 39: D152–D157. https://doi.org/10.1093/nar/gkq1027.

Krek, A., Grün, D., Poy, M.N. et al. (2005). Combinatorial microRNA target predictions. *Nature Genetics* 37: 495–500. https://doi.org/10.1038/ng1536.

Krueger, F. and Andrews, S.R. (2011). Bismark: a flexible aligner and methylation caller for Bisulfite-Seq applications. *Bioinformatics* 27: 1571–1572. https://doi.org/10.1093/bioinformatics/btr167.

Kuan, P.F., Chung, D.J., Pan, G.J. et al. (2011). A statistical framework for the analysis of ChIP-Seq data. *Journal of the American Statistical Association* 106: 891–903. https://doi.org/10.1093/bioinformatics/btt200.

Kumar, S., Chinnusamy, V., and Mohapatra, T. (2018). Epigenetics of modified DNA bases: 5-methylcytosine and beyond. *Frontiers in Genetics* 9: 640. https://doi.org/10.3389/fgene.2018.00640.

Kurdyukov, S. and Bullock, M. (2016). DNA methylation analysis: choosing the right method. *Biology* 5: E3. https://doi.org/10.3390/biology5010003.

Landt, S.G., Marinov, G.K., Kundaje, A. et al. (2012). ChIP-seq guidelines and practices of the ENCODE and modENCODE consortia. *Genome Research* 22: 1813–1831. https://doi.org/10.1101/gr.136184.111.

Langmead, B. and Salzberg, S.L. (2012). Fast gapped-read alignment with Bowtie 2. *Nature Methods* 9: 357–359. https://doi.org/10.1038/nmeth.1923.

Langmead, B., Trapnell, C., Pop, M. et al. (2009). Ultrafast and memory-efficient alignment of short DNA sequences to the human genome. *Genome Biology* 10: R25. https://doi.org/10.1186/gb-2009-10-3-r25.

Latrasse, D., Jégu, T., Li, H. et al. (2017). MAPK-triggered chromatin reprogramming by histone deacetylase in plant innate immunity. *Genome Biology* 18: 131. https://doi.org/10.1186/s13059-017-1261-8.

Lee, S., Fu, F., Xu, S. et al. (2016a). Global regulation of plant immunity by histone lysine methyl transferases. *Plant Cell* 28: 1640–1661. https://doi.org/10.1105/tpc.16.00012.

Lee, B., Baek, J., Park, S. et al. (2016b). deepTarget: end-to-end learning framework for microRNA target prediction using deep recurrent neural networks. *Proceedings of the 7th ACM International Conference on Bioinformatics, Computational Biology, and Health Informatics*, Seattle, WA, USA, 434–442. https://doi.org/10.48550/arXiv.1603.09123

Li, H. and Durbin, R. (2009). Fast and accurate short read alignment with Burrows-Wheeler transform. *Bioinformatics* 25: 1754–1760. https://doi.org/10.1093/bioinformatics/btp324.

Li, H., Ruan, J., and Durbin, R.M. (2008). Mapping short DNA sequencing reads and calling variants using mapping quality scores. *Genome Research* 18: 1851–1858. https://doi.org/10.1101/gr.078212.108.

Li, R., Yu, C., Li, Y. et al. (2009). SOAP2: an improved ultrafast tool for short read alignment. *Bioinformatics* 25: 1966–1967. https://doi.org/10.1093/bioinformatics/btp336.

Li, Q., Brown, J., Huang, H. et al. (2011). Measuring reproducibility of high-throughput experiments. *Annual Applied Statistics* 5: 1752–1779. https://doi.org/10.1214/11-AOAS466.

Li, J., Witten, D.M., Johnstone, I. et al. (2012). Normalization, testing, and false discovery rate estimation for RNA-sequencing data. *Biostatistics* 13: 523–538. https://doi.org/10.1093/biostatistics/kxr031.

Li, X., Jiang, Y., Ji, Z. et al. (2015a). BRHIS1 suppresses rice innate immunity through binding to monoubiquitinated H2A and H2B variants. *EMBO Reports* 16: 1192–1202. https://doi.org/10.15252/embr.201440000.

Li, J., Ma, W., Zeng, P. et al. (2015b). LncTar: a tool for predicting the RNA targets of long noncoding RNAs. *Briefings in Bioinformatics* 16: 806–812. https://doi.org/10.1093/bib/bbu048.

Li, L., Bu, D., and Zhao, Y. (2019). CircRNAwrap - a flexible pipeline for circRNA identification, transcript prediction, and abundance estimation. *FEBS Letters* 593: 1179–1189. https://doi.org/10.1002/1873-3468.13423.

Li, J., Yang, Z., Wang, D., and Li, Z. (2022). WAFNRLTG: a novel model for predicting LncRNA target genes based on weighted average fusion network representation learning method. *Frontiers in Cell Development and Biology* 9: 820342. https://doi.org/10.3389/fcell.2021.820342.

Liang, K. and Keleş, S. (2012a). Normalization of ChIP-seq data with control. *BMC Bioinformatics* 13: 199. https://doi.org/10.1186/1471-2105-13-199.

Liang, K. and Keles, S. (2012b). Detecting differential binding of transcription factors with ChIP-seq. *Bioinformatics* 28: 121–122. https://doi.org/10.1093/bioinformatics/btr605.

Liang, F., Tang, B., Wang, Y. et al. (2014). WBSA: web service for bisulfite sequencing data analysis. *PLoS One* 9: e86707. https://doi.org/10.1371/journal.pone.0086707.

Liao, W.W., Yen, M.R., Ju, E. et al. (2015). MethGo: a comprehensive tool for analyzing whole-genome bisulfite sequencing data. *BMC Genomics* 16 (Suppl 12): S11. https://doi.org/10.1186/1471-2164-16-S12-S11.

Lienhard, M., Grimm, C., Morkel, M. et al. (2014). MEDIPS: genome-wide differential coverage analysis of sequencing data derived from DNA enrichment experiments. *Bioinformatics* 30: 284–286. https://doi.org/10.1093/bioinformatics/btt650.

Lienhard, M., Grasse, S., Rolff, J. et al. (2017). QSEA—modelling of genome-wide DNA methylation from sequencing enrichment experiments. *Nucleic Acids Research* 45: e44. https://doi.org/10.1093/nar/gkw1193.

Lin, M.F., Jungreis, I., and Kellis, M. (2011). PhyloCSF: a comparative genomics method to distinguish protein coding and non-coding regions. *Bioinformatics* 27: i275–i282. https://doi.org/10.1093/bioinformatics/btr209.

Lin, X., Sun, D., Rodriguez, B. et al. (2013). BSeQC: quality control of bisulfite sequencing experiments. *Bioinformatics* 29: 3227–3229. https://doi.org/10.1093/bioinformatics/btt548.

Lisch, D. (2013). How important are transposons for plant evolution? *Nature Reviews Genetics* 14: 49–61. https://doi.org/10.1038/nrg3374.

Liu, B., Yi, J., Sv, A. et al. (2013). QChIPat: a quantitative method to identify distinct binding patterns for two biological ChIP-seq samples in different experimental conditions. *BMC Genomics* 14 (Suppl 8): S3. https://doi.org/10.1186/1471-2164-14-S8-S3.

Liu, Z.W., Shao, C.R., Zhang, C.J. et al. (2014). The SET domain proteins SUVH2 and SUVH9 are required for Pol V occupancy at RNA-directed DNA methylation loci. *PLoS Genetics* 10: e1003948. https://doi.org/10.1371/journal.pgen.1003948.

Liu, Y.C., Li, J.R., Sun, C.H. et al. (2016). CircNet: a database of circular RNAs derived from transcriptome sequencing data. *Nucleic Acids Research* 44: D209–D215. https://doi.org/10.1093/nar/gkv940.

Liu, S.R., Zhou, J.J., Hu, C.G. et al. (2017). MicroRNA-mediated gene silencing in plant defense and viral counter-defense. *Frontiers in Microbiology* 8: 1801. https://doi.org/10.3389/fmicb.2017.01801.

Love, M.I., Huber, W., and Anders, S. (2014). Moderated estimation of fold change and dispersion for RNA-seq data with DESeq2. *Genome Biology* 15: 550. https://doi.org/10.1186/s13059-014-0550-8.

Lunter, G. and Goodson, M. (2011). Stampy: a statistical algorithm for sensitive and fast mapping of Illumina sequence reads. *Genome Research* 21: 936–939. https://doi.org/10.1101/gr.111120.110.

Mahony, S., Edwards, M.D., Mazzoni, E.O. et al. (2014). An integrated model of multiple-condition ChIP-Seq data reveals predeterminants of Cdx2 binding. *PLoS Computational Biology* 10: e1003501. https://doi.org/10.1371/journal.pcbi.1003501.

Mapleson, D., Accinelli, G.G. et al. (2017). KAT: a K-mer analysis toolkit to quality control NGS datasets and genome assemblies. *Bioinformatics* 33: 574–576. https://doi.org/10.1093/bioinformatics/btw663.

Martin, M. (2012). Cutadapt removes adapter sequences from high-throughput sequencing reads. *Bioinformatics in Action* 17: 10–12. https://doi.org/10.14806/ej.17.1.200.

Martin, T.C., Yet, I., Tsai, P.C. et al. (2015). coMET: visualisation of regional epigenome-wide association scan results and DNA co-methylation patterns. *BMC Bioinformatics* 16: 131. https://doi.org/10.1186/s12859-015-0568-2.

Martire, S. and Banaszynski, L.A. (2020). The roles of histone variants in fine-tuning chromatin organization and function. *Nature Reviews Molecular Cell Biology* 21: 522–541. https://doi.org/10.1038/s41580-020-0262-8.

Mas-Ponte, D., Carlevaro-Fita, J., and Palumbo, E. (2017). Lncatlas database for subcellular localization of long noncoding RNAs. *RNA* 23: 1080–1087. https://doi.org/10.1261/rna.060814.117.

Mattick, J.S., Amaral, P.P., Carninci, P. et al. (2023). Long non-coding RNAs: definitions, functions, challenges and recommendations. *Nature Reviews Molecular Cell Biology* 24: 430–447. https://doi.org/10.1038/s41580-022-00566-8.

Matzke, M.A. and Mosher, R.A. (2014). RNA-directed DNA methylation: an epigenetic pathway of increasing complexity. *Nature Reviews Genetics* 15: 394–408. https://doi.org/10.1038/nrg3683.

McLean, C.Y., Bristor, D., Hiller, M. et al. (2010). GREAT improves functional interpretation of cis-regulatory regions. *Nature Biotechnology* 28: 495–501. https://doi.org/10.1038/nbt.1630.

Meissner, A., Gnirke, A., Bell, G.W. et al. (2005). Reduced representation bisulfite sequencing for comparative high-resolution DNA methylation analysis. *Nucleic Acids Research* 33: 5868–5877. https://doi.org/10.1093/nar/gki901.

Meller, B., Kuznicki, D., Arasimowicz-Jelonek, M. et al. (2018). BABA-primed histone modifications in potato for inter-generational resistance to *Phytophthora infestans*. *Frontiers in Plant Science* 9: 1228. https://doi.org/10.3389/fpls.2018.01228.

Memczak, S., Jens, M., Elefsinioti, A. et al. (2013). Circular RNAs are a large class of animal RNAs with regulatory potency. *Nature* 495: 333–338. https://doi.org/10.1038/nature11928.

Mendoza-Parra, M.A., Sankar, M., Walia, M. et al. (2012). POLYPHEMUS: R package for comparative analysis of RNA polymerase II ChIP-seq profiles by non-linear normalization. *Nucleic Acids Research* 40: e30. https://doi.org/10.1093/nar/gkr1205.

Meng, X., Chen, Q., Zhang, P. et al. (2017). CircPro: an integrated tool for the identification of circRNAs with protein-coding potential. *Bioinformatics* 33: 3314–3316. https://doi.org/10.1093/bioinformatics/btx446.

Merino, G.A., Conesa, A., and Fernández, E.A. (2019). A benchmarking of workflows for detecting differential splicing and differential expression at isoform level in human RNA-seq studies. *Briefings in Bioinformatics* 20: 471–481. https://doi.org/10.1093/bib/bbx122.

Merkel, A., Fernandez-Callejo, M., Casals, E. et al. (2018). gemBS—high throughput processing for DNA methylation data from bisulfite sequencing. *Bioinformatics* 35: 737–742. https://doi.org/10.1093/bioinformatics/bty690.

Mette, M.F., Aufsatz, W., Winden, J. et al. (2000). Transcriptional silencing and promoter methylation triggered by double-stranded RNA. *EMBO Journal* 19: 5194–5201. https://doi.org/10.1093/emboj/19.19.5194.

Millán-Zambrano, G., Burton, A., Bannister, A.J. et al. (2022). Histone post-translational modifications—cause and consequence of genome function. *Nature Reviews Genetics* 23: 563–580. https://doi.org/10.1038/s41576-022-00468-7.

Miranda, K.C., Huynh, T., Tay, Y. et al. (2006). A pattern-based method for the identification of microRNA binding sites and their corresponding heteroduplexes. *Cell* 126: 1203–1217. https://doi.org/10.1016/j.cell.2006.07.031.

Mistry, J., Chuguransky, S., Williams, L. et al. (2021). Pfam: the protein families database in 2021. *Nucleic Acids Research* 49: D412–D419. https://doi.org/10.1093/nar/gkaa913.

Montgomery, S.B., Sammeth, M., Gutierrez-Arcelus, M. et al. (2010). Transcriptome genetics using second generation sequencing in a Caucasian population. *Nature* 464: 773–777. https://doi.org/10.1038/nature08903.

Nakato, R. and Sakata, T. (2021). Methods for ChIP-seq analysis: a practical workflow and advanced applications. *Methods* 187: 44–53. https://doi.org/10.1016/j.ymeth.2020.03.005.

Nam, J.W., Rissland, O.S., Koppstein, D. et al. (2014). Global analyses of the effect of different cellular contexts on microRNA targeting. *Molecular Cell* 53: 1031–1043. https://doi.org/10.1016/j.molcel.2014.02.013.

Navarro, L., Dunoyer, P., Jay, F. et al. (2006). A plant miRNA contributes to antibacterial resistance by repressing auxin signaling. *Science* 312: 436–439. https://doi.org/10.1126/science.aae0382.

Neary, J.L. and Carless, M.A. (2020). Methylated DNA immunoprecipitation sequencing (MeDIP-seq): principles and applications. In: *Epigenetics Methods* (ed. T. Tollefsbol), 157–179. Cambridge, MA: Academic Press https://doi.org/10.1016/b978-0-12-819414-00.00009-4.

Newell, R., Pienaar, R., Balderson, B. et al. (2021). ChIP-R: assembling reproducible sets of ChIP-seq and ATAC-seq peaks from multiple replicates. *Genomics* 113: 1855–1866. https://doi.org/10.1016/j.ygeno.2021.04.026.

Ni, P., Nie, F., Zhong, Z. et al. (2023). DNA 5-methylcytosine detection and methylation phasing using PacBio circular consensus sequencing. *Nature Communications* 14: 4054. https://doi.org/10.1038/s41467-023-39784-9.

Niazi, U., Geyer, K.K., Vickers, M.J. et al. (2016). DISMISS: detection of stranded methylation in MeDIP-Seq data. *BMC Bioinformatics* 17: 295. https://doi.org/10.1186/s12859-016-1158-7.

Omony, J., Nussbaumer, T., and Gutzat, R. (2020). DNA methylation analysis in plants: review of computational tools and future perspectives. *Briefings in Bioinformatics* 21: 906–918. https://doi.org/10.1093/bib/bbz039.

Ouyang, W., Zhang, X., Peng, Y. et al. (2021). Rapid and low-input profiling of histone marks in plants using nucleus CUT&Tag. *Frontiers in Plant Science* 12: 634679. https://doi.org/10.3389/fpls.2021.634679.

Pandey, R.V., Pabinger, S., Kriegner, A., and Weinhäusel, A. (2016). ClinQC: a tool for quality control and cleaning of Sanger and NGS data in clinical research. *BMC Bioinformatics* 17: 56. https://doi.org/10.1186/s12859-016-0915-y.

Park, Y. and Wu, H. (2016). Differential methylation analysis for BS-seq data under general experimental design. *Bioinformatics* 32: 1446–1453. https://doi.org/10.1093/bioinformatics/btw026.

Park, S., Min, S., Choi, H., and Yoon, S., (2016). deepMiRGene: deep neural network based precursor microRNA prediction. ArXiv abs/1605.00017: n. pag.

Patel, R.K. and Jain, M. (2012). NGS QC toolkit: a toolkit for quality control of next generation sequencing data. *PLoS One* 7: e30619. https://doi.org/10.1371/journal.pone.0030619.

Pelizzola, M., Koga, Y., Urban, A.E. et al. (2008). MEDME: an experimental and analytical methodology for the estimation of DNA methylation levels based on microarray derived MeDIP-enrichment. *Genome Research* 18: 1652–1659. https://doi.org/10.1101/gr.080721.108.

Peng, C., Han, S., Zhang, H. et al. (2019). RPITER: a hierarchical deep learning framework for ncRNA–protein interaction prediction. *International Journal of Molecular Sciences* 20: 1070. https://doi.org/10.3390/ijms20051070.

Pertea, M., Pertea, G.M., Antonescu, C.M. et al. (2015). StringTie enables improved reconstruction of a transcriptome from RNA-seq reads. *Nature Biotechnology* 33: 290–295. https://doi.org/10.1038/nbt.3122.

Pimentel, H., Bray, N.L., Puente, S. et al. (2017). Differential analysis of RNA-seq incorporating quantification uncertainty. *Nature Methods* 14: 687–690. https://doi.org/10.1038/nmeth.4324.

Price, A.L., Jones, N.C., and Pevzner, P.A. (2005). *De novo* identification of repeat families in large genomes. *Bioinformatics* 21: i351–i358. https://doi.org/10.1093/bioinformatics/bti1018.

Quek, X.C., Thomson, D.W., Maag, J.L. et al. (2015). lncRNAdb v2.0: expanding the reference database for functional long noncoding RNAs. *Nucleic Acids Research* 43: D168–D173. https://doi.org/10.1093/nar/gku988.

Ramírez, F., Dündar, F., Diehl, S. et al. (2014). deepTools: a flexible platform for exploring deep-sequencing data. *Nucleic Acids Research* 42: W187–W191. https://doi.org/10.1093/nar/gku365.

Ramirez-Prado, J.S., Abulfaraj, A.A., Rayapuram, N. et al. (2018a). Plant immunity: from signaling to epigenetic control of defense. *Trends in Plant Science* 23: 833–844. https://doi.org/10.1016/j.tplants.2018.06.004.

Ramirez-Prado, J.S., Piquerez, S.J.M., Bendahmane, A. et al. (2018b). Modify the histone to win the battle: chromatin dynamics in plant–pathogen interactions. *Frontiers in Plant Science* 9: 355. https://doi.org/10.3389/fpls.2018.00355.

Rashid, N.U., Giresi, P.G., Ibrahim, J.G. et al. (2011). ZINBA integrates local covariates with DNA-seq data to identify broad and narrow regions of enrichment, even within amplified genomic regions. *Genome Biology* 12: R67. https://doi.org/10.1186/gb-2011-12-7-r67.

Reczko, M., Maragkakis, M., and Alexiou, P. (2012). Functional microRNA targets in protein coding sequences. *Bioinformatics* 28: 771–776. https://doi.org/10.1093/bioinformatics/bts043.

Riebler, A., Menigatti, M., Song, J.Z. et al. (2014). BayMeth: improved DNA methylation quantification for affinity capture sequencing data using a flexible Bayesian approach. *Genome Biology* 15: R35. https://doi.org/10.1186/gb-2014-15-2-r35.

Robinson, M.D., McCarthy, D.J., and Smyth, G.K. (2010). edgeR: a bioconductor package for differential expression analysis of digital gene expression data. *Bioinformatics* 26: 139–140. https://doi.org/10.1093/bioinformatics/btp616.

Ross-Innes, C.S., Stark, R., Teschendorff, A.E. et al. (2012). Differential oestrogen receptor binding is associated with clinical outcome in breast cancer. *Nature* 481: 389–393. https://doi.org/10.1038/nature10730.

Saripalli, G., Singh, K., Gautam, T. et al. (2020a). Genome-wide analysis of H3K4me3 and H3K27me3 modifications due to *Lr28* for leaf rust resistance in bread wheat (*Triticum aestivum*). *Plant Molecular Biology* 104: 113–136. https://doi.org/10.1007/s11103-020-01029-4.

Saripalli, G., Sharma, C., Gautam, T. et al. (2020b). Complex relationship between DNA methylation and gene expression due to *Lr28* in wheat-leaf rust pathosystem. *Molecular Biology Reports* 47: 1339–1360. https://doi.org/10.1007/s11033-019-05236-1.

Shao, Z., Zhang, Y., Yuan, G.C. et al. (2012). MAnorm: a robust model for quantitative comparison of ChIP-Seq data sets. *Genome Biology* 13: R16. https://doi.org/10.1186/gb-2012-13-3-r16.

Shen, L., Shao, N.Y., Liu, X. et al. (2013). diff Reps: detecting differential chromatin modification sites from ChIP-seq data with biological replicates. *PLoS One* 8: e65598. https://doi.org/10.1371/journal.pone.0065598.

Shin, H., Liu, T., Manrai, A.K. et al. (2009). CEAS: *cis*-regulatory element annotation system. *Bioinformatics* 25: 2605–2606. https://doi.org/10.1093/nar/gkl322.

Singh, K., Saripalli, G., Gautam, T. et al. (2022). BS-seq reveals major role of differential CHH methylation during leaf rust resistance in wheat (*Triticum aestivum* L.). *Molecular Genetics and Genomics* 297: 731–749. https://doi.org/10.1007/s00438-022-01879-1.

Slotkin, R.K. and Martienssen, R. (2007). Transposable elements and the epigenetic regulation of the genome. *Nature Reviews Genetics* 8: 272–285. https://doi.org/10.1038/nrg2072.

Song, Q. and Smith, A.D. (2011). Identifying dispersed epigenomic domains from ChIP-Seq data. *Bioinformatics* 27: 870–871. https://doi.org/10.1093/bioinformatics/btr030.

Song, X., Zhang, N., Han, P. et al. (2016). Circular RNA profile in gliomas revealed by identification tool UROBORUS. *Nucleic Acids Research* 44: e87. https://doi.org/10.1093/nar/gkw075.

Song, L., Fang, Y., Chen, L. et al. (2021). Role of non-coding RNAs in plant immunity. *Plant Communications* 2: 100180. https://doi.org/10.1016/j.xplc.2021.100180.

Spyrou, C., Stark, R., Lynch, A.G. et al. (2009). BayesPeak: Bayesian analysis of ChIP-seq data. *BMC Bioinformatics* 10: 299. https://doi.org/10.1186/1471-2105-10-299.

Stark, A., Brennecke, J., and Bushati, N. (2005). Animal microRNAs confer robustness to gene expression and have a significant impact on 3′UTR evolution. *Cell* 123: 1133–1146. https://doi.org/10.1016/j.cell.2005.11.023.

Stevens, M., Cheng, J.B., Li, D. et al. (2013). Estimating absolute methylation levels at single-CpG resolution from methylation enrichment and restriction enzyme sequencing methods. *Genome Research* 23: 1541–1553. https://doi.org/10.1101/gr.152231.112.

Stockwell, P.A., Chatterjee, A., Rodger, E.J. et al. (2014). DMAP: differential methylation analysis package for RRBS and WGBS data. *Bioinformatics* 30: 1814–1822. https://doi.org/10.1093/bioinformatics/btu126.

Sturm, M., Schroeder, C., and Bauer, P. (2016). SeqPurge: highly-sensitive adapter trimming for paired-end NGS data. *BMC Bioinformatics* 17: 208. https://doi.org/10.1186/s12859-016-1069-7.

Su, Y., Li, H.G., Wang, Y. et al. (2018). Poplar miR472a targeting NBS-LRRs is involved in effective defence against the necrotrophic fungus *Cytospora chrysosperma*. *Journal of Experimental Botany* 69: 5519–5530. https://doi.org/10.1093/jxb/ery304.

Subramanian, A., Tamayo, P., Mootha, V.K. et al. (2005). Gene set enrichment analysis: a knowledge-based approach for interpreting genome-wide expression profiles. *Proceedings of the National Academy of Sciences of the United States of America* 102: 15545–15550. https://doi.org/10.1073/pnas.0506580102.

Sun, P. and Li, G. (2019). CircCode: a powerful tool for identifying circRNA coding ability. *Frontiers in Genetics* 10: 981. https://doi.org/10.3389/fgene.2019.00981.

Sun, L. et al. (2013). Utilizing sequence intrinsic composition to classify protein-coding and long non-coding transcripts. *Nucleic Acids Research* 41: e166. https://doi.org/10.1093/nar/gkt646.

Sun, D., Xi, Y., Rodriguez, B. et al. (2014). MOABS: model based analysis of bisulfite sequencing data. *Genome Biology* 15: R38. https://doi.org/10.1186/gb-2014-15-2-r38.

Sun, X., Zheng, H., and Sui, N. (2018). Regulation mechanism of long non-coding RNA in plant response to stress. *Biochemical and Biophysical Research Communications* 503: 402–407. https://doi.org/10.1016/j.bbrc.2018.07.072.

Szabo, L., Morey, R., Palpant, N.J. et al. (2015). Statistically based splicing detection reveals neural enrichment and tissue-specific induction of circular RNA during human fetal development. *Genome Biology* 16: 126. https://doi.org/10.1186/s13059-015-0690-5.

Taslim, C., Huang, T., and Lin, S. (2011). DIME: R-package for identifying differential ChIP-seq based on an ensemble of mixture models. *Bioinformatics* 27: 1569–1570. https://doi.org/10.1093/bioinformatics/btr165.

Tempel, S. and Tahi, F. (2012). A fast ab-initio method for predicting miRNA precursors in genomes. *Nucleic Acids Research* 40: e80. https://doi.org/10.1093/nar/gks146.

Thakur, R.K., Prasad, P., Bhardwaj, S.C. et al. (2022). Epigenetics of wheat–rust interaction: an update. *Planta* 255: 50. https://doi.org/10.1007/s00425-022-03829-y.

Thrash, A., Arick, I.I.M., and Peterson, D.G. (2018). Quack: a quality assurance tool for high throughput sequence data. *Analytical Biochemistry* 548: 38–43. https://doi.org/10.1016/j.ab.2018.01.028.

Tian, Y., Morris, T.J., Webster, A.P. et al. (2017). ChAMP: updated methylation analysis pipeline for Illumina BeadChips. *Bioinformatics* 33: 3982–3984. https://doi.org/10.1093/bioinformatics/btx513.

Trapnell, C., Williams, B., Pertea, G. et al. (2010). Transcript assembly and quantification by RNA-Seq reveals unannotated transcripts and isoform switching during cell differentiation. *Nature Biotechnology* 28: 511–551. https://doi.org/10.1093/bioinformatics/btx513.

Trapnell, C., Hendrickson, D.G., Sauvageau, M. et al. (2013). Differential analysis of gene regulation at transcript resolution with RNA-seq. *Nature Biotechnology* 31: 46–53. https://doi.org/10.1038/nbt.2450.

Walley, J.W., Shen, Z., McReynolds, M.R. et al. (2018). Fungal-induced protein hyperacetylation in maize identified by acetylome profiling. *Proceedings of the National Academy of Sciences of the United States of America* 115: 210–215. https://doi.org/10.1073/pnas.1717519115.

Wang, C., Xu, J., Zhang, D. et al. (2010a). An effective approach for identification of in vivo protein-DNA binding sites from paired-end ChIP-Seq data. *BMC Bioinformatics* 11: 81. https://doi.org/10.1186/1471-2105-11-81.

Wang, L., Feng, Z., Wang, X. et al. (2010b). DEGseq: an R package for identifying differentially expressed genes from RNA-seq data. *Bioinformatics* 26: 136–138. https://doi.org/10.1093/bioinformatics/btp612.

Wang, K., Singh, D., Zeng, Z. et al. (2010c). MapSplice: accurate mapping of RNA-seq reads for splice junction discovery. *Nucleic Acids Research* 38: e178. https://doi.org/10.1093/nar/gkq622.

Wang, Y., Hu, Q., Wu, Z. et al. (2017). HISTONE DEACETYLASE 6 represses pathogen defence responses in *Arabidopsis thaliana*. *Plant, Cell & Environment* 40: 2972–2986. https://doi.org/10.1111/pce.13047.

Wang, C., Wang, C., Xu, W. et al. (2018). Epigenetic changes in the regulation of *Nicotiana tabacum* response to cucumber mosaic virus infection and symptom recovery through single-base resolution methylomes. *Viruses* 10: 402. https://doi.org/10.3390/v10080402.

Wang, C., Wang, C., Zou, J. et al. (2019). Epigenetics in the plant-virus interaction. *Plant Cell Reports* 38: 1031–1038. https://doi.org/10.1007/s00299-019-02414-0.

Wang, Q., Li, M., Wu, T. et al. (2022a). Exploring epigenomic datasets by ChIPseeker. *Current Protocols* 2: e585. https://doi.org/10.1002/cpz1.585.

Wang, L., Li, J., Guo, B. et al. (2022b). Exonic circular RNAs are involved in Arabidopsis immune response against bacterial and fungal pathogens and function synergistically with corresponding linear RNAs. *Phytopathology* 112: 608–619. https://doi.org/10.1094/PHYTO-09-20-0398-R.

Wang, T., Fowler, J.M., Liu, L. et al. (2023). Direct enzymatic sequencing of 5-methylcytosine at single-base resolution. *Nature Chemical Biology* 19: 1004–1012. https://doi.org/10.1038/s41589-023-01318-1.

Westholm, J.O., Miura, P., Olson, S. et al. (2014). Genome-wide analysis of *Drosophila* circular RNAs reveals their structural and sequence properties and age-dependent neural accumulation. *Cell Reports* 9: 1966–1980. https://doi.org/10.1016/j.celrep.2014.10.062.

Wicker, T., Sabot, F., Hua-Van, A. et al. (2007). A unified classification system for eukaryotic transposable elements. *Nature Reviews Genetics* 8: 973–982. https://doi.org/10.1038/nrg2165.

Wilson, G.A., Dharmi, P., Saito, Y. et al. (2012). Resources for the MeDUSA (Methylated DNA Utility for Sequence Analysis) MeDIP-seq computational analysis pipeline for the identification of differentially methylated regions, and associated methylome data from 18 wild-type and mutant mouse ES, NP and MEF cells. *GigaScience* 1: 11. https://doi.org/10.5524/100035.

Wu, J., Mao, X., Cai, T. et al. (2006). KOBAS server: a web-based platform for automated annotation and pathway identification. *Nucleic Acids Research* 34: W720–W724. https://doi.org/10.1093/nar/gkl167.

Xi, Y. and Wei, L. (2009). BSMAP: whole genome bisulfite sequence MAPping program. *BMC Bioinformatics* 10: 232. https://doi.org/10.1186/1471-2105-10-232.

Xiang, L., Cai, C., Cheng, J. et al. (2018). Identification of circular RNAs and their targets in *Gossypium* under *Verticillium* wilt stress based on RNA-seq. *PeerJournal* 6: e4500. https://doi.org/10.7717/peerj.4500.

Xiao, M.S. and Wilusz, J.E. (2019). An improved method for circular RNA purification using RNase R that efficiently removes linear RNAs containing G-quadruplexes or structured 3′ ends. *Nucleic Acids Research* 47: 8755–8769. https://doi.org/10.1093/nar/gkz576.

Xie, S.S. and Duan, C.G. (2023). Epigenetic regulation of plant immunity: from chromatin codes to plant disease resistance. *aBIOTECH* 17: 1–6. https://doi.org/10.1007/s42994-023-00101-z.

Xing, H., Mo, Y., Liao, W. et al. (2012). Genome-wide localization of protein-DNA binding and histone modification by a Bayesian change-point method with Chip-Seq data. *PLoS Computational Biology* 8: e1002613. https://doi.org/10.1371/journal.pcbi.1002613.

Xu, H., Wei, C.L., Lin, F. et al. (2008). An HMM approach to genome-wide identification of differential histone modification sites from ChIP-seq data. *Bioinformatics* 24: 2344–2349. https://doi.org/10.1093/bioinformatics/btn402.

Xu, J., Liu, S., Yin, P. et al. (2018). MeDEStrand: an improved method to infer genome-wide absolute methylation levels from DNA enrichment data. *BMC Bioinformatics* 19: 540. https://doi.org/10.1186/s12859-018-2574-7.

Yang, G. (2013). Mite digger, an efficient and accurate algorithm for genome wide discovery of miniature inverted repeat transposable elements. *BMC Bioinformatics* 14: 186. https://doi.org/10.1186/1471-2105-14-186.

Yang, X. and Li, L. (2011). miRDeep-P: a computational tool for analyzing the microRNA transcriptome in plants. *Bioinformatics* 27: 2614–2615. https://doi.org/10.1093/bioinformatics/btr430.

Yang, J.H., Li, J.H., Shao, P. et al. (2011). starBase: a database for exploring microRNA–mRNA interaction maps from Argonaute CLIP-Seq and Degradome-Seq data. *Nucleic Acids Research* 39: D202–D209. https://doi.org/10.1093/nar/gkq1056.

Yang, X., Liu, D., Liu, F. et al. (2013). HTQC: a fast quality control toolkit for Illumina sequencing data. *BMC Bioinformatics* 14: 33. https://doi.org/10.1186/1471-2105-14-33.

Yang, L., Mu, X., Liu, C. et al. (2015). Overexpression of potato miR482e enhanced plant sensitivity to *Verticillium dahliae* infection. *Journal of Integrative Plant Biology* 57: 1078–1088. https://doi.org/10.1111/jipb.12348.

Yang, X., Zhang, L., Yang, Y. et al. (2021). miRNA mediated regulation and interaction between plants and pathogens. *International Journal of Molecular Sciences* 22: 2913. https://doi.org/10.3390/ijms22062913.

Yashar, W.M., Kong, G., VanCampen, J. et al. (2022). GoPeaks: histone modification peak calling for CUT&Tag. *Genome Biology* 23: 144. https://doi.org/10.1186/s13059-022-02707-w.

Ye, C., Ji, G., and Liang, C. (2016). *detectMITE*: a novel approach to detect miniature inverted repeat transposable elements in genomes. *Science Reporter* 6: 19688. https://doi.org/10.1038/srep19688.

Yi, H.C., You, Z., Huang, D.S. et al. (2018). A deep learning framework for robust and accurate prediction of ncRNA-protein interactions using evolutionary information. *Molecular Therapy-Nucleic Acids* 11: 337–344. https://doi.org/10.1016/j.omtn.2018.03.001.

Yu, A., Lepere, G., Jay, F. et al. (2013). Dynamics and biological relevance of DNA demethylation in *Arabidopsis* antibacterial defense. *Proceedings of the National Academy of Sciences of the United States of America* 110: 2389–2394. https://doi.org/10.1073/pnas.1211757110.

Yu, Y., Zhou, Y.F., Feng, Y.Z. et al. (2020). Transcriptional landscape of pathogen-responsive lncRNA s in rice unveils the role of ALEX 1 in jasmonate pathway and disease resistance. *Plant Biotechnology Journal* 18: 679–690. https://doi.org/10.1111/pbi.13234.

Yue, X., Xie, Z., Li, M. et al. (2022). Simultaneous profiling of histone modifications and DNA methylation via nanopore sequencing. *Nature Communications* 13: 7939. https://doi.org/10.1038/s41467-022-35650-2.

Zang, C., Schones, D.E., Zeng, C. et al. (2009). A clustering approach for identification of enriched domains from histone modification ChIP-Seq data. *Bioinformatics* 25: 1952–1958. https://doi.org/10.1093/bioinformatics/btp340.

Zhang, Y., Liu, T., Meyer, C.A. et al. (2008). Model-based analysis of ChIP-Seq (MACS). *Genome Biology* 9: R137. https://doi.org/10.1186/gb-2008-9-9-r137.

Zhang, X., Robertson, G., Krzywinski, M. et al. (2011). PICS: probabilistic inference for ChIP-seq. *Biometrics* 67: 151–163. https://doi.org/10.1111/j.1541-0420.2010.01441.x.

Zhang, Y., Lin, Y.H., Johnson, T.D. et al. (2014). PePr: a peak-calling prioritization pipeline to identify consistent or differential peaks from replicated ChIP-Seq data. *Bioinformatics* 30: 2568–2575. https://doi.org/10.1093/bioinformatics/btu372.

Zhang, T., Zhao, Y.L., Zhao, J.H. et al. (2016). Cotton plants export microRNAs to inhibit virulence gene expression in a fungal pathogen. *Nature Plants* 2: 1–6. https://doi.org/10.1038/nplants.2016.153.

Zhang, H., Lang, Z., and Zhu, J.K. (2018). Dynamics and function of DNA methylation in plants. *Nature Reviews Molecular Cell Biology* 19: 489–506. https://doi.org/10.1038/s41580-018-0016-z.

Zhang, X., Ménard, R., Li, Y. et al. (2020a). Arabidopsis SDG8 potentiates the sustainable transcriptional induction of the pathogenesis-related genes *PR1* and *PR2* during plant defense response. *Frontiers in Plant Science* 11: 277. https://doi.org/10.3389/fpls.2020.00277.

Zhang, P., Li, S., and Chen, M. (2020b). Characterization and function of circular RNAs in plants. *Frontiers in Molecular Biosciences* 7: 91. https://doi.org/10.3389/fmolb.2020.00091.

Zhang, H., Guo, H., and Hu, W. (2020c). The emerging role of long non-coding RNAs in plant defense against fungal stress. *International Journal of Molecular Sciences* 21: 2659. https://doi.org/10.3390/ijms21082659.

Zhang, Y., Park, C., Bennett, C. et al. (2021). Rapid and accurate alignment of nucleotide conversion sequencing reads with HISAT-3N. *Genome Research* 31: 1290–1295. https://doi.org/10.1101/gr.275193.120.

Zhang, Y., Waseem, M., Zeng, Z. et al. (2022). MicroRNA482/2118, a miRNA superfamily essential for both disease resistance and plant development. *New Phytologist* 233: 2047–2057. https://doi.org/10.1111/nph.17853.

Zhang, A., Zhang, S., Wang, F. et al. (2023). The roles of microRNAs in horticultural plant disease resistance. *Frontiers in Genetics* 14: 1137471. https://doi.org/10.3389/fgene.2023.1137471.

Zhi, P. and Chang, C. (2021). Exploiting epigenetic variations for crop disease resistance improvement. *Frontiers in Plant Science* 12: 692328. https://doi.org/10.3389/fpls.2021.692328.

Zhou, J.M. and Zhang, Y. (2020). Plant immunity: danger perception and signaling. *Cell* 181: 978–989. https://doi.org/10.1016/j.cell.2020.04.028.

Zhou, Q., Su, X., Wang, A. et al. (2013). QC-chain: fast and holistic quality control method for next-generation sequencing data. *PLoS One* 8: e60234. https://doi.org/10.1371/journal.pone.0060234.

Zhou, Y., Zhu, J., Zhao, M. et al. (2018). Methylation-level inferences and detection of differential methylation with MeDIP-seq data. *PLoS One* 13: e0201586. https://doi.org/10.1371/journal.pone.0201586.

Zhu, L.J., Gazin, C., Lawson, N.D. et al. (2010). ChIPpeakAnno: a bioconductor package to annotate ChIP-seq and ChIP-chip data. *BMC Bioinformatics* 11: 237. https://doi.org/10.1186/1471-2105-11-237.

Zou, B., Yang, D.L., Shi, Z. et al. (2014). Monoubiquitination of histone 2B at the disease resistance gene locus regulates its expression and impacts immune responses in *Arabidopsis*. *Plant Physiology* 65: 309–318. https://doi.org/10.1104/pp.113.227801.

Weblinks

[1] http://www.bioinformatics.babraham.ac.uk/projects/fastqc
[2] http://hannonlab.cshl.edu/fastx_toolkit/
[3] https://github.com/FelixKrueger/TrimGalore
[4] http://www.usadellab.org/cms/?page=trimmomatic
[5] http://cutadapt.readthedocs.io/en/stable/index.html
[6] http://novocraft.com/
[7] https://gensoft.pasteur.fr/docs/Genrich/0.6.1/
[8] https://urgi.versailles.inra.fr/Tools/REPET
[9] https://www.repeatmasker.org/cgi-bin/WEBRepeatMasker
[10] https://alphafold.ebi.ac.uk/

16

The Evolution of Auxin-Binding Protein 1

Siarhei A. Dabravolski[1] *and Stanislav V. Isayenkov*[2,3]

[1] *Department of Biotechnology Engineering, Braude Academic College of Engineering, Karmiel, Israel*
[2] *Institute of Agricultural and Nutritional Sciences, Martin Luther University Halle-Wittenberg, Halle (Saale), Germany*
[3] *Department of Plant Food Products and Biofortification, Institute of Food Biotechnology and Genomics, NAS of Ukraine, Kyiv, Ukraine*

16.1 Abundance of Auxin and Auxin-binding Proteins in Nature

Indole-3-acetic acid (IAA), the main natural form of the plant hormone auxin, is produced ubiquitously in all kingdoms, where auxin is involved in the regulation of:

- bacterial physiology (Matilla et al. 2018), metabolism (Laird et al. 2020), resistance to stresses (Lin et al. 2018; Ahmad et al. 2022), virulence (McClerklin et al. 2018; Kunkel and Johnson 2021), and communication with plants (Duca and Glick 2020);
- algae and microalgae growth (Žižková et al. 2017; Lin et al. 2022);
- fungal physiology, growth, sporulation (Nutaratat et al. 2016; Nicastro et al. 2021), virulence (Leontovyčová et al. 2020), and interaction with microorganisms, other fungi, algae, and plants (Fu et al. 2015);
- virulence of phytopathogenic insects (Yokoyama et al. 2017; Tokuda et al. 2022);
- inflammatory processes in animals and humans (Oliveira et al. 2007; Yokoyama et al. 2017), provide antioxidant and antiapoptotic effects on the renal ischemia-reperfusion injury (Papi et al. 2022) and antidepressive effect (Chen et al. 2022) in the rat model, and anticancer properties in silico simulation (Zarei et al. 2022). The ability of so many organisms from different taxa to synthesize IAA and use it to modulate a wide range of physiological processes suggested that auxin is a pivotal signalling molecule for intra- and interkingdom communication (Duca and Glick 2020; Rico-Jiménez et al. 2022).

However, the regulatory role of IAA and its underlying molecular mechanisms have been studied mostly in the field of plant biology, while information on other organisms is still scarce. For example, the ability to bind auxin was reported for several bacterial sensors, such as chemoreceptor PcpI of *Pseudomonas putida* (Rico-Jiménez et al. 2022), MarR bacterial regulators of auxin catabolism (Conway et al. 2022), solute-binding protein Dde_0634 of *Desulfovibrio desulfuricans* (Vetting et al. 2015), the transcriptional regulator TrpR of

Bioinformatics for Plant Research and Crop Breeding, First Edition. Edited by Jen-Tsung Chen.

Escherichia coli (Herud-Sikimić et al. 2021), and AdmX, a Lys type transcriptional regulator (LTTRs) from *Klebsiella pneumoniae* (Gavira et al. 2023).

Among auxin-binding bacterial sensors, the LTTRs are the best-studied ones. LTTR is formed by an N-terminal helix-turn-helix DNA-binding domain and a C-terminal ligand binding domain (LBD). While LBDs are poorly conserved on the amino acid level, their conservation on a structural level is confirmed with several solved crystal structures (Sainsbury et al. 2009; Koentjoro and Ogawa 2018; Giannopoulou et al. 2021). As was shown in these solved structures, ligand binding caused conformational changes in the LBD which transmitted to the DNA-binding domain and facilitated its binding to DNA (Liu et al. 2021; Soukarieh et al. 2021). So far, only a limited number of ligands for the LBD of LTTRs have been identified. These ligands are represented by fatty, organic and amino acids, aromatic compounds, antibiotics, and various other nutrients (Matilla et al. 2022). For example, the pathway-specific transcriptional activator LTTR AdmX from the rhizosphere isolate *Serratia plymuthica* A153 was shown to recognise IAA, which inhibited the expression of the andrimid gene cluster and suppressed antibiotic production (Matilla et al. 2016). Furthermore, several other pathogenic and beneficial plant-associated bacteria can inhibit the andrimid production in A153 through the synthesis and secretion of IAA (Matilla et al. 2018). However, to the best of our knowledge, there are no data regarding the experimental conformation of any bacterial ABP1 homologous to bind IAA or any other auxin.

Interestingly, two bacterial auxin-binding domains (the solute-binding protein Dde_0634 of *Oleidesulfovibrio alaskensis* and the LBD of *S. plymuthica* A153 LTTR AdmX) belong to the periplasmic binding fold protein superfamily (CL0144). Although many homologous proteins are known in plants, their ability to bind auxin or participate in any kind of auxin-mediated signalling pathways or metabolic reactions is still unknown. Therefore, these hypotheses require further experimental confirmation.

16.2 Auxin in Plants

Auxin, a crucial plant hormone, plays a significant role in regulating a wide range of complex growth and development processes, environmental responses, and various aspects of cell differentiation, division, and elongation (Gomes and Scortecci 2021). Specifically, auxin controls diverse processes in plants, including vascular development and organ patterning (Lavania et al. 2021), light and gravity tropic responses (Lv et al. 2021), root and shoot architecture (Liu et al. 2017; Hou et al. 2021), and adventitious root generation (Li 2021).

IAA serves as the primary natural form of auxin, extensively studied in higher plants, particularly in *Arabidopsis thaliana* (L.) Heynh. The synthesis of auxin mainly occurs through a two-step process from tryptophan. First, tryptophan is converted to indole-3-pyruvic acid (IPyA) by Tryptophan Aminotransferase of Arabidopsis 1 (TAA). Then, IPyA is further converted to IAA by a family of YUCCA flavin-dependent mono-oxygenases (YUCs) (Kriechbaumer et al. 2016). Additionally, an alternative tryptophan-independent auxin biosynthesis pathway involving a cytosolic indole synthase has been reported. However, the enzymology underlying this pathway and its physiological significance remain largely unknown and require further investigation (Wang et al. 2015). Furthermore,

1-naphthalene acetic acid (NAA), a synthetic auxin with high metabolic stability, has been extensively studied and widely used in research (De Klerk et al. 1997).

The cellular responses to auxin are typically dependent on its concentration, which is primarily regulated by auxin biosynthesis and intercellular auxin transport. Over the past few decades, significant research efforts have been dedicated to investigating auxin transporters and their regulation, leading to a comprehensive understanding of both long-distance (phloem flow from shoot to root) and short-distance (cell-to-cell flow facilitated by specific auxin influx/efflux carriers) auxin transport pathways (Semeradova et al. 2020). Several classes of active auxin transporters have been identified and extensively studied, primarily in *A. thaliana*, including pin-formed (PIN) and pin-like (PILS) proteins, P-glycoproteins (belonging to the ATP-binding cassette [ABC] transporter family), and auxin transporter protein 1 (AUX1) and auxin transporter-like protein (LAX) family (Hammes et al. 2022). Moreover, the WALLS ARE THIN1 (WAT1) localized in the tonoplast has been demonstrated to participate in auxin transport and contribute to the maintenance of intracellular auxin homeostasis (Ranocha et al. 2013).

Following changes in cellular auxin concentration, auxin receptors perceive the signal, and this information is transduced through auxin signalling systems to elicit appropriate cellular responses. Currently, three canonical auxin receptor systems have been characterized: SCF^{TIR1} (consisting of an SKP1–cullin-1–F-box complex with transport inhibitor response 1 (TIR1) as the F-box protein), S-phase-kinase-associated protein 2 (SKP2), and ABP1 (Napier 2021; Yu et al. 2022). Additionally, two noncanonical auxin sensing pathways have been identified. In the first mechanism, auxin directly binds to the AUXIN RESPONSE FACTOR (ARF3/ETTIN) transcription factor, influencing chromatin states and interacting with other transcription factors and regulators (Simonini et al. 2016; Kuhn et al. 2020). The second non-canonical mechanism involves TRANSMEMBRANE KINASE 1 (TMK1), a member of the receptor-like kinase family localized in the plasma membrane. TMK1 orchestrates cell expansion and proliferation, lateral root organogenesis, and auxin biosynthesis through various other cellular pathways (Dai et al. 2013; Cao et al. 2019; Wang et al. 2020).

In the subsequent section of this chapter, we focus on ABP1, a protein that has been extensively investigated for a significant period (Tillmann et al. 1989). ABP1 is known to primarily localise to the endoplasmic reticulum and the plasma membrane (Tian et al. 1995; Leblanc et al. 1999). Recently, the developmental functions of ABP1 and the operation of the ABP1-TMK1 signalling module at the cell surface in *A. thaliana* have been comprehensively characterised (Gelová et al. 2021; Friml et al. 2022; Ang and Østergaard 2023). However, the evolutionary aspects of ABP1 remain poorly studied. Consequently, we employ various bioinformatic approaches to dive into the structure-function relationships of ABP in archaea, bacteria, algae, and plants, to enhance our understanding of the overall evolution of ABP1.

16.3 Domain Organisation

Domain architecture is crucial for understanding proteins' functionality and their evolutionary history. Therefore, we evaluated the presence of different domains among identified ABP1 homologous. "Auxin-binding protein" domain (PF02041) belongs to the "Cupin" superfamily, which includes a wide variety of enzymes and non-enzymatic seed storage proteins (Hu et al. 2023). The majority of proteins containing the "auxin-binding protein" domain are single-domain proteins, suggesting that it is the dominant domain architecture specific for these proteins. However, some species have a limited number of other domains, associated with the "auxin-binding protein" domain (Table 16.1).

Thus, 3 proteins from *Tripterygium wilfordii* (Hook.f.) (here and further we are using Uniprot ID if not stated otherwise: A0A7J7CFT4), *Cuscuta campestris* (Yunck.) (A0A484NLZ7) and *Hevea brasiliensis* (Müll.Arg.) (A0A6A6LB20) have N-terminal "auxin-binding protein" domain and C-terminal Glycosyl transferase family 8 domain (PF01501).

Table 16.1 The list of other domains associated with the "auxin-binding protein" domain.

Domain ID	Domain name	Species	Protein ID
PF07762	The protein of unknown function (DUF1618)	*Oryza glumipatula*	A0A0E0BTH8; A0A0E0BTH9
		Oryza nivara	A0A0E0JB44; A0A0E0JB45; A0A0E0JB46
		Oryza meridionalis	A0A0E0FCS0 A0A0E0FCS1
PF01501	Glycosyl transferase family 8	*Cuscuta campestris*	A0A484NLZ7
		Hevea brasiliensis	A0A6A6LB20
		Tripterygium wilfordii	A0A7J7CFT4
PF07883	Cupin domain	*Nocardioides* sp. PD653	A0A1V1W0Q5
		Deltaproteobacteria bacterium	A0A7C3JII1
PF00899	ThiF family domain	*Coptis chinensis*	A0A835I459
PF13428	Tetratricopeptide repeat	*Rhododendron simsii*	A0A834LB65
PF01535	Pentatricopeptide repeat domain	*Prunus yedoensis*	A0A314ZWL6
PF06777	Helical and beta-bridge domain	*Fagus sylvatica*	A0A2N9GW70
PF00657	GDSL-like Lipase/ Acylhydrolase	*Fagus sylvatica*	A0A2N9GED1
PF14223	Gag-polypeptide of LTR copia-type	*Hibiscus syriacus*	A0A6A2Z164
PF05163	DNA damage-inducible family domain	*Singulisphaera* sp. GP187	A0A1N6E577
PF02365	No apical meristem	*Aegilops tauschii*	R7W2T8
PF08637	ATP synthase regulation protein NCA2	*Micractinium conductrix*	A0A2P6VET3
PF00097	Zinc finger, C3HC4 type (RING finger)		

This family includes enzymes associated with lipopolysaccharide biosynthesis and glycogen synthesis via sugar residues transfer to the donor molecules: lipopolysaccharide galactosyltransferase, lipopolysaccharide glucosyltransferase 1 and glycogenin glucosyltransferase (P27128, P27129, and P46976, respectively). GDSL-like Lipase/Acylhydrolase (PF00657) is another interesting domain located to the N-terminal part of the "auxin-binding protein" domain in beech (*Fagus sylvatica* L.) (A0A2N9GED1). Several enzymes of this family were characterized as multifunctional enzymes with esterase (hydrolase activity on ester bonds) and transferase activities. However, among studied enzymes some preferentially function as transferases rather than esterases, while other has acquired transferase activity and lost their reverse esterase activity (Yamashiro et al. 2022).

Furthermore, in monocotyledon species *Aegilops tauschii* (Coss.), "auxin-binding protein" domain was found to be associated with no apical meristem (NAM) domain (PF02365) on its C-terminal part (R7W2T8). NAM proteins are transcription factors, playing a crucial role in plant development processes: responsible for a shoot apical meristem formation, determining positions of meristems and primordial, and regulating the transition between growth by cell division and cell expansion in stamens and petals (Zhong et al. 2016; Phillips et al. 2020). Similarly, the protein (A0A2P6VET3) from *Micractinium conductrix* ([K.Brandt] Pröschold and Darienko) contained N-terminal ATP synthase regulation protein NCA2 domain (PF08637), followed by Zinc finger, C3HC4 type (RING finger) domain (PF00097), and C-terminal "auxin-binding protein" domain. Such domain composition suggested a possible role in the regulation of ATP synthesis (NCA2 domain) (Li et al. 2019) and ubiquitination (RING finger) (Han et al. 2022).

In summary, all the current research on plants ABP1 homologous is conducted on a single-"auxin-binding protein" domain-containing proteins, mostly, on a model species, such as Arabidopsis, rice, corn, and *Physcomitrella*, while the functionality of "auxin-binding protein" domain from proteins with other domain architecture was never studied. Therefore, further research is required to elucidate how other domains affect known properties of the "auxin-binding protein" domain.

16.4 ABP1 Active Sites/Structure/Sequence Analysis

The determined structure of ABP1 derived from maize (*Zea mays* L.) suggests that one auxin-binding site is formed by the interaction of two ABP1 monomers through the dimerisation interface (Figure 16.S1). Furthermore, several amino acids, including Ile22, Leu25, Trp44, Gln46, Thr54, Pro55, Ile130, Phe149, and Trp151, collectively form the binding pocket and establish interactions with the ligand (Woo 2002; Grandits and Oostenbrink 2014). Each subunit of ABP1 coordinates a Zn^{2+} ion through specific histidines (57, 59, and 106) and glutamate (Glu63), and contains a glycosylation site at Asn95. In dicot species such as Arabidopsis, an additional glycosylation site at Asn11 has been identified. Both Zn^{2+} binding and glycosylation have been found to enhance the conformational stability of ABP1 subunits (Woo 2002; Da Costa et al. 2017). Moreover, a C-terminal ER-retention KDEL sequence has been reported to possess protein-stabilising properties (David et al. 2001; Grandits and Oostenbrink 2014) (Figure 16.S1).

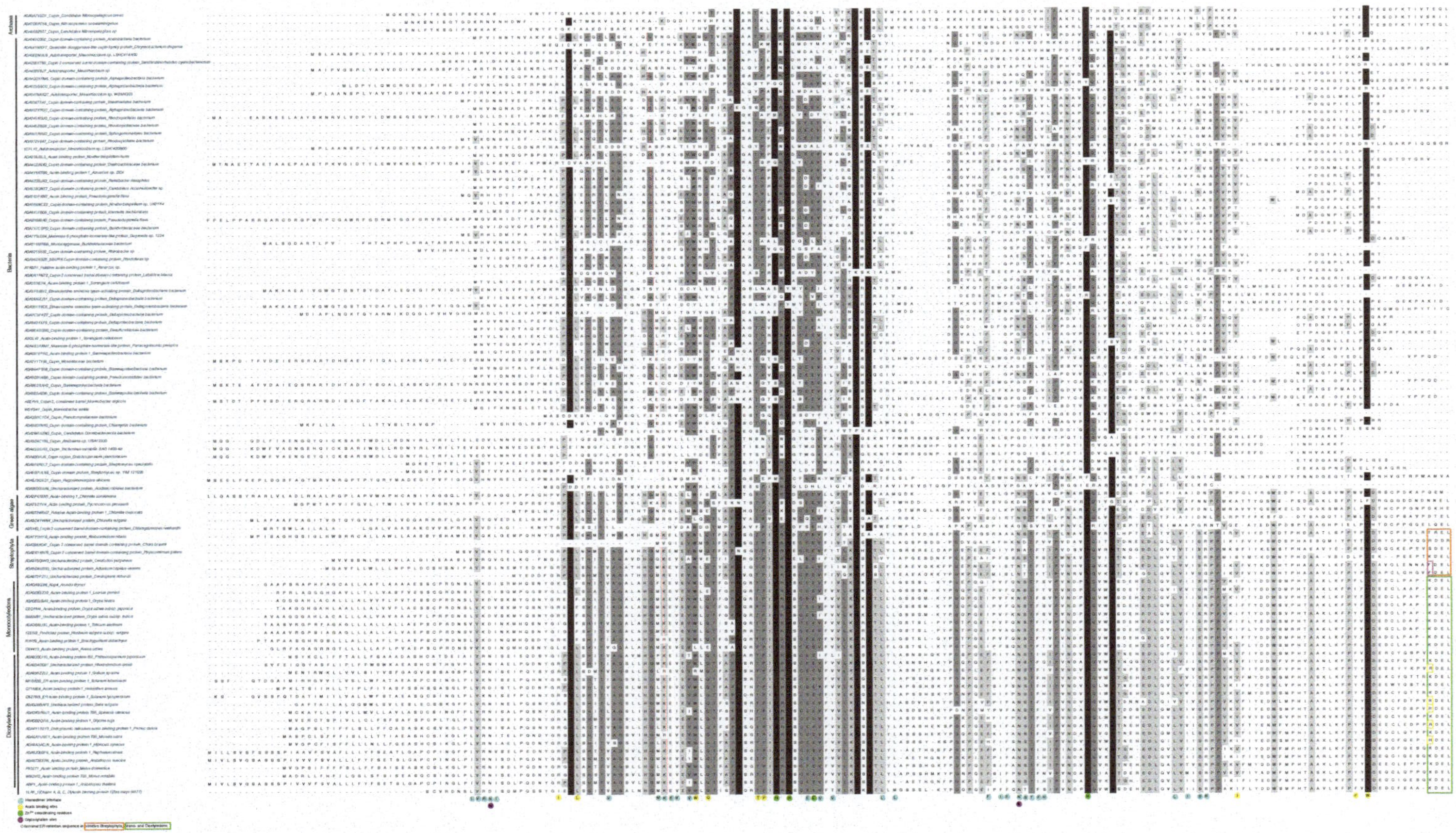

Figure 16.S1 Alignment of the "auxin-binding protein" domain from selected homologues from different taxa. Residues forming the homodimer interface (cyan dots), auxin-binding sites (yellow dots), coordinating Zn^{2+} (green dots), and glycosylation sites (magenta dots) are highlighted. C-terminal ER-retention sequences are depicted with the orange box in primitive Streptophyta and with green box in Mono- and Dicotyledon species; variations in the basic KDEL sequence are depicted with magenta and yellow squares, respectively.

16.4.1 Homodimerisation and Glycosylation Sites

Interestingly, amino acids forming the dimerisation interface are well-conserved in Streptophyta taxa, while in more far-related taxa (green algae, Bacteria, and Archaea) it is rather fragmentary (Figure 16.S1, cyan dots). The same statement can be applied to the major glycosylation site (Asn95). On the other side, the dicot-specific glycosylation site (Asn11) is also presented in some Bacteria (e.g. such as *Rugosimonospora africana* and *Deltaproteobacteria bacterium*), while missing in dicotyledon species *Helianthus annuus* (L.) (Figure 16.S1, magenta dots).

16.4.2 KDEL

The retention of proteins in ER is facilitated through the short C-terminal sequences, which are well-studied in both plant and mammalian systems (KDEL, RDEL, HDEL, DKEL, and KNEL), suggesting that some variations are possible at least in the first two amino acids (Andres et al. 1990; Denecke et al. 1992). Thus, the major ER retention sequence KDEL is widely present in Streptophyta (Figure 16.S1, green box), and less common types of ER retention sequences (such as RDEL and HDEL) have also been identified in several species (such as *Beta vulgaris* L. and *Hibiscus syriacus* L.) (Figure 16.S1, green box, yellow squares).

Interestingly, we found several types of ER-retention sequences in more primitive Streptophyta species (Figure 16.S1, orange box). The classical KDEL sequence was found only in the moss *Ceratodon purpureus* (Hedw.) Brid. Other species contained ER retention sequence with 3 (ferns *Ceratopteris richardii* Brongn. and *Adiantum capillus-veneris* L. – F**DEL**) (Figure 16.S1, orange box, magenta square) and 2 (*Klebsormidium nitens* [Meneghini] Lokhorst – TR**EL**) matching amino acids. Similarly, 3 (*Pycnococcus provasolii* R.R.L.Guillard – N**NEL**, *Chlorella vulgaris* Beijerinck 1890 – **D**T**EL**) and 2 (*Chlamydomonas reinhardtii* P.A.Dang. – LQ**EL**) matching amino acids were found in some green algae. However, the ability of such alternative sequences (F**DEL**, TR**EL**, N**NEL**, **D**T**EL**, and LQ**EL**) to facilitate the retention of secretory plant proteins in the ER was not demonstrated (neither in plant nor in animal systems) and requires experimental confirmation. These results suggested that ABP1 homologous may be retained in the ER through different variations of the classical KDEL sequence not only in vascular plants but also in more primitive plant and green algae species.

16.4.3 Zn-binding Sites

Amino acids coordinating Zn^{2+} ion (histidines 57, 59, and 106) are highly conserved within ABP1 proteins from all defined taxa (Archaea, Bacteria, and Viridiplantae) and are important for the positioning of the ligand. However, the glutamate Glu63 is highly conserved only in Viridiplantae taxa, while in Bacteria and Archaea, the conservation is fragmentary (Figure 16.S1, green dots).

16.4.4 Auxin-binding Sites

Among amino acids, involved in the auxin binding, Leu25, Trp44, Gln46, Thr54, Pro55, and Trp151, were well-conserved in all analysed species (including Archaea, Bacteria, and Viridiplantae). Ile22, Ile130, and Phe149 on the contrary, demonstrated a much lower level of conservation in analysed species (most importantly, including well-studied model plant species from Mono- and Dicotyledon taxa) (Figure 16.S1, yellow dots).

In total, analysis of ABP1 active sites (auxin-binding, Zn^{2+} coordinating, homodimerisation, and glycosylation) suggested that ABP1 from all taxa (Archaea, Bacteria, and Viridiplantae) have all residues necessary for auxin binding, thus could function as a primitive auxin receptor before more specialised proteins of the auxin signalling pathway have evolved.

16.5 ABP1 Evolution

To track the evolutionary origin of the auxin-binding protein, we have searched for the homologous sequences from the UniProt (The UniProt Consortium et al. 2021) (http://www.uniprot.org/), Pfam/InterPro (http://www.ebi.ac.uk/interpro) (El-Gebali et al. 2019; Mitchell et al. 2019), and NCBI databases (http://www.ncbi.nlm.nih.gov/) (Sayers et al. 2022) using both the amino acid-based BLAST and the domain-based search methods (Boratyn et al. 2012). Target proteins were identified based on the presence of the "auxin-binding protein" domain (PF02041) through the NCBI's Conserved Domain Database (Marchler-Bauer et al. 2015). During the sequence search, the truncated, partial, and identical sequences were removed. In the case of multi-domain proteins (which are common for Bacterial and Archaeal taxa), only the sequence of the "auxin-binding protein" domain was used for the alignment and tree reconstruction. In total, 419 sequences were identified (*Viridiplantae*: 355, *Bacteria* 59; and *Archaea* 5). Many bacterial and archaea proteins have been annotated as containing "auxin-binding protein" domain; however, their closer examination demonstrated the presence only of a Cupin domain (PF07883) or other Cupin-related domains.

16.5.1 Phylogenetic Analysis

To understand the evolutionary history of the auxin-binding protein, we reconstruct the phylogenetic trees using both the maximum likelihood and neighbor-joining methods (Felsenstein 1981; Saitou and Nei 1987). In the resulting tree, we can define 7 clades and several sub-clades (Figure 16.1). Clade I was formed by bacteria from different taxa (Cyanobacteria, Alphaproteobacteria, Chlorobi, and others) and two closely related species from Archaea from the TACK group (*Candidatus Nitrosopelagicus brevis* and *Nitrosopumilus cobalaminigenes*). Clade II was composed mostly of proteobacteria (alpha, beta, and gamma) and some other species, such as one Actinomycetota species (*R. africana*) (in clade IIa), or Acidobacteriota (*Acidobacteriota bacterium*) and Myxococcota species (*Sorangium cellulosum*) (in clade IIb). However, clade IIc represented a more interesting set of proteins from diverse taxa, which included different Bacteria, such as Proteobacteria (alpha-, beta-, and delta-), Chlamydiota (*Chlamydiia bacterium*) and Campylobacterota (*Desulfurellaceae bacterium*), and green algae *C. reinhardtii* (P.A.Dang.) and Chlorella (*Chlorella sorokiniana* Shihira & R.W. Krauss and *C. vulgaris* Beijerinck 1890) species.

Further clades were represented only by Eukaryotic organisms. Thus, Clade III was formed mostly by green algae (*K. nitens* (Meneghini) Lokhorst and *P. provasolii* R.R.L.Guillard) and primitive Streptophyta, such as ferns (*A. capillus-veneris* L. and *C. richardii* Brongn.),

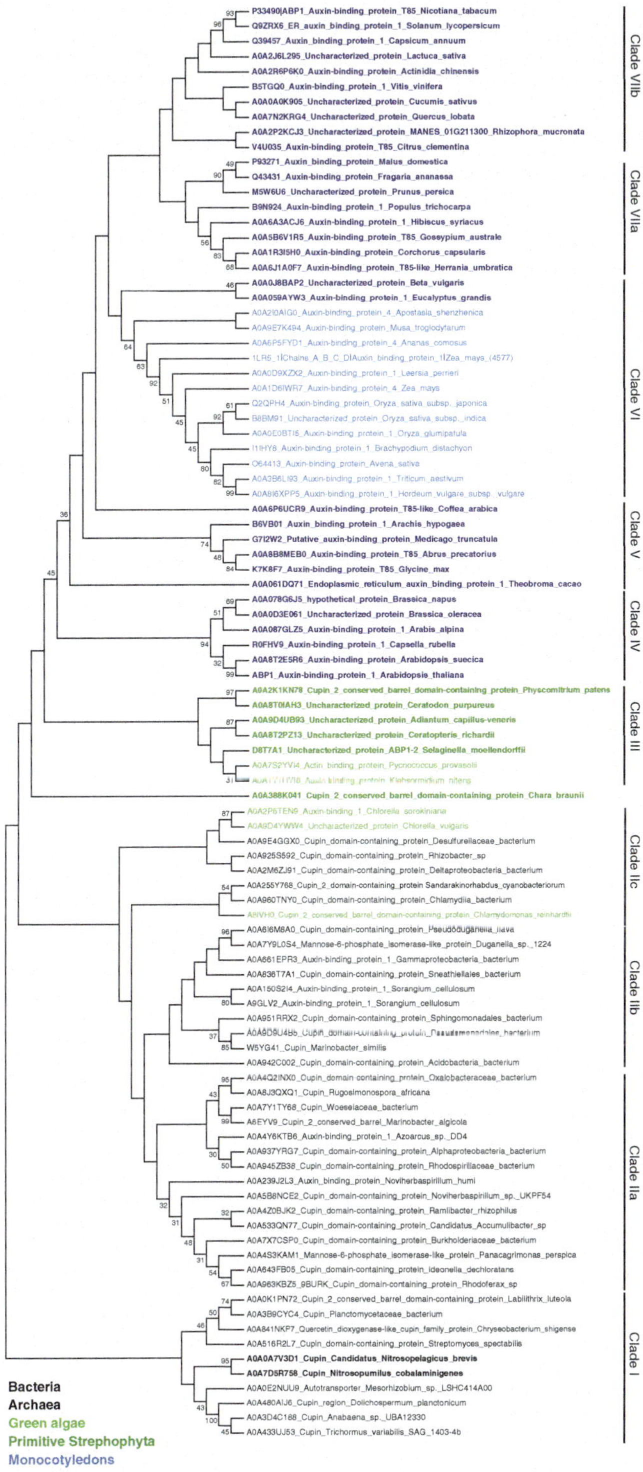

Figure 16.1 Phylogeny estimation of the "auxin-binding protein" domain (PF02041). The Maximum Likelihood method and JTT model were used; 1000 bootstrap replicates. Only branches with bootstrap value >30 are shown. Dicotyledon species are highlighted in blue, Monocotyledon – in cyan, Primitive Strephophyta – in green, Green algae – in light green, Archaea – in bold black, Bacteria – in black.

moss (*Physcomitrella patens* (Hedw.) Bruch & Schimp. and *C. purpureus* (Hedw.) Brid.), spikemoss (*Selaginella moellendorffii* Hieron.). Interestingly, the primitive plant *Chara braunii* (S.G.Gmel.) was located between complex transition clade IIc (formed by proteins from Pro and Eukaryotic organisms) and fully Eukaryotic clade III.

Clade IV was composed of closely related Brassicaceae species (such as *Arabidopsis* sp. and *Brassica* sp.), while clade V combined more diverse species from different high-plant taxa. Here we can name Fabaceae (*Medicago truncatula* Gaertn., *Abrus precatorius* L., *Arachis hypogaea* L. and *Glycine max* (L.) Merr.), Rubiaceae (*Coffea arabica* L.) and Malvaceae (*Theobroma cacao* L.). The following Clade VI was mostly formed by Monocotyledon species (such as *Z. mays* L., *Triticum aestivum* L., *Ananas comosus* (L.) Merr. and others), however, with unexpected inclusion of two dicotyledon species on a separated branch – *B. vulgaris* L. and *Eucalyptus grandis* W.Hill. The Clade VII contained two sub-clades composed of various trees (*Quercus lobata* Née, *Populus trichocarpa* Torr. & A.Gray ex. Hook., *Malus domestica* Borkh., and others), common agricultural and decorative (*Solanum lycopersicum* L., *Cucumis sativus* L., *H. syriacus* L., and others) dicotyledon species.

In general, the reconstructed phylogenetic tree has demonstrated rather low values of bootstrap support in the deep nodes (only nodes with more than 30% are shown). The high values of the bootstrap support can be found only between closely related proteins from bacterial taxa in clade IIa (Alpha-, Beta-, and Gammaproteobacteria), dicotyledon in clades IV and V (Brassicaceae and Fabaceae, respectively), and monocotyledon in clade VI. Besides several highly conserved residues which were discussed in previous sections (dimerisation interface, glycosylation, zinc, and auxin-binding sites), the rest of the sequence, apparently, varies greatly between different taxa. In terms of large-scale evolution, the absence of the "auxin-binding protein" domain among viruses, animal, fungal, and red algae taxa suggested symbiotic gene transfer as the most possible scenario. This suggestion was supported by our phylogenetic tree, where we can trace the "auxin-binding protein" domain from Bacteria to high plants through Green algae and primitive land plants (such as mosses and ferns). Furthermore, the presence of other auxin sensors in bacteria (such as PcpI, MarR, TrpR, and AdmX) may explain why the "auxin-binding protein" domain was found only in a limited number of bacterial taxa.

16.5.2 Structures Modelling and Comparison

Structural models of the "auxin-binding protein" were built with SWISS-MODEL (https://swissmodel.expasy.org/) (Waterhouse et al. 2018). SWISS-MODEL is a fully automated protein homology modelling server, which uses BLAST and HHblits template search algorithms (Camacho et al. 2009; Remmert et al. 2012). The quality and similarity of templates were estimated with the Global Model Quality Estimate (GMQE) (Biasini et al. 2014) and Quaternary Structure Quality Estimate (QSQE) (Bertoni et al. 2017) methods. Predicted 3D structures were refined with the 3D refine tool (https://3drefine.mu.hekademeia.org/) (Bhattacharya et al. 2016) and verified with QMEAN (https://swissmodel.expasy.org/qmean/) a linear combination of six structural descriptors of a given protein (Benkert et al. 2011). iPBA webserver was used to align pdb structures (https://www.dsimb.inserm.fr/dsimb_tools/ipba/index.php) (estimated with RMSD (quality of alignment, calculated from the superimposition of protein pairs based on PB alignment) and normalized score (the dynamic programming alignment score, calculated from the relation of the alignment score to the alignment length)) (Zemla 2003; Tyagi et al. 2006; Tyagi et al. 2008) and PyMol (https://pymol.org/2/) for visualization. Further, all generated models were compared to the original *Z. mays* ABP1 model (1LR5). The similarity of the two structures was defined by a normalised similarity score (the higher the score the higher similarity) and RMSD (with 0 – ideal match). Thus, the alignment of 1LR5 on itself served as a positive control and yielded a maximal score and 0 RMSD (Table 16.2).

Unexpectedly, among all checked structures, a cyanobacterial homologue of ABP1 protein from *Dolichospermum planctonicum* demonstrated the lowest similarity score (with a negative value −56.41). Similarly, a low degree of similarity to the 1LR5 model was shown for the homologues of ABP1 protein from Archaea (*N. cobalaminigenes*) and green algae (*C. reinhardtii*). Interestingly, other Bacterial APB1 homologues from Alphaproteobacteria (*Rhodospirillales bacterium*) and Campylobacterota (*D. bacterium*) demonstrated average and high similarity scores (180.29 and 392.85, respectively). Surprisingly, the similarity of ABP1 homologue from *D. bacterium* was higher than proteins from green algae (*K. nitens*), moss (*Physcomitrium patens*) (with 242.34 and 291.82, respectively) and even homologous

Table 16.2 Quality of generated ABP1 homologues and their alignment with *Zea mays* ABP1 (1LR5).

					Alignment to 1LR5	
Taxa	**Species**	**ID**	**GMQE**	**QMEAN**	**Similarity score**	**RMSD**
Archaea	*Nitrosopumilus cobalaminigenes*	A0A7D5R758	0.96	−0.75	123.18	1.99
Cyanobacteria	*Dolichospermum planctonicum*	A0A480AIJ6	0.95	0.33	−56.41	1.47
Alphaproteobacteria	*Rhodospirillales bacterium*	A0A972V647	0.96	0.74	180.29	1.17
Campylobacterota	*Desulfurellaceae bacterium*	A0A9E4GGX0	0.69	0.33	392.85	0.28
Green algae	*Chlamydomonas reinhardtii*	A8IVH0	0.79	−1.97	98.72	1.7
Green algae	*Klebsormidium nitens*	A0A1Y1HYI8	0.85	−1.49	242.34	1.02
Moss	*Physcomitrium patens*	A0A2K1KN78	0.85	−1.19	291.82	0.7
Eudicotyledon plant	*Arabidopsis thaliana*	P33487	0.87	−3.14	326.77	0.8
Eudicotyledon plant	*Fragaria ananassa*	Q43431	0.87	−2.02	348.65	0.75
Eudicotyledon plant	*Populus trichocarpa*	B9N924	0.88	−0.95	352.66	0.87
Monocotyledon plant	*Musa troglodytarum*	A0A9E7K494	0.89	−0.95	366.01	0.7
Monocotyledon plant	*Zea mays*	1LR5	0.95	1.5	606.73	0.0

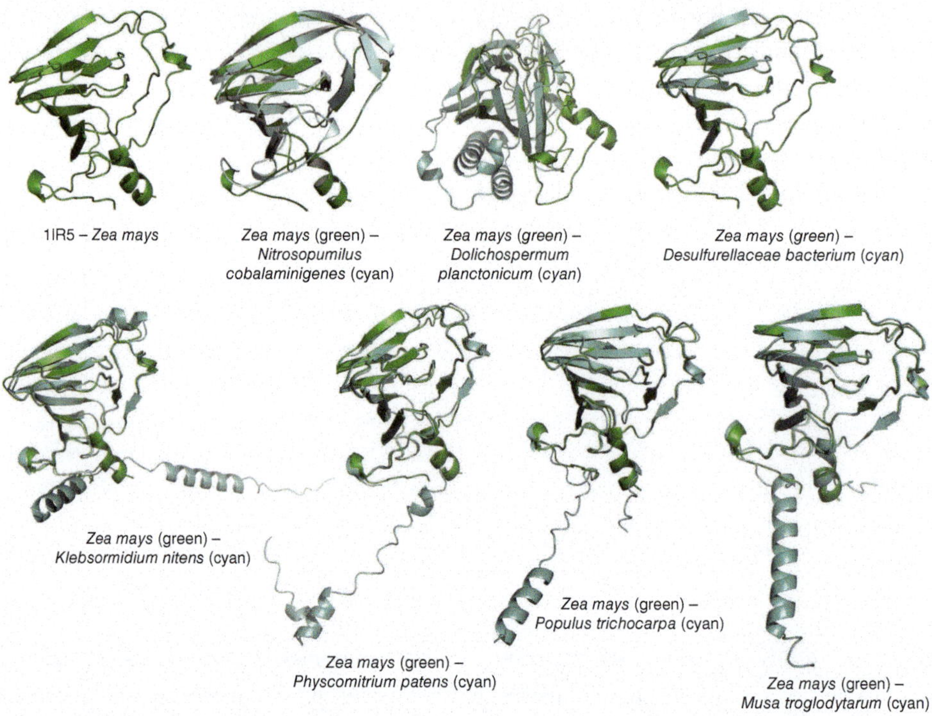

Figure 16.2 Structural alignment of the auxin-binding protein 1 from *Zea mays* (1LR5) (green) with homologous from *Nitrosopumilus cobalaminigenes* (A0A7D5R758), *Dolichospermum planctonicum* (A0A480AIJ6), *Desulfurellaceae bacterium* (A0A9E4GGX0), *Klebsormidium nitens* (A0A1Y1HYI8), *Physcomitrium patens* (A0A2K1KN78), *Populus trichocarpa* (B9N924) and *Musa troglodytarum* (A0A9E7K494) (cyan).

from monocotyledon (*Musa troglodytarum*) and all eudicotyledon plants (*A. thaliana*, *Fragaria ananassa*, and *P. trichocarpa*), which showed high similarity to 1LR5 (with 366.01, 326.77, 348.65 and 352.66, respectively) (Figure 16.2).

The evolution of modern land plants is often explained by the ancient endosymbiotic event when cyanobacteria were engulfed by the last eukaryotic common ancestor (LECA) and gave rise to chloroplasts (Martin et al. 2015). This theory was confirmed and widely accepted by many researchers, thus suggesting a close evolutional relationship in line cyanobacteria → green algae → modern chloroplast (Martin et al. 2002). In particular, many genes have been shown to undergo endosymbiotic transfer events from cyanobacteria to the host nucleus (Keeling 2013). Therefore, the origin of many plant genes can be traced down on the phylogenetic tree from modern land plants to cyanobacteria. However, this is not the case for the "auxin-binding protein." On the reconstructed phylogenetic tree, we can clearly see the closer association between green algae and various Proteobacteria, while Cyanobacteria are located on a separate branch with Archaea ABPs. Also, considering the alignment of the reconstructed models, we can suggest root-associated Proteobacteria as the origin of the "auxin-binding protein." While bacteria of many taxa have "auxin-binding protein" close interaction with plants in the rhizosphere had shaped and unified their structures and, possibly, functions. Therefore, we could not exclude the involvement of another mechanism of gene acquisition, most probably, through the horizontal gene transfer from some *Agrobacterium* species (such as *Agrobacterium tumefaciens* or *Agrobacterium rhizogenes*) (Quispe-Huamanquispe et al. 2015).

In total, results of phylogenetic analysis and alignment of structural models suggested bacterial origin of the "auxin-binding protein." Taking into account the close positioning of the tree and structural similarity between plant and Proteobacterial ABPs, we can speculate that ABP gene was acquired through the endosymbiotic gene transfer from the cyanobacterial ancestor of chloroplast, or, more likely, via horizontal gene transfer from root-associated bacteria such as *Agrobacterium* spp.

16.6 Future Prospective

16.6.1 Role in Archaea

In the introduction section we have reviewed the known function of IAA in all known taxa except Archaea. While animals, fungi, and insects have no ABP1 homologous, they sense IAA with other mechanisms to activate certain signalling pathways and cause desired physiological and/or metabolic responses. However, in the case of Archaea, there are no data regarding any functional significance of IAA and possible involvement of ABP1 homologue in auxin sensing. Future research in this direction would help better understand the evolution of both "auxin-binding protein" and auxin signalling. It appears that the major problem here is the absence of an efficient Archaea cultivation method, thus the current Archaea research relies on indirect analysis (omics approaches) but lacks basic methods used in microbiology (Sun et al. 2020).

16.6.2 Homologous to Bacterial Lys-type Transcriptional Regulator (LTTRs) Auxin-binding Proteins in Plants

Plants have several homologous for the Bacterial auxin-binding proteins (which belong to the Periplasmic binding fold protein superfamily, CL0144) for example, as glutamate receptors. In general, the structural properties, mechanism of action, and functions of this family in plants are well studied. Thus, plant glutamate receptors participate in electrical and Ca^{2+} signalling, and various physiological processes, such as sexual reproduction and wound signalling. Furthermore, recent research suggested that the biological functions of plant glutamate receptors are much wider and include response to pathogens, chemotaxis, regulation of root and pollen development, and many others (Grenzi et al. 2022; Simon et al. 2023). However, the ligand specificity of all glutamate receptors homologous is not well-studied (there are 20 genes in Arabidopsis). Therefore, it would be interesting to check their ability to bind auxin or participate in any kind of auxin-mediated signalling pathways or metabolic reactions.

16.6.3 Function of Other Cupin Domain-containing Proteins in Plants?

Cupin superfamily proteins are widely present in plants and associated with diverse developmental (flowers, roots, leaves, seeds, fruits, and cell wall development) and stress response functions (Hu et al. 2023). At the same time, they can bind different organic molecules (such as benzoic acid, pyroglutamate, acetate, and salicylic acid) (Jain et al. 2016; McPherson 2020), while there is no research approving or disproving their ability to bind auxins. Initial screening to check the ability of plant cupin-domain-containing proteins could be done relatively fast *in silico* with the application of bioinformatic modelling tools, with subsequent wet-lab verification of the most probable candidates.

16.6.4 How Hormone-binding Efficiency/Selectivity has changed Over Evolution?

Changes in the hormone-binding efficiency/selectivity during evolution are another interesting question. Thus, a comparison of the hormone-binding properties between species from different taxa in line of "green lineage" (cyanobacteria/chlorobium → green algae → primitive plants → high plants) would greatly improve our understanding of the role of auxin in land colonisation. Similarly, the comparison of auxin-binding properties between bacteria from different taxa and with different lifestyles (in particular, free-living, plant-associated, and plant pathogens) would help to elucidate the involvement of bacteria in shaping the auxin signalling in plants and provide useful insights into how to protect plants from pathogenic bacteria and facilitate interaction with beneficial bacteria.

16.6.5 IAA Sensors in Fungi, Insects, and Animals?

Interestingly, we did not find any ABP1 homologous in fungi, insects, and animals. It is known that in animals, IPyA (an intermediate in the IAA biosynthesis pathway) acts through the aryl hydrocarbon receptor (Aoki et al. 2018). However, the exact targets for IAA and activated signalling pathways in fungi, insects, and animals are not known. It would be interesting to know if auxin-recognising receptor systems in these organisms are similar (or evolutionally related) to other 4 known plant auxin receptor systems (canonical – SCF^{TIR1} and SKP2, and non-canonical – ARF3/ETTIN and TMK1).

16.6.6 Toxic Effect of IAA on Animals

Several recent publications have reported significant toxicity of IAA on animals. Thus, oral supplementation of rats with 500 mg/kg BW for 14 days resulted in haematological changes (lymphopenia, neutrophilia, leukopenia, and anaemia), a significant increase in the serum levels of Na^+, K^+, Cl– and creatinine, and activities of serum transaminase, creatine kinase-muscle type, creatine kinase-myocardial band, and gamma-glutamyl transferase. Also, levels of gonadotropins, testosterone, and leptin were decreased after treatment (Ismail 2022). Similarly, mice treated with a smaller dose (50 mg/kg BW) for 22 days demonstrated increased reactive oxygen species (ROS) production, apoptosis, chromosome aberrations, DNA damage, and micronuclei. Interestingly, those adverse effects in mice were prevented and alleviated (pre-, post-, and simultaneous treatment) with zinc nanoparticles biosynthesised from *Ochradenus arabicus* (Chaudhary, Hillc. & A.G.Mill.) leaf extracts (Alanazi et al. 2021). However, the exact molecular mechanism underlying these considerable toxic effects was not defined and requires further investigation.

16.6.7 Positive Effects of IPA/IAA on Animals

On the other side, several reports have proved the beneficial effects of IPA and IAA treatments. Thus, IPA application effectively inhibited *cyclooxygenase-2* (*Cox-2*) expression and reduced the severity of UVB-induced IL-1β and IL-6 production, skin lesions, transepithelial water loss, and the augmentation of dermal thickness (Aoki et al. 2014). Also, IPA supplemented diet reduced diarrhea and prevented chronic colonic inflammation, thus suppressing experimental T cell-mediated colitis in mice (Aoki et al. 2018). Both chemicals (IPA and IAA) were reported to alleviate rheumatoid arthritis, an autoimmune and inflammatory disease acting through the aryl hydrocarbon receptor pathway (Huang et al. 2023; Jiang et al. 2023). Finally, IAA protected mice with experimentally induced ankylosing spondylitis, a systemic, chronic, and inflammatory autoimmune disease. Particularly, IAA reduced the production of pro-inflammatory cytokines (tumour necrosis factor α, interleukins 6, 17A, and 23) and promoted the production of the anti-inflammatory cytokine IL-10 (Shen et al. 2022).

16.7 Conclusion

ABP1 is a crucial auxin-binding protein, involved in a wide range of metabolic and physiological responses throughout the life of plants. Based on our results, ABP1 is likely to have evolved from an ancestral bacterial protein from the cupin superfamily. Despite the crucial importance of ABP1 for auxin signalling in plants, the role of ABP1 homologous in Archaea and Bacteria remains a completely unexplored area.

Auxin and its metabolites are ubiquitous signalling molecules, intensively used by organisms from all known taxa (except Archaea – no data). Further research on auxin metabolites and their target receptors should clarify controversial information regarding their toxic and beneficial properties on animals. Such research has wide application in the medical area, in particular, in the treatment and prevention of discussed diseases (such as ankylosing spondylitis and rheumatoid arthritis). Also, practical applications of auxin-related metabolites may be found in agriculture (for example, to modulate communication of plants with pathogenic and/or beneficial bacteria) or the food industry (for example, to modulate microbiota composition and thus, treat various gastrointestinal-related diseases). From the point of evolution, it is interesting which receptors have been involved in auxin binding and signal transduction in animals, and what properties distinguish them from ABP1 and other plant auxin receptors.

References

Ahmad, N., Yasin, D., Bano, F., and Fatma, T. (2022). Ameliorative effects of endogenous and exogenous indole-3-acetic acid on atrazine stressed paddy field cyanobacterial biofertilizer *Cylindrospermum stagnale*. *Scientific Reports* 12: 11175. https://doi.org/10.1038/s41598-022-15415-z.

Alanazi, K.M., Al-kawmani, A.A., Farah, M.A. et al. (2021). Amelioration of indole acetic acid-induced cytotoxicity in mice using zinc nanoparticles biosynthesized with

Ochradenus arabicus leaf extract. *Saudi Journal of Biological Sciences* 28: 7190–7201. https://doi.org/10.1016/j.sjbs.2021.08.015.

Andres, D.A., Dickerson, I.M., and Dixon, J.E. (1990). Variants of the carboxyl-terminal KDEL sequence direct intracellular retention. *The Journal of Biological Chemistry* 265: 5952–5955.

Ang, A.C.H. and Østergaard, L. (2023). Save your TIRs – more to auxin than meets the eye. *New Phytologist* 238: 971–976. https://doi.org/10.1111/nph.18783.

Aoki, R., Aoki-Yoshida, A., Suzuki, C., and Takayama, Y. (2014). Protective effect of indole-3-pyruvate against ultraviolet B-induced damage to cultured HaCaT keratinocytes and the skin of hairless mice. *PLoS One* 9: e96804. https://doi.org/10.1371/journal.pone.0096804.

Aoki, R., Aoki-Yoshida, A., Suzuki, C., and Takayama, Y. (2018). Indole-3-pyruvic acid, an aryl hydrocarbon receptor activator, suppresses experimental colitis in mice. *The Journal of Immunology* 201: 3683–3693. https://doi.org/10.4049/jimmunol.1701734.

Benkert, P., Biasini, M., and Schwede, T. (2011). Toward the estimation of the absolute quality of individual protein structure models. *Bioinformatics* 27: 343–350. https://doi.org/10.1093/bioinformatics/btq662.

Bertoni, M., Kiefer, F., Biasini, M. et al. (2017). Modeling protein quaternary structure of homo- and hetero-oligomers beyond binary interactions by homology. *Scientific Reports* 7: 10480. https://doi.org/10.1038/s41598-017-09654-8.

Bhattacharya, D., Nowotny, J., Cao, R., and Cheng, J. (2016). 3Drefine: an interactive web server for efficient protein structure refinement. *Nucleic Acids Research* 44: W406–W409. https://doi.org/10.1093/nar/gkw336.

Biasini, M., Bienert, S., Waterhouse, A. et al. (2014). SWISS-MODEL: modelling protein tertiary and quaternary structure using evolutionary information. *Nucleic Acids Research* 42: W252–W258. https://doi.org/10.1093/nar/gku340.

Boratyn, G.M., Schäffer, A.A., Agarwala, R. et al. (2012). Domain enhanced lookup time accelerated BLAST. *Biology Direct* 7: 12. https://doi.org/10.1186/1745-6150-7-12.

Camacho, C., Coulouris, G., Avagyan, V. et al. (2009). BLAST+: architecture and applications. *BMC Bioinformatics* 10: 421. https://doi.org/10.1186/1471-2105-10-421.

Cao, M., Chen, R., Li, P. et al. (2019). TMK1-mediated auxin signaling regulates differential growth of the apical hook. *Nature* 568: 240–243. https://doi.org/10.1038/s41586-019-1069-7.

Chen, Y., Tian, P., Wang, Z. et al. (2022). Indole acetic acid exerts anti-depressive effects on an animal model of chronic mild stress. *Nutrients* 14: 5019. https://doi.org/10.3390/nu14235019.

Conway, J.M., Walton, W.G., Salas-González, I. et al. (2022). Diverse MarR bacterial regulators of auxin catabolism in the plant microbiome. *Nature Microbiology* 7: 1817–1833. https://doi.org/10.1038/s41564-022-01244-3.

Da Costa, C.T., Pedebos, C., Verli, H., and Fett-Neto, A.G. (2017). The role of Zn^{2+}, dimerization and N-glycosylation in the interaction of auxin-binding protein 1 (ABP1) with different auxins. *Glycobiology* 27: 1109–1119. https://doi.org/10.1093/glycob/cwx080.

Dai, N., Wang, W., Patterson, S.E., and Bleecker, A.B. (2013). The TMK subfamily of receptor-like kinases in Arabidopsis display an essential role in growth and a reduced sensitivity to auxin. *PLoS One* 8: e60990. https://doi.org/10.1371/journal.pone.0060990.

David, K., Carnero-Diaz, E., Leblanc, N. et al. (2001). Conformational dynamics underlie the activity of the auxin-binding protein, Nt-Abp1. *Journal of Biological Chemistry* 276: 34517–34523. https://doi.org/10.1074/jbc.M102783200.

De Klerk, G.-J., Brugge, J.T., and Marinova, S. (1997). Effectiveness of indoleacetic acid, indolebutyric acid and naphthaleneacetic acid during adventitious root formation *in vitro* in *Malus* 'Jork 9'. *Plant Cell, Tissue and Organ Culture* 49: 39–44. https://doi.org/10.1023/A:1005850222973.

Denecke, J., De Rycke, R., and Botterman, J. (1992). Plant and mammalian sorting signals for protein retention in the endoplasmic reticulum contain a conserved epitope. *The EMBO Journal* 11: 2345–2355. https://doi.org/10.1002/j.1460-2075.1992.tb05294.x.

Duca, D.R. and Glick, B.R. (2020). Indole-3-acetic acid biosynthesis and its regulation in plant-associated bacteria. *Applied Microbiology and Biotechnology* 104: 8607–8619. https://doi.org/10.1007/s00253-020-10869-5.

El-Gebali, S., Mistry, J., Bateman, A. et al. (2019). The Pfam protein families database in 2019. *Nucleic Acids Research* 47: D427–D432. https://doi.org/10.1093/nar/gky995.

Felsenstein, J. (1981). Evolutionary trees from DNA sequences: a maximum likelihood approach. *Journal of Molecular Evolution* 17: 368–376. https://doi.org/10.1007/BF01734359.

Friml, J., Gallei, M., Gelová, Z. et al. (2022). ABP1–TMK auxin perception for global phosphorylation and auxin canalization. *Nature* 609: 575–581. https://doi.org/10.1038/s41586-022-05187-x.

Fu, S.-F., Wei, J.-Y., Chen, H.-W. et al. (2015). Indole-3-acetic acid: a widespread physiological code in interactions of fungi with other organisms. *Plant Signaling & Behavior* 10: e1048052. https://doi.org/10.1080/15592324.2015.1048052.

Gavira, J.A., Rico-Jiménez, M., Ortega, Á. et al. (2023). Emergence of an auxin sensing domain in plant-associated bacteria. *MBio* 14: e03363-22. https://doi.org/10.1128/mbio.03363-22.

Gelová, Z., Gallei, M., Pernisová, M. et al. (2021). Developmental roles of auxin binding protein 1 in *Arabidopsis thaliana*. *Plant Science* 303: 110750. https://doi.org/10.1016/j.plantsci.2020.110750.

Giannopoulou, E., Senda, M., Koentjoro, M.P. et al. (2021). Crystal structure of the full-length LysR-type transcription regulator CbnR in complex with promoter DNA. *The FEBS Journal* 288: 4560–4575. https://doi.org/10.1111/febs.15764.

Gomes, G.L.B. and Scortecci, K.C. (2021). Auxin and its role in plant development: structure, signaling, regulation and response mechanisms. *Journal Of Plant Biology* 23: 894–904. https://doi.org/10.1111/plb.13303.

Grandits, M. and Oostenbrink, C. (2014). Molecular dynamics simulations of the auxin-binding protein 1 in complex with indole-3-acetic acid and naphthalen-1-acetic acid: simulation study of auxin-binding protein 1. *Proteins* 82: 2744–2755. https://doi.org/10.1002/prot.24639.

Grenzi, M., Bonza, M.C., and Costa, A. (2022). Signaling by plant glutamate receptor-like channels: what else! *Current Opinion in Plant Biology* 68: 102253. https://doi.org/10.1016/j.pbi.2022.102253.

Hammes, U.Z., Murphy, A.S., and Schwechheimer, C. (2022). Auxin transporters—a biochemical view. *Cold Spring Harbor Perspectives in Biology* 14: a039875. https://doi.org/10.1101/cshperspect.a039875.

Han, G., Qiao, Z., Li, Y. et al. (2022). RING zinc finger proteins in plant abiotic stress tolerance. *Frontiers in Plant Science* 13: 877011. https://doi.org/10.3389/fpls.2022.877011.

Herud-Sikimić, O., Stiel, A.C., Kolb, M. et al. (2021). A biosensor for the direct visualization of auxin. *Nature* 592: 768–772. https://doi.org/10.1038/s41586-021-03425-2.

Hou, M., Wu, D., Li, Y. et al. (2021). The role of auxin in nitrogen-modulated shoot branching. *Plant Signaling & Behavior* 16: 1885888. https://doi.org/10.1080/15592324.2021.1885888.

Hu, F., Ye, Z., Dong, K. et al. (2023). Divergent structures and functions of the cupin proteins in plants. *International Journal of Biological Macromolecules* 242: 124791. https://doi.org/10.1016/j.ijbiomac.2023.124791.

Huang, T., Cheng, L., Jiang, Y. et al. (2023). Indole-3-pyruvic acid alleviates rheumatoid arthritis via the aryl hydrocarbon receptor pathway. *Annals of Translational Medicine* 11: 213. https://doi.org/10.21037/atm-23-1074.

Ismail, H.T.H. (2022). Assessment toxic effects of exposure to 3-indoleacetic acid via hemato-biochemical, hormonal, and histopathological screening in rats. *Environmental Science and Pollution Research* 29: 90703–90718. https://doi.org/10.1007/s11356-022-22026-8.

Jain, A., Kumar, A., and Salunke, D.M. (2016). Crystal structure of the vicilin from *Solanum melongena* reveals existence of different anionic ligands in structurally similar pockets. *Scientific Reports* 6: 23600. https://doi.org/10.1038/srep23600.

Jiang, Z.-M., Zeng, S.-L., Huang, T.-Q. et al. (2023). Sinomenine ameliorates rheumatoid arthritis by modulating tryptophan metabolism and activating aryl hydrocarbon receptor via gut microbiota regulation. *Science Bulletin* 68: 1540–1555. https://doi.org/10.1016/j.scib.2023.06.027.

Keeling, P.J. (2013). The number, speed, and impact of plastid endosymbioses in eukaryotic evolution. *Annual Review of Plant Biology* 64: 583–607. https://doi.org/10.1146/annurev-arplant-050312-120144.

Koentjoro, M.P. and Ogawa, N. (2018). Structural studies of transcriptional regulation by LysR-type transcriptional regulators in bacteria. *Reviews in Agricultural Science* 6: 105–118. https://doi.org/10.7831/ras.6.105.

Kriechbaumer, V., Botchway, S.W., and Hawes, C. (2016). Localization and interactions between Arabidopsis auxin biosynthetic enzymes in the TAA/YUC-dependent pathway. *Journal of Experimental Botany* 67: 4195–4207. https://doi.org/10.1093/jxb/erw195.

Kuhn, A., Ramans Harborough, S., McLaughlin, H.M. et al. (2020). Direct ETTIN-auxin interaction controls chromatin states in gynoecium development. *eLife* 9: e51787. https://doi.org/10.7554/eLife.51787.

Kunkel, B.N. and Johnson, J.M.B. (2021). Auxin plays multiple roles during plant–pathogen interactions. *Cold Spring Harbor Perspectives in Biology* 13: a040022. https://doi.org/10.1101/cshperspect.a040022.

Laird, T.S., Flores, N., and Leveau, J.H.J. (2020). Bacterial catabolism of indole-3-acetic acid. *Applied Microbiology and Biotechnology* 104: 9535–9550. https://doi.org/10.1007/s00253-020-10938-9.

Lavania, D., Linh, N.M., and Scarpella, E. (2021). Of cells, strands, and networks: auxin and the patterned formation of the vascular system. *Cold Spring Harbor Perspectives in Biology* 13: a039958. https://doi.org/10.1101/cshperspect.a039958.

Leblanc, N., David, K., Grosclaude, J. et al. (1999). A novel immunological approach establishes that the auxin-binding protein, Nt-Abp1, is an element involved in auxin signaling at the plasma membrane. *Journal of Biological Chemistry* 274: 28314–28320. https://doi.org/10.1074/jbc.274.40.28314.

Leontovyčová, H., Trdá, L., Dobrev, P.I. et al. (2020). Auxin biosynthesis in the phytopathogenic fungus *Leptosphaeria maculans* is associated with enhanced transcription of indole-3-pyruvate decarboxylase *LmIPDC2* and tryptophan aminotransferase *LmTAM1*. *Research in Microbiology* 171: 174–184. https://doi.org/10.1016/j.resmic.2020.05.001.

Li, S.-W. (2021). Molecular bases for the regulation of adventitious root generation in plants. *Frontiers in Plant Science* 12: 614072. https://doi.org/10.3389/fpls.2021.614072.

Li, L., Lavell, A., Meng, X. et al. (2019). Arabidopsis DGD1 SUPPRESSOR1 is a subunit of the mitochondrial contact site and cristae organizing system and affects mitochondrial biogenesis. *The Plant Cell* 31: 1856–1878. https://doi.org/10.1105/tpc.18.00885.

Lin, H.-R., Shu, H.-Y., and Lin, G.-H. (2018). Biological roles of indole-3-acetic acid in *Acinetobacter baumannii*. *Microbiological Research* 216: 30–39. https://doi.org/10.1016/j.micres.2018.08.004.

Lin, H., Li, Y., and Hill, R.T. (2022). Microalgal and bacterial auxin biosynthesis: implications for algal biotechnology. *Current Opinion in Biotechnology* 73: 300–307. https://doi.org/10.1016/j.copbio.2021.09.006.

Liu, J., Moore, S., Chen, C., and Lindsey, K. (2017). Crosstalk complexities between auxin, cytokinin, and ethylene in arabidopsis root development: from experiments to systems modeling, and back again. *Molecular Plant* 10: 1480–1496. https://doi.org/10.1016/j.molp.2017.11.002.

Liu, W., Chen, J., Jin, L. et al. (2021). Functional and structural analysis of catabolite control protein C that responds to citrate. *Scientific Reports* 11: 20285. https://doi.org/10.1038/s41598-021-99552-x.

Lv, B., Zhu, J., Kong, X., and Ding, Z. (2021). Light participates in the auxin-dependent regulation of plant growth. *Journal of Integrative Plant Biology* 63: 819–822. https://doi.org/10.1111/jipb.13036.

Marchler-Bauer, A., Derbyshire, M.K., Gonzales, N.R. et al. (2015). CDD: NCBI's conserved domain database. *Nucleic Acids Research* 43: D222–D226. https://doi.org/10.1093/nar/gku1221.

Martin, W., Rujan, T., Richly, E. et al. (2002). Evolutionary analysis of arabidopsis, cyanobacterial, and chloroplast genomes reveals plastid phylogeny and thousands of cyanobacterial genes in the nucleus. *Proceedings of the National Academy of Sciences of the United States of America* 99: 12246–12251. https://doi.org/10.1073/pnas.182432999.

Martin, W.F., Garg, S., and Zimorski, V. (2015). Endosymbiotic theories for eukaryote origin. *Philosophical Transactions of the Royal Society B* 370: 20140330. https://doi.org/10.1098/rstb.2014.0330.

Matilla, M.A., Nogellova, V., Morel, B. et al. (2016). Biosynthesis of the acetyl-CoA carboxylase-inhibiting antibiotic, andrimid in *Serratia* is regulated by Hfq and the LysR-type transcriptional regulator, AdmX: regulation of andrimid biosynthesis. *Environmental Microbiology* 18: 3635–3650. https://doi.org/10.1111/1462-2920.13241.

Matilla, M.A., Daddaoua, A., Chini, A. et al. (2018). An auxin controls bacterial antibiotics production. *Nucleic Acids Research* 46: 11229–11238. https://doi.org/10.1093/nar/gky766.

Matilla, M.A., Velando, F., Martín-Mora, D. et al. (2022). A catalogue of signal molecules that interact with sensor kinases, chemoreceptors and transcriptional regulators. *FEMS Microbiology Reviews* 46: fuab043. https://doi.org/10.1093/femsre/fuab043.

McClerklin, S.A., Lee, S.G., Harper, C.P. et al. (2018). Indole-3-acetaldehyde dehydrogenase-dependent auxin synthesis contributes to virulence of *Pseudomonas syringae* strain DC3000. *PLoS Pathogens* 14: e1006811. https://doi.org/10.1371/journal.ppat.1006811.

McPherson, A. (2020). Binding of benzoic acid and anions within the cupin domains of the vicilin protein canavalin from jack bean (*Canavalia ensiformis*): crystal structures. *Biochemical and Biophysical Research Communications* 524: 268–271. https://doi.org/10.1016/j.bbrc.2020.01.101.

Mitchell, A.L., Attwood, T.K., Babbitt, P.C. et al. (2019). InterPro in 2019: improving coverage, classification and access to protein sequence annotations. *Nucleic Acids Research* 47: D351–D360. https://doi.org/10.1093/nar/gky1100.

Napier, R. (2021). The story of auxin-binding protein 1 (ABP1). *Cold Spring Harbor Perspectives in Biology* 13: a039909. https://doi.org/10.1101/cshperspect.a039909.

Nicastro, R., Raucci, S., Michel, A.H. et al. (2021). Indole-3-acetic acid is a physiological inhibitor of TORC1 in yeast. *PLoS Genetics* 17: e1009414. https://doi.org/10.1371/journal.pgen.1009414.

Nutaratat, P., Srisuk, N., Arunrattiyakorn, P., and Limtong, S. (2016). Indole-3-acetic acid biosynthetic pathways in the basidiomycetous yeast *Rhodosporidium paludigenum*. *Archives of Microbiology* 198: 429–437. https://doi.org/10.1007/s00203-016-1202-z.

Oliveira, D.L., Pugine, S.M.P., Ferreira, M.S.L. et al. (2007). Influence of indole acetic acid on antioxidant levels and enzyme activities of glucose metabolism in rat liver. *Cell Biochemistry and Function* 25: 195–201. https://doi.org/10.1002/cbf.1307.

Papi, S., Ahmadvand, H., Sotoodehnejadnematalahi, F., and Yaghmaei, P. (2022). The protective effects of indole-acetic acid on the renal ischemia-reperfusion injury via antioxidant and antiapoptotic properties in a rat model. *Iranian Journal of Kidney Diseases* 16: 125–134.

Phillips, H.R., Landis, J.B., and Specht, C.D. (2020). Revisiting floral fusion: the evolution and molecular basis of a developmental innovation. *Journal of Experimental Botany* 71: 3390–3404. https://doi.org/10.1093/jxb/eraa125.

Quispe-Huamanquispe, D.G., Gheysen, G., and Kreuze, J.F. (2015). Horizontal gene transfer contributes to plant evolution: the case of *Agrobacterium* T-DNAs. *Frontiers in Plant Science* 2017: 8. https://doi.org/10.3389/fpls.2017.02015.

Ranocha, P., Dima, O., Nagy, R. et al. (2013). Arabidopsis WAT1 Is a vacuolar auxin transport facilitator required for auxin homoeostasis. *Nature Communications* 4: 2625. https://doi.org/10.1038/ncomms3625.

Remmert, M., Biegert, A., Hauser, A., and Söding, J. (2012). HHblits: lightning-fast iterative protein sequence searching by HMM-HMM alignment. *Nature Methods* 9: 173–175. https://doi.org/10.1038/nmeth.1818.

Rico-Jiménez, M., Roca, A., Krell, T., and Matilla, M.A. (2022). A bacterial chemoreceptor that mediates chemotaxis to two different plant hormones. *Environmental Microbiology* 24: 3580–3597. https://doi.org/10.1111/1462-2920.15920.

Sainsbury, S., Lane, L.A., Ren, J. et al. (2009). The structure of CrgA from *Neisseria meningitidis* reveals a new octameric assembly state for LysR transcriptional regulators. *Nucleic Acids Research* 37: 4545–4558. https://doi.org/10.1093/nar/gkp445.

Saitou, N. and Nei, M. (1987). The neighbor-joining method: a new method for reconstructing phylogenetic trees. *Molecular Biology and Evolution* 4: 406–425.

Sayers, E.W., Bolton, E.E., Brister, J.R. et al. (2022). Database resources of the National center for biotechnology information. *Nucleic Acids Research* 50: D20–D26. https://doi.org/10.1093/nar/gkab1112.

Semeradova, H., Montesinos, J.C., and Benkova, E. (2020). All roads lead to auxin: post-translational regulation of auxin transport by multiple hormonal pathways. *Plant Communications* 1: 100048. https://doi.org/10.1016/j.xplc.2020.100048.

Shen, J., Yang, L., You, K. et al. (2022). Indole-3-acetic acid alters intestinal microbiota and alleviates ankylosing spondylitis in mice. *Frontiers in Immunology* 13: 762580. https://doi.org/10.3389/fimmu.2022.762580.

Simon, A.A., Navarro-Retamal, C., and Feijó, J.A. (2023). Merging signaling with structure: functions and mechanisms of plant glutamate receptor ion channels. *Annual Review of Plant Biology* 74: 415–452. https://doi.org/10.1146/annurev-arplant-070522-033255.

Simonini, S., Deb, J., Moubayidin, L. et al. (2016). A noncanonical auxin-sensing mechanism is required for organ morphogenesis in *Arabidopsis*. *Genes & Development* 30: 2286–2296. https://doi.org/10.1101/gad.285361.116.

Soukarieh, F., Mashabi, A., Richardson, W. et al. (2021). Design and evaluation of new quinazolin-4(3*H*)-one derived PqsR antagonists as quorum sensing quenchers in *Pseudomonas aeruginosa*. *ACS Infectious Diseases* 7: 2666–2685. https://doi.org/10.1021/acsinfecdis.1c00175.

Sun, Y., Liu, Y., Pan, J. et al. (2020). Perspectives on cultivation strategies of archaea. *Microbial Ecology* 79: 770–784. https://doi.org/10.1007/s00248-019-01422-7.

The UniProt Consortium, Bateman, A., Martin, M.-J. et al. (2021). UniProt: the universal protein knowledgebase in 2021. *Nucleic Acids Research* 49: D480–D489. https://doi.org/10.1093/nar/gkaa1100.

Tian, H., Klämbt, D., and Jones, A.M. (1995). Auxin-binding protein 1 does not bind auxin within the endoplasmic reticulum despite this being the predominant subcellular location for this hormone receptor. *Journal of Biological Chemistry* 270: 26962–26969. https://doi.org/10.1074/jbc.270.45.26962.

Tillmann, U., Viola, G., Kayser, B. et al. (1989). CDNA clones of the auxin-binding protein from corn coleoptiles (*Zea mays* L.): isolation and characterization by immunological methods. *The EMBO Journal* 8: 2463–2467. https://doi.org/10.1002/j.1460-2075.1989.tb08381.x.

Tokuda, M., Suzuki, Y., Fujita, S. et al. (2022). Terrestrial arthropods broadly possess endogenous phytohormones auxin and cytokinins. *Scientific Reports* 12: 4750. https://doi.org/10.1038/s41598-022-08558-6.

Tyagi, M., Gowri, V.S., Srinivasan, N. et al. (2006). A substitution matrix for structural alphabet based on structural alignment of homologous proteins and its applications. *Proteins* 65: 32–39. https://doi.org/10.1002/prot.21087.

Tyagi, M., de Brevern, A.G., Srinivasan, N., and Offmann, B. (2008). Protein structure mining using a structural alphabet. *Proteins* 71: 920–937. https://doi.org/10.1002/prot.21776.

Vetting, M.W., Al-Obaidi, N., Zhao, S. et al. (2015). Experimental strategies for functional annotation and metabolism discovery: targeted screening of solute binding proteins and unbiased panning of metabolomes. *Biochemistry* 54: 909–931. https://doi.org/10.1021/bi501388y.

Wang, B., Chu, J., Yu, T. et al. (2015). Tryptophan-independent auxin biosynthesis contributes to early embryogenesis in *Arabidopsis*. *Proceedings of the National Academy of Sciences of the United States of America* 112: 4821–4826. https://doi.org/10.1073/pnas.1503998112.

Wang, Q., Qin, G., Cao, M. et al. (2020). A phosphorylation-based switch controls TAA1-mediated auxin biosynthesis in plants. *Nature Communications* 11: 679. https://doi.org/10.1038/s41467-020-14395-w.

Waterhouse, A., Bertoni, M., Bienert, S. et al. (2018). SWISS-MODEL: homology modelling of protein structures and complexes. *Nucleic Acids Research* 46: W296–W303. https://doi.org/10.1093/nar/gky427.

Woo, E.-J. (2002). Crystal structure of auxin-binding protein 1 in complex with auxin. *The EMBO Journal* 21: 2877–2885. https://doi.org/10.1093/emboj/cdf291.

Yamashiro, T., Shiraishi, A., Nakayama, K., and Satake, H. (2022). Key amino acids for transferase activity of GDSL lipases. *IJMS* 23: 15141. https://doi.org/10.3390/ijms232315141.

Yokoyama, C., Takei, M., Kouzuma, Y. et al. (2017). Novel tryptophan metabolic pathways in auxin biosynthesis in silkworm. *Journal of Insect Physiology* 101: 91–96. https://doi.org/10.1016/j.jinsphys.2017.07.006.

Yu, Z., Zhang, F., Friml, J., and Ding, Z. (2022). Auxin signaling: research advances over the past 30 years. *Integrative Plant Biology* 64: 371–392. https://doi.org/10.1111/jipb.13225.

Zarei, M., Rahbar, M.R., and Negahdaripour, M. (2022). Interaction of indole-3-acetic acid with horseradish peroxidase as a potential anticancer agent: from docking to molecular dynamics simulation. *Journal of Biomolecular Structure and Dynamics* 40: 4188–4196. https://doi.org/10.1080/07391102.2020.1854118.

Zemla, A. (2003). LGA: a method for finding 3D similarities in protein structures. *Nucleic Acids Research* 31: 3370–3374. https://doi.org/10.1093/nar/gkg571.

Zhong, J., Powell, S., and Preston, J.C. (2016). Organ boundary NAC-domain transcription factors are implicated in the evolution of petal fusion. *Journal Of Plant Biology* 18: 893–902. https://doi.org/10.1111/plb.12493.

Žižková, E., Kubeš, M., Dobrev, P.I. et al. (2017). Control of cytokinin and auxin homeostasis in cyanobacteria and algae. *Annals of Botany* 119: 151–166. https://doi.org/10.1093/aob/mcw194.

17

Exploring the Potential of Molecular Docking and *In Silico* Studies in Secondary Metabolite and Bioactive Compound Discovery for Plant Research

Amine Elbouzidi[1], Mohamed Taibi[1,2], and Mohamed Addi[1]

[1] *Laboratoire d'Amélioration des Productions Agricoles, Biotechnologie et Environnement (LAPABE), Faculté des Sciences, Université Mohammed Premier, Oujda, Morocco*
[2] *Centre de l'Oriental des Sciences et Technologies de l'Eau et de l'Environnement (COSTEE), Université Mohammed Premier, Oujda, Morocco*

17.1 Introduction

Plants stand as captivating repositories of an astonishing array of chemical compounds, surpassing their conventional stature as mere contributors to growth and development (Wurtzel and Kutchan 2016). Unlike primary metabolites, which are imperative for an organism's fundamental growth, development, and reproductive processes, secondary metabolites hold a different role (Wink 1999; Isah 2019; Yeshi et al. 2022). Their emergence is often triggered by an organism's dynamic interaction with its environment. Alternatively, they might serve as defensive tools against potential predators, thus bolstering the organism's chances of survival (Dewick 2002; Colegate and Molyneux 2007). The synthesis of secondary metabolites finds its origins in fundamental mechanisms such as photosynthesis, glycolysis, and the Krebs cycle. These intricate processes yield biosynthetic intermediates, which in turn culminate in the eventual production of secondary metabolites – commonly recognized as natural substances (Dewick 2002; Kroymann 2011). This botanical treasury of chemicals transcends the rudimentary requirements of plant life, reaching into the realms of complexity and diversity that have long captivated the scientific community (Weng et al. 2012). An intriguing aspect is the seemingly limitless potential for generating novel secondary metabolites, despite the finite array of foundational components available. The core building blocks essential for the synthesis of secondary metabolites predominantly stem from intermediates. Notably, this includes components like Acetyl coenzyme A (acetyl-CoA), shikimic acid, mevalonic acid, and 1-deoxyxylulose-5-phosphate (Dias et al. 2012). These versatile components actively participate in a myriad of biosynthetic pathways, engaging in diverse mechanisms and reactions encompassing alkylation, decarboxylation, aldol, Claisen, and Schiff base formation (do Nascimento and Fett-Neto 2010; Kroymann 2011; Yang et al. 2012; Jan et al. 2021).

Bioinformatics for Plant Research and Crop Breeding, First Edition. Edited by Jen-Tsung Chen.

Plant research has uncovered a treasure trove of natural products, many of which possess remarkable therapeutic, and industrial properties. Secondary metabolites have historically served as a fruitful reservoir of potential compounds for drug development (Haefner 2003; Mishra and Tiwari 2011; Rey-Ladino et al. 2011). Nonetheless, recent endeavors in the exploration and advancement of therapeutics have revealed a waning interest in these natural products (Wurtzel and Kutchan 2016). However, they still offer distinct and diverse chemical structures when compared to standard combinatorial chemistry. This disparity presents an avenue for uncovering new and unique low molecular weight lead compounds. Considering that a mere fraction – less than 10% – of the world's biodiversity has been explored for potential biological activities, there remains a plethora of untapped natural compounds with promising properties. The primary challenge lies in accessing this rich variety of natural chemicals (Yeshi et al. 2022).

The most ancient instances of utilizing natural products for medicinal purposes can be traced back to ancient Mesopotamia around 2600 BCE. During this period, clay tablets inscribed with cuneiform script depicted the use of oils derived from *Cupressus sempervirens* (Cypress) and various *Commiphora* species (myrrh) to address conditions such as coughs, colds, and inflammation. This ancient knowledge continues to influence modern practices (Yeshi et al. 2022). During the following Dark and Middle Ages, monastic institutions in diverse areas, including but not limited to England, Ireland, France, and Germany, safeguarded and upheld Western knowledge. Simultaneously, the Arab world safeguarded and expanded upon Greco-Roman wisdom while incorporating novel herbal resources, including plants from China and India. The eighth century witnessed a notable development with the establishment of the first private pharmacies by the Arabs. Influential figures like Avicenna, a Persian polymath, contributed significantly to the fields of pharmacy and medicine through seminal works such as the *Canon Medicinae* (Yeshi et al. 2022).

These compounds, often isolated from traditional medicinal plants, hold the promise of revolutionizing drug discovery and development. Their intricate interactions with biological systems and their potential to modulate specific cellular processes make them valuable candidates for novel pharmaceutical agents. From anticancer properties to antimicrobial activities, the medicinal potential of secondary metabolites and bioactive compounds has ignited a fervor of research aimed at harnessing these molecules to combat some of the most pressing health challenges of our time. In general, compounds derived from natural origins tend to exhibit distinct structural attributes. These encompass a heightened count of oxygen atoms, an increased presence of chiral centers, amplified steric complexity and molecular rigidity, a greater number of hydrogen bond acceptors and donors, as well as a lower ratio of aromatic ring atoms to overall heavy atoms. Furthermore, a wider spectrum of molecular properties comes into play, spanning features like partition coefficient, molecular mass, and the variety of ring systems (Yeshi et al. 2022). Given these distinctive traits inherent to natural compounds, the task of crafting analogs, be it for enhancing potency and pharmacokinetic properties or diminishing toxicity, presents a formidable challenge for medicinal chemists.

The application of computational techniques to improve and accelerate the drug discovery process rooted in natural products is experiencing a growing trend. This trend is substantiated by various recent chapters in books and review articles (do Nascimento and Fett-Neto 2010; Kroymann 2011; Dias et al. 2012; Weng et al. 2012). These sources delve into the array of techniques encompassing molecular modeling, chemoinformatics (Jan et al. 2021), and machine learning (Yang et al. 2012). These techniques are employed to unravel, identify, and refine potential drug candidates derived from natural sources. They also aid in comprehending the representation of natural products within the chemical space, predicting bioactivity spectra, evaluating factors related to absorption, distribution, metabolism, and excretion (ADME), as well as safety profiles (toxicity). Furthermore, these computational methods are involved in quantifying the similarity between compounds and natural products, which is especially pertinent for engaging in *de novo* design inspired by natural substances.

The primary purpose of this chapter is to provide an in-depth exploration of the role of molecular docking and *in silico* studies in the discovery and characterization of secondary metabolites and bioactive compounds in the context of plant research. These computational techniques have revolutionized the field by allowing researchers to predict and evaluate the interactions between molecules at the molecular level. By simulating these interactions *in silico*, researchers can uncover significant information about the potential bioactivity, binding affinities, and mechanisms of action of various compounds without the need for extensive laboratory work.

17.2 Importance of Structure-based Drug Design from Natural Sources

Comprehending the fundamental factors that govern the interactions between small-molecule ligands and macromolecules is of paramount importance, transcending various domains of research, with particular significance in the realm of drug discovery (Kitchen et al. 2004; Lill 2007; Cooper et al. 2011). One of the most systematic approaches in this domain is structure-based drug design (SBDD), which relies heavily on structural data, often acquired through experimental means or computational homology modeling, of macromolecular targets, commonly referred to as receptors. The central objective of SBDD is the strategic design of ligands possessing precise electrostatic and stereochemical attributes that facilitate robust binding to the receptor, thus eliciting potent biological responses (Kalyaanamoorthy and Chen 2011; Bajad et al. 2021; Aplin et al. 2022).

The advent of three-dimensional macromolecular structures has revolutionized the landscape by enabling a comprehensive exploration of the architectural intricacies within the binding sites. This profound insight encompasses the identification of clefts, cavities, and sub-pockets within the receptor structure. Moreover, it facilitates a meticulous scrutiny of electrostatic characteristics, including charge distribution patterns. Contemporary SBDD techniques empower researchers to craft ligands meticulously tailored to interact optimally with the target receptor, orchestrating the modulation of specific cellular processes with remarkable precision (Mandal and Mandal 2009; Blaney 2012).

At its core, SBDD represents a cyclical and iterative process underpinned by a series of deliberate knowledge acquisition steps (refer to Figure 17.1). It commences with the availability of a known target structure, serving as the foundation for subsequent investigations. *In silico* studies are then conducted to sift through a vast chemical space, identifying potential ligands with binding potential (Temml and Kutil 2021; Radu et al. 2023).

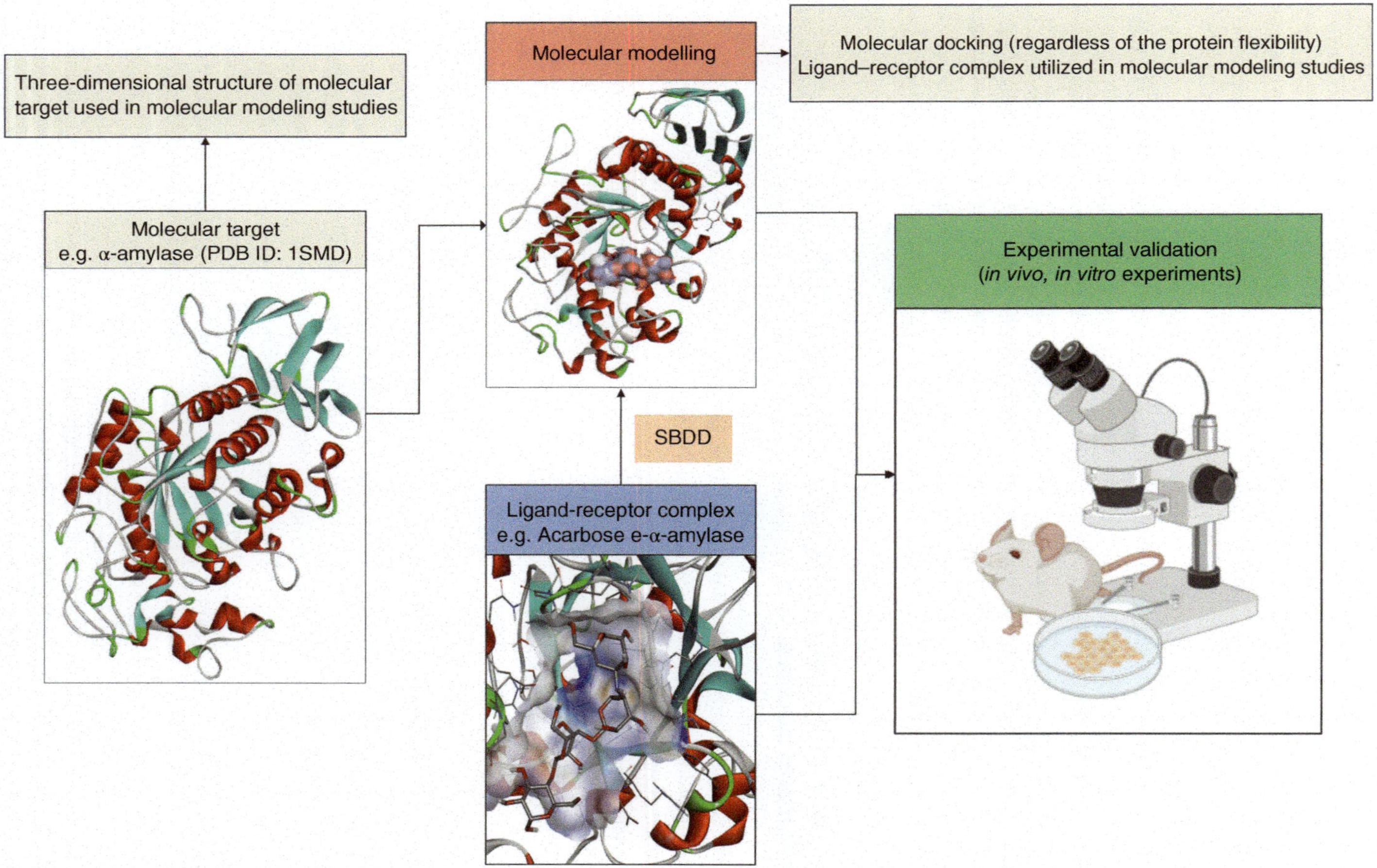

Figure 17.1 Illustration of the process of structure-based drug design (SBDD). It begins with the utilization of the molecular target's three-dimensional structure in molecular modeling investigations. Afterward, potential compounds are synthesized and subjected to experimental assessment. When bioactive small molecules are identified, a ligand–receptor complex structure can be determined. This binding complex is subsequently employed in molecular modeling studies to facilitate the design of new compounds. *Source:* Ferreira et al. (Ferreira et al. 2015)/Adapted from MDPI.

After identifying promising candidates using these molecular modeling procedures, the process of synthesizing these compounds begins. These compounds subsequently undergo rigorous evaluation across multiple biological parameters, including potency, affinity, and efficacy, through an array of experimental platforms (Radu et al. 2023).

Upon the identification of active compounds, the journey enters a crucial phase where the three-dimensional structure of the ligand–receptor complex is unveiled. This structural revelation constitutes a critical juncture in the SBDD process, offering profound insights into various intermolecular facets that underpin molecular recognition. The detailed depiction of ligand–receptor complexes at the structural level proves invaluable in several critical aspects, including the elucidation of binding conformations, characterization of pivotal intermolecular interactions, identification of hitherto unknown binding sites, initiation of mechanistic studies, and the revelation of ligand-induced conformational changes (Kahsai et al. 2011).

In essence, SBDD stands as a multifaceted and dynamic approach with far-reaching implications, extending its influence into the realms of drug discovery and beyond, offering an elegant framework for the rational design of phytomolecules that hold the potential to shape the future of therapeutics and molecular sciences.

17.3 Molecular Docking as a Key Component of SBDD: A Bridge Between Computational and Experimental Approaches

Since its initial introduction in the mid-1970s, docking has established itself as a valuable tool for understanding the interactions between chemical compounds and their molecular targets, as well as for driving progress in the field of drug discovery and development. Notably, there has been a substantial increase in studies utilizing molecular docking for two key purposes:

- Identifying the structural factors essential for effective ligand–receptor binding (Levinthal et al. 1975; DesJarlais et al. 1986; Kitchen et al. 2004).
- The emergence of more precise docking techniques has significantly surged in development since its initial introduction (Levinthal et al. 1975; Kuhl et al. 1984; DesJarlais et al. 1986; Warwicker 1989; Goodsell and Olson 1990; Kitchen et al. 2004).

One of the pioneering and intriguing studies in the realm of docking's application in drug discovery and biology was conducted by Kuntz and colleagues in the first half of the 1980s (Kuntz et al. 1982). In their research, they presented a computational method to investigate potential alignments of ligand–receptor interactions within established structures, including heme-myoglobin/metmyoglobin and thyroxine/prealbumin (Kuntz et al. 1982). Although not the initial investigation to utilize docking for forecasting molecular complex configurations (Levinthal et al. 1975), it introduced a novel approach by utilizing a simplified function incorporating only "hard sphere repulsions" and "hydrogen bonding" to describe protein-ligand interactions, a departure from previous methods (Levinthal et al. 1975; Salemme 1976; Wodak and Janin 1978). Additionally, the researchers were the pioneers in treating the receptor as a fixed, unyielding entity with binding sites denoted as "pockets." Interestingly, this approach effectively anticipated structures closely resembling those found in previously documented X-ray complexes, and it pinpointed protein conformations that could be further refined for energy optimization and the creation of new ligands (Kuntz et al. 1982).

From that point onward, molecular docking has undergone significant advancements, such as the incorporation of flexible algorithms in calculations (see Table 17.1) (Brint and Willett 1987; Böhm 1992; Leach and Kuntz 1992; Eisen et al. 1994). The results of these advancements have reinforced the significance of computer-assisted structure-based drug design in the development of lead compounds (Ferreira et al. 2015).

Considering the potential advantages of this approach, substantial efforts have been channeled into improving docking algorithms and addressing their inherent limitations (Levinthal et al. 1975; DesJarlais et al. 1986; Goodsell and Olson 1990). These limitations

Table 17.1 Major algorithms used in molecular docking technique, characteristics, and examples of most used software and their model of protein flexibility.

No.	Algorithms	Characteristics	Examples of most used software and their protein flexibility model
1.	Matching algorithms	Geometry-based, suitable for virtual screening and database enrichment for its high-speed	DOCK (rigid) (Lorber and Shoichet 1998), Ligand fit (rigid) (Venkatachalam et al. 2003), EUDOC (rigid) (Pang et al. 2001)
2.	Incremental construction	Fragment-based and docking incrementally	FlexX (Induced-fit) (Rarey et al. 1996), DOCK 4.0 (Induced-fit) (Welch et al. 1996), ProPose (rigid) (Seifert et al. 2004), MacDock (Induced-fit) (Fradera et al. 2004), SLIDE (Induced-fit) (Schnecke and Kuhn 1999)
3.	Maximum common substructure search (MCSS)	Fragment-based methods for the *de novo* design	GOLD 3.1 (rigid/Induced-fit) (Jones et al. 1997), HOOK (rigid/Induced-fit) (Eisen et al. 1994)
4.	Monte Carlo	Stochastic search	MCDOCK (rigid) (Liu and Wang 1999), FDS (Induced-fit) (Taylor et al. 2003), AutoDock (rigid) (Goodsell and Olson 1990), PRODOCK (Induced-fit) (Trosset and Scheraga 1999), GLAMDOCK (rigid) (Tietze and Apostolakis 2007), ICM (rigid) (Totrov and Abagyan 1997)
5.	Ligand unbound drug design informatics (LUDI)	Fragment-based methods for the *de novo* design	LUDI (rigid/Induced-fit) (Böhm 1992)
6.	Genetic algorithms	Stochastic search	GOLD 3.1 (rigid/Induced-fit) (Jones et al. 1997), AUTODOCK 4.0 (rigid) (Morris et al. 1998), FLIP DOCK (Induced-fit) (Zhao and Sanner 2007), GAMBLER (rigid) (Pei et al. 2006), PSI-DOCK (rigid) (Pei et al. 2006)

include challenges like the limited exploration of ligand and receptor conformations during pose prediction and the reliance on approximate scoring functions that often produce results inconsistent with experimental binding affinities (Kuhl et al. 1984; Warwicker 1989). Nonetheless, the effectiveness of docking in drug design hinges on the accessibility of crystal structures for the biological targets. To overcome this constraint, various strategies have been employed. These strategies include constructing homology models when 3D structures are lacking, based on highly similar structural templates. Additionally, these approaches can be coupled with molecular dynamics (MD) simulations to further validate and enhance the computational modeling of complexes (Salemme 1976; Wodak and Janin 1978; Kuntz et al. 1982). However, advancements in structural biology and the availability of experimentally determined ligand–target complexes through crystal structures are expected to mitigate this constraint. Additionally, *in silico* techniques, such as molecular dynamics, have been extensively embraced to explore the conformational space of targets, ligands, and ligand–target complexes, thereby offering a more comprehensive insight into their dynamic behavior and improving the results of docking studies (Salemme 1976; Lorber and Shoichet 1998; Venkatachalam et al. 2003). Furthermore, improved virtual screening methods have been developed to enhance the predictions of ligand–target complexes obtained through docking (Rarey et al. 1996; Welch et al. 1996; Pang et al. 2001; Fradera et al. 2004; Seifert et al. 2004). These post-docking refinement and rescoring approaches hold significant promise in the field of drug discovery because of their capacity to yield higher hit rates in virtual screening campaigns and establish stronger correlations with experimental data (Welch et al. 1996; Fradera et al. 2004).

17.4 Molecular Docking and Natural Product Database

The realm of pharmaceutical research is continually seeking novel compounds to address various medical challenges. In this pursuit, bioactive natural products have emerged as valuable assets, serving as initial lead compounds (see Figure 17.2). These compounds hold the potential for optimization through structural modifications, ultimately leading to the development of more potent and effective analogs. Modern medicinal chemistry techniques, including molecular modeling and combinatorial chemistry, play a pivotal role in this transformation. Additionally, the richness of nature's bounty often presents families of structurally related molecules, offering the prospect of extracting valuable insights into structure–activity relationships (SAR). It is important to note that the majority of plant species remain largely unexplored in terms of their biological activities, leaving an abundant source of new therapeutic agents yet to be uncovered. This ongoing exploration of natural sources ensures a steady supply of candidates for rigorous biological screening and the discovery of fresh therapeutic agents. In the contemporary landscape of natural product-based drug discovery, isolated compounds with promising bioactivity undergo detailed SAR studies and molecular modeling processes. This iterative approach aims to design and develop analogs with improved potency, reduced toxicity, and enhanced pharmacokinetic profiles. Furthermore, these investigations may uncover interactions with specific enzymes that can significantly influence the biological activity of the test compounds. Ultimately, the most promising analogs are synthesized in the laboratory and subjected to a battery of

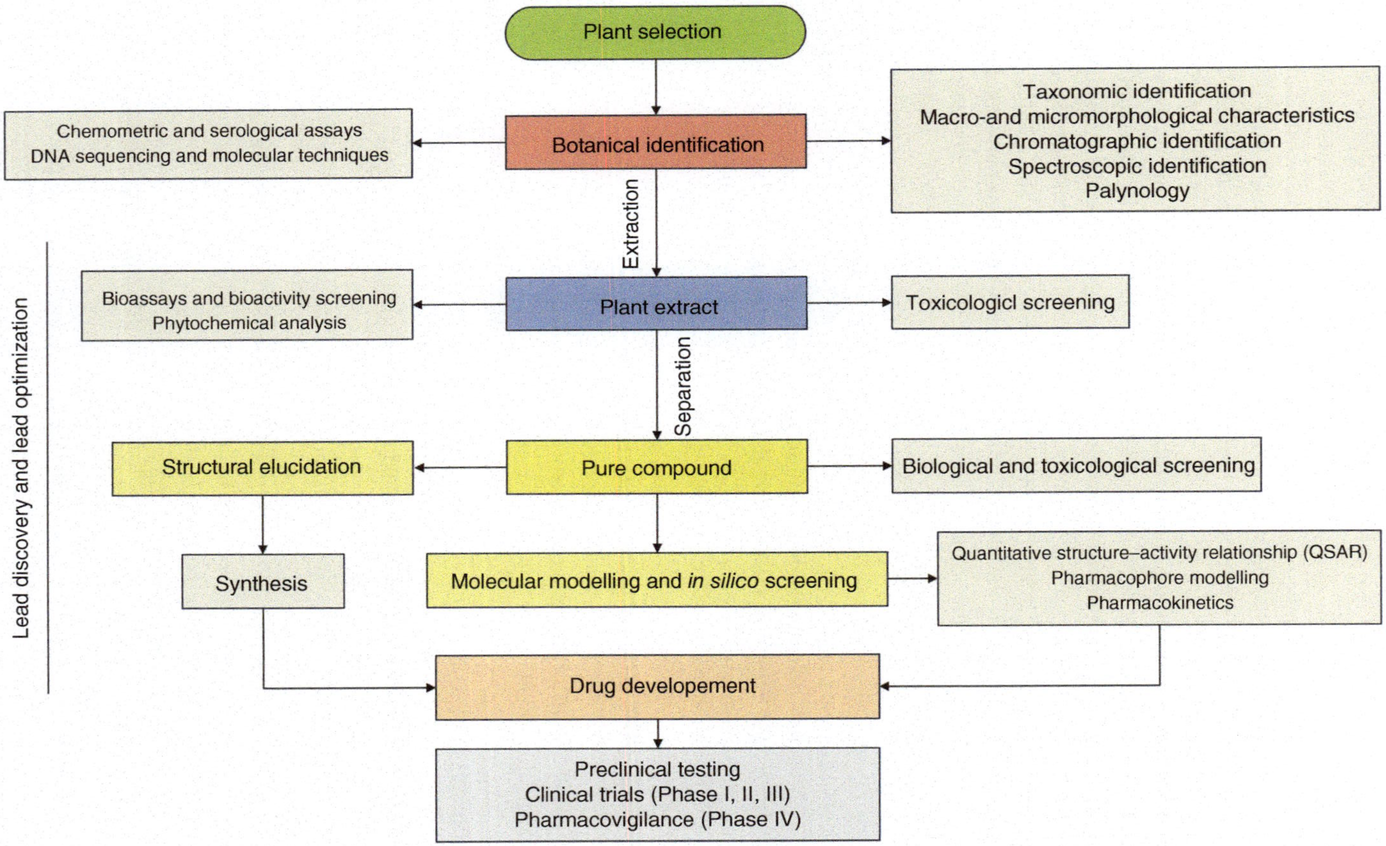

Figure 17.2 The main procedures associated with drug discovery and development of analogs from a naturally isolated lead compound (plant resources).

in vitro and *in vivo* biological assays (Kitchen et al. 2004). To elucidate this comprehensive process further, the subsequent sections delve into each stage.

17.4.1 Constructing and Preparing *In Silico* Ligands

The journey begins with the construction and preparation of *in silico* ligands for molecular modeling procedures (see Figure 17.3). These ligands' 3D structures need to be optimized and provided in PDB format, a prerequisite for subsequent steps. Natural product databases are treasure troves of data, as are repositories like PUBCHEM and ZINC, offering access to structures of known natural compounds in various acceptable formats such as SDF, mol, mol2, and PDB (Sorokina and Steinbeck 2020). However, these structures must undergo rigorous geometry optimization to ensure minimal energy levels. This optimization can occur before the docking process, with tools like AutoDock Vina and Discovery Studio, or it can involve dedicated structure-building and optimization software such as Chimera, Chem 3D Ultra, Avogadro, among others (Liao et al. 2011; Chen et al. 2020).

17.4.2 Preparing the Target

Moving forward, the preparation of the target becomes pivotal. Whether it is proteins (e.g. human serum albumin), receptors (e.g. PPAR-α and PPAR-γ), or enzymes (e.g. cyclooxygenase, topoisomerase II, and protein kinase), their 3D structures are crucial and need to be obtained from the Protein Data Bank (PDB). Subsequently, these structures must undergo geometry optimization and energy minimization. Precisely defining the binding site and calculating standard scores for natural ligands within these target structures forms the bedrock for the subsequent stages.

17.4.3 Docking

The heart of the process lies in docking. Here, the 3D structures of natural products are meticulously docked against the target structures using specialized software. The results are ranked based on binding energy, a critical determinant of molecular interactions. Widely embraced docking tools include AutoDock, AutoDock Vina, Discovery Studio, and MDock . . . (Chen et al. 2020). Successful docking hinges on having optimized 3D structures of both ligands and targets in PDB format. Docking utilizes various search algorithms, with the Lamarckian genetic algorithm being a prime example, employed to pinpoint the optimal binding conformation of the ligand. Post-docking analyses are essential for comprehending intermolecular interactions, providing vital validation for the docking outcomes.

17.4.4 Identifying Hit Molecules

Following the completion of docking simulations, the resultant data undergoes rigorous examination to pinpoint the most promising interactions, primarily based on energy scores. Generally, the top 10 scores are singled out for more detailed analysis via MD simulations. This stage requires two systems: (i) an Apo (uncomplexed) protein/receptor and

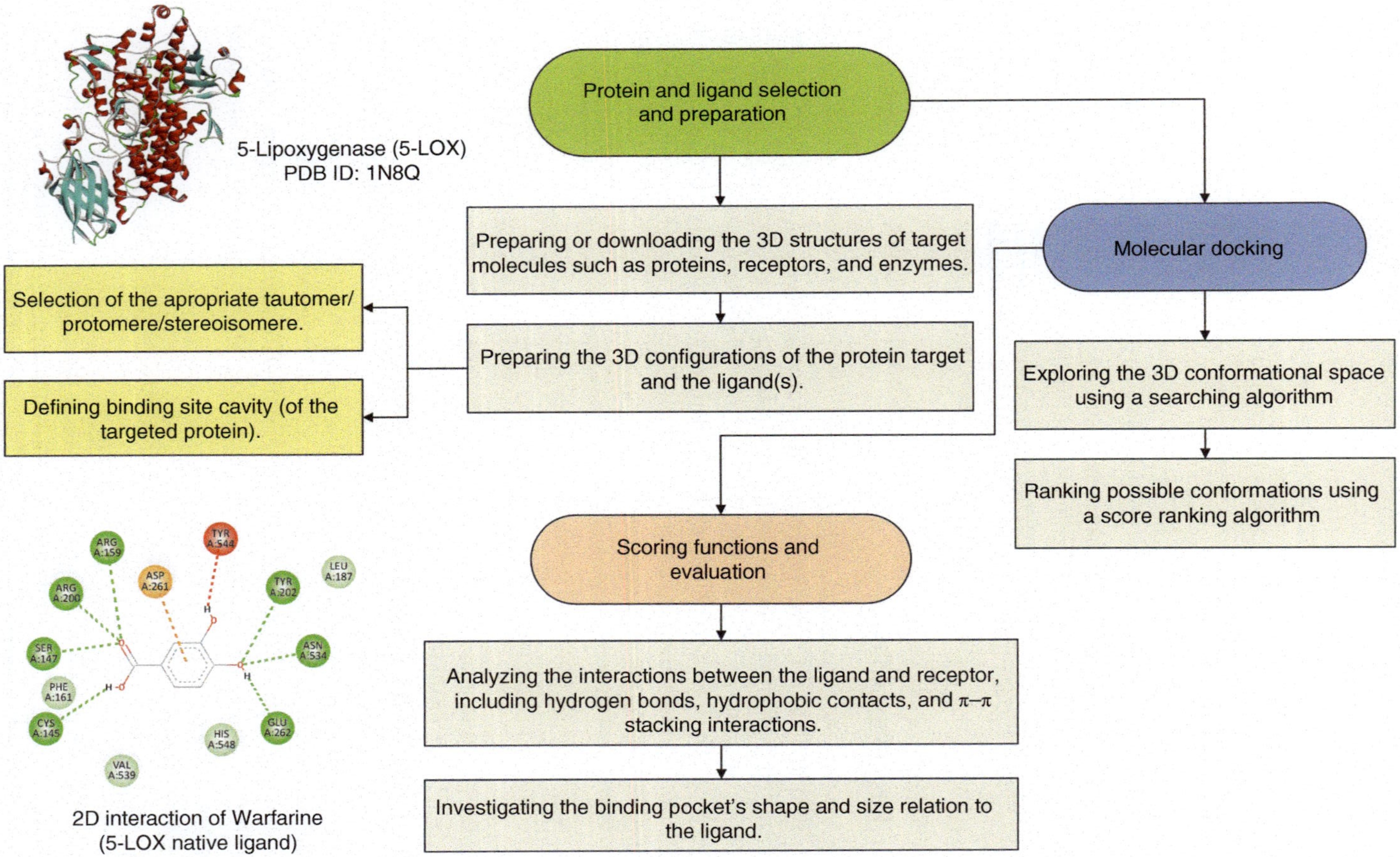

Figure 17.3 Example of step-by-step molecular docking process, including the major steps of protein and ligand selection and preparation, molecular docking, and scoring functions and evaluation.

(ii) protein/receptor complexes with interacting compounds, as initially predicted during the docking study. These comprehensive investigations aim to identify hit molecules demonstrating a strong affinity for the target, relying on ligand ranking and interaction strength.

17.4.5 Optimization of Hits

The optimization phase focuses on refining hit compounds to enhance their affinity for the target. This entails generating a spectrum of analogs to evaluate which exhibit superior affinity. The hits demonstrating the most promising target interactions are taken forward for further development. The next stage involves the assessment of various drug-like properties, including stability, pharmacokinetics, and pharmacodynamics. This is accomplished through the utilization of quantitative structure–activity relationship (QSAR) software (Devillers 2013; Sullivan et al. 2014).

In contemporary drug discovery practices, an alternative approach to the traditional extraction and biological assays of natural products involves the screening of extensive compound libraries derived from natural sources. These libraries house compounds with diverse chemical structures, often maintained by academic institutions and research centers, numbering in the millions. Virtual screening emerges as a powerful tool in this context, capable of identifying potential hits for specific biological activities. This approach establishes robust SAR and streamlines the selection of compounds for subsequent bioassay testing (Bindseil et al. 2001; Stockwell 2004). However, while the natural product database offers structural information for test compounds, validation of newly predicted biological activities requires the availability of these hit molecules for relevant bioassays. These compounds can be synthesized within the laboratory or obtained from commercial entities. Remarkable resources for natural product libraries and databases encompass DNP, Phytopure, ChemSpider, Natural Product Alert, Tim Tech Natural Products, and many others (Gray et al. 2012). This systematic approach ensures a holistic understanding of the intricate process of designing and developing analogs from naturally isolated lead compounds, opening doors to innovative therapeutic solutions.

17.5 Case Studies: Successful Applications of *In Silico* Molecular Docking in Plant Research for Diverse Applications

Molecular docking analysis has risen as a potent tool in the field of drug discovery and in comprehending the mechanisms of action of diverse bioactive products originating from plants. The compounds listed in Table 17.2 exemplify the tremendous potential of this computational technique to unveil promising therapeutic agents. Curcumin, for instance, has long been valued for its antioxidant and anti-inflammatory properties (Hewlings and Kalman 2017). Molecular docking studies elucidated that curcumin can suppress the activity of NF-kB, a key transcription factor in inflammation. This computational insight guided subsequent experimental work, further validating curcumin's therapeutic potential (Hewlings and Kalman 2017). Similarly, taxifolin, an agent with antituberculosis activity, owes its discovery and mechanistic understanding to molecular docking analysis (Kozhikkadan Davis et al. 2018). This computational technique identified that taxifolin

Table 17.2 Examples of the important compounds discovered with different pharmacological activities, from plant origin, and potential mechanisms of action using molecular docking.

Compound/plant origin	Chemical structure	Therapeutic indication	Possible mechanism of action	References
Curcumin/*Curcuma longa* L.	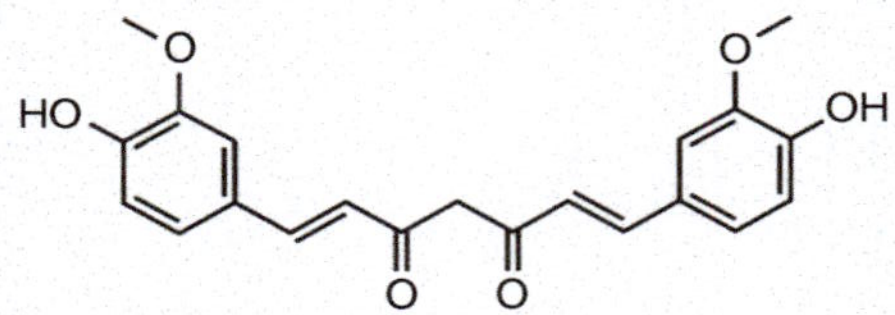	Antioxidant, and anti-inflammatory	Suppression of NF-kB; elimination of reactive oxygen and nitrogen molecules; regulation of GSH, catalase, and SOD functions	(Hewlings and Kalman 2017)
Taxifolin/unknown	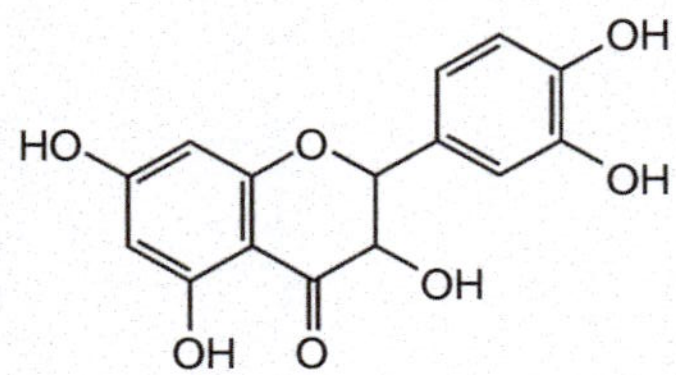	Antituberculosis activity	Inhibition of DNA gyrase and Isoleucyl-tRNA synthetase	(Kozhikkadan Davis et al. 2018)
Naringin/generally found in citrus fruits	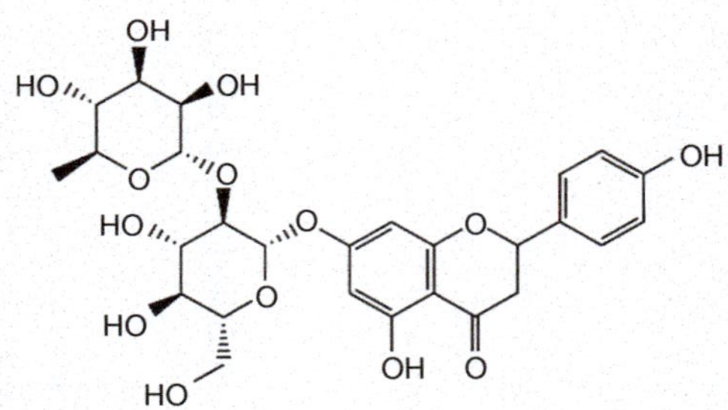	Coronavirus Disease (SARS-CoV-2)	High affinity to the spike protein of SARS-CoV-2	(Jain et al. 2021)

Paclitaxel/*Taxus brevifolia* Nutt.	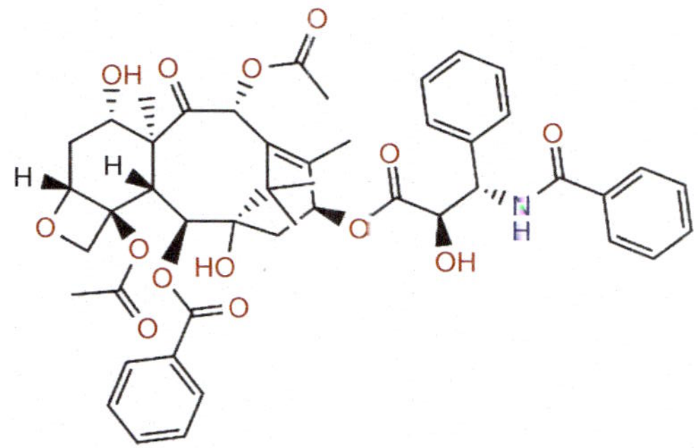	Anticancer activity	Antimitotic potential	(Atanasov et al. 2015)
Aegeline/*Aegle marmelos* L.	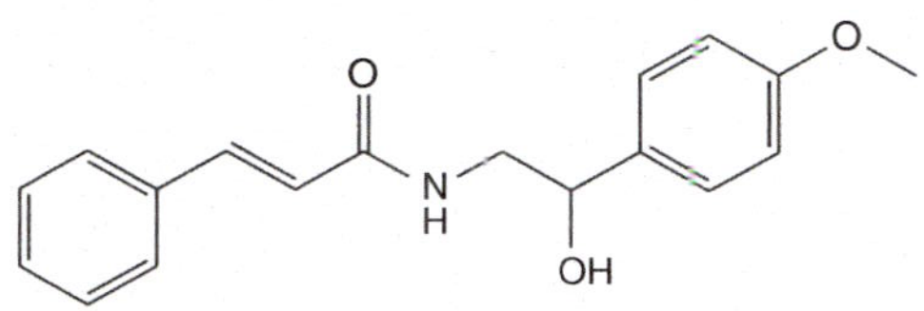	Antidiabetic potential	Inhibition of Glycogen Synthase Kinase-3 Beta	(Prottoy et al. 2019)
α-Terpineol acetate/*Dysphania ambrosioides* (L.) Mosyakin and Clemants	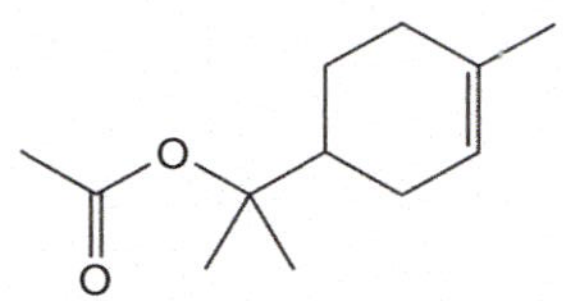	Antioxidant and antibacterial potential	Inhibition of enoyl-acyl carrier protein reductase; elimination of ROS, and radical scavenging potential	(Kandsi et al. 2022)
Mangiferin/*Mangifera indica* L.	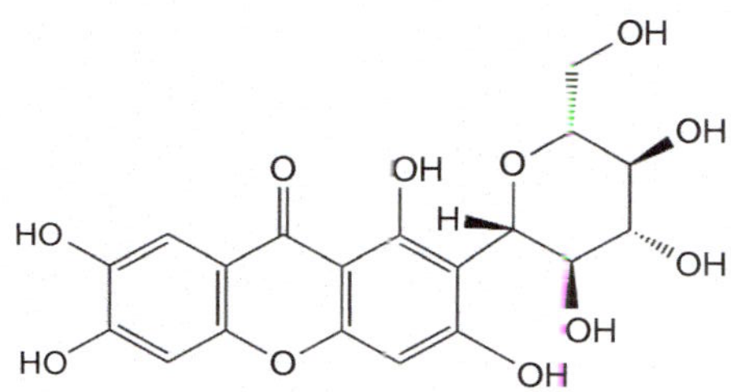	Antidiabetic potential	Inhibition of Glycogen Synthase Kinase-3 Beta	(Prottoy et al. 2019)

(*Continued*)

Table 17.2 (Continued)

Compound/plant origin	Chemical structure	Therapeutic indication	Possible mechanism of action	References
Trimethoxyflavone/*Atriplex halimus* L.	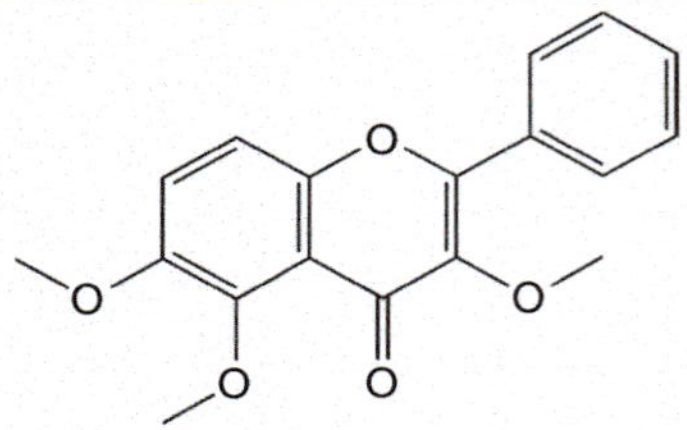	Antioxidant, and anticancer activity	Inhibition of G-protein-coupled receptor subtypes A1 and A2, ATP-binding cassette (ABC) protein transporter inhibition; radical scavenging potential	(Elbouzidi et al. 2022)
Quercetin/*Caralluma europea* (Guss) N.E.Br.	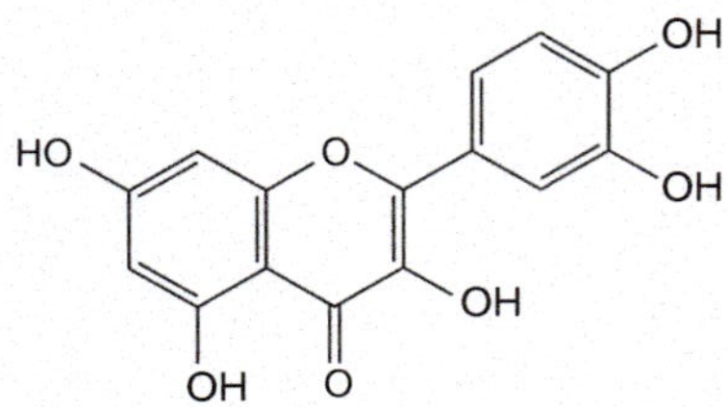	Antioxidant and antidiabetic potential	Antiglycation; inhibitory activity against α-amylase and α-glucosidase enzymes; radical scavenging potential	(Ouassou et al. 2021; Rhizlan et al. 2024)
Capsaicin/*Capsicum annum* L.; *C. minimum* Mill	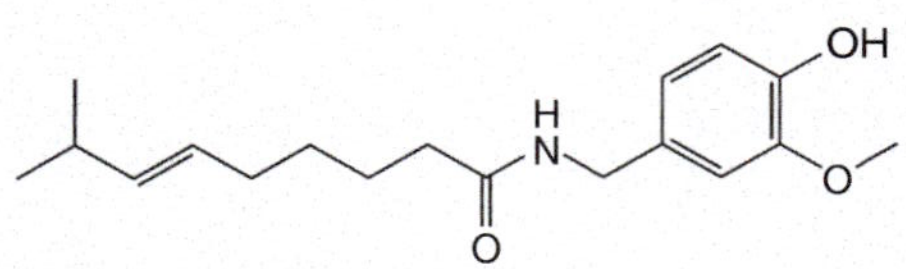	Postherpetic neuralgia and musculoskeletal pain	Stimulates transient receptor potential vanilloid 1 (TRPV1) in sensory neurons	(Sharma et al. 2013)
Galantamine/*Galanthus caucasicus* Grossh	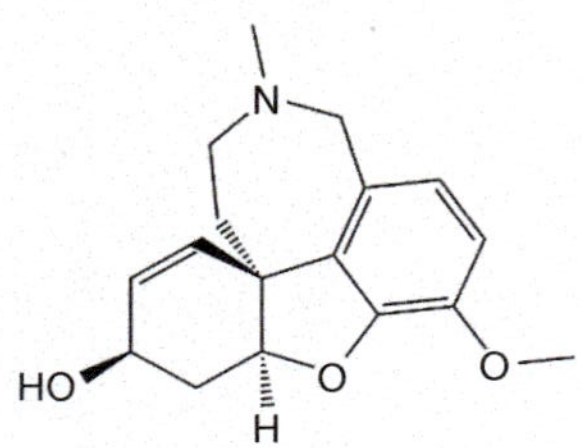	Alzheimer's disease	A reversible inhibitor of acetylcholinesterase; regulates nicotinic acetylcholine receptors (nAChRs)	(Razay and Wilcock 2008)

Masoprocol/*Larrea tridentata* (DC.) Coville	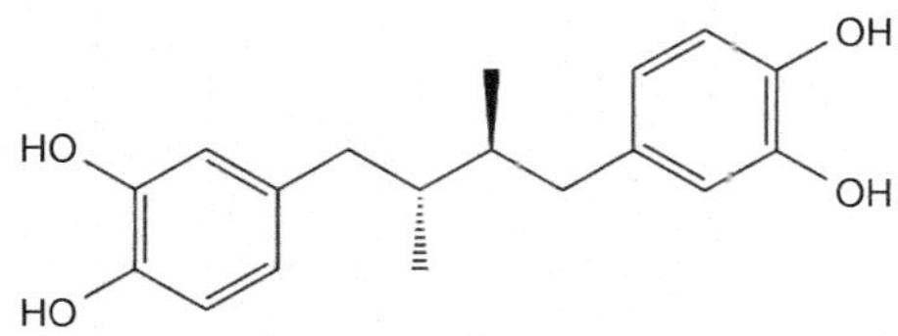	Antineoplastic compound that can be used in cancer chemotherapy	Inhibition of 5-lipoxygenase	(Atanasov et al. 2015)
Rutin/unknown	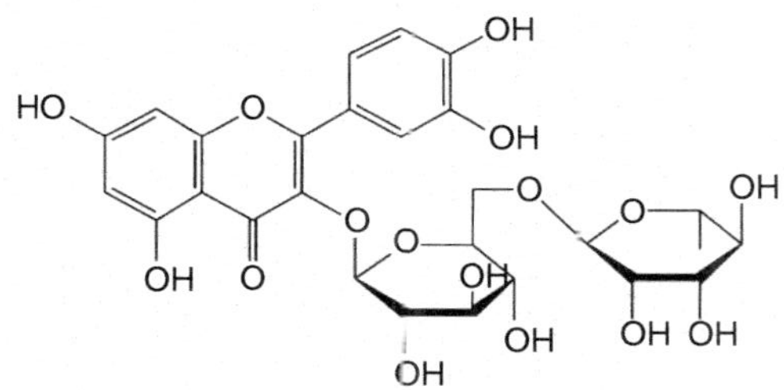	Coronavirus disease (SARS-CoV-2)	Inhibition of ACE-2 enzyme	(Guler et al. 2021)
Gossypetin/*Hibiscus sabdariffa* L.	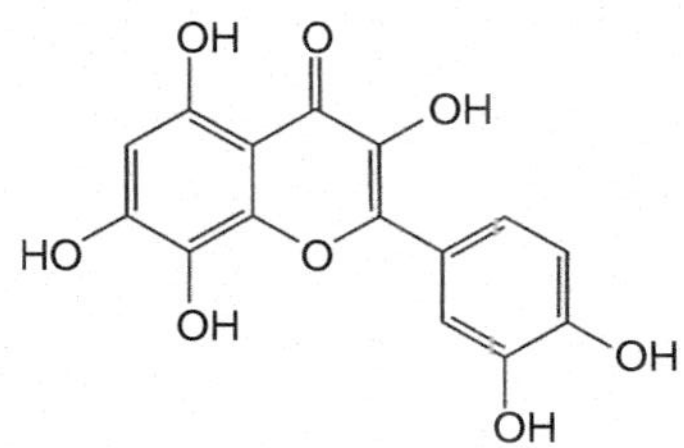	Ebola virus	Inhibition and disruption of the crucial protein structures VP40, VP35, VP30, and VP24	(Raj and Varadwaj 2016)

exerts its effects by inhibiting crucial bacterial proteins, DNA gyrase, and Isoleucyl-tRNA synthetase (Kozhikkadan Davis et al. 2018). These findings hold significant implications for the development of novel antibiotics targeting tuberculosis and other bacterial infections (Kozhikkadan Davis et al. 2018). In the era of the COVID-19 pandemic, naringin, commonly found in citrus fruits, came under scrutiny for its potential to combat the SARS-CoV-2 virus (Ouassou et al. 2020; Jain et al. 2021). Molecular docking analysis played a pivotal role in identifying naringin's high affinity for the spike protein of the virus (Jain et al. 2021). This prediction not only sparked interest in naringin as a potential antiviral agent but also illustrated the utility of computational methods in rapid drug discovery (Jain et al. 2021). Furthermore, paclitaxel, a cornerstone in cancer chemotherapy, owes its discovery to molecular docking (Atanasov et al. 2015). This computational technique elucidated how paclitaxel interferes with microtubules, preventing cancer cell division. Such insights paved the way for its extensive clinical use in cancer treatment (Atanasov et al. 2015). Aegeline, on the other hand, emerged as a potential antidiabetic compound through molecular docking studies (Prottoy et al. 2019). These analyses revealed that aegeline inhibits Glycogen Synthase Kinase-3 Beta, a key player in diabetes. Molecular docking thus provided a foundation for further investigation into aegeline's therapeutic potential (Prottoy et al. 2019). The story continues with α-Terpineol Acetate, a compound found in *Dysphania ambrosioides*, which exhibits antioxidant and antibacterial properties (Kandsi et al. 2022). Molecular docking studies not only identified its interaction with enoyl-acyl carrier protein reductase but also shed light on its radical scavenging potential (Kandsi et al. 2022). This knowledge has opened doors to exploring α-Terpineol Acetate's applications in combating microbial infections and oxidative stress.

Mangiferin, derived from *Mangifera indica* L., also found recognition for its antidiabetic potential through molecular docking (Prottoy et al. 2019). These analyses unveiled how mangiferin interacts with Glycogen Synthase Kinase-3 Beta, providing a molecular rationale for its use in diabetes management (Prottoy et al. 2019). Trimethoxyflavone, originating from *Atriplex halimus* L., intrigued researchers with its antioxidant and anticancer activities (Elbouzidi et al. 2022). Molecular docking elucidated its mechanisms, including the inhibition of G-protein-coupled receptor subtypes and ATP-binding cassette protein transporters, along with its radical scavenging potential (Elbouzidi et al. 2022). These findings have spurred investigations into trimethoxyflavone as a potential anticancer agent. Quercetin, found in *Caralluma europea*, showcased its antioxidant and antidiabetic potential through molecular docking analysis (Ouassou et al. 2021; Rhizlan et al. 2024). It was revealed to exert its effects through antiglycation and inhibitory activity against key enzymes (Ouassou et al. 2021; Rhizlan et al. 2024). These insights have prompted further research into quercetin's applications in managing diabetes and oxidative stress-related conditions. In a different therapeutic context, capsaicin, sourced from *Capsicum annum* and *Capsicum minimum*, was found to alleviate postherpetic neuralgia and musculoskeletal pain by stimulating transient receptor potential vanilloid 1 (TRPV1) in sensory neurons (Sharma et al. 2013). This discovery exemplifies how molecular docking can elucidate mechanisms of pain relief, offering relief to those suffering from these conditions. Finally, galantamine, derived from *Galanthus caucasicus* Grossh., was identified as a valuable compound in Alzheimer's disease treatment. Molecular docking uncovered its role as a reversible inhibitor of acetylcholinesterase and its regulatory effects on nicotinic acetylcholine receptors (nAChRs) (Razay and Wilcock 2008). This knowledge has shaped the development of therapies for Alzheimer's patients.

In the context of infectious diseases, masoprocol from *Larrea tridentata* and gossypetin from *Hibiscus sabdariffa* L. emerged as potential antineoplastic and antiviral agents, respectively, through molecular docking studies (Atanasov et al. 2015; Raj and Varadwaj 2016). Masoprocol inhibits 5-Lipoxygenase, offering promise in cancer chemotherapy (Atanasov et al. 2015), while gossypetin disrupts crucial protein structures in the Ebola virus, opening avenues for antiviral drug development (Raj and Varadwaj 2016).

In conclusion, molecular docking analysis has proven to be an invaluable tool for identifying and understanding the mechanisms of action of bioactive compounds derived from plants. These compounds, with their diverse therapeutic indications, underscore the vital role computational approaches play in modern drug discovery, enabling the rapid exploration of potential therapeutic agents and their mechanisms of action.

17.6 Concluding Remarks and Future Considerations

In conclusion, this chapter has delved into the promising realm of molecular docking and *in silico* studies as invaluable tools for advancing the discovery of secondary metabolites and bioactive compounds in plant research. The increasing demand for novel compounds with therapeutic potential necessitates innovative approaches, and this chapter has underscored the immense potential of molecular docking in streamlining drug discovery development processes. Through a comprehensive review of the principles and methodologies of molecular docking and *in silico* studies, we have highlighted their role in predicting ligand–receptor interactions and optimizing compound design. These methods have already significantly contributed to the identification and validation of secondary metabolites, such as phytochemicals and natural products, which play a pivotal role in various areas of plant research, from pharmacology to crop protection.

Furthermore, the chapter has emphasized the advantages of *in silico* studies in terms of cost-efficiency, speed, and ethical considerations when compared to traditional wet lab approaches. This has profound implications for the sustainable utilization of plant resources, as it offers a means to reduce the need for extensive field collections and minimize the ecological impact. However, it is important to acknowledge that molecular docking and *in silico* studies are not without their limitations. These methods rely heavily on the accuracy of the available molecular models and require validation through experimental assays. Additionally, the selection of appropriate software and parameters can significantly influence the outcomes, underscoring the need for standardization and ongoing research in this field.

In the quest for future advancements in plant research, this chapter delves into several critical areas that demand our attention. These future considerations are pivotal in shaping the trajectory of molecular docking and *in silico* studies within this field.

To embark on this journey into the future of plant research, collaboration emerges as a cornerstone. Drawing expertise from diverse fields, including chemistry, biology, and computational science, becomes imperative. The synergy among these disciplines not only enriches the research process but also validates and applies *in silico* findings with greater precision. In an increasingly interconnected scientific landscape, this multidisciplinary approach will be the bedrock upon which we build our understanding of plant molecular interactions.

References

Aplin, C., Milano, S.K., Zielinski, K.A. et al. (2022). Evolving experimental techniques for structure-based drug design. *The Journal of Physical Chemistry. B* 126: 6599–6607.

Atanasov, A.G., Waltenberger, B., Pferschy-Wenzig, E.-M. et al. (2015). Discovery and resupply of pharmacologically active plant-derived natural products: a review. *Biotechnology Advances* 33: 1582–1614.

Bajad, N.G., Rayala, S., Gutti, G. et al. (2021). Systematic review on role of structure based drug design (SBDD) in the identification of anti-viral leads against SARS-Cov-2. *Current Research in Pharmacology and Drug Discovery* 2: 100026.

Bindseil, K.U., Jakupovic, J., Wolf, D. et al. (2001). Pure compound libraries; a new perspective for natural product based drug discovery. *Drug Discovery Today* 6: 840–847.

Blaney, J. (2012). A very short history of structure-based design: how did we get here and where do we need to go? *Journal of Computer-Aided Molecular Design* 26: 13–14.

Böhm, H.-J. (1992). LUDI: rule-based automatic design of new substituents for enzyme inhibitor leads. *Journal of Computer-Aided Molecular Design* 6: 593–606. https://doi.org/10.1007/BF00126217.

Brint, A.T. and Willett, P. (1987). Algorithms for the identification of three-dimensional maximal common substructures. *Journal of Chemical Information and Computer Sciences* 27: 152–158.

Chen, G., Seukep, A.J., and Guo, M. (2020). Recent advances in molecular docking for the research and discovery of potential marine drugs. *Marine Drugs* 18: 545.

Colegate, S.M. and Molyneux, R.J. (2007). *Bioactive Natural Products: Detection, Isolation, and Structural Determination*. CRC Press ISBN 1420006886.

Cooper, D.R., Porebski, P.J., Chruszcz, M., and Minor, W. (2011). X-ray crystallography: assessment and validation of protein–small molecule complexes for drug discovery. *Expert Opinion on Drug Discovery* 6: 771–782.

DesJarlais, R.L., Sheridan, R.P., Dixon, J.S. et al. (1986). Docking flexible ligands to macromolecular receptors by molecular shape. *Journal of Medicinal Chemistry* 29: 2149–2153.

Devillers, J. (2013). Methods for building QSARs. In: *Computational Toxicology, Volume II* (ed. B. Reisfeld and A.N. Mayeno), 3–27. Totowa, NJ: Humana.

Dewick, P.M. (2002). *Medicinal Natural Products: A Biosynthetic Approach*. Wiley ISBN 0471496413.

Dias, D.A., Urban, S., and Roessner, U. (2012). A historical overview of natural products in drug discovery. *Metabolites* 2: 303–336.

Eisen, M.B., Wiley, D.C., Karplus, M., and Hubbard, R.E. (1994). HOOK: a program for finding novel molecular architectures that satisfy the chemical and steric requirements of a macromolecule binding site. *Proteins: Structure, Function, and Bioinformatics* 19: 199–221.

Elbouzidi, A., Ouassou, H., Aherkou, M. et al. (2022). LC–MS/MS phytochemical profiling, antioxidant activity, and cytotoxicity of the ethanolic extract of *Atriplex halimus* L. against breast cancer cell lines: computational studies and experimental validation. *Pharmaceuticals* 15: 1–25. https://doi.org/10.3390/ph15091156.

Ferreira, L.G., Dos Santos, R.N., Oliva, G., and Andricopulo, A.D. (2015). Molecular docking and structure-based drug design strategies. *Molecules* 20: 13384–13421.

Fradera, X., Kaur, J., and Mestres, J. (2004). Unsupervised guided docking of covalently bound ligands. *Journal of Computer-Aided Molecular Design* 18: 635–650.

Goodsell, D.S. and Olson, A.J. (1990). Automated docking of substrates to proteins by simulated annealing. *Proteins: Structure, Function, and Bioinformatics* 8: 195–202.

Gray, A.I., Igoli, J.O., and Edrada-Ebel, R. (2012). Natural products isolation in modern drug discovery programs. *Natural Products Isolation* 515–534.

Guler, H.I., Tatar, G., Yildiz, O. et al. (2021). Investigation of potential inhibitor properties of ethanolic propolis extracts against ACE-II receptors for COVID-19 treatment by molecular docking study. *Archives of Microbiology* 203: 3557–3564.

Haefner, B. (2003). Drugs from the deep: marine natural products as drug candidates. *Drug Discovery Today* 8: 536–544.

Hewlings, S.J. and Kalman, D.S. (2017). Curcumin: a review of its effects on human health. *Foods* 6: 92.

Isah, T. (2019). Stress and defense responses in plant secondary metabolites production. *Biological Research* 52.

Jain, A.S., Sushma, P., Dharmashekar, C. et al. (2021). *In silico* evaluation of flavonoids as effective antiviral agents on the spike glycoprotein of SARS-CoV-2. *Saudi Journal of Biological Sciences* 28: 1040–1051. https://doi.org/10.1016/j.sjbs.2020.11.049.

Jan, R., Asaf, S., Numan, M. et al. (2021). Plant secondary metabolite biosynthesis and transcriptional regulation in response to biotic and abiotic stress conditions. *Agronomy* 11: 968.

Jones, G., Willett, P., Glen, R.C. et al. (1997). Development and validation of a genetic algorithm for flexible docking. *Journal of Molecular Biology* 267: 727–748.

Kahsai, A.W., Xiao, K., Rajagopal, S. et al. (2011). Multiple ligand-specific conformations of the B2 adrenergic receptor. *Nature Chemical Biology* 7: 692–700.

Kalyaanamoorthy, S. and Chen, Y.-P.P. (2011). Structure-based drug design to augment hit discovery. *Drug Discovery Today* 16: 831–839.

Kandsi, F., Elbouzidi, A., Lafdil, F.Z. et al. (2022). Antibacterial and antioxidant activity of *Dysphania ambrosioides* (L.) mosyakin and clemants essential oils: experimental and computational approaches. *Antibiotics* 11: 482.

Kitchen, D.B., Decornez, H., Furr, J.R., and Bajorath, J. (2004). Docking and scoring in virtual screening for drug discovery: methods and applications. *Nature Reviews Drug Discovery* 3: 935–949. https://doi.org/10.1038/nrd1549.

Kozhikkadan Davis, C., Nasla, K., Anjana, A.K., and Rajanikant, G.K. (2018). Taxifolin as dual inhibitor of Mtb DNA gyrase and isoleucyl-TRNA synthetase: *in silico* molecular docking, dynamics simulation and *in vitro* assays. *Silico Pharmacology* 6: 8. https://doi.org/10.1007/s40203-018-0045-5.

Kroymann, J. (2011). Natural diversity and adaptation in plant secondary metabolism. *Current Opinion in Plant Biology* 14: 246–251.

Kuhl, F.S., Crippen, G.M., and Friesen, D.K. (1984). A combinatorial algorithm for calculating ligand binding. *Journal of Computational Chemistry* 5: 24–34.

Kuntz, I.D., Blaney, J.M., Oatley, S.J. et al. (1982). A geometric approach to macromolecule-ligand interactions. *Journal of Molecular Biology* 161: 269–288. https://doi.org/10.1016/0022-2836(82)90153-X.

Leach, A.R. and Kuntz, I.D. (1992). Conformational analysis of flexible ligands in macromolecular receptor sites. *Journal of Computational Chemistry* 13: 730–748.

Levinthal, C., Wodak, S.J., Kahn, P., and Dadivanian, A.K. (1975). Hemoglobin interaction in sickle cell fibers. I: Theoretical approaches to the molecular contacts. *Proceedings of the National Academy of Sciences* 72: 1330–1334.

Liao, C., Sitzmann, M., Pugliese, A., and Nicklaus, M.C. (2011). Software and resources for computational medicinal chemistry. *Future Medicinal Chemistry* 3: 1057–1085.

Lill, M.A. (2007). Multi-dimensional QSAR in drug discovery. *Drug Discovery Today* 12: 1013–1017.

Liu, M. and Wang, S. (1999). MCDOCK: a Monte Carlo simulation approach to the molecular docking problem. *Journal of Computer-Aided Molecular Design* 13: 435–451.

Lorber, D.M. and Shoichet, B.K. (1998). Flexible ligand docking using conformational ensembles. *Protein Science* 7: 938–950.

Mandal, S. and Mandal, S.K. (2009). Rational drug design. *European Journal of Pharmacology* 625: 90–100.

Mishra, B.B. and Tiwari, V.K. (2011). Natural products: an evolving role in future drug discovery. *European Journal of Medicinal Chemistry* 46: 4769–4807.

Morris, G.M., Goodsell, D.S., Halliday, R.S. et al. (1998). Automated docking using a Lamarckian genetic algorithm and an empirical binding free energy function. *Journal of Computational Chemistry* 19: 1639–1662.

do Nascimento, N.C. and Fett-Neto, A.G. (2010). Plant secondary metabolism and challenges in modifying its operation: an overview. In: *Plant Secondary Metabolism Engineering, Methods in Molecular Biology* (ed. A.G. Fett-Neto), 1–13. Totowa, NJ: Humana.

Ouassou, H., Kharchoufa, L., Bouhrim, M. et al. (2020). The pathogenesis of coronavirus disease 2019 (COVID-19): evaluation and prevention. *Journal of Immunology Research* 2020: 1357983. https://doi.org/10.1155/2020/1357983.

Ouassou, H., Bouhrim, M., Kharchoufa, L. et al. (2021). *Caralluma europaea* (Guss) N.E.Br.: a review on ethnomedicinal uses, phytochemistry, pharmacological activities, and toxicology. *Journal of Ethnopharmacology* 273: 113769. https://doi.org/10.1016/j.jep.2020.113769.

Pang, Y., Perola, E., Xu, K., and Prendergast, F.G. (2001). EUDOC: a computer program for identification of drug interaction sites in macromolecules and drug leads from chemical databases. *Journal of Computational Chemistry* 22: 1750–1771.

Pei, J., Wang, Q., Liu, Z. et al. (2006). PSI-DOCK: towards highly efficient and accurate flexible ligand docking. *Proteins: Structure, Function, and Bioinformatics* 62: 934–946.

Prottoy, N.I., Sarkar, B., Ullah, A. et al. (2019). Molecular docking and pharmacological property analysis of antidiabetic agents from medicinal plants of Bangladesh against type II diabetes: a computational approach. *PharmaTutor* 7: 6–15.

Radu, A.-F., Bungau, S.G., Negru, A.P. et al. (2023). Novel potential janus kinase inhibitors with therapeutic prospects in rheumatoid arthritis addressed by *in silico* studies. *Molecules* 28: 4699.

Raj, U. and Varadwaj, P.K. (2016). Flavonoids as multi-target inhibitors for proteins associated with ebola virus: *in silico* discovery using virtual screening and molecular docking studies. *Interdisciplinary Sciences: Computational Life Sciences* 8: 132–141.

Rarey, M., Kramer, B., Lengauer, T., and Klebe, G. (1996). A fast flexible docking method using an incremental construction algorithm. *Journal of Molecular Biology* 261: 470–489.

Razay, G. and Wilcock, G.K. (2008). Galantamine in Alzheimer's disease. *Expert Review of Neurotherapeutics* 8: 9–17.

Rey-Ladino, J., Ross, A.G., Cripps, A.W. et al. (2011). Natural products and the search for novel vaccine adjuvants. *Vaccine* 29: 6464–6471.

Rhizlan, A., Amine, E., Ouassou, H. et al. (2024). *In vitro* study on antioxidant and antiglycation activities, and molecular docking of Moroccan medicinal plants for diabetes. *Current Traditional Medicine* 10: 1–14.

Salemme, F.R. (1976). An hypothetical structure for an intermolecular electron transfer complex of cytochromes c and B5. *Journal of Molecular Biology* 102: 563–568.

Schnecke, V. and Kuhn, L.A. (1999). Database screening for HIV protease ligands: the influence of binding-site conformation and representation on ligand selectivity. *Proceedings of the ISMB*, 242–251.

Seifert, M.H.J., Schmitt, F., Herz, T., and Kramer, B. (2004). ProPose: a docking engine based on a fully configurable protein–ligand interaction model. *Journal of Molecular Modeling* 10: 342–357.

Sharma, S.K., Vij, A.S., and Sharma, M. (2013). Mechanisms and clinical uses of capsaicin. *European Journal of Pharmacology* 720: 55–62.

Sorokina, M. and Steinbeck, C. (2020). Review on natural products databases: where to find data in 2020. *Journal of Cheminformatics* 12: 20.

Stockwell, B.R. (2004). Exploring biology with small organic molecules. *Nature* 432: 846–854.

Sullivan, K.M., Manuppello, J.R., and Willett, C.E. (2014). Building on a solid foundation: SAR and QSAR as a fundamental strategy to reduce animal testing. *SAR and QSAR in Environmental Research* 25: 357–365.

Taylor, R.D., Jewsbury, P.J., and Essex, J.W. (2003). FDS: flexible ligand and receptor docking with a continuum solvent model and soft-core energy function. *Journal of Computational Chemistry* 24: 1637–1656.

Temml, V. and Kutil, Z. (2021). Structure-based molecular modeling in SAR analysis and lead optimization. *Computational and Structural Biotechnology Journal* 19: 1431–1444.

Tietze, S. and Apostolakis, J. (2007). GlamDock: development and validation of a new docking tool on several thousand protein–ligand complexes. *Journal of Chemical Information and Modeling* 47: 1657–1672.

Totrov, M. and Abagyan, R. (1997). Flexible protein–ligand docking by global energy optimization in internal coordinates. *Proteins: Structure, Function, and Bioinformatics* 29: 215–220.

Trosset, J. and Scheraga, H.A. (1999). PRODOCK: software package for protein modeling and docking. *Journal of Computational Chemistry* 20: 412–427.

Venkatachalam, C.M., Jiang, X., Oldfield, T., and Waldman, M. (2003). LigandFit: a novel method for the shape-directed rapid docking of ligands to protein active sites. *Journal of Molecular Graphics & Modelling* 21: 289–307.

Warwicker, J. (1989). Investigating protein-protein interaction surfaces using a reduced stereochemical and electrostatic model. *Journal of Molecular Biology* 206: 381–395.

Welch, W., Ruppert, J., and Jain, A.N. (1996). Hammerhead: fast, fully automated docking of flexible ligands to protein binding sites. *Chemistry & Biology* 3: 449–462.

Weng, J.-K., Philippe, R.N., and Noel, J.P. (2012). The rise of chemodiversity in plants. *Science* 336: 1667–1670.

Wink, M. (1999). *Biochemistry of Plant Secondary Metabolism*, vol. 2. CRC Press ISBN 0849340853.

Wodak, S.J. and Janin, J. (1978). Computer analysis of protein-protein interaction. *Journal of Molecular Biology* 124: 323–342.

Wurtzel, E.T. and Kutchan, T.M. (2016). Plant metabolism, the diverse chemistry set of the future. *Science* 353: 1232–1236. https://doi.org/10.1126/science.aad2062.

Yang, C., Fang, X., Wu, X. et al. (2012). Transcriptional regulation of plant secondary metabolism F. *Journal of Integrative Plant Biology* 54: 703–712.

Yeshi, K., Crayn, D., Ritmejerytė, E., and Wangchuk, P. (2022). Plant secondary metabolites produced in response to abiotic stresses has potential application in pharmaceutical product development. *Molecules* 27: 313.

Zhao, Y. and Sanner, M.F. (2007). FLIPDock: docking flexible ligands into flexible receptors. *Proteins: Structure, Function, and Bioinformatics* 68: 726–737.

18

Exploring Secondary Metabolites in Plants Through Bioinformatics

Sneha Murmu[1], *Ritwika Das*[1], *Bharati Pandey*[2], *Soumya Sharma*[1], *and Mohammad Samir Farooqi*[1]

[1] *Division of Agricultural Bioinformatics, ICAR-Indian Agricultural Statistics Research Institute, New Delhi, India*
[2] *Animal Biotechnology Division, ICAR-National Dairy Research Institute, Karnal, Haryana, India*

18.1 Introduction

Plants synthesize a diverse array of organic substances that play supportive roles in their growth, development, and reproductive processes. These compounds, referred to as secondary metabolites, are small organic molecules with molecular weights typically below 3000 Da, evolving from primary metabolites throughout the plant's life cycle (Twaij and Hasan 2022). They are often produced via various metabolic pathways and serve as vital defense mechanisms in the natural environment, protecting plants against diverse abiotic and biotic stresses, including drought, salinity, heat, UV radiation, pathogens, insects, microbes, and predators (Kumar et al. 2023).

Plant secondary metabolites accumulate at higher levels under stressful conditions, serving as signaling molecules to enhance the expression of defense-related genes. For example, in *Fusarium oxysporum*, α-tomatine, a potent antifungal saponin, triggers the activation of phosphotyrosine kinase and G-protein pathways, inducing a burst of reactive oxygen species (ROS) and elevating intracellular Ca^{2+} levels (Zaynab et al. 2018). Rice plants combat *Magnaporthe oryzae*-induced rice blast by producing various diterpenoid phytoalexins (Azizi et al. 2019). Maize responds to *Rhizopus microspores* and *Colletotricum graminicola* infections by synthesizing substantial quantities of kauralexins, a type of diterpenoid phytoalexin (Schmelz et al. 2011). *Fusarium graminearum* infection in maize leads to the production of sesquiterpenoid phytoalexins, such as zealexin, which confer resistance against *R. microspores* and *Aspergillus flavus*. Moreover, numerous secondary metabolites in plants exhibit antibacterial activities (Huffaker et al. 2011).

Plant secondary metabolites, including caffeine, have been revealed to exhibit potent antibacterial properties against various plant pathogenic bacteria such as *Ralstonia solancearum, Dickeya solani, Clavibacter michiganesis, Pectobacterium atrosepticum, Pectobacterium carotovorum*, and *Xanthomonas campestris* (Uche-Okereafor et al. 2019).

Bioinformatics for Plant Research and Crop Breeding, First Edition. Edited by Jen-Tsung Chen.

Moreover, in the case of brassica, glucosinolate and phenolic substances have been identified to confer resistance against bacterial infections, as discussed in the study conducted by Velasco et al. (2013). Additionally, *in vitro* studies have demonstrated that methanol extracts of *Polyalthia longifolia*, *Terminalia chebula*, and *Zingiber officinale*, along with chloroform extracts of *Z. officinale*, exhibit substantial efficacy against *X. campestris* pv. *campestris*.

Numerous secondary metabolites, such as phenolics, alkaloids, and flavonoids, exhibit potent antiviral properties (Reichling 2018). Traditional Chinese herbs contain approximately 18,500 alkaloid compounds, including 7-deoxytrans-dihydronarciclasine, which have been observed to demonstrate antiviral effects (Erharuyi et al. 2014). In a related context, the compound Bruceine-D, identified in *Brassica javanica* extract, shows inhibitory effects against PVY, CMV, and TMV (Shen et al., 2015). The analogs of the antiviral protein (APV) found in *Bougainvillaea spectabilis* and *Prosopis chilensis* effectively combat viral infections in sunflower and cowpea (Gurjar et al. 2012). Apart from their defensive roles, plant secondary metabolites also participate in the regulation of plant growth and development. Glucosinolate compounds, for instance, regulate various aspects of plant physiology, including stomatal opening, circadian clock regulation, and flowering time. Similarly, externally applied flavonoids exert a regulatory influence on auxin transport in plants.

For many years, secondary metabolites have been a plentiful supply of therapeutic substances. Many of the traditional medicines used around the world are derived from plants. Drugs for various illnesses such as pain, malaria, and cancer have been developed from secondary metabolites such as alkaloids (such as morphine and quinine), terpenoids, and artemisinin. Resveratrol and other polyphenolic chemicals in plants, such as flavonoids, have antioxidant characteristics that are good for the immune system against oxidative damage. A lot of herbal medicines treat various illnesses by utilizing the therapeutic qualities of plant secondary metabolites. Secondary metabolites have highly significant economic importance. These are utilized as organic pesticides, herbicides, and fungicides in agriculture such as neem oil. Terpenoids are important ingredients in many perfumes, cosmetics, and flavors, i.e., aroma of mint, lavender, etc. A secondary metabolite used for coloring textiles is indigo, which is obtained from the indigo plant. A number of culinary spices, including turmeric, cinnamon, and ginger, include secondary metabolites that give food flavor and scent.

18.2 Classification of Plant Secondary Metabolites

Types of plant secondary metabolites vary according to plant species, plant parts as well as types of stress. For instance, leaves typically produce larger concentrations of phytoalexins which is an antimicrobial stress suppressor compound. Secondary metabolites, which are produced by biosynthetic modifications such as methylation, glycosylation, and hydroxylation, are typically the byproduct of primary metabolites. As compared to primary metabolites, secondary metabolites have more complex molecular structural makeup and side chains. Around 20,000 plant secondary metabolites have been isolated and characterized which can be classified into four major classes, viz., terpenes (composed mainly of carbon and hydrogen), phenolics (composed of simple sugars and benzene rings), N-containing compounds, and S-containing compounds (Satish et al. 2020) (Figure 18.1).

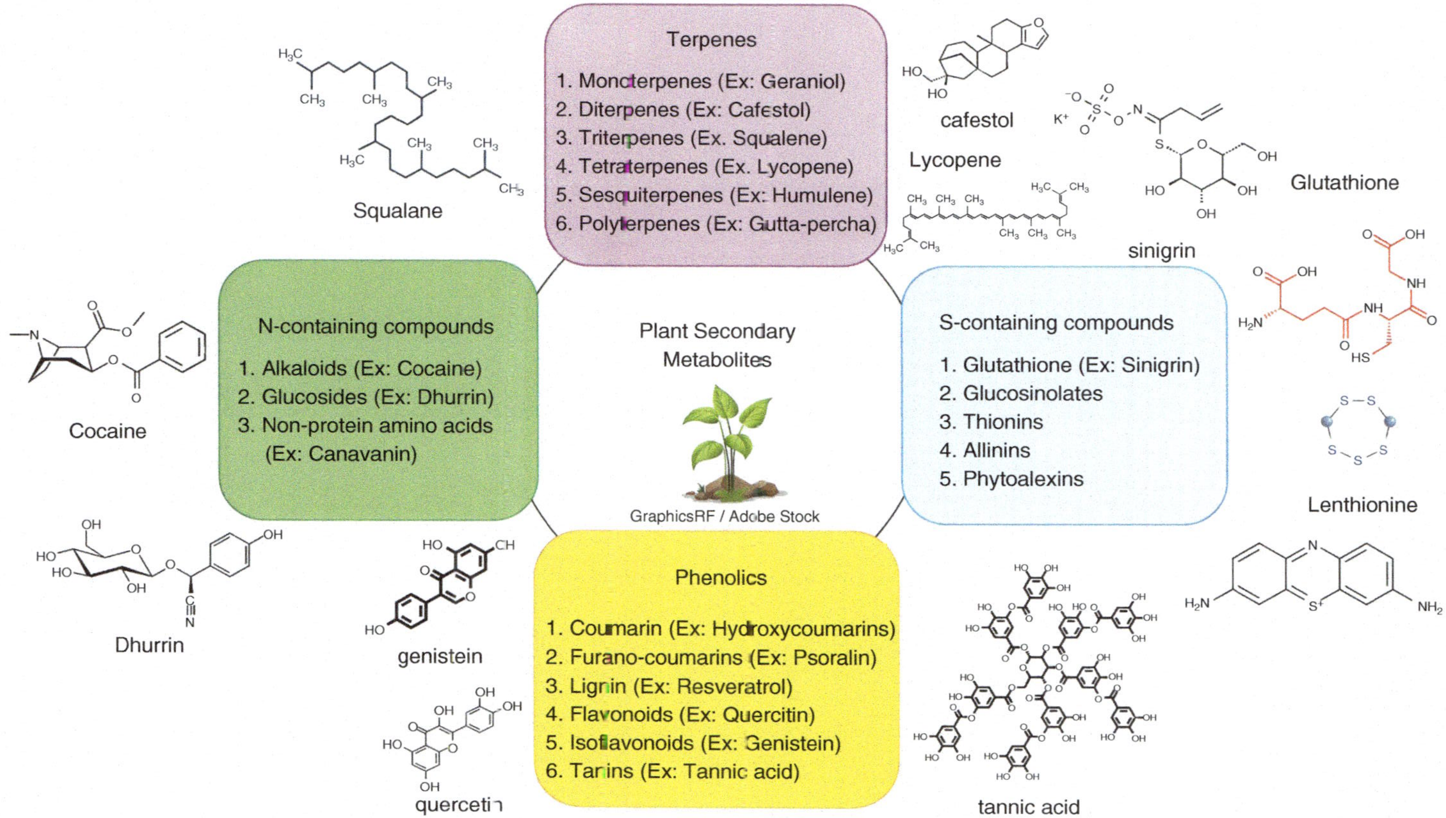

Figure 18.1 Classification of plant secondary metabolites.

18.2.1 Terpenes

A diverse array of terpenes constitutes this group of complex and large metabolites, with over 35,000 of them identified to date (Twaij and Hasan 2022). Terpenes consist of isoprene units, which can be modified through cyclization reactions, rendering them easily identifiable. These compounds are categorized based on the number of isoprene units present in their carbon framework. Terpenoids are bioactive metabolites exhibiting properties such as antitubercular, anticancer, anxiolytic, and mutagenic activities.

18.2.2 Phenolics

More than 4000 distinct flavonoids with plant origins have been identified. These are mostly found as glycosides in the leaves, flowers, stems, and roots of green plants. Chalcones, flavones, flavanols, flavanones, anthocyanins, and isoflavones are important flavonoids that are often magnificently colored because they are made up of two benzene rings. Only plants and microbes can utilize phenylpropanoids, which are derived from the shikimate pathway and produce important aromatic amino acids like phenylalanine and tyrosine.

18.2.3 N-containing Compounds

A wide variety of complex and sizable metabolites form this collection of terpenes, with more than 35,000 of them recognized as of the latest findings (Twaij and Hasan 2022). Terpenes are comprised of isoprene units that can be altered via cyclization reactions, making them easily recognizable. These compounds are classified based on the quantity of isoprene units found in their carbon structure. Terpenoids are active biological metabolites demonstrating characteristics such as antitubercular, anticancer, anxiolytic, and mutagenic activities.

18.2.4 S-containing Compounds

Thio-compounds, or secondary metabolites of plants that include sulfur, are important for plant defense and other biological activities. Their chemical makeup includes sulfur, frequently in the form of thiol (–SH) or disulfide (–S–S–) linkages. Example: glucosinolates, which contribute to the distinctive flavors and health advantages of cruciferous vegetables like broccoli and mustard greens, and alliin, which is present in garlic. These chemicals can repel herbivores and have positive health effects on human beings.

18.3 Secondary Metabolites Pathways in Plants

Numerous categories of secondary metabolites, such as alkaloids and flavonoids, play diverse biological roles. Alkaloids, in particular, exhibit widespread pharmacological and recreational effects. These secondary metabolites are synthesized as byproducts of various primary metabolite biosynthetic pathways. The phenylpropanoid pathway in plants produces phenolic chemicals, which encompass a range of structural components and contribute to the vivid colors of flowers, fruits, and vegetables. Nitrogen-based organic compounds are derived from the shikimic acid pathway and consist of a heterocyclic ring of aromatic amino acids (Maeda and Dudareva 2012). Tryptophan and phenylalanine, the aromatic amino acids, serve as common metabolic precursors for both phenolic compounds and alkaloids (Twaij and Hasan 2022). The subsequent section outlines several significant biosynthetic pathways associated with plant secondary metabolites.

18.3.1 Shikimic Acid Pathway

According to Herrmann and Weaver (1999) and Maeda and Dudareva (2012), the shikimic acid pathway involves a series of enzymatic reactions that lead to the production of shikimic acid. This compound serves as a crucial precursor for the biosynthesis of aromatic amino acids, including phenylalanine, tyrosine, and tryptophan. The pathway begins with the conversion of erythrose-4-phosphate and phosphoenolpyruvate (PEP) into shikimic acid, also known as 3-deoxy-D-arabinoheptulosonate-7-phosphate (DAHP), through several metabolic transformations. Subsequently, DAHP is further metabolized into chorismate, which plays a pivotal role in the synthesis of phenylalanine, tyrosine, and tryptophan.

18.3.2 Terpene Biosynthesis Pathway

Terpene biosynthesis in plants constitutes a multifaceted pathway leading to the production of a diverse array of terpenoid compounds. The process commences with the generation of two essential isoprenoid precursors, isopentenyl diphosphate (IPP) and dimethylallyl diphosphate (DMAPP), which are biosynthesized through distinct routes. The mevalonate pathway (MVA) operates within the cytoplasm of plant cells, and the enzyme 3-hydroxy-3-methylglutaryl-CoA reductase (HMGR) facilitates the conversion of HMG-CoA to mevalonate (Rohmer and Rohmer 1999). In contrast, the 2-C-methyl-D-erythritol 4-phosphate pathway (MEP) occurs in plastids, with DMAPP being generated through the catalytic action of enzymes 1-deoxy-D-xylulose 5-phosphate synthase (DXS) and 1-deoxy-D-xylulose 5-phosphate reductoisomerase (DXR) (Eisenreich et al. 1998). Subsequent to the formation of these precursors, an assortment of linear or cyclic terpenoid compounds, including monoterpenes, diterpenes, tetraterpenes, and sesquiterpenes, is synthesized within plant cells, facilitated by various critical enzymes such as prenyltransferases (Davis and Croteau 2000; Tholl 2015).

18.3.3 Phenylpropanoid Pathway

Phenolic compounds, including lignin, flavonoids, tannins, and phenolic acids, are synthesized through the phenylpropanoid pathway (Vogt 2010). This metabolic route commences with the conversion of phenylalanine to cinnamic acid facilitated by the enzyme phenylalanine ammonia-lyase (PAL). Various enzymatic processes then facilitate the transformation of cinnamic acid into diverse phenolic compounds, such as sinapic acid, ferulic acid, caffeic acid, and *p*-coumaric acid. Additionally, the phenylpropanoid pathway gives rise to flavonoids (Winkel-Shirley 2002) and lignin (Boerjan et al. 2003), serving critical functions in plant pigmentation, shielding against UV radiation, and providing structural support to cells.

18.3.4 Alkaloid Biosynthesis Pathway

According to Dey et al. (2020), certain alkaloids are synthesized from common amino acid precursors such as tryptophan, tyrosine, or lysine. These amino acids serve as the building blocks for the formation of diverse alkaloids through a series of enzymatic reactions. For instance, nicotine and tropane are alkaloids that originate from the transformation of tryptophan. Similarly, tryptophan contributes to the synthesis of indole alkaloids like ajmalicine and vincristine, involving intricate enzymatic pathways. Furthermore, tyrosine is involved in the production of dopamine and ephedrine, as highlighted by Dey et al. (2020). Additionally, ornithine or arginine precursors play a role in the biosynthesis of pyrrolizidine alkaloids, as described in the same study.

18.3.5 Glucosinolate Biosynthesis Pathway

Glucosinolates represent crucial sulfur-containing secondary metabolites in plants, synthesized through multiple enzymatic steps from diverse amino acid precursors. These compounds can be classified into different groups, including aliphatic, indole, or aromatic types. For instance, sinigrin, classified as an aliphatic glucosinolate, originates from amino acids such as methionine and cysteine (Kliebenstein 2008). Tryptophan acts as a precursor for the production of indole glucosinolates like glucobrassicin (Mikkelsen et al. 2000). On the other hand, aromatic glucosinolates such as gluconasturtiin are derived from aromatic amino acids like phenylalanine and tyrosine (Textor et al. 2004).

18.3.6 Glycoside Biosynthesis Pathway

Glycosides represent a diverse class of plant secondary metabolites, encompassing compounds such as flavonoids, alkaloids, and cyanogenic substances. Their synthesis involves the utilization of sugar nucleotides, for example, UDP-glucose or UDP-rhamnose, as suppliers of activated sugars during the glycosylation process. The production of these sugar nucleotides relies on enzymatic reactions using precursor monosaccharides (Schnurr et al. 2006). Aglycones, the specific targets for glycosylation, may undergo modifications or activation steps to prepare them for the glycosylation process. The transfer of sugar moieties from activated sugar donors to the aglycones is facilitated by glycosyltransferase enzymes, resulting in the formation of glycosides. Subsequent to the initial glycosylation event (Reily et al. 2019), glycosides can undergo further modifications such as acylation or methylation, leading to the production of a diverse range of structurally distinct compounds (Alseekh et al. 2020).

18.4 Mining of Omics Data

Omics technologies, including genomics, transcriptomics, proteomics, and metabolomics, have revolutionized biological research. They provide comprehensive insights into the genetic, transcriptional, and metabolic processes within organisms. When applied to secondary metabolite research, these data sources offer valuable information about the biosynthesis, regulation, and distribution of these compounds.

18.4.1 Genomics

Elucidating the genetic underpinnings of secondary metabolite production in plants entails the exploration of genomic data for secondary metabolite identification. Analyzing genomic data facilitates the discovery of the underlying genes, enzymes, and regulatory elements implicated in secondary metabolite biosynthesis. A concise overview of the process for mining plant secondary metabolites through genomic data analysis is presented below.

18.4.1.1 Genomic Data Retrieval

In the exploration of secondary metabolites in a plant, the genomic sequence of the specific plant species is frequently accessible in publicly accessible databases. Notably, the following repositories serve as valuable sources for obtaining genomic sequences:

- The National Center for Biotechnology Information's GenBank (https://www.ncbi.nlm.nih.gov/) serves as a comprehensive reservoir housing genomic sequences of diverse organisms, including various plant species.
- Ensembl Plants (https://plants.ensembl.org/index.html) offers researchers access to annotated genomes and associated information for numerous plant species.
- Phytozome (https://phytozome-next.jgi.doe.gov/) stands as an essential resource for comparative plant genomics, providing a wealth of genomic data pertaining to a diverse array of plant species.

18.4.1.2 Genome Annotation

Genome annotation refers to the identification and characterization of various elements within a genome, encompassing regulatory regions, noncoding sequences, and the localization of genes potentially implicated in the synthesis of these compounds. The process involves the following key steps:

- **Gene prediction tools**: Utilizing specialized tools, such as AUGUSTUS (Stanke and Morgenstern 2005), GeneMark (Lukashin and Borodovsky 1998), and MAKER (Cantarel et al. 2008), enables the prediction of gene locations and structures within the genomic sequence. These tools employ algorithms that consider diverse factors, including coding regions, exon–intron boundaries, and protein-coding potential, to pinpoint potential gene candidates.
- **Functional annotation**: Once genes are predicted, the subsequent phase involves assigning functions to these genes. This is accomplished by comparing the predicted gene sequences with known protein databases using the widely employed tool, BLAST (Basic Local Alignment Search Tool) (https://blast.ncbi.nlm.nih.gov/Blast.cgi). By identifying similarities to known genes, the functions of the newly identified genes can be inferred.
- **Identification of potential secondary metabolite genes**: Within the identified genes, a critical task is identifying those genes potentially linked to the biosynthesis of secondary metabolites. These may include genes encoding enzymes such as polyketide synthases (PKS), terpene synthases (TPS) (Wawrzyn et al. 2015), and other enzymes known to be associated with secondary metabolite production.

- **Gene clusters**: Secondary metabolite genes often exhibit clustering within the genome, where these clusters comprise multiple genes involved in the synthesis, regulation, and transport of secondary metabolites. Detecting and characterizing these clusters constitute a pivotal aspect of genome annotation, particularly in the context of secondary metabolite exploration.

18.4.1.3 Comparative Genomics

Analysis of comparative genomics among closely related plant species can uncover shared secondary metabolite biosynthesis genes and gene clusters, offering valuable insights into the evolutionary trajectories of these pathways.

- Opt for one or more intimately related plant species to the focal species. Such closely related species are more inclined to exhibit similar secondary metabolite biosynthesis pathways.
- Access the genomic data of the chosen species from openly accessible databases or sequencing initiatives, ensuring the data's comprehensive annotation.
- Perform orthology analysis to pinpoint orthologous genes between the target species and the closely related counterparts. Orthologs, derived from a common ancestor, generally retain analogous functionalities across different species.
- Employ comparable tools and methodologies for genome annotation of the closely related species, ensuring consistency in gene prediction and annotation, similar to those employed for the target species.
- Leverage tools like MultiGeneBlast (Medema et al. 2013), OrthoFinder (Emms and Kelly 2015), or BLASTp to discern conserved gene clusters linked to secondary metabolite biosynthesis. These tools facilitate the identification of gene groups prevalent in both the target and related species, indicating potential involvement in analogous metabolic pathways.
- Functionally annotate the identified genes and clusters utilizing tools like InterProScan (Jones et al. 2014) or by conducting BLAST searches against established protein databases. This process aids in attributing putative functions to the conserved genes.
- Construct phylogenetic trees for the identified conserved genes to analyze their evolutionary relationships. Phylogenetic scrutiny provides insights into the extent of divergence or conservation exhibited by these genes across different evolutionary timeframes.

18.4.1.4 Differential Expression Analysis

Differential analysis of gene expression is crucial for identifying genes that demonstrate significant upregulation or downregulation under specific experimental conditions compared to control conditions. Commonly employed tools for this purpose include:

- **DESeq2**: This extensively utilized R package facilitates the analysis of differential gene expression from RNA-Seq data (Love et al., 2014).
- **EdgeR**: Another R package designed for differential expression analysis, particularly suitable for scenarios involving small sample sizes or rare events (Robinson et al. 2010).
- **limma-voom**: This package is well-suited for analyzing both microarray and RNA-Seq data, accommodating diverse experimental designs (Costa-Silva et al. 2017).

Additionally, clustering genes based on their expression profiles can aid in identifying groups of genes exhibiting similar expression patterns across various samples or experimental conditions, potentially associated with similar biological processes, including secondary metabolite biosynthesis.

18.4.1.5 Functional Annotation

Annotate the differentially expressed genes (DEGs) based on their functions and domains using tools such as InterProScan or Blast2GO. This process aids in attributing potential functions to the DEGs.

- Perform functional enrichment analysis on the DEGs to detect overrepresented gene ontology (GO) terms (Du Plessis et al. 2011) or KEGG pathways (Kanehisa and Goto 2000). This facilitates the comprehension of the biological functions and pathways associated with the identified genes.
- Choose genes that demonstrate significant differential expression and hold biological relevance to secondary metabolite biosynthesis. Strong candidates are those with heightened expression levels in tissues or conditions conducive to secondary metabolite production.
- Substantiate the role of candidate genes through experimental methods, such as gene knockout, overexpression, or other functional assays, to confirm their participation in secondary metabolite biosynthesis.

18.4.1.6 Co-expression Analysis

Conduct a co-expression network analysis to pinpoint gene clusters linked with the DEGs of interest. This approach aids in unraveling potential regulatory networks and gene clusters implicated in the biosynthesis of secondary metabolites. Weighted gene co-expression network analysis (WGCNA) by Li et al. (2018) is a frequently employed tool for this specific analysis.

18.4.2 Transcriptomic Data

Mining of secondary metabolites in plants using transcriptomic data involves analyzing the RNA expression profiles of genes to identify those involved in secondary metabolite biosynthesis. Transcriptomics provides insights into which genes are actively transcribed under different conditions, tissues, or developmental stages, allowing researchers to link gene expression to secondary metabolite production. Here's a step-by-step guide on how to mine secondary metabolites in plants using transcriptomic data.

18.4.2.1 Data Acquisition

RNA sequencing (RNA-Seq) data is commonly used for transcriptomic analysis and can be retrieved from public databases, such as the Gene Expression Omnibus (GEO) or Sequence Read Archive (SRA).

18.4.2.2 Data Preprocessing

Preprocess the transcriptomic data to clean, filter, and normalize it. This step is essential to remove noise and ensure data quality. It includes quality control, trimming adapter sequences, and filtering out low-quality reads. Tools like FastQC (https://www.bioinformatics.babraham.ac.uk/index.html) and Trimmomatic (Bolger et al. 2014) are often used for these tasks.

18.4.2.3 RNA-Seq Data Alignment

The first step in transcriptomic annotation is to align or map the RNA-Seq reads to a reference genome or, in the absence of a reference genome, perform a *de novo* assembly. The goal is to determine the precise locations of RNA transcripts within the genome.

- **Reference genome alignment:** If a well-annotated reference genome for plant species is available, map the RNA-Seq reads to the reference genome using bioinformatics tools. This process associates reads with specific genomic coordinates, enabling the identification of exons, introns, and splice junctions.
- ***De novo* assembly**: In cases where there is no reference genome or for nonmodel organisms, *de novo* assembly tools like Trinity or SOAPdenovo-Trans can be used to assemble RNA-Seq reads into contigs, transcripts, and potentially full-length genes. This assembly step generates a set of potential transcripts based solely on the RNA-Seq data.
- **Transcript assembly:** After alignment or de novo assembly, the next step is to assemble the aligned or assembled reads into transcript models. This involves identifying splice junctions, exons, and introns to reconstruct complete transcripts.

18.4.2.4 Annotation of Transcriptomic Data

Annotation of transcriptomic data is a crucial step in analyzing RNA-Seq data to identify and characterize genes and their isoforms. This annotation process helps to understand which genes are being expressed and how they contribute to various biological processes, including secondary metabolite biosynthesis. Some commonly used tools and pipelines for gene annotation include:

- **StringTie**: StringTie is a popular tool that estimates transcript abundances and generates annotation files based on the assembled transcripts. It can provide information on gene expression levels and isoform variation (Pertea et al. 2015).
- **Cufflinks/Cuffmerge/Cuffcompare**: The Cufflinks suite is a set of tools that aids in transcriptome assembly, merging of assemblies from multiple samples, and comparison with reference annotations. Cuffcompare helps assess the quality of transcript assembly (Ghosh and Chan 2016).
- **HISAT2-StringTie-Ballgown pipeline**: This pipeline combines HISAT2 for alignment and StringTie for transcript assembly, followed by the Ballgown package for visualization and analysis of differential expression (Liu et al. 2022).

Follow steps 4–6 as mentioned in the genomic analysis protocol.

18.4.3 Proteomics Data

Mining proteomics data for secondary metabolites involves a systematic analysis of protein expression and functional information to identify proteins associated with secondary metabolite biosynthesis, regulation, and functions.

18.4.3.1 Data Collection

Data collection in proteomics involves obtaining relevant proteomics datasets that contain information about protein abundances, posttranslational modifications, and protein–protein interactions. Mass spectrometry such as liquid chromatography–mass spectrometry (LC-MS/MS) or matrix-assisted laser desorption/ionization (MALDI-MS) is the primary technology used to generate proteomics data. It allows for the identification and quantification of proteins in biological samples.

18.4.3.2 Data Repositories and Format Conversion

- Retrieve proteomics datasets from proteomics data repositories and databases that host publicly available. These resources may contain data from previous studies related to secondary metabolites. Examples of proteomics data repositories include PRIDE (Jones et al. 2006), PeptideAtlas (Deutsch et al. 2008), and ProteomeXchange (Vizcaíno et al. 2014).
- Convert raw mass spectrometry data files into standardized formats such as mzML or mzXML for compatibility with proteomics analysis software.

18.4.3.3 Quality Control

Evaluate the quality of the proteomics data by checking for issues like data completeness, mass spectrometry quality, and sample integrity. MaxQuant (Tyanova et al. 2016) provides quality control metrics and visualization tools to assess the quality of mass spectrometry data. Skyline is a software tool that allows for visual inspection of mass spectrometry data to identify issues or anomalies (Adams et al. 2020).

18.4.3.4 Data Preprocessing

Normalize protein abundance values to account for variations in protein abundance across samples, ensuring that meaningful comparisons can be made. Proteome Discoverer (Orsburn 2021) includes built-in normalization methods for label-free quantification (LFQ) data. MSstats is an R package (Choi et al. 2014) for label-free quantitative proteomics that includes normalization functions.

18.4.3.5 Protein Identification and Quantification

- Protein identification tools such as MaxQuant, Proteome Discoverer, or Mascot are commonly used.
- Quantify protein abundance using LFQ or isotope labeling techniques like SILAC (stable isotope labeling by amino acids in cell culture) or iTRAQ (isobaric tags for relative and absolute quantitation).

18.4.3.6 Differential Expression Analysis

Differential expression analysis in proteomics involves comparing protein expression levels between different experimental conditions to identify proteins that are significantly upregulated or downregulated under conditions relevant to secondary metabolite production.

- Perform t-tests or ANOVA to compare protein expression levels between different conditions.
- Allows for the calculation of p-values and false discovery rates (FDR) to assess the significance of differential expression.

18.4.4 Metabolomics Data

Mining metabolomic data for secondary metabolites involves the systematic analysis of metabolites present in biological samples to identify those associated with secondary metabolite biosynthesis and regulation. Here's a step-by-step guide on how to conduct such mining:

18.4.4.1 Data Acquisition

- **Metabolomic data**: Obtain metabolomic data through techniques such as liquid chromatography–mass spectrometry (LC-MS), gas chromatography–mass spectrometry (GC-MS), or nuclear magnetic resonance (NMR) spectroscopy.
- **Sample collection**: Collect samples from conditions or tissues where secondary metabolite production is expected or has been observed. This could involve different growth stages, environmental stressors, or genetic modifications.
- **Public databases**: Publicly available metabolomics databases and repositories like MetaboLights, Metabolomics Workbench, or the Human Metabolome Database (HMDB) for reference datasets.

18.4.4.2 Feature Identification

- **Metabolite annotation**: Assign putative metabolite identities to the detected features by comparing mass-to-charge ratios (*m/z*), retention times, or spectra to reference databases like HMDB, KEGG, or MassBank.
- **Statistical significance**: Assess the statistical significance of feature identifications to prioritize those most likely associated with secondary metabolite pathways.

18.4.4.3 Differential Analysis

- **Comparative analysis**: Compare metabolite abundances between different experimental conditions, e.g., comparing samples with and without secondary metabolite production.
- **Statistical tests**: Use statistical tests such as *t*-tests, ANOVA, or nonparametric tests to identify metabolites that show significant differences in abundance.

18.4.4.4 Functional Annotation

- **Pathway analysis**: Perform pathway enrichment analysis using tools like MetaboAnalyst (https://www.metaboanalyst.ca/) or MetaboMiner (Xia et al. 2008) to identify metabolic pathways enriched with differentially regulated metabolites.
- **Biosynthetic pathway mapping**: Map metabolites to known biosynthetic pathways related to secondary metabolites using databases like KEGG.
- Create visualizations such as heatmaps, volcano plots, and pathway maps to represent metabolomic data and highlight significant changes in metabolite abundance. Use software like Cytoscape (https://cytoscape.org/) to visualize metabolic pathways and networks.

18.4.5 Multi-omics Integration

The integration of multi-omics data has emerged as a transformative approach in the realm of secondary metabolite identification. By combining genomics, transcriptomics, proteomics, and metabolomics datasets, researchers gain a comprehensive and multidimensional view of the intricate processes governing secondary metabolite biosynthesis. This holistic approach allows for the simultaneous exploration of gene expression patterns, protein-level interactions, and metabolite profiles, offering valuable insights into the regulation, biosynthetic pathways, and distribution of secondary metabolites. Multi-omics integration not only enables the identification of candidate genes and enzymes involved in biosynthesis but also facilitates the elucidation of complex regulatory networks governing secondary metabolite production. Moreover, it aids in the discovery of novel metabolites and their potential bioactivities. The synergy between these omics disciplines enhances the accuracy and efficiency of secondary metabolite identification, paving the way for innovative applications in drug discovery, agriculture, and biotechnology while deepening our understanding of the intricate world of natural bioactive compounds.

18.5 Bioinformatics Tools for Analysis of Secondary Metabolites and Pathways

Over the past decade, the field of metabolomics has produced a vast volume of metabolic data through the utilization of a diverse array of analytical instruments. To address the challenges posed by this abundance of data, various web-based software tools have been developed. These tools are specifically tailored for the purposes of processing raw metabolomics data, conducting data mining and assessment, facilitating data interpretation, performing statistical analyses, and enabling the mathematical modeling of metabolomic networks. Their primary objective is to aid in the identification of secondary metabolites and associated pathways. In the subsequent section, we have classified several widely utilized tools and programs into two distinct categories. The first group pertains to data processing and annotation, while the second group is focused on pathway analysis.

18.5.1 Data Processing and Annotation

Numerous software applications are available for the computational analysis of extensive spectral data generated by diverse analytical instruments in metabolomics. These online tools are utilized for tasks such as raw data processing, valuable information extraction, and integration of metabolite datasets. Typically, the gathered data goes through several processing stages, including baseline shift correction, background noise reduction, peak detection and alignment, and ultimately, mass spectra deconvolution (Table 18.1). Several bioinformatics tools have been specifically developed for data preprocessing, including XCMS (accessible at https://xcmsonline.scripps.edu), METLIN (available at http://metlin.scripps.edu), AMDIS (automated mass spectral deconvolution and identification system), MeltDB, MetaboAnalys, MetAlign, MZmine 2, and AnalyzerPro, tailored for different analytical techniques (Table 18.1). For example, XCMS is an online bioinformatics platform that streamlines the direct upload of raw data, assisting users in data processing and statistical analysis (Tautenhahn et al. 2012). In LC-MS experiments, XCMS has been adapted for automated data transfer, leading to reduced data processing time and improved efficiency within an online framework (Montenegro-Burke et al. 2017). On the other hand, METLIN is an online database widely used in studies related to plant stress response metabolic

Table 18.1 Available bioinformatics and statistical tools for metabolite identification.

Database	Primary function	Data input	Website (URL, Accessed on 23 September 2023)	References
AllCSS	Metabolite prediction and annotation	DTIM-MS TWIM-MS	http://allccs.zhulab.cn/	Zhou et al. (2020)
AMDIS	Data processing	GC-MS	http://www.amdis.net/	Behrends et al. (2011)
BinBase	Metabolite annotation	GC-MS	https://fiehnlab.ucdavis.edu/projects/binbase-setup	Skogerson et al. (2011)
FiehnLib	Metabolic profiling	GC-qTOF-MS	https://fiehnlab.ucdavis.edu/projects/fiehnlib	Kind et al. (2009)
GMDB	Metabolite annotation	MALDI-TOF	https://jcggdb.jp/rcmg/glycodb/Ms_ResultSearch	(Kameyama et al. 2005)
GNPS	Data processing, visualization, and metabolite annotation	GC-MS-EILC-MS	https://gnps.ucsd.edu/ProteoSAFe/static/gnps-splash.jsp	Wang et al. (2016)
KEGG	Metabolic models	—	http://www.genome.jp/kegg/	Aoki and Kanehisa (2005)
KNApSAcK	Metabolite database	FT/ICR-MS	http://kanaya.naist.jp/KNApSAcK/	Hall et al. (2002)
MarVis	Metabolite annotation	LC-MS	http://marvis.gobics.de/	(Kaever et al. 2015)
MassBase	Metabolite annotation	MS	http://webs2.kazusa.or.jp/massbase/	Ara et al. (2021)
MAVEN	Data processing	LC-MS	https://maven.apache.org/	Clasquin et al. (2012)
MeltDB 2.0	Data processing	GC-MS & LC-MS	https://meltdb.cebitec.uni-bielefeld.de	Kessler et al. (2013)
MetaboAnalyst	Statistical analysis	GC-MS & LC-MS	www.metaboanalyst.ca/	Xia et al. (2009)
Metabox	Analysis workflow	MS	https://github.com/kwanjeeraw/metabox	Wanichthanarak et al. (2017)
MetAlign	Data processing and statistical analysis	GC-MS & LC-MS	www.metalign.nl	Lommen and Kools (2012)
MetAssign	Data annotation	LC-MS	http://mzmatch.sourceforge.net/	Daly et al. (2014)
MetFrag	Metabolite annotation	MS	https://ipb-halle.github.io/MetFrag/	Ruttkies et al. (2016)

Table 18.1 (Continued)

Database	Primary function	Data input	Website (URL, Accessed on 23 September 2023)	References
MetiTree	Data annotation	MS	http://www.metitree.nl/	Rojas-Cherto et al. (2012)
METLIN	Metabolite annotation	LC-MS & MS/MS	https://metlin.scripps.edu/	Smith et al. (2005)
MMCD	Metabolite annotation	MS	https://www.g6g-softwaredirectory.com/bio/metabolomics/dbs-kbs/20670-Univ-Madison-WI-MMCD.php	Cui et al. (2008)
Mzcloud	Metabolite annotation	MS/MS & MSn	https://www.mzcloud.org/	Mistrik et al. (2013)
MZmine2	Data processing	LC-MS	http://mzmine.github.io/	Pluskal et al. (2010)
NIST	Metabolite annotation	GC-MS, LC-MS & MS/MS	http://www.nist.gov/srd/nist1a.cfm	Rumble et al. (1992)
PRIMe	Metabolite annotation	GC-MS, LC-MS & CE-MS	http://prime.psc.riken.jp/	Sakurai et al. (2013)
XCMS	Data processing	GC-MS, LC-MS & MS2	https://xcmsonline.scripps.edu	Tautenhahn et al. (2012)

profiling. It proves particularly valuable for the profiling of specific plant metabolites and streamlines the processes of data processing, mining, and annotation (Smith et al. 2005). MeltDB (https://meltdb.cebitec.uni-bielefeld.de) serves as a crucial web-based platform for conducting data evaluation, processing, and statistical analysis in the field of plant metabolomics (Kessler et al. 2013). Moreover, the online platform MetaboAnalyst offers a versatile enrichment analysis tool with various topological and visualization capabilities (Xia and Wishart 2016). The Global Natural Product Social Molecular Networking (GNPS) platform (http://gnps.ucsd.edu) is a web-based mass spectrometry (MS/MS) resource designed for metabolite processing and annotation (Wang et al. 2016). GNPS plays a pivotal role in identifying and uncovering metabolites across the entire data lifecycle, ranging from data acquisition and analysis to post-publication stages (Wang et al. 2016). Lastly, MZmine 2 is a publicly accessible data processing module renowned for its support of high-resolution spectral analysis. It proves suitable for both targeted and nontargeted metabolomic investigations and excels in handling substantial datasets efficiently (Pluskal et al. 2010). Various computational web-based tools, statistical software, and online bioinformatics resources are commonly employed for data analysis in plant metabolomics (Table 18.1).

18.5.2 Pathway Analysis

Pathway analysis strives to merge biochemical knowledge with gathered metabolomics data to pinpoint metabolite patterns that correspond to metabolic pathways (Pérez-Alonso et al. 2018). Metabolic pathways can be perceived as assemblages of metabolites connected by shared biological processes and linked through one or more enzymatic reactions. All-inclusive metabolic pathway databases, such as KEGG (Kanehisa et al. 2017), MetaCyc (Caspi et al. 2007), AraCyc (Mueller et al. 2003), and the small molecule pathway database (SMPDB) (Jewison et al. 2014), encompass a wide array of metabolic pathways (Table 18.2). Diverse software tools, including metabolite set enrichment analysis (MSEA), MPEA, IMPaLA, MBRole, VANTED, MetaboAnalyst, Paintomics, ProMeTra, Metscape2, and MetaMapRR, facilitate statistical and other metabolite enrichment analyses (Table 18.2). MSEA methods can be systematically categorized into over-representation (ORA), single-sample profiling (SSP), and quantitative enrichment analysis (QEA) (Xia and Wishart 2010a). Metscape2 (Karnovsky et al. 2012), an extension of the well-known Cytoscape software (Kohl et al. 2011), enables the visualization of metabolites, genes, and pathways within the context of metabolic networks. Furthermore, platform-independent online resources like Paintomics (García-Alcalde et al. 2011), ProMeTra (Neuweger et al. 2009), and MetaMapRR (Grapov et al. 2015) are also accessible.

Table 18.2 Databases and tools for metabolite enrichment analysis and pathway visualization.

Database	Description	Website (URL, Accessed on 23 Sept 2023)	References
AraCyc	Provides a curated database of metabolic pathways and enzymes in *Arabidopsis thaliana*	https://www.plantcyc.org/typeofpublication/aracyc	Mueller et al. (2003)
Cytoscape	A versatile software for visualizing and analyzing molecular interaction networks	http://www.cytoscape.org/	Kohl et al. (2011)
IMPaLA	Enables the integrative analysis of metabolomic data for pathway enrichment and visualization	http://impala.molgen.mpg.de	Kamburov et al. (2011)
iPath	Facilitates the exploration and visualization of various biological pathways and networks	http://pathways.embl.de/	Letunic et al. (2008)
KEGG	Offers a comprehensive resource for understanding high-level functions and utilities of biological systems	http://www.genome.jp/kegg/	Kanehisa et al. (2017)
MapMan	Allows the user to display and analyze large datasets in the context of biological pathways and processes	http://mapman.gabipd.org/web/guest/mapman	Thimm et al. (2004)

Table 18.2 (Continued)

Database	Description	Website (URL, Accessed on 23 Sept 2023)	References
MBRole	Provides a platform for exploring the roles of metabolites in various biological processes and pathways	http://csbg.cnb.csic.es/mbrole2/	Chagoyen and Pazos (2011)
Metabolonote	A database for storing and sharing information related to metabolomics experiments and data	http://metabolonote.kazusa.or.jp/	Ara et al. (2015)
MetaCrop	Offers information on crop plant metabolism and pathways, aiding in the understanding of plant biochemistry	http://metacrop.ipk-gatersleben.de	Schreiber et al. (2012)
MetaCyc	Provides a comprehensive collection of metabolic pathways and enzymes from various organisms	http://www.metacyc.org	Caspi et al. (2007)
MetPA	Facilitates the pathway analysis of metabolomic data, enabling the identification of significant pathways	-	Xia and Wishart (2010b)
MetaboAnalyst	Allows comprehensive metabolomic data analysis, including statistical, functional, and pathway analysis	http://www.metaboanalyst.ca	Xia and Wishart (2016)
PathwayExplorer	Enables the exploration and visualization of biological pathways and associated molecular interactions	http://genome.tugraz.at/pathwayexplorer/pathwayexplorer_description.shtml	Mlecnik et al. (2005)
SMPDB	Offers a comprehensive, high-quality, and manually curated database of small molecule pathways found in humans	http://www.smpdb.ca	Jewison et al. (2014)
WikiPathways	A collaborative platform that allows the community to curate and share biological pathways in a wiki-based format	http://wikipathways.org	Kelder et al. (2012)

CE-MS refers to capillary electrophoresis-mass spectrometry, DTIM-MS stands for drift tube ion mobility–mass spectrometry, EI is used for electrospray ionization, FTICR-MS denotes Fourier transform ion cyclotron resonance-mass spectrometry, GC-TOF-MS represents gas chromatography–time-of-flight mass spectrometry, GC-MS is for gas chromatography–mass spectrometry, HPLC stands for high-performance liquid chromatography, LC-MS refers to liquid chromatography–mass spectrometry, MALDI-TOF represents matrix-assisted laser desorption/ionization-time of flight, and TWIM-MS is traveling wave ion mobility–mass spectrometry.

18.6 Conclusion

In conclusion, the realm of plant secondary metabolites stands as a vast and intricate landscape, rich with compounds that hold great promise in various fields. Bioinformatics emerges as an indispensable guiding light in our journey to explore and harness the potential of these natural treasures. This chapter has illuminated the multifaceted role of bioinformatics, showcasing its ability to decipher intricate pathways, predict and characterize the key players, and unveil the regulatory mechanisms that govern secondary metabolites in plants. From genomics to transcriptomics, from data mining to integration, bioinformatics has empowered us with a holistic understanding of these bioactive compounds. As we move forward, this powerful toolkit continues to lead us toward innovative applications in medicine, industry, and agriculture. It not only deepens our knowledge but also offers practical solutions for a sustainable and bioactive-rich future.

References

Adams, K.J., Pratt, B., Bose, N. et al. (2020). Skyline for small molecules: a unifying software package for quantitative metabolomics. *Journal of Proteome Research* 19 (4): 1447–1458.

Alseekh, S., de Souza, L.P., Benina, M., and Fernie, A.R. (2020). The style and substance of plant flavonoid decoration; towards defining both structure and function. *Phytochemistry* 174: 112347. https://doi.org/10.1016/j.phytochem.2020.112347.

Aoki, K.F. and Kanehisa, M. (2005). Using the KEGG database resource. *Current Protocols in Bioinformatics* 11 (1): 1–12.

Ara, T., Enomoto, M., Arita, M. et al. (2015). Metabolonote: a wiki-based database for managing hierarchical metadata of metabolome analyses. *Frontiers in Bioengineering and Biotechnology* 3: 38.

Ara, T., Sakurai, N., Suzuki, H. et al. (2021). MassBase: a large-scaled depository of mass spectrometry datasets for metabolome analysis. *Plant Biotechnology* 38 (1): 167–171.

Azizi, P., Osman, M., Hanafi, M.M. et al. (2019). Adaptation of the metabolomics profile of rice after *Pyricularia oryzae* infection. *Plant Physiology and Biochemistry* 144: 466–479. https://doi.org/10.1016/j.plaphy.2019.10.014.

Behrends, V., Tredwell, G.D., and Bundy, J.G. (2011). A software complement to AMDIS for processing GC-MS metabolomic data. *Analytical Biochemistry* 415 (2): 206–208.

Boerjan, W., Ralph, J., and Baucher, M. (2003). Lignin biosynthesis. *Annual Review of Plant Biology* 54 (1): 519–546. https://doi.org/10.1146/annurev.arplant.54.031902.134938.

Bolger, A.M., Lohse, M., and Usadel, B. (2014). Trimmomatic: a flexible trimmer for Illumina sequence data. *Bioinformatics* 30 (15): 2114–2120.

Cantarel, B.L., Korf, I., Robb, S.M. et al. (2008). MAKER: an easy-to-use annotation pipeline designed for emerging model organism genomes. *Genome Research* 18 (1): 188–196.

Caspi, R., Foerster, H., Fulcher, C.A. et al. (2007). The MetaCyc Database of metabolic pathways and enzymes and the BioCyc collection of Pathway/Genome Databases. *Nucleic Acids Research* 36 (suppl_1): D623–D631.

Chagoyen, M. and Pazos, F. (2011). MBRole: enrichment analysis of metabolomic data. *Bioinformatics* 27 (5): 730–731.

Choi, M., Chang, C.Y., Clough, T. et al. (2014). MSstats: an R package for statistical analysis of quantitative mass spectrometry-based proteomic experiments. *Bioinformatics* 30 (17): 2524–2526.

Clasquin, M.F., Melamud, E., and Rabinowitz, J.D. (2012). LC-MS data processing with MAVEN: a metabolomic analysis and visualization engine. *Current Protocols in Bioinformatics* 37 (1): 14–11.

Costa-Silva, J., Domingues, D., and Lopes, F.M. (2017). RNA-Seq differential expression analysis: an extended review and a software tool. *PLoS One* 12 (12): e0190152.

Cui, Q., Lewis, I.A., Hegeman, A.D. et al. (2008). Metabolite identification via the madison metabolomics consortium database. *Nature Biotechnology* 26 (2): 162–164.

Daly, R., Rogers, S., Wandy, J. et al. (2014). MetAssign: probabilistic annotation of metabolites from LC–MS data using a Bayesian clustering approach. *Bioinformatics* 30 (19): 2764–2771.

Davis, E.M. and Croteau, R. (2000). Cyclization enzymes in the biosynthesis of monoterpenes, sesquiterpenes, and diterpenes. In: *Biosynthesis* (ed. F.J. Leeper and J.C. Vederas), 53–95. Berlin, Heidelberg: Springer https://doi.org/10.1007/3-540-48146-X_2.

Deutsch, E.W., Lam, H., and Aebersold, R. (2008). PeptideAtlas: a resource for target selection for emerging targeted proteomics workflows. *EMBO Reports* 9 (5): 429–434.

Dey, P., Kundu, A., Kumar, A. et al. (2020). Analysis of alkaloids (indole alkaloids, isoquinoline alkaloids, tropane alkaloids). In: *Recent Advances in Natural Products Analysis* (ed. A.S. Silva, S.F. Nabavi, M. Saeedi, and S.M. Nabavi), 505–567. Elsevier https://doi.org/10.1016/B978-0-12-816455-6.00015-9.

Du Plessis, L., Škunca, N., and Dessimoz, C. (2011). The what, where, how and why of gene ontology—a primer for bioinformaticians. *Briefings in Bioinformatics* 12 (6): 723–735.

Eisenreich, W., Schwarz, M., Cartayrade, A. et al. (1998). The deoxyxylulose phosphate pathway of terpenoid biosynthesis in plants and microorganisms. *Chemistry & Biology* 5 (9): R221–R233. https://doi.org/10.1016/S1074-5521(98)90002-3.

Emms, D.M. and Kelly, S. (2015). OrthoFinder: solving fundamental biases in whole genome comparisons dramatically improves orthogroup inference accuracy. *Genome Biology* 16 (1): 1–14.

Erharuyi, O., Falodun, A., and Langer, P. (2014). Medicinal uses, phytochemistry and pharmacology of *Picralima nitida* (Apocynaceae) in tropical diseases: a review. *Asian Pacific Journal of Tropical Medicine* 7 (1): 1–8. https://doi.org/10.1016/S1995-7645(13)60182-0.

García-Alcalde, F., García-López, F., Dopazo, J., and Conesa, A. (2011). Paintomics: a web based tool for the joint visualization of transcriptomics and metabolomics data. *Bioinformatics* 27 (1): 137–139.

Ghosh, S. and Chan, C.K.K. (2016). Analysis of RNA-Seq data using TopHat and Cufflinks. In: *Plant Bioinformatics: Methods and Protocols* (ed. D. Edwards), 339–361. New York: Humana.

Grapov, D., Wanichthanarak, K., and Fiehn, O. (2015). MetaMapR: pathway independent metabolomic network analysis incorporating unknowns. *Bioinformatics* 31 (16): 2757–2760.

Gurjar, M.S., Ali, S., Akhtar, M., and Singh, K.S. (2012). Efficacy of plant extracts in plant disease management. *Agricultural Sciences* 3 (3): 425–433. https://doi.org/10.4236/as.2012.33050.

Hall, R., Beale, M., Fiehn, O. et al. (2002). Plant metabolomics: the missing link in functional genomics strategies.

Herrmann, K.M. and Weaver, L.M. (1999). The shikimate pathway. *Annual Review of Plant Physiology and Plant Molecular Biology* 50: 473–503. https://doi.org/10.1146/annurev.arplant.50.1.473.

Huffaker, A., Dafoe, N.J., and Schmelz, E.A. (2011). ZmPep1, an ortholog of Arabidopsis elicitor peptide 1, regulates maize innate immunity and enhances disease resistance. *Plant Physiology* 155 (3): 1325–1338. https://doi.org/10.1104/pp.110.166710.

Jewison, T., Su, Y., Disfany, F.M. et al. (2014). SMPDB 2.0: big improvements to the Small Molecule Pathway Database. *Nucleic Acids Research* 42 (D1): D478–D484.

Jones, P., Côté, R.G., Martens, L. et al. (2006). PRIDE: a public repository of protein and peptide identifications for the proteomics community. *Nucleic Acids Research* 34 (suppl_1): D659–D663.

Jones, P., Binns, D., Chang, H.Y. et al. (2014). InterProScan 5: genome-scale protein function classification. *Bioinformatics* 30 (9): 1236–1240.

Kaever, A., Landesfeind, M., Feussner, K. et al. (2015). MarVis-pathway: integrative and exploratory pathway analysis of non-targeted metabolomics data. *Metabolomics* 11: 764–777.

Kamburov, A., Cavill, R., Ebbels, T.M. et al. (2011). Integrated pathway-level analysis of transcriptomics and metabolomics data with IMPaLA. *Bioinformatics* 27 (20): 2917–2918.

Kameyama, A., Kikuchi, N., Nakaya, S. et al. (2005). A strategy for identification of oligosaccharide structures using observational multistage mass spectral library. *Analytical Chemistry* 77 (15): 4719–4725.

Kanehisa, M. and Goto, S. (2000). KEGG: kyoto encyclopedia of genes and genomes. *Nucleic Acids Research* 28 (1): 27–30.

Kanehisa, M., Furumichi, M., Tanabe, M. et al. (2017). KEGG: new perspectives on genomes, pathways, diseases and drugs. *Nucleic Acids Research* 45 (D1): D353–D361.

Karnovsky, A., Weymouth, T., Hull, T. et al. (2012). Metscape 2 bioinformatics tool for the analysis and visualization of metabolomics and gene expression data. *Bioinformatics* 28 (3): 373–380.

Kelder, T., Van Iersel, M.P., Hanspers, K. et al. (2012). WikiPathways: building research communities on biological pathways. *Nucleic Acids Research* 40 (D1): D1301–D1307.

Kessler, N., Neuweger, H., Bonte, A. et al. (2013). MeltDB 2.0–advances of the metabolomics software system. *Bioinformatics* 29 (19): 2452–2459.

Kind, T., Wohlgemuth, G., Lee, D.Y. et al. (2009). FiehnLib: mass spectral and retention index libraries for metabolomics based on quadrupole and time-of-flight gas chromatography/mass spectrometry. *Analytical Chemistry* 81 (24): 10038–10048.

Kliebenstein, D.J. (2008). A role for gene duplication and natural variation of gene expression in the evolution of metabolism. *PLoS One* 3 (3): e1838. https://doi.org/10.1371/journal.pone.0001838.

Kohl, M., Wiese, S., and Warscheid, B. (2011). Cytoscape: software for visualization and analysis of biological networks. In: *Data Mining in Proteomics: From Standards to Applications* (ed. M. Hamacher, M. Eisenacher, and C. Stephan), 291–303. Totowa, NJ: Humana.

Kumar, S., Korra, T., Thakur, R. et al. (2023). Role of plant secondary metabolites in defence and transcriptional regulation in response to biotic stress. *Plant Stress* 8: 100154. https://doi.org/10.1016/j.stress.2023.100154.

Letunic, I., Yamada, T., Kanehisa, M., and Bork, P. (2008). iPath: interactive exploration of biochemical pathways and networks. *Trends in Biochemical Sciences* 33 (3): 101–103.

Li, J., Zhou, D., Qiu, W. et al. (2018). Application of weighted gene co-expression network analysis for data from paired design. *Scientific Reports* 8 (1): 622.

Liu, X., Zhao, J., Xue, L. et al. (2022). A comparison of transcriptome analysis methods with reference genome. *BMC Genomics* 23 (1): 1–15.

Lommen, A. and Kools, H.J. (2012). MetAlign 3.0: performance enhancement by efficient use of advances in computer hardware. *Metabolomics* 8: 719–726.

Love, M.I., Huber, W., and Anders, S. (2014). Moderated estimation of fold change and dispersion for RNA-seq data with DESeq2. *Genome Biology* 15 (12): 1–21.

Lukashin, A.V. and Borodovsky, M. (1998). GeneMark. hmm: new solutions for gene finding. *Nucleic Acids Research* 26 (4): 1107–1115.

Maeda, H. and Dudareva, N. (2012). The shikimate pathway and aromatic amino acid biosynthesis in plants. *Annual Review of Plant Biology* 63: 73–105. https://doi.org/10.1146/annurev-arplant-042811-105439.

Medema, M.H., Takano, E., and Breitling, R. (2013). Detecting sequence homology at the gene cluster level with MultiGeneBlast. *Molecular Biology and Evolution* 30 (5): 1218–1223.

Mikkelsen, M.D. et al. (2000). Cytochrome P450 CYP79B2 from Arabidopsis catalyzes the conversion of tryptophan to indole-3-acetaldoxime, a precursor of indole glucosinolates and indole-3-acetic acid. *Journal of Biological Chemistry* 275 (43): 33712–33717. https://doi.org/10.1074/jbc.M001667200.

Mistrik, R., Lutisan, J., Huang, Y. et al. (2013). mzCloud: a key conceptual shift to understand 'Who's who'in untargeted metabolomics. *A Key Conceptual Shift to Understand'Who's Who'in Untargeted Metabolomics. Metabolomics Society 2013 Conference*, Glasgow.

Mlecnik, B., Scheideler, M., Hackl, H. et al. (2005). PathwayExplorer: web service for visualizing high-throughput expression data on biological pathways. *Nucleic Acids Research* 33 (suppl_2): W633–W637.

Montenegro-Burke, J.R., Aisporna, A.E., Benton, H.P. et al. (2017). Data streaming for metabolomics: accelerating data processing and analysis from days to minutes. *Analytical Chemistry* 89 (2): 1254–1259.

Mueller, L.A., Zhang, P., and Rhee, S.Y. (2003). AraCyc: a biochemical pathway database for Arabidopsis. *Plant Physiology* 132 (2): 453–460.

Neuweger, H., Persicke, M., Albaum, S.P. et al. (2009). Visualizing post genomics data-sets on customized pathway maps by ProMeTra–aeration-dependent gene expression and metabolism of *Corynebacterium glutamicum* as an example. *BMC Systems Biology* 3 (1): 1–14.

Orsburn, B.C. (2021). Proteome discoverer—a community enhanced data processing suite for protein informatics. *Proteomes* 9 (1): 15.

Pérez-Alonso, M.M., Carrasco-Loba, V., and Pollmann, S. (2018). Advances in plant metabolomics. *Annual Plant Reviews Online* 1 (2): 557–588.

Pertea, M., Pertea, G.M., Antonescu, C.M. et al. (2015). StringTie enables improved reconstruction of a transcriptome from RNA-seq reads. *Nature Biotechnology* 33 (3): 290–295.

Pluskal, T., Castillo, S., Villar-Briones, A., and Orešič, M. (2010). MZmine 2: modular framework for processing, visualizing, and analyzing mass spectrometry-based molecular profile data. *BMC Bioinformatics* 11 (1): 1–11.

Reichling, J. (2018). Plant–microbe interactions and secondary metabolites with antibacterial, antifungal and antiviral properties. In: *Annual Plant Reviews Online* (ed. M. Wink), 214–347. Wiley https://doi.org/10.1002/9781119312994.apr0420.

Reily, C., Stewart, T.J., Renfrow, M.B., and Novak, J. (2019). Glycosylation in health and disease. *Nature Reviews Nephrology* 15 (6): 346–366. https://doi.org/10.1038/s41581-019-0129-4.

Robinson, M.D., McCarthy, D.J., and Smyth, G.K. (2010). edgeR: a Bioconductor package for differential expression analysis of digital gene expression data. *Bioinformatics* 26 (1): 139–140.

Rohmer, M. and Rohmer, M. (1999). The discovery of a mevalonate-independent pathway for isoprenoid biosynthesis in bacteria, algae and higher plants. *Natural Product Reports* 16 (5): 565–574. https://doi.org/10.1039/a709175c.

Rojas-Cherto, M., van Vliet, M., Peironcely, J.E. et al. (2012). MetiTree: a web application to organize and process high-resolution multi-stage mass spectrometry metabolomics data. *Bioinformatics* 28 (20): 2707–2709.

Rumble, J.R. Jr., Bickham, D.M., and Powell, C.J. (1992). The NIST X-ray photoelectron spectroscopy database. *Surface and Interface Analysis* 19 (1–12): 241–246.

Ruttkies, C., Schymanski, E.L., Wolf, S. et al. (2016). MetFrag relaunched: incorporating strategies beyond *in silico* fragmentation. *Journal of Cheminformatics* 8 (1): 1–16.

Sakurai, T., Yamada, Y., Sawada, Y. et al. (2013). PRIMe update: innovative content for plant metabolomics and integration of gene expression and metabolite accumulation. *Plant and Cell Physiology* 54 (2): e5.

Satish, L., Shamili, S., Yolcu, S. et al. (2020). Biosynthesis of secondary metabolites in plants as influenced by different factors. In: *Plant-derived Bioactives* (ed. M.K. Swamy), 61–100. Singapore: Springer Singapore https://doi.org/10.1007/978-981-15-1761-7_3.

Schmelz, E.A., Kaplan, F., Huffaker, A. et al. (2011). Identity, regulation, and activity of inducible diterpenoid phytoalexins in maize. *Proceedings of the National Academy of Sciences of the United States of America* 108 (13): 5455–5460. https://doi.org/10.1073/pnas.1014714108.

Schnurr, J.A., Storey, K.K., Jung, H.J.G. et al. (2006). UDP-sugar pyrophosphorylase is essential for pollen development in Arabidopsis. *Planta* 224 (3): 520–532. https://doi.org/10.1007/s00425-006-0240-1.

Schreiber, F., Colmsee, C., Czauderna, T. et al. (2012). MetaCrop 2.0: managing and exploring information about crop plant metabolism. *Nucleic Acids Research* 40 (D1): D1173–D1177.

Shen, E., Zou, J., Hubertus Behrens, F. et al. (2015). Identification, evolution, and expression partitioning of miRNAs in allopolyploid *Brassica napus*. *Journal of Experimental Botany* 66 (22): 7241–7253. https://doi.org/10.1093/jxb/erv420.

Skogerson, K., Wohlgemuth, G., Barupal, D.K., and Fiehn, O. (2011). The volatile compound BinBase mass spectral database. *BMC Bioinformatics* 12: 1–15.

Smith, C.A., O'Maille, G., Want, E.J. et al. (2005). METLIN: a metabolite mass spectral database. *Therapeutic Drug Monitoring* 27 (6): 747–751.

Stanke, M. and Morgenstern, B. (2005). AUGUSTUS: a web server for gene prediction in eukaryotes that allows user-defined constraints. *Nucleic Acids Research* 33 (suppl_2): W465–W467.

Tautenhahn, R., Patti, G.J., Rinehart, D., and Siuzdak, G. (2012). XCMS online: a web-based platform to process untargeted metabolomic data. *Analytical Chemistry* 84 (11): 5035–5039.

Textor, S., Bartram, S., Kroymann, J. et al. (2004). Biosynthesis of methionine-derived glucosinolates in *Arabidopsis thaliana*: recombinant expression and characterization of methylthioalkylmalate synthase, the condensing enzyme of the chain-elongation cycle. *Planta* 218 (6): 1026–1035. https://doi.org/10.1007/s00425-003-1184-3.

Thimm, O., Bläsing, O., Gibon, Y. et al. (2004). MAPMAN: a user-driven tool to display genomics data sets onto diagrams of metabolic pathways and other biological processes. *The Plant Journal* 37 (6): 914–939.

Tholl, D. (2015). Biosynthesis and biological functions of terpenoids in plants. In: *Biotechnology of Isoprenoids* (ed. J. Schrader and J. Bohlmann), 63–106. Cham: Springer https://doi.org/10.1007/10_2014_295.

Twaij, B.M. and Hasan, M.N. (2022). Bioactive secondary metabolites from plant sources: types, synthesis, and their therapeutic uses. *International Journal of Plant Biology* 13 (1): 4 14. https://doi.org/10.3390/ijpb13010003.

Tyanova, S., Temu, T., and Cox, J. (2016). The MaxQuant computational platform for mass spectrometry-based shotgun proteomics. *Nature Protocols* 11 (12): 2301–2319.

Uche-Okereafor, N. et al. (2019). Antibacterial activities of crude secondary metabolite extracts from pantoea species obtained from the stem of *Solanum mauritianum* and their effects on two cancer cell lines. *International Journal of Environmental Research and Public Health* 16 (4): 602. https://doi.org/10.3390/ijerph16040602.

Velasco, P., Lema, M., Francisco, M. et al. (2013). *In vivo* and *in vitro* effects of secondary metabolites against *Xanthomonas campestris* pv. *campestris*. *Molecules (Basel, Switzerland)* 18 (9): 11131–11143. https://doi.org/10.3390/molecules180911131.

Vizcaíno, J.A., Deutsch, E.W., Wang, R. et al. (2014). ProteomeXchange provides globally coordinated proteomics data submission and dissemination. *Nature Biotechnology* 32 (3): 223–226.

Vogt, T. (2010). Phenylpropanoid biosynthesis. *Molecular Plant* 3 (1): 2–20. https://doi.org/10.1093/mp/ssp106.

Wang, M., Carver, J.J., Phelan, V.V. et al. (2016). Sharing and community curation of mass spectrometry data with Global Natural Products Social Molecular Networking. *Nature Biotechnology* 34 (8): 828–837.

Wanichthanarak, K., Fan, S., Grapov, D. et al. (2017). Metabox: a toolbox for metabolomic data analysis, interpretation and integrative exploration. *PLoS One* 12 (1): e0171046.

Wawrzyn, G.T., Held, M.A., Bloch, S.E., and Schmidt-Dannert, C. (2015). Genome mining for fungal secondary metabolic gene clusters. In: *Biosynthesis and Molecular Genetics of Fungal Secondary Metabolites*, vol. 2 (ed. S. Zeilinger, J.-F. Martín, and C. García-Estrada), 43–65. New York: Springer.

Winkel-Shirley, B. (2002). Biosynthesis of flavonoids and effects of stress. *Current Opinion in Plant Biology* 5 (3): 218–223. https://doi.org/10.1016/S1369-5266(02)00256-X.

Xia, J. and Wishart, D.S. (2010a). MSEA: a web-based tool to identify biologically meaningful patterns in quantitative metabolomic data. *Nucleic Acids Research* 38 (suppl_2): W71–W77.

Xia, J. and Wishart, D.S. (2010b). MetPA: a web-based metabolomics tool for pathway analysis and visualization. *Bioinformatics* 26 (18): 2342–2344.

Xia, J. and Wishart, D.S. (2016). Using MetaboAnalyst 3.0 for comprehensive metabolomics data analysis. *Current Protocols in Bioinformatics* 55 (1): 14–10.

Xia, J., Bjorndahl, T.C., Tang, P., and Wishart, D.S. (2008). MetaboMiner–semi-automated identification of metabolites from 2D NMR spectra of complex biofluids. *BMC Bioinformatics* 9 (1): 1–16.

Xia, J., Psychogios, N., Young, N., and Wishart, D.S. (2009). MetaboAnalyst: a web server for metabolomic data analysis and interpretation. *Nucleic Acids Research* 37 (suppl_2): W652–W660.

Zaynab, M., Fatima, M., Abbas, S. et al. (2018). Role of secondary metabolites in plant defense against pathogens. *Microbial Pathogenesis* 124: 198–202. https://doi.org/10.1016/j.micpath.2018.08.034.

Zhou, Z., Luo, M., Chen, X. et al. (2020). Ion mobility collision cross-section atlas for known and unknown metabolite annotation in untargeted metabolomics. *Nature Communications* 11 (1): 4334.

19

Understanding Plant Secondary Metabolism Using Bioinformatics Tools: Recent Advances and Prospects

Dola Mukherjee[1] *and Ashutosh Mukherjee*[2]

[1] *Amity Institute of Biotechnology, Amity University, Kolkata, West Bengal, India*
[2] *Department of Botany, Vivekananda College, Kolkata, West Bengal, India*

19.1 Introduction

Secondary or specialized metabolites in plants have the potential to treat various infections, health disorders, and illnesses. Although they are often being replaced by other synthetic drugs, these natural products from higher plants are major sources of different industries such as pharmaceutical, agrochemical, flavor, and aroma. Medicinal and aromatic plants have contributed to almost 80% world's population to achieve their primary health care requirements (Laloo et al. 2021). The therapeutic properties of these plants are due to strong phyto-constituents (Zengin et al. 2022). Secondary metabolites, produced from primary metabolites, are created by biosynthetic modifications such as glycosylation, methylation, and hydroxylation. Secondary metabolites are not directly involved in plant growth and development. Compared to primary metabolites, secondary metabolites are structurally more complex (Twaij and Hasan 2022). In plants, secondary metabolites play different functions such as mediating relationships between organisms, response to environmental challenges, and protecting plants against herbivores and pest attacks and infections (Divekar et al. 2022). Additionally, plant microbiomes participate in many of the above-mentioned processes either directly or indirectly by influencing the metabolism in the plant system. According to reports, plants can affect their microbiome by secreting diverse metabolites, and the microbiome can affect the host plant's metabolome (Pang et al. 2021). Although plant secondary metabolites are abundant, they can be broadly classified as (i) phenolic compounds which are composed of simple sugars and benzene rings, (ii) terpenoids, (iii) sulfur-containing compounds, and (iv) nitrogen-containing compounds (Barut et al. 2022; Twaij and Hasan 2022) (Figure 19.1).

A brief description of various secondary metabolites is given below.

i) **Phenolics**: Only plants and microbes can utilize phenylpropanoids, which are derived from the shikimate pathway and produce important aromatic amino acids like phenylalanine and tyrosine. Plants form a component of the food chain so that humans can

Bioinformatics for Plant Research and Crop Breeding, First Edition. Edited by Jen-Tsung Chen.

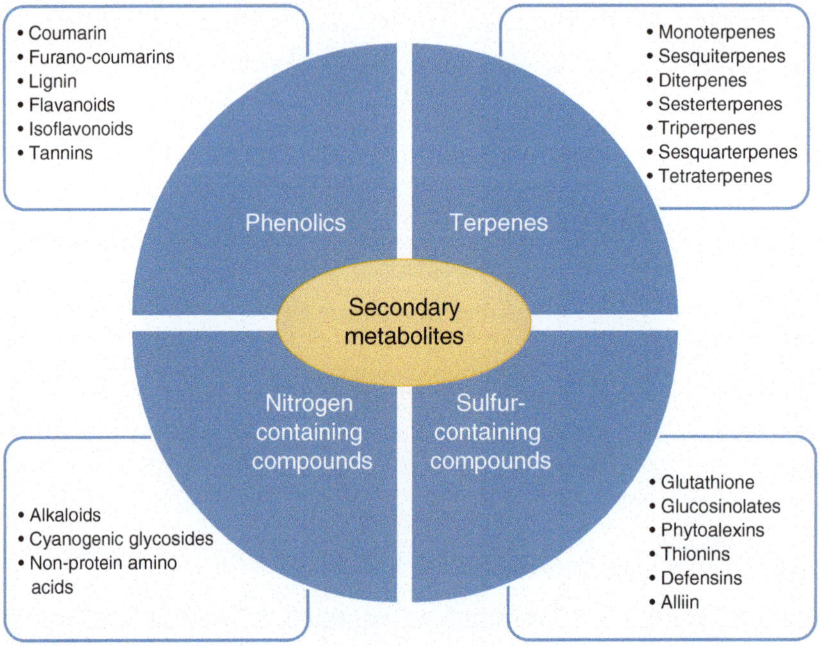

Figure 19.1 Major types of secondary metabolites.

receive vital amino acids as they are mostly needed by animals and humans (Twaij and Hasan 2022). Based on the chemical structures, phenolic compounds can further be divided into subgroups such as coumarin, furanocoumarins, flavonoids, and isoflavonoids (Jan et al. 2021). Soluble phenolic compounds are typically synthesized in the endoplasmic reticulum and then stored in vacuoles. Soluble phenolic compounds conjugate with the molecules of the cell wall through ester and glycosidic bonds to form bound phenolic compounds (Jan et al. 2021).

ii) **Terpenes**: With at least 50,000 distinct terpenes identified thus far, this is a structurally varied and sizable and lipid soluble metabolite group (Jan et al. 2021). Terpenes are easily recognized since they are made up of isoprene units that can be modified by cyclization processes. The quantity of isoprene units in their carbon skeleton determines the classification into several groups. Terpenoids are metabolites that have mutagenic, anxiolytic, anti-tubercular, and anticancer properties (Garcia-Brugger et al. 2006).

iii) **Sulfur-containing compounds**: These are relatively a small group of plant secondary metabolites with about 200 metabolites (Jan et al. 2021). These include the familiar glucosinolates and their breakdown products, including thiocyanates, isothiocyanates, epithionitriles, and oxazolidinethiones (Venditti and Bianco 2020). Allinin, glycosphingolipid, glutathione, some phytoalexins, defensins, and thionins are some of the sulfur-containing secondary metabolites that take part in the plant defense system (Mazid et al. 2011; Jan et al. 2021).

iv) **Nitrogen-containing compounds**: Alkaloids, cyanogenic glycosides and non-protein amino acids are the main categories of N-containing compounds (Twaij and Hasan 2022). Of these, alkaloids contain basic nitrogen atoms and are structurally diverse (Barut et al. 2022). Since they are usually soluble in aqueous solutions, they can easily be extracted in water upon protonating the nitrogen (Twaij and Hasan 2022). These substances include some of the most well-known substances, including caffeine, nicotine, cocaine, and morphine (Heinrich et al. 2021). These substances are recognized for their hallucinogenic, analgesic, and anxiolytic properties and frequently affect the central nervous system physiologically (Dewick 2002; Twaij and Hasan 2022). Alkaloids make up nearly 50% of plant-derived medications while being a relatively minor class of metabolites (Oksman-Caldentey and Inzé 2004).

The various secondary metabolites can be synthesized via a variety of biochemical pathways beginning with photosynthesis, such as acetate-malonate for phenols and alkaloids; shikimic acid for phenols, tannins, and aromatic alkaloids; pentose for glycosides, polysaccharides, mevalonic acid, and methyl erythrose phosphate for terpenoids, alkaloids, and steroids (Trenchard et al. 2015; Shirley et al. 2019). However, many plant secondary metabolites remain uncharacterized (Lou et al. 2021). Although these are of huge industrial potential, their chemical synthesis is often unachievable or if achievable, often cost-prohibitive due to their structural complexity (Chemler and Koffas 2008). So, uncovering the biosynthetic pathways as well as characterization of the enzymes involved in these pathways are crucial for further biotechnological approaches aimed toward their *in vitro* synthesis, manipulation, and engineering (Wiltschi et al. 2020). Recent advances in the use of omics technologies have paved the way for a better understanding of plant secondary metabolism. The myriad approaches of genomics, transcriptomics, proteomics, and metabolomics have allowed a holistic approach to harness the power of natural products (Roychowdhury et al. 2023). Recent genome-level studies are helping in identifying the responsible genes for the production of such a wide and diverse range of bioactive metabolites. Understanding these networks of genes, proteins, and metabolites can lead to increased production of important specialized metabolites using both conventional and advanced plant biotechnology techniques.

19.2 Secondary Metabolic Gene Clusters

A biosynthetic gene cluster may be described as the presence of two or more nonhomologous genes nearby on the same chromosome and are engaged in the same biosynthetic pathway for producing a specialized metabolite or a set of similar metabolites (Medema et al. 2015). In bacteria and filamentous fungi, the genes responsible for producing specific metabolites are often grouped for synchronized expression. The clustered metabolic genes are responsible for the production of toxins in filamentous fungi and antibiotics in actinomycetes. Although previously it was thought that genes forming plant metabolic pathways were not clustered, several recent studies report that plants have clustered metabolic genes. Similar to prokaryotic gene clustering, plants also have gene clusters that encode enzymes involved in the manufacture of specific metabolites. Throughout the plant kingdom, over 20 gene clusters responsible for the production of various metabolites have already been identified. It is now assumed that gene clusters have evolved through duplication and neofunctionalization of primary metabolic pathway genes (Qi et al. 2006). These gene clusters are usually subjected to strict regulation, often at the nucleosome level. The size of the plant metabolic gene clusters usually ranges between 35 kb and several hundred kb (Bharadwaj et al. 2021). Gene clusters consist of two parts: primarily the genes that are responsible for determining the class of metabolites and secondarily one or two genes to produce metabolic diversity by modulating metabolite scaffold (Schneider et al. 2016; Nützmann et al. 2020). The architectural and evolutionary features of fungal and plant gene clusters are somewhat similar, although the idea of horizontal gene transfer (HGT) of clusters in fungi is not found in plants (Slot and Rokas 2010; Nützmann et al. 2018). Furthermore, gene clusters for producing both primary and secondary metabolites are found in filamentous fungi, while in plants, mostly specialized metabolite-producing gene clusters are found. It has been revealed that fungal gene clusters contain pathway-specific transcriptional regulators and transporters, a feature lacking in plant systems (Rokas et al. 2020). However, transporters and coordinately activated regulatory genes are seen in a few plant gene clusters (Darbani et al. 2016; Hen-Avivi et al. 2016; Shen et al. 2021). Multifaceted gene clusters are quite common in fungal origin. This comprises multigene clusters for an entire biosynthetic pathway, the occurrence of two gene clusters on the same chromosome, and intertwined clusters for the production of different metabolites (Yu et al. 2000; Bradshaw et al. 2013; Wiemann et al. 2013). Recent studies on the evolution of gene clusters suggest that plants recruited genes by duplicating genes involved in primary metabolism and assigning them new functions. As a result, they were able to reprogram their functional traits to adapt to different ecological niches (Qi et al. 2006; Field and Osbourn 2008; Krokida et al. 2013). Some of the metabolic gene clusters in plants and their features are listed in Table 19.1.

19.3 Sequencing Techniques and Analytical Tools for Plant Metabolomics Study

DNA sequencing technologies paved a long way since the first generation sequencing techniques, i.e. enzymatic sequencing (Sanger et al. 1977) and chemical analysis (Maxam and Gilbert 1977). With the advent of next-generation sequencing (NGS) technologies, sequencing became fast, cost-effective, and high-throughput compared to the previous techniques. NGS includes 454 Roche Diagnostics, Illumina/Solexa, SOLiD, and ion torrent sequencing (Younessi-Hamzekhanlu et al. 2022). These technologies are regularly being used for genomic and transcriptomic sequencing of many plants including medicinally important plants. The NGS methods especially Illumina HiSeq have been employed for the sequencing of many medicinal plants including *Allium sativum* (Sun et al. 2020), *Aloe vera* (Jaiswal et al. 2021), *Azadirachta indica* (Krishnan et al. 2012), *Catharanthus roseus* (Kellner et al. 2015), and so on (Cheng et al. 2021). Now, third-generation sequencing techniques are employed for genome sequencing. Examples of these new technologies include PacBio (Pacific Biosciences) sequencing, which uses a single-molecule, real-time (SMRT) approach, and Oxford Nanopore sequencing (ONT; based on nanopore technology) (Cheng et al. 2021; Satam et al. 2023). These third-generation sequencing technologies have been used for genome sequencing of *Juglans regia* (Marrano et al. 2020), *Panax notoginseng* (Jiang et al. 2021), *Tripterygium wilfordii* (Tu et al. 2020), etc.

Table 19.1 List of some Secondary metabolic gene clusters present in plants.

Sl	Species	Metabolites	Compound type	Chromosome	Features
1	*Arabidopsis thaliana*	Thalianol (Field et al. 2011)	Terpenes	5	Plant dwarfism is caused by to unregulated expression of cluster genes which can affect the microbiota of the roots
		Marneral (Field et al. 2011)	Triterpenes	5	Plant dwarfism is caused by to unregulated expression of cluster genes
2	*Avena strigosa*	Avenacin (Mugford et al. 2013)	Triterpenes	1	Defense against pathogens
3	*Cumis sativas*	Cucurbitacins (Shang et al. 2014)	Triterpenes	6	Insect restraint property. Medicinal property
4	*Ricinus communisis*	Casbenes (King et al. 2014, 2016)	Diterpenes	1	Medicinal property. Used for the treatment of cancer and HIV
5	*Jatropha curcus*	Casbenes (King et al. 2014, 2016)	Diterpenes	1	Medicinal property. Used for the treatment of cancer and HIV
6	*Zea mays*	2,4-dihydroxy-7-methoxy-1,4-benzoxazin-3-one (DIMBOA) (Krokida et al. 2013)	Hydroxamic acid	4	Plant defense activity
7	*Oryza sativa*	Momilactones (Shimura et al., 2007; Wang et al. 2011)	Diterpenes	4	Anti-fungal and insect restraint properties
		Oryzalides (Swaminathan et al. 2009; Wu et al. 2011)	Diterpenes	2	Plant defense activity
		Oryzalides (Swaminathan et al. 2009; Wu et al. 2011)	Diterpenes	2	Herbivore deterrent attributes

(*Continued*)

Table 19.1 (Continued)

Sl	Species	Metabolites	Compound type	Chromosome	Features
8	*Lotus japonicus*	Linamarin (Takos et al. 2011)	Cyanogenic glucosides	3	Herbivore deterrent attributes
		Lotuastralin Takos et al. (2011)	Cyanogenic glucosides	3	Herbivore deterrent attributes
		20-Hydroxy betulinic acid (Krokida et al. 2013)	Triterpene	3	Take part in plant development and nodule formation
9	*Sorghum bicolor*	Dhurrin (Takos et al. 2011)	Cyanogenic glucosides	1	Herbivore deterrent attributes
10	*Solanum lycopersicum*	α-tomatine (Itkin et al. 2013)	Steroidal glycoalkaloid	7	Demonstrates toxicity and anti-pathogenic properties toward humans
		Lycosantalonol (Matsuba et al. 2013, 2015; Zhou and Pichersky 2020)	Mono-terpenes	8	Role in pollinator attraction and defense against herbivores as well as fungal infections
		β-phellandrene (Matsuba et al. 2013, 2015; Zhou and Pichersky 2020)	Di-terpenes		Role in pollinator attraction and defense against herbivores as well as fungal infections
		Falcarindiol (Jeon et al. 2020)	Fatty acid gene cluster	12	Induced by pathogen attack and involved in defense response
11	*Solanum tuberosum*	α- solanine (Itkin et al. 2013)	Steroidal glycoalkaloid	7	Exhibits anti-pathogenic activities and toxicity toward human
12	*Papaver somniferum*	Noscapaine (Guo et al. 2018)	Benzylisoquinoline	11	High-value secondary metabolite
		Thebaine (Guo et al. 2018)	Benzylisoquinoline	11	High-value secondary metabolite
13	*Capsicim annum*	Capsidiol (Lee et al. 2017)	Sesquiterpene	2,12	Pathogen induced production

The study of plant metabolomics employs several advanced and high-throughput separatory techniques such as liquid chromatography (LC), gas chromatography (GC), mass spectrometry (MS), and nuclear magnetic resonance (NMR) for detecting, identifying, and evaluating the complex nature of plant metabolome (Gao et al. 2018; Adeniji et al. 2020). GC and LC are often associated with MS and these are the two most utilized tools for this purpose. LC-MS are used for the metabolite profiling of compounds with higher molecular weight, polar, and thermo-labile compounds such as glucosinolates, phenolics, and vitamins. On the other hand, GC-MS is suitable for profiling volatile compounds, organic acids, amino acids, low molecular weight hydrocarbons, etc. (Razzaq et al. 2019; Alawiye and Babalola 2021). Other tools are high-performance liquid chromatography (HPLC) and ultra-high-performance liquid chromatography (UHPLC), High-performance thin layer chromatography (HPTLC), Fourier transform ion-cyclotron resonance mass spectrometry (FI-ICR-MS), direct-infusion mass spectrometry (DIMS) and capillary electrophoresis–mass spectrometry (CE-MS) (Mashabela et al. 2022).

19.4 Bioinformatics Tools for the Elucidation of Secondary Metabolism in Plants

With the advancement of new technologies, an enormous amount of data is being generated every day from many organisms. With the rapid accumulation of these multi-omics data, bioinformatics plays a vital role in facilitating systems-level understanding of metabolism by integrating different types of data. Bioinformatics thus provides crucial mechanisms to analyze bulk information generated from these high-throughput techniques. Many databases provide genome-scale data which can be accessed freely for further studies. The best-known databases for gene and genome and transcriptome level information are NCBI GenBank (Benson et al. 2013; Sayers et al. 2022), RefSeq (O'Leary et al. 2016), NCBI-GEO (Edgar et al. 2002; Barrett et al. 2013), DDBJ (Mashima et al. 2017), ENA (Amid et al. 2020), etc. However, for plant metabolic data analysis, there are several other important databases to study genomic, transcriptomic, proteomic, and metabolomic data (Table 19.2).

19.5 Medicinal Plant Genome and/or Metabolome Databases

There are many databases dedicated to medicinal plants. These databases try to integrate crucial information regarding medicinal plants, their metabolites, and sequence data. Some of them are discussed below.

19.5.1 1 K Medicinal Plant Genome Database

Su et al (2022) developed the 1 K Medicinal Plant Genome Database (http://www.herbgenome.com/) which is a comprehensive source of genomes as well as metabolites of medicinal plants. As of now, the database contains information about more than 100 species and it aims to collect data for 1000 medicinal plants. It contains many types of data such as ploidy level and karyotype, genome size, number of genes, vernacular name in

Table 19.2 List of some appropriate databases for the study of plant secondary metabolism.

Type of database	Database	Description	References
Plant genome data	Ensembl Plants	An integrative resource database showing several sequenced genome-scale data of plant species	(Bolser et al. 2016; Yates et al. 2022)
	Gramene	It is an open-source, curated online resource for model organisms and crops. It helps scientists working in the fields of pathway analysis and comparative genomics in plants	(Tello-Ruiz et al. 2022)
	Phytozome	A comparative genomics portal for plants includes gene families that are related and represent the contemporary descendants of ancestral genes that are assembled at significant phylogenetic nodes	(Goodstein et al. 2012)
	PlantCARE	PlantCARE is a database of plant cis-acting regulatory components, including enhancers and repressors	(Lescot et al. 2002)
	TAIR	Information regarding model plant *Arabidopsis thaliana* is made available through a comprehensive online resource called the Arabidopsis Information Resource (TAIR)	(Berardini et al. 2015)
	PlantGDB	The PlantGDB database provides the groundwork for identifying gene sets that are either common to all plants or unique to a particular species by incorporating a number of bioinformatics tools that support gene prediction and cross-species comparisons	(Duvick et al. 2008)
	PLAZA	PLAZA is a platform for functional, comparative, and evolutionary plant genomics	(Van Bel et al. 2022)
	Plant rDNA database	A comprehensive online resource with open access, the Plant rDNA database provides details about the positions, dimensions, and quantity of the 18S-5.8S-26S (35S) and 5S ribosomal DNA loci	(Garcia et al. 2014)
Plant metabolite data	Golm Metabolome Database (GMD)	The Golm Metabolome Database is publicly accessible. Mass spectral libraries, metabolite profiling experiments, and supplementary data and resources are made available to the public by GMD	(Kopka et al. 2005)
	Metabolomics Workbench	It is an open repository for experimental data and metabolomics metadata from a variety of species, including plants. Data from nuclear magnetic resonance spectroscopy (NMR) and mass spectrometry (MS) are available here	(Sud et al. 2016)
	Plant Metabolite Database	It is a freely accessible plant metabolite database that contains many ESI-MSn spectra from the public mass spectrometry library. Several types of data analysis like Principal Component Analysis can be performed	(Fu et al. 2017; Liu et al. 2019)
	PubChem	PubChem is a freely accessible repository of data regarding chemical compounds and their biological actions	(Kim et al. 2023)

Plant proteome data	PeptideAtlas	An extensive collection of peptides found in tandem mass spectrometry proteomics experiments is compiled in the publicly available PeptideAtlas database	(Desiere et al. 2006; van Wijk et al. 2021)
	ProteomicsDB	ProteomicsDB is a multi-omics and multi-organism resource for life science research. It includes proteomics, transcriptomics, and phenomics data	(Samaras et al. 2020)
	PPDB	Plant Proteome Database (PPDB) contains curated information about protein function, protein properties, and subcellular localization, as well as experimental data from in-house proteome and mass spectrometry analysis. Information regarding *Zea mays* and *Arabidopsis thaliana* is included in the database	(Sun et al. 2009)
	PRIDE	Protein and peptide identifications along with the corresponding expression values, posttranslational modifications, and supporting mass spectra evidence are all included in the centralized, standards-compliant PRoteomics IDEntifications (PRIDE) Archive database for mass spectrometry proteomics data	(Perez-Riverol et al. 2019)
Metabolic network data	KEGG	The database resource KEGG aids in comprehending the higher-level operations and features of the biological system. Enzymes, genes, chemical compounds, and metabolic pathways are all covered	(Kanehisa and Goto 2000; Kanehisa 2019; Kanehisa et al. 2023)
	PMN (Plant Metabolic Network) and PlantCyc	For a number of plant species, the PMN gives users access to information about enzymes, pathways, etc. that has been manually curated or computationally predicted. Information on common and distinct metabolic pathways found in more than 500 plant species is available through PlantCyc.	(Hawkins et al. 2021)
	BioCyc	BioCyc is a set of 20,042 Pathway/Genome Databases (PGDBs) with software tools for exploring thousands of microbes and model eukaryotes. This database has a focus on microbes, but it also includes some plant genomes	(Karp et al. 2019)
	MetaCyc	It is an assembled database of metabolic pathways from every domain of life that has been experimentally curated. Together with related data like genes, enzymes, reactions, and metabolites, it includes information on primary and secondary metabolic pathways	(Caspi et al. 2020)
	BRENDA	An extensive database covering every facet of enzymes, such as their structure, role in illness, and use	(Chang et al. 2021)
	PlantReactome	A comprehensive database that covers every aspect of enzymes, including their use, structure, and role in disease. It includes intermediary metabolism, signaling, transcriptional regulation, and developmental pathways	(Naithani et al. 2020)
	M-CSA	A database on the mechanisms of enzyme reactions. For hundreds of enzymes, it offers annotation on the protein, cofactors, catalytic residues, and reaction mechanisms	(Ribeiro et al. 2018)
	Rhea	A repository of carefully selected chemical and transport reactions with biological significance	(Bansal et al. 2022)

Chinese, and some important metabolite and their KEGG entry number. It includes BLAST (Mount 2007), JBrowse (Buels et al. 2016), and other tools such as primer design. This database contains information on many important medicinal plants like *Abies alba* (Pinaceae), *Arachis hypogaea* (Leguminosae), and *J. regia* (Juglandaceae). For some species, genomic data, GFF (general feature format) files along with coding sequences and protein sequences are also available.

19.5.2 TCMPG

Traditional Chinese Medicine Plant Genome (Meng et al. 2022) (TCMPG; http://cbcb.cdutcm.edu.cn/TCMPG/) is another database developed for medicinal plants. This database includes information about 160 plants with 195 corresponding genomes (some plants *Actinidia chinensis* and *A. hypogaea* have more than one annotated genome). Also, plants are grouped hierarchically and graphically to give a holistic view of the genomic information from a taxonomic perspective. Also, genome size-wise classification is provided. The database provides SSR finder, synteny viewer, and HmmSearch along with BLAST and JBrowse.

19.5.3 IGTCM

Integrative Genome Database of Traditional Chinese Medicines (Ye et al. 2023) (IGTCM; http://yeyn.group:96/) is an integrated database dealing with traditional Chinese medicine (TCM) genomes and other vital information. The database contains genomes of 83 medicinal plants such as *Abrus precatorius*, *Handroanthus impetiginosus*, and *Papaver somniferum* to name a few. The database contains six components such as information regarding the herb, ingredient, gene, protein, RNA, and cds (coding sequence). Moreover, several genomic and proteomic features like GC percentage, enzyme EC number, gene ontology, KEGG pathway information, 2D structures, molecular formulae, and SMILE information of the key ingredients are also available. The database is also linked to BLAST and JBrowse.

19.5.4 IMPPAT

Indian Medicinal Plants, Phytochemistry And Therapeutics (Mohanraj et al. 2018) (IMPPAT; https://cb.imsc.res.in/imppat/) is a comprehensive phytochemical database for the study of Indian medicinal plants. IMPPAT 2.0 (Vivek-Ananth et al. 2023), the current version is the largest digital database regarding Indian medicinal plants and their phytochemicals until now. As of now, the database contains information about 4010 medicinal plants, 17,967 phytochemicals, and 27,365 associations between phytochemicals and human target proteins. The search option is very robust, i.e. one can search by plants using scientific names or Indian vernacular names of a particular plant or by phytochemical name. An advanced search option is also given using which one can get information by searching phytochemical information such as Molecular weight, hydrogen bond acceptors (HBA) or donors (HBD), Lipinski RO5 violation, and so on. However, genomic or pathway information is not included in this database.

19.5.5 GPGD

Global Pharmacopoeia Genome Database (Liao et al. 2022) (GPGD; http://www.gpgenome.com/) is a comprehensive and robust database for medicines derived from eight international pharmacopeias (China, USA, Europe, Japan, Korea, India, Egypt, and Brazil). It provides data regarding DNA barcodes, genomic and organellar DNA sequences, and transcriptome sequences. Various tools such as BLAST, species identification by DNA barcode, genome browser, and sequence extraction are also available.

19.5.6 CMAUP

Collective Molecular Activities of Useful Plants (Zeng et al. 2019) (CMAUP; https://bidd.group/CMAUP/about.html) provide collective molecular activities of more than 5000 useful plants including more than 2500 medicinal plants from various nations/regions. It also provides data regarding the human target proteins, gene ontology, and KEGG pathways associated with these plants.

19.6 Automation of Natural Product Detection by Identification of Metabolic Gene Cluster

Natural products from plants and microbial sources are well elucidated through bioassay-guided identification, and screening of extracts leading to the discovery of novel therapeutic compounds. Advanced computer power along with quicker and affordable genome sequencing technologies have combined to provide for more logical and foresighted quests. The first complete genome sequence of the bacteria *Haemophilus influenza* (Fleischmann et al. 1995) was published in the mid-1990s, just about the time a potent graphic user interface (GUI) in computer technology was introduced (Li et al. 2009). This interdisciplinary association between molecular biology and computational technology marked the beginning of whole-genome analysis as a method for drug discovery by looking for durable protein targets (Li et al. 2009). It was later realized that genome scanning may be used to find beneficial secondary metabolites including terpenoids, polyketides, and nonribosomal peptides (NRPs) (Van Lanen and Shen 2006; Li et al. 2009). Actually, computational approaches such as multiple sequence analysis and structure analysis through crystallographic analysis tools enabled scientists to clarify the features and functions of key enzymes associated with the assembly and tailoring of natural products (Li et al. 2009). Simplified molecular input line entry specification (SMILES) (Weininger 1988), a common letter coding for molecules, has helped organize chemical structures among numerous large chemical databases such as Zinc (Sterling and Irwin 2015), NCBI (Sayers et al. 2022), and PubChem (Kim et al. 2023). A software program named ClustScan (Starcevic et al. 2008) produces SMILES by scanning NRPS/PKS (polyketide synthases) clusters. Yet another software, NP.searcher (Li et al. 2009), can also identify and decode DNA of possible gene clusters associated with the biosynthesis of natural products.

Secondary metabolite prediction tools such as NP.searcher are extremely useful. It also enables the search engine closer to the prediction of molecules for lead optimization toward

Table 19.3 List of some automation tools for secondary metabolic gene cluster prediction.

Sl.	Name	Description	References
1	antiSMASH and PlantiSMASH	It is a tool for quickly identifying, annotating, and analyzing gene clusters engaged in the biosynthesis of secondary metabolites, from both bacterial and fungal genome sequences PlantiSMASH is an extension of antiSMASH. It automatically identifies potential plant biosynthesis gene clusters	(Medema et al. 2011; Blin et al. 2023)
2	NP.searcher	By reading a hidden Markov model of DNA, this web-based, open-source program can assemble a variety of small molecules made up of four acyl-CoA extender units incorporated into polyketides, twenty standard amino acids, and twenty-two additional chain-elongation intermediates used in nonribosomal peptide systems	(Li et al. 2009)
3	PhytoClust	This program looks through plant genomes to find potential candidate metabolic gene clusters. It enables mining for new clusters in addition to searching for recognized gene cluster types	(Töpfer et al. 2017)
4	PlantClusterFinder	It uses genomic location information with reaction and enzyme-related data for identifying metabolic gene clusters in plants	(Schläpfer et al. 2017)

drug discovery. Along with the progression of SMILES analysis for various biological activities, the advancement of docking software permits SMILES as input and dock these with proteins. This promises to be a valuable new tool for the research and development of pharmaceuticals. The predicted natural products in SMILES format can be utilized to build both 2D and 3D representations in various chemical software. These SMILES-generated 3D models can be further utilized to dock with proteins using well-established methods. Thus, several bioinformatic tools are already available for the identification of natural products by identifying metabolic gene clusters. Some tools for the identification of plant metabolic gene clusters are listed in Table 19.3.

19.7 The Big Data and Systems Biology Approach

The advancement of high-throughput data generated by different omics tools and the expansion of computer and software tool capabilities made it possible to make biology a data-intensive science. Thus, biology has gradually become a big data science and it is incorporating ideas from various other scientific disciplines such as mathematics, statistics, physics, and obviously, chemistry (Altaf-Ul-Amin et al. 2014). Currently, the combination of such diverse information can lead to a better understanding of a cell or an organism at the system level. This is the systems biology approach with the help of big data generated from different omics technologies. The integration of different omics data is especially vital for higher organisms such as plants (Sampaio et al. 2022). Different systems biology approaches like constraint-based modeling (CBM) and genome-scale metabolic models (GSMMs) are used for the study of metabolism (Sampaio et al. 2022). To get a holistic view of biological systems, CBM performs modeling of the internal components of those systems and the interactions that result from them using both computational and mathematical models (Feist et al. 2009). GSMMs are genome annotations derived *in silico* metabolic flux models representing all the metabolic reactions that take place within an organism (Sampaio et al. 2022). The multi-tissue model generally consists of various copies of a GSMM which are linked by inter-tissue exchange reactions (Sampaio et al. 2022). Additionally, tissue-specific omics data can be used to define the constraints for each tissue model for improving the flux predictions (de Oliveira Dal'Molin and Nielsen 2018; Shaw and Cheung 2020). Constraint-based methods can then be used to simulate GSSMs (Sampaio et al. 2022). One such approach is flux balance analysis (FBA) which is used for predicting the metabolic phenotype of an organism in diverse conditions (Orth et al. 2010). Flux variability analysis (FVA), one more constraint-based approach, computes each reaction's maximum and minimum flux for a particular set of constraints (Mahadevan and Schilling 2003). Dynamic flux balance analysis (dFBA; an extension of FBA) represents the rates of uptake or excretion by assuming that intracellular metabolites are at steady state, but exchange metabolites and the entire biomass are constrained with dynamic equations (Mahadevan et al. 2002).

19.8 Application of Machine Learning in Plant Secondary Metabolism

Since higher organisms like plants are complex organisms with a variety of cells and tissues, they must integrate various omics data in metabolic models. Thus, generic models may result in wrong interpretations, as some reactions or pathways are tissue or condition-specific. In nonmodel organisms with poorly characterized metabolism, this becomes more difficult (Sampaio et al.2022). As the omics datasets are huge, heterogeneous, and complex, sophisticated and advanced computation approaches are needed to process them. Machine learning (ML) is now being extensively employed to process, investigate, and integrate several omics data and to extract biological knowledge from these data (Misra et al. 2019). Presently, various databases systematically store and manage a variety of omics data that may be retrieved efficiently for the application of ML algorithms (Kim et al. 2020). ML algorithms can identify patterns, select relevant features, and draw inferences from the observed data without making any biological assumptions (Zampieri et al. 2019; Rana et al. 2020).

The following sections will familiarize important concepts of ML which are important in the context of plant secondary metabolism study. The roles of these algorithms in modern plant research are also discussed simultaneously.

19.9 Artificial Intelligence (AI)

AI refers to certain computer systems that can enhance human intelligence for perceiving and analyzing certain data packets and drawing inferences from them to take appropriate actions responding to them (Bini 2018). Even though artificial intelligence (AI) has been around for nearly a century, it is recently gaining popularity. The major reason behind this is the availability of large data sets, i.e. the big data, advancement of computer technologies, and increase in computational speed and development of modern algorithms developed over the years (Ali et al. 2023). A branch of AI is ML which develops algorithms used for predictive modeling or executing various tasks from multidimensional datasets (Camacho et al. 2018).

19.10 Machine Learning (ML)

ML is a branch of computer science where machines, i.e. computers, are programmed to discover hidden nonobvious patterns from data (Chicco 2017; Camacho et al. 2018). ML verifies resemblances among different databases by recognizing characteristic regularities between input data and output results and it has been used in natural product research (Seetharam et al. 2019). The ability to automatically identify patterns in data is the main reason for gaining popularity of ML particularly when the data is very large, there are exceptions to the general cases and the expert knowledge is incomplete or inaccurate (Yip et al. 2013). Presently, Bioinformatics and computational biology fulfils all these parameters (Chicco 2017) and thus, ML is becoming popular day by day in the fields of omics data and bioinformatics.

Different ML algorithms have already been deployed to interpret large metabolic datasets and to develop tools for studying cellular metabolism (Costello and Martin 2018; Cuperlovic-Culf 2018; Antonakoudis et al. 2020). The primary function of ML is accumulating experimental knowledge through the use of large data sets and a multifaceted mixture of algorithms. Based on the labeling of training samples, ML algorithms are generally classified into the following three types (Yan and Wang 2022):

a) **supervised learning (SL)** which makes use of a labeled training dataset (input data that is already tagged with the correct output) for model training and based on that predicts the output for an unlabeled test data set,
b) **unsupervised learning (UL)** which utilizes an unlabeled data set for training a model as well as identifying relationships within the data,
c) **semi-supervised learning (SSL)** in which the training data is made up of both labeled and unlabeled samples. Based on both these labeled and unlabeled samples in the training data, models make predictions or decisions.

SL can be utilized for designing new compounds and pathways by utilizing the plethora of big omics data. These models can be used in predicting reaction outcomes to derive reactants from the products, assigning known enzymes for orphan reactions, predicting the toxicity of pathway intermediates, and evaluating the feasibility of reactions (Lim et al. 2023). Support vector machine (SVM) is one category of supervised ML algorithm that has been applied to classify some plants using economic uses of plants and corresponding metabolite groups (Liu et al. 2017). SVM has also been used to predict the antibacterial ability of several plant secondary metabolites (Karim et al. 2022). SVM has also been used to discover the missing link enzymes of the plant benzylisoquinoline alkaloid pathway (Vavricka et al. 2022). SVM and random forest (RF) algorithms have been used to distinguish secondary metabolic genes from general (primary) metabolic genes in *Arabidopsis thaliana* genome (Moore et al. 2019). UL and SSL algorithms have been applied in the areas of data clustering, dimensionality reduction (DR), gene regulatory network (GRN) inference, single-cell omics data, and cross-species prediction. In plants, large-scale omics research is often hindered due to the dearth of annotated genes and pathways. SSL and transfer learning strategies have become suitable approaches for prediction jobs. These strategies consider conservation of genes and pathways among the evolutionarily related species (Yan and Wang 2022).

ML and CBM approaches are applied in combination to study fluxomics, and multi-omics research and to generate constraints-based models and fluxomics. The flux distribution is anticipated by CBM and is further analyzed by different ML methods. On the other hand, diverse omics data can be included in GSMMs' constraints to generate context-specific models to make accurate flux predictions. These predicted fluxes can be combined with omics data to be analyzed jointly by ML algorithms (Sampaio et al. 2022). A series of ML models have been utilized for predicting enzyme turnover numbers by employing data on network properties, structural features of enzymes, biochemical information as well as assay conditions (Heckmann et al. 2018). The predicted enzyme turnover numbers have then been utilized for simulating the two diverse forms of a GSMM, viz. metabolic modeling with enzyme kinetics (MOMENT) (Adadi et al. 2012) and a GSMM of metabolism and macromolecular expression (ME) (Lerman et al. 2012). These resulted in a more precise prediction of quantitative proteome allocation compared to earlier approaches (Kim et al. 2020). ML algorithms such as support vector regression (SVR), regression decision tree, lasso regression, k-nearest neighbors regression as well as ensemble methods have been used to explain the rates of changes in metabolic concentrations associated with isopentenol and limonene pathway using the time-series data of metabolome and proteome as the input (Costello and Martin 2018). ML has also been used in classifying angiosperms based on sulfur-containing compounds (Abdullah-Zawawi et al. 2022).

19.11 Deep Learning (DL)

Deep learning (DL) is an extension of ML that involves the processing and validation of large training datasets between input and output units (Calderaro et al. 2022). Thus, DL is gaining popularity for efficient processing of large quantities of data and identifying patterns concealed deep inside biological data. DL employs different structural network models for analyzing different types of data and for different applications. The primary models of DL include convolutional neural network (CNN), Elman recurrent neural network (RNN), long short-term memory (LSTM), and generative adversarial network (GAN) (Ma et al. 2023).

19.11.1 Convolutional Neural Network (CNN)

CNN is a deep feedforward neural network with the features of local connection and weight-sharing that employs a stack of convolution layers for the extraction of features (Ma et al. 2023). CNN along with SVM has been applied to process HPLC-MS fingerprints that have been applied to distinguish different plant parts from the plant family Apiaceae (Turova et al. 2022). A deep learning model called GTC has been developed using CNN and graph transformer networks. This has successfully been used in predicting plant secondary metabolic pathways by using the KEGG database, PMN, and SMILE information (Bao et al. 2023). DeepEC is an EC (Enzyme Commission) number prediction tool with a protein sequence as input. DeepEC employs three independent CNNs for different classification tasks and a homology analysis tool (Ryu et al. 2019).

19.11.2 Recurrent Neural Network (RNN)

Elman RNN or simply the RNN is especially important for processing time-dependent sequential data (Ma et al. 2023). RNN has been applied for the identification of time-delayed GRNs which will be helpful for reverse engineering (Kordmahalleh et al. 2017). Recently, a phytochemical library has been developed from Korean medicinal plants and this database has been used for identifying phytochemicals useful against SARS-CoV-2 virus using RNN (Hossain et al. 2023).

19.11.3 Long Short-Term Memory (LSTM)

LSTM is a type of RNN and can preserve long-term dependencies by using internal feedback (Le et al. 2022). LSTM has been used to help develop a catalog of software tools named MSCAT (Metabolomics Software CATalog) (Dekermanjian et al. 2021). LSTM has also been used to ease the task of literature review relating to metabolomics study (Yeung et al. 2022). LSTM is effective for the study of time-series data and also for interpreting cumulative changes, and thus, LSTM is adapted to intricate crop and climatic patterns (Moon et al. 2019).

19.11.4 Generative Adversarial Network (GAN)

GAN learns the generative model of data distribution through adversarial methods and is used in several fields including omics data (Lan et al. 2020). GAN in combination with near-infrared (NIR) spectra has been used to identify two plant species namely *Cuminum cyminum* (cumin) and *Foeniculum vulgare* (fennel) from different regions (Yang et al. 2021). This is important as these plants collected from different regions vary in the content of lipids, phenols, and proteins as well as essential oil composition. Another study successfully achieved rapid identification of the *F. vulgare* (fennel) from different origins with the help of different ML algorithms including GAN (Zuo et al. 2022).

19.11.5 Artificial Neural Networks (ANNs)

ANNs are computer programs that are biologically inspired and designed to simulate similar to how the human brain processes information (Agatonovic-Kustrin and Beresford 2000). ANNs collect their knowledge not from programming, but by identifying the patterns and relationships in data and are trained or learn through the experience. An ANN is formed from hundreds of single units, artificial neurons, or processing elements (PE), connected with coefficients (weights), which constitute the neural structure and are organized in layers (Agatonovic-Kustrin and Beresford 2000). ANNs are part of an extensive family of ML algorithms that try to learn rules or conditions from data samples and sometimes can be automatically improved through the huge amount of available data (Basheer and Hajmeer 2000; Pomyen et al. 2020). ANN has successfully been used for the prediction of the secondary metabolite content of plants in response to soil factors and the environment (Ray et al. 2020). ANN prediction model has been employed for studying the effect of soil nutrients and environmental parameters on podophyllotoxin content in *Podophyllum hexandrum* (Alam and Naik 2009), curcumin content (Akbar et al. 2016), and yield of essential oil (Akbar et al. 2018) in *Curcuma longa* and coronarin D content in *Hedychium coronarium* (Ray et al. 2020).

A concise diagram showing the relationship between different omics data and computational approaches including ML and DL for the study of plant secondary metabolism is given in Figure 19.2.

19.12 Conclusion and Future Perspective

Plant specialized metabolism evolved independently from the other eukaryotes. Moreover, as these metabolites evolved in relation to environmental interaction, there are a plethora of these metabolites and many of them are yet to be discovered. Although they are classified in some major groups such as terpenoids, phenolics alkaloids, and sulfur-containing metabolites, there are often poor overlaps among similar types of pathways in many plant species. Thus, the study of only some model plants is not enough to understand this type of metabolism. A holistic approach is needed not only for a better understanding of the specialized metabolism in plants but also to produce industrially important metabolites. Improvements in sequencing technologies as well as faster and more efficient computational tools like ML and DL are enabling our understanding of metabolic pathways in many medicinal plants more rapidly than ever. Coupled with metabolomics knowhows and sequencing technologies, these advanced bioinformatic tools are showing us the complex and versatile nature of plant specialized metabolic pathways and their variations. Moreover, biotechnology companies are also increasingly using ML techniques for the advancement of production systems. Thus, we expect that more sophisticated computational tools will help to unleash the full spectrum of secondary metabolism in plants as well as speed up the field of systems metabolic engineering in the future.

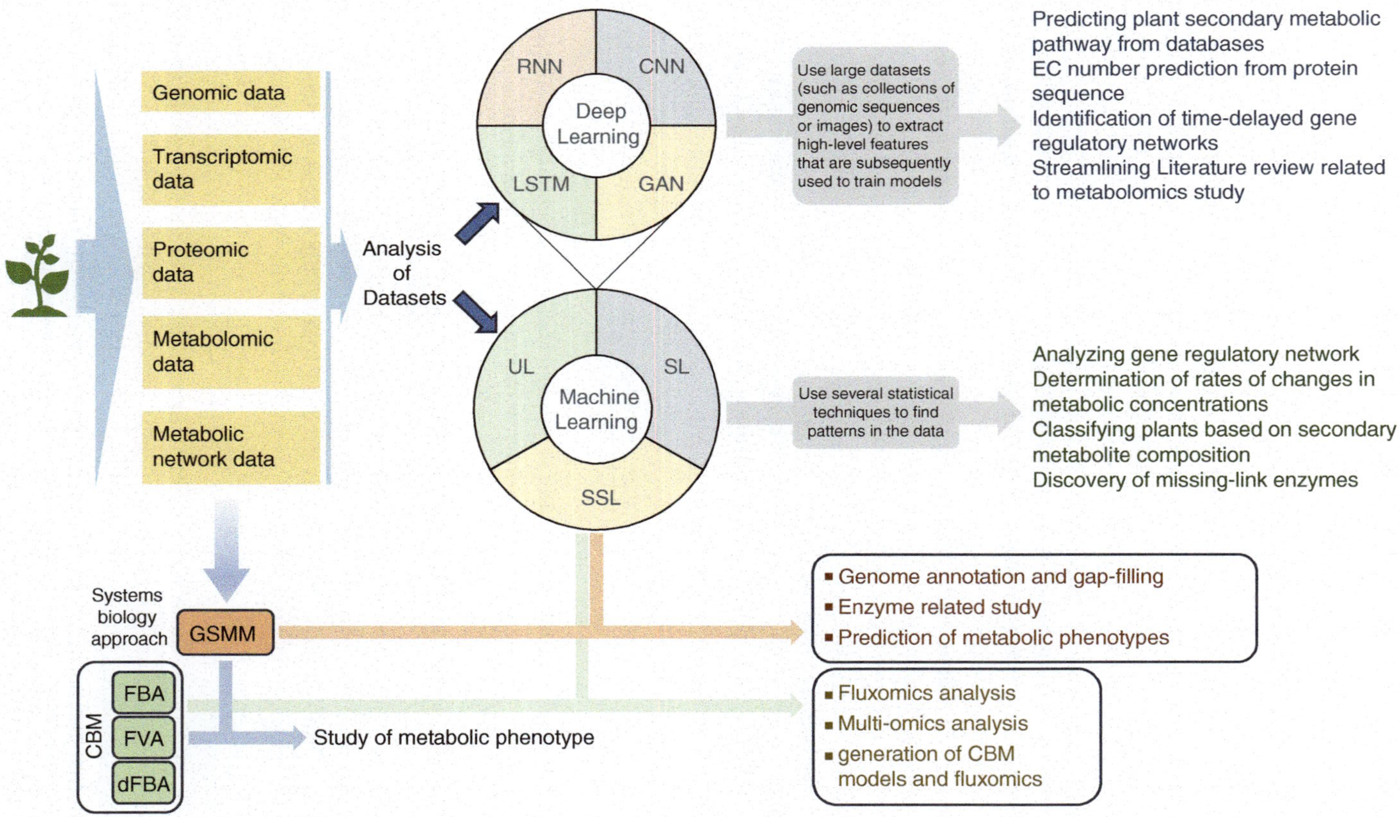

Figure 19.2 Application of ML, DL, and other computational approaches in the study of plant secondary metabolism with the help of different omics approaches. See text for details.

References

Abdullah-Zawawi, M.-R., Govender, N., Karim, M.B. et al. (2022). Chemoinformatics-driven classification of Angiosperms using sulfur-containing compounds and machine learning algorithm. *Plant Methods* 18 (1): 118. https://doi.org/10.1186/s13007-022-00951-6.

Adadi, R., Volkmer, B., Milo, R. et al. (2012). Prediction of microbial growth rate versus biomass yield by a metabolic network with kinetic parameters. *PLoS Computational Biology* 8 (7): e1002575. https://doi.org/10.1371/journal.pcbi.1002575.

Adeniji, A.A., Babalola, O.O., and Loots, D.T. (2020). Metabolomic applications for understanding complex tripartite plant-microbes interactions: strategies and perspectives. *Biotechnology Reports (Amsterdam, Netherlands)* 25: e00425. https://doi.org/10.1016/j.btre.2020.e00425.

Agatonovic-Kustrin, S. and Beresford, R. (2000). Basic concepts of artificial neural network (ANN) modeling and its application in pharmaceutical research. *Journal of Pharmaceutical and Biomedical Analysis* 22 (5): 717–727. https://doi.org/10.1016/S0731-7085(99)00272-1.

Akbar, A., Kuanar, A., Joshi, R.K. et al. (2016). Development of prediction model and experimental validation in predicting the curcumin content of turmeric (*Curcuma longa* L.). *Frontiers in Plant Science* 7: 01507. https://doi.org/10.3389/fpls.2016.01507.

Akbar, A., Kuanar, A., Patnaik, J. et al. (2018). Application of Artificial Neural Network modeling for optimization and prediction of essential oil yield in turmeric (*Curcuma longa* L.). *Computers and Electronics in Agriculture* 148: 160–178. https://doi.org/10.1016/j.compag.2018.03.002.

Alam, M.A. and Naik, P.K. (2009). Impact of soil nutrients and environmental factors on podophyllotoxin content among 28 *Podophyllum hexandrum* populations of Northwestern Himalayan Region using linear and nonlinear approaches. *Communications in Soil Science and Plant Analysis* 40 (15–16): 2485–2504. https://doi.org/10.1080/00103620903111368.

Alawiye, T.T. and Babalola, O.O. (2021). Metabolomics: current application and prospects in crop production. *Biologia* 76 (1): 227–239. https://doi.org/10.2478/s11756-020-00574-z.

Ali, F., Sarfraz, S., Hameed, A. et al. (2023). Big data revolution and machine learning to solve genetic mysteries in crop breeding. In: *Sustainable Agriculture in the Era of the OMICs Revolution* (ed. C.S. Prakash, S. Fiaz, M.A. Nadeem, et al.), 83–101. Cham: Springer.

Altaf-Ul-Amin, M., Afendi, F.M., Kiboi, S.K. et al. (2014). Systems biology in the context of big data and networks. *BioMed Research International* 2014: 428570. https://doi.org/10.1155/2014/428570.

Amid, C., Alako, B.T.F., Kadhirvelu, V.B. et al. (2020). The European nucleotide archive in 2019. *Nucleic Acids Research* 48 (D1): D70–D76. https://doi.org/10.1093/nar/gkz1063.

Antonakoudis, A., Barbosa, R., Kotidis, P. et al. (2020). The era of big data: genome-scale modelling meets machine learning. *Computational and Structural Biotechnology Journal* 18: 3287–3300. https://doi.org/10.1016/j.csbj.2020.10.011.

Bansal, P., Morgat, A., Axelsen, K.B. et al. (2022). Rhea, the reaction knowledgebase in 2022. *Nucleic Acids Research* 50 (D1): D693–D700. https://doi.org/10.1093/nar/gkab1016.

Bao, H., Zhao, J., Zhao, X. et al. (2023). Prediction of plant secondary metabolic pathways using deep transfer learning. *BMC Bioinformatics* 24 (1): 348. https://doi.org/10.1186/s12859-023-05485-9.

Barrett, T., Wilhite, S.E., Ledoux, P. et al. (2013). NCBI GEO: archive for functional genomics data sets—update. *Nucleic Acids Research* 41 (D1): D991–D995. https://doi.org/10.1093/nar/gks1193.

Barut, M., Nadeem, M.A., Akgür, Ö. et al. (2022). Medicinal and aromatic plants in the omics era: application of plant breeding andbiotechnology for plant secondary metabolite production. *Turkish Journal of Agriculture and Forestry* 46 (2): 182–203. https://doi.org/10.55730/1300-011X.2970.

Basheer, I.A. and Hajmeer, M. (2000). Artificial neural networks: fundamentals, computing, design, and application. *Journal of Microbiological Methods* 43 (1): 3–31. https://doi.org/10.1016/S0167-7012(00)00201-3.

Benson, D.A., Cavanaugh, M., Clark, K. et al. (2013). GenBank. *Nucleic Acids Research* 41 (D1): D36–D42. https://doi.org/10.1093/nar/gks1195.

Berardini, T.Z., Reiser, L., Li, D. et al. (2015). The Arabidopsis information re*source:* making and mining the "gold standard" annotated reference plant genome. *Genesis* 53 (8): 474–485. https://doi.org/10.1002/dvg.22877.

Bharadwaj, R., Kumar, S.R., Sharma, A. et al. (2021). Plant metabolic gene clusters: evolution, organization, and their applications in synthetic biology. *Frontiers in Plant Science* 12: 697318. https://doi.org/10.3389/fpls.2021.697318.

Bini, S.A. (2018). Artificial intelligence, machine learning, deep learning, and cognitive computing: what do these terms mean and how will they impact health care? *The Journal of Arthroplasty* 33 (8): 2358–2361. https://doi.org/10.1016/j.arth.2018.02.067.

Blin, K., Shaw, S., Augustijn, H.E. et al. (2023). antiSMASH 7.0: new and improved predictions for detection, regulation, chemical structures and visualisation. *Nucleic Acids Research* 51 (W1): W46–W50. https://doi.org/10.1093/nar/gkad344.

Bolser, D., Staines, D.M., Pritchard, E. et al. (2016). Ensembl plants: integrating tools for visualizing, mining, and analyzing plant genomics data. In: *Plant Bioinformatics: Methods and Protocols*, Methods in Molecular Biology (ed. D. Edwards), 115–140. New York: Springer https://doi.org/10.1007/978-1-4939-3167-5_6.

Bradshaw, R.E., Slot, J.C., Moore, G.G. et al. (2013). Fragmentation of an aflatoxin-like gene cluster in a forest pathogen. *New Phytologist* 198 (2): 525–535. https://doi.org/10.1111/nph.12161.

Buels, R., Yao, E., Diesh, C.M. et al. (2016). JBrowse: a dynamic web platform for genome visualization and analysis. *Genome Biology* 17 (1): 66. https://doi.org/10.1186/s13059-016-0924-1.

Calderaro, J., Seraphin, T.P., Luedde, T. et al. (2022). Artificial intelligence for the prevention and clinical management of hepatocellular carcinoma. *Journal of Hepatology* 76 (6): 1348–1361. https://doi.org/10.1016/j.jhep.2022.01.014.

Camacho, D.M., Collins, K.M., Powers, R.K. et al. (2018). Next-generation machine learning for biological networks. *Cell* 173 (7): 1581–1592. https://doi.org/10.1016/j.cell.2018.05.015.

Caspi, R., Billington, R., Keseler, I.M. et al. (2020). The MetaCyc database of metabolic pathways and enzymes - a 2019 update. *Nucleic Acids Research* 48 (D1): D445–D453. https://doi.org/10.1093/nar/gkz862.

Chang, A., Jeske, L., Ulbrich, S. et al. (2021). BRENDA, the ELIXIR core data resource in 2021: new developments and updates. *Nucleic Acids Research* 49 (D1): D498–D508. https://doi.org/10.1093/nar/gkaa1025.

Chemler, J.A. and Koffas, M.A.G. (2008). Metabolic engineering for plant natural product biosynthesis in microbes. *Current Opinion in Biotechnology* 19 (6): 597–605. https://doi.org/10.1016/j.copbio.2008.10.011.

Cheng, Q.-Q., Ouyang, Y., Tang, Z.-Y. et al. (2021). Review on the development and applications of medicinal plant genomes. *Frontiers in Plant Science* 12: 791219. https://doi.org/10.3389/fpls.2021.791219.

Chicco, D. (2017). Ten quick tips for machine learning in computational biology. *Biodata Mining* 10 (1): 35. https://doi.org/10.1186/s13040-017-0155-3.

Costello, Z. and Martin, H.G. (2018). A machine learning approach to predict metabolic pathway dynamics from time-series multiomics data. *NPJ Systems Biology and Applications* 4 (1): 19. https://doi.org/10.1038/s41540-018-0054-3.

Cuperlovic-Culf, M. (2018). Machine learning methods for analysis of metabolic data and metabolic pathway modeling. *Metabolites* 8 (1): 4. https://doi.org/10.3390/metabo8010004.

Darbani, B., Motawia, M.S., Olsen, C.E. et al. (2016). The biosynthetic gene cluster for the cyanogenic glucoside dhurrin in *Sorghum bicolor* contains its co-expressed vacuolar MATE transporter. *Scientific Reports* 6 (1): 37079. https://doi.org/10.1038/srep37079.

Dekermanjian, J., Labeikovsky, W., Ghosh, D. et al. (2021). MSCAT: a machine learning assisted catalog of metabolomics software tools. *Metabolites* 11 (10): 678. https://doi.org/10.3390/metabo11100678.

Desiere, F., Deutsch, E.W., King, N.L. et al. (2006). The PeptideAtlas project. *Nucleic Acids Research* 34 (suppl_1): D655–D658. https://doi.org/10.1093/nar/gkj040.

Dewick, P.M. (2002). *Medicinal Natural Products: A Biosynthetic Approach*. Wiley.

Divekar, P.A., Narayana, S., Divekar, B.A. et al. (2022). Plant secondary metabolites as defense tools against herbivores for sustainable crop protection. *International Journal of Molecular Sciences* 23 (5): 2690. https://doi.org/10.3390/ijms23052690.

Duvick, J., Fu, A., Muppirala, U. et al. (2008). PlantGDB: a resource for comparative plant genomics. *Nucleic Acids Research* 36 (suppl_1): D959–D965. https://doi.org/10.1093/nar/gkm1041.

Edgar, R., Domrachev, M., and Lash, A.E. (2002). Gene Expression Omnibus: NCBI gene expression and hybridization array data repository. *Nucleic Acids Research* 30 (1): 207–210. https://doi.org/10.1093/nar/30.1.207.

Feist, A.M., Herrgård, M.J., Thiele, I. et al. (2009). Reconstruction of biochemical networks in microorganisms. *Nature Reviews Microbiology* 7 (2): 129–143. https://doi.org/10.1038/nrmicro1949.

Field, B. and Osbourn, A.E. (2008). Metabolic diversification—independent assembly of operon-like gene clusters in different plants. *Science* 320 (5875): 543–547. https://doi.org/10.1126/science.1154990.

Field, B., Fiston-Lavier, A.-S., Kemen, A. et al. (2011). Formation of plant metabolic gene clusters within dynamic chromosomal regions. *Proceedings of the National Academy of Sciences, USA* 108 (38): 16116–16121. https://doi.org/10.1073/pnas.1109273108.

Fleischmann, R.D., Adams, M.D., White, O. et al. (1995). Whole-genome random sequencing and assembly of *Haemophilus influenzae* Rd. *Science* 269 (5223): 496–512. https://doi.org/10.1126/science.7542800.

Fu, H.-Y., Guo, X.-M., Zhang, Y.-M. et al. (2017). AntDAS: automatic data analysis strategy for UPLC–QTOF-based nontargeted metabolic profiling analysis. *Analytical Chemistry* 89 (20): 11083–11090. https://doi.org/10.1021/acs.analchem.7b03160.

Gao, Y., Chen, Y., Yue, X. et al. (2018). Development of simultaneous targeted metabolite quantification and untargeted metabolomics strategy using dual-column liquid chromatography coupled with tandem mass spectrometry. *Analytica Chimica Acta* 1037: 369–379. https://doi.org/10.1016/j.aca.2018.08.042.

Garcia, S., Gálvez, F., Gras, A. et al. (2014). Plant rDNA database: update and new features. *Database* 2014: bau063. https://doi.org/10.1093/database/bau063.

Garcia-Brugger, A., Lamotte, O., Vandelle, E. et al. (2006). Early signaling events induced by elicitors of plant defenses. *Molecular Plant-Microbe Interactions* 19 (7): 711–724. https://doi.org/10.1094/MPMI-19-0711.

Goodstein, D.M., Shu, S., Howson, R. et al. (2012). Phytozome: a comparative platform for green plant genomics. *Nucleic Acids Research* 40 (D1): D1178–D1186. https://doi.org/10.1093/nar/gkr944.

Guo, L., Winzer, T., Yang, X. et al. (2018). The opium poppy genome and morphinan production. *Science* 362 (6412): 343–347. https://doi.org/10.1126/science.aat4096.

Hawkins, C., Ginzburg, D., Zhao, K. et al. (2021). Plant metabolic network 15: a resource of genome-wide metabolism databases for 126 plants and algae. *Journal of Integrative Plant Biology* 63 (11): 1888–1905. https://doi.org/10.1111/jipb.13163.

Heckmann, D., Lloyd, C.J., Mih, N. et al. (2018). Machine learning applied to enzyme turnover numbers reveals protein structural correlates and improves metabolic models. *Nature Communications* 9 (1): 5252. https://doi.org/10.1038/s41467-018-07652-6.

Heinrich, M., Mah, J., and Amirkia, V. (2021). Alkaloids used as medicines: structural phytochemistry meets biodiversity—an update and forward look. *Molecules* 26 (7): 1836. https://doi.org/10.3390/molecules26071836.

Hen-Avivi, S., Savin, O., Racovita, R.C. et al. (2016). A metabolic gene cluster in the wheat W1 and the barley *Cer-cqu* loci determines β-diketone biosynthesis and glaucousness. *The Plant Cell* 28 (6): 1440–1460. https://doi.org/10.1105/tpc.16.00197.

Hossain, A., Rahman, M.E., Rahman, M.S. et al. (2023). Identification of medicinal plant-based phytochemicals as a potential inhibitor for SARS-CoV-2 main protease (M^{pro}) using molecular docking and deep learning methods. *Computers in Biology and Medicine* 157: 106785. https://doi.org/10.1016/j.compbiomed.2023.106785.

Itkin, M., Heinig, U., Tzfadia, O. et al. (2013). Biosynthesis of antinutritional alkaloids in solanaceous crops is mediated by clustered genes. *Science* 341 (6142): 175–179. https://doi.org/10.1126/science.1240230.

Jaiswal, S.K., Mahajan, S., Chakraborty, A. et al. (2021). The genome sequence of *Aloe vera* reveals adaptive evolution of drought tolerance mechanisms. *iScience* 24 (2): 102079. https://doi.org/10.1016/j.isci.2021.102079.

Jan, R., Asaf, S., Numan, M. et al. (2021). Plant secondary metabolite biosynthesis and transcriptional regulation in response to biotic and abiotic stress conditions. *Agronomy* 11 (5): 968. https://doi.org/10.3390/agronomy11050968.

Jeon, J.E., Kim, J.-G., Fischer, C.R. et al. (2020). A pathogen-responsive gene cluster for highly modified fatty acids in tomato. *Cell* 180 (1): 176–187. https://doi.org/10.1016/j.cell.2019.11.037.

Jiang, Z., Tu, L., Yang, W. et al. (2021). The chromosome-level reference genome assembly for *Panax notoginseng* and insights into ginsenoside biosynthesis. *Plant Communications* 2 (1): 100113. https://doi.org/10.1016/j.xplc.2020.100113.

Kanehisa, M. (2019). Toward understanding the origin and evolution of cellular organisms. *Protein Science: A Publication of the Protein Society* 28 (11): 1947–1951. https://doi.org/10.1002/pro.3715.

Kanehisa, M. and Goto, S. (2000). KEGG: kyoto encyclopedia of genes and genomes. *Nucleic Acids Research* 28 (1): 27–30. https://doi.org/10.1093/nar/28.1.27.

Kanehisa, M., Furumichi, M., Sato, Y. et al. (2023). KEGG for taxonomy-based analysis of pathways and genomes. *Nucleic Acids Research* 51 (D1): D587–D592. https://doi.org/10.1093/nar/gkac963.

Karim, M.B., Kanaya, S., and Altaf-Ul-Amin, M. (2022). Antibacterial activity prediction of plant secondary metabolites based on a combined approach of graph clustering and deep neural network. *Molecular Informatics* 41 (7): 2100247. https://doi.org/10.1002/minf.202100247.

Karp, P.D., Billington, R., Caspi, R. et al. (2019). The BioCyc collection of microbial genomes and metabolic pathways. *Briefings in Bioinformatics* 20 (4): 1085–1093. https://doi.org/10.1093/bib/bbx085.

Kellner, F., Kim, J., Clavijo, B.J. et al. (2015). Genome-guided investigation of plant natural product biosynthesis. *The Plant Journal* 82 (4): 680–692. https://doi.org/10.1111/tpj.12827.

Kim, G.B., Kim, W.J., Kim, H.U. et al. (2020). Machine learning applications in systems metabolic engineering. *Current Opinion in Biotechnology* 64: 1–9. https://doi.org/10.1016/j.copbio.2019.08.010.

Kim, S., Chen, J., Cheng, T. et al. (2023). PubChem 2023 update. *Nucleic Acids Research* 51 (D1): D1373–D1380. https://doi.org/10.1093/nar/gkac956.

King, A.J., Brown, G.D., Gilday, A.D. et al. (2014). Production of bioactive diterpenoids in the Euphorbiaceae depends on evolutionarily conserved gene clusters. *The Plant Cell* 26 (8): 3286–3298. https://doi.org/10.1105/tpc.114.129668.

King, A.J., Brown, G.D., Gilday, A.D. et al. (2016). A cytochrome P450-mediated intramolecular carbon–carbon ring closure in the biosynthesis of multidrug-resistance-reversing lathyrane diterpenoids. *ChemBioChem* 17 (17): 1593–1597. https://doi.org/10.1002/cbic.201600316.

Kopka, J., Schauer, N., Krueger, S. et al. (2005). GMD@CSB.DB: the Golm Metabolome Database. *Bioinformatics* 21 (8): 1635–1638. https://doi.org/10.1093/bioinformatics/bti236.

Kordmahalleh, M.M., Sefidmazgi, M.G., Harrison, S.H. et al. (2017). Identifying time-delayed gene regulatory networks via an evolvable hierarchical recurrent neural network. *Biodata Mining* 10 (1): 29. https://doi.org/10.1186/s13040-017-0146-4.

Krishnan, N.M., Pattnaik, S., Jain, P. et al. (2012). A draft of the genome and four transcriptomes of a medicinal and pesticidal angiosperm *Azadirachta indica*. *BMC Genomics* 13 (1): 464. https://doi.org/10.1186/1471-2164-13-464.

Krokida, A., Delis, C., Geisler, K. et al. (2013). A metabolic gene cluster in *Lotus japonicus* discloses novel enzyme functions and products in triterpene biosynthesis. *New Phytologist* 200 (3): 675–690. https://doi.org/10.1111/nph.12414.

Laloo, D., Kalita, J.M., and Prasad, S.K. (2021). Molecular docking studies of plant-derived bioactive compounds: a comprehensive *in silico* standardization approach. In: *Evidence Based Validation of Traditional Medicines: A Comprehensive Approach* (ed. S.C. Mandal,

R. Chakraborty, and S. Sen), 371–404. Singapore: Springer https://doi.org/10.1007/978-981-15-8127-4_19.

Lan, L., You, L., Zhang, Z. et al. (2020). Generative adversarial networks and its applications in biomedical informatics. *Frontiers in Public Health* 8: 00164. https://doi.org/10.3389/fpubh.2020.001648.

Le, V.-T., Tran-Trung, K., and Hoang, V.T. (2022). A comprehensive review of recent deep learning techniques for human activity recognition. *Computational Intelligence and Neuroscience* 2022: 8323962. Available at: https://doi.org/10.1155/2022/8323962.

Lee, H.-A., Kim, S., Kim, S. et al. (2017). Expansion of sesquiterpene biosynthetic gene clusters in pepper confers nonhost resistance to the Irish potato famine pathogen. *New Phytologist* 215 (3): 1132–1143. https://doi.org/10.1111/nph.14637.

Lerman, J.A., Hyduke, D.R., Latif, H. et al. (2012). *In silico* method for modelling metabolism and gene product expression at genome scale. *Nature Communications* 3 (1): 929. https://doi.org/10.1038/ncomms1928.

Lescot, M., Déhais, P., Thijs, G. et al. (2002). PlantCARE, a database of plant cis-acting regulatory elements and a portal to tools for *in silico* analysis of promoter sequences. *Nucleic Acids Research* 30 (1): 325–327. https://doi.org/10.1093/nar/30.1.325.

Li, M.H.T., Ung, P.M.U., Zajkowski, J. et al. (2009). Automated genome mining for natural products. *BMC Bioinformatics* 10: 185. https://doi.org/10.1186/1471-2105-10-185.

Liao, B., Hu, H., Xiao, S. et al. (2022). Global Pharmacopoeia Genome Database is an integrated and mineable genomic database for traditional medicines derived from eight international pharmacopoeias. *Science China Life Sciences* 65 (4): 809–817. https://doi.org/10.1007/s11427-021-1968-7.

Lim, P.K., Julca, I., and Mutwil, M. (2023). Redesigning plant specialized metabolism with supervised machine learning using publicly available data sources. *Computational and Structural Biotechnology Journal* 21: 1639–1650. https://doi.org/10.1016/j.csbj.2023.01.013.

Liu, K., Abdullah, A.A., Huang, M. et al. (2017). Novel approach to classify plants based on metabolite-content similarity. *BioMed Research International* 2017: e5296729. https://doi.org/10.1155/2017/5296729.

Liu, P., Zhou, H., Zheng, Q. et al. (2019). An automatic UPLC-HRMS data analysis platform for plant metabolomics. *Plant Biotechnology Journal* 17 (11): 2038–2040. https://doi.org/10.1111/pbi.13180.

Lou, Y.-R., Anthony, T.M., Fiesel, P.D. et al. (2021). It happened again: convergent evolution of acylglucose specialized metabolism in black nightshade and wild tomato. *Science Advances* 7 (46): eabj8726. https://doi.org/10.1126/sciadv.abj8726.

Ma, S., Liu, J., Li, W. et al. (2023). Machine learning in TCM with natural products and molecules: current status and future perspectives. *Chinese Medicine* 18 (1): 1–17. https://doi.org/10.1186/s13020-023-00741-9.

Mahadevan, R. and Schilling, C.H. (2003). The effects of alternate optimal solutions in constraint-based genome-scale metabolic models. *Metabolic Engineering* 5 (4): 264–276. https://doi.org/10.1016/j.ymben.2003.09.002.

Mahadevan, R., Edwards, J.S., and Doyle III, F.J. (2002). Dynamic flux balance analysis of diauxic growth in *Escherichia coli*. *Biophysical Journal* 83 (3): 1331–1340. https://doi.org/10.1016/S0006-3495(02)73903-9.

Marrano, A., Britton, M., Zaini, P.A. et al. (2020). High-quality chromosome-scale assembly of the walnut (*Juglans regia* L.) reference genome. *GigaScience* 9 (5): giaa050. https://doi.org/10.1093/gigascience/giaa050.

Mashabela, M.D., Masamba, P., and Kappo, A.P. (2022). Metabolomics and chemoinformatics in agricultural biotechnology research: complementary probes in unravelling new metabolites for crop improvement. *Biology* 11 (8): 1156. https://doi.org/10.3390/biology11081156.

Mashima, J., Kodama, Y., Fujisawa, T. et al. (2017). DNA Data Bank of Japan. *Nucleic Acids Research* 45 (D1): D25–D31. https://doi.org/10.1093/nar/gkw1001.

Matsuba, Y., Nguyen, T.T.H., Wiegert, K. et al. (2013). Evolution of a complex locus for terpene biosynthesis in *Solanum*. *The Plant Cell* 25 (6): 2022–2036. https://doi.org/10.1105/tpc.113.111013.

Matsuba, Y., Zi, J., Jones, D. et al. (2015). Biosynthesis of the diterpenoid lycosantalonol via nerylneryl diphosphate in *Solanum lycopersicum*. *PLoS One* 10 (3): e0119302. https://doi.org/10.1371/journal.pone.0119302.

Maxam, A.M. and Gilbert, W. (1977). A new method for sequencing DNA. *Proceedings of the National Academy of Sciences, USA* 74 (2): 560–564. https://doi.org/10.1073/pnas.74.2.560.

Mazid, M., Khan, T.A., and Mohammad, F. (2011). Role of secondary metabolites in defense mechanisms of plants. *Biology and Medicine* 3 (2): 232–249.

Medema, M.H., Blin, K., Cimermancic, P. et al. (2011). antiSMASH: rapid identification, annotation and analysis of secondary metabolite biosynthesis gene clusters in bacterial and fungal genome sequences. *Nucleic Acids Research* 39 (suppl_2): W339–W346. https://doi.org/10.1093/nar/gkr466.

Medema, M.H., Kottmann, R., Yilmaz, P. et al. (2015). Minimum Information about a Biosynthetic Gene cluster. *Nature Chemical Biology* 11 (9): 625–631. https://doi.org/10.1038/nchembio.1890.

Meng, F., Tang, Q., Chu, T. et al. (2022). TCMPG: an integrative database for traditional Chinese medicine plant genomes. *Horticulture Research* 9: uhac060. https://doi.org/10.1093/hr/uhac060.

Misra, B.B., Langefeld, C.D., Olivier, M. et al. (2019). Integrated omics: tools, advances and future approaches. *Journal of Molecular Endocrinology* 62 (1): R21–R45. https://doi.org/10.1530/jme-18-0055.

Mohanraj, K., Karthikeyan, B.S., Vivek-Ananth, R.P. et al. (2018). IMPPAT: a curated database of Indian Medicinal Plants, Phytochemistry And Therapeutics. *Scientific Reports* 8 (1): 4329. https://doi.org/10.1038/s41598-018-22631-z.

Moon, T., Ahn, T.I., and Son, J.E. (2019). Long short-term memory for a model-free estimation of macronutrient ion concentrations of root-zone in closed-loop soilless cultures. *Plant Methods* 15 (1): 59. https://doi.org/10.1186/s13007-019-0443-7.

Moore, B.M., Wang, P., Fan, P. et al. (2019). Robust predictions of specialized metabolism genes through machine learning. *Proceedings of the National Academy of Sciences, USA* 116 (6): 2344–2353. https://doi.org/10.1073/pnas.1817074116.

Mount, D.W. (2007). Using the basic local alignment search tool (BLAST). *Cold Spring Harbor Protocols* 2007 (7): pdb.top17. https://doi.org/10.1101/pdb.top17.

Mugford, S.T., Louveau, T., Melton, R. et al. (2013). Modularity of plant metabolic gene clusters: a trio of linked genes that are collectively required for acylation of triterpenes in oat. *The Plant Cell* 25 (3): 1078–1092. https://doi.org/10.1105/tpc.113.110551.

Naithani, S., Gupta, P., Preece, J. et al. (2020). Plant Reactome: a knowledgebase and resource for comparative pathway analysis. *Nucleic Acids Research* 48 (D1): D1093–D1103. https://doi.org/10.1093/nar/gkz996.

Nützmann, H.-W., Scazzocchio, C., and Osbourn, A. (2018). Metabolic gene clusters in eukaryotes. *Annual Review of Genetics* 52: 159–183. https://doi.org/10.1146/annurev-genet-120417-031237.

Nützmann, H.-W., Doerr, D., Ramírez-Colmenero, A. et al. (2020). Active and repressed biosynthetic gene clusters have spatially distinct chromosome states. *Proceedings of the National Academy of Sciences, USA* 117 (24): 13800–13809. https://doi.org/10.1073/pnas.1920474117.

O'Leary, N.A., Wright, M.W., Brister, J.R. et al. (2016). Reference sequence (RefSeq) database at NCBI: current status, taxonomic expansion, and functional annotation. *Nucleic Acids Research* 44 (Database issue): D733–D745. https://doi.org/10.1093/nar/gkv1189.

Oksman-Caldentey, K.-M. and Inzé, D. (2004). Plant cell factories in the post-genomic era: new ways to produce designer secondary metabolites. *Trends in Plant Science* 9 (9): 433–440. https://doi.org/10.1016/j.tplants.2004.07.006.

de Oliveira Dal'Molin, C.G. and Nielsen, L.K. (2018). Plant genome-scale reconstruction: from single cell to multi-tissue modelling and omics analyses. *Current Opinion in Biotechnology* 49: 42–48. https://doi.org/10.1016/j.copbio.2017.07.009.

Orth, J.D., Thiele, I., and Palsson, B.Ø. (2010). What is flux balance analysis? *Nature Biotechnology* 28 (3): 245–248. https://doi.org/10.1038/nbt.1614.

Pang, Z., Chen, J., Wang, T. et al. (2021). Linking plant secondary metabolites and plant microbiomes: a review. *Frontiers in Plant Science* 12: 621276. https://doi.org/10.3389/fpls.2021.621276.

Perez-Riverol, Y., Csordas, A., Bai, J. et al. (2019). The PRIDE database and related tools and resources in 2019: improving support for quantification data. *Nucleic Acids Research* 47 (D1): D442–D450. https://doi.org/10.1093/nar/gky1106.

Pomyen, Y., Wanichthanarak, K., Poungsombat, P. et al. (2020). Deep metabolome: applications of deep learning in metabolomics. *Computational and Structural Biotechnology Journal* 18: 2818–2825. https://doi.org/10.1016/j.csbj.2020.09.033.

Qi, X., Bakht, S., Qin, B. et al. (2006). A different function for a member of an ancient and highly conserved cytochrome P450 family: from essential sterols to plant defense. *Proceedings of the National Academy of Sciences, USA* 103 (49): 18848–18853. https://doi.org/10.1073/pnas.0607849103.

Rana, P., Berry, C., Ghosh, P. et al. (2020). Recent advances on constraint-based models by integrating machine learning. *Current Opinion in Biotechnology* 64: 85–91. https://doi.org/10.1016/j.copbio.2019.11.007.

Ray, A., Halder, T., Jena, S. et al. (2020). Application of artificial neural network (ANN) model for prediction and optimization of coronarin D content in *Hedychium coronarium*. *Industrial Crops and Products* 146: 112186. https://doi.org/10.1016/j.indcrop.2020.112186.

Razzaq, A., Sadia, B., Raza, A. et al. (2019). Metabolomics: a way forward for crop improvement. *Metabolites* 9 (12): 303. https://doi.org/10.3390/metabo9120303.

Ribeiro, A.J.M., Holliday, G.L., Furnham, N. et al. (2018). Mechanism and Catalytic Site Atlas (M-CSA): a database of enzyme reaction mechanisms and active sites. *Nucleic Acids Research* 46 (D1): D618–D623. https://doi.org/10.1093/nar/gkx1012.

Rokas, A., Mead, M.E., Steenwyk, J.L. et al. (2020). Biosynthetic gene clusters and the evolution of fungal chemodiversity. *Natural Product Reports* 37 (7): 868–878. https://doi.org/10.1039/C9NP00045C.

Roychowdhury, R., Das, S.P., Gupta, A. et al. (2023). Multi-omics pipeline and omics-integration approach to decipher plant's abiotic stress tolerance responses. *Genes* 14 (6): 1281. https://doi.org/10.3390/genes14061281.

Ryu, J.Y., Kim, H.U., and Lee, S.Y. (2019). Deep learning enables high-quality and high-throughput prediction of enzyme commission numbers. *Proceedings of the National Academy of Sciences, USA* 116 (28): 13996–14001. https://doi.org/10.1073/pnas.1821905116.

Samaras, P., Schmidt, T., Frejno, M. et al. (2020). ProteomicsDB: a multi-omics and multi-organism resource for life science research. *Nucleic Acids Research* 48 (D1): D1153–D1163. https://doi.org/10.1093/nar/gkz974.

Sampaio, M., Rocha, M., and Dias, O. (2022). Exploring synergies between plant metabolic modelling and machine learning. *Computational and Structural Biotechnology Journal* 20: 1885–1900. https://doi.org/10.1016/j.csbj.2022.04.016.

Sanger, F., Nicklen, S., and Coulson, A.R. (1977). DNA sequencing with chain-terminating inhibitors. *Proceedings of the National Academy of Sciences, USA* 74 (12): 5463–5467. https://doi.org/10.1073/pnas.74.12.5463.

Satam, H., Joshi, K., Mangrolia, U. et al. (2023). Next-generation sequencing technology: current trends and advancements. *Biology* 12 (7): 997. https://doi.org/10.3390/biology12070997.

Sayers, E.W., Bolton, E.E., Brister, J.R. et al. (2022). Database resources of the national center for biotechnology information. *Nucleic Acids Research* 50 (D1): D20–D26. https://doi.org/10.1093/nar/gkab1112.

Schläpfer, P., Zhang, P., Wang, C. et al. (2017). Genome-wide prediction of metabolic enzymes, pathways, and gene clusters in plants. *Plant Physiology* 173 (4): 2041–2059. https://doi.org/10.1104/pp.16.01942.

Schneider, L.M., Adamski, N.M., Christensen, C.E. et al. (2016). The *Cer-cqu* gene cluster determines three key players in a β-diketone synthase polyketide pathway synthesizing aliphatics in epicuticular waxes. *Journal of Experimental Botany* 67 (9): 2715–2730. https://doi.org/10.1093/jxb/erw105.

Seetharam, K., Kagiyama, N., and Sengupta, P.P. (2019). Application of mobile health, telemedicine and artificial intelligence to echocardiography. *Echo Research and Practice* 6 (2): R41–R52. https://doi.org/10.1530/ERP-18-0081.

Shang, Y., Ma, Y., Zhou, Y. et al. (2014). Biosynthesis, regulation, and domestication of bitterness in cucumber. *Science* 346 (6213): 1084–1088. https://doi.org/10.1126/science.1259215.

Shaw, R. and Cheung, C.Y.M. (2020). Multi-tissue to whole plant metabolic modelling. *Cellular and Molecular Life Sciences* 77: 489–495. https://doi.org/10.1007/s00018-019-03384-y.

Shen, S., Peng, M., Fang, H. et al. (2021). An *Oryza*-specific hydroxycinnamoyl tyramine gene cluster contributes to enhanced disease resistance. *Science Bulletin* 66 (23): 2369–2380. https://doi.org/10.1016/j.scib.2021.03.015.

Shimura, K., Okada, A., Okada, K. et al. (2007). Identification of a biosynthetic gene cluster in rice for momilactones. *Journal of Biological Chemistry* 282 (47): 34013–34018. https://doi.org/10.1074/jbc.M703344200.

Shirley, H.J., Koyioni, M., Muncan, F. et al. (2019). Synthesis of lamellarin alkaloids using orthoester-masked α-keto acids. *Chemical Science* 10 (15): 4334–4338. https://doi.org/10.1039/C8SC05678A.

Slot, J.C. and Rokas, A. (2010). Multiple *GAL* pathway gene clusters evolved independently and by different mechanisms in fungi. *Proceedings of the National Academy of Sciences, USA* 107 (22): 10136–10141. https://doi.org/10.1073/pnas.0914418107.

Starcevic, A., Zucko, J., Simunkovic, J. et al. (2008). *ClustScan*: an integrated program package for the semi-automatic annotation of modular biosynthetic gene clusters and *in silico* prediction of novel chemical structures. *Nucleic Acids Research* 36 (21): 6882–6892. https://doi.org/10.1093/nar/gkn685.

Sterling, T. and Irwin, J.J. (2015). ZINC 15 – ligand discovery for everyone. *Journal of Chemical Information and Modeling* 55 (11): 2324–2337. https://doi.org/10.1021/acs.jcim.5b00559.

Su, X., Yang, L., Wang, D. et al. (2022). 1 K Medicinal Plant Genome Database: an integrated database combining genomes and metabolites of medicinal plants. *Horticulture Research* 9: uhac075. https://doi.org/10.1093/hr/uhac075.

Sud, M., Fahy, E., Cotter, D. et al. (2016). Metabolomics workbench: an international repository for metabolomics data and metadata, metabolite standards, protocols, tutorials and training, and analysis tools. *Nucleic Acids Research* 44 (D1): D463–D470. https://doi.org/10.1093/nar/gkv1042.

Sun, Q., Zybailov, B., Majeran, W. et al. (2009). PPDB, the Plant Proteomics Database at Cornell. *Nucleic Acids Research* 37 (Database issue): D969–D974. https://doi.org/10.1093/nar/gkn654.

Sun, X., Zhu, S., Li, N. et al. (2020). A chromosome-level genome assembly of garlic (*Allium sativum*) provides insights into genome evolution and allicin biosynthesis. *Molecular Plant* 13 (9): 1328–1339. https://doi.org/10.1016/j.molp.2020.07.019.

Swaminathan, S., Morrone, D., Wang, Q. et al. (2009). CYP76M7 is an *ent*-cassadiene C11α-hydroxylase defining a second multifunctional diterpenoid biosynthetic gene cluster in rice. *The Plant Cell* 21 (10): 3315–3325. https://doi.org/10.1105/tpc.108.063677.

Takos, A.M., Knudsen, C., Lai, D. et al. (2011). Genomic clustering of cyanogenic glucoside biosynthetic genes aids their identification in *Lotus japonicus* and suggests the repeated evolution of this chemical defence pathway. *The Plant Journal* 68 (2): 273–286. https://doi.org/10.1111/j.1365-313X.2011.04685.x.

Tello-Ruiz, M.K., Jaiswal, P., and Ware, D. (2022). Gramene: a resource for comparative analysis of plants genomes and pathways. *Methods in Molecular Biology (Clifton, N.J.)* 2443: 101–131. https://doi.org/10.1007/978-1-0716-2067-0_5.

Töpfer, N., Fuchs, L.-M., and Aharoni, A. (2017). The PhytoClust tool for metabolic gene clusters discovery in plant genomes. *Nucleic Acids Research* 45 (12): 7049–7063. https://doi.org/10.1093/nar/gkx404.

Trenchard, I.J., Siddiqui, M.S., Thodey, K. et al. (2015). *De novo* production of the key branch point benzylisoquinoline alkaloid reticuline in yeast. *Metabolic Engineering* 31: 74–83. https://doi.org/10.1016/j.ymben.2015.06.010.

Tu, L., Su, P., Zhang, Z. et al. (2020). Genome of *Tripterygium wilfordii* and identification of cytochrome P450 involved in triptolide biosynthesis. *Nature Communications* 11 (1): 971. https://doi.org/10.1038/s41467-020-14776-1.

Turova, P., Stavrianidi, A., Svekolkin, V. et al. (2022). Analysis of primary liquid chromatography mass spectrometry data by neural networks for plant samples classification. *Metabolites* 12 (10): 993. https://doi.org/10.3390/metabo12100993.

Twaij, B.M. and Hasan, M.N. (2022). Bioactive secondary metabolites from plant sources: types, synthesis, and their therapeutic uses. *International Journal of Plant Biology* 13 (1): 4–14. https://doi.org/10.3390/ijpb13010003.

Van Bel, M., Silvestri, F., Weitz, E.M. et al. (2022). PLAZA 5.0: extending the scope and power of comparative and functional genomics in plants. *Nucleic Acids Research* 50 (D1): D1468–D1474. https://doi.org/10.1093/nar/gkab1024.

Van Lanen, S.G. and Shen, B. (2006). Microbial genomics for the improvement of natural product discovery. *Current Opinion in Microbiology* 9 (3): 252–260. https://doi.org/10.1016/j.mib.2006.04.002.

Vavricka, C.J., Takahashi, S., Watanabe, N. et al. (2022). Machine learning discovery of missing links that mediate alternative branches to plant alkaloids. *Nature Communications* 13 (1): 1405. https://doi.org/10.1038/s41467-022-28883-8.

Venditti, A. and Bianco, A. (2020). Sulfur-containing secondary metabolites as neuroprotective agents. *Current Medicinal Chemistry* 27 (26): 4421–4436. https://doi.org/10.2174/0929867325666180912105036.

Vivek-Ananth, R.P., Mohanraj, K., Sahoo, A.K. et al. (2023). IMPPAT 2.0: an enhanced and expanded phytochemical atlas of Indian medicinal plants. *ACS Omega* 8 (9): 8827–8845. https://doi.org/10.1021/acsomega.3c00156.

Wang, Q., Hillwig, M.L., and Peters, R.J. (2011). CYP99A3: functional identification of a diterpene oxidase from the momilactone biosynthetic gene cluster in rice. *The Plant Journal* 65 (1): 87–95. https://doi.org/10.1111/j.1365-313X.2010.04408.x.

Weininger, D. (1988). SMILES, a chemical language and information system. 1. Introduction to methodology and encoding rules. *Journal of Chemical Information and Computer Sciences* 28 (1): 31–36. https://doi.org/10.1021/ci00057a005.

Wiemann, P., Guo, C.-J., Palmer, J.M. et al. (2013). Prototype of an intertwined secondary-metabolite supercluster. *Proceedings of the National Academy of Sciences, USA* 110 (42): 17065–17070. https://doi.org/10.1073/pnas.1313258110.

van Wijk, K.J., Leppert, T., Sun, Q. et al. (2021). The Arabidopsis PeptideAtlas: harnessing worldwide proteomics data to create a comprehensive community proteomics resource. *The Plant Cell* 33 (11): 3421–3453. https://doi.org/10.1093/plcell/koab211.

Wiltschi, B., Cernava, T., Dennig, A. et al. (2020). Enzymes revolutionize the bioproduction of value-added compounds: from enzyme discovery to special applications. *Biotechnology Advances* 40: 107520. https://doi.org/10.1016/j.biotechadv.2020.107520.

Wu, Y., Hillwig, M.L., Wang, Q. et al. (2011). Parsing a multifunctional biosynthetic gene cluster from rice: biochemical characterization of CYP71Z6 & 7. *FEBS Letters* 585 (21): 3446–3451. https://doi.org/10.1016/j.febslet.2011.09.038.

Yan, J. and Wang, X. (2022). Unsupervised and semi-supervised learning: the next frontier in machine learning for plant systems biology. *The Plant Journal* 111 (6): 1527–1538. https://doi.org/10.1111/tpj.15905.

Yang, B., Chen, C., Chen, F. et al. (2021). Identification of cumin and fennel from different regions based on generative adversarial networks and near infrared spectroscopy.

Spectrochimica Acta Part A: Molecular and Biomolecular Spectroscopy 260: 119956. https://doi.org/10.1016/j.saa.2021.119956.

Yates, A.D., Allen, J., Amode, R.M. et al. (2022). Ensembl Genomes 2022: an expanding genome resource for non-vertebrates. *Nucleic Acids Research* 50 (D1): D996–D1003. https://doi.org/10.1093/nar/gkab1007.

Ye, Y.-N., Liang, D.-F., Yi, J.-H. et al. (2023). IGTCM: an integrative genome database of traditional Chinese medicine plants. *The Plant Genome* 16 (2): e20317. https://doi.org/10.1002/tpg2.20317.

Yeung, C.S., Beck, T., and Posma, J.M. (2022). MetaboListem and TABoLiSTM: two deep learning algorithms for metabolite named entity recognition. *Metabolites* 12 (4): 276. https://doi.org/10.3390/metabo12040276.

Yip, K.Y., Cheng, C., and Gerstein, M. (2013). Machine learning and genome annotation: a match meant to be? *Genome Biology* 14 (5): 205. https://doi.org/10.1186/gb-2013-14-5-205.

Younessi-Hamzekhanlu, M., Ozturk, M., Jafarpour, P. et al. (2022). Exploitation of next generation sequencing technologies for unraveling metabolic pathways in medicinal plants: a concise review. *Industrial Crops and Products* 178: 114669. https://doi.org/10.1016/j.indcrop.2022.114669.

Yu, J., Chang, P.-K., Bhatnagar, D. et al. (2000). Cloning of a sugar utilization gene cluster in *Aspergillus parasiticus. Biochimica et Biophysica Acta (BBA)-Gene Structure and Expression* 1493 (1–2): 211–214. https://doi.org/10.1016/S0167-4781(00)00148-2.

Zampieri, G., Vijayakumar, S., Yaneske, E. et al. (2019). Machine and deep learning meet genome-scale metabolic modeling. *PLoS Computational Biology* 15 (7): e1007084. https://doi.org/10.1371/journal.pcbi.1007084.

Zeng, X., Zhang, P., Wang, Y. et al. (2019). CMAUP: a database of collective molecular activities of useful plants. *Nucleic Acids Research* 47 (D1): D1118–D1127. https://doi.org/10.1093/nar/gky965.

Zengin, G., Mahomoodally, M.F., Yıldıztugay, E. et al. (2022). Chemical composition, biological activities and *in silico* analysis of essential oils of three endemic prangos species from Turkey. *Molecules* 27 (5): 1676. https://doi.org/10.3390/molecules27051676.

Zhou, F. and Pichersky, E. (2020). The complete functional characterisation of the terpene synthase family in tomato. *New Phytologist* 226 (5): 1341–1360. https://doi.org/10.1111/nph.16431.

Zuo, E., Sun, L., Yan, J. et al. (2022). Rapidly detecting fennel origin of the near-infrared spectroscopy based on extreme learning machine. *Scientific Reports* 12 (1): 13593. https://doi.org/10.1038/s41598-022-17810-y.

20

An Appraisal of Flavonoids Through Bioinformatics

Manoj Kumar Mishra[1] *and Vibha Pandey*[2]

[1] *Department of Biotechnology, Faculty of Engineering, Rama University, Kanpur, Uttar Pradesh, India*
[2] *Molecular Biology and Biotechnology Division, CSIR-National Botanical Research Institute, Lucknow, Uttar Pradesh, India*

20.1 Overview of Flavonoids

Flavonoids are known to be a diverse group of plant secondary metabolites found in plants and plant-derived products, particularly they belong to a class of secondary metabolites with a polyphenolic structure (Mishra et al. 2021). They have a variety of beneficial biochemical and antioxidant effects that are linked to various chronic diseases such as diabetes, cancer, Alzheimer's disease (AD), atherosclerosis, and so on (Zhang et al. 2017; Panche et al. 2016; Kumar and Pandey 2013). Flavonoids have a wide range of health-promoting properties and are used in a variety of nutraceutical, pharmaceutical, medicinal, and cosmetic applications. This is due to their ability to modulate important cellular enzyme functions as well as their antioxidative, anti-inflammatory, antimutagenic, and anticarcinogenic properties (Zhang et al. 2017). Flavonoids play a variety of biological activities in plants, i.e. they act as growth regulators and enhance tolerance against oxidative stress (Burda and Oleszek 2001). It has long been known that flavonoids are synthesized at specific locations in plants and are responsible for the color and fragrance of flowers as well as for attracting pollinators to fruits and, as a consequence fruit dispersion to aid in seed and spore germination, as well as the growth and development of seedlings (Panche et al. 2016; Wang et al. 2000). Flavonoids protect plants from various biotic and abiotic stresses and serve as special UV filters, signaling molecules, allopathic substances, phytoalexins, detoxifying agents, and antimicrobial defensive compounds (Mishra et al. 2017, 2021, 2022; Burda and Oleszek 2001). Flavonoids have protective effects against frost hardiness, and drought resistance, and may function to help plants adapt to heat and tolerate freezing temperatures (Mishra et al. 2013, 2017, 2021). It was shown that functional gene silencing in plants was associated with flavonoid biosynthesis and that the early advances in floral genetics were primarily due to mutation techniques having an impact on flower colors derived from flavonoids (Dixon and Pasinetti 2010). The health benefits of flavonoids have been attributed to both humans and animals, and the current focus is on chemoprevention and disease treatment. There are currently approximately 6000 flavonoids that contribute to the

Bioinformatics for Plant Research and Crop Breeding, First Edition. Edited by Jen-Tsung Chen.

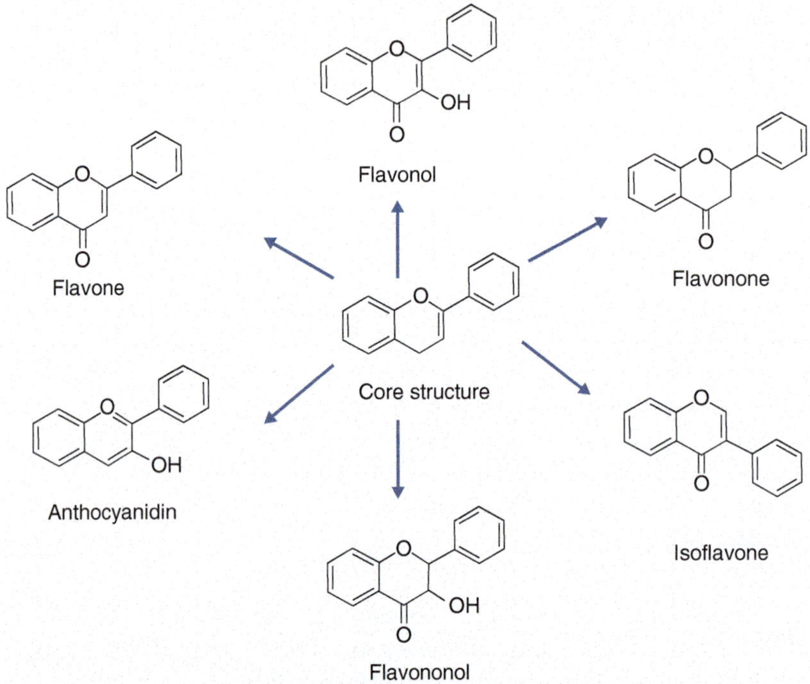

Figure 20.1 Chemical structure of different flavonoids.

vibrant fruit pigment, fruit-derived products, vegetables, and medicinal plants. Flavones, flavonols, flavanones, isoflavones, biflavanones, catechins, and anthocyanidins are examples of flavonoids (Figure 20.1) (Dixon and Pasinetti 2010; Ray et al. 2008). The identification and prediction of potential biological activities of flavonoids by artificial intelligence and chalcone synthase (CHS) have been attempted to discuss in this chapter.

20.2 Identification of Flavonoid Biosynthetic Genes and Enzymes by Computational Tools

Flavonoids in fruits and vegetables are difficult to identify or profile because their complex structures, distribution, and concentrations in plants vary greatly (Zhang et al. 2017; Panche et al. 2016). Traditionally, a number of spectroscopic technologies, including IR, UV, NMR, and MS, can be used to determine the structure of flavonoids. In the last few years, various computational prediction methods such as molecular docking, virtual screening, and other artificial intelligence tools evolved for the identification, structure analysis, and prediction of the properties of flavonoids (Karydas et al. 2020; Deng et al. 2018; Djeradi et al., 2014). Predicting flavonoids computationally can be useful for drug discovery, understanding their biological activities, and designing new compounds with desired properties. Here are some commonly used computational approaches and tools for predicting flavonoids.

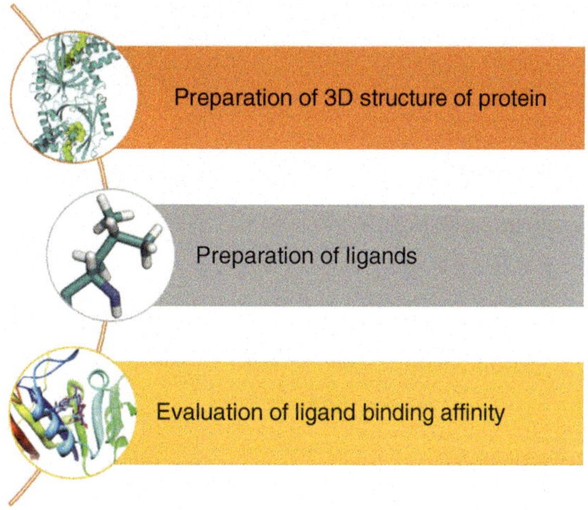

Figure 20.2 Three steps of molecular docking for drug development.

20.2.1 Chemoinformatics and Molecular Docking

Cheminformatics involves the use of chemical databases and computational techniques to analyze chemical structures and properties of flavonoids. It is accompanied by chemical experiment results that are used to predict some obscure chemical phenomena (Al-Jumaili and Al Hdeethi 2021). Molecular docking is a bioinformatics modeling technique that involves bonding two or more molecules to create a stable structure. Molecular docking can predict the three-dimensional properties of molecules based on their binding properties. It is used to predict the binding affinity of flavonoids with precise biological targets, such as enzymes or receptors (Moulishankar and Lakshmanan 2020). Figure 20.2 depicts the three steps in drug development that involve molecular docking. Software tools like AutoDock 4 and AutoDock Vina 1 are open-source programs that can be used for molecular docking studies. The source code is available at https://github.com/ccsb-scripps/AutoDock-Vina (Eberhardt et al. 2021).

20.2.2 Virtual Screening

Virtual screening is a technique used to screen large chemical databases to identify potential flavonoids with specific biological activities (Lavecchia and Di Giovanni 2013). In virtual screening molecular dynamic simulations of the flavonoid–DNA complexes showed that the flavonoids interacted with DNA through hydrogen bonding, hydrophobic interaction, and π-stacking (Oliveira et al. 2022). Molecular docking and machine learning methods can be combined in virtual screening workflows to prioritize candidate compounds for further experimental testing (Figure 20.3). Molecular dynamic simulations suggested that the in virtual screening flavonoid–DNA complexes. Several virtual screening methods are available for the identification of flavonoids and active biomolecules.

20.2.2.1 Virtual High-throughput Screening

This method is used to identify flavonoids from large collections of virtual compound libraries. This is accomplished by employing optimal parameters that compute the target receptor's complementarity with compounds.

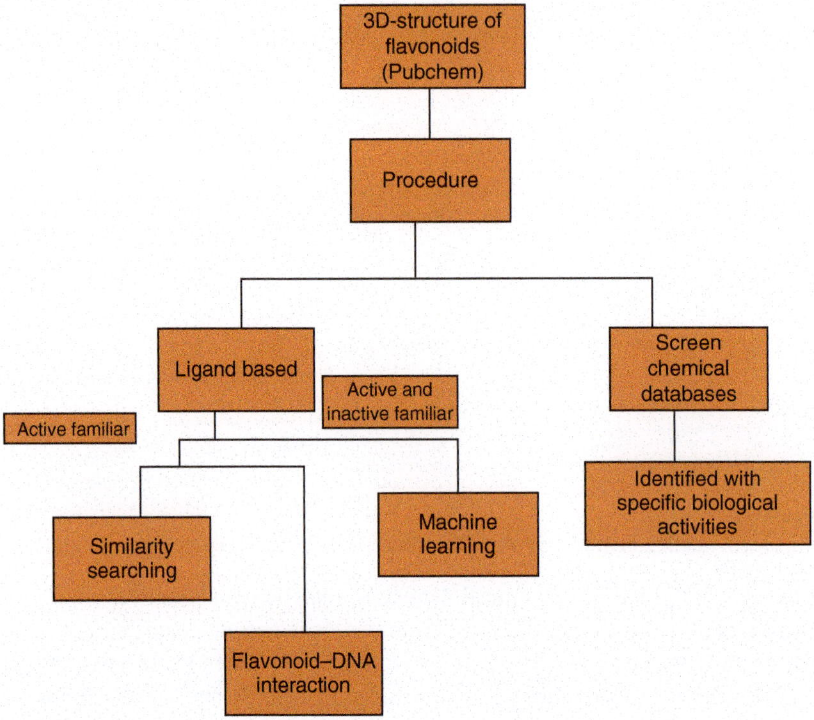

Figure 20.3 Virtual screening of flavonoids with Machine learning.

20.2.2.2 Active Site Prediction

By virtually screening a compound against numerous flavonoids or flavones that represent different active site types, the active site of the compound can be predicted. The comprehensive effects of compound active site prediction include the identification of potential targets, the prognosis of harmful effects, and the identification of therapeutic agents.

20.2.2.3 Structure-based Screening

In this method, the estimation of the affinity likelihood that the ligand will bind to targets is combined with real internal coordinate docking.

20.2.2.4 Ligand-based Screening

A screening technique known as "ligand-based screening" makes use of the collective knowledge held in sets of ligands produced from one or more high-affinity scaffolds, like hydrogen bond donors and acceptors.

20.2.2.5 Machine Learning Techniques

Because machine learning techniques are based on high-dimensional data processing capabilities, they are suitable for virtual screening of large compound libraries to classify molecules as active or inactive or to rank them according to their levels of activity.

20.2.2.6 Virtual Combinatorial Libraries

All techniques for the virtual screening of already-developed compound libraries are included in this virtual screening design. Structure-based docking and scoring as well as similarity-based compound clustering techniques are the most crucial methods.

20.2.3 Flavon Q

Flavon Q, an automated data processing tool, allows field-acquired expertise to be transferred into data analysis and facilitates flavonoid research (Zhang et al. 2017). It is intended for the automated analysis of data from the flavone and flavonol glycoside subclasses of flavonoids. Flavon Q employs ultra-high performance liquid chromatography-diode array detection, which is capable of data format conversion, peak detection, flavone and flavonol glycoside peak extraction, identification of flavone and flavonol glycosides, and quantitative results. By examining data from samples with mixed flavonoid standards and plant samples, Flavon Q was validated. Due to the enormous diversity of flavonoids and the dearth of commercially available standards, profiling different flavonoids in natural products can occasionally be difficult.

20.2.4 Flavon Q-2.0

FlavonQ-2.0 is an updated software program designed specifically for the analysis of flavonoids (Zhang et al. 2017; Harnly et al. 2006). Flavon Q-2.0 has made several significant advancements over its predecessor, Flavon Q. Like Flavon Q, Flavon Q-2.0 includes all of the functions needed to detect chromatographic peaks, integrate peak areas, interpret MS spectra, and generate qualitative and quantitative results. Flavon Q-2.0's significant advancements are:

1) Different classes of flavonoids such as flavone/flavonol, flavan/flavan/flavanol, flavanone/flavanonol, isoflavone, anthocyanidins, and hydroxycinnamic acids (nonflavonoids) can all be analyzed by Flavon Q-2.0.
2) The chemometric pattern recognition method is used by the Flavon Q-2.0 program to categorize the different classes of flavonoids. Under this program, flavonoids are first classified using UV–vis spectra in a step-by-step manner (known as step-wise classification). After that individual flavonoids in each classified class are identified by their mass spectra (Zhang et al. 2017).
3) SIMICA and FOAM are well-known modeling techniques, but they can also be used in classification mode (Wabuyele and Harrington 1996). Because modeling methods rely on feature similarities within each independent class, the test sample may belong to none of the classes in the training sets (Zhang et al. 2017).
4) The outcome of the aforementioned step is correlated with the peak's HRAM/MSn spectra, and a tentative identification is made using a search of the internal flavonoid database.

20.2.5 Bioinformatics Tools

Bioinformatics tools can be used to analyze the biological pathways and networks influenced by flavonoids. This can help predict their potential roles in different physiological processes. Tools like Cytoscape and pathway analysis software can be valuable in this regard (Bindea et al. 2009). It's important to note that computational predictions should be validated through experimental studies to confirm the accuracy of the results. Additionally, the choice of computational method or tool may depend on the specific research question and available resources. Integrating multiple computational approaches can often provide more robust predictions of flavonoid properties and activities (Žuvela et al. 2018).

20.3 Prediction of the Potential Biological Activities of Flavonoids Based on Their Chemical Structure

20.3.1 Quantitative Structure–Activity Relationship (QSAR)

In QSAR modeling, predictors have physicochemical properties of chemicals or hypothetical molecular descriptors; response variables in QSAR modeling could be the biological activity of chemical components. Based on their chemical structures, QSAR models are used to predict the biological activity of flavonoids (Žuvela et al. 2018). These models investigate the relationship between a compound's chemical structure and its biological activity. Hence, quantitative structure–activity relationship (QSAR) models have been developed to estimate and predict the antioxidant activity of flavonoids using descriptors derived from the chemical structure and various physicochemical parameters (Tao et al. 2023). The use of regression and classification methods to build QSAR models relating molecular descriptors and experimental antioxidant capacity allows for the prediction of antioxidant capacity of other flavonoids whose antioxidant capacity is unknown experimentally; this may lay the theoretical groundwork for extracting and synthesizing flavonoids with high antioxidant capacity (Tao et al. 2023; Djeradi et al. 2014). Several QSAR models have been created using various techniques, such as multiple linear regression (MLR), partial least squares (PLS), artificial neural networks (ANN), and support vector regression (SVR), to more accurately predict antioxidant capacity (Gramatica 2013). It has been established that the antioxidant activity of flavonoids as determined by *in vitro* chemical experiments is closely related to the Fukui indices, zero-order connectivity index, molar refractivity, and electrotopological state (Djeradi et al. 2014).

A QSAR is a mathematical model in the following form:

$$\text{Modelling activity} = f\left(\text{physiochemical properties and / or structural properties}\right) + \text{error}$$

The following are the main steps in QSAR:

1) Data set selection and extraction of structural/empirical descriptors,
2) variable selection,
3) model construction, and
4) validation evaluation

QSAR modeling software, such as Cheminformatics tools or software packages like RDKit and Chem Des, can be used for this purpose.

20.3.2 Machine Learning and Data Mining

Machine learning is an artificial intelligence subfield dealing with the ways in which machines learn from data. Additionally, it involves using computers and mathematical algorithms to make predictions about the future using data-driven learning (Kusumiyati and Asikin 2023). It is thought to be a multistep process that includes the (selection, preprocessing, transformation, data mining, interpretation, and evaluation) depicted in (Figure 20.4) (Kavakiotis et al. 2017). Machine learning algorithms can be trained on a dataset of known flavonoids and their biological activities to predict the properties of new compounds. Features extracted from the chemical structures can be used as input for machine learning models (Kusumiyati and Asikin 2023). Kusumiyati and Asikin (2023), combined near-infrared spectroscopy (NIRS) which can predict organic samples in solid, liquid, and gaseous forms, with machine learning to predict total flavonoid content. Python libraries like scikit-learn and deep learning frameworks like Tensor Flow and PyTorch can be employed for this purpose.

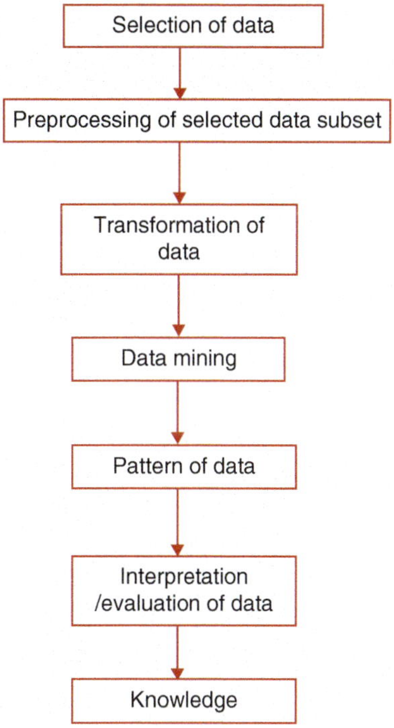

Figure 20.4 Important steps involved in the machine learning process.

20.3.3 Molecular Dynamics Simulation

Molecular dynamics simulations can be used to study the dynamic behavior of flavonoids in a biological environment. These simulations provide insights into the interactions between flavonoids and biomolecules over time. Software packages like GROMACS or AMBER are commonly used for molecular dynamics simulations.

20.4 Chalcone Synthase

Chalcone synthase (CHS; EC. 2.3.1.74) is the first identified enzyme of the flavonoid biosynthesis pathway and gatekeeper of the anthocyanin pathway. This enzyme is widely distributed in Cryptogamous plants and has been confirmed to retain a conserved nature with 60% homologous sequences among angiosperms. Characteristics of CHS have been studied in countless plants including, *Alpinia oxyphylla*, *Arabidopsis thaliana*, *Boesenbergia rotunda*, *Coleus forskohlii*, *Curcuma longa*, *Garbera hybrida*, *Grewia asiatica*, *Juglans regia*, *Oncidium Gower Ramsay*, *Oryza sativa*, *Malus domestica*, *Medicago sativa*, *Rheum emodi*,

Triticum aestivum, and *Zea mays*. High similarity among CHS (type III polyketide synthase) of different plants has been proved with two exons and one intron in gene structure, while the two active sites and highly conserved catalytic center of, Cys-His-Asn-Phe in protein structure. The first step of the phenylpropanoid pathway is catalyzed by CHS to produce naringenin chalcone. The catalytic process involved decarboxylation, condensation, and cyclization, utilizing three molecules of malonyl-CoA molecules along with one molecule of p-coumaroyl-CoA molecule. Produced chalcone can serve as a substrate further in the pathway to synthesize tannins, flavones, flavonols, isoflavones, resveratrol, anthocyanins, etc. These ultimate products have significant biological and physiological functions in environmental adaptability, floral pigments, UV protectants, plant resistance, insect repellents, etc. (Ahmad et al. 2023; Akram et al. 2023; Liu et al. 2023; Mahajan et al. 2023).

Flavonoid biosynthetic pathway presented a model to establish the evolutionary understanding of genes/enzymes as well as their relative importance in structural or regulatory processes, in the context of an ecological perspective. Plenty of studies revealed that CHSs were encoded by a multigene family. The number and distribution of CHS genes varied among the different species of plants along with different expression and regulation mechanisms, which indicate diverse events during its evolution process.

Therefore, it is evident to study the characteristics of CHS genes and enzymes using different aspects of molecular, biochemical, and physiological assets. *In silico* analysis provides the basic foundation to create a hypothetical approach, in order to achieve the detailed knowledge of genes and proteins. Such analysis has been very helpful in confirming the evolutionary status of the CHS gene and the involvement of CHS enzyme active site amino acids to complete the enzymatic reaction.

20.5 Sequence Retrieval

In silico analyses of genes/proteins depends upon the availability of sequences. The complete sequence of genes and/or enzymes is a must in order to start any *in silico* analysis. These sequences can be retrieved from online databases like NCBI, or databases designated to specific plants. If the sequences are not available, gene amplification, cloning, and sequencing will be required to achieve the goal.

20.5.1 Through Cloning

CHS from *Ginkgo biloba* (GbCHS; Pang et al. 2004; Xu et al. 2007), *Zingiber officinale* (ZoCHS; Ghosh and Mandi 2015), *C. forskohlii* (CfCHS; Awasthi et al. 2016); *Dysoxylum gotadhora* (DgCHS; Mahajan et al. 2023) was cloned and sequenced using cDNA from leaf tissues. The conserved domain and gDNA helped in the amplification and assembly of CHS from *Rhododendron* × *hybridum* (Jia et al. 2023).

20.5.2 Online Search

EST sequences of *Sorghum bicolor* were used to extract eight virtual CHS, using BLASTN and assembled using the SeqManII program of Lasergene software, followed by sequencing (Lo et al. 2002). The complete sequence of two CHS from *Scutellaria baicalensis* was obtained from GenPept by (Lei et al. 2009). Four CHS protein sequences of *Solanum tuberosum* were obtained from Swiss-Prot (Gupta et al. 2014). The protein sequence of the OsCHS gene family (OsCHS1-OsCHS27) in *O. sativa* previously reported by Han et al. (2017) was retrieved from the iTAK (Ahmad et al. 2023). Nine CHS from *Linum usitatissimum* (LuCHS; Eom and Hyun 2016), 14-CHS from *Z. mays* (ZmCHS; Han et al. 2016), 11-CHS from *Zostera marina* (ZomaCHS; Ma et al. 2021); 27-CHS from *O. sativa* (OsCHS; Ahmad et al. 2023) were identified using CHS protein query against genome databases. Using a few databases and online tools (Pfam, NCBI, SMART, expasy) 87 wheat CHS (TaCHS) proteins have been set aside, using 5-CHS protein sequences of barley, maize, and *Arabidopsis* (Liu et al. 2023). NCBI database was used to find a total of 22 (2-complete and 20-partial) CHS1 sequences that belong to the family Zingiberaceae (16 species), Costaceae (5 species), and Poaceae (1 species) (Akram et al. 2023).

20.6 Localization

Localization of SbCHS in the cytoplasmic matrix has been confirmed by TargetP 1.1 Server, where flavonoid synthesis takes place. Also, the hydrophilic nature and absence of transmembrane topological structure were confirmed by TMHMM Server v.2.0 and ProtFun 2.2 Server to confirm catalytic function (Lei et al. 2009). The cytoplasmic existence of LuCHS was confirmed using Plant-mPLoc web services (Eom and Hyun 2016). Among 27 OsCHS proteins, 21 resided in the cytoplasm, 5 (OsCHS01, OsCHS02, OsCHS03, OsCHS08, and OsCHS27) in chloroplast and only one (OsCHS22) in mitochondria (Ahmad et al. 2023).

20.7 Homology Search

Mostly an intron is present between two exons of *SbCHS* (Lo et al. 2002) and ZmCHS (Han et al. 2016) similar to barley, soybean, and *Arabidopsis* as well as RhCHS (180, 990 bp exon and 870 bp intron; Jia et al. 2023), TaCHS (Liu et al. 2023). Intron varies from 150 to 1000 bp in length. BLASTN search detected homology between 8-CHS of *S. bicolor* and *Z. mays*, with two conserved reasons having 87–90% identity. Sequences of SbCHS 1–8 at nucleotide (77–79%) and protein (81–82%) levels exposed their similarity (Lo et al. 2002). LuCHS genes were detected with the presence of 1–3 exons, where the existence of the first intron was sited between first and second nucleotide (Eom and Hyun 2016). ZomaCHS mostly have one or two exons, but can have up to four exons (Ma et al. 2021) There are variations in lengths and numbers of exons and introns among OsCHS (Ahmad et al. 2023).

20.8 Conserved Domain

Two conserved domains, the malonyl-CoA binding site, and 6-motifs, including the N-glycosylation motif were confirmed through ScanProsite and CDD (Lei et al. 2009). Transferase activity was associated with two conserved C-terminal and N-terminal domains were present in 14-ZmCHS (Han et al. 2016) and 11-ZomaCHS (Ma et al. 2021) represented by two and six as well as four and three conserved motifs, respectively. PLN03170,

PLN03171, and PLN03173 superfamily domains were confirmed in 2-, 19-, and 2-OsCHS genes, along with PLN03169 functional and CHS-Like conserved domains in 2- and 1-OsCHS, respectively (Ahmad et al. 2023). Also, the MEME online server established five motif distributions in 22-OsCHS genes, followed by 4-motif in 4-OsCHS and 3-motifs in OsCHS22 (Ahmad et al. 2023), while out of 15 motifs of 87-TaCHS, 8 motifs were detected in TaCHS of clade I–III (Liu et al. 2023). Three out of 6-TaCHS of Clad-I contain 13 motifs, while the other three contain 14 motifs. Among clad-II 26-TaCHS, one- motif was absent in 17-TaCHS. Alignment of CHS from 10 plants (*Rhododenron× hybridum*, *Rhododendron simsii*[three amino acids], *Vaccinium myrtillus*, *Camellia sinensis*, *Camellia nitidissima*, *M. domestica*, *Prunus armeniaca*, and *Rosa chinensis*) pronounced presence of five (I–V) conserved domains in RhCHS (Jia et al. 2023).

20.9 Sequence Alignment and Phylogeny

Sequence alignment and phylogenetic evaluation is the most common yet significant analysis, which unveils the similarities and dissimilarities among nucleotides/proteins together with their evolutionary behavior.

20.9.1 Active Site

Alignment of SbCHS 1–8, GbCHS, and StCHS revealed the position of conserved amino acids Cys^{164}, Phe^{215}, His^{303}, Asn^{336}, and Pro^{375}, defining the active site of CHS. Among these four residues, Cys, His, and Asn are catalytic triads of the enzyme and are positioned at the lower side of hydrophobic pockets. Any mutation in these amino acids will cause dysfunction (Lo et al. 2002; Pang et al. 2004; Lei et al. 2009; Mallika et al. 2011; Gupta et al. 2014; Eom and Hyun 2016; Ma et al. 2021; Jia et al. 2023; Mahajan et al. 2023). Residue Met^{137}, Ile^{254}, and Phe^{265} can be declared conserved among angiosperms. Apart from these Cys^{65}-Cys^{89}- Cys^{135}- Cys^{169}- Cys^{195}- Cys^{347} were also found to be conserved (Mallika et al. 2011). Cys is present in the highly conserved "MMYQQGCFAGGTVLR" region of CHS (Lo et al. 2002). RhCHS was discovered with the active site "RLMLYQQGCFAGGTVLR" in conserved sequence II (position 156–172) and the signature motif "GVLFGFGPGLT" in conserved sequence V (position 368–380). RhCHS was also noted with Ser^{133}, Glu^{192}, Thr^{194}, Thr^{197}, and Ser^{338} as CoA-binding sites (Jia et al. 2023). Other conserved motifs in DbCHS are FGFGPGL, GNMSSA, WIAHPGGPA, LFGDG, and MMYQGCF (Mahajan et al. 2023).

20.9.2 Phylogeny

Pang et al. 2004 used the software Vector NTIe Suite 8 and Clustal W for multi-alignment of GbCHS, showing more than 80% similarity with gymnosperm and angiosperm. However, GbCHS was found closer to the gymnosperm. CHS showed wide distribution, though SbCHS 1–7 clustered together, while CHS8 remains outside the cluster (Lo et al. 2002). MEGA 2.0 was used by Pang et al. (2004) for phylogenetic tree construction, confirming the sharing of the same family signature sequences and motifs by GbCHS, closer to gymnosperms (same cluster). Lei et al. (2009) utilized Vector NTI Suite 8 software for multiple sequence alignment with 16 other plant CHS. Phylogeny by Ma et al. (2021) distributed ZomaCHS into three clades, where clade-I and clade-II validated with other plants CHS while clade-III is a cluster of no-CHA from plants. ClustalX aligned SbCHS sequences with other plants' CHS while constructing a phylogenetic tree using MEGA3 and PHYLIP. There was no clear evolutionary pattern spotted among 22-CHS of Zingiberaceae, Costaceae, and Poaceae to confirm the conserved nature of CHS within the Commelinids clade. The phylogenetic tree divided these CHS into Clade-I (10 species from Zingiberaceae and 1 from Costaceae) and two Subclusters of Clade-II (with 3 species from Zingiberaceae and 1 from Costacea. Single species of Poaceae used in the study belonged to Subcluster-II (Akram et al. 2023). Phylogenetic relationships among CHS of *O. sativa*, *A. thaliana*, and *Capsicum annuum* classified them into 4-subgroups. Subgroup-I contained 6 OsCHS proteins, while subgroup-II, III, and IV contained 2-, 8-, and 11-OsCHS proteins, respectively (Ahmad et al. 2023). The phylogenetic tree of 32 CHS (Genebank) including RhCHS clustered them into 8 groups, where *Rhododendron* CHS clade was closer to *Vaccinium* CHS as they belong to the same family *Ericaceae*, which is closely related to Camellia family clad (Jia et al. 2023). CHS from wheat (TaCHS) clustered into three major groups (Clade I, II, and III) with 6, 26, and 55 CHS proteins, respectively. Though clade-III only consists of TaCHS genes, in clade-I and clade-II TaCHS genes were clustered with 7-CHS-like genes and 5CHS genes barley, maize, and *Arabidopsis* (Liu et al. 2023). The phylogenetic tree of DbCHS with CHS proteins from bacteria, fungi, monocots, and dicots displayed >80% similarity in higher plants. Therefore, DbCHS is associated with dicotyledonous plants, and most closely to the CHS from *Dimocarpus longan* (Mahajan et al. 2023).

Members of the plant CHS superfamily share 80% similarity, suggesting a conservative nature. CHS from angiosperms and gymnosperms grouped together in the phylogenetic tree. CHS proteins can be regarded as an ancestor of the plant type III PKS superfamily, as CHS from fungi, bryophytes and plants cluster sequentially (Mallika et al. 2011). There were seven single nucleotide polymorphisms observed among ZoCHS from 17 different genotypes (Ghosh and Mandi 2015). ZmCHS genes were classified into 4-classes during the phylogenetic study (Han et al. 2016). Tandem, proximal, and transposed duplication were observed in ZomaCHS (Ma et al. 2021). Three mutations (deletion, insertion, and substitution) along with 18.23% of SNPs with low polymorphism in the conserved region of the CHS gene were detected while analyzing nucleotide similarities and dissimilarities among 22-CHS of Zingiberaceae, Costaceae, and Poaceae (Akram et al. 2023).

Therefore, in order to complement with diverse requirements of flavonoid function, plant CHS genes have also evolved diversification with different expression patterns, with the possibility of matching ancestors. Also, a single amino acid change can transform CHS proteins into non-CHS proteins.

20.10 Chromosome Location

ZmCHS were located unevenly on 6 (1, 2, 3, 4, 5, 7) out of 10 chromosomes of maize. Gene duplication can be considered the driving force in the evolution of CHS genes (Han et al. 2016). Chromosomal distribution of OsCHS genes represented that, chromosomes one, four, five, eleven, seven, and ten with 1-, 2-, 4-, 6-, 7- and 6- OsCHS genes, respectively

(Ahmad et al. 2023). Similarly, TaCHS were disseminated among 19-wheat chromosomes, with a rather even distribution of 29, 29, and 27-TaCHS among A-, B- and D- sub-genomes (Liu et al. 2023).

20.11 Characterization

A high extent of polar residues (https://resources.qiagenbioinformatics.com/manuals/clcsequenceviewer/current/index.php?manual=User_interface.html) confirmed their hydrophilic character of cytosolic (TargetP software) StCHS proteins. Though, acidic in nature, the high portion of acidic and basic residue supports proper salt bridge formation (Gupta et al. 2014). Chalcone isomerase (CHI) was the most interactive among the 7-identified interactive proteins with OsCHS, using the STRING database. Like CHS, CHI is also a key and stress-responsive enzyme of the flavonoid biosynthesis pathway (Ahmad et al. 2023).

20.12 Three-dimensional Structure

A better understanding of the working characteristics of any enzyme can be well-defined through its structure, which can be created *in silico* using available software. SWISS-MODEL was used by Pang et al. 2004 to create a 3-D structure of GbCHS to reveal a symmetric dimerization of two homomers, with 11 α-helix and 15 β-sheets of each monomer. A study of protein structure by Lei et al. (2009) used GOR4, Swiss-Modeling, and WebLab Viewer-Lite 4.2 to conclude that random coils and α-helixes as main machinery while β-strands intersperse in the whole proteins by sharing a proportion of 45.38%, 31.54% and 23.08% in SbCHS. The most appropriate modeled structure of StCHS (SOPMA tool) was evaluated and validated for structure and stereochemistry (Rampage, Procheck, Errat, Ramachandran plot 2, and Vadar) after calculating minimized energy and refining using a 3Drefine server. The generated model has been submitted to the Protein Model Data Base. The modeled structure was abundant in α-helices as compared to β-sheets (Gupta et al. 2014). SWISS-MODEL-based structure of ZoCHS shared 81.09% identity with CHS of alfalfa (Ghosh and Mandi 2015). Schrödinger Suite-based protein model of CfCHS was analyzed by procheck server to confirm the most favorable region with >90.3% residues (Awasthi et al. 2016). Validation of the secondary structure of ZmCHS confirmed its similarity with other plant CHSs (Han et al. 2016). The three-dimensional structure of RhCHS was predicted by Jia et al. (2023) using 5UC5, as a model template using SWISS-MODEL. Secondary structure prediction revealed that DbCHS protein contains α-helix (40%) and random coil (31.39%) predominantly. Extended strand and β-turn had 17.72% and 10.89% prevalence (Mahajan et al. 2023). The closest homolog found is type III PKS in *O. sativa* (PDB ID 4YJY) and was used as a template to model the structure of DbCHS using SWISSMODEL. Ramachandran plot (PROCHECK) detected DbCHS with 94%, 5.7%, 0.3%, and 0.0% amino acid in the most favorable, additional allowed, generously allowed, and disallowed regions, respectively to confirm its appropriateness of modeled structure (Mahajan et al. 2023).

20.13 Docking

Each subunit of homodimeric CHS protein can perform condensation reactions, independently. Docking simulation proper placement of cinnamic acid- CoA-esters and its derivatives in the binding pocket of CfCHS protein model. Protein–ligand binding suggested the involvement of Phe266 through π–π interaction as well as Csy164, Lys270, and Asn337 by H-bond during enzyme–substrate complex formation (Awasthi et al. 2016). Therefore, even a single unit of DsCHS utilizes coumaric acid (one molecule) and malonic acid (three molecules) substrates to produce naringenin chalcone and ultimately its isomerized form naringenin. Docking simulation showed a 10 nm grid around substrates and catalytic centers with (Cys164, His303, and Asn336) of DbCHS (Mahajan et al. 2023).

20.14 Conclusion and Prospects

Flavonoids are more powerful molecules that contain antioxidant activity and affect the quality and stability of foods by acting as flavor enhancers. Due to the increased attention received by flavonoids, this chapter provides information about various flavonoids and their potential biological activity prediction by bioinformatics tools using artificial intelligence technologies such as cheminformatics, virtual screening, molecular docking, and machine learning programs. The study of flavonoids is complicated due to the diversity of different molecular structures and the lack of data on bioavailability. Therefore, analytical techniques need to be improved to allow more data to be collected on absorption and excretion. Several reports have emphasized that molecular docking studies are needed to identify potential molecules of flavonoids for their use in the treatment of various diseases in the human health system. Future research should focus on how flavonoids interact with receptor molecules to treat both acute and chronic illnesses. QSAR model for the antioxidant activity of flavonoids was developed using descriptors. The QSAR model for the antioxidant activity of flavonoids was developed using descriptors. Therefore, QSARs can be expected to help facilitate the future design of additional substituted flavonoids with good antioxidant activity. More research is needed to discover new flavonoid compounds from nature's gift so that they can replace the use of synthetic drugs which are harmful to the body. In this context, there is a need for research and development programs involving computational biology, artificial intelligence studies along with *in vivo* research that will give an optimistic and secure picture of the future.

References

Ahmad, S., Jeridi, M., Siddiqui, S. et al. (2023). Genome-wide identification, characterization, and expression analysis of the shalcone synthase gene family in *Oryza sativa* under abiotic stresses. *Plant Stress* 9: 100201.

Akram, S., Zulkifly, S., Khamis, S., and Ghani, N.I.A. (2023). Preliminary *in silico* analysis of *CHS1* gene in commelinids clade: family Zingiberaceae, Costaceae, and Poaceae. *Pertanika Journal of Tropical Agricultural Science* 46 (2): 425–437.

Al-Jumaili, M.H.A. and Al Hdeethi, M.K.Y. (2021). Study of selected flavonoid structures and their potential activity as breast anticancer agents. *Cancer Informatics* 20: 1–6.

Awasthi, P., Mahajan, V., Jamwal, V.L. et al. (2016). Cloning and expression analysis of chalcone synthase gene from *Coleus forskohlii*. *Journal of Genetics* 95: 647–657.

Bindea, G., Mlecnik, B., Hackl, H. et al. (2009). ClueGO: a Cytoscape plug-in to decipher functionally grouped gene ontology and pathway annotation networks. *Bioinformatics* 25 (8): 1091–1093.

Burda, S. and Oleszek, W. (2001). Antioxidant and antiradical activities of flavonoids. *Journal of Agricultural and Food Chemistry* 49 (6): 2774–2779.

Deng, Y., Li, C., Li, H., and Lu, S. (2018). Identification and characterization of flavonoid biosynthetic enzyme genes in *Salvia miltiorrhiza* (Lamiaceae). *Molecules* 23 (6): 1467.

Dixon, R.A. and Pasinetti, G.M. (2010). Flavonoids and isoflavonoids: from plant biology to agriculture and neuroscience. *Plant Physiology* 154 (2): 453–457.

Djeradi, H., Rahmouni, A., and Cheriti, A. (2014). Antioxidant activity of flavonoids: a QSAR modeling using Fukui indices descriptors. *Journal of Molecular Modeling* 20: 1–9.

Eberhardt, J., Santos-Martins, D., Tillack, A.F., and Forli, S. (2021). AutoDock Vina 1.2. 0: new docking methods, expanded force field, and python bindings. *Journal of Chemical Information and Modeling* 61 (8): 3891–3898.

Eom, S.H. and Hyun, T.K. (2016). Genome-wide identification and transcriptional expression analysis of chalcone synthase in flax (*Linum usitatissimum* L.). *Gene Reports* 5: 51–56.

Ghosh, S. and Mandi, S.S. (2015). SNP in chalcone synthase gene is associated with variation of 6-gingerol content in contrasting landraces of *Zingiber officinale*. Roscoe. *Gene* 566 (2): 184–188.

Gramatica, P. (2013). On the development and validation of QSAR models. In: *Computational Toxicology: Volume II* (ed. B. Reisfeld and A.N. Mayeno), 499–526. Totowa, NJ: Humana.

Gupta, P.S.S., Banerjee, S., Mondal, S. et al. (2014). An insight into the structure and function of chalcone synthase from sequence of *Solanum tuberosum*. *Journal of Advanced Bioinformatics Applications and Research* 5 (1): 8–22.

Han, Y., Ding, T., Su, B., and Jiang, H. (2016). Genome-wide identification, characterization and expression analysis of the chalcone synthase family in maize. *International Journal of Molecular Sciences* 17 (2): 161.

Han, Y., Cao, Y., Jiang, H., and Ding, T. (2017). Genome-wide dissection of the chalcone synthase gene family in *Oryza sativa*. *Molecular Breeding* 37: 1–12.

Harnly, J.M., Doherty, R.F., Beecher, G.R. et al. (2006). Flavonoid content of US fruits, vegetables, and nuts. *Journal of Agricultural and Food Chemistry* 54 (26): 9966–9977.

Jia, Y.H., He, F., Shen, Z.L. et al. (2023). Molecular characterization of a chalcone synthase gene RhCHS from *Rhododendron× hybridum* Hort. *Gene* 857: 147176.

Karydas, C., Iatrou, M., Kouretas, D. et al. (2020). Prediction of antioxidant activity of cherry fruits from UAS multispectral imagery using machine learning. *Antioxidants* 9 (2): 156.

Kavakiotis, I., Tsave, O., Salifoglou, A. et al. (2017). Machine learning and data mining methods in diabetes research. *Computational and Structural Biotechnology Journal* 15: 104–116.

Kumar, S. and Pandey, A.K. (2013). Chemistry and biological activities of flavonoids: an overview. *The Scientific World Journal* 2013: 162750.

Kusumiyati, K. and Asikin, Y. (2023). Machine learning-based prediction of total phenolic and flavonoid in horticultural products. *Open Agriculture* 8 (1): 20220163.

Lavecchia, A. and Di Giovanni, C. (2013). Virtual screening strategies in drug discovery: a critical review. *Current Medicinal Chemistry* 20 (23): 2839–2860.

Lei, W., Sun, M., Luo, K.M. et al. (2009). Compute simulation to characterize structure and function of chalcone synthase from *Scutellaria baicalensis* Georgi. *Molecular Biology* 43: 1008–1013.

Liu, Y., Bai, J., Yuan, S. et al. (2023). Characterization and expression analysis of chalcone synthase gene family members suggested their roles in the male sterility of a wheat temperature-sensitive genic male sterile (TGMS) line. *Gene* 888: 147740.

Lo, C., Coolbaugh, R.C., and Nicholson, R.L. (2002). Molecular characterization and *in silico* expression analysis of a chalcone synthase gene family in *Sorghum bicolor*. *Physiological and Molecular Plant Pathology* 61 (3): 179–188.

Ma, M., Zhong, M., Zhang, Q. et al. (2021). Phylogenetic implications and functional disparity in the chalcone synthase gene family of common Seagrass *Zostera marina*. *Frontiers in Marine Science* 8: 760902.

Mahajan, V., Chouhan, R., Jamwal, V.L. et al. (2023). A wound inducible chalcone synthase gene from *Dysoxylum gotadhora* (DbCHS) regulates flavonoid biosynthesis. *Physiology and Molecular Biology of Plants* 29: 959–969.

Mallika, V., Sivakumar, K.C., and Soniya, E.V. (2011). Evolutionary implications and physicochemical analyses of selected proteins of type III polyketide synthase family. *Evolutionary Bioinformatics* 7: 41–53.

Mishra, M.K., Chaturvedi, P., Singh, R. et al. (2013). Overexpression of WsSGTL1 gene of *Withania somnifera* enhances salt tolerance, heat tolerance and cold acclimation ability in transgenic Arabidopsis plants. *PLoS One* 8 (4): e63064.

Mishra, M.K., Srivastava, M., Singh, G. et al. (2017). Overexpression of *Withania somnifera* SGTL1 gene resists the interaction of fungus *Alternaria brassicicola* in *Arabidopsis thaliana*. *Physiological and Molecular Plant Pathology* 97: 11–19.

Mishra, M.K., Pandey, S., Niranjan, A., and Misra, P. (2021). Comparative analysis of phenolic compounds from wild and *in vitro* propagated plant *Thalictrum foliolosum* and antioxidant activity of various crude extracts. *Chemical Papers* 75 (9): 4873–4885.

Mishra, M.K., Tiwari, S., Srivastava, M. et al. (2022). Ectopic expression of WsSGTL3. 1 gene in *Arabidopsis thaliana* confers enhanced resistance to *Pseudomonas syringae*. *Journal of Plant Growth Regulation* 41 (4): 1871–1886.

Moulishankar, A. and Lakshmanan, K. (2020). Data on molecular docking of naturally occurring flavonoids with biologically important targets. *Data in Brief* 29: 105243.

Oliveira, A.P.S., Lima, D.R., Bezerra, L.L. et al. (2022). Virtual screening of flavonoids from *Chamaecrista* genus: ADME and pharmacokinetic properties, interactions of flavonoid–DNA complex by molecular docking and molecular dynamics. *Journal of Biomolecular Structure and Dynamics* 41: 7677–7685.

Panche, A.N., Diwan, A.D., and Chandra, S.R. (2016). Flavonoids: an overview. *Journal of Nutritional Science* 5: e47.

Pang, Y., Shen, G.A., Liu, C. et al. (2004). Molecular cloning and sequence analysis of a novel chalcone synthase cDNA from *Ginkgo biloba*. *DNA Sequence* 15 (4): 283–290.

Ray, S., Sengupta, C., and Roy, K. (2008). QSAR modeling for lipid peroxidation inhibition potential of flavonoids using topological and structural parameters. *Open Chemistry* 6 (2): 267–276.

Tao, Y., Zhang, H., and Wang, Y. (2023). Revealing and predicting the relationship between the molecular structure and antioxidant activity of flavonoids. *LWT* 174: 114433.

Wabuyele, B.W. and Harrington, P.D.B. (1996). Fuzzy optimal associative memory for background prediction of near-infrared spectra. *Applied Spectroscopy* 50 (1): 35–42.

Wang, H., Provan, G.J., and Helliwell, K. (2000). Tea flavonoids: their functions, utilisation and analysis. *Trends in Food Science & Technology* 11 (4–5): 152–160.

Xu, F., Cheng, S., Cheng, S. et al. (2007). Time course of expression of chalcone synthase gene in *Ginkgo biloba*. *Journal of Plant Physiology and Molecular Biology* 33 (4): 309.

Zhang, M., Sun, J., and Chen, P. (2017). Development of a comprehensive flavonoid analysis computational tool for ultrahigh-performance liquid chromatography-diode array detection-high-resolution accurate mass-mass spectrometry data. *Analytical Chemistry* 89 (14): 7388–7397.

Žuvela, P., David, J., and Wong, M.W. (2018). Interpretation of ANN-based QSAR models for prediction of antioxidant activity of flavonoids. *Journal of Computational Chemistry* 39 (16): 953–963.

21

Golden Opportunities: Harnessing Bioinformatics to Revolutionize Plant Research and Unleash the Power of Golden Rice in Crop Breeding

Poulami Majumder

Department of Biotechnology, Maulana Abul Kalam Azad University of Technology, Kolkata, West Bengal, India

21.1 Introduction

In the annals of agricultural science and human nutrition, few breakthroughs have garnered as much attention as the development of Golden Rice. Envisioned as a formidable weapon in the fight against vitamin A deficiency, a condition that afflicts millions, particularly in developing nations, the Golden Rice represents the synergistic fusion of traditional plant breeding techniques with the might of modern bioinformatics (Al-Babili and Beyer 2005). This confluence of biology and computational science has heralded a new era in plant research. This interdisciplinary field has made it possible to analyze vast and complex biological datasets, especially those associated with plant genomics. As the world's population swells, ensuring food security becomes not just about quantity but also quality. Herein lies the power of bioinformatics; it does not just augment crop yields but also aids in improving the nutritional profile of the crops themselves. Golden rice is a quintessential manifestation of this potential. By engineering rice, the staple food of over half the world's population, to produce and accumulate provitamin A (beta-carotene) in its grains, scientists have unveiled a potential remedy to combat micronutrient deficiencies, particularly vitamin A deficiency which affects hundreds of thousands of children globally (Beyer 2010). However, the path to developing Golden Rice was not straightforward. It is a detailed understanding of rice genomics, an exploration of genes responsible for beta-carotene synthesis, and their precise integration into the rice genome without affecting its inherent attributes. The tools and resources offered by bioinformatics have played an indispensable role in this journey. Sequence databases, as vast repositories of genomic information, paved the way for identifying target genes and potential pathways for beta-carotene synthesis. Functional genomics tools offered insights into gene expression patterns, ensuring that beta-carotene synthesis and accumulation were optimal. Integrative approaches, encompassing the spectrum of systems biology, provided a holistic view, ensuring that the introduction of new genes did not inadvertently disrupt other vital metabolic pathways (Tang et al. 2009). However, bioinformatics is not only about data analysis, but it also

Bioinformatics for Plant Research and Crop Breeding, First Edition. Edited by Jen-Tsung Chen.

stands at the crossroads of multiple disciplines – genetics, molecular biology, computer science, mathematics, and even statistics. This multidisciplinary nature has also enabled the field to offer tools and methodologies that push the boundaries of what is possible in plant breeding. The powerful gene-editing tool, CRISPR-Cas9, has its roots deeply embedded in bioinformatics. From the early days of its development using Agrobacterium-mediated transformation to the latest versions that leverage CRISPR-Cas9 for precision breeding, the journey of Golden Rice offers a case study of how bioinformatics can revolutionize crop breeding. Golden rice, in its successive iterations, has greatly benefited from such advancements. Besides gene editing, bioinformatics has also contributed immensely to other breeding strategies. Marker-assisted selection (MAS), which involves using specific DNA sequences as "markers" to select desirable traits, has its foundation in the genomics databases and analysis tools that bioinformatics offers. Similarly, genomic selection is a more comprehensive approach than MAS and also relies on the data-rich environment that bioinformatics provides. Speed breeding, another method, greatly benefits from insights derived from genomic data, ensuring that the right traits are selected in the shortest possible time. But the journey of Golden Rice, despite its promise, has not been without challenges. And many of these are not just scientific but also sociopolitical. Ethical considerations about genetically modifying a staple crop, public perceptions fueled by misinformation, stringent regulatory landscapes, and the overarching issue of accessibility, have often overshadowed the tangible benefits that Golden Rice offers. By providing transparent and open-access data, building predictive models that demonstrate the safety and efficacy of genetically modified (GM) crops, and collaborating with interdisciplinary teams that include sociologists, policymakers, and educators. In essence, bioinformatics has transitioned from being a support tool to a central pillar in plant research and crop breeding. Whether it is decoding the intricate rice genome, enhancing the drought resistance of maize, or improving the nutritional profile of tomatoes, bioinformatics stands at the forefront, driving innovation. With the ever-increasing global food demand and the looming challenges of climate change, bioinformatics will undoubtedly play a pivotal role in ensuring food security and sustainability for future generations. In a world grappling with the twin challenges of climate change and burgeoning populations, the tale of Golden Rice and bioinformatics offers hope, a testament to human ingenuity and the relentless pursuit of a better, healthier future for all (Paine et al. 2005).

In this chapter, we will delve deep into the fascinating intersection of bioinformatics and Golden Rice. We will chart the journey of Golden Rice from its conceptualization to its current iterations, focusing on the pivotal role that bioinformatics has played. Through the lens of Golden Rice, we will also explore the larger narrative of how bioinformatics is shaping the future of crop breeding and, by extension, global nutrition. Golden rice, fortified with the power of bioinformatics, holds the golden promise of turning this vision into a reality. Golden rice is not just another variety of rice; it is a symbol of what modern agriculture can achieve when science, vision, and humanitarian needs converge. Its significance in global agriculture is multifaceted, encompassing health, economic, environmental, and societal dimensions. While debates around its adoption and broader implications continue, there is no denying that Golden Rice represents a promising stride toward leveraging biotechnology for the greater good of humanity.

21.1.1 An Overview of Bioinformatics in Plant Research and Crop Breeding

Bioinformatics, the fusion of biology and computational technology, has significantly revolutionized plant research and crop breeding in recent decades (Wesseler and Zilberman 2014). By enabling scientists to delve deep into the intricacies of plant genomes, understand gene functions, and optimize crop traits, bioinformatics has emerged as a linchpin in modern agronomy (Stein et al. 2006). In Figure 21.1, a schematic diagram is used to illustrate the theme of this work.

- **Plant genomic databases:** These databases serve as vast repositories of plant genomic sequences. For example, the PLAZA platform provides a comprehensive comparative genomics environment, while Phytozome hosts genomes from diverse plant families. These resources facilitate the exploration of evolutionary relationships, identification of conserved genes, and discovery of novel genes responsible for specific traits.

Figure 21.1 A schematic Illustrated diagram showing the fusion of bioinformatics and rice and showing the golden rice as the final product. *Source:* ChatGPT 4.0

- **Functional genomics and transcriptomics:** Bioinformatics tools have made it feasible to understand gene function by analyzing their expression patterns under various conditions. RNA-seq data analysis, for instance, provides insights into which genes are expressed under drought conditions or which ones respond to pest attacks. By understanding these pathways, researchers can devise strategies to bolster plant resilience and improve yield.
- **Proteomics and metabolomics:** Just as transcriptomics deals with RNA, proteomics focuses on protein profiles, and metabolomics explores metabolic compounds. Software tools like MaxQuant or MetaboAnalyst enable scientists to dissect the protein profiles and metabolic pathways, respectively. Understanding these can be pivotal, especially when the objective is to enhance certain metabolites (like flavonoids in berries or essential oils in herbs).
- **Genetic engineering and molecular breeding:** Advances in bioinformatics have also informed genetic engineering techniques. CRISPR-Cas9, one of the most transformative tools in recent times, relies heavily on bioinformatic algorithms to predict precise genomic edits. Likewise, bioinformatics is crucial in MAS, where DNA markers associated with desirable traits are used to streamline the breeding process. With MAS, breeders can more confidently select plants that have the desired traits, thereby significantly accelerating the breeding timeline.
- **Genome-wide association studies (GWAS):** In plant breeding, identifying genes or genetic markers linked to desirable traits can be a game-changer. GWAS is a method used to identify these genetic markers across the genome. Bioinformatics software and platforms help manage and analyze vast datasets involved in GWAS, highlighting associations between specific genetic variants and observed traits.
- **Predictive modelling:** One of the standout advantages of bioinformatics is predictive modeling. Leveraging machine learning and statistical models, researchers can pr
- edict outcomes based on genomic data. For instance, predicting how a particular crop variety might respond to future climate change scenarios or foreseeing potential disease resistance based on genetic markers.
- **Integration with systems biology:** As plant research advances, there is a growing trend toward a systems biology approach, where the objective is to understand the plant as an integrated and interacting network of genes, proteins, and biochemical reactions. Bioinformatics tools enable this holistic view, integrating diverse data types to provide a comprehensive picture of plant physiology and metabolism.

21.1.2 The Significance of Golden Rice in Global Agriculture

This GM rice variety is engineered to produce provitamin A (beta-carotene) in its endosperm, giving the grains their characteristic golden hue (Wu et al. 2021). The significance of Golden Rice extends far beyond its color; its potential impacts span from global nutrition to the future dynamics of crop breeding (Harmon 2013).

- **Addressing micronutrient deficiency:** The primary aim behind the development of Golden Rice was to combat vitamin A deficiency (VAD) prevalent in many parts of the world. VAD can lead to severe visual impairment, a compromised immune system, and even death. Given that rice is a staple for more than half the global population, especially in regions where VAD is rampant, Golden Rice presents a direct solution to this public health challenge.

- **Proof-of-concept for biofortification:** Golden rice stands as a testament to the power of biofortification, the process of enhancing the nutritional profile of crops through natural breeding techniques or biotechnology. It showcases the possibility of tailoring our staple crops to address specific nutritional deficiencies, thus paving the way for other biofortified crops targeting different nutrients.
- **Economic implications:** Improving public health through nutrition has economic implications. Healthier populations mean reduced medical costs and increased productivity. By preventing VAD-associated complications, Golden Rice can potentially reduce healthcare burdens on struggling economies and improve overall societal well-being.
- **Sustainable solution:** Supplemental interventions like vitamin capsules or syrups are often used to combat VAD. However, these interventions require recurrent logistical arrangements and financial commitments. Golden rice, on the other hand, offers a sustainable, once-off solution. Once farmers adopt Golden Rice, communities can reap its benefits year after year without the need for recurring external interventions.
- **Fostering acceptance of GMOs:** Golden rice, by virtue of its humanitarian goals, provides a compelling narrative for the potential benefits of genetically modified organisms (GMOs). Given the widespread skepticism surrounding GMOs, Golden Rice serves as an ambassador for the positive, life-saving possibilities that biotechnological advancements in agriculture can offer.
- **Shaping future agricultural research:** The success of Golden Rice has spurred further research into genetically modifying crops for various purposes, from nutrition enhancement to resistance against pests or environmental stresses. It has set a precedent for public and private sector investments in similar bioengineering endeavors.
- **Ethical and social implications:** Golden rice's journey from labs to fields has initiated numerous discussions on the ethical and social aspects of GMOs. It forces societies, policymakers, and scientists to ponder questions like: Who gets access to such technologies? How do we ensure GMOs are safe? And how do we balance potential benefits against ecological or socio-economic concerns?

21.1.3 Objective and Scope of the Chapter

To provide a comprehensive overview of the history and evolution of bioinformatics in relation to plant research, emphasizing its transformative impact and detailing the rapid advancements in plant genomics and crop breeding techniques, illustrating how they revolutionize modern agriculture.

- Enumerate and delve into the vast array of bioinformatics tools and resources pivotal for plant research, spanning from sequence databases to integrative systems biology, and explore the cutting-edge genetic engineering and breeding strategies applied to the development and enhancement of Golden Rice, including the novel applications of CRISPR-Cas9, MAS, and speed breeding.
- Provide an in-depth case study on Golden Rice, encompassing its early development, the integration of bioinformatics for gene identification, and the tangible improvements

seen in its agronomic and nutritional quality and highlight the multifaceted challenges faced in the journey of Golden Rice from labs to fields, discussing ethical quandaries, regulatory hurdles, and public perceptions.
- Offer insights into the future, shedding light on ensuring Golden Rice's accessibility to the most vulnerable populations and the boundless potential of bioinformatics in heralding further breakthroughs in crop breeding.

21.2 Background

Bioinformatics, the fusion of biology and computational analysis, has evolved tremendously since its rudimentary applications in the 1970s. Initially, it was primarily employed for sequence alignment and database searches, aiding in understanding the genetic code of various organisms. However, its potential became even more pronounced with the inception of the Human Genome Project in the late 20th century, which paved the way for comprehensive genomic analyses in various organisms, including plants (De Steur et al. 2022). In the realm of plant research, bioinformatics became indispensable for navigating and interpreting the enormous data sets generated from plant genomes. One such groundbreaking application was the mapping and modification of rice genes to produce Golden Rice. Understanding the metabolic pathways of rice, scientists identified genes from daffodils and later from maize that could produce beta-carotene when inserted into rice. Bioinformatics tools facilitated the identification of these specific genes, predicted their behavior in rice, and optimized their expression levels. Moreover, as sequencing technologies became more sophisticated and cheaper, bioinformatics adapted to manage and make sense of the mounting data (Beyer et al. 2002). For Golden Rice, this meant refining the initial prototypes to improve carotenoid levels and ensuring the stability of these modifications over multiple generations. In essence, without the advances in bioinformatics, the precise genetic modifications and subsequent validations required to produce and optimize Golden Rice would have been exceptionally challenging, if not impossible. Golden Rice stands as a testament to the power of biotechnological innovation in addressing global health challenges. Since its inception, this GM rice has been specifically designed to produce beta-carotene, a precursor of vitamin A. Traditional rice varieties, despite being a staple for half the world's population, lack this vital nutrient. In regions where rice is predominantly consumed, VAD has been rampant, leading to severe health complications, including blindness and increased susceptibility to infectious diseases. Golden Rice, thus, emerged as a beacon of hope for millions. Its development and potential to mitigate micronutrient deficiencies showcase how bioinformatics and modern plant breeding techniques can harmoniously blend to produce solutions that not only address food security but also nutrition quality (Dubock 2014a,b). The rise of plant genomics has ushered in a new era for crop breeding, catalyzing significant improvements in yield, resilience, and nutritional quality. Leveraging the latest scientific innovations, researchers and breeders have been able to identify and harness genetic variations in plants with unprecedented accuracy and speed.

- **Sequencing technologies and genomic resources:** One of the most impactful developments in plant genomics has been the advent and continuous improvement of next-generation sequencing (NGS) technologies. NGS allows for rapid and cost-effective sequencing of entire genomes, paving the way for the establishment of reference genomes for many plant species. With these reference genomes, researchers can identify genetic markers, genes, and pathways responsible for specific traits, facilitating their manipulation in crop breeding programs.

- **Genome-wide association studies (GWAS):** GWAS are used to identify genetic markers associated with specific traits by analyzing the genomes of a large number of individuals in a particular species. For crops, this means recognizing markers related to desirable traits like drought tolerance, disease resistance, or yield. Once identified, these markers can be utilized in breeding programs to screen and select plants with the desired genetic makeup.
- **Genomic selection (GS):** Building on traditional breeding methodologies, GS uses statistical models to predict the performance of hybrid crops based on their genomic data. This allows breeders to select the best possible candidates without having to physically cultivate and test each one, considerably speeding up the breeding process.
- **CRISPR-Cas9 and genome editing:** One of the most transformative tools in modern biology is the CRISPR-Cas9 genome editing system. It allows for precise modifications in the DNA sequence, enabling breeders to introduce or remove specific traits with unparalleled accuracy. For example, if a gene responsible for disease susceptibility in a crop is identified, it can be "turned off" or replaced using CRISPR, leading to disease-resistant varieties.
- **High-throughput phenotyping:** Complementing genomics, high-throughput phenotyping uses sensors, drones, and artificial intelligence to collect and analyze data on plant characteristics (phenotypes) in real-time. This technology assists breeders in correlating genetic information with observable traits, streamlining the selection process in breeding programs.
- **Synthetic biology and bioengineering:** Beyond mere tweaks in the genome, synthetic biology aims to design and construct entirely new biological parts, devices, and systems. In the context of plant breeding, this might mean engineering plants to produce novel compounds or exhibit traits that are not found in nature. Such advancements could revolutionize industries like biofuels, pharmaceuticals, and more.
- **Multi-omics approaches:** While genomics focuses on DNA sequences, other "omics" technologies study various aspects of cellular processes, such as transcriptomics (RNA), proteomics (proteins), and metabolomics (metabolites). Integrating data from these diverse sources gives a more holistic view of plant biology, enabling breeders to understand how changes at the genetic level translate to overall plant function and physiology.
- **Systems biology and predictive modeling:** Using computational models, systems biology aims to predict how changes in one part of a plant's biology might affect the organism as a whole. These models, refined by real-world data, assist breeders in anticipating how genetic modifications might manifest in the overall plant phenotype. As the world struggles with the dual challenges of a growing population and changing climate, these tools offer hope (Dwiningsih 2010). From the precision of CRISPR-Cas9 to the holistic insights offered by multi-omics approaches, modern science empowers breeders with a toolkit of unprecedented depth and versatility. With these advances, the goal of producing crops that are more nutritious, resilient, and sustainable is not just a vision but a rapidly approaching reality (Stein et al. 2008).

21.3 Bioinformatics Tools and Resources for Plant Research

21.3.1 Sequence Databases and Repositories

Given the significant potential of Golden Rice to address global micronutrient deficiencies, a plethora of bioinformatics tools, sequence databases, and repositories have been developed or adapted to support its research and development. These resources not only house critical genetic information but also foster collaboration and ensure transparency among researchers, enabling rapid advancements in the field.

- **GenBank:** GenBank, maintained by the National Center for Biotechnology Information (NCBI), is the primary repository for nucleotide sequences and their protein translations. Any research associated with the genetic makeup of Golden Rice has likely deposited sequences in GenBank, making it a crucial first stop for researchers looking into the rice's genetic data.
- **Rice Genome Annotation Project (RGAP):** This is a comprehensive resource dedicated specifically to rice genomics. The database contains not only the complete sequence of the rice genome but also detailed annotations. Researchers focusing on Golden Rice can glean insights into the specific genes involved in beta-carotene production and their context within the broader rice genome.
- **Phytozome:** A hub for green plant genomics, Phytozome provides access to numerous plant genomes, including rice. With its comparative genomics features, researchers can contrast the genomes of Golden Rice with other rice varieties and plants, enhancing the understanding of the genetic modifications made.
- **Oryzabase:** An integrated science database for rice, Oryzabase contains detailed genetic, genomic, and morphological information about rice. It is a comprehensive resource that allows researchers to link the phenotypic traits of Golden Rice (like beta-carotene enrichment) with its underlying genotypic data.
- **PLAZA:** While not exclusive to rice, PLAZA is a platform for plant comparative genomics. The database helps researchers understand the evolutionary lineage of specific genes and can be instrumental in tracing the origin and adaptation of the genes introduced into Golden Rice.
- **IRRI's Rice Knowledge Bank:** The International Rice Research Institute (IRRI) has been a leading institution in rice research. Their Rice Knowledge Bank houses a wealth of information, including some focused on GM rice varieties like Golden Rice.
- **SRA (Sequence Read Archive):** Also managed by NCBI, the SRA is a repository that stores raw sequencing data. As Golden Rice underwent various stages of development and optimization, researchers would have sequenced different prototypes. These raw sequences, potentially deposited in SRA, can provide a granular look at the genetic evolution of Golden Rice.
- **KEGG (Kyoto Encyclopedia of Genes and Genomes):** While KEGG encompasses much more than plants, its metabolic pathways database can be vital for researchers looking into the biochemical processes that Golden Rice's beta-carotene production involves.

The bioinformatics landscape for Golden Rice, as showcased by these databases and repositories, is both expansive and intricate (Dawe et al. 2002). These resources not only provide the raw genetic data but also contextualize it within broader biological systems, enabling researchers to develop a holistic understanding of Golden Rice's biology (Swamy et al. 2019). The availability of such comprehensive databases is a testament to the global scientific community's commitment to leveraging genomics for societal good.

21.3.2 Sequence Alignment and Phylogenetic Analysis

Sequence alignment and phylogenetic analysis are foundational tools in the realm of bioinformatics and evolutionary biology. They enable researchers to understand the evolutionary relationships among different organisms and discern the functional and structural properties of genes and proteins.

- **Sequence alignment:** It is the process of arranging sequences (DNA, RNA, or protein) to identify regions of similarity. These similarities could be the result of functional, structural, or evolutionary relationships between the sequences. There are four types of sequence alignment:
  - **Pairwise alignment:** This involves comparing two sequences to each other. It can further be divided into:
  - **Global alignment (e.g., Needleman–Wunsch algorithm):** Where the entirety of both sequences is aligned.
  - **Local alignment (e.g., Smith–Waterman algorithm):** Where the most similar subregions of sequences are aligned, even if the entire sequence does not align well.
  - **Multiple sequence alignment (MSA):** This involves comparing more than two sequences simultaneously. Tools like CLUSTAL, MUSCLE, and T-Coffee are widely used for MSA.

  Alignments help in several areas, such as:
  - Predicting the function of newly discovered genes based on the similarity to known genes.
  - Inferring evolutionary processes and identifying conservation among species.
  - Designing experiments to probe functional elements in genes and proteins.
- **Phylogenetic analysis:** Phylogenetic analysis seeks to understand the evolutionary relationships among a set of species or genes. It involves the construction of a phylogenetic tree, where the branches represent an evolutionary lineage.
  - **Phylogenetic trees:** These are graphical representations that depict the evolutionary relationships among entities. The branching points, or nodes, represent common ancestors, and the length of branches can represent the genetic distance or time since divergence.
    - **Tree construction methods:** Multiple methods exist for constructing phylogenetic trees. Some of the most common include:
    - **Distance-based methods:** These use genetic distance matrices as input. Examples include neighbor-joining and UPGMA.
    - **Character-based methods:** These use actual sequence data. Examples include maximum parsimony and maximum likelihood.
    - **Bayesian methods:** These use a combination of likelihood methods and Bayes theorem to infer phylogenies.

- Phylogenetic analyses aid in:
  - Tracing the origin and spread of infectious diseases.
  - Classifying organisms and understanding biodiversity.
  - Predicting the function of genes and proteins based on evolutionary conservation.
  - Understanding the adaptive evolution of organisms in response to changing environments.

21.3.2.1 Applications in Plant Genomics

For crops like Golden Rice, sequence alignment and phylogenetic analyses are essential. Aligning the sequences of modified genes with those from other plants can identify the best candidates for genetic engineering. Phylogenetic analysis, meanwhile, can trace the evolutionary origin of these genes, offering insights into their natural function and potential implications when introduced into rice. Together, sequence alignment and phylogenetic analysis provide a comprehensive view of the molecular evolution of life (Greedy 2018). In an era where genomic data is exponentially increasing, these tools are critical for converting raw sequence information into meaningful biological insights (Stone and Glover 2017).

21.3.3 Functional Genomics and Transcriptomics Tools

While sequencing provides the raw blueprint of an organism's genome, functional genomics and transcriptomics dive deeper, exploring how genes give rise to function and how they are expressed over time and in response to various stimuli. When it comes to Golden Rice, understanding the functional landscape of its genome and its gene expression patterns is crucial for optimizing its nutritional benefits and ensuring its safety. Here are some essential tools and methodologies in this domain:

- **Functional Genomics**
  - **Gene knockout and overexpression:** One of the direct ways to decipher a gene's function is to either disable (knockout) or amplify (overexpress) it and then observe the resulting phenotypic changes. In the context of Golden Rice, understanding the function of genes involved in the beta-carotene synthesis pathway can be crucial.
  - **Reporter gene assays:** These assays involve attaching a "reporter" gene (like one that codes for a fluorescent protein) to a gene of interest. When the gene of interest is active, the reporter gene is also expressed, serving as a visual indicator of gene activity.
  - **T-DNA insertional mutagenesis:** This technique uses the bacterium Agrobacterium tumefaciens to introduce a piece of DNA (T-DNA) into the plant genome. Wherever the T-DNA inserts, it can disrupt gene function, providing clues about the role of affected genes.
- **Transcriptomics**

  Transcriptomics involves studying the transcriptome, which is the complete set of RNA transcripts produced by the genome under specific conditions.
  - **RNA Sequencing (RNA-Seq):** This NGS method quantifies and identifies RNA molecules in a sample. For Golden Rice, RNA-Seq can pinpoint which genes are active during different growth stages and under various environmental conditions, shedding light on when and where beta-carotene synthesis is most active.

 - **Microarrays:** Before the dominance of RNA-Seq, microarrays were the go-to method for transcriptomics. These involve attaching thousands of probes to a solid surface and then hybridizing them with labeled RNA from the sample. Though less popular now, they have been instrumental in early transcriptomic studies on rice.
 - **Differential expression analysis:** This computational approach compares the gene expression levels between different conditions or tissues. In the context of Golden Rice, it could be used to contrast expression levels in modified versus traditional rice varieties to ascertain the effects of genetic modifications.
 - **Quantitative real-time PCR (qRT-PCR):** Often used in conjunction with RNA-Seq, qRT-PCR quantifies specific RNA molecules, verifying the results from broader transcriptomic studies.
 - **Applications in Golden Rice:** For Golden Rice, functional genomics tools can be employed to fine-tune the metabolic pathway for beta-carotene synthesis. If a particular gene in the pathway is not functioning as intended, techniques like gene knockout or overexpression can be used to investigate and optimize its role.

On the transcriptomics front, tools like RNA-Seq offer insights into the broader impacts of introducing beta-carotene synthesis. By comparing Golden Rice's transcriptome to that of traditional rice, researchers can ensure that the modifications do not inadvertently affect other essential genes or pathways. Functional genomics and transcriptomics tools provide a holistic understanding of the dynamic relationship between Golden Rice's genome and its phenotype (Mayer 2005). These techniques and methodologies, combined with advancements in bioinformatics, ensure that Golden Rice is not only nutritious but also safe and sustainable for broader agricultural practices (Datta et al. 2007).

- Proteomics and Metabolomics Tools

  Proteomics and metabolomics are crucial disciplines that follow the flow of biological information from genes to proteins and metabolites, respectively. These studies give researchers insights into the actual molecular machinery of cells and the end products of cellular function. In relation to Golden Rice, they can provide deeper insights into the actual protein products and metabolic changes resulting from genetic modifications.
- Proteomics

  Proteomics deals with the large-scale study of proteins, especially their functions and structures.
  - **Mass spectrometry (MS):** One of the core techniques in proteomics, MS identifies and quantifies proteins based on their mass-to-charge ratios. Coupled with liquid chromatography (LC-MS/MS), it can dissect complex protein mixtures, identifying and quantifying thousands of proteins in a single run.
  - **Two-dimensional gel electrophoresis (2D-GE):** This technique separates proteins based on their isoelectric point and molecular weight. Once separated, proteins can be identified and quantified, often using MS.
  - **Protein microarrays:** Analogous to DNA microarrays, these are surfaces to which a vast array of proteins are attached. They can be used for various applications, including protein–protein interaction studies, enzyme substrate identification, and antibody binding studies.

  - **Protein–protein interaction networks:** Techniques like yeast two-hybrid screening or affinity purification followed by mass spectrometry can identify interaction partners for a protein of interest. These interactions can help decipher the role of proteins within cellular pathways.
- Metabolomics
  Metabolomics focuses on the complete set of metabolites within a cell or organism, giving insights into the end products of gene expression.
  - **Gas chromatography–mass spectrometry (GC-MS) and liquid chromatography–mass spectrometry (LC-MS):** Both are staple techniques in metabolomics. They can separate and quantify small molecules, providing insights into the metabolic status of a sample.
  - **Nuclear magnetic resonance (NMR) spectroscopy:** Another vital tool in metabolomics, NMR can identify and quantify metabolites based on their magnetic properties.
  - **Metabolic pathway analysis:** Using tools like MetaboAnalyst, researchers can map identified metabolites onto known metabolic pathways, helping to visualize changes in metabolic flow resulting from genetic modifications or environmental changes.

21.3.4 Applications in Golden Rice

In the context of Golden Rice, proteomic tools can monitor the levels of proteins related to the introduced beta-carotene synthesis pathway, ensuring that they are produced in the right amounts and are not adversely affecting other cellular pathways (Zhu et al. 2018). Metabolomic studies are even more direct, examining the end products of the genetic modifications. In Golden Rice, metabolomics can quantify the levels of beta-carotene and its precursors, as well as other metabolites in the grain. This ensures that the rice not only produces beta-carotene but also retains its nutritional content without accumulating potentially harmful by-products. Proteomics and metabolomics tools provide a direct window into the machinery and end products of the cellular processes in Golden Rice. By understanding the protein landscape and metabolic changes that result from genetic modifications, researchers can ensure that Golden Rice is nutritionally enhanced while maintaining its overall health benefits and safety (Moghissi et al. 2016).

21.3.4.1 Integration of Omics Data for Systems Biology

Systems biology is an integrative discipline that seeks to understand biological systems holistically, viewing them as more than just the sum of their parts. It relies heavily on the synthesis of diverse datasets from various "omics" disciplines, including genomics, transcriptomics, proteomics, metabolomics, and more. The integration of these datasets provides a multi-dimensional view of biological processes, offering insights into how genes, transcripts, proteins, and metabolites interact in the context of the entire system.

1) Multi-omics Data Integration
   Combining data from various omics platforms can be a daunting task due to the inherent complexity and high dimensionality of the datasets.
   - **Data preprocessing:** Each omics platform produces data in different formats, scales, and quality. Hence, preprocessing steps such as normalization, missing value imputation, and batch effect correction are crucial.

- **Dimensionality reduction:** Tools like principal component analysis (PCA) and t-distributed stochastic neighbor embedding (t-SNE) can reduce the dimensionality of data, allowing for easier visualization and analysis.
- **Network-based integration:** Biological processes are interconnected. Networks (such as protein–protein interaction or metabolic pathways) can be used to integrate multi-omics data, highlighting key nodes or hubs in a system.

2) **Systems Biology Tools and Platforms**
   - **Cytoscape:** A software platform for visualizing molecular interaction networks and integrating these with gene expression profiles and other state data.
   - **Genome2D:** A tool for visualizing and interpreting integrated omics data in the context of metabolic pathways.
   - **MAGPIE:** Stands for Model Automated Genomics and Proteomics Integrated Environment. It is a platform for building predictive models based on integrated omics datasets.
3) **Systems Biology Models**

   By combining various omics datasets, researchers can develop holistic models that simulate biological processes.
   - **Constraint-based models:** These use the stoichiometry of reactions in a system (often metabolic reactions) to predict feasible steady-state flux distributions.
   - **Kinetic models:** These rely on enzyme kinetics and detail the rate of reactions. When combined with proteomic data, they can predict how changes in enzyme concentration affect system behavior.
   - **Agent-based models:** These simulate individual entities (e.g., cells or proteins) as agents with defined rules, offering insights into emergent behaviors from complex interactions.

For a GMO like Golden Rice, integrated systems biology can be particularly insightful. By combining transcriptomic data (which genes are turned on/off), proteomic data (which proteins are produced and in what amounts), and metabolomic data (which end products are created), researchers can develop comprehensive models detailing how the introduction of beta-carotene synthesis impacts rice at various levels (Figure 21.2). This can offer insights into potential unintended effects, optimization strategies, and a deeper understanding of rice biology in general. The integration of omics data for systems biology represents a paradigm shift in how we approach biological research. Instead of focusing on individual components, systems biology acknowledges the inherent complexity and interconnectivity of biological systems (Kettenburg et al. 2018). Through a combination of cutting-edge techniques, computational tools, and integrative thinking, researchers can unravel the multifaceted intricacies of life, paving the way for innovations in fields like biotechnology, medicine, and agriculture (Mallikarjuna Swamy et al. 2021).

21.4 Genetic Engineering and Breeding Strategies for Golden Rice

Golden Rice stands as a testament to the power of modern biotechnology in addressing global nutritional challenges. Central to its development has been genetic engineering, which enabled the insertion of genes responsible for beta-carotene (provitamin A)

Figure 21.2 Illustration of a digital dashboard displaying various bioinformatics tools for plant research. Icons represent sequence databases, functional genomics tools, proteomics techniques, and systems biology integration. Each icon is connected by digital pathways, emphasizing the integrated nature of bioinformatics. In the foreground, a magnified view of Golden Rice grain, with its DNA structure highlighted and annotated, symbolizing the targeted research and advancements. *Source:* ChatGPT 4.0

synthesis into rice, a crop that naturally lacks this vital nutrient in its grains. Utilizing a set of genes primarily from daffodils and later versions with genes from maize, scientists engineered rice plants to produce beta-carotene in the endosperm – the edible part of the grain. However, the journey of Golden Rice from the lab to the fields also involved traditional breeding techniques. Once the initial GM rice was developed, breeding strategies were employed to cross this rice with high-yielding local varieties, ensuring that the beta-carotene-producing trait was combined with other desired agronomic traits. This integrative approach ensures that Golden Rice is not only nutritious but also tailored to meet the agronomic and cultural needs of the regions where VAD is prevalent (Corrigan et al. 2009).

21.4.1 CRISPR-Cas9 and Other Gene-editing Technologies

Gene editing has revolutionized modern biology, providing precise tools to modify the DNA of organisms. Among these tools, CRISPR-Cas9 has emerged as the most prominent, but there are several other notable systems. Below, we delve into CRISPR-Cas9 and its peers in the gene-editing realm:

a) **CRISPR-Cas9:** The CRISPR-Cas9 system originates from a bacterial defense mechanism against viral invaders. It was first observed in bacteria, where they used the system to store snippets of viral DNA (in CRISPR arrays) and then employ the Cas9 protein to target and cleave matching viral DNA during subsequent invasions. In the lab, researchers have harnessed this mechanism by designing synthetic RNA sequences (guide RNAs) that can target specific locations in an organism's genome. When introduced into cells alongside the Cas9 protein, precise cuts can be made at desired genetic locations. These cuts can then be naturally repaired by the cell, leading to gene deletions, or researchers can introduce desired sequences to be incorporated. CRISPR-Cas9 has a vast array of applications, from basic biological research to therapeutic genome editing, agricultural modifications, and more.
b) **Zinc finger nucleases (ZFNs):** ZFNs are proteins that can be designed to target specific DNA sequences. They function by fusing a DNA-binding domain (the zinc finger) with a DNA-cleaving domain. Before CRISPR's rise, ZFNs were among the primary tools for gene editing, especially in therapeutic contexts. However, designing specific ZFNs can be labor-intensive and expensive.
c) **Transcription activator-like effector nucleases (TALENs):** Similar to ZFNs, TALENs are fusions of a DNA-binding domain and a DNA-cleaving domain. The DNA-binding domain is derived from proteins secreted by certain plant pathogenic bacteria. TALENs can be designed to target a wider range of sequences compared to ZFNs, and they have a relatively higher specificity. However, like ZFNs, their design can be more labor-intensive than CRISPR systems.
d) **Base editors:** Evolved from the CRISPR-Cas9 system, base editors can make specific point mutations (change one base to another) without inducing double-strand breaks. They function by fusing a modified Cas protein with another enzyme that can convert one DNA base into another. Base editors offer a more subtle way of editing genes, allowing for the precise alteration of DNA without causing significant disruptions.
e) **Prime editing:** Prime editing is a newer form of gene editing that combines aspects of CRISPR-Cas9 and base editing. It uses a modified form of Cas9 and a specially designed RNA to directly copy edits into a target DNA site. This technique is considered more accurate and versatile than other CRISPR methods and can be used to insert, delete, or replace DNA bases with high precision.

While CRISPR-Cas9 has garnered the limelight in recent years, the world of gene editing is rich and diverse. Each method offers its advantages, limitations, and specificities, and the choice of one over another depends on the particular requirements of the task at hand. As our understanding and technology progress, it is likely that we will continue to see innovations and improvements in this dynamic field, driving forward the frontiers of genetics and biomedicine (Regis 2019).

21.4.2 Marker-assisted Selection (MAS)

MAS is a modern breeding technique that employs molecular markers to identify and select desirable genetic traits or alleles in plants and animals. Instead of waiting for the phenotype (observable traits) to manifest, breeders can make selection decisions much earlier in the development process using the genetic markers associated with those traits. This approach has the potential to accelerate the breeding process and enhance the precision of breeding programs.

Molecular markers: These are short DNA sequences with known locations on chromosomes. Their presence (or absence) can be associated with particular phenotypic traits.

Genotyping: The process of determining the genetic makeup of an individual at specific marker sites.

Identification of markers: The first step is identifying molecular markers that are linked to the desired trait. This is typically done through population studies where the inheritance of markers and traits are observed across generations.

Screening for markers: Once markers are identified, breeding populations are screened for these markers using various molecular techniques.

Selection: Plants or animals showing the marker for the desired trait are selected for further breeding.

21.4.2.1 Advantages of MAS

Speed: Traditional breeding relies on phenotypic selection, which often requires waiting for the organism to grow to a certain stage or even waiting for specific seasons. MAS can expedite the selection process.

Precision: MAS can reduce the influence of environmental factors, leading to more accurate selection.

Complex traits: For traits controlled by multiple genes, MAS offers a way to select multiple markers simultaneously.

Cost-effective: In the long run, by reducing the number of generations and field evaluations, MAS can be more cost-effective than conventional breeding for certain traits.

21.4.2.2 Limitations

Initial investment: Setting up MAS requires significant initial investment in terms of technology, equipment, and expertise.

Trait complexity: If a trait is controlled by many genes and the genetic architecture is not well understood, MAS can be challenging.

Linkage drag: Sometimes, the selected marker may be linked to undesired traits, which can also get selected unintentionally.

21.4.2.3 Applications

MAS has been employed in a range of agricultural contexts:

Disease resistance: For many crops, MAS has been used to select for resistance against specific diseases.

Quality traits: In crops like rice and wheat, MAS has been used to enhance grain quality.

Abiotic stress: With changing climate patterns, MAS is increasingly being used to select for drought or salinity tolerance in crops.

The MAS represents the confluence of traditional breeding techniques with modern molecular biology. While it is not a panacea for all breeding challenges, when used judiciously and in combination with other tools, MAS offers a powerful approach to developing the next generation of high-performing, resilient crops and livestock.

21.4.3 Genomic Selection and Genome-wide Association Studies (GWAS)

The advent of high-throughput sequencing technologies has ushered in a new era of genetics, enabling researchers to probe deeper into the genome's intricacies (Tang et al. 2012). Two techniques that stand out in this realm are GS and GWAS. Both approaches exploit the vast wealth of genomic data but serve different purposes within the spectrum of genetic analysis and breeding.

a) **Genomic Selection (GS):**

   Concept: GS is a method of predicting the potential performance of an individual based on its entire genomic profile rather than a few specific markers. This approach utilizes data from many individuals to build a predictive model.

   Procedure: The process typically begins with the phenotypic measurement of a training population and their genotypic data. Using this, a model is built to predict the phenotypic value based on the genomic information. This model is then applied to a selection candidate population that lacks phenotypic measurements but has genotypic data, to predict their performance.

   Advantages: GS allows breeders to evaluate potential breeding candidates at a much earlier developmental stage, greatly speeding up the breeding process. It is especially powerful for traits that are expensive, difficult, or time-consuming to measure.

b) **Genome-Wide Association Studies (GWAS):**

   Concept: GWAS is employed to identify genetic variants [usually single nucleotide polymorphisms (SNPs)] associated with specific traits or diseases in populations.

   Procedure: GWAS involves scanning the genomes of many individuals (both with and without a specific trait or condition) to identify genetic markers that correlate with the trait of interest. If a genetic variant occurs more frequently in individuals with the condition, it is likely linked to the trait.

   Advantages: GWAS has the potential to unveil the genetic underpinnings of complex traits or diseases, guiding researchers toward genes of interest or pathways involved in the trait's manifestation.

   Limitations:

   Multiple testing: Since GWAS evaluates many genetic variants simultaneously, it requires stringent statistical corrections to account for multiple testing, which might lower the power of the study.

   Population stratification: Differences in allele frequencies in subpopulations can result in spurious associations. Proper experimental design and statistical methods are required to mitigate this.

c) **Relationship Between GS and GWAS:**

   While GS focuses on prediction and GWAS on understanding the genetic basis of traits, both can inform each other. For example:

Insights from GWAS can be used to enhance the predictive models in GS by giving more weight to significant markers.

GS can identify potential candidate genes or regions, which can then be investigated in depth using GWAS.

Both GS and GWAS represent advanced tools in the geneticist's arsenal. While GS is steering the course of modern breeding by allowing predictions based on vast swathes of genomic data, GWAS continues to unearth genetic variants associated with critical traits, deepening our understanding of biology. As sequencing technologies evolve and computational methods become more sophisticated, the integration and application of GS and GWAS will only grow in importance, promising breakthroughs in fields ranging from agriculture to medicine.

21.4.4 Speed Breeding and High-throughput Phenotyping

In the pursuit of accelerating the development of improved plant varieties, two methodologies – speed breeding and high-throughput phenotyping – have emerged as transformative tools. While the former expedites the plant's growth cycle, the latter aids in the rapid assessment of plant traits, together contributing to a more efficient breeding process.

a) **Speed Breeding:**

   Concept: Speed breeding is a method to achieve faster plant growth, thereby reducing the time it takes for a plant to progress from germination to maturity. The result is multiple plant generations in a single year.

   Procedure: Controlled Environment: Speed breeding typically uses growth chambers or glasshouses where environmental conditions like light, temperature, and humidity are controlled.

   Extended photoperiod: By extending the daylight hours to up to 22 hours and optimizing light spectra, plants are tricked into growing faster.

   Advantages:

   Rapid generations: Achieving up to 6 wheat generations per year (as opposed to the typical 2) can significantly speed up breeding and genetic research.

   Versatility: The technique has been successfully applied to major crops including wheat, barley, chickpea, and pea.

b) **High-throughput phenotyping (HTP):**

   Concept: HTP is an advanced method for rapidly measuring plant traits (phenotypes) using automated systems and state-of-the-art imaging technologies.

   Procedure:

   Sensors and platforms: High-resolution cameras, multispectral and hyperspectral imaging devices, and even drones are used to capture data on plant growth, architecture, and responses to environmental factors.

   Data analysis: The massive datasets generated require advanced computational tools, including machine learning algorithms, to process, analyze, and correlate with genetic data.

Advantages:

Scale: Traditional phenotyping is time-consuming and often subjective. HTP allows for the objective measurement of thousands of plants in a short time.

Precision: High-resolution imaging captures nuanced differences between plants, allowing for more precise trait selection.

Integration with genomics: Combining HTP data with genomic data can assist in linking phenotypes to specific genetic markers or regions, aiding MAS and genomic selection.

c) **Integration of Speed Breeding and HTP:**

When combined, speed breeding and high-throughput phenotyping offer a potent combination for modern plant breeding:

Rapid development and assessment: Breeders can not only grow plants faster but also assess their traits quickly.

Iterative process: The combination allows for a more iterative breeding approach, where feedback from one generation can be rapidly integrated into the selection and breeding of the next.

The challenges of the 21st century – including climate change, growing global populations, and evolving pests and diseases – demand innovative solutions for crop improvement. Speed breeding and high-throughput phenotyping represent the cutting edge of plant breeding technology, offering the potential to quickly develop crops that are not only higher-yielding but also more resilient and nutritious. As these techniques become more refined and accessible, they are set to redefine the landscape of agricultural research and production.

21.5 Case Study: Development and Improvement of Golden Rice

The urgent pursuit of sustainable solutions to global malnutrition has motivated scientific innovation and collaboration across disciplines and continents. Standing as a paragon of such efforts is the story of Golden Rice, a biofortified crop engineered to address a pressing public health issue: VAD. This case study delves into the genesis, evolution, and refinement of Golden Rice, tracing its journey from the drawing boards of visionary scientists to the fields of farmers in rice-consuming nations. Beyond its technical and scientific facets, the tale of Golden Rice encapsulates the broader challenges of melding biotechnology with traditional agriculture, navigating regulatory and societal landscapes, and ensuring that cutting-edge science translates to tangible health benefits for the global populace (Dubock 2019). As we unpack this compelling narrative, we also gain insights into the dynamic interplay of science, policy, public perception, and the overarching goal of leveraging modern biotechnology for the greater good (Potrykus 2010).

21.5.1 The Initial Development of Golden Rice (GR1 and GR2)

The story of Golden Rice, marked by innovation, collaboration, and the pursuit of a humanitarian mission, began in the late 20th century. The initial development of Golden Rice (GR1 and GR2) stands as a groundbreaking chapter in the quest to address VAD, a widespread public health issue in many developing countries. Here is an overview of the key milestones and developments:

Early 1990s: The concept of Golden Rice was conceived by two renowned plant scientists, Dr. Ingo Potrykus from the Swiss Federal Institute of Technology and Dr. Peter Beyer from the University of Freiburg in Germany. Their vision was to create a GM rice variety that could produce beta-carotene, a precursor of Vitamin A, in its grains.

Genetic engineering of GR1: In 1999, the prototype, known as Golden Rice 1 (GR1), was developed. It involved the insertion of two foreign genes – crtI from daffodils and psy from bacteria – into the rice genome. These genes allowed the rice plant to produce beta-carotene.

21.5.1.1 Challenges and Regulatory Hurdles

Technical challenges: GR1, while a groundbreaking achievement, had limitations. It produced relatively low levels of beta-carotene, and there were concerns about the feasibility of achieving sufficient nutrient intake through normal rice consumption.

Regulatory hurdles: The development of GMOs raised concerns among regulatory authorities, particularly in countries where rice is a dietary staple. Approval for field trials and cultivation of GR1 proved to be a protracted and complex process.

The development of GR2: Early 2000s: Responding to the limitations of GR1, scientists continued their work, leading to the development of Golden Rice 2 (GR2). GR2 incorporated additional genetic modifications, resulting in significantly increased beta-carotene levels in rice grains.

Field trials and global collaboration: In 2004, field trials of GR2 began in several countries, including the Philippines, to assess its agronomic performance and safety.

Collaboration: The development of Golden Rice involved an international collaboration between public and private institutions, as well as the support of organizations like the Rockefeller Foundation and the Bill & Melinda Gates Foundation.

21.5.1.2 Challenges and Continued Development

Scientific challenges: The evolution of Golden Rice faced technical challenges related to optimizing beta-carotene production and addressing concerns regarding yield, taste, and nutrient stability.

Regulatory challenges: The approval process for GM crops remained a significant hurdle, with differing regulatory standards and public perceptions across countries.

The initial development of Golden Rice, from the creation of GR1 to the enhancements leading to GR2, marked the beginning of a promising journey to combat VAD through biotechnology. However, as the case study unfolds, it becomes evident that the path from scientific innovation to widespread adoption and impact is paved with scientific, regulatory, and societal complexities, which continued to shape the course of this remarkable endeavor.

21.6 Bioinformatics-guided Identification of Target Genes for Provitamin A Enhancement

The quest to alleviate global malnutrition through the biofortification of staple crops has led to innovative approaches that leverage bioinformatics as a powerful tool for identifying target genes. Provitamin A enhancement, exemplified by the development of Golden Rice, illustrates how bioinformatics plays a pivotal role in this endeavor.

a) **Genomic Data Mining**

Genomic databases: Bioinformaticians tap into extensive genomic databases to access the genetic information of the target crop, such as rice.

Sequence alignment: Bioinformatics tools perform sequence alignments to identify genes in the crop's genome that are homologous to known genes involved in provitamin A synthesis.

b) **Comparative Genomics:**

Orthologous genes: By comparing the genomes of the target crop with other plants known to produce high levels of provitamin A, bioinformatic analyses identify orthologous genes with similar functions.

c) **Predictive Modeling:**

***In silico* expression analysis:** Bioinformatics allows for the prediction of gene expression patterns by analyzing promoter regions and regulatory elements. This helps identify genes that are likely to be active in the desired tissues (e.g., endosperm for rice).

d) **Functional Annotation:**

Gene ontology (GO) analysis: Bioinformatics tools categorize genes based on their potential functions and assign Gene Ontology terms. This helps prioritize genes likely to be involved in provitamin A synthesis.

e) **Pathway Analysis:**

Metabolic pathway analysis: Bioinformatics enables the reconstruction of metabolic pathways involved in provitamin A biosynthesis. Genes that participate in these pathways are prime candidates for modification.

f) **Sequence Variation Analysis:**

SNPs: Bioinformatics tools identify natural genetic variations (SNPs) in the target crop's genes, which can be leveraged to enhance provitamin A content.

g) **Functional Genomics and Transcriptomics:**

RNA-Seq: High-throughput sequencing technologies, guided by bioinformatics, are used to analyze gene expression patterns in response to various conditions, providing insights into gene regulation.

h) **Integration of Multi-Omics Data:**

Integrative bioinformatics: Bioinformatics integrates data from genomics, transcriptomics, proteomics, and metabolomics to construct comprehensive models of gene networks involved in provitamin A synthesis.

i) **Candidate Gene Selection:**

Prioritization: Bioinformatics-driven analyses prioritize candidate genes based on their expression levels, sequence conservation, and potential to enhance provitamin A content. The bioinformatics-guided identification of target genes for provitamin A enhancement exemplifies how modern biotechnology converges with computational biology. By harnessing the power of genomics, comparative genomics, predictive modeling, and multi-omics integration, researchers can pinpoint genes and regulatory elements that hold the promise of transforming staple crops into nutrient-rich biofortified varieties. This approach, as showcased by Golden Rice, has the potential to address pressing global nutritional challenges while exemplifying the synergy between scientific innovation and computational prowess.

21.6.1.1 Employing CRISPR-Cas9 and Other Breeding Strategies for Golden Rice Improvement

Golden Rice, initially developed to combat VAD, represents a pioneering example of genetic modification in crop breeding. However, ongoing research and development seek to further enhance its nutritional value, agronomic performance, and societal acceptance. Employing CRISPR-Cas9 and other breeding strategies, scientists are refining Golden Rice in the following ways:

a) **CRISPR-Cas9 for Precision Editing:**
   - **Targeted trait enhancement:** CRISPR-Cas9 is employed to precisely edit the rice genome, allowing for the enhancement of specific traits, such as increased beta-carotene production in Golden Rice.
   - **Reduced off-target effects:** Advanced CRISPR-Cas9 techniques and bioinformatics tools minimize off-target effects, ensuring the safety and predictability of genetic modifications.
   - **Nutrient stability:** Researchers are utilizing CRISPR-Cas9 to stabilize the beta-carotene content in Golden Rice, making it more resilient to environmental factors and storage conditions.

b) **Genomic Selection (GS):**
   - **Improved breeding efficiency:** GS combines genotypic and phenotypic data to predict an individual's genetic potential. It accelerates the selection of plants with enhanced nutritional traits while considering their overall performance.
   - **Multi-trait selection:** GS allows for the simultaneous selection of multiple traits, optimizing not only beta-carotene content but also other agronomically important characteristics.

c) **Marker-assisted selection (MAS):**
   - **Selection for specific alleles:** MAS, guided by molecular markers, enables the selection of rice plants with desired alleles associated with increased beta-carotene production.
   - **Combining traits:** MAS can be used to combine the beta-carotene trait with other traits like disease resistance or high yield, ensuring a well-rounded crop.

d) **High-throughput phenotyping (HTP):**
   - **Efficient trait evaluation:** HTP technologies, including automated imaging and data analysis, streamline the evaluation of thousands of rice plants, making it faster and more accurate to identify those with enhanced beta-carotene content.

e) **Traditional Breeding Strategies:**
   - **Hybridization:** Traditional crossbreeding techniques are employed to introduce genetic diversity from local rice varieties, potentially improving beta-carotene traits while preserving important local characteristics.

- **Backcrossing:** Backcrossing with locally adapted varieties helps to transfer the beta-carotene trait into widely cultivated rice varieties.

f) **Regulatory Compliance and Societal Engagement:**
- **Safety and regulation:** Researchers and breeders are working closely with regulatory authorities to ensure that all genetic modifications comply with safety standards and regulations.
- **Public acceptance:** Engaging with stakeholders, including farmers and consumers, is essential to gain acceptance and trust in GM crops.

As Golden Rice undergoes continuous improvement, these breeding strategies, including CRISPR-Cas9-mediated precision editing, genomics-driven selection, and traditional breeding methods, synergize to create a crop that not only addresses VAD but also meets the diverse needs of farmers and consumers worldwide. By balancing scientific innovation with safety and societal acceptance, Golden Rice exemplifies the potential of biotechnology to contribute to global food security and nutrition.

21.6.2 Assessing the Agronomic Performance and Nutritional Quality of Improved Golden Rice Lines

Assessing the agronomic performance and nutritional quality of improved Golden Rice lines is crucial to ensuring that the crop not only delivers the desired nutritional benefits but also meets the practical requirements of farmers and consumers. Here is an assessment of both aspects:

a) **Agronomic Performance:**

Yield and growth characteristics: One of the primary concerns when improving Golden Rice is to maintain or enhance its yield and growth characteristics compared to conventional rice varieties. This includes assessing factors such as grain yield per hectare, plant height, flowering time, and resistance to pests and diseases. Improved lines should ideally match or outperform existing varieties in these agronomic traits.

Environmental adaptability: Golden Rice should exhibit adaptability to diverse environmental conditions and regions where rice is a staple crop. It should grow well in different soil types, temperature ranges, and rainfall patterns.

Pest and disease resistance: Given the potential risks of increased susceptibility to pests and diseases due to genetic modifications, it is essential to assess the resistance of improved Golden Rice lines to common rice pathogens and pests.

Agronomic practices: Farmers' ability to cultivate Golden Rice using standard agronomic practices, such as planting, fertilization, and irrigation, should be evaluated. Compatibility with existing farming methods is critical for adoption.

Stability across generations: Agronomic traits should remain stable across generations to ensure that farmers can rely on consistent performance year after year.

b) **Nutritional Quality:**

Beta-carotene content: The primary nutritional goal of Golden Rice is to provide a significant source of provitamin A (beta-carotene). Improved lines should exhibit higher and more stable levels of beta-carotene in the edible parts of the rice grains.

Nutrient stability: It is essential to assess the stability of beta-carotene content during storage and cooking. Golden Rice should retain its nutritional value even after processing and storage.

Taste and sensory properties: Nutritional improvements should not compromise the taste, aroma, or overall sensory qualities of Golden Rice. Consumer acceptance is vital, and any undesirable changes in taste or texture must be addressed.

Digestibility: The digestibility and bioavailability of beta-carotene in Golden Rice should be evaluated to ensure that the provitamin A is effectively absorbed by the human body.

Micronutrient profile: Apart from beta-carotene, it is essential to assess other micronutrients and anti-nutritional factors to ensure that the overall nutritional profile of Golden Rice is balanced and safe for consumption.

c) **Field Trials and Real-world Conditions:**

Field trials conducted in diverse geographical locations and under different growing conditions are crucial to assess both agronomic performance and nutritional quality. These trials should involve farmers and communities to obtain feedback on the practicality and acceptability of improved Golden Rice lines.

d) **Monitoring and Continuous Improvement:**

The assessment of agronomic performance and nutritional quality should be an ongoing process, with continuous monitoring and refinement of the crop. Researchers, breeders, and agricultural extension services should work together to address any issues that arise and ensure that improved Golden Rice lines meet the needs of both farmers and consumers.

In conclusion, the successful development and adoption of improved Golden Rice lines require a balanced approach that considers both agronomic performance and nutritional quality. By rigorously evaluating these aspects and continuously engaging with stakeholders, Golden Rice can fulfill its potential as a powerful tool in addressing VAD and improving global nutrition (Depositario et al. 2009).

e) **Challenges and Future Perspectives**

The journey of Golden Rice, from its conceptualization to its development and improvement, has been marked by remarkable achievements, but it also faces several challenges and holds vast potential for the future (Ilg et al. 2010).

Challenges:

- **Regulatory hurdles:** Regulatory approvals for GM crops can be time-consuming and complex, varying across countries (Potrykus 2012). Overcoming these regulatory hurdles is essential to ensure the widespread cultivation and distribution of improved Golden Rice (Fuchs and Glaab 2011).
- **Societal acceptance:** Public perception of GMOs remains a challenge. Engaging with the public, addressing concerns, and building trust are ongoing efforts needed to gain acceptance.
- **Maintaining agronomic performance:** While enhancing nutritional quality, it is crucial to ensure that agronomic performance is not compromised. Maintaining high yields, pest resistance, and adaptability to diverse environments remains a significant challenge.
- **Nutrient stability:** Ensuring that the beta-carotene content remains stable during storage, cooking, and processing is vital to provide consistent nutritional benefits to consumers.

- **Access and distribution:** Ensuring that improved Golden Rice reaches the communities that need it most, especially in remote and marginalized areas, requires robust distribution channels and equitable access (Biswas et al. 2021).

f) **Future Perspectives:**
- **Continuous research and improvement:** Ongoing research will focus on further enhancing the nutritional quality of Golden Rice while addressing any challenges that arise. Advanced breeding techniques, such as CRISPR-Cas9, will play a pivotal role in achieving these objectives (Gayen et al. 2015).
- **Global adoption:** Expanding the cultivation and adoption of improved Golden Rice varieties across regions with VAD remains a priority. Collaboration with governments, NGOs, and local communities is essential for successful implementation (Glover et al. 2020).
- **Diverse varieties:** Developing diverse Golden Rice varieties tailored to different geographical regions and local preferences will broaden its acceptance and impact.
- **Synergy with other nutritional solutions:** Integrating Golden Rice into broader nutrition and food security programs, along with other interventions like vitamin supplementation and dietary diversification, can maximize its impact on reducing malnutrition.
- **Nutrient-enriched crops:** The success of Golden Rice has spurred research into nutrient enrichment of other staple crops, such as cassava, maize, and wheat, offering the potential to combat various nutrient deficiencies.
- **Sustainability:** Sustainable agricultural practices, including crop rotation, agroecological approaches, and organic farming, should be considered to ensure the long-term viability of improved Golden Rice cultivation.
- **Education and awareness:** Continued efforts to educate farmers, consumers, and policymakers about the benefits and safety of Golden Rice are essential for its acceptance and utilization (Krawinkel 2007).

In conclusion, improved Golden Rice stands as a symbol of scientific innovation and humanitarian efforts to address global malnutrition (Chong 2003). While challenges persist, the future of Golden Rice holds immense promise as researchers, policymakers, and communities collaborate to unlock its potential to improve nutrition, enhance food security, and transform the lives of millions around the world (Yang et al. 2014).

21.7 Ethical and IP Issues

21.7.1 Ethical Considerations and Public Perception of Genetically Modified Crops

The introduction of GM crops has raised important ethical considerations and generated a wide range of public perceptions and attitudes. These issues touch upon concerns related to safety, environmental impact, ownership of genetic resources, and global food security.

a) **Environmental Impact:**

Ethical Concerns:

Unintended consequences: Critics worry about the unintended environmental consequences of GM crops, such as the potential harm to nontarget organisms, soil health, and ecosystem stability.

Gene flow: The transfer of modified genes from GM crops to wild or non-GM varieties through pollen flow raises questions about genetic contamination and its ecological implications.

Public Perception:

Divergent Views: Public opinion varies widely, with some individuals expressing concerns about environmental risks, while others perceive GM crops as a way to reduce the environmental impact of agriculture through reduced pesticide use and enhanced resource efficiency (Wesseler and Zilberman 2017).

b) **Food Safety:**

Ethical Concerns:

Human health: Ethical debates center on the safety of GM foods for human consumption, particularly potential allergenicity and long-term health effects.

Labeling: The right to know whether a product contains GM ingredients is considered an ethical issue, and labeling laws are implemented differently in various countries.

Public Perception:

Mixed views: Public perceptions of food safety are influenced by factors like media coverage, cultural beliefs, and trust in regulatory agencies. While some are concerned, others view GM foods as rigorously tested and safe.

c) **Ownership and Access:**

Ethical Concerns:

Corporate control: Critics argue that the consolidation of GM seed patents and intellectual property (IP) rights in the hands of a few large corporations can limit access to technology and seeds for small-scale farmers in developing countries.

Dependency: Concerns are raised about developing countries becoming dependent on multinational corporations for GM seeds and technologies.

Public Perception:

Fairness and equity: Public sentiment often aligns with ethical concerns regarding the equitable distribution of benefits and access to GM crop technologies, especially in the context of global agricultural systems.

d) **Global Food Security:**

Ethical Concerns:

Distribution of benefits: Ensuring that GM crop technologies benefit marginalized and food-insecure populations is an ethical imperative (Bollinedi et al. 2019).

Addressing nutritional needs: Ethical questions arise regarding the focus on crops like Golden Rice, engineered to address specific nutrient deficiencies, and the allocation of research resources to such projects.

Public Perception:

Balancing needs: Public perception regarding GM crops' role in global food security varies, with some seeing them as a promising tool and others emphasizing the importance of holistic approaches to food access and nutrition.

e) **Ethical Oversight:**

Ethical Concerns:

Ethical oversight: Ethical oversight and governance of GM crop development, including transparency, stakeholder engagement, and accountability, are essential to address concerns and build trust.

Public Perception:

Trust in institutions: Public perception often hinges on trust in regulatory agencies, scientific expertise, and the transparency of decision-making processes.

In navigating the ethical considerations and public perception of GM crops, it is crucial to foster open dialogue, engage with diverse stakeholders, and prioritize robust risk assessments and safety measures. Striking a balance between technological innovation, ethical principles, and societal concerns is essential to harness the potential benefits of GM crops while addressing the complex issues they raise.

21.7.2 Regulatory and Intellectual Property Issues

The development and commercialization of GM crops bring forth intricate regulatory frameworks and IP concerns that impact various stakeholders, including biotechnology companies, farmers, consumers, and regulatory authorities (Gayen et al. 2016). Here is an overview of the key issues in each domain:

a) **Regulatory Issues:**
   - **Safety assessments:** Regulatory authorities in different countries require comprehensive safety assessments to evaluate the potential risks associated with GM crops. These assessments encompass environmental, human health, and ecological considerations (Wesseler et al. 2014).
   - **Transboundary movement:** The global trade in GM crops necessitates the harmonization of regulatory standards across nations. Issues arise when countries have divergent regulatory approaches, leading to trade disputes and market access challenges (Al-Babili et al. 2006).
   - **Labeling and traceability:** Regulations regarding the labeling of GM products vary worldwide. Ethical and consumer right-to-know considerations underlie labeling requirements, but they can also affect trade and consumer choice.
   - **Environmental impact:** Regulatory authorities scrutinize the environmental impact of GM crops, including potential effects on nontarget organisms, soil ecosystems, and the emergence of herbicide-resistant weeds or insect pests.
   - **Human health:** GM crops intended for human consumption undergo rigorous safety assessments to ensure that they are safe to eat and do not pose allergenic or toxic risks.
   - **Coexistence:** The coexistence of GM crops with conventional or organic agriculture presents challenges in terms of minimizing gene flow and contamination.

b) **Intellectual Property Issues:**
   - **Patents and ownership:** Biotechnology companies often hold patents on GM crop technologies, which grant them exclusive rights to use, make, sell, and license these crops. This can lead to concerns about corporate control and limited access for small-scale farmers.
   - **Licensing and royalties:** Licensing agreements and royalty fees can affect the affordability of GM seeds for farmers, especially in developing countries. These issues raise questions about equitable access to technology.
   - **Seed saving and farmer rights:** Restrictions on seed saving and replanting GM seeds can be contentious. Farmers' rights to save and exchange seeds are weighed against IP protection.

- **Biopiracy:** Concerns about biopiracy arise when genetic resources from developing countries are used to develop GM crops without appropriate compensation or benefit-sharing agreements.
- **Innovation and public goods:** The tension between protecting IP rights and ensuring the development of public goods, such as GM crops designed to address malnutrition or food security, underscores ethical dilemmas in IP protection.
- **Technology transfer:** Issues related to the transfer of GM crop technology to developing countries involve both IP considerations and the need for capacity building and infrastructure development (Datta and Datta 2020).

Addressing these regulatory and IP issues involves a delicate balancing act between fostering innovation, ensuring food security, protecting public and environmental health, and respecting the rights and interests of various stakeholders. As the use of GM crops continues to grow worldwide, efforts to harmonize regulations, encourage technology sharing, and establish equitable benefit-sharing mechanisms will play a pivotal role in shaping the future of GM crop development and adoption.

21.7.3 Ensuring Accessibility and Adoption of Golden Rice in Target Communities

The successful introduction and utilization of Golden Rice, especially in regions with VAD, requires a multifaceted approach that addresses accessibility, acceptance, and sustainability. Here are strategies to ensure its accessibility and adoption in target communities:

a) **Collaborative Partnerships:**
   Engage with local communities: Collaborate closely with local communities, including farmers, nutritionists, and health workers, to understand their needs, concerns, and preferences regarding Golden Rice.
   Government support: Seek support and partnership with local governments to facilitate regulatory approvals, distribution networks, and extension services.
b) **Education and Awareness:**
   Community education: Conduct educational programs to inform target communities about the nutritional benefits of Golden Rice, its safety, and its role in combating VAD.
   Training for farmers: Provide training to farmers on cultivation practices, including planting, harvesting, and seed preservation techniques specific to Golden Rice.
c) **Affordable Access:**
   Subsidies and incentives: Explore options for making Golden Rice seeds and technology more affordable, such as subsidies or incentives for farmers to grow it.
   Local seed production: Encourage local seed production to reduce the cost of seeds and ensure a consistent supply.
d) **Distribution Networks:**
   Establish distribution channels: Develop efficient distribution networks to deliver Golden Rice seeds to remote and underserved areas, working with local agricultural cooperatives and organizations.
   Access for vulnerable groups: Ensure that Golden Rice is accessible to vulnerable groups, including smallholder farmers and marginalized communities (Qiu 2012).
e) **Sustainable Farming Practices:**
   Integrated agriculture: Promote the integration of Golden Rice into diversified farming systems to enhance resilience and food security.

Conservation agriculture: Emphasize conservation agriculture practices that improve soil health and reduce environmental impact.

f) **Addressing Cultural and Taste Preferences:**

Cultural sensitivity: Respect local dietary habits and cultural preferences, and work to ensure that Golden Rice aligns with traditional recipes and tastes.

g) **Monitoring and Evaluation:**

Data collection: Establish a robust monitoring and evaluation system to assess the impact of Golden Rice adoption on nutrition and health outcomes in target communities (Dubock 2017).

Feedback mechanism: Encourage feedback from farmers and consumers to continuously improve Golden Rice varieties and address any challenges that arise.

h) **Community Ownership:**

Participatory approach: Involve local communities in decision-making processes and give them a sense of ownership in the development and promotion of Golden Rice.

i) **Addressing Gender Disparities:**

Gender-inclusive approaches: Recognize the role of women in agriculture and nutrition and implement gender-inclusive strategies that empower women farmers.

j) **Continuous Research and Improvement:**

Research investment: Commit to ongoing research and development to improve Golden Rice varieties, addressing not only Vitamin A but also other nutritional needs and agronomic challenges.

k) **Global Collaboration:**

International cooperation: Collaborate with international organizations, research institutions, and NGOs to leverage resources, expertise, and knowledge sharing.

By implementing these strategies, stakeholders can work together to ensure that Golden Rice becomes a sustainable and impactful solution for combating VAD and improving the overall nutrition and health of target communities.

As bioinformatics continues to advance, it will empower breeders with the tools and knowledge needed to develop crops that are not only more productive and resilient but also tailored to meet evolving nutritional and environmental challenges (Grusak 2005). The integration of bioinformatics with traditional breeding methods represents a powerful synergy that promises to shape the future of agriculture and food security (Bollinedi et al. 2017).

21.8 Conclusion

Through the lens of the chapter, we revisited the transformative role of bioinformatics in the domain of plant research and crop breeding, with Golden Rice being a salient example. This GM crop not only exemplifies a successful application of bioinformatics in enhancing nutritional qualities but also serves as a beacon of hope against global micronutrient deficiencies. We dissected various tools and resources available through bioinformatics, from sequence databases to integrative systems biology approaches, and how these assets have been pivotal in developing, refining, and improving Golden Rice. Furthermore, we brought to the forefront the challenges and considerations that have been part and parcel of Golden Rice's journey, such as ethical concerns, public perception issues, and regulatory hurdles.

The chapter elucidated that bioinformatics is not merely a supportive field but is central to present and future breakthroughs in plant research and crop breeding. It has transitioned into a linchpin that not only decodes complex genomes but also facilitates the enhancement of nutritional profiles in crops. The narrative of Golden Rice showcased bioinformatics as an essential player in realizing the vision of fortified and nutritionally superior crops. As technologies continue to evolve, bioinformatics will likely remain integral to developing innovative solutions, as seen through the application of CRISPR-Cas9, MAS, genomic selection, and speed breeding in Golden Rice. The future indeed holds more possibilities where bioinformatics will continue driving innovation in crop breeding, paving the way for not only high yields but also nutritionally enriched food sources.

Golden Rice's development and the overarching role of bioinformatics have profound implications for global food security and nutrition. With an ever-increasing global population and the persistent challenge of malnutrition, particularly in developing nations, the need for crops that are both high-yielding and nutritionally enhanced is imperative. Golden Rice represents a tangible solution to these challenges, addressing the pervasive issue of VAD. The lessons learned and successes achieved through the case of Golden Rice offer a blueprint for the development of other GM crops, each with the potential to address different nutritional deficiencies and food security challenges. Furthermore, the approach taken with Golden Rice, guided and supported by bioinformatics, could serve as a model for future endeavors in the field of agricultural biotechnology, each contributing toward the creation of a world where access to nutritious food is not a privilege but a right for all. The journey and the narrative of Golden Rice indeed provide optimism and direction, illuminating the path for leveraging biotechnology for humanity's greater good. Through the integration of bioinformatics, there is an opportunity to meet global food demands sustainably, efficiently, and nutritiously, heralding a future where food security and optimal nutrition are accessible realities for populations worldwide. In summing up, the interplay of bioinformatics and the development of crops like Golden Rice underscores a promising trajectory for addressing the complex challenges of global food security and nutrition. From unfolding the genetic mysteries of plants to facilitating the inception and improvement of nutritionally superior crops, bioinformatics has proven to be an indispensable ally. The case of Golden Rice is illustrative, providing not only a framework for future scientific endeavors but also hope for millions worldwide afflicted by nutritional deficiencies. As we move forward, it is evident that the tandem of bioinformatics and diligent research will continue to unveil solutions, with the potential to bring about significant positive changes to global agriculture, health, and nutrition. The golden opportunities lying at this intersection of technology and biology are indeed vast and beckoning, offering a vista of a future where hunger and malnutrition could well be relics of the past.

Declaration

Author used ChatGPT 4.0 for initial drafting the article.

References

Al-Babili, S. and Beyer, P. (2005). Golden Rice–five years on the road–five years to go? *Trends in Plant Science* 10 (12): 565–573.

Al-Babili, S., Hoa, T.T., and Schaub, P. (2006). Exploring the potential of the bacterial carotene desaturase CrtI to increase the β-carotene content in Golden Rice. *Journal of Experimental Botany* 57 (4): 1007–1014.

Beyer, P. (2010). Golden Rice and 'Golden' crops for human nutrition. *New Biotechnology* 27 (5): 478–481.

Beyer, P., Al-Babili, S., Ye, X. et al. (2002). Golden rice: introducing the β-carotene biosynthesis pathway into rice endosperm by genetic engineering to defeat vitamin A deficiency. *The Journal of Nutrition* 132 (3): 506S–510S.

Biswas, P.S., Swamy, B.M., Kader, M.A. et al. (2021). Development and field evaluation of near-isogenic lines of GR2-EBRRI dhan29 golden rice. *Frontiers in Plant Science* (12): 619739.

Bollinedi, H., Prabhu, K.V., Singh, N.K. et al. (2017). Molecular and functional characterization of GR2-R1 event based backcross derived lines of Golden Rice in the genetic background of a mega rice variety Swarna. *PloS One* 12 (1): e0169600.

Bollinedi, H., Dhakane-Lad, J., Krishnan, S.G. et al. (2019). Kinetics of β-carotene degradation under different storage conditions in transgenic Golden Rice® lines. *Food Chemistry* 278: 773–779.

Chong, M. (2003). Acceptance of golden rice in the Philippine 'rice bowl'. *Nature Biotechnology* 21 (9): 971–972.

Corrigan, J.R., Depositario, D.P., Nayga, R.M. Jr. et al. (2009). Comparing open-ended choice experiments and experimental auctions: an application to golden rice. *American Journal of Agricultural Economics* 91 (3): 837–853.

Datta, S.K. and Datta, K. (2020). *Golden Rice. The Future of Rice Demand: Quality Beyond Productivity*, 135–147. Cham: Springer.

Datta, S.K., Datta, K., Parkhi, V. et al. (2007). Golden rice: introgression, breeding, and field evaluation. *Euphytica* 154: 271–278.

Dawe, D., Robertson, R., and Unnevehr, L. (2002). Golden rice: what role could it play in alleviation of vitamin A deficiency? *Food Policy* 27 (5–6): 541–560.

De Steur, H., Stein, A.J., and Demont, M. (2022). From Golden Rice to Golden Diets: how to turn its recent approval into practice. *Global Food Security* 32: 100596.

Depositario, D.P., Nayga, R.M. Jr., Wu, X., and Laude, T.P. (2009). Effects of information on consumers' willingness to pay for golden rice. *Asian Economic Journal* 23 (4): 457–476.

Dubock, A. (2014a). The politics of golden rice. *GM Crops & Food* 5 (3): 210–222.

Dubock, A. (2014b). The present status of Golden Rice. *Journal of Huazhong Agricultural University* 33 (6): 69–84.

Dubock, A. (2017). An overview of agriculture, nutrition and fortification, supplementation and biofortification: Golden Rice as an example for enhancing micronutrient intake. *Agriculture & Food Security* 6 (1): 59.

Dubock, A. (2019). *Golden Rice: To Combat Vitamin A Deficiency for Public Health*. London, UK: IntechOpen.

Dwiningsih, Y. (2010). Golden Rice: Vitamin A in the Grain. *BioS Biology Popular Magazine* 1 (1).

Fuchs, D. and Glaab, K. (2011). Material power and normative conflict in global and local agrifood governance: the lessons of 'Golden Rice' in India. *Food Policy* 36 (6): 729–735.

Gayen, D., Ali, N., Sarkar, S.N. et al. (2015). Down-regulation of lipoxygenase gene reduces degradation of carotenoids of golden rice during storage. *Planta* 242: 353–363.

Gayen, D., Ghosh, S., Paul, S. et al. (2016). Metabolic regulation of carotenoid-enriched golden rice line. *Frontiers in Plant Science* 7: 1622.

Glover, D., Kim, S.K., and Stone, G.D. (2020). Golden Rice and technology adoption theory: a study of seed choice dynamics among rice growers in the Philippines. *Technology in Society* (60): 101227.

Greedy, D. (2018). Golden Rice is safe to eat, says FDA. *Nature Biotechnology* 36 (7): 559.

Grusak, M.A. (2005). Golden Rice gets a boost from maize. *Nature Biotechnology* 23 (4): 429–430.

Harmon, A. (2013 Aug 24). Golden rice: lifesaver. *The New York Times* 23.

Ilg, A., Yu, Q., Schaub, P. et al. (2010). Overexpression of the rice carotenoid cleavage dioxygenase 1 gene in Golden Rice endosperm suggests apocarotenoids as substrates in planta. *Planta* 232: 691–699.

Kettenburg, A.J., Hanspach, J., Abson, D.J., and Fischer, J. (2018). From disagreements to dialogue: unpacking the Golden Rice debate. *Sustainability Science* 13: 1469–1482.

Krawinkel, M.B. (2007). What we know and don't know about Golden Rice. *Nature Biotechnology* 25 (6): 623.

Mallikarjuna Swamy, B.P., Marundan, S. Jr., Samia, M. et al. (2021). Development and characterization of GR2E Golden rice introgression lines. *Scientific Reports* 11 (1): 2496.

Mayer, J.E. (2005). The golden rice controversy: useless science or unfounded criticism? *BioScience* 55 (9): 726–727.

Moghissi, A.A., Pei, S., and Liu, Y. (2016). Golden rice: scientific, regulatory and public information processes of a genetically modified organism. *Critical Reviews in Biotechnology* 36 (3): 535–541.

Paine, J.A., Shipton, C.A., Chaggar, S. et al. (2005). Improving the nutritional value of Golden Rice through increased pro-vitamin A content. *Nature Biotechnology* 23 (4): 482–487.

Potrykus, I. (2010). Lessons from the 'Humanitarian Golden Rice' project: regulation prevents development of public good genetically engineered crop products. *New Biotechnology* 27 (5): 466–472.

Potrykus, I. (2012). "Golden Rice", a GMO-product for public good, and the consequences of GE-regulation. *Journal of Plant Biochemistry and Biotechnology* 21 (Suppl 1): 68–75.

Qiu, J. (2012). China sacks officials over Golden Rice controversy. *Nature* 10.

Regis, E. (2019). *Golden Rice: The Imperiled Birth of a GMO Superfood*. Johns Hopkins University Press.

Stein, A.J., Sachdev, H.P., and Qaim, M. (2006). Potential impact and cost-effectiveness of Golden Rice. *Nature Biotechnology* 24 (10): 1200–1201.

Stein, A.J., Sachdev, H.P., and Qaim, M. (2008). Genetic engineering for the poor: Golden Rice and public health in India. *World Development* 36 (1): 144–158.

Stone, G.D. and Glover, D. (2017). Disembedding grain: Golden Rice, the Green Revolution, and heirloom seeds in the Philippines. *Agriculture and Human Values* 34 (1): 87–102.

Swamy, B.M., Samia, M., Boncodin, R. et al. (2019). Compositional analysis of genetically engineered GR2E "Golden Rice" in comparison to that of conventional rice. *Journal of Agricultural and Food Chemistry* 67 (28): 7986–7994.

Tang, G., Qin, J., Dolnikowski, G.G. et al. (2009). Golden Rice is an effective source of vitamin A. *The American Journal of Clinical Nutrition* 89 (6): 1776–1783.

Tang, G., Hu, Y., Yin, S.A. et al. (2012). Retracted: β-Carotene in Golden Rice is as good as β-carotene in oil at providing vitamin A to children. *The American Journal of Clinical Nutrition* 96 (3): 658–664.

Wesseler, J. and Zilberman, D. (2014). The economic power of the Golden Rice opposition. *Environment and Development Economics* 19 (6): 724–742.

Wesseler, J. and Zilberman, D. (2017). Golden Rice: no progress to be seen. Do we still need it? *Environment and Development Economics* 22 (2): 107–109.

Wesseler, J.H., Kaplan, S., and Zilberman, D. (2014). The cost of delaying approval of golden rice. *Agricultural and Resource Economics Update* 17 (3): 1–3.

Wu, F., Wesseler, J., Zilberman, D. et al. (2021). Allow golden rice to save lives. *Proceedings of the National Academy of Sciences* 118 (51): e2120901118.

Yang, J., Xu, K., and Rodriguez, L. (2014). The rejection of science frames in the news coverage of the golden rice experiment in Hunan, China. *Health, Risk & Society* 16 (4): 339–354.

Zhu, Q., Zeng, D., Yu, S. et al. (2018). From golden rice to aSTARice: bioengineering astaxanthin biosynthesis in rice endosperm. *Molecular Plant* 11 (12): 1440–1448.

22

Going Wild: Genomics of Forest Plants and the Future of Crop Improvement

Ajinkya Bharatraj Patil[1], *Debojyoti Kar*[2], *Sourav Datta*[2], *and Nagarjun Vijay*[1]

[1] *Computational Evolutionary Genomics Lab, Department of Biological Sciences, IISER Bhopal, Bhauri, Madhya Pradesh, India*
[2] *Plant Cell and Developmental Biology Lab, Department of Biological Sciences, IISER Bhopal, Bhauri, Madhya Pradesh, India*

22.1 Introduction

Of the more than 300 thousand plant species known to humans, only a fourth (~75,000) have seen extensive utilization for their benefits. Yet, only 1% of these plant species have been cultivated or seen domestication. Only a fraction (0.1%) of these domesticated species have seen widespread global cultivation. Such a focus on a small set of plants results from properties such as shorter harvest time, resilience to drought, and stress that determine the ease of cultivation. Early attempts to systematically exploit plants and organized cultivation have evaluated various species and selected those suited for human consumption. Subsequently, trade relations between different parts of the world and the suitability of these plants to diverse climates have been major factors in determining the plants that are widely cultivated today. Many plant species have failed to establish themselves as staple food despite their high utility and past popularity. The reasons for such limited use of plant species range from longer generation times to habitat-specific adaptations which prevent their widespread cultivation. Hence, plants amenable to traditional breeding practices have been the focus of human capitalization.

The idea of plant breeding dates back to almost 10,000 years ago when humans started cultivating plants, giving rise to the concept of agriculture. Conventionally, plant breeding was based on observed variation and selection of individuals with a superior phenotype that outperformed the original landraces. One classic example of breeding for more desirable traits is the derivation of maize from teosinte. Crop improvement involved crossing both parents and selecting the useful phenotype that combines the best characteristics. This idea of trait inheritance through generations and the laws governing them was proposed by Sir Gregor Mendel in 1865. The laws of heredity established by Mendel by studying peas paved the way for modern plant breeding technologies. He provided the concepts of alleles, alternate forms of a gene, each inherited from one of the two parents. It was also clear that one allele is dominant over the other (recessive), and the effect of the dominant allele is reflected in the phenotype. The understanding of random segregation of gametes

Bioinformatics for Plant Research and Crop Breeding, First Edition. Edited by Jen-Tsung Chen.

and the mechanism of phenotypic trait inheritance established by Mendel was a major milestone in using genetics for crop improvement. These approaches have solved the problem of selecting only superior phenotypes in cultivars, resulting in lower line stability and adaptability (Breseghello and Coelho, 2013). However, the findings of Mendel were suitable for particularly unlinked genes, and this selective breeding favored a better selection of qualitative traits.

These early efforts to link phenotypic diversity with genes using the segregation of traits allowed the construction of linear maps containing trait-specific loci. Following the discovery of DNA as hereditary material, advances in genetics allowed the identification of genetic changes responsible for causing specific qualitative phenotypes. Identification of DNA loci associated with a phenotype with continuous variation resulted in the definition of quantitative trait loci (QTLs). Mapping QTLs of varied quantitative phenotypes in diverse crop species became popular with the availability of genomic resources. The QTL-based studies mainly identified large genomic segments encompassing multiple genes, which resulted in limited resolution. High-throughput genotyping approaches that compared a large number of individuals with distinct phenotypes to pinpoint specific polymorphisms responsible for a particular trait. Such genome-wide association studies (GWAS) allowed for the fine-scale mapping of qualitative and quantitative phenotypic traits. However, these approaches are limited to phenotypic variation available within a species, pinpointing differences in traits between the populations or cultivars. Moreover, most quantitative traits are considered polygenic, and these approaches are not efficient in mapping such traits. Furthermore, the statistical power required to identify the genetic basis of phenotypes with low effect sizes makes these methods very expensive. Although all the techniques discussed above have led to advances in plant breeding and crop improvement, these methods have limited utility in characterizing phenotypic variation between species. Hence, bioinformatic approaches using population genomic datasets allow evaluating within and between species variation, identifying candidate gene loci responsible for phenotypes at fine scale, introgression between species, and adaptation mechanisms involving selection and demography. The ability to perform such varied analyses provides an in-depth knowledge of phenotype-associated variation and their mechanisms.

Precious plant products such as saffron calyx, specific types of timber, alkaloids, etc., are highly sought after and are expensive due to their scarcity resulting from our inability to grow these plants outside their native range. The advances in genetic tools and the availability of genomic resources have meant that such exotic plants may be tamed to be tolerant of non-native habitats and subjected to worldwide cultivation. Such genomic invigoration of plant species using the latest advances in biology has the potential to reclaim various plant phenotypes that were missed out in the initial round of plant domestication. The use of metabolic engineering to develop novel bio-processes using "platform cell factories" has been possible for producing artemisinic acid and opioids. Commonly used cell factories are genetically well-characterized yeast and bacterial systems. These approaches have traditionally identified a molecule of interest and subsequently incorporated the pathway to produce this molecule in cell factories. Earlier approaches to identifying the metabolites and the pathways involved in their production were not always successful. The availability of large genomic datasets has made it increasingly feasible to identify pathways involved in producing diverse metabolites. The advent of sophisticated genome editing tools such as clustered regularly interspaced short palindromic repeats (CRISPR) and their application in eukaryotic systems has also allowed for the specific manipulation of plant systems. We contemplate that the next phase of crop improvement can benefit from the knowledge gleaned from genomic analysis of forest-adapted and hard-to-cultivate plants. It would be possible to engineer complex phenotypes in plant systems by incorporating entire pathways responsible for their production. For example, comparative genomics of closely related species can help identify the genetic architecture responsible for distinct phenotypes. The possibility of changes in gene families, gain or loss of specific genes involved in certain metabolic pathways, changes in interactions, regulatory rewiring, structural variation, other genetic polymorphism, repeat insertion, etc., and their impact on the phenotypes can be revealed. Attaining specific phenotypes will be possible once the molecular mechanisms responsible for the natural traits are characterized. Genomic manipulation of target plant species to incorporate diverse phenotypes from distant taxa can improve cultivation ability and ensure climate resilience.

Reducing the cost of genomic technologies has allowed its use to address diverse questions in non-model plant species. Such studies have revealed the varied genetic mechanisms involved in generating similar phenotypes. Millennia of natural experimentation has devised multiple ways to produce an astounding diversity, providing a ready, well-tested toolkit to achieve distinct phenotypes. Genomic analysis has been instrumental in elucidating these diverse mechanisms throughout the tree of life. In this chapter, we highlight the utility of applying whole-genomic approaches in leveraging the phenotypic diversity of plant species. Based on a literature survey of a diverse set of genomic studies, we enumerate three major directions to understand the genetic basis of phenotypic diversity (see Figure 22.1). (i) Role of repetitive elements in generating phenotypic diversity. (ii) Utility of population genomics to identify candidate loci in attaining natural phenotypes and

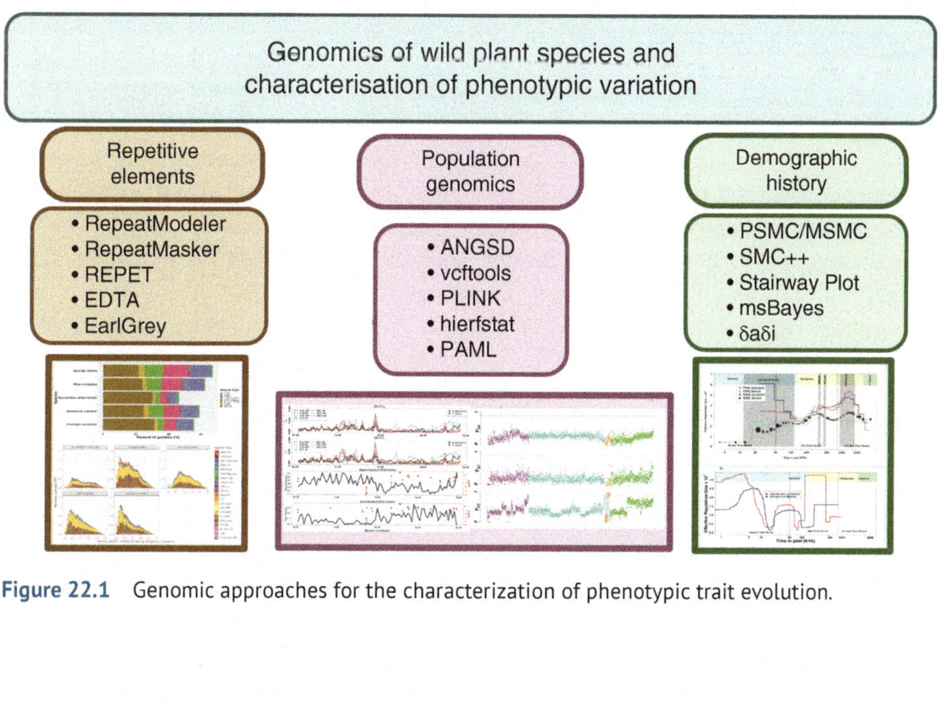

Figure 22.1 Genomic approaches for the characterization of phenotypic trait evolution.

adaptations. (iii) Understand the role of stochastic demographic processes to identify the effects of bottlenecks in these populations and their response to climatic fluctuations.

22.2 Repetitive Genomic Elements as Phenotypic Trait Variation Machinery

The repetitive elements constitute an abundant part of the genomes of eukaryotes, especially plants. Historically known as "Junk" or "Useless" DNA sequences, they occupy a significant, if not maximum, part of the genomes of plants. The repetitive elements are primarily divided into two main classes called tandem repeats and dispersed repeats. There has been much research on tandem repeats in the early part of the discovery due to their implications for human diseases. The seminal work by Barbara McClintock on the repetitive elements and their effect on the regulation of kernel color of Maize gave Transposable elements a new identity. Since then, transposons have been implicated in multiple phenotypes in various plant species. The transposons have risen as nature's machinery of creating variation. The repetitive elements, especially transposons, are notorious DNA sequences capable of introducing rapid changes or mutations in the genome, which gives rise to different phenotypes on which selection can be acted upon. The radiation of angiosperms and diversification of various plant phenotypes can be attributed to such genomic elements, which introduced such rapid phenotypic changes, eventually fueling the adaptation to their local environments. The domestication of multiple plants has now been implicated in the transposable elements and their mediated accelerated evolution.

The influence of transposons on phenotypic traits has been widespread across all plant taxa. The transposon-mediated genomic evolution is not only relevant for wild plants, but it also drives the domestication of crops. The polyploidization events and transposon accumulation influence each other and consequently contribute to genome architecture changes and size (Vicient and Casacuberta 2017). Larger polyploid genomes such as maize, wheat, and other monocot crop species have a more significant proportion of transposable elements and have played an essential part in their evolution and domestication (Giraud et al. 2021). A well-known example is Maize, which is derived from its wild relative Teosinte. The domesticated Maize has multiple transposon insertions regulating its expression and domestication by modifying its apical dominance (Studer et al. 2011) and changes to its photoperiod activity to adapt to varied landscapes and climates (Yang et al. 2013; Huang et al. 2017). These and multiple other traits were influenced by transposons such as kernel morphology and anthocyanin content, seed dispersal, and spreading such other phenotypes resulted in the domesticated highly cultivated corn (Feschotte et al. 2002; Lisch 2013). The Sunflower (*Helianthus annus*) comprises mainly retrotransposons, predominantly LTR-Gypsy, and shows ongoing active Gypsy activity in the genome (Staton et al. 2012; Badouin et al. 2017). The transcriptional and chromatin regulation was observed due to transposon insertion in the sunflower genome, further proving transposon-mediated trait acquisition in sunflowers (Gagliardi et al. 2019). Transposon-influenced trait evolution is not restricted to larger genomes. In contrast, Foxtail millet (*Setaria italica*) with a smaller genome (~420 Mb) has shaped its phenotype driven by transposon insertions (Wang et al. 2012). Foxtail millet has lost its seed-shattering phenotype due to transposon insertion near *the sh1* gene, and this phenotype was subsequently selected and domesticated (Liu et al. 2022). One of the essential staple diets of the world, rice, is also shaped by transposable elements, and its influence has led to its multiple varieties, phenotypes, and widespread domestication (Akakpo et al. 2020). The differences in transposon content in various rice strains led to a theory of multiple domestication centers of Rice in Australasian countries (Civán et al. 2015; Carpentier et al. 2019), although the theory is opposed by others (Huang and Han 2015) the involvement of transposons in the domestication of rice is undeniable in all studies. Other widely cultivated crops like wheat (Zhao et al. 2023), barley (Kashino-Fujii et al. 2018), wrinkled pea (Bhattacharyya et al. 1990), etc., have multiple phenotypes influenced by the activity of transposons. Not only in grain and oil seed crops, but the activity of transposons has been crucial in many fruit and flowering dicotyledons, too. The tomato (*Solanum lycopersicum*) pan-genome analysis revealed multiple causal traits associated with TE insertion polymorphisms (TIPs) in various cultivars (Domínguez et al. 2020). The variety of grapevine phenotypes such as Cabernet, Chardonnay, Ruby Okayama, etc., are due to retrotransposon insertion and subsequent modification in *the Gret1* gene, which gives various colors and flavors to grapes (Kobayashi et al. 2004; Lisch 2013). The blood orange phenotype in *Citrus sinensis* and their different cultivars are due to the insertion of Copia-like transposons inserted within *the Ruby* locus, which control *Ruby* expression and lead to differing levels of anthocyanins in these oranges (Butelli et al. 2012). The control of anthocyanin content in Apples (*Malus domestica*) is also controlled by a redTE transposon insertion, which mediated the control of MYB1 expression, which imparts red-colored phenotypes of apples (Zhang et al. 2019b). The effective control of expression and their respective phenotypes prove that transposons are the candidates to leverage phenotype selection and crop improvement through genetic engineering. Leveraging the abundance of wild plants and their desirable phenotypes could be possible by examining their genomic abundance of repetitive elements and their annotation. Identification of repetitive elements in wild plants' genomic resources, followed by their classification and characterization of sequences influencing coding sequences of important gene candidates, would be the way to leverage this information. Hence, considering the involvement of transposons in the phenotypic control of definitive traits in plants, it becomes imperative to identify, annotate, and characterize repetitive sequences with every sequenced genome.

Next-generation sequencing (NGS) technologies have boosted genomic information for diverse species. With so many genomes getting sequenced, characterizing the repetitive landscape in these genomes is crucial. Repeat annotation pipelines like RepeatModeler (Flynn et al. 2020) and RepeatMasker (Tarailo-Graovac and Chen 2009), along with RebBase libraries, have been a constant standard for identifying and characterizing repeats for genome assemblies for a decade now. Nevertheless, there have been constant improvements in such pipelines in terms of the use of unassembled data for identification as well as integration of multiple software to use all kinds of evidence to have holistic and nonredundant identification repeat annotation (Kojima et al. 2016; Ou and Jiang 2018). Estimating the abundance of repetitive elements gives us a quantitative idea of change occurring between species, but placing them on a temporal scale allows us to identify species-specific insertions. These species-specific insertions are more likely to be causal elements for its phenotypic evolution (Feschotte et al. 2002). In estimating divergence from the consensus sequence to the target, the Kimura-2 parameter distance can be calculated, an index of the temporal scale of insertions of transposons. In our study of the Wild Jack (*Artocarpus hirsutus*) genome, a closely related species of Jackfruit (*Artocapus heterophyllus*) endemic to the Western Ghats, we ventured to find out phenotypic differences between closely related species and their phenotypic variation (Patil et al. 2022). Jackfruit can be considered a semi-domesticated species; its strikingly larger fruits and low-hanging fruits with reticulated branching tree structure define it as a unique fruiting plant. Breadfruit (*Artocarpus altilis*) is another domesticated plant that originated from the wild relatives of New Guinea and has mid-sized green-colored fruits with moderately high trees. In contrast, the Wild Jack is a typical forest-adapted species with long trunks reaching above 80 meters with apical small-sized, yellow-colored fruiting phenotype. We compared the repetitive sequence abundances between Breadfruit (*A. altilis*), Jackfruit (*A. heterophyllus*), and Wild Jack (*Artocarpus hirustus*). We found that even though Breadfruit and Wild Jack share similar repetitive landscapes and abundances, in contrast, Jackfruit has more than ~130 Mbp more LTR-Copia transposon abundance than the other two species. The Kimura-2 divergence estimates explained that these LTR-Copia sequences had been accumulated in Jackfruit after post-divergence from the other two species, which further establishes the importance of Copia transposons in attaining these striking phenotypes in Jackfruit. Further studies can exploit the experimental resources to explore the presence of LTR Copia sequences in the vicinity of coding sequences to pin down genes responsible for these diverse phenotypes of Jackfruit. Such characterization can be leveraged to obtain highly nectar-rich fruits with high flavonoid and protein-rich foods.

Although identification, annotation, and classification of repetitive elements in eukaryotic genomes is crucial, there are many challenges in executing it efficiently. Repetitive elements, by nature, consist of repeating sequences which can be extremely hard to capture and sequence. The short-read sequencing techniques proved to be sub-optimal in conquest to sequence complex repetitive genomic regions, leading to underestimation and adding to missing information that impedes understanding crucial ongoing regulatory mechanisms in organisms. The advent of long-read sequencing techniques like Oxford Nanopore, PacBio, optical mapping, HiC chromatin capture techniques, etc., have led to the generation of highly contiguous and nearly complete genome assemblies. These highly contiguous assemblies are proving to be highly useful in identifying intact repetitive genomic regions and their abundance throughout the genome and helping characterize the mechanisms of trait-specific variation. As a blessing in disguise, recent highly contiguous haplotype resolved assemblies efficiently identify the repetitive elements responsible, and the characterization of downstream structural variants present in pangenomes is being explored efficiently. Structural variations identified within and between species have taken phenotype-associated associations to the next level. Now, scenarios of domestication, artificial selection, and invasiveness potential of the species and their mechanisms are being characterized efficiently. For instance, The haplotype resolved assembly of Potato (*Solanum tuberosum*) was able to identify that the duplications and translocations were highly enriched for the transposons like LTR Gypsy and Copia and had an essential role in introducing these variations. In contrast, the inversions were independent of the TE activity (Sun et al. 2022). The Soybean pan-genome study elaborated that the structural variations in the genomes of all relatives and cultivars are due to repetitive genomic elements (Liu et al. 2020). Further, they associated gene expression changes due to these structural variations and their association with the critical agronomic traits which played an essential part in domesticated variants. Apart from such important agronomic traits, some traits, such as adaptation to varied and divergent climates, play a central role in some plant biology, such as invasive species. Invasive species rapidly adapt to the novel environment by punctuated changes and accumulating changes quickly to be able to shape their genomes accordingly. *Ambrosia artemisiifolia* is an invasive species of weeds native to Northern America but has now spread through Europe. A phased genome assembly of this weed reported around 40% of retroelements in their genomes and identification of region-specific haploblocks enriching for flowering-time related genes (Battlay et al. 2023). In addition to these haploblocks, population genomic analysis identified flowering regulator ELF3 (Early Flowering 3) as putatively differentiated loci potentially under selection. Such population genetic analyses provide an in-depth understanding of the changes that occurred at the base level, their subsequent fixation, and their relevance to the adaptation to the local environment.

22.3 Naturally Acquired Traits Can Be Effectively Characterized Using Population Genomics

Domestication of crop plants generally includes artificially selecting favorable traits over multiple generations. Due to artificial selection, these crops are vulnerable to multiple pathogens/diseases, which are nonexistent in local habitats but are abundant in alien landscapes. This vulnerability to diverse plant pathogens and diseases restricts crops from being cultivated all over the globe. Additionally, climatic parameters such as temperature, humid ity, etc., are highly distinct from different landscapes worldwide. The endemicity of plants to specific environmental factors disallows cultivation in varied climatic conditions due to the inability to show tolerance to heat/cold, drought, osmotic pressures, etc. Identifying genes or genomic regions involved in adaptation to alien environments becomes crucial at such a juncture. Invasive plant species and plants with broader distribution scales are great candidates for such exploration. As discussed previously, the invasive Ragweed native to North America could invade landscapes throughout Europe, and the gene *ELF3* was found to be highly differentiated between North American and European populations. Differentiation at *ELF3* loci allowed this weed to adapt and proliferate in the alien environment. Disease-resistance proteins such as NB-LRR domain-containing proteins have been implicated in protecting plants from biotic and abiotic stresses (McHale et al. 2006). NB-LRR proteins provide multiple types of resistance in diverse plant species and can be exploited to provide resistance in focal plants (Lee and Yeom 2015). However, apart from generalized proteins and factors providing resistance to plants, multiple genomic loci can be involved in attaining adaptation due to independent selection pressures and mechanisms in different species (Shinozaki and Yamaguchi-Shinozaki 2022). Identifying such loci involved in providing tolerance to various species of plants would result in diverse possibilities in attaining a specific phenotype, and such exploration of loci includes analyzing genomes of wild plants for signatures of natural selection.

Exploration of naturally selected candidate loci in a particular plant species, its genetic variation, and differentiation across multiple populations provides a framework for the adaptation phenomenon. Such exploration of population-specific variation within a species employs various population genetic analyses and metrics, allowing adequate characterization of genomic loci responsible for adaptation. For instance, the domestication of wheat has been an exciting story as to how two species, *Triticum* and *Aegilops*, were instrumental in rapid domestication and acquiring the adaptation potential. The recent study on population genomics of wheat genomes identified the gene *Ppd-D1* as instrumental in acquiring cultivation potential worldwide due to its early-flowering phenotype (Zhao et al. 2023). Such candidate loci with signatures of selection for acclimatization to newer environments can be found and introduced in other endemic plants to enhance their cultivation potential.

Population genomics is a field of genomics that employs distribution-wide genomic sampling from multiple populations, mainly of the same species and characterizes population-specific clustering, admixture, diversity, divergence, and differentiation leading up to the identification of the adaptive genomic loci. The analysis helps identify genetically distinct populations, diversity levels within them, inbreeding and runs of homozygosity (ROH), amount of gene flow and introgression between them, levels of divergence and differentiation between them, along with genomic loci experiencing different forms of selective pressures such as purifying, balancing, etc. Such in-depth analysis of populations can elucidate the diverse phenomenon of the evolutionary history of the species, its populations, and the dynamics between them. We need to calculate specific metrics from the population-level genomic datasets to make sense of the population genomic analysis. The first set of metrics can be categorized into within-population metrics, which include estimation of genetic diversity, ROH, etc. The relevant parameters to these within-population metrics are – θ_W (Watterson's estimator) and θ_π (pairwise estimator or nucleotide diversity), used to estimate genetic diversity within a population. θ_W is calculated as the total number of polymorphic sites across all the samples normalized by the sample size, giving an idea of total allelic diversity in the population. Whereas θ_π which is calculated by taking the average of the pairwise nucleotide differences across all possible sites in the samples of a population, which gives an idea of per-site differences. Both the metrics are derivatives of the coalescent theory, which translates to $\theta \sim N_e\mu$ (effective population size × mutation rate) and should be equal under constant mutation rate and effective population size. However, in real data, differences in effective population sizes usually introduce differences in these metrics, allowing us to identify the causes. The greater genetic variation in populations is considered chiefly better due to the possibility of adaptive potential. In contrast, differences in levels of diversity at the same loci can hint toward fixation as one of the scenarios. θ_W is skewed due to minor or rare alleles, whereas θ_π shows the variation that major or high-frequency alleles capture (Booker et al. 2017).

Tajima's D is another within-population parameter that calculates differences between pairwise diversity and Watterson's estimator, i.e. $(\theta_W - \theta_\pi)$ normalized by their standard deviation (Tajima 1989). When the data deviates from the neutral model, Tajima's D shows a skew to either positive or negative values. The neutrality can be affected by genetic hitchhiking/selective sweeps or demographic processes such as drift or bottleneck. The skew of Tajima's D to negative values indicates either an influx of alleles from more individuals, i.e.

an increase in population size or directional positive selection. In contrast, the skew to positive values of Tajima's D indicates either population contraction/bottleneck or balancing or purifying selection acting on the genomic loci under consideration. Although Tajima's D is great at identifying deviation from neutrality, it is not helpful in distinguishing differences in changes due to selection or demography. Another parameter, Fay and Wu's H, is also a test for deviation from neutrality but is more influenced by selection than demographic forces. If we compare Tajima's D values with Fay and Wu's H, we can distinguish the influence of demography from selection, and the causal mechanism can be identified. While detecting positive selection in the dataset, including ancestral states of the alleles becomes important to decide the direction of change and the propensity of the positive selection. Even though Fay and Wu's H can show changes due to selection, another metric called Fu and Li's D can effectively integrate the allelic states, i.e. ancestral allele to derived allele, which allows effective calculation of propensity of the positive selective forces acting on the sites (Fu and Li 1993).

Apart from changes in diversity among populations, detecting linkage disequilibrium (LD) is also crucial to identify the extent of linked selection between loci and their effects due to recombination and other factors (Cutter and Payseur 2013). Detection of ROH is also an analysis that informs about the history and relatedness of the samples included in the analysis (Gibson et al. 2006). Apart from these metrics important for within-population analysis, haplotype statistics are a robust type of analysis that incorporates information from different haplotypes and the extent of LD acting upon them, which informs about changes in selective pressures (Fariello et al. 2013). There are multiple haplotype statistics such as EHH (extended haplotype homozygosity), iHH (integrated EHH), iHS (integrated haplotype score), etc. EHH provides the likelihood of the same origin of two polymorphisms considered, and iHH informs the extent of haplotype length and effect. In contrast, iHS indicates that the genomic loci have undergone positive selection recently.

Within-population statistics can inform a few characteristics about the population in focus. However, the most exciting questions can be asked when divergent populations, mostly with diverse traits, are compared. In such an inter-population genomics framework, we use specific statistics to characterize differentiation, gene flow, clustering, admixture, and identification of adaptive loci. The first statistic that is highly useful in defining differentiation between populations is F_{ST} (fixation index), which is used extensively in population genomics while investigating adaptive loci in populations and speciation. F_{ST} is calculated as a normalized difference between allele frequencies of individuals between and within the population. The index varies between 0 and 1, where 0 indicates similar levels of segregating alleles between populations. In contrast, a value of 1 for a genomic position indicates fixation of an allele in one of the populations. Due to its more straightforward interpretation, this statistic was used widely but is vulnerable to diversity levels and will show bias in such cases. Multiple standardized versions have been introduced to robustly calculate the F_{ST} for identifying differentiated loci between populations to overcome the bias in estimates of F_{ST}. One such statistic is PBS (population branch statistic), which includes information on ancestral alleles to identify lineage-specific selection events by calculating a change in allele frequency for each population separately from the ancestral sequence. Using PBS along with F_{ST} provides robustness to identifying differentiating loci across the populations (Wolf and Ellegren 2016).

Another metric used widely in characterizing between-population relationships is D_{XY}, an important statistic in determining the extent of population divergence. D_{XY} is estimated by calculating total segregating polymorphisms across populations and is proportional to the genomic variation in ancestral sequence, divergence time from the ancestor, and the mutation rate (Charlesworth 1998). In contrast to differentiation statistics, D_{XY} is not vulnerable to within-population genetic diversity differences and, hence, can effectively detect changes in gene flow. The between-population haplotype statistic XP-EHH is calculated as a ratio of integrated haplotype scores (iES: integrated EHHS) of two populations contrasted (Sabeti et al. 2007). The XP-EHH is a haplotype-aware metric that identifies genomic loci under positive selection while integrating recombination information and diversity between the populations. The genomic loci experiencing selection pressures could be acted upon the existing standing variation or be introgressed from some other species or diverged population of the same species, providing adaptive potential. Most recently, one of the most interesting phenomena in the adaptation genomics has been "genomic introgression." Maize has extensive introgressed genomic regions from ancestral species, which experienced strong selective pressures and provided adaptation by modulating traits such as flowering time, macrohairs, cold tolerance, and altitude adaptation (Calfee et al. 2021). The introgressed genomic region has been instrumental in adaptation to higher altitudes and achieving wider distribution for Maize (Gonzalez-Segovia et al. 2019). Reports of large genomic regions introgressed in multiple domesticated species, such as Potato, Grapevine, Ryegrass, etc., have been instrumental in adapting to newer environments and attaining important phenotypes (Blanco-Pastor 2022). Detection of introgression is based on the estimation of D-statistic from the four-species genomic test, also known as the ABBA-BABA test (Martin et al. 2015). Introgressed loci can be wrongly identified and must be carefully addressed to distinguish from the effect of standing variation (Owens et al. 2021). Although D-statistics has been highly applicable in finding evidence of signatures of introgression, more robust methods such as D_3 statistics have been developed, providing better implementation (Hahn and Hibbins 2019).

Many contemporary software and packages are available, and multiple ongoing modifications and improvements are being introduced in population genomics. VCFTOOLS (Danecek et al. 2011) and PLINK (Purcell et al. 2007) are the classical suites of tools that allow the calculation of most of the population genomic statistics and related analysis. Many technical challenges were at hand in these contemporary tools, like the effect of low coverage, genome assembly errors and fragmentations, variant calling challenges, etc. Further improvements to overcome many limitations, easy implementation and parallelization, led to the development of tools like ANGSD (Korneliussen et al. 2014), pixy (Korunes and Samuk 2021), hierfstat (Goudet 2005), BayeScEnv (de Villemereuil and Gaggiotti 2015), Ldhat (Auton and McVean 2007), mlRho (Haubold et al. 2010), PopLDdecay (Zhang et al. 2019a), SweepFinder (Huber et al. 2016), TreeMix (Pickrell and Pritchard 2012), etc. which help in analyzing and calculating most of the population genomic parameters. Although all these parameters are efficient in identifying signatures of selection, expansion in population sizes leads to the accumulation of mutation and their fixation irrespective of selective pressures, the phenomenon also known as "Allele surfing" (Gralka et al. 2016). Many demographic processes leave similar footprints as selection across the genome and can be wrongly identified as signatures of selective pressures. To delineate the effects of demography from selective pressures, we also need to study the demographic history of the focal species and its populations (Muers 2009).

22.4 Demographic Forces Are Crucial in Shaping Genome Dynamics

Detecting signatures of selection at genomic loci has seen a lot of promise and improvements over time for robust detection. However, the influx of population genomic data from diverse sets of organisms has illuminated different forces acting on the genome apart from selection, such as demographic events. Detection of selective pressures includes analyzing changes in diversity at localized genomic loci between populations, and the increased diversity in one population gives a putative signal of selection. However, the influx of variation at particular genomic loci in a population could also be due to demographic processes such as population expansion or bottleneck events in one population in consideration (Jensen et al. 2005). The genomic loci distributed throughout the genome have independent coalescence history and have a specific TMRCA (time to most recent common ancestor) of their own. The genomic segments with a constant TMRCA are separated by the ancestral recombination events, which allows the delineation of independently evolving loci (Li and Durbin 2011). The difficulty in distinguishing the causes of differences in levels of diversity is because of the use of one-dimensional diversity data while neglecting the effects of recombination on the genomic loci. Hence, including recombination events information, such as ancestral recombination graph (ARG), becomes imperative while estimating demography and selection. ARG involves estimating genomic variation with Coalescence events while estimating recombination breaks across genomic loci post-divergence of the population in consideration (Rasmussen et al. 2014). ARG-based statistics such as recombination rate, Leaf Trace, TMRCA, pi (average distance between two leaves), popsize (effective population size for a local tree), Relative TMRCA half-time (RTH), etc., help in identifying putative selection signatures while distinguishing from the demographic processes (Hubisz and Siepel 2020). Coalescent theory has been instrumental in estimating demographic processes across genomes, and there have been multiple implementations to estimate Ne (Effective population size). Estimating temporal dynamics of historical changes in Ne of a species allows for identifying evolutionary history, including genetic drift, expansions, and bottlenecks, which led up to current distribution and levels of genomic variation. The historical demographic processes shape genomes, leading to changes in phenotypic traits through stochastic processes in response to climate and environmental conditions (Nadachowska-Brzyska et al. 2022).

Wild relatives of domesticated crops comparatively have higher levels of genomic variation, which works as a reservoir to adapt to various biotic and abiotic changes in their environments. The domestication process often involves artificial selection and repeated inbreeding for particular traits reduces genetic diversity (Gaut et al. 2018). More than not, domesticated plants are subsampled parts that expand to more significant numbers through cultivation, leading to extensive population bottleneck. Due to the smaller effective population size of the cultivated population, these populations experience extensive effects of random genetic drift, and multiple unrelated genomic loci observe fixation of deleterious alleles and repetitive elements (Lefébure et al. 2017; Wang et al. 2017). Fixation of unrelated deleterious traits makes these populations vulnerable to various biotic and abiotic factors. Hence, it becomes crucial to study comparative demographic processes involved in naturally occurring wild relatives of domesticated crops to understand the diversity and effects of random genetic drift in response to biotic and abiotic factors (Garnier et al. 2018). Moreover, forest plants with longer generation times are indicators of climate fluctuations, and studying their demographic histories informs us of the responses to biotic and abiotic changes in their habitats. In our study of the *Mesua ferrea* genome, we estimated comparative demographic histories of tropical forest plants and Neotropical ones (Patil et al. 2021). We observed population size contractions during extensive mid-Pleistocene transition (MPT) glaciations in the forest plants distributed in tropical regions, in contrast to neotropical plants. Evidence from multiple forest plant species showing similar trends of bottlenecks with similar distributions further establishes the importance of studying plants with longer generation times in understanding older environmental fluctuations, their responses, and changes in genomic variation. Analyzing the genomic demographic histories across closely related species distributed at varied habitats informs critical responses to climate change and long-term environmental changes. Comparing closely related species can further establish specific lineage-specific factors other than abiotic changes, such as co-evolutionary species interactions and differences in the habitat-specific pathogen load (Bai et al. 2018). Such micro-habitat changes affecting inter-specific interactions, altitudinal variation, and other adaptation-specific changes recorded in genomes are effectively elucidated through demographic history estimations.

Estimating demographic histories has been sought in genetics for a long time, and multiple theoretical frameworks and implementations have been used until now. Approximate Bayesian computation (ABC)-based methods were popular in marker-based studies with reduced representation sequencing data from multiple populations. The most popular ABC-based software implementations include δaδi, ABCtoolbox, fastsimcoal, ONeSAMP, etc. ABC-based methods have seen multiple improvements in terms of amounts of data; the inclusion of multiple variables and parameters, such as PopSizeABC, MMC-ABC, DILS, etc., have shown much more promise in these tools (Nadachowska-Brzyska et al. 2022). After the advent of NGS, a tremendous amount of whole genome sequencing datasets became available and allowed efficient analysis of total neutral variation. Tools developed for analyzing reduced representation data were inefficient for larger genomic datasets that needed an advent in the field. Heng Li's implementation of the hidden Markov model (HMM) based estimation of historical Ne using Coalescence theory, pairwise sequential Markovian coalescent (PSMC), was a much-needed efficient software (Li and Durbin 2011). PSMC efficiently handled the genomic dataset from a single individual to estimate historical Ne and allowed investigation of changes in population size between populations, including divergence. Although PSMC proved ideal in multiple applications, the effect of population structure induced differences in the estimations and needed to be addressed using robust species sampling. Methodological improvements in SMC-based methods to include multiple genomes led to the development of important tools such as MSMC, MSMC2, SMC++, and eSMC (Spence et al. 2018). These methodological advances robustly estimate temporal demographic estimations for a particular species while considering population structure and inter-population variation. A set of tools that uses population genomic data by estimating site frequency spectrum (SFS), such as Stairway Plot and its modification Stairway Plot 2, are developed to estimate more recent changes in the demographic history. Demographic modeling and estimating demographic history from empirical data provide efficient scenarios of population dynamics within species.

Initial methods developed to analyze demographic processes in the organisms were largely dependent on the TE polymorphisms due to the nature of these sequences (Biémont 2010). TE dynamics are largely shaped by demographic processes and selection, allowing crucial modification in the genetic architecture (Lockton et al. 2008; Bourgeois and Boissinot 2019). Recent genomic evidence shows an influx of TE sequences during domestication, which allowed rapid adaptation and provided important phenotypes to be propagated through the populations (Cai et al. 2022). Hence, neglecting repetitive sequences such as TEs while estimating demographic history might demonstrate false scenarios of the populations in consideration. Nevertheless, while using whole genome-based demographic estimations, mostly SMC-based methods mask the repetitive DNA due to their notorious mutation profiles (Slotkin 2018). Suppose the repetitive elements are profiled independently and assessed for divergent mutation rates. In that case, we can efficiently include the important repetitive element families while estimating demographic histories and recovering the real historical population dynamics scenario. With this motivation, we developed a pipeline which profiles all the repetitive families and their effects on the demographic history estimations using PSMC (Patil and Vijay 2021). Our study elucidated that most repetitive elements do not have divergent mutational profiles and should be included in the demographic history estimations to recover real demographic scenarios for species. Efficient computational analysis of demographic forces acting on the genome will illuminate crucial mechanisms and clues to improve our knowledge of plant breeding approaches.

Phenotypic plasticity is the ability of an organism to modify its phenotypes in response to abiotic or environmental changes. This ability to respond to variability in an organism's habitat and climate is regarded as the most important characteristic due to selection and demographic forces that can effectively lead to adaptation (Chevin et al. 2013). It has been proposed that plasticity can be "Adaptive" or "Nonadaptive," in which only adaptive plasticity can provide raw material to reach phenotypic optimum and attain fitness, unlike nonadaptive plasticity. Nevertheless, adaptive plasticity in a certain context could probably be nonadaptive and become adaptive due to historical selection events in response to certain environmental cues (Ghalambor et al. 2007). In a study on Trinidadian guppies (*Poecilia reticulata*) residing in a predator-prone population, after an artificial change to a predator-free habitat evolved parallel changes in expression in a similar direction to the existing predator-free population. These changes in gene expression illuminate the involvement of plasticity in rapid adaptation in response to changes in the environment. Moreover, it concludes that the particular environmental change will determine the effect of directional selection, and in some cases, adaptive plasticity defines boundaries for evolution, whereas nonadaptive plasticity harbors the ability to potentiate evolution through increasing strength of directional selection (Ghalambor et al. 2015). Since plasticity is the ability to show rapid changes to the environment, these are more likely due to epigenetic mechanisms and regulatory elements of the genome. Regulatory elements mediate the changes in expression by controlling the exonic structure through alternative splicing, methylating genic regions or demethylation, and many other mechanisms. Understanding the regulome of organisms through an epigenetic lens will be crucial in understanding the mechanisms of the plastic nature of genomic architecture.

22.5 Conclusion

The challenges in breeding non-model species hitherto impeded the elucidation of the mechanisms involved in their striking phenotypes. The influx of genomic information has allowed the investigation of mechanisms of such diverse phenotypic traits. Mechanisms involved in developing these phenotypes could be due to different evolutionary forces leading to changes in the genetic information. The effective analysis of genomes would include the analysis of repeat dynamics, changes in gene content, structural variation, signatures of selection, and the effects of demography. This book chapter explores using genomic analysis in plant research to demystify trait evolution, which can be applied to future crop improvement programs.

References

Akakpo, R. et al. (2020). The impact of transposable elements on the structure, evolution, and function of the rice genome. *New Phytologist* 226 (1): 44–49. https://doi.org/10.1111/NPH.16356.

Auton, A. and McVean, G. (2007). Recombination rate estimation in the presence of hotspots. *Genome Research* 17 (8): 1219. https://doi.org/10.1101/GR.6386707.

Badouin, H. et al. (2017). The sunflower genome provides insights into oil metabolism, flowering and Asterid evolution. *Nature* 546 (7656): 148–152. https://doi.org/10.1038/nature22380.

Bai, W.N. et al. (2018). Demographically idiosyncratic responses to climate change and rapid Pleistocene diversification of the walnut genus *Juglans* (Juglandaceae) revealed by whole-genome sequences. *New Phytologist* 217: 1726–1736. https://doi.org/10.1111/nph.14917.

Battlay, P. et al. (2023). Large haploblocks underlie rapid adaptation in the invasive weed *Ambrosia artemisiifolia*. *Nature Communications* 14 (1): 1–15. https://doi.org/10.1038/s41467-023-37303-4.

Bhattacharyya, M.K. et al. (1990). The wrinkled-seed character of pea described by Mendel is caused by a transposon-like insertion in a gene encoding starch-branching enzyme. *Cell* 60 (1): 115–122. https://doi.org/10.1016/0092-8674(90)90721-P.

Biémont, C. (2010). Anecdotal, historical and critical commentaries on genetics: a brief history of the status of transposable elements: from junk DNA to major players in evolution. *Genetics* 186 (4): 1085. https://doi.org/10.1534/GENETICS.110.124180.

Blanco-Pastor, J.L. (2022). Alternative modes of introgression-mediated selection shaped crop adaptation to novel climates. *Genome Biology and Evolution* 14 (8): https://doi.org/10.1093/GBE/EVAC107.

Booker, T.R., Jackson, B.C., and Keightley, P.D. (2017). Detecting positive selection in the genome. *BMC Biology* 15 (1): 98. https://doi.org/10.1186/S12915-017-0434-Y.

Bourgeois, Y. and Boissinot, S. (2019). On the population dynamics of junk: a review on the population genomics of transposable elements. *Genes* 10 (6): 419. https://doi.org/10.3390/GENES10060419.

Breseghello, F. and Coelho, A.S.G. (2013). Traditional and modern plant breeding methods with examples in rice (*Oryza sativa* L.). *Journal of Agricultural and Food Chemistry* 61 (35): 8277–8286. https://doi.org/10.1021/jf305531j.

Butelli, E. et al. (2012). Retrotransposons control fruit-specific, cold-dependent accumulation of anthocyanins in blood oranges. *The Plant Cell* 24 (3): 1242–1255. https://doi.org/10.1105/TPC.111.095232.

Cai, X. et al. (2022). Transposable element insertion: a hidden major source of domesticated phenotypic variation in *Brassica rapa*. *Plant Biotechnology Journal* 20 (7): 1298–1310. https://doi.org/10.1111/PBI.13807.

Calfee, E. et al. (2021). Selective sorting of ancestral introgression in maize and teosinte along an elevational cline. *PLoS Genetics* 17 (10): e1009810. https://doi.org/10.1371/JOURNAL.PGEN.1009810.

Carpentier, M.C. et al. (2019). Retrotranspositional landscape of Asian rice revealed by 3000 genomes. *Nature Communications* 10 (1): 24. https://doi.org/10.1038/S41467-018-07974-5.

Charlesworth, B. (1998). Measures of divergence between populations and the effect of forces that reduce variability. *Molecular Biology and Evolution* 15 (5): 538–543. https://doi.org/10.1093/OXFORDJOURNALS.MOLBEV.A025953.

Chevin, L.M., Collins, S., and Lefèvre, F. (2013). Phenotypic plasticity and evolutionary demographic responses to climate change: taking theory out to the field. *Functional Ecology* 27 (4): 967–979. https://doi.org/10.1111/J.1365-2435.2012.02043.X.

Civán, P. et al. (2015). Three geographically separate domestications of Asian rice. *Nature Plants* 1: 15164. https://doi.org/10.1038/NPLANTS.2015.164.

Cutter, A.D. and Payseur, B.A. (2013). Genomic signatures of selection at linked sites: unifying the disparity among species. *Nature Reviews Genetics* 14 (4): 262–274. https://doi.org/10.1038/nrg3425.

Danecek, P. et al. (2011). The variant call format and VCFtools. *Bioinformatics* 27 (15): 2156–2158. https://doi.org/10.1093/BIOINFORMATICS/BTR330.

Domínguez, M. et al. (2020). The impact of transposable elements on tomato diversity. *Nature Communications* 11 (1): 1–11. https://doi.org/10.1038/s41467-020-17874-2.

Fariello, M.I. et al. (2013). Detecting signatures of selection through haplotype differentiation among hierarchically structured populations. *Genetics* 193 (3): 929–941. https://doi.org/10.1534/GENETICS.112.147231.

Feschotte, C., Jiang, N., and Wessler, S.R. (2002). Plant transposable elements: where genetics meets genomics. *Nature Reviews Genetics* 3 (5): 329–341. https://doi.org/10.1038/nrg793.

Flynn, J.M. et al. (2020). RepeatModeler2 for automated genomic discovery of transposable element families. *Proceedings of the National Academy of Sciences of the United States of America* 117 (17): 9451–9457. https://doi.org/10.1073/pnas.1921046117.

Fu, Y.X. and Li, W.H. (1993). Statistical tests of neutrality of mutations. *Genetics* 133 (3): 693–709. https://doi.org/10.1093/GENETICS/133.3.693.

Gagliardi, D. et al. (2019). Dynamic regulation of chromatin topology and transcription by inverted repeat-derived small RNAs in sunflower. *Proceedings of the National Academy of Sciences of the United States of America* 116 (35): 17578–17583. https://doi.org/10.1073/pnas.1903131116.

Garnier, E. et al. (2018). Plant demographic and functional responses to management intensification: a long-term study in a Mediterranean rangeland. *Journal of Ecology* 106 (4): 1363–1376. https://doi.org/10.1111/1365-2745.12996.

Gaut, B.S. et al. (2018). Demography and its effects on genomic variation in crop domestication. *Nature Plants* 4 (8): 512–520. https://doi.org/10.1038/S41477-018-0210-1.

Ghalambor, C.K. et al. (2007). Adaptive versus non-adaptive phenotypic plasticity and the potential for contemporary adaptation in new environments. *Functional Ecology* 21 (3): 394–407. https://doi.org/10.1111/J.1365-2435.2007.01283.X.

Ghalambor, C.K. et al. (2015). Non-adaptive plasticity potentiates rapid adaptive evolution of gene expression in nature. *Nature* 525 (7569): 372–375. https://doi.org/10.1038/nature15256.

Gibson, J., Morton, N.E., and Collins, A. (2006). Extended tracts of homozygosity in outbred human populations. *Human Molecular Genetics* 15 (5): 789–795. https://doi.org/10.1093/HMG/DDI493.

Giraud, D. et al. (2021). Gene and transposable element expression evolution following recent and past polyploidy events in *Spartina* (Poaceae). *Frontiers in Genetics* 12: 589160. https://doi.org/10.3389/fgene.2021.589160.

Gonzalez-Segovia, E. et al. (2019). Characterization of introgression from the teosinte *Zea mays* ssp. *mexicana* to Mexican highland maize. *PeerJ* 7: e6815. https://doi.org/10.7717/PEERJ.6815.

Goudet, J. (2005). hierfstat, a package for r to compute and test hierarchical F-statistics. *Molecular Ecology Notes* 5 (1): 184–186. https://doi.org/10.1111/j.1471-8286.2004.00828.x.

Gralka, M. et al. (2016). Allele surfing promotes microbial adaptation from standing variation. *Ecology Letters* 19 (8): 889–898. https://doi.org/10.1111/ELE.12625.

Hahn, M.W. and Hibbins, M.S. (2019). A three-sample test for introgression. *Molecular Biology and Evolution* 36 (12): 2878–2882. https://doi.org/10.1093/MOLBEV/MSZ178.

Haubold, B., Pfaffelhuber, P., and Lynch, M. (2010). mlRho – a program for estimating the population mutation and recombination rates from shotgun-sequenced diploid genomes. *Molecular Ecology* 19 (Suppl 1): 277–284. https://doi.org/10.1111/J.1365-294X.2009.04482.X.

Huang, X. and Han, B. (2015). Rice domestication occurred through single origin and multiple introgressions. *Nature Plants* 2 (1): 15207. https://doi.org/10.1038/nplants.2015.207.

Huang, C. et al. (2017). ZmCCT9 enhances maize adaptation to higher latitudes. *Proceedings of the National Academy of Sciences of the United States of America* 115 (2): E334–E341. https://doi.org/10.1073/pnas.1718058115.

Huber, C.D. et al. (2016). Detecting recent selective sweeps while controlling for mutation rate and background selection. *Molecular Ecology* 25 (1): 142–156. https://doi.org/10.1111/MEC.13351.

Hubisz, M. and Siepel, A. (2020). Inference of ancestral recombination graphs using ARGweaver. *Methods in Molecular Biology (Clifton, N.J.)* 2090: 231–266. https://doi.org/10.1007/978-1-0716-0199-0_10.

Jensen, J.D. et al. (2005). Distinguishing between selective sweeps and demography using DNA polymorphism data. *Genetics* 170 (3): 1401–1410. https://doi.org/10.1534/GENETICS.104.038224.

Kashino-Fujii, M. et al. (2018). Retrotransposon insertion and DNA methylation regulate aluminum tolerance in European Barley Accessions. *Plant Physiology* 178 (2): 716–727. https://doi.org/10.1104/PP.18.00651.

Kobayashi, S., Goto-Yamamoto, N., and Hirochika, H. (2004). Retrotransposon-induced mutations in grape skin color. *Science (New York, N.Y.)* 304 (5673): 982. https://doi.org/10.1126/SCIENCE.1095011.

Kojima, K. et al. (2016). Short tandem repeat number estimation from paired-end reads for multiple individuals by considering coalescent tree. *BMC Genomics* 17 (Suppl 5): 494. https://doi.org/10.1186/S12864-016-2821-0.

Korneliussen, T.S., Albrechtsen, A., and Nielsen, R. (2014). ANGSD: analysis of next generation sequencing data. *BMC Bioinformatics* 15 (1): 356.

Korunes, K.L. and Samuk, K. (2021). pixy: unbiased estimation of nucleotide diversity and divergence in the presence of missing data. *Molecular Ecology Resources* 21 (4): 1359–1368. https://doi.org/10.1111/1755-0998.13326.

Lee, H.A. and Yeom, S.I. (2015). Plant NB-LRR proteins: tightly regulated sensors in a complex manner. *Briefings in Functional Genomics* 14 (4): 233–242. https://doi.org/10.1093/BFGP/ELV012.

Lefébure, T. et al. (2017). Less effective selection leads to larger genomes. *Genome Research* 27 (6): 1016–1028. https://doi.org/10.1101/GR.212589.116.

Li, H. and Durbin, R. (2011). Inference of human population history from individual whole-genome sequences. *Nature* 475 (7357): 493–496. https://doi.org/10.1038/nature10231.

Lisch, D. (2013). How important are transposons for plant evolution? *Nature Reviews Genetics* 14 (1): 49–61. https://doi.org/10.1038/NRG3374.

Liu, Y. et al. (2020). Pan-genome of wild and cultivated soybeans. *Cell* 182 (1): 162–176.e13. https://doi.org/10.1016/J.CELL.2020.05.023.

Liu, H. et al. (2022). Transposon insertion drove the loss of natural seed shattering during foxtail millet domestication. *Molecular Biology and Evolution* 39 (6): https://doi.org/10.1093/MOLBEV/MSAC078.

Lockton, S., Ross-Ibarra, J., and Gaut, B.S. (2008). Demography and weak selection drive patterns of transposable element diversity in natural populations of *Arabidopsis lyrata*. *Proceedings of the National Academy of Sciences of the United States of America* 105 (37): 13965–13970. https://doi.org/10.1073/pnas.0804671105.

Martin, S.H., Davey, J.W., and Jiggins, C.D. (2015). Evaluating the use of ABBA–BABA statistics to locate introgressed loci. *Molecular Biology and Evolution* 32 (1): 244–257. https://doi.org/10.1093/MOLBEV/MSU269.

McHale, L. et al. (2006). Plant NBS-LRR proteins: adaptable guards. *Genome Biology* 7 (4): https://doi.org/10.1186/GB-2006-7-4-212.

Muers, M. (2009). Separating demography from selection. *Nature Reviews Genetics* 10 (5): 280–281. https://doi.org/10.1038/nrg2581.

Nadachowska-Brzyska, K., Konczal, M., and Babik, W. (2022). Navigating the temporal continuum of effective population size. *Methods in Ecology and Evolution* 13 (1): 22–41. https://doi.org/10.1111/2041-210X.13740.

Ou, S. and Jiang, N. (2018). LTR_retriever: a highly accurate and sensitive program for identification of long terminal repeat retrotransposons. *Plant Physiology* 176 (2): 1410–1422. https://doi.org/10.1104/pp.17.01310.

Owens, G.L. et al. (2021). Standing variation rather than recent adaptive introgression probably underlies differentiation of the texanus subspecies of *Helianthus annuus*. *Molecular Ecology* 30 (23): 6229–6245. https://doi.org/10.1111/MEC.16008.

Patil, A.B. and Vijay, N. (2021). Repetitive genomic regions and the inference of demographic history. *Heredity* 127 (2): 151–166. https://doi.org/10.1038/s41437-021-00443-8.

Patil, A.B. et al. (2021). The genome sequence of *Mesua ferrea* and comparative demographic histories of forest trees. *Gene* 769: 145214. https://doi.org/10.1016/j.gene.2020.145214.

Patil, A.B. et al. (2022). Jack of all trades: genome assembly of Wild Jack and comparative genomics of Artocarpus. *Frontiers in Plant Science* 13: 1029540. https://doi.org/10.3389/FPLS.2022.1029540.

Pickrell, J.K. and Pritchard, J.K. (2012). Inference of population splits and mixtures from genome-wide allele frequency data. *PLoS Genetics* 8 (11): e1002967. https://doi.org/10.1371/JOURNAL.PGEN.1002967.

Purcell, S. et al. (2007). PLINK: a tool set for whole-genome association and population-based linkage analyses. *American Journal of Human Genetics* 81 (3): 559–575. https://doi.org/10.1086/519795.

Rasmussen, M.D. et al. (2014). Genome-wide inference of ancestral recombination graphs. *PLoS Genetics* 10 (5): e1004342. https://doi.org/10.1371/JOURNAL.PGEN.1004342.

Sabeti, P.C. et al. (2007). Genome-wide detection and characterization of positive selection in human populations. *Nature* 449 (7164): 913–918. https://doi.org/10.1038/NATURE06250.

Shinozaki, K. and Yamaguchi-Shinozaki, K. (2022). Functional genomics in plant abiotic stress responses and tolerance: from gene discovery to complex regulatory networks and their application in breeding. *Proceedings of the Japan Academy. Series B, Physical and Biological Sciences* 98 (8): 470–492. https://doi.org/10.2183/PJAB.98.024.

Slotkin, R.K. (2018). The case for not masking away repetitive DNA. *Mobile DNA* 9 (1): 1–4. https://doi.org/10.1186/s13100-018-0120-9.

Spence, J.P. et al. (2018). Inference of population history using coalescent HMMs: review and outlook. *Current Opinion in Genetics & Development* 53: 70–76. https://doi.org/10.1016/J.GDE.2018.07.002.

Staton, S.E. et al. (2012). The sunflower (*Helianthus annuus* L.) genome reflects a recent history of biased accumulation of transposable elements. *The Plant Journal: For Cell and Molecular Biology* 72 (1): 142–153. https://doi.org/10.1111/J.1365-313X.2012.05072.X.

Studer, A. et al. (2011). Identification of a functional transposon insertion in the maize domestication gene tb1. *Nature Genetics* 43 (11): 1160–1163. https://doi.org/10.1038/ng.942.

Sun, H. et al. (2022). Chromosome-scale and haplotype-resolved genome assembly of a tetraploid potato cultivar. *Nature Genetics* 54 (3): 342–348. https://doi.org/10.1038/S41588-022-01015-0.

Tajima, F. (1989). Statistical method for testing the neutral mutation hypothesis by DNA polymorphism. *Genetics* 123 (3): 585–595. https://doi.org/10.1093/GENETICS/123.3.585.

Tarailo-Graovac, M. and Chen, N. (2009). Using RepeatMasker to identify repetitive elements in genomic sequences. *Current Protocols in Bioinformatics* 4 (SUPPL. 25): bi0410s25. https://doi.org/10.1002/0471250953.BI0410S25.

Vicient, C.M. and Casacuberta, J.M. (2017). Impact of transposable elements on polyploid plant genomes. *Annals of Botany* 120 (2): 195–207. https://doi.org/10.1093/AOB/MCX078.

de Villemereuil, P. and Gaggiotti, O.E. (2015). A new FST-based method to uncover local adaptation using environmental variables. *Methods in Ecology and Evolution* 6 (11): 1248–1258. https://doi.org/10.1111/2041-210X.12418.

Wang, J. et al. (2012). Genome sequence of foxtail millet (*Setaria italica*) provides insights into grass evolution and biofuel potential. *Nature Biotechnology* 30 (6): 549–554. https://doi.org/10.1038/NBT.2195.

Wang, L. et al. (2017). The interplay of demography and selection during maize domestication and expansion. *Genome Biology* 18 (1): 215. https://doi.org/10.1186/S13059-017-1346-4.

Wolf, J.B.W. and Ellegren, H. (2016). Making sense of genomic islands of differentiation in light of speciation. *Nature Reviews Genetics* 18 (2): 87–100. https://doi.org/10.1038/nrg.2016.133.

Yang, Q. et al. (2013). CACTA-like transposable element in ZmCCT attenuated photoperiod sensitivity and accelerated the postdomestication spread of maize. *Proceedings of the National Academy of Sciences of the United States of America* 110 (42): 16969–16974. https://doi.org/10.1073/PNAS.1310949110.

Zhang, C. et al. (2019a). PopLDdecay: a fast and effective tool for linkage disequilibrium decay analysis based on variant call format files. *Bioinformatics (Oxford, England)* 35 (10): 1786–1788. https://doi.org/10.1093/BIOINFORMATICS/BTY875.

Zhang, L. et al. (2019b). A high-quality apple genome assembly reveals the association of a retrotransposon and red fruit colour. *Nature Communications* 10 (1): 1494. https://doi.org/10.1038/S41467-019-09518-X.

Zhao, X. et al. (2023). Population genomics unravels the Holocene history of bread wheat and its relatives. *Nature Plants* 9 (3): 403–419. https://doi.org/10.1038/S41477-023-01367-3.

Index

Note: **Bold** page numbers refer to tables and *Italics* page numbers refer to figures.

d

e

h

q